T0257438

Breast Cancer

From Fundamental Biology to Therapeutic Strategies

A subject collection from *Cold Spring Harbor Perspectives in Medicine*

OTHER SUBJECT COLLECTIONS FROM *COLD SPRING HARBOR PERSPECTIVES IN MEDICINE*

Retinal Disorders: Approaches to Diagnosis and Treatment, Second Edition

Combining Human Genetics and Causal Inference to Understand Human Disease and Development

Lung Cancer: Disease Biology and Its Potential for Clinical Translation

Influenza: The Cutting Edge

Leukemia and Lymphoma: Molecular and Therapeutic Revolution

Addiction, Second Edition

Hepatitis C Virus: The Story of a Scientific and Therapeutic Revolution

The PTEN Family

Metastasis: Mechanism to Therapy

Genetic Counseling: Clinical Practice and Ethical Considerations

Bioelectronic Medicine

Function and Dysfunction of the Cochlea: From Mechanisms to Potential Therapies

Next-Generation Sequencing in Medicine

Prostate Cancer

RAS and Cancer in the 21st Century

Enteric Hepatitis Viruses

Bone: A Regulator of Physiology

Multiple Sclerosis

SUBJECT COLLECTIONS FROM *COLD SPRING HARBOR PERSPECTIVES IN BIOLOGY*

Synthetic Biology and Greenhouse Gases

Wound Healing: From Bench to Bedside

The Endoplasmic Reticulum, Second Edition

Sex Differences in Brain and Behavior

Regeneration

The Nucleus, Second Edition

Auxin Signaling: From Synthesis to Systems Biology, Second Edition

Stem Cells: From Biological Principles to Regenerative Medicine

Heart Development and Disease

Cell Survival and Cell Death, Second Edition

Calcium Signaling, Second Edition

Engineering Plants for Agriculture

Protein Homeostasis, Second Edition

Translation Mechanisms and Control

Cytokines

Circadian Rhythms

Immune Memory and Vaccines: Great Debates

Cell–Cell Junctions, Second Edition

Breast Cancer

From Fundamental Biology to Therapeutic Strategies

A subject collection from *Cold Spring Harbor Perspectives in Medicine*

EDITED BY

Jane E. Visvader
The Walter and Eliza Hall Institute of Medical Research

Jeffrey M. Rosen
Baylor College of Medicine

Samuel Aparicio
University of British Columbia

COLD SPRING HARBOR LABORATORY PRESS
Cold Spring Harbor, New York • www.cshlpress.org

Breast Cancer: From Fundamental Biology to Therapeutic Strategies

A subject collection from *Cold Spring Harbor Perspectives in Medicine*
Articles online at www.perspectivesinmedicine.org

Executive Editor	Richard Sever
Project Supervisor	Barbara Acosta
Permissions Administrator	Carol Brown
Production Editor	Diane Schubach
Production Manager/Cover Designer	Denise Weiss
Publisher	John Inglis

Front cover artwork: Figure legend, TP53$^{-/-}$ syngeneic GEM model recapitulates human breast cancer luminal-like subtype. Imaging mass cytometry depicts an enriched tumor immune microenvironment within the tumor of a luminal-like GEM model. (White: F4/80; red: Ki67; green: Vimentin; magenta: CD8; yellow: S100AS8/9; aqua: PanCK; blue: DAPI.) Cover provided by Diego A. Pedroza and the Cytometry and Cell Sorting Core, Baylor College of Medicine, Houston, TX.

Library of Congress Cataloging-in-Publication Data

Names: Visvader, Jane E., editor. | Rosen, Jeffrey M., editor. | Aparicio, Samuel, 1963- editor.

Title: Breast cancer: from fundamental biology to therapeutic strategies / edited Jane E. Visvader, Walter and Eliza Hall Institute, Jeffrey M. Rosen, Baylor College of Medicine, and Samuel Aparicio, University of British Columbia.

Description: Cold Spring Harbor, New York : Cold Spring Harbor Laboratory Press, [2024] | Series: Cold Spring Harbor perspectives in medicine | "A subject collection from Cold Spring Harbor perspectives in medicine"-Title page. | Includes bibliographical references and index. |

Summary: "Breast cancer is the most common malignancy in women worldwide. This volume reviews our current understanding of the molecular, pathological, and biomechanical characteristics of breast cancer, the risk factors involved, and how these features influence disease progression and metastasis. Recent advances in the detection, disease monitoring, and treatment of breast cancer are also discussed, as are the models and technologies (e.g., organoids and 3D imaging) that are moving the field forward"-- Provided by publisher.

Identifiers: LCCN 2023034428 (print) | LCCN 2023034429 (ebook) | ISBN 9781621824664 (hardcover) | ISBN 9781621824671 (epub)

Subjects: LCSH: Breast–Cancer. | Breast–Cancer–Genetic aspects. | Breast–Cancer–Treatment.

Classification: LCC RC280.B8 B672224 2024 (print) | LCC RC280.B8 (ebook) | DDC 616.99/449042–dc23/eng/20230817

LC record available at https://lccn.loc.gov/2023034428

LC ebook record available at https://lccn.loc.gov/2023034429

Contents

Contents

Preface

Quite remarkably, breast cancer was first described in 1600 BC on Egyptian papyrus, where treatment of eight patients by cauterization was documented. Today, breast cancer remains the most common cancer in women worldwide and the incidence continues to increase in most countries. The sobering statistics of more than 685,000 deaths from breast cancer alone in 2020 underscores the need for improved strategies for prevention, earlier detection, tumor stratification, and optimized therapies. Over the past two decades, we have witnessed an exponential increase in our understanding of breast cancer from molecular and cellular insights to clinical trials. The revolutions in genomics, imaging, and computation have enabled a deep understanding of genetic drivers, breast tumor heterogeneity, and cellular complexity at unprecedented resolution. The ability to decipher stromal and immune components within the complex tumor microenvironment (TME) has led to an enhanced appreciation of the role of noncarcinoma cells in tumor progression. Concurrently, there have been significant advances across the board in histopathology, risk factor identification, germline genetic testing, new model systems for breast tumor initiation and progression, and an enormous increase in the number of clinical trials.

In this volume, we have brought together prominent experts in the field to provide their perspectives on important new developments and future directions on a range of topics central to breast cancer. Where possible, we attempted to align two independent experts to collaborate on writing a chapter to provide their complementary expertise. However, this was not intended to be a comprehensive review and we acknowledge that no single volume can do complete justice to this rapidly advancing field. Accordingly, we apologize to investigators whose work was not cited. With emphasis on translational research, our distinguished authors have sought to address some of the most poignant questions in the cancer field. It is now well recognized that breast cancer is not a single disease but displays profound heterogeneity in pathology, genomic/epigenetic alterations, gene expression, and the TME. Untangling these factors is crucial as they directly impinge on clinical behavior, response to therapy, and tumor relapse.

The volume begins with chapters on the pathology and epidemiology of breast cancer and genetic risk factors. Pathology continues to play a critical role in breast cancer, despite the "omics" era of tumor analysis. More than 20 histological breast cancer subtypes based on morphology and immunohistochemical biomarker expression have been identified. Kos et al. describe common and rare histological subtypes and highlight how rare (and difficult-to-treat) subtypes can be recognized by trained pathologists and linked to gene and biomarker expression. In terms of understanding the etiology of breast cancer, Terry and Colditz summarize advances from epidemiological studies and their implications for breast cancer risk reduction. They focus on modifiable risk factors for breast cancer risk reduction and population prevention. In complementary work, Yadav et al. discuss the increasing number of pathogenic variants in cancer predisposition genes contributing to breast and ovarian cancer, and the evolution of multigene panels and germline genetic testing parameters required for improved risk assessment.

The transformative leaps in our understanding of the molecular landscape of breast cancer are discussed in chapters on ductal carcinoma in situ (DCIS) (Behbod et al.) and breast cancer heterogeneity (intratumoral and intertumoral) through the prism of gene expression profiling (Swarbrick et al.), epigenetics (Bahl et al.), and proteomics (Lei et al.). The chapter on DCIS also summarizes the latest developments in the biology and progression of DCIS and underscores the need to identify early biomarkers of long-term risk and mortality, and measures to avoid overtreatment. With the rapid development of single-cell technologies and its application to breast cancer, we now have a deeper understanding of heterogeneity within the tumor ecosystem and the inherent variation across the different intrinsic subtypes. Swarbrick et al. discuss developments in prognostic gene expression assays used in the clinic and cover recent insights into the entire tumor ecosystem from single-cell molecular profiling. Beyond mutations, chromatin variants are emerging as important drivers of breast cancer. The function and causes of chromatin variants (polymorphisms in chromatin states) and their impact on intertumoral and intratumoral heterogeneity are reviewed by Bahl et al. Proteomics approaches provide another tier of data for a more informed view of breast cancer. It is becoming clear that when combined with genomics (the new area of proteogenomics) and transcriptomics, there is the power to identify more accurate biomarkers and treatment options.

The processes of metastasis and dormancy still pose many challenges as underscored by the current perspectives by Nolan et al. and Dalla et al., respectively. Nolan et al. describe mechanisms governing organ-specific metastasis of breast cancer cells, with emphasis on breast tumor cell–niche interactions and coevolution processes that occur at secondary sites. As cancer patients usually succumb to death through metastasis, a much deeper understanding of the metastatic niche is required, with the possibility of targeting TMEs as metastasis-specific therapies. In the case of breast cancers that relapse after several years, often in bone, tumor dormancy provides a mechanism to explain delayed recurrence in a subset of patients. Dalla et al. integrate clinical data with recent findings on mechanisms leading to acquisition of the dormant state or release from this state at the metastatic site and discuss a framework to address residual tumor cell dormancy and prevent tumor recurrence in patients at high risk.

The modeling of breast cancer using state-of-the-art models has allowed the field to elucidate the functions of genes and pathways in a more physiological context and to establish preclinical platforms. For human tissue, there has been intense activity to derive patient-derived xenograft (PDX) and organoid tumor models. In recognition of the importance of cell context and the microenvironment, organoid cultures have been developed to mimic the architecture of normal tissue and tumors. Muthuswamy and Brugge discuss the importance of organoid models to interrogate cancer biology, drug response and resistance, and highlight the need for better recapitulating interactions between tumor cells and the microenvironment. In the chapter by Lewis and Caldos, the power and limitations of PDX models as experimental platforms for drug treatment and development are highlighted. Although large-scale studies are not generally possible given the cost and time required to generate these models, they serve a crucial role in determining the consistency between in vivo PDX studies and patient cohorts. Bu and Li describe the power of genetically engineered mouse models, and the most recent advances in intraductal virus injections, precision editing using in vivo CRISPR/Cas9 technology, and the development of somatic rat models to better recapitulate certain subtypes of breast cancer.

Adding a further dimension for deconvoluting the tumor ecosystem, revolutionary advances in 3D imaging and spatial mapping technologies have paved the way for investigating cellular context. Multidimensional imaging is being used to conduct large-scale analyses of breast tissue to view the spatial distribution of tumor cells in their native TME at single-cell resolution. Rios et al. describe how cells can be tracked in real time during tumor progression and treatment using sophisticated imaging to provide high-definition data on tumor cell types, cell movement, and behavior. The chapter by Ali and West reviews cutting-edge spatial imaging technologies (both spatial transcriptomics and proteomics), and their role in understanding spatial context and heterogeneity in tumor biology. These platforms have provided insights into novel subtypes of cells within tumors, tissue features, and cellular neighborhoods. Although the full clinical significance of multiplex imaging technology is yet to be realized, it will be critical to determine ways in which high-resolution spatial data can be translated into clinical practice.

It is becoming increasingly evident that the immune system is an important determinant of clinical outcome. The relatively poor response to immune checkpoint inhibitors by the majority of breast cancers has warranted a more informed understanding of the immune system. Toward this end, and complementing the chapters on spatial imaging, Quail et al. discuss immune heterogeneity between breast cancer subtypes, the immune response against breast cancer, and the utility of preclinical models to decipher tumor clearance and immune evasion. Emens and Loi then describe how the type of immune infiltrate can significantly impact cancer outcomes and the exciting progress that has been made in breast cancer immunotherapy over the last decade. Indeed, specific therapies have recently become the new standard of care for triple-negative breast cancer (TNBC) in different settings. Moving forward, they underscore the need to clarify how best to use immunotherapy in breast cancer and to incorporate it into strategies for more tailored management of patients.

In the final chapter, "Emerging Therapies for Breast Cancer" Goel and Chandarlapaty discuss other emerging therapies and cover the principles that underpin advances for both early-stage and advanced breast cancer patients. This chapter includes recent developments in targeting key drivers, anti-cell-cycle machinery therapies, and synthetic lethal targets, together with advances in anti-estrogen therapy. It is thus clear that the enormous growth in our understanding of cancer biology over the past 20 years is changing the disease trajectory across all breast tumor subtypes and is moving us closer toward precision medicine. Nonetheless, there remain major gaps such as targetable molecular pathways for TNBC.

The editors wish to thank all authors for their time and effort in contributing to this volume and providing their collective wisdom. We are particularly grateful for their contributions, which were commenced during the throes of the pandemic and the challenges it posed for many. We are grateful to Barbara Acosta for her dedication and patience in putting this book together, and to Richard Sever and Maria Smit for conceiving this volume and bringing it to light. We also thank colleagues at CSH Laboratory Press for their excellent editorial skills during this undertaking. Collectively, we hope these chapters will provide a timely overview of many exciting discoveries and debates in the field of breast cancer that will further inform the next research priorities and future generation of therapeutics.

Jane E. Visvader
Jeffrey M. Rosen
Samuel Aparicio

Breast Cancer Histopathology in the Age of Molecular Oncology

Zuzana Kos,[1,2] Torsten O. Nielsen,[1,3] and Anne-Vibeke Laenkholm[4,5]

[1]Department of Pathology and Laboratory Medicine, University of British Columbia, Vancouver, British Columbia V6T 1Z4, Canada

[2]BC Cancer Vancouver Centre, Vancouver, British Columbia V5Z 4E6, Canada

[3]Molecular and Advanced Pathology Core, Vancouver, British Columbia V6H 3Z6, Canada

[4]Department of Surgical Pathology, Zealand University Hospital, 4000 Roskilde, Denmark

[5]Department of Clinical Medicine, Faculty of Health and Medical Sciences, University of Copenhagen, 2200 Copenhagen, Denmark

Correspondence: zuzana.kos@bccancer.bc.ca

For more than a century, microscopic histology has been the cornerstone for cancer diagnosis, and breast carcinoma is no exception. In recent years, clinical biomarkers, gene expression profiles, and other molecular tests have shown increasing utility for identifying the key biological features that guide prognosis and treatment of breast cancer. Indeed, the most common histologic pattern—invasive ductal carcinoma of no special type—provides relatively little guidance to management beyond triggering grading, biomarker testing, and clinical staging. However, many less common histologic patterns can be recognized by trained pathologists, which in many cases can be linked to characteristic biomarker and gene expression patterns, underlying mutations, prognosis, and therapy. Herein we describe more than a dozen such histomorphologic subtypes (including lobular, metaplastic, salivary analog, and several good prognosis special types of breast cancer) in the context of their molecular and clinical features.

Breast cancer is the most commonly diagnosed cancer worldwide, yet, as we have come to understand in the last decades, it is not a single disease entity. Different pathways of pathogenesis give rise to distinct tumors with differing biological properties, behaviors, and prognoses, which can be defined by microscopic histomorphology, protein biomarkers, RNA expression profiles, and DNA alterations. Additional clinicopathological factors that are prognostic in early breast cancer include age, lymph node status, tumor size, histological subtype, grade, lymphovascular invasion, and biomarker expression—especially estrogen receptor (ER) and human epidermal growth factor receptor 2 (HER2) status (Table 1). The most common histological types of breast cancer are invasive ductal carcinoma (invasive breast carcinoma of no special type) and invasive lobular carcinoma, but there more than 20 recognized histological spe-

Table 1. Landscape of breast cancer

Clinical subtype	Molecular subtype		Typical pathological grade	IHC surrogate (gene signature)	Risk of recurrence	Sites of metastasis (distribution) (Kennecke et al. 2010)
	Intrinsic	TNBC				
ER⁺ HER2⁻	Luminal A		Low	ER⁺/PR (>20%)/HER2⁻/Ki67 low[a] (gene expression signature low risk)	Lowest risk, continuous over time	Bone (67%) >> liver (29%), lung (24%) > brain (8%)
	Luminal B		Intermediate/ high	ER⁺/HER2⁻/PR <20% and/or Ki67 high[a] (gene expression signature high risk)	Higher risk than LumA, continuous over time	Bone (71%) >> liver (32%), lung (30%) > brain (11%)
HER2⁺	Luminal B		High	ER⁺/PR any/HER2⁺/Ki67 any (usually high)	High risk, early recurrence (highest risk within 2 yr)	Bone (65%) > liver (44%), lung (37%) > brain (15%)
	HER2-E		High	ER⁻/PR⁻/HER2⁺/Ki67 any (usually high)	High risk, early recurrence (highest risk within 2 yr)	Bone (60%) > lung (47%), liver (46%), brain (29%)
ER⁻ HER2⁻ (TNBC)	Basal-like	BL1 BLIS, BL2 BLIA, IM MES, BLIS, MSL	High	ER⁻/PR⁻/HER2⁻/Ki67 any (usually high)/CK5/6⁺ and/or EGFR⁺ and/or SOX10⁺	Highest risk, early recurrence (highest risk within 2 yr)	Lung (43%), bone (39%), brain (25%), liver (21%)
	Claudin low	M MES, BLIS, MSL	Intermediate/ high	ER⁻/PR⁻/HER2⁻/Ki67 any/E-cadherin⁻		(No consensus reported)
	Luminal A/B, HER2E	LAR	Intermediate/ high	ER⁻/PR⁻/HER2⁻/AR⁺/Ki67 any	High risk, early recurrence (highest risk within 2 yr)	Bone > lung, liver > brain

(TNBC) Triple-negative breast cancer, (IHC) immunohistochemistry, (ER) estrogen receptor, (PR) progesterone receptor, (HER2) human epidermal growth factor receptor 2, (HER2-E) HER2-enriched, (BL1) basal-like 1, (BL2) basal-like 2, (BLIS) basal-like immune-suppressed, (BLIA) basal-like immune-activated, (IM) immunomodulatory, (M) mesenchymal, (MES) mesenchymal, (MSL) mesenchymal stem-like, (LAR) luminal androgen receptor, (CK5/6) cytokeratin 5/6, (EGFR) epidermal growth factor receptor, (SOX10) SRY-box transcription factor 10, (AR) androgen receptor, (LumA) luminal A.

[a]The cut-off point between "low" and "high" values is not universally defined, although values >13% specifically distinguishes Luminal B from A (Cheang et al. 2009). The St. Gallen 2021 Consensus Panel majority opinion supported tumors with Ki67 <5% do not receive chemotherapy, whereas tumors with Ki67 > 30% receive chemotherapy (Burstein et al. 2021).

Cite this article as *Cold Spring Harb Perspect Med* doi: 10.1101/cshperspect.a041647

Table 2. Histological subtypes of breast cancer

Histological subtype	Most common biomarker subtype[a]	Most common gene expression subtype	Most common mutations	Biological behavior/ prognosis
Invasive ductal—no special type (IDC-NST)	ER⁺HER2⁻	Luminal	PIK3CA, TP53, GATA3, ERBB2, CCND1	Intermediate
Invasive lobular	ER⁺HER2⁻	Luminal A	CDH1, PIK3CA	Intermediate
Tubular; cribriform	ER⁺HER2⁻	Luminal A		Good
Mucinous	ER⁺HER2⁻	Luminal A		Good
Invasive micropapillary	ER⁺HER2⁻	Luminal	PIK3CA	Intermediate
Apocrine	TNBC	LAR	PIK3CA, PTEN, AKT	Intermediate
Squamous	TNBC	Basal (BL2)	TERT	Aggressive
Low-grade adenosquamous	TNBC	Basal	PIK3CA	Good
Fibromatosis-like	TNBC	Claudin-low	PIK3CA, TERT	Good
Spindle cell	TNBC	Claudin-low, BL2, mesenchymal	TERT	Aggressive
Metaplastic-heterologous mesenchymal	TNBC	Basal	TP53	Aggressive
Adenoid cystic	TNBC	Basal	MYB-NFIB	Good
Secretory	TNBC	Basal	ETV6-NTRK3	Good

(TNBC) Triple-negative breast cancer, (LAR) luminal androgen receptor.

[a]HER2⁺ tumors constitute a significant minority of cases with IDC-NST, invasive micropapillary, and apocrine histology, whereas TNBC makes up an additional minority of IDC-NST and invasive micropapillary cancers.

cial types of cancer that have distinct morphology and behavior (International Agency for Research on Cancer 2019). Breast cancer treatment is determined based on the combined evaluation of these various tumor factors and may include surgery, radiotherapy and systemic treatment (chemotherapy, endocrine therapy, targeted drugs, and immunotherapy) according to the patient's individual risk. This work first reviews the key standard immunohistochemical (IHC) biomarkers and gene expression subtypes of breast cancer, which lay the context for our subsequent description of the main histologic categories of breast cancer currently recognized by the pathology community (Table 2), with a focus on their morphological, phenotypical, and molecular characterizations and relationship to prognosis.

BREAST CANCER SUBTYPING BY PROTEIN AND RNA EXPRESSION

Clinical Biomarkers in Breast Cancer

Regardless of histology, pathologists evaluate several protein biomarkers on every new case of breast cancer. These are typically assessed by IHC, in their morphologic context.

Estrogen Receptor

ER belongs to a family of nuclear hormone receptors that act as ligand-activated transcription factors, with signaling reflecting a balance between two opposing receptors—ERα (the clinically measured form) and ERβ (Heldring et al. 2007). ER is assessed by IHC as the percentage of carcinoma cells showing nuclear staining with anti-ERα antibody and is reported as positive when >10% of cells stain for ER, and as low positive in the 1%–10% range. Up to 80% of breast cancers express ER but rates vary considerably depending on country and race/ethnicity (Kong et al. 2020; Popli et al. 2021). ER⁺ cancers derive substantial benefit from endocrine therapy, including reduced local, contralateral, and distant recurrence rates and reduced 15-year breast cancer mortality (Early Breast Cancer Trialists' Collaborative Group 2005; Early Breast Cancer Trialists' Collaborative Group et al. 2011) with higher levels of ER expression predicting greater benefit (Early Breast Cancer Trialists' Collaborative Group 1998). Conversely, ER positivity confers a degree of chemoresistance as ER$^+$ cancers are less likely than ER$^-$ cancers to achieve complete pathological response from neoadjuvant chemotherapy (Cortazar et al. 2014).

Progesterone Receptor

Progesterone receptor (PR) is also a member of the steroid hormone receptor superfamily and is a downstream effector of ER. PR signaling potentiates mammary gland proliferation and may act as a driver for breast cancer (Lange 2008; Knutson et al. 2012). Findings from the Women's Health Initiative randomized trial showed that hormone replacement therapy containing estrogen and progesterone (but not estrogen alone) significantly increased the risk of developing invasive breast cancer (Chlebowski et al. 2003; LaCroix et al. 2011). PR is assessed by immunohistochemistry, with ≥1% nuclear staining of tumor cells considered as positive. In ER$^+$ breast cancer, PR adds no predictive information regarding benefit from endocrine treatment (Early Breast Cancer Trialists' Collaborative Group et al. 2011; Early Breast Cancer Trialists' Collaborative Group 2015). The importance of PR is in prognosis (Allison et al. 2020), where it has long been recognized that ER$^+$/PR$^+$ cancers have the most favorable prognosis (Pichon et al. 1980). PR > 20% has been suggested as a criterion distinguishing luminal A from luminal B breast carcinoma (Prat et al. 2012).

Human Epidermal Growth Factor Receptor 2

Human epidermal growth factor receptor 2 (HER2) is a member of the HER family of transmembrane receptor tyrosine kinases that regulate cell growth, survival, differentiation, migration, and other cellular responses through several signaling pathways, including RAS/RAF/MEK/ERK and PI3K/AKT/mTOR (Wieduwilt and Moasser 2008). Prior to the development of anti-HER2 targeted therapy, HER2$^+$ breast cancers carried the worst prognosis (Cheang et al. 2008). The development and clinical implementation of targeted HER2 therapy has transformed

the course of HER2$^+$ disease and rendered HER2 status a strong predictive biomarker (for review, see Swain et al. 2023). HER2 protein overexpression and gene amplification are strongly correlated and are clinically assessed by IHC and in situ hybridization (ISH) assays, respectively. HER2 IHC is scored as 0 (no staining or ≤10% of cells showing faint/barely perceptible incomplete membranous staining), 1+ (>10% of cells showing faint/barely perceptible incomplete membranous staining), 2+ (>10% of cells showing weak to moderate complete membranous staining), or 3+ (>10% of cells showing strong, complete membranous staining). A HER2$^+$ result for first-generation HER2-targeted drugs consists of either 3+ IHC staining or gene amplification by ISH (HER2/CEP17 ratio ≥2 with average HER2 signals/cell ≥4 or average HER2 signals/cell ≥6 with 2+ IHC staining) (Wolff et al. 2023). Until recently, IHC 0 and 1+ results have been reported as negative, and IHC 2+ results reported as equivocal with the specimen reflexed to ISH testing. In 2022, the publication of the DESTINY-Breast04 trial of a HER2 antibody drug conjugate showed improved survival in patients with cancers showing 1+ and 2+/ISH negative results (termed in the trial protocol as HER2-low) (Modi et al. 2022). There are now a number of trials targeting HER2-low and HER2-ultralow cancers (Venetis et al. 2022). In the 2023 HER2 testing in breast cancer update, the American Society of Clinical Oncology–College of American Pathologists did not endorse HER2 low as a new category but stressed the importance of reporting IHC score for drug eligibility (Wolff et al. 2023).

Ki67

Ki67 is a chromatin protein expressed in all cycling cells and is the most commonly used IHC marker of proliferation. Proliferation genes (including MKI67, encoding Ki67) are important components of many commercially available multigene prognostic signatures used to estimate an individual patient's risk of breast cancer recurrence (Tian et al. 2010; Wallden et al. 2015). By IHC, Ki67 index is expressed as the percentage of invasive breast cancer cells with nuclear staining and is undeniably prognostic, with higher Ki67 proliferation rates associated with worse long-term survival (de Azambuja et al. 2007; Inwald et al. 2013; Petrelli et al. 2015). In the neoadjuvant chemotherapy setting, Ki67 predicts for pathological complete response (de Azambuja et al. 2007; Luporsi et al. 2012; Denkert et al. 2013), and higher Ki67 in residual tumor tissue is associated with worse long-term survival (van Minckwitz et al. 2013; Ács et al. 2017). In neoadjuvant endocrine therapy trials, on-treatment Ki67 assessment in postmenopausal women provides information on long-term survival and identifies patients that may benefit from additional therapy (Goncalves et al. 2012; Smith et al. 2020). Low Ki67 has been incorporated into the surrogate IHC definition for differentiating luminal A from luminal B tumors. Nevertheless, controversy surrounding Ki67 remains, particularly related to its reproducibility, as well as cut points that would justify chemotherapy or not (Nielsen et al. 2020).

Triple-Negative Breast Cancer

The term triple-negative breast cancer (TNBC) refers to tumors that are ER⁻, PR⁻, and HER2⁻. This phenotype accounts for ∼10%–15% of breast cancers in Western populations and 27% of breast cancers across African countries, with a high of 46% in West Africa (Hercules et al. 2022). TNBC represents a heterogeneous group of histological subtypes, the majority being high-grade invasive ductal carcinomas of no special type and of basal-like molecular subtype (Perou et al. 2000). At primary diagnosis, patients with TNBC tend to be younger and at risk of carrying a pathogenic germline mutation such as *BRCA1/2* (Shimelis et al. 2018). *BRCA* testing in TNBC can identify patients eligible for adjuvant poly (ADP-ribose) polymerase (PARP) inhibitor therapy (Tutt et al. 2021; Emens and Loi 2023), risk-reducing surgery, and genetic counseling. TNBC is also more likely to be immunogenic and is currently the only subtype of breast cancer eligible for immune checkpoint inhibitor therapy (Karn et al. 2020; Schmid et al. 2020). There are also, however, some rare special type TNBCs that are recognized to be clinically

very different cancers with favorable prognosis (Cserni et al. 2021), underscoring the importance of taking the histological subtype classification into consideration during treatment decision-making.

Gene Expression Subtypes

Beyond the biomarker-defined clinical subtypes, breast cancers are also classified by gene expression profiling into their "intrinsic" molecular subtypes (luminal A, luminal B, basal-like, HER2-enriched, and the later addition claudin-low) (Parker et al. 2009; Prat et al. 2010). The relationship between clinical, intrinsic molecular, and IHC surrogate subtype definitions is presented in Table 1. IHC surrogates show only moderate agreement with intrinsic molecular subtype and should not be considered interchangeable (Viale et al. 2018; Holm et al. 2020).

Specific to TNBC, six molecular subtypes have been described (basal-like 1 [BL1], basal-like 2 [BL2], immunomodulatory [IM], mesenchymal [M], mesenchymal stem-like [MSL], and luminal androgen receptor [LAR]) (Lehmann et al. 2011), which were then refined into four core subtypes BL1, BL2, M, and LAR (Lehmann et al. 2016). Other groups have also reported four stable triple-negative subtypes, described as basal-like immune-suppressed (BLIS), basal-like immune-activated (BLIA), mesenchymal (MES), and LAR (Table 3; Burstein et al. 2015).

Intrinsic Molecular Subtypes

Luminal A

Luminal A breast cancers represent the group with the best prognosis, characterized by expression of genes activated by ER signaling pathways typically seen in luminal epithelial cells of breast glands. There is low expression of genes related to cell proliferation (Perou et al. 2000; Sørlie et al. 2001). Consequently, they show high level expression of ER and PR, are negative for HER2, and have a low Ki67 proliferation rate. They are pathologically low-grade tumors with lower incidence of relapse and higher survival rates than any other molecular type. Risk of relapse, al-

Table 3. Molecular subtypes of triple-negative breast cancer (TNBC)—relationship between different signatures

TNBCtype-4 (Lehmann et al. 2016)	TNBCtype (Lehmann et al. 2011)	TNBC subtypes (Burstein et al. 2015)

(LAR) Luminal androgen receptor, (NSL) mesenchymal stem-like, (M) mesenchymal, (BL1) basal-like 1, (IM) immunomodulatory, (BL2) basal-like 2, (MES) mesenchymal, (BLIS) basal-like immune-suppressed, (BLIA) basal-like immune-activated.

though low, is continuous over time and, if it occurs, is most frequently to the bone. Even in the setting of relapse, they have longer survival (Kennecke et al. 2010).

Luminal B

Luminal B cancers show increased expression of proliferation genes, are higher grade, and are more aggressive than luminal A tumors (Perou et al. 2000; Sørlie et al. 2001). In clinical parlance, luminal B is often used to refer to ER^+ $HER2^-$ cancers that show low-level or negative PR and/or high Ki67 proliferation rate. In the intrinsic molecular subtyping data, the luminal B cluster includes most of the ER^+HER2^+ tumors (which are clinically often referred to as luminal-HER2). Most are of high or intermediate histological grade. Risk of relapse is continuous over time in $HER2^-$ luminal B tumors, whereas recurrences in $HER2^+$ cases occur early. The most common site of metastasis is bone, but with higher rates of lung and liver metastasis than luminal A tumors, and of brain metastases for $HER2^+$ tumors (Table 1; Kennecke et al. 2010).

HER2-Enriched

HER2-enriched breast cancers are characterized by high HER2 pathway signaling, overexpression of proliferation-related genes, and low expression of luminal genes. Most tumors in this group are positive for HER2 and negative for ER and PR (Parker et al. 2009; Prat and Perou 2011). They

are high-grade, aggressive tumors with a fast growth rate. Their prognosis has improved significantly after the introduction of HER2-targeted therapies (Wang and Xu 2019). Although bone metastases are still most common, the HER2-enriched intrinsic subtype confers a higher rate of brain, lung, and liver metastases (Kennecke et al. 2010).

Basal-Like

Basal-like breast cancers are particularly aggressive tumors characterized by genes ordinarily expressed by the outer layer basal or myoepithelial cells of breast glands. Most are triple-negative but stain with basal IHC markers such as high-molecular-weight basal keratins (CK5, CK14), EGFR, nestin, and SOX10. They are more prevalent in younger women and those of African ancestry. Basal-like breast cancers have the highest rate of metastases, relapsing early and with a predilection for lung and brain (Kennecke et al. 2010). They are also the most responsive to chemotherapeutic regimens, with the highest pathological complete response rates out of all of the subgroups (Prat et al. 2015).

Claudin-Low

Claudin-low tumors are characterized by low expression of tight junction proteins claudin 3, 4, and 7, E-cadherin, and luminal differentiation markers and enrichment for cancer stem-cell-like features, epithelial-to-mesenchymal transi-

Cite this article as *Cold Spring Harb Perspect Med* doi: 10.1101/cshperspect.a041647

tion marker, and immune response genes. The majority are triple-negative and histologically show a high frequency of metaplastic- and medullary-like features. Claudin-low tumors have a response rate to neoadjuvant chemotherapy that falls between basal-like and luminal tumors (Prat et al. 2010).

Triple-Negative Molecular Subtypes

Luminal Androgen Receptor

LAR is the only subtype uniformly identified in the major triple-negative subtype profiles with excellent concordance (Lehmann et al. 2011, 2016; Burstein et al. 2015). LAR breast cancers are characterized by androgen, ER, prolactin, and ErbB4 signaling, despite negative ERα IHC. They show high expression of androgen receptor, which is strongly positive by IHC. Histologically, these tumors often have apocrine histology, and they show poor response to neoadjuvant chemotherapy compared to other triple-negative cancers.

Basal-Like Subtypes

Lehmann's basal-like subtypes are highly proliferative tumors showing high $MKI67$ expression and Ki67 IHC labeling rates of >70% (Lehmann et al. 2011). The BL1 subtype is characterized by enrichment of cell cycle and cell division pathways accompanied by elevated DNA damage response pathways. The BL2 subtype involves growth factor signaling pathways, glycolysis, and gluconeogenesis; is uniquely enriched in growth factor receptors such as EGFR, MET, and EPHA2; and has features suggestive of basal/myoepithelial origin, including higher expression of TP63 and MME (CD10) (Lehmann et al. 2011). Burstein's BLIS subtype is characterized by down-regulation of B-cell, T-cell, and natural killer (NK) cell function, and low expression of genes related to antigen presentation, immune cell differentiation and innate and adaptive immune cell communication. In contrast the BLIA subtype shows up-regulation of genes related to B-cell, T-cell, and NK cell function and high expression of STAT genes. This group encompasses Lehmann's immunomodulatory

group. Unsurprisingly, the immune-rich BLIA subgroup has the best outcomes, whereas the BLIS group has the worst (Burstein et al. 2015).

Immunomodulatory Subtype

IM subtype breast cancers express genes encoding immune antigens, cytokines, and core immune signal transduction pathways. These transcripts have been determined to originate from infiltrating lymphocytes rather than a distinct carcinoma cell profile. As might be expected given the prognostic significance of tumor infiltrating lymphocytes in TNBC, the IM subtype shows the best overall and relapse-free survival (Lehmann et al. 2016). Therefore, although IM is more a descriptor of the immune state of a tumor rather than an independent subtype, an IM profile is strongly prognostic. As mentioned above, BLIA encompasses the IM group.

Mesenchymal and Mesenchymal Stem-Like Subtypes

M and MSL subtypes are enriched in pathways associated with epithelial-to-mesenchymal transition and cell motility, with MSL also having deceased expression of proliferation-related genes. A portion of these tumors represent the claudin-low subtype. The MSL group has subsequently been shown to represent tumors with substantial tumor-associated mesenchymal cells rather than representing an independent subtype (Lehmann et al. 2016); nevertheless, Burstein's MES subtype incorporates most of the MSL and some of the M tumors, and is characterized by pathways related to cell cycle, mismatch repair, DNA damage, and hereditary breast cancer signaling pathways, as well as high expression of genes normally seen in osteocytes and adipocytes including growth factors such as IGF-1 (Burstein et al. 2015).

BREAST CANCER SUBTYPING BY HISTOMORPHOLOGY

Invasive Ductal Carcinoma, No Special Type

One reason why histology, important as it is for the initial recognition of malignancy, has less impact on prognosis and management than biomarker expression is that the majority of breast cancers fall into the histological category of invasive ductal carcinoma, no special type (IDC-NST). These are a heterogeneous group of cancers, encompassing all clinical and molecular subgroups, united only by the fact that they do not fall into one of the histological special type categories. Invasive ductal carcinomas most frequently harbor alterations in *TP53*, *PIK3CA*, *GATA3*, *ERBB2*, and *CCND1* (Consortium et al. 2017), but rates differ depending on subgroup. ER^+ cancers have a high prevalence of *PIK3CA* mutations, whereas TNBC have high rates of *TP53* mutations. Like all breast cancers, invasive ductal carcinomas are clinically defined by biomarker groups into three main categories: ER^+HER2^-, $HER2^+$, and TNBC. Histo-

logically, they are most defined by grade, which is a descriptor of differentiation and aggressiveness (Fig. 1). Additional pathological prognostic indicators include tumor size, lymphovascular invasion, the numbers of lymph nodes involved, and surgical margin status. Treatment of invasive ductal carcinoma depends on a combination of clinicopathological factors, the most clinically important of which are the standard IHC biomarkers. In early-stage ER^+HER2^- cancers, other factors such as grade, Ki67, and prognostic gene expression assays come into play.

Within the category of IDC-NST are a number of descriptive morphological patterns, including sebaceous, glycogen-rich clear cell, lipid-rich, oncocytic, melanotic, choriocarcinomatous, pleomorphic, with osteoclastic-like giant cells, with

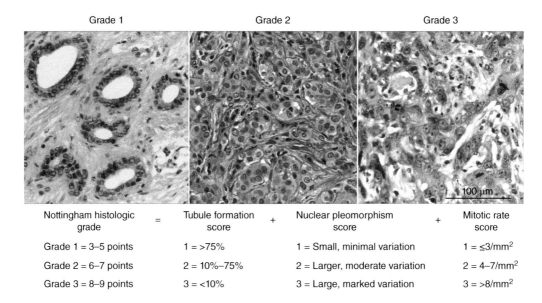

Nottingham histologic grade	=	Tubule formation score	+	Nuclear pleomorphism score	+	Mitotic rate score
Grade 1 = 3–5 points		1 = >75%		1 = Small, minimal variation		$1 = \leq 3/mm^2$
Grade 2 = 6–7 points		2 = 10%–75%		2 = Larger, moderate variation		$2 = 4–7/mm^2$
Grade 3 = 8–9 points		3 = <10%		3 = Large, marked variation		$3 = >8/mm^2$

Figure 1. Pathologic grading of breast cancer. Shown are different grades of invasive ductal carcinoma, no special type. Grade combines three characteristics: tubule formation, nuclear pleomorphism, and mitotic rate. Tubule formation is assessed at low power throughout the whole tumor, counting only structures with cells polarized around a central lumen, with cutoffs at 75% and 10% of tumor area to determine score. Nuclear pleomorphism is assessed at high power in the area of the tumor showing highest pleomorphism and is assessed in relation to normal breast glands. Score 1 nuclei are similar in size to normal glandular epithelial cells and show minimal variation in size, an even chromatin pattern, and inconspicuous nucleoli. Score 2 nuclei are larger (1.5×–2× the size of normal epithelium), with mild to moderate variation and small nucleoli. Score 3 nuclei are even larger, showing marked variation in size and shape, vesicular chromatin, and one or multiple prominent nucleoli. Mitotic score is assessed at high power in the most mitotically active area. Mitotic counts are expressed as the number of mitoses per 10 high power fields (or mm^2 for digital analysis). The values from the three scores are added to produce a final score, which is assigned a grade: Grade 1 (low-grade, well-differentiated) = 3–5 points; Grade 2 (intermediate-grade, moderately differentiated) = 6 or 7 points; Grade 3 (high-grade, poorly differentiated = 8 or 9 points).

neuroendocrine differentiation and medullary pattern (Fig. 2). Of these, only IDC with medullary pattern is clinically relevant. IDC with medullary pattern (previously called medullary and atypical medullary carcinoma) refers to well-circumscribed high-grade tumors with solid syncytial growth pattern and abundant tumor infiltrating lymphocytes (TILs). These tumors show multiple adverse prognostic factors, including significant nuclear pleomorphism, high mitotic/Ki67 rates, *TP53* mutations, TNBC phenotype, and basal-like molecular subtype; nevertheless, they have a favorable prognosis compared to other TNBC, because of the high content of TILs (which have shown level 1B evidence as a positive prognostic factor in TNBC) (Loi et al. 2022; Emens and Loi 2023). In the TNBC molecular classifications, these tumors fit the IM state and cluster as BL2/BLIA.

Invasive Lobular Carcinoma

Invasive lobular carcinoma (ILC) represents ∼8%–14% of breast cancers and is characterized by cellular discohesion. In the classic form, ILC consists of small tumor cells arranged in single file or as single tumor cells (International Agency for Research on Cancer 2019). Variants of ILC are described based on growth pattern (e.g., solid, alveolar, tubulolobular) or cytology (e.g., pleomorphic, histiocytoid/apocrine, signet ring cell) (Fig. 3; Christgen et al. 2021). The evident cellular discohesion of ILC is due to loss of E-cadherin in >90% of tumors through somatic *CDH1* mutations and *CDH1* loss of heterozygosity (LOH) (Desmedt et al. 2016). Loss of E-cadherin can be visualized by IHC documenting a negative or aberrant staining pattern. Loss of 16q, where *CDH1* is located, is seen in nearly all ILCs. Gain of 1q and 16p and *PIK3CA* mutations are also frequent. Compared to IDC, ILCs are enriched for mutations targeting PTEN, TBX3, and FOXA1 (Ciriello et al. 2015). The majority of ILCs are classic type, which are almost invariably ER+HER2− and predominantly luminal A molecular subtype (Lænkholm et al. 2020). Classic ILC is hormonally responsive and relatively chemoresistant, rarely achieving pathological complete response (pCR) to neoadjuvant che-

motherapy (Cortazar and Geyer 2015; Provenzano et al. 2015). The solid variant tends to be higher grade and show an increased mitotic count; the majority are still ER+HER2− but are more often luminal B molecular subtype. Apocrine and most histiocytoid ILCs, conversely, are ER−, and when also pleomorphic may be HER2+. The rare pleomorphic variant is characterized by prominent nuclear pleomorphism and a high mitotic rate; it is high grade (grade 3) and may be ER− and/or HER2+. The growth patterns of ILCs, which tend to be permeative and elicit minimal tissue response, mean they are more often mammographically occult, present at larger size and higher stage, and have a higher rate of resection margin positivity than IDC-NST. Compared to IDCs, ILCs also show a higher frequency of multifocality and bilaterality. Additionally, ILC demonstrates a different metastatic pattern than IDC, with unusual metastatic sites, including the gastrointestinal tract, retroperitoneum, peritoneal surfaces, meninges, and reproductive organs (Ferlicot et al. 2004). Patients with ILC experience late recurrence, and although ILC initially presents with better outcome as compared to IDC, the prognosis appears to be worse during long-term follow-up (Pestalozzi et al. 2008).

Tubular and Cribiform Carcinoma

These tumors (Fig. 4A,B) are rare, accounting for ∼1%–2% and ∼0.5%–1% of breast carcinomas, respectively. Tubular carcinoma is characterized by tubules and glands lined by a single layer of small neoplastic luminal cells with loss of the myoepithelial cell layer, set within a desmoplastic stroma. To fulfill the diagnostic criteria for tubular carcinoma, >90% of the tumor should present the tubular morphology. Cribriform carcinoma is characterized by multilayer uniform tumor cells presenting with a net-like pattern (International Agency for Research on Cancer 2019). Tubular carcinomas have a low frequency of genomic alterations, the most frequent of which are loss of 16q and gain of 1q (Waldman et al. 2001). Both subtypes are low-grade tumors, per definition are grade 1, ER+HER2− with luminal A molecular subtype and favorable prognosis (Weigelt et al. 2008; Liu et al. 2021). In fact, the

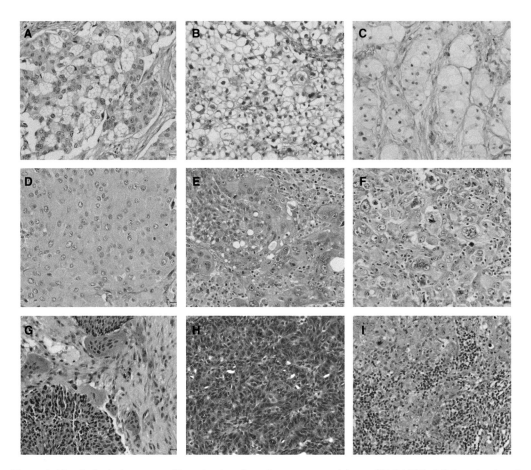

Figure 2. Morphological patterns of invasive ductal carcinoma, no special type (IDC-NST). (*A*) Invasive ductal carcinoma, no special type, with sebaceous pattern showing sebaceous-appearing cells with lipid-containing clear, vacuolated cytoplasm. (*B*) Invasive ductal carcinoma, no special type, with glycogen-rich clear cell pattern showing tumor cells with abundant clear cytoplasm containing glycogen. (*C*) Invasive ductal carcinoma, no special type, with lipid-rich pattern showing large, polygonal tumor cells with abundant foamy or multivacuolated cytoplasm due to neutral lipids. (*D*) Invasive ductal carcinoma, no special type, with oncocytic pattern showing tumor cells with abundant eosinophilic granular cytoplasm and round nuclei with prominent nucleoli. There is considerable morphological overlap with apocrine carcinoma; however, IDCs with oncocytic pattern are negative for androgen receptor (AR) and positive for antimitochondrial antibody. (*E*) Invasive ductal carcinoma, no special type, with choriocarcinomatous pattern showing multinucleated tumor cells wrapping around monocytic tumor cells, mimicking the biphasic growth pattern of choriocarcinoma. (*F*) Invasive ductal carcinoma, no special type, with pleomorphic pattern showing wildly pleomorphic, bizarre tumor cells. (*G*) Invasive ductal carcinoma, no special type, with osteoclastic-like giant cells showing typical invasive ductal carcinoma with hypervascular stroma containing extravasated red blood cells and several osteoclast-like giant cells that are related to histiocytes and not part of the carcinoma itself. (*H*) Invasive ductal carcinoma, no special type, with neuroendocrine differentiation showing cells with scant cytoplasm arranged in rosette-like pattern but with insufficient neuroendocrine histologic features and neuroendocrine marker expression to be classified as a neuroendocrine carcinoma or neuroendocrine tumor of the breast. (*I*) Invasive ductal carcinoma, no special type, with medullary pattern showing syncytial sheets of highly pleomorphic cells with abundant mitoses and a dense tumor infiltrating lymphocyte response.

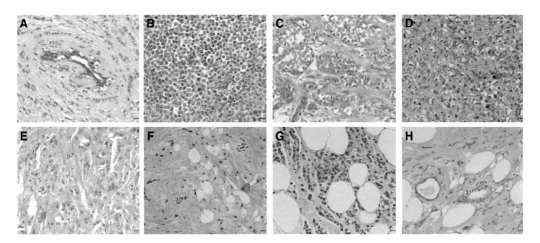

Figure 3. Morphologic patterns of invasive lobular carcinoma. (*A*) Invasive lobular carcinoma, classic type showing single-file and single-cell infiltration with minimal tissue reaction and displaying characteristic "targetoid" arrangement around a benign duct. (*B*) Invasive lobular carcinoma with solid growth pattern showing sheets of dyscohesive cells. (*C*) Invasive lobular carcinoma with alveolar pattern showing globular aggregates of more than 20 tumor cells separated by thin bands of collagenous fibrosis, which may mimic lobular carcinoma in situ. (*D*) Invasive lobular carcinoma, pleomorphic type (pleomorphic lobular carcinoma) showing pleomorphic cells with increased mitoses. (*E*) Apocrine differentiation can be seen in all variants of invasive lobular carcinoma, particularly the pleomorphic type as seen here. (*F*) Histiocytoid invasive lobular carcinoma showing cells with low- to intermediate-grade nuclei and abundant pale eosinophilic cytoplasm, which resemble histiocytes or (if more apocrine) granular cell tumors. (*G*) Invasive lobular carcinoma with signet ring cells showing intracytoplasmic mucin with peripherally displaced nuclei. (*H*) The tubulolobular pattern is composed of an admixture of tubular and typical lobular growth pattern.

prognosis of tubular carcinoma is better than grade 1 IDC-NST, and patients with pure tubular carcinoma have overall survival rates similar to the general population (Rakha et al. 2009). Cribriform carcinoma shows similar behavior and prognosis to tubular carcinoma.

Mucinous Carcinoma

Mucinous carcinomas present as well circumscribed tumors consisting of irregular islands of tumor cells located in abundant extracellular mucin (Fig. 4C). They can be divided into type A with abundant mucin and low cellularity and type B with high cellularity and frequent neuroendocrine differentiation (Weigelt et al. 2009a). Pure mucinous carcinoma includes a >90% mucinous component and accounts for ∼2% of breast cancers. They are mainly ER^+HER2^- with luminal A molecular subtype (Laenkholm et al. 2018) and favorable prognosis (Bae et al. 2011; Lian et al. 2020). Up-regulation of *MUC2*, *SEC16A*, and *CRACR2A*

is a common feature of carcinomas with mucinous differentiation across various primary sites of origin (Nguyen et al. 2021). Despite the ER^+ phenotype, the level of *PIK3CA* mutations is low compared to ER^+ IDC-NST.

Invasive Micropapillary Carcinoma

This is a rare tumor characterized by an inside-out growth pattern visualized by IHC for epithelial membrane antigen (EMA) with staining of the part of the cell membrane facing the surrounding stroma (Li et al. 2006). Pure micropapillary carcinoma (Fig. 4D) has to include >90% of the inside-out growth pattern (International Agency for Research on Cancer 2019). This subtype is, in general, ER^+ but the TNBC phenotype has been described (Nassar et al. 2001). A variable proportion of tumors are reported to be $HER2^+$, but in contrast to other tumor types may show moderate to strong incomplete U-shaped basolateral membranous HER2 IHC

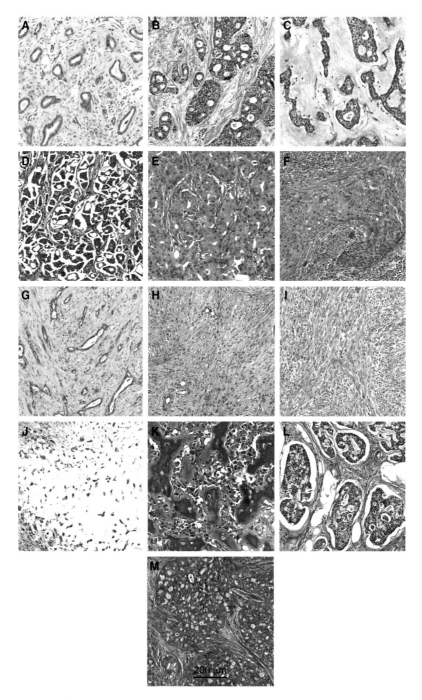

Figure 4. Special types of invasive breast cancer. (*A*) Tubular carcinoma. (*B*) Cribiform carcinoma. (*C*) Mucinous carcinoma. (*D*) Invasive micropapillary carcinoma. (*E*) Carcinoma with apocrine differentiation (apocrine carcinoma). (*F*) Metaplastic squamous cell carcinoma. (*G*) Low-grade adenosquamous carcinoma. (*H*) Fibromatosis-like metaplastic carcinoma. (*I*) Metaplastic spindle cell carcinoma. (*J*) Metaplastic chondroid matrix-producing carcinoma (metaplastic carcinoma with heterologous mesenchymal differentiation). (*K*) Metaplastic osseus matrix-producing carcinoma (metaplastic carcinoma with heterologous mesenchymal differentiation). (*L*) Adenoid cystic carcinoma. (*M*) Secretory carcinoma.

staining, which requires reflex HER2 fluorescence in situ hybridization (FISH) (Perron et al. 2020). The mutational landscape is comparable to the luminal A/B molecular subtype (Thennavan et al. 2021). At primary diagnosis these tumors often show lymphovascular invasion and positive axillary lymph node status associated with higher rates of locoregional recurrence (Wu et al. 2017). There is, however, no difference in overall survival as compared to IDC (Wu et al. 2017).

Apocrine Carcinoma

Around 1% of malignant breast tumors are classified as apocrine carcinomas. They are characterized by large, round or irregular, fairly pleomorphic nuclei with prominent nucleoli (Fig. 4E; International Agency for Research on Cancer 2019). The surrounding granular eosinophilic cytoplasm expresses GCDFP-15 by IHC. Apocrine carcinoma express nuclear staining for androgen receptor, which is mandatory for this subtype classification. Mutations in PIK3CA/PTEN/AKT are common; TP53 and BRAF/KRAS are also seen (Vranic et al. 2015). Apocrine carcinomas are typically ER⁻ and may show HER2 overexpression or alternatively triple-negative phenotype; they are HER2-enriched or luminal by intrinsic molecular subtype (Vranic et al. 2015; Thennavan et al. 2021) and LAR by TNBC molecular classification. The prognosis of apocrine carcinoma depends on clinicopathological factors (Zhang et al. 2017).

Metaplastic Carcinoma

Metaplastic carcinoma is an umbrella term used to describe a range of rare cancers with differing morphology and behavior. They contain, as a unifying feature, a neoplastic component other than breast glandular epithelium. This category contains both highly aggressive as well as some of the best prognosis tumors, so studies combining all into a single metaplastic carcinoma category not surprisingly show differing results regarding prognosis. Most are TNBC and basal-like or claudin-low molecular subtypes (Weigelt et al. 2009b), BL2 or mesenchymal by gene expression profile

(Lehmann et al. 2016). The high-grade subtypes are generally regarded to be more aggressive and less chemoresponsive than other triple-negative carcinomas (Wong et al. 2021). As a whole, mutations in the Wnt/β-catenin pathway (CTNNB1, APC, and WISP3) are common in metaplastic carcinomas (Hayes et al. 2008). These tumors also harbor a high frequency of TP53 and TERT promoter mutations and are more enriched for PIK3CA/PIK3R1 and Ras-Map kinase pathway aberrations compared to other triple-negative carcinomas, although this varies among the subtypes described below (Krings and Chen 2018).

Metaplastic Squamous Cell Carcinoma

Metaplastic squamous cell carcinoma is comprised of cells with squamous differentiation, with variable degrees of cytological atypia and keratinization (Fig. 4F) and is histologically indistinguishable from squamous cell carcinoma of any other site. Tumors can be solid or can be cystic with a cavity lined by atypical squamous cells. Squamous carcinomas of the breast can be pure but are more often mixed with invasive ductal carcinoma (sometimes called high-grade adenosquamous carcinoma) or other metaplastic components (usually spindle cells). The diagnosis of pure metaplastic squamous cell carcinoma requires exclusion of primary cutaneous or metastatic squamous cell carcinoma from another site, such as lung. Tumors are usually triple-negative, basal-like intrinsic subtype, BL2 by TNBC gene classification, and enriched for TERT promoter mutations (Krings and Chen 2018). Metaplastic squamous cell carcinoma is an aggressive tumor and treated the same as other high-grade TNBC.

Low-Grade Adenosquamous Carcinoma

Low-grade adenosquamous carcinoma (LGASC) is comprised of infiltrative well-formed syringomatous like glands with bland cytology and squamoid nests within a desmoplastic stroma (Fig. 4G). Lymphoid aggregates are often present at the periphery. LGASC may arise in association with benign sclerosing lesions. These breast cancers are usually triple-negative and basal-like but actually have an excellent prognosis. The tumors show a high rate of *PIK3CA* mutations but not *TP53* mutations, another contrast from conventional TNBCs (Bataillon et al. 2018). Although these tumors may be locally aggressive and recur with incomplete excision, metastases are exceedingly rare (Hoeven et al. 1993). Because of the excellent prognosis, this entity is particularly important to distinguish from the other high-grade metaplastic carcinomas.

Fibromatosis-Like Metaplastic Carcinoma

Fibromatosis-like metaplastic carcinoma is a variant composed of cytologically bland spindle cells morphologically resembling fibromatosis, within variably collagenous stroma (Fig. 4H). Fibromatosis-like metaplastic carcinoma must always be considered in the differential of any bland spindle cell proliferation in the breast. The diagnosis rests on the presence of positive staining for cytokeratin and/or p63, as an in situ or more conventional invasive component is generally lacking. The tumors are typically triple-negative and claudin-low (Rito et al. 2014). *PIK3CA* and *TERT* promoter mutations are common (Zhong et al. 2022); *TP53* mutations are not seen (Krings and Chen 2018). Similar to LGASC, fibromatosis-like metaplastic carcinoma may arise in association with sclerosing lesions (Gobbi et al. 2003) and show high rates of local recurrence but low risk of metastases (Gobbi et al. 1999; Sneige et al. 2001).

Spindle Cell Carcinoma

Spindle cell carcinoma of the breast (Fig. 4I) is characterized by atypical spindle cells arranged in any of several microscopic patterns, including long or short fascicles, storiform, herringbone, or interwoven architectures. Most spindle cell carcinomas are high-grade, typically triple-negative, claudin-low intrinsic subtype and BL2 or mesenchymal by TNBC classification. The spindle cell component may be admixed with conventional invasive ductal carcinoma or other metaplastic elements like squamous epithelial morphology. Some tumors, however, may present with a purely spindle cell pattern, in which diagnosis as spindle cell carcinoma rather than a sarcoma is rendered by IHC evidence of epithelial/myoepithelial differentiation (e.g., keratins, p63, CD10, SMA, or S100) or presence of an associated in situ component (Adem et al. 2002; Dunne et al. 2003; Leibl et al. 2005); although in a minority of cases such evidence of epithelial differentiation is lacking (Rakha et al. 2021). Spindle cell carcinomas often lack *TP53* mutations (Krings and Chen 2018) but are enriched for *TERT* promoter mutations (Krings and Chen 2018). Spindle cell carcinoma is an aggressive malignancy, with metastatic spread (most commonly to lung) often occurring in the absence of nodal involvement (Khan et al. 2003; Carter et al. 2006).

Metaplastic Carcinoma with Heterologous Mesenchymal Differentiation

Metaplastic carcinoma with heterologous mesenchymal differentiation incorporates chondroid, sometimes osseous, or rarely rhabdoid (or other) mesenchymal components (Fig. 4J,K). The tumors showing chondroid or osseous differentiation are also called matrix-producing carcinomas. Although rare pure tumors exist, most often there is an admixed carcinomatous component present (invasive ductal or squamous cell carcinoma), which aids in the diagnosis. In purely mesenchymal tumors, immunohistochemical stains are helpful to identify epithelial/myoepithelial differentiation (e.g., keratins, p63, CD10, SMA, or S100). Most chondroid matrix–producing carcinomas are positive for SOX10 (Cimino-Mathews et al. 2013). Metaplastic carcinomas with mesenchymal differentiation are usually triple-negative. Chondroid matrix–producing carcinomas are basal-like and enriched for *TP53* mutations and lack PI-3 kinase, Ras-Map kinase, or *TERT* promoter aberrations (Krings and Chen 2018). SNAI1, BCL2L11 protein, and Akt1 pathway activity is increased compared to nonmesenchymal metaplastic carcinomas (McQuerry et al. 2019). Metaplastic carcinomas with heterologous mesenchymal differentiation are aggressive and treated akin to high-grade TNBCs.

Salivary Gland Type Tumors

Homologous tumors occur in the breast and salivary gland, showing similar histology and molecular alterations but distinctly different behavior and prognoses. In the malignant category are adenoid cystic carcinoma, secretory carcinoma, acinic cell carcinoma, mucoepidermoid carcinoma, and polymorphous adenocarcinoma. Exceedingly rare as a group, only adenoid cystic and secretory carcinomas are seen with any real frequency.

Adenoid Cystic Carcinoma

Breast adenoid cystic carcinoma (AdCC) is a rare histological subtype accounting for 0.1%–1% of breast cancers (Marchiò et al. 2010). AdCC (Fig. 4L) is comprised of a dual population of cells, both epithelial (luminal) and myoepithelial (basal), arranged in distinctive architectural patterns (cribriform, tubular, trabecular, and solid). AdCC is characterized most commonly by *MYB-NFIB* fusions (Persson et al. 2009) or less frequently by other *MYB* or *MYBL1* rearrangements (including *MYBL1-ACTN1*, *MYBL1-NFIB*, or *MYB* amplification) (Kim et al. 2018). These breast cancers are usually triple-negative and basal-like by gene expression. In contrast to most triple-negative basal-like cancers, classical AdCC is low-grade, has a low Ki67 proliferation rate, and a good prognosis. There is also a solid basaloid variant of AdCC, however, which is high-grade, shows a high Ki67 proliferation rate, and displays more aggressive clinical behavior (Slodkowska et al. 2020).

Secretory Carcinoma

Secretory carcinoma of the breast accounts for <1% of breast cancers. Originally described in children and named "juvenile carcinoma," secretory carcinoma can occur at any age (Horowitz et al. 2012). The histological appearance is very distinctive with vacuolated tumor cells arranged in microcystic, solid, and tubular architectures containing intracellular and extracellular secretions (Fig. 4M). The *ETV6-NTRK3* fusion is pathognomonic. Most tumors are triple-negative and basal-like but can also show low-level ER expression. Secretory carcinomas lack *TP53* mutations or the complex genomic alterations (high mutational burden, frequent copy number aberrations) seen in conventional triple-negative

basal-like breast cancers (Castillo et al. 2015; Krings et al. 2017). Further in contrast to most TNBCs, the majority of secretory carcinomas are low-grade and have an excellent prognosis, even in the setting of nodal involvement or metastases (Horowitz et al. 2012). High-grade variants have been reported, which do show more aggressive behavior (Castillo et al. 2015). Identifying the histology as secretory carcinoma is of particular importance as patients with recurrent or metastatic tumors are eligible for treatment with TRK inhibitors (Drilon et al. 2018, 2017; Doebele et al. 2020).

CONCLUDING REMARKS

In the age of molecular oncology, it remains critical for pathologists to be able to recognize the many and varying possible histologies of breast cancers to identify invasive carcinomas and distinguish them from benign and in situ disease (the latter being described in a separate monograph; Behbod et al. 2023). But having done so, in part because so many breast cancers have a generic histology (invasive ductal carcinoma of no special type), subsequent treatment decisions often rely less on histology than they do on other tumor-related factors (the most important of which are biomarker status and stage). In IDC and ILC, morphologic variants (Figs. 2 and 3) have limited clinical impact (outside rare IDC-medullary and ILC pleomorphic types). Relevant additional levels of information can then be layered on, with several validated RNA expression and DNA mutation tests available that predict behavior and response to therapy. However, multiple distinct histological special types can be quickly and efficiently recognized by trained pathologists employing standard light microscopy (Fig. 4), which can often be linked to a specific set of molecular drivers, clinical behaviors, and consequent treatment decisions. Thus, even in this molecular era characterized by continuing advances in the "omics" of tumors, classical histological features and morphological subtypes remain important in diagnosing breast cancer, guiding subsequent molecular workup, and efficiently optimizing treatment.

ACKNOWLEDGMENTS

We thank Samuel Leung and Jamie Yu for assistance in preparing figures and references. T.O.N. is supported by grants from the Canadian Cancer Society (grant 706768) and the Cancer Research Society (grant 944513).

REFERENCES

Ács B, Zámbó V, Vízkeleti L, Szász AM, Madaras L, Szentmártoni G, Tőkés T, Molnár BÁ, Molnár IA, Vári Kakas S, et al. 2017. Ki-67 as a controversial predictive and prognostic marker in breast cancer patients treated with neoadjuvant chemotherapy. *Diagn Pathol* 12: 20. doi:10.1186/s13000-017-0608-5

Adem C, Reynolds C, Adlakha H, Roche PC, Nascimento AG. 2002. Wide spectrum screening keratin as a marker of metaplastic spindle cell carcinoma of the breast: an immunohistochemical study of 24 patients. *Histopathol* 40: 556–562. doi:10.1046/j.1365-2559.2002.01417.x

Allison KH, Hammond MEH, Dowsett M, McKernin SE, Carey LA, Fitzgibbons PL, Hayes DF, Lakhani SR, Chavez-MacGregor M, Perlmutter J, et al. 2020. Estrogen and progesterone receptor testing in breast cancer: ASCO/CAP guideline update. *J Clin Oncol* 38: 1346–1366. doi:10.1200/JCO.19.02309

Bae SY, Choi MY, Cho DH, Lee JE, Nam SJ, Yang JH. 2011. Mucinous carcinoma of the breast in comparison with invasive ductal carcinoma: clinicopathologic characteristics and prognosis. *J Breast Cancer* 14: 308–313. doi:10.4048/jbc.2011.14.4.308

Bataillon G, Fuhrmann L, Girard E, Menet E, Laé M, Capovilla M, Treilleux I, Arnould L, Penault-Llorca F, Rouzier R, et al. 2018. High rate of *PIK3CA* mutations but no *TP53* mutations in low-grade adenosquamous carcinoma of the breast. *Histopathol* 73: 273–283. doi:10.1111/his.13514

Behbod F, Chen JH, Thompson A. 2023. Human ductal carcinoma in situ: advances and future perspectives. *Cold Spring Harb Perspect Med*: a041319. doi:10.1101/cshperspect.a041319

Burstein MD, Tsimelzon A, Poage GM, Covington KR, Contreras A, Fuqua SAW, Savage MI, Osborne CK, Hilsenbeck SG, Chang JC, et al. 2015. Comprehensive genomic analysis identifies novel subtypes and targets of triple-negative breast cancer. *Clin Cancer Res* 21: 1688–1698. doi:10.1158/1078-0432.CCR-14-0432

Burstein HJ, Curigliano G, Thürlimann B, Weber WP, Poortmans P, Regan MM, Senn HJ, Winer EP, Gnant M; Panelists of the St Gallen Consensus Conference. 2021. Customizing local and systemic therapies for women with early breast cancer: the St. Gallen International Consensus Guidelines for treatment of early breast cancer. *Ann Oncol* 32: 1216–1235. doi:10.1016/j.annonc.2021.06.023

Carter MR, Hornick JL, Lester S, Fletcher CDM. 2006. Spindle cell (sarcomatoid) carcinoma of the breast. *Am J Surg Pathol* 30: 300–309. doi:10.1097/01.pas.0000184809.27735.a1

Castillo MD, Chibon F, Arnould L, Croce S, Ribeiro A, Perot G, Hostein I, Geha S, Bozon C, Garnier A, et al. 2015. Secretory breast carcinoma. *Am J Surg Pathol* 39: 1458–1467. doi:10.1097/PAS.0000000000000487

Cheang MCU, Voduc D, Bajdik C, Leung S, McKinney S, Chia SK, Perou CM, Nielsen TO. 2008. Basal-Like breast cancer defined by five biomarkers has superior prognostic value than triple-negative phenotype. *Clin Cancer Res* 14: 1368–1376. doi:10.1158/1078-0432.CCR-07-1658

Cheang MC, Chia SK, Voduc D, Gao D, Leung S, Snider J, Watson M, Davies S, Bernard PS, Parker JS, et al. 2009. Ki67 index, HER2 status, and prognosis of patients with luminal B breast cancer. *J Natl Cancer Inst* 101: 736–750. doi:10.1093/jnci/djp082

Chlebowski RT, Hendrix SL, Langer RD, Stefanick ML, Gass M, Lane D, Rodabough RJ, Gilligan MA, Cyr MG, Thomson CA, et al. 2003. Influence of estrogen plus progestin on breast cancer and mammography in healthy postmenopausal women: the women's health initiative randomized trial. *J Am Med Assoc* 289: 3243–3253. doi:10.1001/jama.289.24.3243

Christgen M, Cserni G, Floris G, Marchio C, Djerroudi L, Kreipe H, Derksen PWB, Vincent-Salomon A. 2021. Lobular breast cancer: histomorphology and different concepts of a special spectrum of tumors. *Cancers (Basel)* 13: 3695. doi:10.3390/cancers13153695

Cimino-Mathews A, Subhawong AP, Elwood H, Warzecha HN, Sharma R, Park BH, Taube JM, Illei PB, Argani P. 2013. Neural crest transcription factor Sox10 is preferentially expressed in triple negative and metaplastic breast carcinomas. *Hum Pathol* 44: 959–965. doi:10.1016/j.humpath.2012.09.005

Ciriello G, Gatza ML, Beck AH, Wilkerson MD, Rhie SK, Pastore A, Zhang H, McLellan M, Yau C, Kandoth C, et al. 2015. Comprehensive molecular portraits of invasive lobular breast cancer. *Cell* 163: 506–519. doi:10.1016/j.cell.2015.09.033

Consortium TAPG, André F, Arnedos M, Baras AS, Baselga J, Bedard PL, Berger MF, Bierkens M, Calvo F, Cerami E, et al. 2017. AACR project GENIE: powering precision medicine through an international consortium. *Cancer Discov* 7: 818–831. doi:10.1158/2159-8290.CD-17-0151

Cortazar P, Geyer CE. 2015. Pathological complete response in neoadjuvant treatment of breast cancer. *Ann Surg Oncol* 22: 1441–1446. doi:10.1245/s10434-015-4404-8

Cortazar P, Zhang L, Untch M, Mehta K, Costantino JP, Wolmark N, Bonnefoi H, Cameron D, Gianni L, Valagussa P, et al. 2014. Pathological complete response and long-term clinical benefit in breast cancer: the CTNeoBC pooled analysis. *Lancet* 384: 164–172. doi:10.1016/S0140-6736(13)62422-8

Cserni G, Quinn CM, Foschini MP, Bianchi S, Callagy G, Chmielik E, Decker T, Fend F, Kovács A, van Diest PJ, et al. 2021. Triple-negative breast cancer histological subtypes with a favourable prognosis. *Cancers (Basel)* 13: 5694. doi:10.3390/cancers13225694

de Azambuja E, Cardoso F, de Castro G, Colozza M, Mano MS, Durbecq V, Sotiriou C, Larsimont D, Piccart-Gebhart MJ, Paesmans M. 2007. Ki-67 as prognostic marker in early breast cancer: a meta-analysis of published studies involving 12,155 patients. *Br J Cancer* 96: 1504–1513. doi:10.1038/sj.bjc.6603756

Denkert C, Loibl S, Müller BM, Eidtmann H, Schmitt WD, Eiermann W, Gerber B, Tesch H, Hilfrich J, Huober J, et al. 2013. Ki67 levels as predictive and prognostic parameters in pretherapeutic breast cancer core biopsies: a translational investigation in the neoadjuvant GeparTrio trial. *Ann Oncol* **24:** 2786–2793. doi:10.1093/annonc/mdt350

Desmedt C, Zoppoli G, Gundem G, Pruneri G, Larsimont D, Fornili M, Fumagalli D, Brown D, Rothé F, Vincent D, et al. 2016. Genomic characterization of primary invasive lobular breast cancer. *J Clin Oncol* **34:** 1872–1881. doi:10.1200/JCO.2015.64.0334

Doebele RC, Drilon A, Paz-Ares L, Siena S, Shaw AT, Farago AF, Blakely CM, Seto T, Cho BC, Tosi D, et al. 2020. Entrectinib in patients with advanced or metastatic NTRK fusion-positive solid tumours: integrated analysis of three phase 1–2 trials. *Lancet Oncol* **21:** 271–282. doi:10.1016/S1470-2045(19)30691-6

Drilon A, Siena S, Ou SHI, Patel M, Ahn MJ, Lee J, Bauer TM, Farago AF, Wheler JJ, Liu SV, et al. 2017. Safety and antitumor activity of the multitargeted pan-TRK, ROS1, and ALK inhibitor entrectinib: combined results from two phase I trials (ALKA-372-001 and STARTRK-1). *Cancer Discov* **7:** 400–409. doi:10.1158/2159-8290.CD-16-1237

Drilon A, Laetsch TW, Kummar S, DuBois SG, Lassen UN, Demetri GD, Nathenson M, Doebele RC, Farago AF, Pappo AS, et al. 2018. Efficacy of larotrectinib in TRK fusion–positive cancers in adults and children. *N Engl J Med* **378:** 731–739. doi:10.1056/NEJMoa1714448

Dunne B, Lee AHS, Pinder SE, Bell JA, Ellis IO. 2003. An immunohistochemical study of metaplastic spindle cell carcinoma, phyllodes tumour and fibromatosis of the breast. *Hum Pathol* **34:** 1009–1015. doi:10.1053/S0046-8177(03)00414-3

Early Breast Cancer Trialists' Collaborative Group. 1998. Tamoxifen for early breast cancer: an overview of the randomised trials. *Lancet* **351:** 1451–1467. doi:10.1016/S0140-6736(97)11423-4

Early Breast Cancer Trialists' Collaborative Group. 2005. Effects of chemotherapy and hormonal therapy for early breast cancer on recurrence and 15-year survival: an overview of the randomised trials. *Lancet* **365:** 1687–1717. doi:10.1016/S0140-6736(05)66544-0

Early Breast Cancer Trialists' Collaborative Group. 2015. Aromatase inhibitors versus tamoxifen in early breast cancer: patient-level meta-analysis of the randomised trials. *Lancet* **386:** 1341–1352. doi:10.1016/S0140-6736(15)61074-1

Early Breast Cancer Trialists' Collaborative Group; Davies C, Godwin J, Gray R, Clarke M, Cutter D, Darby S, McGale P, Pan HC, Taylor C, et al. 2011. Relevance of breast cancer hormone receptors and other factors to the efficacy of adjuvant tamoxifen: patient-level meta-analysis of randomized trials. *Lancet* **378:** 771–784. doi:10.1016/S0140-6736(11)60993-8

Emens LA, Loi S. 2023. Immunotherapy approaches for breast cancer patients in 2023. *Cold Spring Harb Perspect Med* **13:** a041332. doi:10.1101/cshperspect.a041332

Ferlicot S, Vincent-Salomon A, Médioni J, Genin P, Rosty C, Sigal-Zafrani B, Fréneaux P, Jouve M, Thiery JP, Sastre-Garau X. 2004. Wide metastatic spreading in infiltrating lobular carcinoma of the breast. *Eur J Cancer* **40:** 336–341. doi:10.1016/j.ejca.2003.08.007

Gobbi H, Simpson JF, Borowsky A, Jensen RA, Page DL. 1999. Metaplastic breast tumors with a dominant fibromatosis-like phenotype have a high risk of local recurrence. *Cancer* **85:** 2170–2182. doi:10.1002/(SICI)1097-0142(19990515)85:10<2170::AID-CNCR11>3.0.CO;2-X

Gobbi H, Simpson JF, Jensen RA, Olson SJ, Page DL. 2003. Metaplastic spindle cell breast tumors arising within papillomas, complex sclerosing lesions, and nipple adenomas. *Mod Pathol* **16:** 893–901. doi:10.1097/01.MP.0000085027.75201.B5

Goncalves R, Ma C, Luo J, Suman V, Ellis MJ. 2012. Use of neoadjuvant data to design adjuvant endocrine therapy trials for breast cancer. *Nat Rev Clin Oncol* **9:** 223–229. doi:10.1038/nrclinonc.2012.21

Hayes MJ, Thomas D, Emmons A, Giordano TJ, Kleer CG. 2008. Genetic changes of Wnt pathway genes are common events in metaplastic carcinomas of the breast. *Clin Cancer Res* **14:** 4038–4044. doi:10.1158/1078-0432.CCR-07-4379

Heldring N, Pike A, Andersson S, Matthews J, Cheng G, Hartman J, Tujague M, Ström A, Treuter E, Warner M, et al. 2007. Estrogen receptors: how do they signal and what are their targets. *Physiol Rev* **87:** 905–931. doi:10.1152/physrev.00026.2006

Hercules SM, Alnajar M, Chen C, Mladjenovic SM, Shipeolu BA, Perkovic O, Pond GR, Mbuagbaw L, Blenman KR, Daniel JM. 2022. Triple-negative breast cancer prevalence in Africa: a systematic review and meta-analysis. *BMJ Open* **12:** e055735. doi:10.1136/bmjopen-2021-055735

Hoeven KHV, Drudis T, Cranor ML, Erlandson RA, Rosen PP. 1993. Low-grade adenosquamous carcinoma of the breast. *Am J Surg Pathol* **17:** 248–258. doi:10.1097/00000478-199303000-00005

Holm J, Yu NYL, Johansson A, Ploner A, Hall P, Lindström LS, Czene K. 2020. Concordance of immunohistochemistry based and gene expression-based subtyping in breast cancer. *JNCI Cancer Spectr* **5:** pkaa087. doi:10.1093/jncics/pkaa087

Horowitz DP, Sharma CS, Connolly E, Gidea-Addeo D, Deutsch I. 2012. Secretory carcinoma of the breast: results from the survival, epidemiology and end results database. *Breast* **21:** 350–353. doi:10.1016/j.breast.2012.02.013

International Agency for Research on Cancer. 2019. *Breast tumours, WHO classification of tumours*, 5th ed. World Health Organization, Lyon, France.

Inwald EC, Klinkhammer-Schalke M, Hofstädter F, Zeman F, Koller M, Gerstenhauer M, Ortmann O. 2013. Ki-67 is a prognostic parameter in breast cancer patients: results of a large population based cohort of a cancer registry. *Breast Cancer Res Treat* **139:** 539–552. doi:10.1007/s10549-013-2560-8

Karn T, Denkert C, Weber KE, Holtrich U, Hanusch C, Sinn BV, Higgs BW, Jank P, Sinn HP, Huober J, et al. 2020. Tumour mutational burden and immune infiltration as independent predictors of response to neoadjuvant immune checkpoint inhibition in early TNBC in GeparNuevo. *Ann Oncol* **31:** 1216–1222. doi:10.1016/j.annonc.2020.05.015

Kennecke H, Yerushalmi R, Woods R, Cheang MCU, Voduc D, Speers CH, Nielsen TO, Gelmon K. 2010. Metastatic behavior of breast cancer subtypes. *J Clin Oncol* **28**: 3271–3277. doi:10.1200/JCO.2009.25.9820

Khan HN, Wyld L, Dunne B, Lee AHS, Pinder SE, Evans AJ, Robertson JFR. 2003. Spindle cell carcinoma of the breast: a case series of a rare histological subtype. *Eur J Surg Oncol (EJSO)* **29**: 600–603. doi:10.1016/S0748-7983(03)00107-0

Kim J, Geyer FC, Martelotto LG, Ng CK, Lim RS, Selenica P, Li A, Pareja F, Fusco N, Edelweiss M, et al. 2018. *MYBL1* rearrangements and *MYB* amplification in breast adenoid cystic carcinomas lacking the *MYB–NFIB* fusion gene. *J Pathol* **244**: 143–150. doi:10.1002/path.5006

Knutson TP, Daniel AR, Fan D, Silverstein KA, Covington KR, Fuqua SA, Lange CA. 2012. Phosphorylated and sumoylation-deficient progesterone receptors drive proliferative gene signatures during breast cancer progression. *Breast Cancer Res* **14**: R95. doi:10.1186/bcr3211

Kong X, Liu Z, Cheng R, Sun L, Huang S, Fang Y, Wang J. 2020. Variation in breast cancer subtype incidence and distribution by race/ethnicity in the United States from 2010 to 2015. *JAMA Netw Open* **3**: e2020303. doi:10.1001/jamanetworkopen.2020.20303

Krings G, Chen YY. 2018. Genomic profiling of metaplastic breast carcinomas reveals genetic heterogeneity and relationship to ductal carcinoma. *Mod Pathol* **31**: 1661–1674. doi:10.1038/s41379-018-0081-z

Krings G, Joseph NM, Bean GR, Solomon D, Onodera C, Talevich E, Yeh I, Grenert JP, Hosfield E, Crawford ED, et al. 2017. Genomic profiling of breast secretory carcinomas reveals distinct genetics from other breast cancers and similarity to mammary analog secretory carcinomas. *Mod Pathol* **30**: 1086–1099. doi:10.1038/modpathol.2017.32

LaCroix AZ, Chlebowski RT, Manson JE, Aragaki AK, Johnson KC, Martin L, Margolis KL, Stefanick ML, Brzyski R, Curb JD, et al. 2011. Health outcomes after stopping conjugated equine estrogens among postmenopausal women with prior hysterectomy: A randomized controlled trial. *J Am Med Assoc* **305**: 1305–1314. doi:10.1001/jama.2011.382

Laenkholm AV, Jensen MB, Eriksen JO, Buckingham W, Ferree S, Nielsen TO, Ejlertsen B. 2018. The ability of PAM50 risk of recurrence score to predict 10-year distant recurrence in hormone receptor-positive postmenopausal women with special histological subtypes. *Acta Oncol* **57**: 44–50. doi:10.1080/0284186X.2017.1403044

Lænkholm AV, Jensen MB, Eriksen JO, Roslind A, Buckingham W, Ferree S, Nielsen T, Ejlertsen B. 2020. Population-based study of prosigna-PAM50 and outcome among postmenopausal women with estrogen receptor-positive and HER2-negative operable invasive lobular or ductal breast cancer. *Clin Breast Cancer* **20**: e423–e432. doi:10.1016/j.clbc.2020.01.013

Lange CA. 2008. Challenges to defining a role for progesterone in breast cancer. *Steroids* **73**: 914–921. doi:10.1016/j.steroids.2007.12.023

Lehmann BD, Bauer JA, Chen X, Sanders ME, Chakravarthy AB, Shyr Y, Pietenpol JA. 2011. Identification of human triple-negative breast cancer subtypes and preclinical models for selection of targeted therapies. *J Clin Investig* **121**: 2750–2767. doi:10.1172/JCI45014

Lehmann BD, Jovanović B, Chen X, Estrada MV, Johnson KN, Shyr Y, Moses HL, Sanders ME, Pietenpol JA. 2016. Refinement of triple-negative breast cancer molecular subtypes: implications for neoadjuvant chemotherapy selection. *PLoS ONE* **11**: e0157368. doi:10.1371/journal.pone.0157368

Leibl S, Gogg-Kammerer M, Sommersacher A, Denk H, Moinfar F. 2005. Metaplastic breast carcinomas: Are they of myoepithelial differentiation?: immunohistochemical profile of the sarcomatoid subtype using novel myoepithelial markers. *Am J Surg Pathol* **29**: 347–353. doi:10.1097/01.pas.0000152133.60278.d2

Li Y, Kaneko M, Sakamoto DG, Takeshima Y, Inai K. 2006. The reversed apical pattern of MUC1 expression is characteristics of invasive micropapillary carcinoma of the breast. *Breast Cancer (Auckl)* **13**: 58–63. doi:10.2325/jbcs.13.58

Lian W, Zheng J, Chen D. 2020. Different prognosis by subtype in the early mucinous breast cancer: a SEER population-based analysis. *Transl Cancer Res* **9**: 5969–5978. doi:10.21037/tcr-20-1237

Liu J, Zheng X, Han Z, Lin S, Han H, Xu C. 2021. Clinical characteristics and overall survival prognostic nomogram for invasive cribriform carcinoma of breast: a SEER population-based analysis. *BMC Cancer* **21**: 168. doi:10.1186/s12885-021-07895-5

Loi S, Salgado R, Adams S, Pruneri G, Francis PA, Lacroix-Triki M, Joensuu H, Dieci MV, Badve S, Demaria S, et al. 2022. Tumour infiltrating lymphocyte stratification of prognostic staging of early-stage triple negative breast cancer. *NPJ Breast Cancer* **8**: 3. doi:10.1038/s41523-021-00362-1

Luporsi E, André F, Spyratos F, Martin P-M, Jacquemier J, Penault-Llorca F, Tubiana-Mathieu N, Sigal Zafrani B, Arnould L, Gompel A, et al. 2012. Ki-67: level of evidence and methodological considerations for its role in the clinical management of breast cancer: analytical and critical review. *Breast Cancer Res Treat* **132**: 895–915. doi:10.1007/s10549-011-1837-z

Marchiò C, Weigelt B, Reis-Filho JS. 2010. Adenoid cystic carcinomas of the breast and salivary glands (or "The strange case of Dr Jekyll and Mr Hyde" of exocrine gland carcinomas). *J Clin Pathol* **63**: 220–228. doi:10.1136/jcp.2009.073908

McQuerry JA, Jenkins DF, Yost SE, Zhang Y, Schmolze D, Johnson WE, Yuan Y, Bild AH. 2019. Pathway activity profiling of growth factor receptor network and stemness pathways differentiates metaplastic breast cancer histological subtypes. *BMC Cancer* **19**: 881. doi:10.1186/s12885-019-6052-z

Modi S, Jacot W, Yamashita T, Sohn J, Vidal M, Tokunaga E, Tsurutani J, Ueno NT, Prat A, Chae YS, et al. 2022. Trastuzumab deruxtecan in previously treated HER2-low advanced breast cancer. *N Engl J Med* **387**: 9–20. doi:10.1056/NEJMoa2203690

Nassar H, Wallis T, Andea A, Dey J, Adsay V, Visscher D. 2001. Clinicopathologic analysis of invasive micropapillary differentiation in breast carcinoma. *Mod Pathol* **14**: 836–841. doi:10.1038/modpathol.3880399

Nguyen B, Sanchez-Vega F, Fong CJ, Chatila WK, Boroujeni AM, Pareja F, Weigelt B, Sotiriou C, Larsimont D, Reis-Filho JS, et al. 2021. The genomic landscape of carcinomas with mucinous differentiation. *Sci Rep* **11**: 9478. doi:10.1038/s41598-021-89099-2

Nielsen TO, Leung SCY, Rimm DL, Dodson A, Acs B, Badve S, Denkert C, Ellis MJ, Fineberg S, Flowers M, et al. 2020. Assessment of Ki67 in breast cancer: updated recommendations from the international Ki67 in breast cancer working group. *JNCI J Natl Cancer Inst* **113**: 808–819. doi:10.1093/jnci/djaa201

Parker JS, Mullins M, Cheang MCU, Leung S, Voduc D, Vickery T, Davies S, Fauron C, He X, Hu Z, et al. 2009. Supervised risk predictor of breast cancer based on intrinsic subtypes. *J Clin Oncol* **27**: 1160–1167. doi:10.1200/JCO.2008.18.1370

Perou CM, Sørlie T, Eisen MB, van de Rijn M, Jeffrey SS, Rees CA, Pollack JR, Ross DT, Johnsen H, Akslen LA, et al. 2000. Molecular portraits of human breast tumours. *Nature* **406**: 747–752. doi:10.1038/35021093

Perron M, Wen HY, Hanna MG, Brogi E, Ross DS. 2020. HER2 immunohistochemistry in invasive micropapillary breast carcinoma: complete assessment of an incomplete pattern. *Arch Pathol Lab Med* **145**: 979–987. doi:10.5858/arpa.2020-0288-OA

Persson M, Andrén Y, Mark J, Horlings HM, Persson F, Stenman G. 2009. Recurrent fusion of *MYB* and *NFIB* transcription factor genes in carcinomas of the breast and head and neck. *Proc Natl Acad Sci* **106**: 18740–18744. doi:10.1073/pnas.0909114106

Pestalozzi BC, Zahrieh D, Mallon E, Gusterson BA, Price KN, Gelber RD, Holmberg SB, Lindtner J, Snyder R, Thürlimann B, et al. 2008. Distinct clinical and prognostic features of infiltrating lobular carcinoma of the breast: combined results of 15 international breast cancer study group clinical trials. *J Clin Oncol* **26**: 3006–3014. doi:10.1200/JCO.2007.14.9336

Petrelli F, Viale G, Cabiddu M, Barni S. 2015. Prognostic value of different cut-off levels of Ki-67 in breast cancer: a systematic review and meta-analysis of 64,196 patients. *Breast Cancer Res Treat* **153**: 477–491. doi:10.1007/s10549-015-3559-0

Pichon MF, Pallud C, Brunet M, Milgrom E. 1980. Relationship of presence of progesterone receptors to prognosis in early breast cancer. *Cancer Res* **40**: 3357–3360.

Popli P, Gutterman EM, Omene C, Ganesan S, Mills D, Marlink R. 2021. Receptor-defined breast cancer in five east African countries and its implications for treatment: systematic review and meta-analysis. *JCO Glob Oncol* **7**: 289–301. doi:10.1200/GO.20.00398

Prat A, Perou CM. 2011. Deconstructing the molecular portraits of breast cancer. *Mol Oncol* **5**: 5–23. doi:10.1016/j.molonc.2010.11.003

Prat A, Parker JS, Karginova O, Fan C, Livasy C, Herschkowitz JI, He X, Perou CM. 2010. Phenotypic and molecular characterization of the claudin-low intrinsic subtype of breast cancer. *Breast Cancer Res* **12**: R68. doi:10.1186/bcr2635

Prat A, Cheang MCU, Martín M, Parker JS, Carrasco E, Caballero R, Tyldesley S, Gelmon K, Bernard PS, Nielsen TO, et al. 2012. Prognostic significance of progesterone receptor–positive tumour cells within immunohistochemically defined luminal A breast cancer. *J Clin Oncol* **31**: 203–209. doi:10.1200/JCO.2012.43.4134

Prat A, Fan C, Fernández A, Hoadley KA, Martinello R, Vidal M, Viladot M, Pineda E, Arance A, Muñoz M, et al. 2015. Response and survival of breast cancer intrinsic subtypes following multi-agent neoadjuvant chemotherapy. *BMC Med* **13**: 303. doi:10.1186/s12916-015-0540-z

Provenzano E, Bossuyt V, Viale G, Cameron D, Badve S, Denkert C, MacGrogan G, Penault-Llorca F, Boughey J, Curigliano G, et al. 2015. Standardization of pathologic evaluation and reporting of postneoadjuvant specimens in clinical trials of breast cancer: recommendations from an international working group. *Mod Pathol* **28**: 1185–1201. doi:10.1038/modpathol.2015.74

Rakha EA, Lee AHS, Evans AJ, Menon S, Assad NY, Hodi Z, Macmillan D, Blamey RW, Ellis IO. 2009. Tubular carcinoma of the breast: further evidence to support its excellent prognosis. *J Clin Oncol* **28**: 99–104. doi:10.1200/JCO.2009.23.5051

Rakha EA, Quinn CM, Foschini MP, Martin MM, Dabbs DJ, Lakhani S, Varga Z, Pinder SE, Schmitt FC, Reis Filho JS, et al. 2021. Metaplastic carcinomas of the breast without evidence of epithelial differentiation: a diagnostic approach for management. *Histopathology* **78**: 759–771. doi:10.1111/his.14290

Rito M, Schmitt F, Pinto AE, André S. 2014. Fibromatosis-like metaplastic carcinoma of the breast has a laudin-low immunohistochemical phenotype. *Virchows Arch* **465**: 185–191. doi:10.1007/s00428-014-1603-9

Schmid P, Cortes J, Pusztai L, McArthur H, Kümmel S, Bergh J, Denkert C, Park YH, Hui R, Harbeck N, et al. 2020. Pembrolizumab for early triple-negative breast cancer. *N Engl J Med* **382**: 810–821. doi:10.1056/NEJMoa1910549

Shimelis H, LaDuca H, Hu C, Hart SN, Na J, Thomas A, Akinhanmi M, Moore RM, Brauch H, Cox A, et al. 2018. Triple-negative breast cancer risk genes identified by multigene hereditary cancer panel testing. *JNCI J Natl Cancer Inst* **110**: 855–862. doi:10.1093/jnci/djy106

Slodkowska E, Xu B, Kos Z, Bane A, Barnard M, Zubovits J, Iyengar P, Faragalla H, Turbin D, Williams P, et al. 2020. Predictors of outcome in mammary adenoid cystic carcinoma. *Am J Surg Pathol* **44**: 214–223. doi:10.1097/PAS.0000000000001378

Smith I, Robertson J, Kilburn L, Wilcox M, Evans A, Holcombe C, Horgan K, Kirwan C, Mallon E, Sibbering M, et al. 2020. Long-term outcome and prognostic value of Ki67 after perioperative endocrine therapy in postmenopausal women with hormone-sensitive early breast cancer (POETIC): an open-label, multicentre, parallel-group, randomised, phase 3 trial. *Lancet Oncol* **21**: 1443–1454. doi:10.1016/S1470-2045(20)30458-7

Sneige N, Yaziji H, Mandavilli SR, Perez ER, Ordonez NG, Gown AM, Ayala A. 2001. Low-grade (fibromatosis-like) spindle cell carcinoma of the breast. *Am J Surg Pathol* **25**: 1009–1016. doi:10.1097/00000478-200108000-00004

Sørlie T, Perou CM, Tibshirani R, Aas T, Geisler S, Johnsen H, Hastie T, Eisen MB, van de Rijn M, Jeffrey SS, et al. 2001. Gene expression patterns of breast carcinomas distinguish tumour subclasses with clinical implications. *Proc Natl Acad Sci* **98**: 10869–10874. doi:10.1073/pnas.191367098

Swain SM, Shastry M, Hamilton E. 2023. Targeting HER2-positive breast cancer: advances and future directions.

Nat Rev Drug Discov **22**: 101–126. doi:10.1038/s41573-022-00579-0

Thennavan A, Beca F, Xia Y, Garcia-Recio S, Allison K, Collins LC, Tse GM, Chen YY, Schnitt SJ, Hoadley KA, et al. 2021. Molecular analysis of TCGA breast cancer histologic types. *Cell Genom* **1**: 100067. doi:10.1016/j.xgen.2021.100067

Tian S, Roepman P, van't Veer LJ, Bernards R, Snoo FD, Glas AM. 2010. Biological functions of the genes in the mammaprint breast cancer profile reflect the hallmarks of cancer. *Biomark Insights* **5**: BMI.S6184. doi:10.4137/BMI.S6184

Tutt ANJ, Garber JE, Kaufman B, Viale G, Fumagalli D, Rastogi P, Gelber RD, de Azambuja E, Fielding A, Balmaña J, et al. 2021. Adjuvant olaparib for patients with *BRCA1*- or *BRCA2*-mutated breast cancer. *N Engl J Med* **384**: 2394–2405. doi:10.1056/NEJMoa2105215

Venetis K, Crimini E, Sajjadi E, Corti C, Guerini-Rocco E, Viale G, Curigliano G, Criscitiello C, Fusco N. 2022. HER2 low, ultra-low, and novel complementary biomarkers: expanding the spectrum of HER2 positivity in breast cancer. *Front Mol Biosci* **9**: 834651. doi:10.3389/fmolb.2022.834651

Viale G, de Snoo FA, Slaets L, Bogaerts J, van 't Veer L, Rutgers EJ, Piccart-Gebhart MJ, Stork-Sloots L, Glas A, Russo L, et al. 2018. Immunohistochemical versus molecular (BluePrint and MammaPrint) subtyping of breast carcinoma. Outcome results from the EORTC 10041/BIG 3-04 MINDACT trial. *Breast Cancer Res Treat* **167**: 123–131. doi:10.1007/s10549-017-4509-9

von Minckwitz G, Schmitt WD, Loibl S, Müller BM, Blohmer JU, Sinn BV, Eidtmann H, Eiermann W, Gerber B, Tesch H, et al. 2013. Ki67 measured after neoadjuvant chemotherapy for primary breast cancer. *Clin Cancer Res* **19**: 4521–4531. doi:10.1158/1078-0432.CCR-12-3628

Vranic S, Marchiò C, Castellano I, Botta C, Scalzo MS, Bender RP, Payan-Gomez C, di Cantogno LV, Gugliotta P, Tondat F, et al. 2015. Immunohistochemical and molecular profiling of histologically defined apocrine carcinomas of the breast. *Hum Pathol* **46**: 1350–1359. doi:10.1016/j.humpath.2015.05.017

Waldman FM, Hwang ES, Etzell J, Eng C, DeVries S, Bennington J, Thor A. 2001. Genomic alterations in tubular breast carcinomas. *Hum Pathol* **32**: 222–226. doi:10.1053/hupa.2001.21564

Wallden B, Storhoff J, Nielsen T, Dowidar N, Schaper C, Ferree S, Liu S, Leung S, Geiss G, Snider J, et al. 2015. Development and verification of the PAM50-based Pro-

signa breast cancer gene signature assay. *BMC Méd Genom* **8**: 54. doi:10.1186/s12920-015-0129-6

Wang J, Xu B. 2019. Targeted therapeutic options and future perspectives for HER2-positive breast cancer. *Signal Transduct Target Ther* **4**: 34. doi:10.1038/s41392-019-0069-2

Weigelt B, Horlings H, Kreike B, Hayes M, Hauptmann M, Wessels L, de Jong D, de Vijver MV, Veer LV, Peterse J. 2008. Refinement of breast cancer classification by molecular characterization of histological special types. *J Pathol* **216**: 141–150. doi:10.1002/path.2407

Weigelt B, Geyer FC, Horlings HM, Kreike B, Halfwerk H, Reis-Filho JS. 2009a. Mucinous and neuroendocrine breast carcinomas are transcriptionally distinct from invasive ductal carcinomas of no special type. *Mod Pathol* **22**: 1401–1414. doi:10.1038/modpathol.2009.112

Weigelt B, Kreike B, Reis-Filho JS. 2009b. Metaplastic breast carcinomas are basal-like breast cancers: A genomic profiling analysis. *Breast Cancer Res Treat* **117**: 273–280. doi:10.1007/s10549-008-0197-9

Wieduwilt MJ, Moasser MM. 2008. The epidermal growth factor receptor family: biology driving targeted therapeutics. *Cell Mol Life Sci* **65**: 1566–1584. doi:10.1007/s00018-008-7440-8

Wolff AC, Somerfield MR, Dowsett M, Hammond MEH, Hayes DF, McShane LM, Saphner TJ, Spears PA, Allison KH. 2023. Human epidermal growth factor receptor 2 testing in breast cancer: ASCO College of American pathologists guideline update. *J Clin Oncol* **41**: 3867–3872. doi:10.1200/JCO.22.02864

Wong W, Brogi E, Reis-Filho JS, Plitas G, Robson M, Norton L, Morrow M, Wen HY. 2021. Poor response to neoadjuvant chemotherapy in metaplastic breast carcinoma. *NPJ Breast Cancer* **7**: 96. doi:10.1038/s41523-021-00302-z

Wu Y, Zhang N, Yang Q. 2017. The prognosis of invasive micropapillary carcinoma compared with invasive ductal carcinoma in the breast: a meta-analysis. *BMC Cancer* **17**: 839. doi:10.1186/s12885-017-3855-7

Zhang N, Zhang H, Chen T, Yang Q. 2017. Dose invasive apocrine adenocarcinoma has worse prognosis than invasive ductal carcinoma of breast: evidence from SEER database. *Oncotarget* **8**: 24579–24592. doi:10.18632/oncotarget.15597

Zhong S, Zhou S, Li A, Lv H, Li M, Tang S, Xu X, Shui R, Yang W. 2022. High frequency of *PIK3CA* and *TERT* promoter mutations in fibromatosis-like spindle cell carcinomas. *J Clin Pathol* **75**: 477–482. doi:10.1136/jclinpath-2020-207071

Epidemiology and Risk Factors for Breast Cancer: 21st Century Advances, Gaps to Address through Interdisciplinary Science

Mary Beth Terry[1] and Graham A. Colditz[2]

[1]Department of Epidemiology, Mailman School of Public Health, Columbia University, Chronic Disease Unit Leader, Department of Epidemiology, Herbert Irving Comprehensive Cancer Center, Associate Director, New York, New York 10032, USA

[2]Division of Public Health Sciences, Department of Surgery, Washington University School of Medicine and Alvin J. Siteman Cancer Center at Washington University School of Medicine and Barnes-Jewish Hospital in St Louis, St. Louis, Missouri 63110, USA

Correspondence: mt146@cumc.columbia.edu; colditzg@wustl.edu

Research methods to study risk factors and prevention of breast cancer have evolved rapidly. We focus on advances from epidemiologic studies reported over the past two decades addressing scientific discoveries, as well as their clinical and public health translation for breast cancer risk reduction. In addition to reviewing methodology advances such as widespread assessment of mammographic density and Mendelian randomization, we summarize the recent evidence with a focus on the timing of exposure and windows of susceptibility. We summarize the implications of the new evidence for application in risk stratification models and clinical translation to focus prevention-maximizing benefits and minimizing harm. We conclude our review identifying research gaps. These include: pathways for the inverse association of vegetable intake and estrogen receptor (ER)-ve tumors, prepubertal and adolescent diet and risk, early life adiposity reducing lifelong risk, and gaps from changes in habits (e.g., vaping, binge drinking), and environmental exposures.

Breast cancer, the most common cancer in women worldwide, has been increasing annually in most countries over the last few decades, and the increasing trends cannot be fully explained by changes in fertility (Lima et al. 2021). In the United States, breast cancer incidence has been increasing in most age groups except for ages 55–69 yr (Kehm et al. 2019b) (annual percent change [APC] and 95% confidence interval for women ages 25–39 = 0.53 [0.29, 0.78]; for ages 40–54 = 0.32 [0.07, 0.57]; for ages 55–69 = 0.14 [−0.18, 0.46]; and for ages 70–84 = 0.95 [0.48, 1.41]). The increase in women under age 40 yr is notable as this increase cannot be attributed to population based breast cancer screening by mammography, which is not recommended in women under age 40 yr. There remain major health disparities in breast cancer incidence and mortality in the United States with the greatest cancer burden seen in Non-Hispanic Black women (NHB) compared with Non-Hispanic White (NHW) women with higher incidence rates of one of the most aggressive molecular subtypes of breast cancer (triple-negative breast cancer [TNBC]) and higher mortality rates overall seen in NHB and Hispanic women (Li et al. 2003; Shoemaker et al. 2018; Acheampong et al. 2020). Although much rarer, in absolute incidence, breast cancer in men has also been increasing every year since 1975 (APC = 0.67 [0.49–0.85]). This review focuses on advances from epidemiological studies over the past two decades summarizing the evidence to help inform scientific discovery and clinical and public health translation. We specifically focus on modifiable risk factors for breast cancer risk reduction (and population prevention), but do not include chemoprevention as that is covered in the literature by Brown (2022). We discuss methodological advances in epidemiological approaches, summarize the evidence regarding modifiable risk factors, discuss implications of the new epidemiological evidence risk stratification and clinical translation, and, finally, suggest areas for interdisciplinary science to address identified gaps.

METHODOLOGICAL ADVANCES

Advances in the evidence base from epidemiological methods include (1) molecular epidemiologic evidence linking exposures to measurable biomarkers as well as genomic markers as instruments and modifiers to assess causality and mechanisms; (2) evidence from tumor pathology examining risk factors by molecular subtype of breast cancer; and (3) evidence regarding breast tissue density as measured by mammograms and other methods has changed how we think about risk and risk assessment. Further, in the last two decades, advances from life course studies have provided new perspectives on windows of susceptibility (WOS), periods when the breast tissue is more susceptible to carcinogens. As the WOS framework has major implications for understanding breast biology as well as potential windows to optimize risk reduction in public health and clinical interventions, we discuss the epidemiological evidence, where available, for each risk factor construct as it pertains to WOS.

Molecular Epidemiology

First introduced in the 1980s (Perera and Weinstein 1982), the last two decades have seen an acceleration of molecular epidemiological studies particularly as high-throughput assays have made such studies possible. Molecular epidemiological studies include studies that integrate blood, urine, and/or saliva collected in affected and unaffected individuals to measure exposures as well as potential mediators and modifiers. Biospecimens have also been essential in measuring exposures such as nutrients and environmental chemicals. Use of biospecimens to measure potential mediators between exposures and outcomes, including epigenetic changes, markers of DNA damage like DNA adduct formation, and metabolomic profiles, provide insight on potential mechanistic pathways as well as additional evidence of causality. Advances in genomics over the last two decades has permitted both gene–environmental interaction studies using genetic variants in breast cancer susceptibility genes as well as the use of genetic variants as proxies for exposure using a methodology called Mendelian randomization (MR) (Schatzkin et al. 2009; Swerdlow et al. 2016; Yarmolinsky et al. 2018). The concept behind MR is to select genetic variant(s) as instruments to proxy for exposure constructs. Such instruments are by design uncorrelated with other confounders in an observational study. As most risk factors for cancer cannot be ethically randomized, the MR approach combined with other epidemiological evidence helps establish causality from observational epidemiology. Below we summarize how large-scale molecular epidemiologic studies have provided important evidence to assess causality by each major risk factor category considered.

Tumor Pathology and Molecular Subtypes

The classification of breast cancer based on molecular subtype of the tumors has transformed clinical care (see Swarbrick et al. 2022). The most common molecular subtype is luminal A (estrogen receptor–positive [ER$^+$] and progesterone receptor–positive [PR$^+$], HER2-negative,

Cite this article as *Cold Spring Harb Perspect Med* doi: 10.1101/cshperspect.a041317

and low levels of the protein Ki-67). The rarest subtype is HER2-enriched breast cancer (ER-, PR-, and HER2-positive). The second and third in prevalence are TNBC or basal-like breast cancer (ER-, PR-, HER2-negative) and luminal B (ER$^+$, HER2-negative, and PR$^-$ and/or high levels of Ki-67), but the ranking depends on age and race/ethnicity (Acheampong et al. 2020). Although major advances have delineated different treatment patterns, the epidemiology has been less clear on whether factors map to specific molecular subtypes. For example, while studies of alcohol intake generally support consistency with luminal A tumors, many other risk factors like breast density relate to most molecular subtypes. A key challenge in understanding whether or not there are etiologically distinct breast cancers based on molecular subtypes is that many large epidemiological studies typically may only have data on ER and PR status, with less complete data for other markers that help define the molecular subtypes (e.g., HER2, Ki-67). However, data that do exist support important etiologic differences between cancers characterized by ER and PR status (e.g., alcohol, see below). Further, many epidemiological studies are comprised mainly of individuals diagnosed with postmenopausal breast cancer and may have insufficient numbers of individuals with other molecular subtypes to examine etiological heterogeneity.

Breast Tissue Density

Another key advance has been the ability to do large studies of breast tissue density. Classification of breast tissue density first made by radiologists starting with qualitative and visual assessments of patterns (Wolfe 1976) to semi-automated methods (e.g., Byng et al. 1994) to fully automated methods (e.g., Keller et al. 2012) has allowed epidemiologists to study how classification of breast tissue density affects risk and whether breast tissue density mediates the effects between exposures and breast cancer. The automation of breast tissue measures combined with machine learning methods also supports the utility of mammography for rapid risk assessment in the clinic (e.g., Corbex et al. 2014;

Tehranifar et al. 2021). It is now established that individuals with denser breast tissue (more epithelial and stromal tissue relative to adipose tissue) are at higher risk of breast cancer. The majority of the states in the United States now have passed breast density notification laws so that individuals are informed after a screening mammogram that they may be at higher risk of breast cancer based on their breast tissue density. As reviewed below, risk factors, with the exception of exogenous hormones, have not consistently mapped to mammographic breast density, but research is ongoing as to whether some risk factors are mediated by breast density (Rice et al. 2016; Burton et al. 2017; Ward et al. 2022). Large genome-wide association studies (GWAS) support that some genetic variants associated with breast cancer are also associated with mammographic breast density, but also that there are novel genetic variants for the breast density phenotype itself (e.g., Chen et al. 2022).

Windows of Susceptibility (WOS)

Another major advance has been our understanding of when the breast tissue is more susceptible to carcinogens (Colditz and Frazier 1995; Colditz et al. 2014; Terry et al. 2019b). For example, accumulating data, both through animal studies and human studies, support that there are particular WOS including the prenatal, pubertal, pregnancy, and lactation, and the menopausal windows where there are stronger and more consistent signals from environmental chemical exposures on breast cancer risk (Terry et al. 2019b). The breast changes in form and function during these WOS, increasing the vulnerability of the breast cells to effects from carcinogens. Despite this recognition, most epidemiological studies collect information on exposures outside of WOS. For example, a systematic review of different classes of environmental chemicals and compounds summarizing over 150 studies on breast cancer over 10 yr (2006–2016) found that only 11% of the studies measured exposures during WOS (Rodgers et al. 2018). Evidence that supports stronger and more consistent associations for exposures within WOS than outside of WOS may provide powerful insight to understanding breast biology as well as to focus risk reduction programs to specific WOS, which may be more feasible if interventions are needed for shorter periods of time.

MODIFIABLE RISK FACTORS

Two major trends in breast cancer epidemiological studies occurred over the past two decades. First, many more large cohort studies have been conducted examining breast cancer risk factors to complement evidence from case-control studies. The increase in cohort studies, including pooled analysis of individual participant data, increases the ability to rule out bias in observational studies. This is because consistency in evidence from cohort studies and case-control studies makes it unlikely that selection and information biases that operate differently in these designs can account for the observed effects. Both study designs still may face unmeasured or residual confounding but, as mentioned above, there are other advances that are being used to rule out confounding including MR studies and other types of molecular epidemiological studies that can examine causal mechanisms. In addition, over the last two decades, there has been much more focus on identifying and establishing causality for risk factors that are more amenable to modification as many of the established breast cancer risk factors related to reproductive timing (age at menarche, age at menopause) and reproductive decisions (e.g., pregnancy and parity) are less amenable to modification. We also have not included discussion of breastfeeding, which may be amenable to modification, particularly if work leave policies support breastfeeding, as it was identified as a factor that reduces breast cancer risk over a half century ago. What has become clearer, over the last two decades, however, is that breastfeeding reduces the risk of more aggressive molecular phenotypes including ER$^-$ and TNBC cancer (Millikan et al. 2008; Work et al. 2014; Islami et al. 2015; John et al. 2018) in addition to reducing ER$^+$ breast cancers. Thus, we focus our review on the following categories of risk factors: exogenous hormone use, diet, adiposity, weight gain and physical activity, alcohol and tobacco

use, and environmental chemical use. In summarizing the evidence, we draw on systematic reviews and hallmark studies related to each topic, with a particular focus on what we have learned from the methodological advances including the WOS framework.

Exogenous Hormone Use—Hormone-Replacement Therapy

The largest randomized trial of population-based breast cancer prevention—the Women's Health Initiative (WHI) —was stopped early in 2002 because of the higher coronary heart disease rates in women taking estrogen plus progestin versus placebo (Fletcher and Colditz 2002; Rossouw et al. 2002). The decision to stop the trial after an average follow-up of 5.2 yr (planned duration, 8.5 yr) was made when these results met predetermined levels of harm as defined in the trial protocol. In addition to the increased risk of coronary heart disease, there was a higher risk of breast cancer, stroke, and pulmonary embolism, and a decreased risk of colorectal cancer risk. After the trial results were published, there was a large decrease in hormone-replacement therapy (HRT) prescribed in women in midlife and a subsequent decrease in postmenopausal breast cancer rates at a population level (Chlebowski et al. 2020). This population level decline showed the profound effect that modifying individual risk factors can have on population level incidence. But it also revealed major health disparities as the decline in incidence was primarily seen in white and higher-income women based on hormone therapy prescription patterns (Krieger 2008). Use of hormone therapy has also been observed to increase breast tissue density, and, thus, breast density in its role as both a risk factor for breast cancer and a potential mediator has been determined through the HRT studies (for review, see Azam et al. 2020).

Although studies over the last two decades specific to newer formulations of hormonal contraceptives used during premenopausal years have been less consistent than the associations seen with the higher dose formulations used when oral contraceptives (OCs) were first introduced (e.g., use before 1975; see Work et al. 2014), the evidence has supported modest increases in breast cancer risk for both OCs as well as with use of levonorgestrel intrauterine devices (for review, see Zürcher et al. 2022). A challenge, however, from the epidemiological literature is that most women use multiple types of hormonal contraceptives throughout their lives and therefore it is difficult to assess the risk of one type of hormonal contraceptive versus the other. OCs have opposing effects on breast versus ovarian cancer risk (e.g., Ferris et al. 2014), with very consistent inverse associations seen with duration of OC use and lower risk of ovarian cancer.

It should be noted that menopausal hormone therapy and hormonal contraceptive use have different indications for use and, unlike the constructs we review below, should not be considered as a modifiable risk factor that can be universally recommended for all individuals. It should also be noted that different types of exogenous hormone use may affect cancer risk differently. For example, unopposed menopausal hormone therapy in the WHI and other studies has been shown to reduce colorectal cancer use, and selected types of hormonal contraceptive use have been consistently associated with reduced risk of ovarian cancer. Thus, recommendations regarding exogenous hormone use should be considered with respect to their primary indication as well as an individual's risk of other cancers (e.g., colorectal and ovarian).

Diet

Dietary factors have long been studied as possible modifiers of breast cancer risk. One of the earliest hypotheses, based initially on international ecologic studies, was that level of intake of total dietary fat was positively associated with risk. Further investigations, including the WHI randomized controlled study (Prentice et al. 2007), have ultimately found no link for dietary fat change in midlife. A range of more contemporary research methods has made it possible for nuanced and detailed study of the associations between diet and breast cancer.

Overall, evidence suggests that a prudent, largely plant-based diet may lower the risk of breast cancer, while a typical Western diet, in-cluding larger amounts of red and processed meat, refined grains, and full-fat dairy, may increase risk. A meta-analysis by Xiao et al. (2019) of 14 cohort studies and 18 case-control studies found that a largely plant-based diet was associated with a 14% decreased risk of premenopausal breast cancer, with slightly larger reductions in ER/PR$^+$ disease. There was a 9% reduction in disease risk with the prudent diet among postmenopausal women, with a 32% reduction in ER/PR$^-$ disease.

Among individual components of an overall diet, caloric balance and overnutrition are some of the most important in relation to breast cancer risk. Discussed in detail below, adiposity and weight gain have an established association with risk, contributing to the overall burden of disease.

Other individual dietary components with links to breast cancer include alcohol (discussed below), vegetables, carotenoids, dairy, and calcium. Apart from alcohol, which is a well-established risk factor (see below), these components have relatively consistent associations with breast cancer, but a convincing level of evidence is currently lacking.

There is little evidence of an association between overall breast cancer risk and intake of vegetables. But the American Institute for Cancer Research (AICR) Continuous Update Project (CUP) (WCRF 2018) showed a 21% decreased risk for ER/PR$^-$ disease with each increase in 200 g of non-starchy vegetables per day (WCRF 2018). And the Pooling Project showed a 12% reduction for ER$^-$ disease with each increase in 300 g per day of vegetables (Jung et al. 2013). Carotenoids, which are found in high levels in many types of fruits and vegetables, have also been inversely associated with breast cancer risk. CUP meta-analyses have found reduced breast cancer risk with increasing serum levels of total carotenoids, β-carotene, and lutein (WCRF 2018). Reductions in risk associated with carotenoids may be larger in ER$^-$ disease. There is also some evidence suggesting that early life intake of vegetable protein and soy may be associated with a lower risk of breast cancer and benign breast disease (Su et al. 2010; Sahni and Parsad 2013; Berkey et al. 2020).

Intake of both dairy products and calcium has been investigated as a potential dietary factor that may reduce risk of breast cancer. A CUP meta-analysis found a small risk reduction with each 200 g per day increase in dairy intake for premenopausal, but not postmenopausal, disease. Milk intake alone showed no association, a finding confirmed by MR studies. Overall calcium intake including dairy and other sources of calcium, however, was inversely associated with both premenopausal and postmenopausal breast cancer.

Although some MR studies are now being reported on components of diet such as milk and coffee, micronutrient and metabolomic markers remain to be more thoroughly evaluated. Like adult diet and cancer risk, we need to align measurement of exposure with the biologic timing in the course of cancer development; studies of adult diet and mammographic density also are challenged by timing. The majority of dietary studies to date with the exception of alcohol show no relation with mammographic density.

As noted above, suggestions for childhood and adolescent diet and lifestyle have key roles in childhood development and cancer risk provide strong evidence for WOS. Greater linear growth and attained height is directly related to pre- and postmenopausal breast cancer (WCRF 2018). Factors related to attained height include milk consumption (Berkey et al. 2009; de Beer 2012), high animal fat intake, and low vegetable fiber intake (Berkey et al. 2000). In addition, studies documenting peak height growth velocity is positively related to lifetime risk of breast cancer (Berkey et al. 1999; Ahlgren et al. 2004), drawing further attention to the importance of understanding pathways and opportunities for prevention.

Adiposity, Weight Gain, and Physical Activity

Energy imbalance plays a key role in breast cancer risk, with the direction of effect varying depending on stage of life. Overall, adulthood adiposity is well established as a protective factor for premenopausal breast cancer and a risk factor for postmenopausal disease (van den Brandt et al. 2000; WCRF 2018). CUP meta-analyses have found a 7% decreased risk for each 5 kg/m^2 increase in overall adulthood body mass index (BMI) for premenopausal disease, and a 12% increased risk for postmenopausal disease (WCRF 2018).

Weight gain in adulthood is positively associated with postmenopausal but not premenopausal breast cancer. CUP meta-analyses found a 6% increased risk of postmenopausal disease with each 5 kg gained (van den Brandt et al. 2000; WCRF 2018). When stratified by hormone receptor status, positive associations with weight gain only remained for receptor-positive disease (ER$^+$ PR$^+$), and when stratified by menopausal hormone therapy use, positive associations only remained for never and former users.

Adiposity in young adulthood (18–30 yr), youth, and adolescence has a durable inverse association not only with risk of premenopausal disease but also reaching into postmenopausal breast cancer (Rosner et al. 2017). CUP meta-analyses have found an 18% decreased risk for each 5 kg/m^2 increase in young adult BMI despite menopausal status. This is confirmed by MR studies (Hao et al. 2022), although the MR studies have not been consistent for adult BMI and breast cancer risk, likely due to potential bias in these adult BMI MR studies. Transcriptomics studies suggest pathways for early life adiposity to influence risk though these need confirmation and better understanding of any modifiable pathways (Wang et al. 2021).

Early adiposity is directly related to lower mammographic density is one possible mechanism for this protection, although other pathways suggested by transcriptomic studies include four gene sets (two gene sets down-regulated and two gene sets up-regulated) that were affected by early life body size in estrogen receptor–positive (ER$^+$) tumors and 15 gene sets (14 gene sets down-regulated and one gene set up-regulated) in ER-negative (ER$^-$) tumors. Higher early life size was associated with down-regulation of cellular proliferation pathways (MYC targets) in both ER$^+$ and ER$^-$ tumors (Wang et al. 2021).

While adiposity does have an established protective effect on breast cancer, particularly before menopause, it is important to place this into a broader context, both for breast cancer and other diseases. In North America and across much of Europe, ~80% of breast cancer cases are diagnosed after menopause (Heer et al. 2020). So, despite the dual effect that overweight and obesity have on risk, adiposity has a substantial negative impact on the burden of breast cancer overall. And this health burden grows substantially greater when considering other conditions related to overweight and obesity, including cardiovascular disease, metabolic syndrome, physical disability, and multiple other cancers (Lauby-Secretan et al. 2016).

Physical activity can include a wide range of activities, from specific bouts of exercise to occupational tasks to household chores. Vigorous physical activity, such as running or fast cycling, is inversely associated with premenopausal breast cancer, while both vigorous and total overall physical activity are inversely associated with postmenopausal disease. Comparing highest versus lowest levels of vigorous physical activity, CUP meta-analyses found risk reductions of 17% for premenopausal breast cancer and 10% for postmenopausal disease. For highest versus lowest total physical activity levels, risk reduction was 13% for postmenopausal disease. There is some evidence that level of physical activity in youth and the teenage years is inversely associated with later-adulthood breast cancer risk (Monninkhof et al. 2007; Niehoff et al. 2017). The reduction in breast cancer risk associated with physical activity consistently reported in the general population has also been found in higher risk women, including those with pathogenic variants (PVs) in BRCA1 or BRCA2 or a family history (Kehm et al. 2020). More study, though, is needed to fully describe this potential association and how it may impact prevention messaging and programs (Bucy et al. 2022).

In addition to other potential mechanisms, physical activity can play an important role in energy balance and weight maintenance, which can impact multiple factors associated with breast cancer risk.

Alcohol and Tobacco Smoking

One of the earliest identified modifiable factors for breast cancer was alcohol with sufficient evidence in 1987 to recommend limited intake, based on results of 17 studies, in a *New England Journal of Medicine* (*NEJM*) editorial accompanying two prospective studies (Graham 1987). This recommendation is further supported by meta-analyses (Longnecker 1994) and an International Agency for Research on Cancer (IARC) review of evidence supporting a causal effect of alcohol on breast cancer (IARC Working Group on the Evaluation of Carcinogenic Risks to Humans 2007: Lyon France 2010). Yet there has been slower uptake in translation of this information to the public (Strebel and Terry 2021), and many questioned the extent to which the modest relative associations could be explained by bias (Ellison et al. 2001). The modest size coupled with the known measurement issues of underreporting of alcohol made many epidemiologists hesitant to say there was causal association. As more data emerged from cohort studies (Smith-Warner et al. 1998; Brennan et al. 2010), the consistency in the modest associations from moderate alcohol intake and breast cancer risk enabled selection and information bias to be ruled out. Advances from molecular epidemiologic studies provided further evidence regarding causality by reducing the possibility of confounding (McDonald et al. 2013). For example, high-throughput genotyping enabled both gene–environment interaction studies with alcohol metabolizing genotypes (e.g., Terry et al. 2006; Barrdahl et al. 2017) as well as MR using genetic variants in alcohol metabolizing genes as instruments (e.g., Ong et al. 2021); these studies support that the association with breast cancer risk is unlikely to be explained by confounding. The association between alcohol intake and breast cancer was stronger with a longer interval from menarche to first pregnancy (Liu et al. 2013). Large-scale metabolomic studies also help support a causal association between alcohol and breast cancer by demonstrating an alcohol-based metabolomic signature (e.g., Playdon et al. 2017).

Studies of alcohol intake and subtype of breast cancer have consistently supported an association with ER$^+$ cancers, and luminal A specifically when measured. This pattern is also generally supported when examining the association between alcohol intake in women with known PVs in *BRCA1* or *BRCA2* genes who are more likely to have other cancers that are ER⁻ and other molecular subtypes (McGuire et al. 2006). Some studies support that there is a positive association between alcohol intake and mammographic breast density (for reviews, see McDonald et al. 2013; Lester et al. 2022), but the direction of the association may depend on the timing of alcohol intake, as intake earlier in life may be associated with lower breast density.

In contrast, increased breast cancer risk from tobacco smoking in women with PV in BRCA1 or BRCA2 have been reported particularly for smoking prior to first pregnancy (Breast Cancer Family Registry et al. 2008; Terry et al. 2011; Zeinomar et al. 2019). Higher risk has also been reported for individuals who both smoke and drink (Zeinomar et al. 2019). MR-based studies have also supported a causal association between cigarette intake (Dimou et al. 2021; Park et al. 2021) and breast cancer risk.

Studies of alcohol and tobacco exposure during WOS provide evidence of the complex role timing may have as both alcohol and tobacco can operate as carcinogens, and both can affect endogenous hormone levels. For example, prenatal exposure to alcohol has been associated with altered steroid hormone levels (Stevens et al. 2005), and prenatal exposure to maternal tobacco smoke has been associated with daughter's decreased breast density (Terry et al. 2011). Yet cumulative exposure to passive smoking exposure starting with household smoke exposure in childhood has been associated with increased breast cancer risk (Luo et al. 2011).

A new area of inquiry is related to whether binge drinking (defined by number of drinks within a short time interval) is independently related to breast cancer over and beyond regular use. For example, in a large U.S. Sister Study cohort, binge drinking was independently associated with breast cancer risk even after controlling for regular alcohol intake (White et al. 2015, 2017). As discussed below, this is a promising area of inquiry as it may open up new areas of targeted prevention where campaigns focus on reducing binge drinking particularly in younger women rather than complete abstinence. In par-

ticular, as alcohol and its metabolites have been demonstrated to mutate stem cells (Garaycoechea et al. 2018), it will be important to know whether the impact of alcohol can be reduced through slower intake when endogenous DNA repair mechanisms can repair any damage.

Environmental Exposures

In addition to individual-level lifestyle factors, there have been accumulating data for many different environmental chemicals and breast cancer risk from both case-control and cohort studies over the last two decades. These include studies on dioxins, polychlorinated biphenyls (PCBs), organochlorine pesticides (OCPs) (including dichlorodiphenyltrichloroethane [DDT]), polycyclic aromatic hydrocarbons (PAHs), air pollution, bisphenol A (BPA), phthalates, nonylphenols, hair dyes and straighteners, perfluoroalkyl substances (PFAS), flame retardants, drinking water, organic solvents, and heavy metals (for review, see Larsson et al. 2015; Hiatt and Brody 2018; Rodgers et al. 2018). One of the first and most comprehensive studies of environmental exposures and breast cancer was the Long Island Breast Cancer Study Project (LIBCSP), which was a large population-based, case-control study examining environmental exposures measured by questionnaire, blood, urine, dust, soil, water, and geographic modeling (Gammon et al. 2002a). Methods to measure environmental exposures through these alternative non-questionnaire approaches really advanced the field as apart from occupationally exposed individuals, it is often very challenging to remember past exposure to selected environmental exposures and measurement error from biomarker measurements is most likely to be a nondifferential error. Although the LIBCSP did observe associations between reported pesticide exposure and breast cancer risk (Teitelbaum et al. 2007) as measured by questionnaires, as it was a case-control study it remains possible that the association was driven by recall bias. The LIBCSP did not observe associations with OCPs as measured by blood biomarkers (Gammon et al. 2002c); the study did find associations with PAH biomarkers (Gammon et al. 2002b). The difference in findings between blood biomarker measures of pesticides and questionnaire measures may be because of reporting bias but also could be because questionnaires ask about lifetime exposure, and biomarkers typically only measure exposure within the last few months or years (depending on the half-life of the chemical) if using blood biomarkers and/or the last few hours or days if using urinary biomarkers.

Despite the large increase in epidemiological studies using blood, tissue, and urinary biomarkers of exposures over the last two decades, most studies collected samples at the time of diagnosis if a case-control study, or outside of any specific WOS if a cohort study. For example, a systematic review of different classes of environmental chemicals and compounds summarizing more than 150 studies on breast cancer over 10 yr (2006–2016) found that only 11% of the studies measured exposures during WOS (Rodgers et al. 2018). The data outside of WOS have been less consistent as many environmental chemical exposures are very common and experienced frequently by most women throughout their life. It remains challenging to observe consistent signals for such ubiquitous exposures when the exposure is measured without considering susceptibility. For example, in the few studies that have measured environmental chemicals during WOS, the evidence is much more consistent (see Table 1; Terry et al. 2019b). Some of these same chemicals may affect breast density (e.g., White et al. 2019; Krigbaum et al. 2020), particularly in women with a breast cancer family history (e.g., McDonald et al. 2020).

Further, there is a much greater consistency in the evidence with environmental chemicals when underlying susceptibility of cancer is considered. Many epidemiological cohort studies are not enriched for participants at increased absolute risk of cancer, which limits power to detect a signal and identify gene–environment interactions (Terry et al. 2016). Cancer gene discovery uses the approach of enrichment to identify genes that are relevant to all but would be difficult to identify in average risk cohorts. In a similar way, enriched cohorts have been used to examine whether modifiable factors have a different relative risk based on underlying absolute

risk (Terry et al. 2016, 2019a; Hopper et al. 2018; Kehm et al. 2019a, 2020; Zeinomar et al. 2019); this approach is also relevant for environmental exposures (Zeinomar et al. 2020). In a systematic review, 68 publications from 36 unique studies revealed a much more consistent pattern when examining breast cancer risk with the following environmental chemical compounds: PAH, air pollution, OCPs and metabolites, PCBs, metals, solvents, fungicides and other pesticides, PFAS, personal care products, phthalates, dioxins, and polybrominated diphenyl ethers (PBDEs) (Zeinomar et al. 2020). Specifically, when either examining studies of individuals at higher risk because of their cancer family history, or stratifying based on early-onset and/or genetic variants related to poorer DNA repair, >74% of studies supported an association with a specific environmental exposure and breast cancer risk.

As we know that more aggressive phenotypes like ER$^-$ and TNBC are more likely to be diagnosed earlier, these studies of enriched cohorts point to a need for interdisciplinary studies regarding environmental exposures and breast cancer risk as environmental exposures also are inequitably distributed. For example, prenatal exposure to some chemicals found in personal care products, including hair products, is associated with earlier menarche (McDonald et al. 2018; Harley et al. 2019)—a key risk factor for breast cancer. Furthermore, emerging evidence supports an association between hair product use and breast cancer risk and breast cancer clinicopathology (Llanos et al. 2017; Eberle et al. 2020; Rao et al. 2022), particularly among NHB women who use more hair products per capita than any other racial and ethnic group. Environmental exposures are thus a plausible driver of breast cancer disparities, especially given that the cancer burden is higher in the same neighborhoods that have higher exposure to environmental contaminants (Ruiz et al. 2018; Richmond-Bryant et al. 2020; Fong and Bell 2021), in households and individuals using products high in chemical exposures (Duty et al. 2005; James-Todd et al. 2011, 2012), and with heightened vulnerability to deleterious health effects of these environmental toxicants (Hicken et al. 2012; Alcala et al. 2019). Further, a largely

Table 1. Types of exposures supported by epidemiological evidence from breast cancer studies that relate to windows of susceptibility (WOS)

Studies	Prenatal	Premenarche	Menarche to first birth	Pregnancy window	Menopausal window
Exogenous hormone use					Hormone-replacement therapy
Diet		Vegetable protein	Animal proteins		
Adiposity and weight gain		Lifelong reduction in risk with higher adiposity			
Physical activity			Protective benefit, strongest for premenopausal breast cancer		
Alcohol and tobacco	Tobacco exposure related to lower density	Passive exposure to household smoker combined with later life passive smoker exposure may increase risk	Some evidence with tobacco intake of increased effect, alcohol increases risk		
Environmental chemical exposures	Regional total suspended particulates; dichlorodiphenyltrichloroethane (DDT)	DDT		Perfluoroalkyl carboxylic acid (PFCA), perfluorosulfonic acid (PFSA), perfluorooctanesulfonamide (PFOSA), polychlorinated biphenyl (PCB)	Cadmium

unexplored area is how these same chemical exposures may affect and/or interfere with hormonal cancer treatment (see Table 2).

Male Breast Cancer

As outlined in the introduction, increasing incidence of breast cancer has been seen in women in the United States and around the world (for review, see Chen et al. 2020). Less epidemiological research exists for male breast cancer particularly with respect to cohort studies, but many of the factors discussed above likely also play a role in male breast cancer (for review, see Zheng and Leone 2022). The correlation between male and female breast cancer incidence increasing over time also points to factors other than reproductive factors in driving these trends. Nevertheless, there are some important differences including that most male breast cancer is hormone receptor positive and BRCA2 mutations are more common (Gucalp et al. 2019).

IMPLICATIONS FOR RISK STRATIFICATION AND CLINICAL TRANSLATION

The increasing number of identified risk factors over the past two decades as well as the growing evidence of the importance of high, intermediate, and low penetrant genetic variants in breast cancer susceptibility genes (see Couch 2022) has increased the potential to move to precision prevention. In particular, hundreds of low penetrant genetic variants have been combined into a single polygenic risk score (PRS). New statistical models to integrate all these types of data including mammographic breast images, standard risk factors, and PRS are being developed. The ever-increasing volume of data and the repeated assessments provide opportunities for dynamic risk assessment and establishment of points for interventions.

Breast cancer risk prediction models have largely been divided into models using detailed family history pedigrees (e.g., Claus [Claus et al. 1991], BOADICEA [Antoniou et al. 2004], and BRCA-PRO [Berry et al. 2002]) and models using risk factor data with summary measures of family history (e.g., GAIL/BCRAT [Gail and Benichou 1994]). There are models that have integrated family pedigree data with risk factor data (e.g., IBIS [Brentnall and Cuzick 2020]/TC and BOADICIA [Carver et al. 2021]). These models are updated with additional risk factor data including alcohol and body weight (e.g., Pfeiffer et al. 2013). Other models add PRS plus mammographic density plus established risk factors (Rosner et al. 2021), and the combination is being evaluated in an implementation study in Canada (Brooks et al. 2021). Models that rely on mammographic density, however (e.g., Breast Cancer Surveillance Consortium [BCSC] model [Tice et al. 2019]), are limited to using only in populations that are already screened with mammography and thus are not as applicable for finding young high-risk women and particularly women who may be at risk of TNBC. Also, currently, none of the models have environmental chemical exposures included in the model and other major risk factors that may be driving the major health disparities in the breast cancer burden.

A key limitation of the models is that they have largely been developed and validated in a primarily white population of European ancestry with few exceptions (e.g., CARE model extended from the Breast Cancer Risk Assessment Tool [BCRAT] [Gail et al. 2007]). There are many considerations when translating models developed in one population and applying them to another population including the use of clinical PRS scores that have largely been developed in white populations of European ancestry (e.g., Liu et al. 2021). There has also been growing interest in using more than breast density from the mammographic image (Gastounioti et al. 2022). There remains debate whether machine learning approaches with mammographic images and limited epidemiological and clinical data (e.g., Yala et al. 2019, 2021, 2022) versus mammogram-based models with limited risk factors data and without costly genetic data (Jiang et al. 2021; Rosner et al. 2021) will prove to have more clinical utility. Determining the model performance to decide best practices for providing simple, rapid, and scalable risk assessment at the clinic is necessary to

Table 2. Identified gaps in evidence for interdisciplinary research and translation across basic, population, and clinical sciences for breast cancer reduction

Studies	Gaps in evidence to address by basic science	Gaps in evidence to address by population science	Gaps in evidence to address by clinical science
Exogenous hormone use		Use of newer hormonal methods and risk; risk stratification based on other outcomes	Implementation of short-term hormone-replacement therapy (HRT) and/or HRT based on risk stratification (e.g., higher risk of colorectal cancer than breast cancer)
Diet	Relationship to estrogen receptor–negative (ER⁻) tumors for vegetables	Data regarding prepubertal and adolescent diet and breast cancer risk	Implementation of the importance of nutrition earlier in life
Adiposity and weight gain	Pathways for protective effect of early life adiposity with consideration of height growth; is it associated with height growth?	Can adult weight loss or lifestyle interventions achieve the same benefit as early life adiposity? Is the consistent, negative association with childhood and adolescent body size driven by a lack of adjustment for height growth?	Implementation of weight-management programs earlier in adulthood
Physical activity		How long after interventions or increases in activity does benefit persist?	How to implement programs; interventions based on adolescence
Alcohol	Agents to block alcohol effect on breast cancer risk?	More data on binge drinking	No clear evidence that alcohol after diagnosis matters but does binge drinking?
Smoking	What mechanisms explain the dual effect, and can knowledge of the mechanisms help with targeted prevention for windows of susceptibility (WOS)?	How does vaping effect breast density and cancer risk	Can clinical tobacco cessation programs reduce cancer burden in higher risk individuals
Environmental chemicals	Mechanistic evidence for modifiers of the impact of (endocrine-disrupting chemicals [EDCs], metals and other chemicals, specific evidence regarding ER⁻ and triple-negative breast cancer [TNBC] specifically)	Additional data needed in higher risk populations and specific to WOS	Studies of the impact of EDCs on hormonal cancer treatments

ensure identification of those in greatest need of more frequent screening, identification and treatment, and less radiation exposure from overscreening and unnecessary treatments.

CONCLUDING REMARKS: NEXT STEPS UNANSWERED QUESTIONS—NEED FOR INTERDISCIPLINARY RESEARCH

Although there have been great advances in understanding the etiology of breast cancer, there remain major gaps. We believe that these gaps need to be addressed through interdisciplinary work across basic, population, and clinical sciences. We have outlined major gaps in Table 2 that we believe should be prioritized based on our review of the epidemiological evidence but recognize that this selection is illustrative rather than a complete list. In addition to gaps for specific risk factor constructs, there remain fundamental gaps in understanding whether the etiological pathways from benign lesions to ductal carcinoma in situ to invasive lesions are distinct. In addition to improving public health and clinical translation, we need more studies that collect repeated measures of biomarkers and exposures to allow evaluation of changes over time. Often our inferences are based on single biomarker measures. This may be changed in the future with the availability of electronic health record (EHR) data with clinical biomarker data stored centrally. Epidemiological studies have also largely focused on studying the first primary cancer, although emerging data show that PRS is also related to risk of second breast cancer. Again, with the increased potential of merging large clinical databases across regions and with the increasing integration of risk factor data into the EHR, future studies may more easily integrate required data from the clinical treatment of the first cancer as well as risk factor data.

What is clear, however, is that there remains an urgent need in investing in interdisciplinary studies of the etiology of TNBC including biologic pathways and specific risk factors that have not been as well studied including environmental chemicals, markers of stress, and adolescent risk factors. Epidemiological studies primarily have focused on producing evidence related to

HR$^+$ cancers and investigation and translation of risk stratification to individuals of different cultures, literacy, genders, settings, and with different access to services is urgently needed. We also recognize that in addition to observational studies, there is a major gap in evidence-based interventions to address the different breast cancer subtypes and particularly as relevant for WOS.

ACKNOWLEDGMENTS

We thank the Breast Cancer Research Foundation for their continued support (BCRF-21-028 and BCRF 21-143 to G.A.C. and M.B.T., respectively).

REFERENCES

*Reference is also in this subject collection.

Acheampong T, Kehm RD, Terry MB, Argov EL, Tehranifar P. 2020. Incidence trends of breast cancer molecular subtypes by age and race/ethnicity in the US from 2010 to 2016. *JAMA Netw Open* **3:** e2013226. doi:10.1001/jama networkopen.2020.13226

Ahlgren M, Melbye M, Wohlfahrt J, Sørensen TI. 2004. Growth patterns and the risk of breast cancer in women. *N Engl J Med* **351:** 1619–1626. doi:10.1056/NEJMoa 040576

Alcala E, Brown P, Capitman JA, Gonzalez M, Cisneros R. 2019. Cumulative impact of environmental pollution and population vulnerability on pediatric asthma hospitalizations: a multilevel analysis of CalEnviroScreen. *Int J Environ Res Public Health* **16:** 2683. doi:10.3390/ijer ph16152683

Antoniou AC, Pharoah PP, Smith P, Easton DF. 2004. The BOADICEA model of genetic susceptibility to breast and ovarian cancer. *Br J Cancer* **91:** 1580–1590. doi:10.1038/sj .bjc.6602175

Azam S, Jacobsen KK, Aro AR, Lynge E, Andersen ZJ. 2020. Hormone replacement therapy and mammographic density: a systematic literature review. *Breast Cancer Res Treat* **182:** 555–579. doi:10.1007/s10549-020-05744-w

Barrdahl M, Rudolph A, Hopper JL, Southey MC, Broeks A, Fasching PA, Beckmann MW, Gago-Dominguez M, Castelao JE, Guénel P, et al. 2017. Gene–environment interactions involving functional variants: results from the Breast Cancer Association Consortium. *Int J Cancer* **141:** 1830–1840. doi:10.1002/ijc.30859

Berkey CS, Frazier AL, Gardner JD, Colditz GA. 1999. Adolescence and breast carcinoma risk. *Cancer* **85:** 2400–2409. doi:10.1002/(SICI)1097-0142(19990601)85:11<24 00::AID-CNCR15>3.0.CO;2-O

Berkey CS, Gardner JD, Frazier AL, Colditz GA. 2000. Relation of childhood diet and body size to menarche and adolescent growth in girls. *Am J Epidemiol* **152:** 446–452. doi:10.1093/aje/152.5.446

Berkey CS, Colditz GA, Rockett HR, Frazier AL, Willett WC. 2009. Dairy consumption and female height growth: prospective cohort study. *Cancer Epidemiol Biomarkers Prev* **18:** 1881–1887. doi:10.1158/1055-9965.EPI-08-1163

Berkey CS, Tamimi RM, Willett WC, Rosner B, Hickey M, Toriola AT, Frazier AL, Colditz GA. 2020. Adolescent alcohol, nuts, and fiber: combined effects on benign breast disease risk in young women. *NPJ Breast Cancer* **6:** 61. doi:10.1038/s41523-020-00206-4

Berry DA, Iversen ES Jr, Gudbjartsson DF, Hiller EH, Garber JE, Peshkin BN, Lerman C, Watson P, Lynch HT, Hilsenbeck SG, et al. 2002. BRCAPRO validation, sensitivity of genetic testing of *BRCA1/BRCA2*, and prevalence of other breast cancer susceptibility genes. *J Clin Oncol* **20:** 2701–2712. doi:10.1200/JCO.2002.05.121

Breast Cancer Family Registry; Kathleen Cuningham Consortium for Research into Familial Breast Cancer (Australasia); Ontario Cancer Genetics Network. 2008. Smoking and risk of breast cancer in carriers of mutations in BRCA1 or BRCA2 aged less than 50 years. *Breast Cancer Res Treat* **109:** 67–75. doi:10.1007/s10549-007-9621-9

Brennan SF, Cantwell MM, Cardwell CR, Velentzis LS, Woodside JV. 2010. Dietary patterns and breast cancer risk: a systematic review and meta-analysis. *Am J Clin Nutr* **91:** 1294–1302. doi:10.3945/ajcn.2009.28796

Brentnall AR, Cuzick J. 2020. Risk models for breast cancer and their validation. *Stat Sci* **35:** 14–30. doi:10.1214/19-STS729

Brooks JD, Nabi HH, Andrulis IL, Antoniou AC, Chiquette J, Després P, Devilee P, Dorval M, Droit A, Easton DF, et al. 2021. Personalized risk assessment for prevention and early detection of breast cancer: integration and implementation (PERSPECTIVE I&I). *J Pers Med* **11:** 511. doi:10.3390/jpm11060511

* Brown P. 2022. Prevention in breast cancer. *Cold Spring Harb Perspect Med* doi:10.1101/cshperspect.a041345

Bucy AM, Valencia CI, Howe CL, Larkin TJ, Conard KD, Anderlik EW, Valdivi SI, Bea JW. 2022. Physical activity in young BRCA carriers and reduced risk of breast cancer. *Am J Prev Med* **63:** 837–845. doi:10.1016/j.amepre.2022.04.022

Burton A, Maskarinec G, Perez-Gomez B, Vachon C, Miao H, Lajous M, López-Ridaura R, Rice M, Pereira A, Garmendia ML, et al. 2017. Mammographic density and ageing: a collaborative pooled analysis of cross-sectional data from 22 countries worldwide. *PLoS Med* **14:** e1002335. doi:10.1371/journal.pmed.1002335

Byng JW, Boyd NF, Fishell E, Jong RA, Yaffe MJ. 1994. The quantitative analysis of mammographic densities. *Phys Med Biol* **39:** 1629–1638. doi:10.1088/0031-9155/39/10/008

Carver T, Hartley S, Lee A, Cunningham AP, Archer S, Babb de Villiers C, Roberts J, Ruston R, Walter FM, Tischkowitz M, et al. 2021. Canrisk tool—a web interface for the prediction of breast and ovarian cancer risk and the likelihood of carrying genetic pathogenic variants. *Cancer Epidemiol Biomarkers Prev* **30:** 469–473. doi:10.1158/1055-9965.EPI-20-1319

Chen Z, Xu L, Shi W, Zeng F, Zhuo R, Hao X, Fan P. 2020. Trends of female and male breast cancer incidence at the

global, regional, and national levels, 1990–2017. *Breast Cancer Res Treat* **180:** 481–490. doi:10.1007/s10549-020-05561-1

Chen H, Fan S, Stone J, Thompson DJ, Douglas J, Li S, Scott C, Bolla MK, Wang Q, Dennis J, et al. 2022. Genome-wide and transcriptome-wide association studies of mammographic density phenotypes reveal novel loci. *Breast Cancer Res* **24:** 27. doi:10.1186/s13058-022-01524-0

Chlebowski RT, Aragaki AK, Anderson GL, Prentice RL. 2020. Forty-year trends in menopausal hormone therapy use and breast cancer incidence among postmenopausal black and white women. *Cancer* **126:** 2956–2964. doi:10.1002/cncr.32846

Claus EB, Risch N, Thompson WD. 1991. Genetic analysis of breast cancer in the cancer and steroid hormone study. *Am J Hum Genet* **48:** 232–242.

Colditz GA, Frazier AL. 1995. Models of breast cancer show that risk is set by events of early life: prevention efforts must shift focus. *Cancer Epidemiol Biomarkers Prev* **4:** 567–571.

Colditz GA, Bohlke K, Berkey CS. 2014. Breast cancer risk accumulation starts early: prevention must also. *Breast Cancer Res Treat* **145:** 567–579. doi:10.1007/s10549-014-2993-8

Corbex M, Bouzbid S, Boffetta P. 2014. Features of breast cancer in developing countries, examples from North Africa. *Eur J Cancer* **50:** 1808–1818. doi:10.1016/j.ejca.2014.03.016

* Couch F. 2022. Genetic predisposition and breast cancer. *Cold Spring Harb Perspect Med* doi:10.1101/cshperspect.a041318

de Beer H. 2012. Dairy products and physical stature: a systematic review and meta-analysis of controlled trials. *Econ Hum Biol* **10:** 299–309. doi:10.1016/j.ehb.2011.08.003

Dimou N, Yarmolinsky J, Bouras E, Tsilidis KK, Martin RM, Lewis SJ, Gram IT, Bakker MF, Brenner H, Figueiredo JC, et al. 2021. Causal effects of lifetime smoking on breast and colorectal cancer risk: Mendelian randomization study. *Cancer Epidemiol Biomarkers Prev* **30:** 953–964. doi:10.1158/1055-9965.EPI-20-1218

Duty SM, Ackerman RM, Calafat AM, Hauser R. 2005. Personal care product use predicts urinary concentrations of some phthalate monoesters. *Environ Health Perspect* **113:** 1530–1535. doi:10.1289/ehp.8083

Eberle CE, Sandler DP, Taylor KW, White AJ. 2020. Hair dye and chemical straightener use and breast cancer risk in a large US population of black and white women. *Int J Cancer* **147:** 383–391. doi:10.1002/ijc.32738

Ellison RC, Zhang Y, McLennan CE, Rothman KJ. 2001. Exploring the relation of alcohol consumption to risk of breast cancer. *Am J Epidemiol* **154:** 740–747. doi:10.1093/aje/154.8.740

Ferris JS, Daly MB, Buys SS, Genkinger JM, Liao Y, Terry MB. 2014. Oral contraceptive and reproductive risk factors for ovarian cancer within sisters in the breast cancer family registry. *Br J Cancer* **110:** 1074–1080. doi:10.1038/bjc.2013.803

Fletcher SW, Colditz GA. 2002. Failure of estrogen plus progestin therapy for prevention. *JAMA* **288:** 366–368. doi:10.1001/jama.288.3.366

Fong KC, Bell ML. 2021. Do fine particulate air pollution (PM2.5) exposure and its attributable premature mortality differ for immigrants compared to those born in the United States? *Environ Res* **196:** 110387. doi:10.1016/j.envres.2020.110387

Gail MH, Benichou J. 1994. Validation studies on a model for breast cancer risk. *J Natl Cancer Inst* **86:** 573–575. doi:10.1093/jnci/86.8.573

Gail MH, Costantino JP, Pee D, Bondy M, Newman L, Selvan M, Anderson GL, Malone KE, Marchbanks PA, McCaskill-Stevens W, et al. 2007. Projecting individualized absolute invasive breast cancer risk in African American women. *J Natl Cancer Inst* **99:** 1782–1792. doi:10.1093/jnci/djm223

Gammon MD, Neugut AI, Santella RM, Teitelbaum SL, Britton JA, Terry MB, Eng SM, Wolff MS, Stellman SD, Kabat GC, et al. 2002a. The Long Island Breast Cancer Study Project: description of a multi-institutional collaboration to identify environmental risk factors for breast cancer. *Breast Cancer Res Treat* **74:** 235–254. doi:10.1023/A:1016387020854

Gammon MD, Santella RM, Neugut AI, Eng SM, Teitelbaum SL, Paykin A, Levin B, Terry MB, Young TL, Wang LW, et al. 2002b. Environmental toxins and breast cancer on Long Island. I: Polycyclic aromatic hydrocarbon DNA adducts. *Cancer Epidemiol Biomarkers Prev* **11:** 677–685.

Gammon MD, Wolff MS, Neugut AI, Eng SM, Teitelbaum SL, Britton JA, Terry MB, Levin B, Stellman SD, Kabat GC, et al. 2002c. Environmental toxins and breast cancer on Long Island. II: Organochlorine compound levels in blood. *Cancer Epidemiol Biomarkers Prev* **11:** 686–697.

Garaycoechea JI, Crossan GP, Langevin F, Mulderrig L, Louzada S, Yang F, Guilbaud G, Park N, Roerink S, Nik-Zainal S, et al. 2018. Alcohol and endogenous aldehydes damage chromosomes and mutate stem cells. *Nature* **553:** 171–177. doi:10.1038/nature25154

Gastounioti A, Desai S, Ahluwalia VS, Conant EF, Kontos D. 2022. Artificial intelligence in mammographic phenotyping of breast cancer risk: a narrative review. *Breast Cancer Res* **24:** 14. doi:10.1186/s13058-022-01509-z

Graham S. 1987. Alcohol and breast cancer. *N Engl J Med* **316:** 1211–1213. doi:10.1056/NEJM198705073161908

Gucalp A, Traina TA, Eisner JR, Parker JS, Selitsky SR, Park BH, Elias AD, Baskin-Bey ES, Cardoso F. 2019. Male breast cancer: a disease distinct from female breast cancer. *Breast Cancer Res Treat* **173:** 37–48. doi:10.1007/s10549-018-4921-9

Hao Y, Xiao J, Liang Y, Wu X, Zhang H, Xiao C, Zhang L, Burgess S, Wang N, Zhao X, et al. 2022. Reassessing the causal role of obesity in breast cancer susceptibility: a comprehensive multivariable Mendelian randomization investigating the distribution and timing of exposure. *Int J Epidemiol* dyac143. doi:10.1093/ije/dyac143

Harley KG, Berger KP, Kogut K, Parra K, Lustig RH, Greenspan LC, Calafat AM, Ye X, Eskenazi B. 2019. Association of phthalates, parabens and phenols found in personal care products with pubertal timing in girls and boys. *Hum Reprod* **34:** 109–117. doi:10.1093/humrep/dey337

Heer E, Harper A, Escandor N, Sung H, McCormack V, Fidler-Benaoudia MM. 2020. Global burden and trends in premenopausal and postmenopausal breast cancer: a population-based study. *Lancet Glob Health* **8:** e1027–e1037. doi:10.1016/S2214-109X(20)30215-1

Hiatt RA, Brody JG. 2018. Environmental determinants of breast cancer. *Annu Rev Public Health* **39:** 113–133. doi:10.1146/annurev-publhealth-040617-014101

Hicken MT, Gee GC, Morenoff J, Connell CM, Snow RC, Hu H. 2012. A novel look at racial health disparities: the interaction between social disadvantage and environmental health. *Am J Public Health* **102:** 2344–2351. doi:10.2105/AJPH.2012.300774

Hopper JL, Dite GS, MacInnis RJ, Liao Y, Zeinomar N, Knight JA, Southey MC, Milne RL, Chung WK, Giles GG, et al. 2018. Age-specific breast cancer risk by body mass index and familial risk: prospective family study cohort (ProF-SC). *Breast Cancer Res* **20:** 132. doi:10.1186/s13058-018-1056-1

IARC Working Group on the Evaluation of Carcinogenic Risks to Humans (2007: Lyon France). 2010. *Alcohol consumption and ethyl carbamate.* International Agency for Research on Cancer, WHO Press, Lyon, France.

Islami F, Liu Y, Jemal A, Zhou J, Weiderpass E, Colditz G, Boffetta P, Weiss M. 2015. Breastfeeding and breast cancer risk by receptor status—a systematic review and meta-analysis. *Ann Oncol* **26:** 2398–2407. doi:10.1093/annonc/mdv379

James-Todd T, Terry MB, Rich-Edwards J, Deierlein A, Senie R. 2011. Childhood hair product use and earlier age at menarche in a racially diverse study population: a pilot study. *Ann Epidemiol* **21:** 461–465. doi:10.1016/j.annepidem.2011.01.009

James-Todd T, Senie R, Terry MB. 2012. Racial/ethnic differences in hormonally active hair product use: a plausible risk factor for health disparities. *J Immigr Minor Health* **14:** 506–511. doi:10.1007/s10903-011-9482-5

Jiang S, Cao J, Colditz GA, Rosner B. 2021. Predicting the onset of breast cancer using mammogram imaging data with irregular boundary. *Biostatistics* kxab032. doi:10.1093/biostatistics/kxab032

John EM, Hines LM, Phipps AI, Koo J, Longacre TA, Ingles SA, Baumgartner KB, Slattery ML, Wu AH. 2018. Reproductive history, breast-feeding and risk of triple negative breast cancer: the Breast Cancer Etiology in Minorities (BEM) study. *Int J Cancer* **142:** 2273–2285. doi:10.1002/ijc.31258

Jung S, Spiegelman D, Baglietto L, Bernstein L, Boggs DA, van den Brandt PA, Buring JE, Cerhan JR, Gaudet MM, Giles GG, et al. 2013. Fruit and vegetable intake and risk of breast cancer by hormone receptor status. *J Natl Cancer Inst* **105:** 219–236. doi:10.1093/jnci/djs635

Kehm RD, Hopper JL, John EM, Phillips KA, MacInnis RJ, Dite GS, Milne RL, Liao Y, Zeinomar N, Knight JA, et al. 2019a. Regular use of aspirin and other non-steroidal anti-inflammatory drugs and breast cancer risk for women at familial or genetic risk: a cohort study. *Breast Cancer Res* **21:** 52. doi:10.1186/s13058-019-1135-y

Kehm RD, Yang W, Tehranifar P, Terry MB. 2019b. 40 years of change in age- and stage-specific cancer incidence rates in US women and men. *JNCI Cancer Spectr* **3:** pkz038. doi:10.1093/jncics/pkz038

Kehm RD, Genkinger JM, MacInnis RJ, John EM, Phillips KA, Dite GS, Milne RL, Zeinomar N, Liao Y, Knight JA, et al. 2020. Recreational physical activity is associated with reduced breast cancer risk in adult women at high risk for breast cancer: a cohort study of women selected for familial and genetic risk. *Cancer Res* **80:** 116–125. doi:10.1158/0008-5472.CAN-19-1847

Keller BM, Nathan DL, Wang Y, Zheng Y, Gee JC, Conant EF, Kontos D. 2012. Estimation of breast percent density in raw and processed full field digital mammography images via adaptive fuzzy c-means clustering and support vector machine segmentation. *Med Phys* **39:** 4903–4917. doi:10.1118/1.4736530

Krieger N. 2008. Hormone therapy and the rise and perhaps fall of US breast cancer incidence rates: critical reflections. *Int J Epidemiol* **37:** 627–637. doi:10.1093/ije/dyn055

Krigbaum NY, Cirillo PM, Flom JD, McDonald JA, Terry MB, Cohn BA. 2020. In utero DDT exposure and breast density before age 50. *Reprod Toxicol* **92:** 85–90. doi:10.1016/j.reprotox.2019.11.002

Larsson SC, Orsini N, Wolk A. 2015. Urinary cadmium concentration and risk of breast cancer: a systematic review and dose-response meta-analysis. *Am J Epidemiol* **182:** 375–380. doi:10.1093/aje/kwv085

Lauby-Secretan B, Scoccianti C, Loomis D, Grosse Y, Bianchini F, Straif K; International Agency for Research on Cancer Handbook Working Group. 2016. Body fatness and cancer—viewpoint of the IARC Working Group. *N Engl J Med* **375:** 794–798. doi:10.1056/NEJMsr1606602

Lester SP, Kaur AS, Vegunta S. 2022. Association between lifestyle changes, mammographic breast density, and breast cancer. *Oncologist* **27:** 548–554. doi:10.1093/oncolo/oyac084

Li CI, Malone KE, Daling JR. 2003. Differences in breast cancer stage, treatment, and survival by race and ethnicity. *Arch Intern Med* **163:** 49–56. doi:10.1001/archinte.163.1.49

Lima SM, Kehm RD, Terry MB. 2021. Global breast cancer incidence and mortality trends by region, age-groups, and fertility patterns. *EClinicalMedicine* **38:** 100985. doi:10.1016/j.eclinm.2021.100985

Liu Y, Colditz GA, Rosner B, Berkey CS, Collins LC, Schnitt SJ, Connolly JL, Chen WY, Willett WC, Tamimi RM. 2013. Alcohol intake between menarche and first pregnancy: a prospective study of breast cancer risk. *J Natl Cancer Inst* **105:** 1571–1578. doi:10.1093/jnci/djt213

Liu C, Zeinomar N, Chung WK, Kiryluk K, Gharavi AG, Hripcsak G, Crew KD, Shang N, Khan A, Fasel D, et al. 2021. Generalizability of polygenic risk scores for breast cancer among women with European African, and Latinx ancestry. *JAMA Netw Open* **4:** e2119084.

Llanos AAM, Rabkin A, Bandera EV, Zirpoli G, Gonzalez BD, Xing CY, Qin B, Lin Y, Hong CC, Demissie K, et al. 2017. Hair product use and breast cancer risk among African American and white women. *Carcinogenesis* **38:** 883–892. doi:10.1093/carcin/bgx060

Longnecker MP. 1994. Alcoholic beverage consumption in relation to risk of breast cancer: meta-analysis and review. *Cancer Causes Control* **5:** 73–82. doi:10.1007/BF01830729

Luo J, Margolis KL, Wactawski-Wende J, Horn K, Messina C, Stefanick ML, Tindle HA, Tong E, Rohan TE. 2011. Association of active and passive smoking with risk of breast cancer among postmenopausal women: a prospec-

tive cohort study. *BMJ* **342:** d1016. doi:10.1136/bmj.d1016

McDonald JA, Goyal A, Terry MB. 2013. Alcohol intake and breast cancer risk: weighing the overall evidence. *Curr Breast Cancer Rep* **5:** 208–221. doi:10.1007/s12609-013-0114-z

McDonald JA, Tehranifar P, Flom JD, Terry MB, James-Todd T. 2018. Hair product use, age at menarche and mammographic breast density in multiethnic urban women. *Environ Health* **17:** 1. doi:10.1186/s12940-017-0345-y

McDonald JA, Cirillo PM, Tehranifar P, Krigbaum NY, Engmann NJ, Cohn BA, Terry MB. 2020. In utero DDT exposure and breast density in early menopause by maternal history of breast cancer. *Reprod Toxicol* **92:** 78–84. doi:10.1016/j.reprotox.2019.08.009

McGuire V, John EM, Felberg A, Haile RW, Boyd NF, Thomas DC, Jenkins MA, Milne RL, Daly MB, Ward J, et al. 2006. No increased risk of breast cancer associated with alcohol consumption among carriers of BRCA1 and BRCA2 mutations ages <50 years. *Cancer Epidemiol Biomarkers Prev* **15:** 1565–1567. doi:10.1158/1055-9965.EPI-06-0323

Millikan RC, Newman B, Tse CK, Moorman PG, Conway K, Dressler LG, Smith LV, Labbok MH, Geradts J, Bensen JT, et al. 2008. Epidemiology of basal-like breast cancer. *Breast Cancer Res Treat* **109:** 123–139. doi:10.1007/s10549-007-9632-6

Monninkhof EM, Elias SG, Vlems FA, van der Tweel I, Schuit AJ, Voskuil DW, van Leeuwen FE; TFPAC. 2007. Physical activity and breast cancer: a systematic review. *Epidemiology* **18:** 137–157. doi:10.1097/01.ede.0000251167.75581.98

Niehoff NM, White AJ, Sandler DP. 2017. Childhood and teenage physical activity and breast cancer risk. *Breast Cancer Res Treat* **164:** 697–705. doi:10.1007/s10549-017-4276-7

Ong JS, Derks EM, Eriksson M, An J, Hwang LD, Easton DF, Pharoah PP, Berchuck A, Kelemen LE, Matsuo K, et al. 2021. Evaluating the role of alcohol consumption in breast and ovarian cancer susceptibility using population-based cohort studies and two-sample Mendelian randomization analyses. *Int J Cancer* **148:** 1338–1350. doi:10.1002/ijc.33308

Park HA, Neumeyer S, Michailidou K, Bolla MK, Wang Q, Dennis J, Ahearn TU, Andrulis IL, Anton-Culver H, Antonenkova NN, et al. 2021. Mendelian randomisation study of smoking exposure in relation to breast cancer risk. *Br J Cancer* **125:** 1135–1145. doi:10.1038/s41416-021-01432-8

Perera FP, Weinstein IB. 1982. Molecular epidemiology and carcinogen-DNA adduct detection: new approaches to studies of human cancer causation. *J Chronic Dis* **35:** 581–600. doi:10.1016/0021-9681(82)90078-9

Pfeiffer RM, Park Y, Kreimer AR, Lacey JV Jr, Pee D, Greenlee RT, Buys SS, Hollenbeck A, Rosner B, Gail MH, et al. 2013. Risk prediction for breast, endometrial, and ovarian cancer in white women aged 50 y or older: derivation and validation from population-based cohort studies. *PLoS Med* **10:** e1001492. doi:10.1371/journal.pmed.1001492

Playdon MC, Ziegler RG, Sampson JN, Stolzenberg-Solomon R, Thompson HJ, Irwin ML, Mayne ST, Hoover

RN, Moore SC. 2017. Nutritional metabolomics and breast cancer risk in a prospective study. *Am J Clin Nutr* **106:** 637–649. doi:10.3945/ajcn.116.150912

Prentice RL, Thomson CA, Caan B, Hubbell FA, Anderson GL, Beresford SA, Pettinger M, Lane DS, Lessin L, Yasmeen S, et al. 2007. Low-fat dietary pattern and cancer incidence in the Women's Health Initiative dietary modification randomized controlled trial. *J Natl Cancer Inst* **99:** 1534–1543. doi:10.1093/jnci/djm159

Rao R, McDonald JA, Barrett ES, Greenberg P, Teteh DK, Montgomery SB, Qin B, Lin Y, Hong CC, Ambrosone CB, et al. 2022. Associations of hair dye and relaxer use with breast tumor clinicopathologic features: findings from the women's circle of health study. *Environ Res* **203:** 111863. doi:10.1016/j.envres.2021.111863

Rice MS, Bertrand KA, VanderWeele TJ, Rosner BA, Liao X, Adami HO, Tamimi RM. 2016. Mammographic density and breast cancer risk: a mediation analysis. *Breast Cancer Res* **18:** 94. doi:10.1186/s13058-016-0750-0

Richmond-Bryant J, Mikati I, Benson AF, Luben TJ, Sacks JD. 2020. Disparities in distribution of particulate matter emissions from US coal-fired power plants by race and poverty status after accounting for reductions in operations between 2015 and 2017. *Am J Public Health* **110:** 655–661. doi:10.2105/AJPH.2019.305558

Rodgers KM, Udesky JO, Rudel RA, Brody JG. 2018. Environmental chemicals and breast cancer: an updated review of epidemiological literature informed by biological mechanisms. *Environ Res* **160:** 152–182. doi:10.1016/j.envres.2017.08.045

Rosner B, Eliassen AH, Toriola AT, Chen WY, Hankinson SE, Willett WC, Berkey CS, Colditz GA. 2017. Weight and weight changes in early adulthood and later breast cancer risk. *Int J Cancer* **140:** 2003–2014. doi:10.1002/ijc.30627

Rosner B, Tamimi RM, Kraft P, Gao C, Mu Y, Scott C, Winham SJ, Vachon CM, Colditz GA. 2021. Simplified breast risk tool integrating questionnaire risk factors, mammographic density, and polygenic risk score: development and validation. *Cancer Epidemiol Biomarkers Prev* **30:** 600–607. doi:10.1158/1055-9965.EPI-20-0900

Rossouw JE, Anderson GL, Prentice RL, LaCroix AZ, Kooperberg C, Stefanick ML, Jackson RD, Beresford SA, Howard BV, Johnson KC, et al. 2002. Risks and benefits of estrogen plus progestin in healthy postmenopausal women: principal results from the Women's Health Initiative randomized controlled trial. *JAMA* **288:** 321–333. doi:10.1001/jama.288.3.321

Ruiz D, Becerra M, Jagai JS, Ard K, Sargis RM. 2018. Disparities in environmental exposures to endocrine-disrupting chemicals and diabetes risk in vulnerable populations. *Diabetes Care* **41:** 193–205. doi:10.2337/dc16-2765

Sahni K, Parsad D. 2013. Stability in vitiligo: is there a perfect way to predict it? *J Cutan Aesthet Surg* **6:** 75–82. doi:10.4103/0974-2077.112667

Schatzkin A, Abnet CC, Cross AJ, Gunter M, Pfeiffer R, Gail M, Lim U, Davey Smith G. 2009. Mendelian randomization: how it can—and cannot—help confirm causal relations between nutrition and cancer. *Cancer Prev Res (Phila)* **2:** 104–113. doi:10.1158/1940-6207.CAPR-08-0070

Shoemaker ML, White MC, Wu M, Weir HK, Romieu I. 2018. Differences in breast cancer incidence among young women aged 20–49 years by stage and tumor characteristics, age, race, and ethnicity, 2004–2013. *Breast Cancer Res Treat* **169:** 595–606. doi:10.1007/s10549-018-4699-9

Smith-Warner SA, Spiegelman D, Yaun SS, van den Brandt PA, Folsom AR, Goldbohm RA, Graham S, Holmberg L, Howe GR, Marshall JR, et al. 1998. Alcohol and breast cancer in women: a pooled analysis of cohort studies. *JAMA* **279:** 535–540. doi:10.1001/jama.279.7.535

Stevens RG, Cohen RD, Terry MB, Lasley BL, Siiteri P, Cohn BA. 2005. Alcohol consumption and serum hormone levels during pregnancy. *Alcohol* **36:** 47–53. doi:10.1016/j.alcohol.2005.07.006

Strebel J, Terry MB. 2021. Alcohol, binge drinking, and cancer risk: accelerating public health messaging through countermarketing. *Am J Public Health* **111:** 812–814. doi:10.2105/AJPH.2021.306233

Su X, Tamimi RM, Collins LC, Baer HJ, Cho E, Sampson L, Willett WC, Schnitt SJ, Connolly JL, Rosner BA, et al. 2010. Intake of fiber and nuts during adolescence and incidence of proliferative benign breast disease. *Cancer Causes Control* **21:** 1033–1046. doi:10.1007/s10552-010-9532-7

* Swarbrick A, Fernandez-Martinez A, Perou CM. 2022. Gene expression profiling to decipher breast cancer inter- and intra-tumor heterogeneity. *Cold Spring Harb Perspect Med* doi:10.1101/cshperspect.a041320

Swerdlow DI, Kuchenbaecker KB, Shah S, Sofat R, Holmes MV, White J, Mindell JS, Kivimaki M, Brunner EJ, Whittaker JC, et al. 2016. Selecting instruments for Mendelian randomization in the wake of genome-wide association studies. *Int J Epidemiol* **45:** 1600–1616. doi:10.1093/ije/dyw088

Tehranifar P, Wei Y, Terry MB. 2021. Less is more—ways to move forward for improved breast cancer risk stratification. *Cancer Epidemiol Biomarkers Prev* **30:** 587–589. doi:10.1158/1055-9965.EPI-20-1627

Teitelbaum SL, Gammon MD, Britton JA, Neugut AI, Levin B, Stellman SD. 2007. Reported residential pesticide use and breast cancer risk on Long Island, New York. *Am J Epidemiol* **165:** 643–651. doi:10.1093/aje/kwk046

Terry MB, Gammon MD, Zhang FF, Knight JA, Wang Q, Britton JA, Teitelbaum SL, Neugut AI, Santella RM. 2006. ADH3 genotype, alcohol intake and breast cancer risk. *Carcinogenesis* **27:** 840–847. doi:10.1093/carcin/bgi285

Terry MB, Schaefer CA, Flom JD, Wei Y, Tehranifar P, Liao Y, Buka S, Michels KB. 2011. Prenatal smoke exposure and mammographic density in mid-life. *J Dev Orig Health Dis* **2:** 340–352. doi:10.1017/S2040174411000614

Terry MB, Phillips KA, Daly MB, John EM, Andrulis IL, Buys SS, Goldgar DE, Knight JA, Whittemore AS, Chung WK, et al. 2016. Cohort profile: the breast cancer prospective family study cohort (ProF-SC). *Int J Epidemiol* **45:** 683–692. doi:10.1093/ije/dyv118

Terry MB, Daly MB, Phillips KA, Ma X, Zeinomar N, Leoce N, Dite GS, MacInnis RJ, Chung WK, Knight JA, et al. 2019a. Risk-reducing oophorectomy and breast cancer risk across the spectrum of familial risk. *J Natl Cancer Inst* **111:** 331–334. doi:10.1093/jnci/djy182

Terry MB, Michels KB, Brody JG, Byrne C, Chen S, Jerry DJ, Malecki KMC, Martin MB, Miller RL, Neuhausen SL, et al. 2019b. Environmental exposures during windows of susceptibility for breast cancer: a framework for prevention research. *Breast Cancer Res* **21**: 96. doi:10.1186/s13058-019-1168-2

Tice JA, Bissell MCS, Miglioretti DL, Gard CC, Rauscher GH, Dabbous FM, Kerlikowske K. 2019. Validation of the breast cancer surveillance consortium model of breast cancer risk. *Breast Cancer Res Treat* **175**: 519–523. doi:10.1007/s10549-019-05167-2

van den Brandt PA, Spiegelman D, Yaun SS, Adami HO, Beeson L, Folsom AR, Fraser G, Goldbohm RA, Graham S, Kushi L, et al. 2000. Pooled analysis of prospective cohort studies on height, weight, and breast cancer risk. *Am J Epidemiol* **152**: 514–527. doi:10.1093/aje/152.6.514

Wang J, Peng C, Guranich C, Heng YJ, Baker GM, Rubadue CA, Glass K, Eliassen AH, Tamimi RM, Polyak K, et al. 2021. Early-life body adiposity and the breast tumor transcriptome. *J Natl Cancer Inst* **113**: 778–784. doi:10.1093/jnci/djaa169

Ward SV, Burton A, Tamimi RM, Pereira A, Garmendia ML, Pollan M, Boyd N, Dos-Santos-Silva I, Maskarinec G, Perez-Gomez B, et al. 2022. The association of age at menarche and adult height with mammographic density in the international consortium of mammographic density. *Breast Cancer Res* **24**: 49. doi:10.1186/s13058-022-01545-9

WCRF. 2018. Diet, nutrition, physical activity and cancer: a global perspective. Continuous Update Project Expert Report 2018, World Cancer Research Fund/American Institute for Cancer Research, London. https://www.wcrf.org/wp-content/uploads/2021/02/Breast-cancer-report.pdf

White A, Castle IJ, Chen CM, Shirley M, Roach D, Hingson R. 2015. Converging patterns of alcohol use and related outcomes among females and males in the United States, 2002 to 2012. *Alcohol Clin Exp Res* **39**: 1712–1726. doi:10.1111/acer.12815

White AJ, DeRoo LA, Weinberg CR, Sandler DP. 2017. Lifetime alcohol intake, binge drinking behaviors, and breast cancer risk. *Am J Epidemiol* **186**: 541–549. doi:10.1093/aje/kwx118

White AJ, Weinberg CR, O'Meara ES, Sandler DP, Sprague BL. 2019. Airborne metals and polycyclic aromatic hydrocarbons in relation to mammographic breast density. *Breast Cancer Res* **21**: 24. doi:10.1186/s13058-019-1110-7

Wolfe JN. 1976. Risk for breast cancer development determined by mammographic parenchymal pattern. *Cancer* **37**: 2486–2492. doi:10.1002/1097-0142(197605)37:5<2486::AID-CNCR2820370542>3.0.CO;2-8

Work ME, John EM, Andrulis IL, Knight JA, Liao Y, Mulligan AM, Southey MC, Giles GG, Dite GS, Apicella C, et al. 2014. Reproductive risk factors and oestrogen/progesterone receptor-negative breast cancer in the breast cancer family registry. *Br J Cancer* **110**: 1367–1377. doi:10.1038/bjc.2013.807

Xiao Y, Xia J, Li L, Ke Y, Cheng J, Xie Y, Chu W, Cheung P, Kim JH, Colditz GA, et al. 2019. Associations between dietary patterns and the risk of breast cancer: a systematic review and meta-analysis of observational studies. *Breast Cancer Res* **21**: 16. doi:10.1186/s13058-019-1096-1

Yala A, Lehman C, Schuster T, Portnoi T, Barzilay R. 2019. A deep learning mammography-based model for improved breast cancer risk prediction. *Radiology* **292**: 60–66. doi:10.1148/radiol.2019182716

Yala A, Mikhael PG, Strand F, Lin G, Smith K, Wan YL, Lamb L, Hughes K, Lehman C, Barzilay R. 2021. Toward robust mammography-based models for breast cancer risk. *Sci Transl Med* **13**: eaba4373. doi:10.1126/scitranslmed.aba4373

Yala A, Mikhael PG, Strand F, Lin G, Satuluru S, Kim T, Banerjee I, Gichoya J, Trivedi H, Lehman CD, et al. 2022. Multi-institutional validation of a mammography-based breast cancer risk model. *J Clin Oncol* **40**: 1732–1740. doi:10.1200/JCO.21.01337

Yarmolinsky J, Wade KH, Richmond RC, Langdon RJ, Bull CJ, Tilling KM, Relton CL, Lewis SJ, Davey Smith G, Martin RM. 2018. Causal inference in cancer epidemiology: what is the role of Mendelian randomization? *Cancer Epidemiol Biomarkers Prev* **27**: 995–1010. doi:10.1158/1055-9965.EPI-17-1177

Zeinomar N, Knight JA, Genkinger JM, Phillips KA, Daly MB, Milne RL, Dite GS, Kehm RD, Liao Y, Southey MC, et al. 2019. Alcohol consumption, cigarette smoking, and familial breast cancer risk: findings from the prospective family study cohort (ProF-SC). *Breast Cancer Res* **21**: 128. doi:10.1186/s13058-019-1213-1

Zeinomar N, Oskar S, Kehm RD, Sahebzeda S, Terry MB. 2020. Environmental exposures and breast cancer risk in the context of underlying susceptibility: a systematic review of the epidemiological literature. *Environ Res* **187**: 109346. doi:10.1016/j.envres.2020.109346

Zheng G, Leone JP. 2022. Male breast cancer: an updated review of epidemiology, clinicopathology, and treatment. *J Oncol* **2022**: 1734049.

Zürcher A, Knabben L, Janka H, Stute P. 2022. Influence of the levonorgestrel-releasing intrauterine system on the risk of breast cancer: a systematic review. *Arch Gynecol Obstet* doi:10.1007/s00404-022-06640-y

Germline Genetic Testing for Hereditary Breast and Ovarian Cancer: Current Concepts in Risk Evaluation

Siddhartha Yadav,[1] Fergus J. Couch,[2] and Susan M. Domchek[3]

[1]Department of Oncology, Mayo Clinic, Rochester, Minnesota 55905, USA

[2]Department of Laboratory Medicine and Pathology, Mayo Clinic, Rochester, Minnesota 55901, USA

[3]Basser Center for BRCA, Abramson Cancer Center, University of Pennsylvania, Philadelphia, Pennsylvania 19104, USA

Correspondence: yadav.siddhartha@mayo.edu; couch.fergus@mayo.edu; susan.domchek@pennmedicine.upenn.edu

Our understanding of hereditary breast and ovarian cancer has significantly improved over the past two decades. In addition to *BRCA1/2*, pathogenic variants in several other DNA-repair genes have been shown to increase the risks of breast and ovarian cancer. The magnitude of cancer risk is impacted not only by the gene involved, but also by family history of cancer, polygenic risk scores, and, in certain genes, pathogenic variant type or location. While estimates of breast and ovarian cancer risk associated with pathogenic variants are available, these are predominantly based on studies of high-risk populations with young age at diagnosis of cancer, multiple primary cancers, or family history of cancer. More recently, breast cancer risk for germline pathogenic variant carriers has been estimated from population-based studies. Here, we provide a review of the field of germline genetic testing and risk evaluation for hereditary breast and ovarian cancers in high-risk and population-based settings.

As detailed by Lynch et al. (2004) reports of familial clustering of cancer date back to the sixteenth century. For hereditary breast and ovarian cancer, major advances came in the 1990s when *BRCA1* and *BRCA2* (*BRCA1/2* hereafter) were identified and then cloned (Hall et al. 1990; Narod et al. 1991; Miki et al. 1994; Wooster et al. 1995; Tavtigian et al. 1996). Over the past two decades, there has been significant progress in our understanding of familial breast and ovarian cancer. Pathogenic variants (PVs) in multiple genes in addition to *BRCA1/2* have been identified as contributing to the risk of breast or ovarian cancer. Germline genetic testing criteria and multigene panels for the identification of women at risk of breast or ovarian cancer have also evolved significantly. With an expansion of the number of risk-associated genes with variable penetrance and differing cancer risk, risk assessment and management in carriers of germline PVs have become more nuanced. This review will highlight the current understanding of the risk of breast and/or ovarian cancer with germline PVs in cancer predisposition genes, the nuances in interpreting the cancer risk, and the controversies in the genetic testing criteria for identifying women at risk of breast and/or ovarian cancer.

GERMLINE GENETIC TESTING FOR BREAST AND OVARIAN CANCER RISK

The criteria for which germline genetic testing for breast and ovarian cancer risk is recommended have been in continuous evolution. When first offered commercially, genetic testing was time-consuming, expensive, limited to *BRCA1* and *BRCA2*, had unclear clinical utility, and often resulted in variants of uncertain significance (VUS) (the latter of which often caused confusion and anxiety in patients and providers alike). Over time, evidence has accumulated that knowledge of genetic risk can lead to improved early detection and risk reduction strategies. The number of genes associated with breast and ovarian cancer risk expanded, sequencing technology has improved, and costs have lowered. There are multiple guidelines for the consideration of genetic testing, which vary by professional societies, payors, and countries (National Collaborating Centre for Cancer 2013; Forbes et al. 2019; Manahan et al. 2019; Owens et al. 2019; Marmolejo et al. 2021; Pujol et al. 2021; National Comprehensive Cancer Network 2023). Although in some scenarios there is a universal agreement (e.g., all individuals with high-grade epithelial ovarian cancer should undergo genetic testing), for breast cancer the guidelines vary. For example, while the American Society of Breast Surgeons (ASBrS) (Manahan et al. 2019) advocates that all women with breast cancer should be offered genetic testing, the National Comprehensive Cancer Network (NCCN) sees the value of genetic testing as low in non-Ashkenazi Jewish women over 65 with estrogen receptor (ER)-positive breast cancer and no concerning family history (National Comprehensive Cancer Network 2023). Several prior probability models exist to identify patients who are at high risk of being PV carriers; however, patients may meet the threshold for germline genetic testing by one criterion but not by another, which can lead to confusion for patients and clinicians. In part, the confusion is due to the fact that the majority of these guidelines and models are not optimized for the detection of germline PVs in moderate-penetrance genes (Yadav and Couch 2019). Table 1 lists some of the commonly used criteria for germline genetic testing in clinical practice. Separate criteria exist for Li–Fraumeni syndrome, hereditary diffuse gastric cancer, and Lynch syndrome (Table 1; Li et al. 1988; Park et al. 2000; Chompret et al. 2001; Kaurah et al. 2007; Blair et al. 2020; National Comprehensive Cancer Network 2022), all of which are associated with an increased risk of breast and/or ovarian cancer, classically in the setting of other specific features. The identification of a PV in a gene such as *TP53* or *CDH1* in an individual without a personal or family history consistent with the associated syndrome is another area of uncertainty and ongoing research (Katona et al. 2020; Kratz et al. 2021).

Germline Genetic Testing Criteria in Women with Breast and/or Ovarian Cancer

All women with epithelial ovarian, peritoneal or fallopian tube cancer, as well as their relatives, should be considered for germline genetic testing. However, uptake of genetic testing is low (Childers et al. 2017; Weinmann et al. 2022) as less than one-third of women with ovarian cancer undergo genetic testing (Childers et al. 2017; Kurian et al. 2019). The undertesting of women with ovarian cancer is of serious concern and several factors, including racial disparities and access to testing, have been implicated (Kurian et al. 2019). These factors need to be investigated further to improve the overall rate of germline genetic testing in women with ovarian cancer.

The criteria for germline genetic testing in women with breast cancer are controversial. The NCCN recommends testing based on age at diagnosis, pathology, ethnicity, family history of cancer, presence of metastatic disease, and potential for use of PARP inhibitors (National Comprehensive Cancer Network 2023), whereas the ASBrS recommends germline genetic testing for all women with breast cancer (Manahan et al. 2019). The ASBrS recommendations were based on articles that demonstrated that a significant proportion of PV carriers, up to 50%, were

Table 1. List of commonly used guidelines and prior probability models for germline genetic testing

Testing criteria	Description	Link to criteria
NCCN testing criteria for high-penetrance breast and/or ovarian cancer susceptibility genes (National Comprehensive Cancer Network 2023)	Designed for the detection of *BRCA1, BRCA2 PALB2, CDH1, PTEN, TP53* among women with or at risk of HBOC	https://www.nccn.org/professionals/physician_gls/pdf/genetics_bop.pdf
ASBrS consensus guideline on genetic testing for hereditary breast cancer (Manahan et al. 2019)	Focuses on detection of germline mutations in women with or at risk of breast cancer	https://www.breastsurgeons.org/docs/statements/Consensus-Guideline-on-Genetic-Testing-for-Hereditary-Breast-Cancer.pdf
USPSTF recommendation statement for *BRCA*-related cancer (Owens et al. 2019)	Testing criteria for *BRCA*-related cancers	https://www.uspreventiveservicestaskforce.org/Page/Document/RecommendationStatementFinal/brca-related-cancer-risk-assessment-genetic-counseling-and-genetic-testing1
NICE guidelines on germline genetic testing (National Collaborating Centre for Cancer 2013)	Guidelines for germline genetic testing in the United Kingdom	https://www.nice.org.uk/guidance/cg164/chapter/Recommendations#genetic-testing
Classic Li–Fraumeni syndrome criteria (Li et al. 1988)	Original criteria for identifying patients with Li–Fraumeni syndrome and germline *TP53* mutation	https://www.lfsassociation.org/what-is-lfs/lfs-critieria
Chompret criteria (Chompret et al. 2001)	Designed to identify affected families beyond the classic criteria for Li–Fraumeni syndrome and germline *TP53* mutation	https://www.lfsassociation.org/what-is-lfs/lfs-critieria
Testing criteria for Cowden syndrome/PTEN hamartoma tumor syndrome (National Comprehensive Cancer Network 2023)	Major and minor criteria for the detection of germline *PTEN* mutations	https://www.nccn.org/professionals/physician_gls/pdf/genetics_bop.pdf
IGCLC hereditary diffuse gastric cancer clinical practice guidelines (Blair et al. 2020)	Testing criteria for detection of hereditary diffuse gastric cancer and germline *CDH1* mutations	https://hereditarydiffusegastriccancer.org/wp-content/uploads/2022/09/HDGC-IGCLC-Guideline-2020.pdf
Lynch syndrome criteria (National Comprehensive Cancer Network 2022)	Criteria for the evaluation of Lynch syndrome	https://www.nccn.org/professionals/physician_gls/pdf/genetics_colon.pdf

(NCCN) National Comprehensive Cancer Network, (ASBrS) American Society of Breast Surgeons, (HBOC) hereditary breast and ovarian cancer, (IGCLC) International Gastric Cancer Linkage Consortium, (USPSTF) United States Preventive Services Taskforce, (NICE) National Institute for Health and Care Excellence.

missed by NCCN germline genetic testing criteria (Yang et al. 2018a; Beitsch et al. 2019). However, concerns have been raised about selection biases in these studies, as well as the fact that the majority of PVs that were "missed" were genes either not associated with breast cancer (e.g.,

MUTYH) or genes associated with moderate risks of breast cancer (e.g., *CHEK2* and *ATM*) (Pal et al. 2019). It is well established that several breast cancer patient characteristics, specifically triple-negative breast cancer (TNBC); young-onset breast cancer; Ashkenazi Jewish descent;

and personal or family history of ovarian cancer, pancreatic cancer, or metastatic prostate cancer, are associated with PVs in high penetrance genes such as *BRCA1*, *BRCA2*, and *PALB2*. These factors are incorporated in NCCN guidelines. In the study by Beitsch et al. (2019), 0.8% of those without TNBC or young age of onset had PV in these high-risk genes. In another study of a hospital-based cohort of women with breast cancer, it was demonstrated that testing all women with breast cancer meeting NCCN criteria or diagnosed under the age of 65 may achieve greater than 98% sensitivity for PVs in *BRCA1/2* and greater than 90% sensitivity for other breast cancer predisposition gene PVs without the need to test everyone (Yadav et al. 2020a).

As thresholds for testing continue to fall and be modified, it becomes more important than ever to consider the selection of genes on panels. As will be detailed below, many genes initially included on "breast cancer panels" do not appear to be associated with breast cancer risk, including *RAD50*, *MRE11*, and *NBN*. In addition, it is very important that clinicians appropriately communicate with patients the differing risk between, for example, a *BRCA1* PV and a *CHEK2* PV (the former with lifetime risk of breast cancer, often triple negative, of up to 70% and a lifetime risk of ovarian cancer up to 45%; the latter with a lifetime risk of breast cancer, almost always ER-positive, of up to 25%, with no associated ovarian cancer risk). Detecting a *BRCA1* or *BRCA2* PV may be lifesaving, as risk-reducing oophorectomy has been shown to significantly decrease ovarian cancer-specific and all-cause mortality (Domchek et al. 2010; Finch et al. 2014). In addition, adjuvant olaparib has been shown to confirm an overall survival benefit in *BRCA1/2* PV carriers with high-risk breast cancer (Tutt et al. 2021; Geyer et al. 2022).

Germline Genetic Testing Criteria in Women without a Personal History of Cancer

Among women without a personal history of cancer, germline genetic testing is guided by personal and family history of cancer (Manahan et al. 2019; National Comprehensive Cancer Network 2023). Testing of an affected family member initially is preferred, but testing may be considered in unaffected women if affected family members are not available for testing. All individuals with a family history of epithelial ovarian cancer, male breast cancer, pancreatic cancer, or metastatic prostate cancer qualify for germline genetic testing (National Comprehensive Cancer Network 2023). In women with family history of cancer, age at diagnosis of cancer in family members, number of affected family members, type and features of cancers in the family, and Ashkenazi Jewish ancestry is typically used to assess the need for genetic testing (National Comprehensive Cancer Network 2023). In addition, breast cancer probability models, as recommended by the United States Preventive Services Task Force (USPSTF), are useful to identify women who may benefit from testing in this context (Owens et al. 2019).

As the cost of genetic testing has gone down, there is an increased interest in population-wide genetic testing of all unaffected women (King et al. 2014). In the Ashkenazi Jewish population, universal genetic testing has been demonstrated to be cost-effective (Manchanda et al. 2015), and such screening is now offered to Ashkenazi Jewish women in Israel. Similar decision-analysis models have also shown that universal genetic testing of all women in the general population might be cost-effective (Manchanda et al. 2018). In the United States currently, many women who meet traditional criteria for genetic testing have not been tested and there remains significant racial disparities in testing (Cragun et al. 2017; Kurian et al. 2019). Many studies are ongoing to determine optimal implementation strategies. There are no guidelines in the United States at this time that recommend population screening, although the NCCN states that genetic testing can be considering in those of Ashkenazi Jewish ancestry without any additional risk factors (National Comprehensive Cancer Network 2023).

Germline Genetic Testing Modality and the Role of Multigene Panel Testing

Multigene panels have, for the most part, replaced sequential single-gene testing for evaluation of inherited cancer risk. In 2015, multigene panels were ordered in approximately two-thirds of cases undergoing genetic testing after a diagnosis of breast cancer, whereas *BRCA1/2* testing was ordered in only one-third of cases (Kurian et al. 2018). Although multigene panels can identify additional patients at increased risk for hereditary breast and ovarian cancer (HBOC), the panels are also criticized for the inclusion of genes unrelated to the phenotype at hand (e.g., *SDHA*, *VHL*), genes of unclear clinical utility (e.g., *MRE11*, *NBN*) that often result in high rates of VUS, which increase as the number of genes on the panel increases (Hall et al. 2014; Robson 2014; Axilbund 2016; Thompson et al. 2016; Robson and Domchek 2019). To address some of the challenges with risk interpretation and high rates of VUS with multigene panels, it is recommended that genetic testing should be performed in coordination with professionals with expertise and experience in cancer genetics and consideration be given to the selection of genes included on a panel.

PREVALENCE OF GERMLINE PVs IN CANCER PREDISPOSITION GENES

The frequency of germline PVs in cancer predisposition genes among women at risk depends on several factors, including personal history of cancer, family history, cancer subtype, race/ethnicity, and the selection criteria for patients undergoing testing.

Prevalence of Germline PVs among Women with Breast and/or Ovarian Cancer

Germline PVs in *BRCA1/2* are detected in ~10%–15% of all women with high-grade serous ovarian cancer (Ramus et al. 2007; Walsh et al. 2011; Zhang et al. 2011; Alsop et al. 2012; Song et al. 2014; Norquist et al. 2016; LaDuca et al. 2019). In addition, multigene panels have identified PVs in DNA-repair genes other than *BRCA1/2*, including *RAD51C*, *RAD51D*, *BARD1*, *BRIP1*, etc., in ~5% of women with ovarian cancer (Walsh et al. 2011; Song et al. 2015; Norquist et al. 2016; Lilyquist et al. 2017; Eoh et al. 2018; LaDuca et al. 2019). PVs in mismatch repair genes are also observed in ~1% of

patients with ovarian cancer (Pal et al. 2012; Lilyquist et al. 2017; LaDuca et al. 2019).

Among women with breast cancer referred for genetic testing, ~10% are found to carry a germline PV in cancer predisposition genes (Kurian et al. 2014; Tung et al. 2015; Couch et al. 2017; LaDuca et al. 2019); however, these studies are impacted by ascertainment bias. Recently, two large studies (Dorling et al. 2021; Hu et al. 2021) have provided more precise estimates of the prevalence of germline PVs in breast cancer patients in the general population. Hu et al. investigated 32,247 women with breast cancer (case patients) and 32,544 unaffected women (controls) from population-based studies in the Cancer Risk Estimates Related to Susceptibility (CARRIERS) Consortium. PVs in 12 established breast cancer–predisposition genes (*ATM*, *BARD1*, *BRCA1*, *BRCA2*, *CDH1*, *CHEK2*, *NF1*, *PTEN*, *PALB2*, *RAD51C*, *RAD51D*, and *TP53*) were detected in 5.03% of case patients and in 1.63% of controls. PVs in *BRCA1*, *BRCA2*, and *PALB2* were identified in 2.6% of breast cancer patients. In a parallel study, which included both clinically ascertained and population-based cases and controls, the Breast Cancer Association Consortium examined 60,466 women with breast cancer and 53,461 controls. In the population-based subset of 48,826 patients and 50,703 controls, 5.6% had a PV in one of these 12 genes. These two studies have also provided estimates of relative risk associated with PVs in these genes. It is possible that PVs in other genes confer risks, but as they are rare, larger studies will need to be conducted. Importantly, in both of these studies (Dorling et al. 2021; Hu et al. 2021), PVs in *BRIP1*, *MRE11*, *RAD50*, and *NBN* were not associated with breast cancer, including the *NBN* Slavic founder PV 657del5 (Seemanová et al. 2007). While testing for PVs in *BRIP1* is still relevant due to significant association with ovarian cancer (Suszynska et al. 2020), testing for *MRE11*, *RAD50*, and *NBN* has no clinical utility at this time.

Prevalence of Germline PVs by Breast Cancer Subtype

The prevalence of germline PVs also differs by the specific subtype of breast cancer with the highest frequency, ~15% in breast cancer predisposition genes in women with TNBC undergoing genetic testing (Shimelis et al. 2018; Hu et al. 2020). In addition, differences in relative frequencies of different breast cancer predisposition genes have been observed by the breast cancer subtype. For instance, BRCA1/2, BARD1, PALB2, RAD51C, and RAD51D PVs are enriched in women with ER-negative breast cancer, and TP53 and CHEK2 PVs in women with HER2-positive breast cancer, compared to women with ER-positive/HER2-negative breast cancer (Hu et al. 2020; Mavaddat et al. 2022). Similarly, PVs in ATM or CHEK2 are significantly enriched in women with ER-positive breast cancer compared to ER-negative breast cancer (Hu et al. 2020; Mavaddat et al. 2022). In addition, BRCA1 PVs are less frequent and CDH1 PVs are more common in women with invasive lobular carcinoma compared to women with invasive ductal carcinoma (Yadav et al. 2021). However, the overall frequency of CDH1 PVs, even among women with invasive lobular cancer, is low (0.2%–0.5%) (Yadav et al. 2021). In addition, there is significant ambiguity in the appropriate management of carriers of CDH1 PVs who do not have a personal or family history of gastric cancer (Blair et al. 2020). Hence, testing for CDH1 PVs may not be necessary for all women with invasive lobular cancer due to its low prevalence and challenges in management of gastric cancer risk.

Racial and Ethnic Differences in the Frequency of Germline PVs

Racial and ethnic differences in the frequency of germline PVs are known to exist. In particular, ~2.5% of unaffected women of Ashkenazi Jewish ancestry are carriers of germline PVs in BRCA1/2 (Rubinstein 2004). In Ashkenazi Jewish women with breast cancer, ~10% are found to carry germline PVs in BRCA1/2, which increases to 20%–30% for those diagnosed at a younger age (Abeliovich et al. 1997; Rubinstein 2004). This is primarily due to three founder PVs in BRCA1/2 (c.68_69delAG and c.5266dupC in BRCA1, and c.5946delT in BRCA2) in women with Ashkenazi Jewish ancestry (Levy-Lahad

et al. 1997). Several other founder PVs in breast cancer predisposition genes have been described in other races/ethnicities, which may change the relative frequency of germline PVs in specific populations. The CHEK2_c.1100delC is found in ~1%–2% of non-Hispanic White women of European ancestry with breast cancer (CHEK2 Breast Cancer Case-Control Consortium 2004). This PV is not commonly seen in women of other races/ethnicities with breast cancer (Caswell-Jin et al. 2018; Yadav et al. 2020b).

The frequency of germline PVs in minority populations is an area of ongoing research. Initial studies reported differences in the relative frequencies of PVs in several genes compared to non-Hispanic Whites (Caswell-Jin et al. 2018; Yadav et al. 2020b). For instance, BRCA1/2, RAD51C, and RAD51D PVs are noted to be more common in minority populations (Palmer et al. 2020; Yadav et al. 2020b). More recently, in the population-based CARRIERS study, among 3946 Black women and 25,287 non-Hispanic White women with breast cancer, there was no statistically significant difference by race in the combined prevalence of PVs in the 12 breast cancer susceptibility genes evaluated (5.65% in Black vs. 5.06% in non-Hispanic White women; $P = 0.12$). The prevalence of PVs in CHEK2 was higher in non-Hispanic White than Black patients (1.29% vs. 0.38%; $P < 0.001$), whereas Black patients had a higher prevalence of PVs in BRCA2 (1.80% vs. 1.24%; $P = 0.005$) and PALB2 (1.01% vs. 0.40%; $P < 0.001$). For ER-negative breast cancer, the prevalence of PVs was 10.3% and not different except for PALB2, which was higher in Black women. In women diagnosed before age 50 yr, there was no difference in overall prevalence of PVs in Black versus non-Hispanic White women (8.83% vs. 10.04%; $P = 0.25$), and among individual genes, only CHEK2 PV prevalence differed by race. After adjustment for age at diagnosis, the standardized prevalence ratio of PVs in non-Hispanic White relative to Black women was 1.08 (95% CI, 1.02–1.14), and there was no longer a statistically significant difference in BRCA2 PV prevalence (Palmer et al. 2020; Domchek et al. 2021). Since minority populations have been understudied, the rates of VUS are known to be higher in minorities compared to non-Hispanic Whites (Caswell-Jin et al. 2018; Yadav et al. 2020b).

Prevalence of Germline PVs among Men with Breast Cancer

Male relatives of women with inherited breast or ovarian cancer syndromes are at risk of breast cancer. Among men with breast cancer undergoing multigene panel testing, the frequency of germline PVs is noted to be ~18% (Pritzlaff et al. 2017). *BRCA2* is the most common PV noted in this population, detected in 11% of men, followed by germline *CHEK2* PVs in ~4% (Pritzlaff et al. 2017). Due to the high frequency of germline PVs in *BRCA1/2* and other genes, all men with breast cancer are recommended to undergo germline genetic testing (National Comprehensive Cancer Network 2023).

Prevalence of Germline PVs among Women without a Personal History of Cancer in the General Population

In women without a personal history of cancer, the frequency of germline PVs is dependent on several factors, including race/ethnicity and family history. Among unaffected women of Ashkenazi Jewish ancestry, the frequency of germline PVs in *BRCA1/2* is approximately one in 40 (Rubinstein 2004). Recently, two population-based case-control studies including unaffected women in the general population who were not selected for family history, the frequency of germline PVs in cancer predisposition genes was 1.6% (Dorling et al. 2021; Hu et al. 2021), with the prevalence of *BRCA1*, *BRCA2*, and *PALB2* of 0.47% and 0.49%, respectively, which is consistent with other estimates (Maxwell et al. 2016).

INTERPRETING RISK OF BREAST AND/OR OVARIAN CANCER IN PV CARRIERS

Germline PVs in *BRCA1/2* and several other DNA-repair genes have been identified to be associated with an increased risk of breast and/or ovarian cancer (Norquist et al. 2016; Kurian et al.

2017; LaDuca et al. 2019; Dorling et al. 2021; Hu et al. 2021). Germline PVs in *BRCA1* is associated with ~10- to 15-fold increase of ovarian cancer, while PVs in *BRCA2* is associated with a 5- to 10-fold increase compared to women in the general population (Kurian et al. 2017; Lilyquist et al. 2017; LaDuca et al. 2019). The cumulative ovarian cancer risk by age 70–80 yr is estimated to be ~40%–50% for *BRCA1* and 15%–20% for *BRCA2* PV carriers (Kuchenbaecker et al. 2017a; Chen et al. 2020). Apart from *BRCA1/2*, PVs in *ATM, BRIP1, MLH1, MSH2, MSH6, PALB2, RAD51C,* and *RAD51D* have also been noted to be associated with an increased risk of ovarian cancer (Table 2; Ramus et al. 2015; Kurian et al. 2017; Lilyquist et al. 2017; Domchek and Robson 2019; LaDuca et al. 2019), although with varied level of risk. For example, the absolute risk of ovarian cancer associated with PV in *ATM* may be less than 3% (Norquist et al. 2016; Kurian et al. 2017; Lilyquist et al. 2017), a level at which risk-reducing salpingo-oophorectomy is not routinely recommended; for *PALB2*, the lifetime risk is ~5% (Yang et al. 2020; Song et al. 2021), whereas for *RAD51C, RAD51D,* and *BRIP1*, it is closer to 10% (Ramus et al. 2015; Song et al. 2015; Domchek and Robson 2019). There is insufficient or conflicting evidence for associations between germline PVs in several other genes such as *BARD1, NBN,* and *PMS2* (Domchek and Robson 2019), and further studies are needed to clarify the role of these PVs in ovarian cancer. PVs in *CHEK2, MRE11, RAD50,* and *RAD51B* are not associated with increased ovarian cancer risk (Song et al. 2015; Norquist et al. 2016; Lilyquist et al. 2017).

Genes associated with increased risk of breast cancer can be divided into two categories: high-risk and moderate-risk genes. Germline PVs in the high-risk genes confer a fourfold or higher risk of breast cancer, while PVs in the moderate-risk genes confer approximately two- to fourfold risk of breast cancer, compared to women in the general population. *BRCA1, BRCA2, PALB2, TP53, PTEN,* and *CDH1* are considered to be high-risk genes, whereas *ATM* and *CHEK2* are considered moderate-risk genes for breast cancer risk (Couch et al. 2017; Dorling et al. 2021; Hu et al. 2021; Yadav et al. 2021). The

Table 2. Estimates of breast and/or ovarian cancer risk associated with germline PVs in cancer predisposition genes

Gene	Breast cancer risk and OR[a] or lifetime risk[b]	Ovarian cancer risk and OR[a] or lifetime risk[b]
ATM	Moderately increased risk; lifetime risk: 20%–25% (Dorling et al. 2021; Hu et al. 2021)	Likely increased risk; OR: 1.9–2.2 (Kurian et al. 2017; Lilyquist et al. 2017; LaDuca et al. 2019)
BARD1	Moderately increased risk of TNBC; OR for TNBC: 2.5 (Hu et al. 2021)	Unknown; unlikely to be associated with increased risk (Ramus et al. 2015; Kurian et al. 2017; Lilyquist et al. 2017; LaDuca et al. 2019)
BRCA1	High-risk of breast cancer; lifetime risk: 50% (Dorling et al. 2021; Hu et al. 2021)	Increased risk; lifetime risk: 40%–50% (Kuchenbaecker et al. 2017a; Chen et al. 2020)
BRCA2	High-risk of breast cancer; lifetime risk: 45%–50% (Dorling et al. 2021; Hu et al. 2021)	Increased risk; lifetime risk: 15%–20% (Kuchenbaecker et al. 2017a; Chen et al. 2020)
BRIP1	Unlikely to be associated with breast cancer risk (Dorling et al. 2021; Hu et al. 2021)	Increased risk; OR: 2.6–4.9 (Ramus et al. 2015; Kurian et al. 2017; Lilyquist et al. 2017; LaDuca et al. 2019)
CHEK2	Moderately increased risk; lifetime risk: 20%–25% (Dorling et al. 2021; Hu et al. 2021)	No increased risk (Kurian et al. 2017; Lilyquist et al. 2017; LaDuca et al. 2019)
CDH1	High risk of breast cancer, specifically lobular breast cancer; lifetime risk: 40%–60% (Hansford et al. 2015; Roberts et al. 2019)	Unknown
Lynch syndrome genes (*MLH1*, *MSH2*, *MSH6*, and *PMS2*)	Unknown—possible increased risk especially for *MSH6* (Dorling et al. 2021; Hu et al. 2021)	Increased risk, especially in *MLH1*, *MSH2*, and *MSH6* carriers; lifetime risk for *MLH1*: 11%–20%; lifetime risk for *MSH2*: 15%–24%; OR for *MSH6*: 1.9–5.0; unclear risk for *PMS2* (Domchek and Robson 2019)
MRE11A	Risk not increased (Dorling et al. 2021; Hu et al. 2021)	Risk not increased (Norquist et al. 2016; Kurian et al. 2017; LaDuca et al. 2019)
NBN	No evidence of increased risk overall (Dorling et al. 2021; Hu et al. 2021); possible moderately increased risk for individuals with 657del5 variant (Steffen et al. 2006; Zhang et al. 2013)	Unknown; unlikely to be associated with increased risk
NF1	Moderately increased risk of breast cancer in the setting of clinical NF1; OR: 3.2 (LaDuca et al. 2019)	Risk not increased (LaDuca et al. 2019)
PALB2	High risk of breast cancer; lifetime risk: 35%–50% (Dorling et al. 2021; Hu et al. 2021)	Increased risk; lifetime risk: 5% (Yang et al. 2020; Song et al. 2021)
PTEN	High risk of breast cancer; OR: 8.8 (LaDuca et al. 2019); lifetime risk: 85% (Tan et al. 2012)	Unknown
RAD50	Risk not increased (Dorling et al. 2021; Hu et al. 2021)	Risk not increased (Norquist et al. 2016; LaDuca et al. 2019)

Continued

Table 2. *Continued*

Gene	Breast cancer risk and OR[a] or lifetime risk[b]	Ovarian cancer risk and OR[a] or lifetime risk[b]
RAD51C	Increased risk of TNBC; OR for TNBC: 2.2–5.7 (Dorling et al. 2021; Hu et al. 2021)	Increased risk; OR: 5–8 (Kurian et al. 2017; Lilyquist et al. 2017; LaDuca et al. 2019)
RAD51D	Increased risk of TNBC; OR for TNBC: 3.9–6.0 (Dorling et al. 2021; Hu et al. 2021)	Increased risk; OR: 3–6 (Kurian et al. 2017; Lilyquist et al. 2017; LaDuca et al. 2019)
TP53	High risk of breast cancer; OR: 4.3–5.3 (Kurian et al. 2017; LaDuca et al. 2019)	Unknown—conflicting evidence (Kurian et al. 2017; LaDuca et al. 2019); high risk of multiple other cancers

(OR) Odds ratio, (TNBC) triple-negative breast cancer.
[a]Compared to unaffected women in the reference population. Estimates are approximate ratios.
[b]Lifetime risk by age 70 to 80. Estimates are approximate percentages.

lifetime risk (up to age 80) of breast cancer from population-based studies is estimated to be ~50% for *BRCA1* and *BRCA2*, 35% for *PALB2*, and 25% for *ATM* and *CHEK2* PV carriers (Hu et al. 2021).

The risk of breast cancer with PVs in mismatch repair genes (*MLH1*, *MSH2*, *MSH6*, and *PMS2*) has not been established: some studies have shown an association with breast cancer risk, while others have not (Harkness et al. 2015; Couch et al. 2017; Goldberg et al. 2017; Kurian et al. 2017; LaDuca et al. 2019; Dorling et al. 2021; Hu et al. 2021). Among the mismatch repair genes, PVs in *MSH6* were observed to be twofold enriched in breast cancer cases compared to matched controls in two large studies recently, suggesting a potential association with breast cancer risk (Dorling et al. 2021; Hu et al. 2021). However, the association between PVs in *MSH6* and breast cancer risk needs to be further investigated and, at present, there is insufficient evidence to suggest that PVs in mismatch repair genes are associated with an increased risk of breast cancer. Case-control studies have also demonstrated that germline PVs in *NBN*, *RAD50*, and *MRE11A* are not associated with an increased risk of breast cancer (Couch et al. 2017; Kurian et al. 2017; Dorling et al. 2021; Hu et al. 2021). Germline PVs in *BRIP1*, *RAD51C*, and *RAD51D* are associated with an increased risk of TNBC, and their associations with other subtypes of breast cancer have not been established (Shimelis et al. 2018; Hu et al. 2020). Similarly, the *FANCM* p. Arg658ter truncating variant might also be as-

sociated with an increased risk of TNBC only (Figlioli et al. 2019).

SPECIAL CONSIDERATIONS IN INTERPRETING CANCER RISK IN PV CARRIERS

Incorporating Effects of Age and Family History in Cancer Risk Assessment in PV Carriers

The residual or remaining risk of breast or ovarian cancer in PV carriers at any given age tends to be highest in women under the age of 50, who have not yet undergone menopause. As the cost and accessibility to genetic testing continue to improve, many women over the age of 65 are undergoing germline genetic testing. Among PV carriers over the age of 65, the remaining lifetime risk of breast cancer up to age 85 is ~20% for *BRCA1/2*, 15% for *PALB2* and *CHEK2*, and 10% for *ATM* (Boddicker et al. 2021). In addition to the effects of age, family history of cancer can further help discern breast or ovarian cancer risk among germline PV carriers. Family history may be an indicator of other underlying genetic factors that may modify the absolute risk of breast and/or ovarian cancer. For instance, breast cancer risk for *PALB2* PV carriers by age 70 has been estimated at 33% for those with family history of but 58% for those with two or more first-degree relatives with breast cancer (Antoniou et al. 2014). Hence, it is important to incorporate the effects of age and family history in breast or ovarian cancer risk assessment in PV carriers to guide discussions on cancer-risk management options.

Contralateral Breast-Cancer Risk and Risk of Other Cancers in PV Carriers

In carriers of *BRCA1/2* PVs with breast cancer, the risk of contralateral breast cancer is estimated to be ~25% to 40% (Metcalfe et al. 2004; Kuchenbaecker et al. 2017a). Therefore, in women with newly diagnosed breast cancer presurgical genetic testing may inform surgical options, such as bilateral mastectomies to reduce the risk of subsequent contralateral disease (Hartmann et al. 2001; van Sprundel et al. 2005; Kaas et al. 2010; Metcalfe et al. 2014). A similar approach may be applied to women with germline PVs in other high-risk genes. Presurgical genetic testing in women with breast cancer is feasible and has been demonstrated to reduce the need for additional surgeries (Lokich et al. 2014; Chiba et al. 2016; Yadav et al. 2017; Yadav et al. 2018). The risk of contralateral breast cancer is approximately twofold higher among carriers of truncating variants in *CHEK2* including c.1100delC, but the contralateral breast cancer risk is not elevated for the *CHEK2* c.470T > C (p.Ile157Thr) variant (Yadav et al. 2023). *ATM* PV carriers do not have an increased risk of contralateral breast cancer (Weitzel et al. 2021; Yadav et al. 2023). Among *PALB2* PV carriers with breast cancer, compared to noncarriers, the risk of contralateral breast cancer may be significantly higher in women with ER-negative tumor, but the risk may not be elevated in *PALB2* PV carriers with ER-positive breast cancer (Yadav et al. 2023). However, these findings need to be investigated further due to the small number of PV carriers in the study (Yadav et al. 2023). In addition to the risk of breast and ovarian cancer, germline PVs in several genes are also associated with an increased risk of other cancers, including melanoma and pancreatic cancer in both men and women, and prostate cancer in men (Cybulski et al. 2007; Mersch et al. 2015; Yadav et al. 2019; Chen et al. 2022). A comprehensive evaluation of the risk of breast and other cancers in PV carriers is essential to guide risk management strategies.

Interpreting VUS Results

Women who are found to have a VUS result in the absence of a pathogenic or likely PV are managed similar to those receiving negative results. The rate of VUS has increased after the introduction of multigene panels and is dependent on the number of genes included in the panels (Slavin et al. 2018). In addition, minority populations are known to have high VUS rates because they are relatively understudied and are underrepresented in the majority of reference genetic databases (Caswell-Jin et al. 2018; Yadav et al. 2020b). VUS results can lead to anxiety and unwarranted interventions when the risk is not interpreted correctly (Hamilton et al. 2009; O'Neill et al. 2009; Eccles et al. 2015; Welsh et al. 2017). Variant reclassification is an area of ongoing research and innovative novel methods for faster reclassification of variants are now being explored through incorporation of functional, structural, and clinical data (Findlay et al. 2018; Boonen et al. 2019; Richardson et al. 2021; Hu et al. 2022).

Risk by Specific Variants and Genomic Regions in PV Carriers

Specific variants or regions within a gene may confer different risk of breast and/or ovarian cancer. For instance, breast and ovarian cancer risks vary by type and location of *BRCA1/2* PVs and separate breast and ovarian cancer cluster regions have been identified within the *BRCA1/2* genes that confer differential breast or ovarian cancer risks (Rebbeck et al. 2015). In addition, specific germline PVs such as the *ATM* c.7271T > G is associated with a high risk of breast cancer (lifetime risk of ~70%), and carriers of this PV should be managed differently than other *ATM* carriers (van Os et al. 2016). Similarly, risk management of low penetrance variants in *BRCA1/2* or *CHEK2* should also be different compared to other usual variants in these genes (Al-Mulla et al. 2009; Han et al. 2013). For instance, the *CHEK2* c.470T > C (p.Ile157Thr) and c.1283C > T (p.Ser428Phe) are both classified as PVs but the risk of breast cancer in carriers is only marginally increased (1.2- to 1.4-fold) compared to noncarriers (Kilpivaara et al. 2004; Laitman et al. 2022). In contrast, the truncating *CHEK2* c.1100delC variant confers an approximately twofold increased risk of breast cancer in carriers compared to noncarriers (CHEK2 Breast Cancer Case-Control Consortium 2004). It is important to recognize these differences at the variant level in a gene to interpret the cancer risk accurately.

Integrating Polygenic Risk Scores in Risk Estimation

There has been increased interest in polygenic risk scores (PRS) to further stratify the level of cancer risk in PV carriers and those without a germline PV. PRS scores are derived from a collection of several single-nucleotide polymorphisms associated with slightly increased risk of cancer, and separate risk scores exist for evaluation of the risk of breast and ovarian cancers (Mavaddat et al. 2015, 2019; Black et al. 2018; Rudolph et al. 2018; Yang et al. 2018b). These scores can help identify a subset of patients with germline PVs who are at higher risk of breast and/or ovarian cancer than expected from the germline PV (Kuchenbaecker et al. 2017b; Muranen et al. 2017; Gao et al. 2019; Lakeman et al. 2021), including men (Barnes et al. 2022). On the flip side, they can also help identify women who may not be at high risk of breast or ovarian cancer despite carrying a germline PV. This may provide an individualized assessment of risk to inform management of cancer risk. Clinical tests using PRS in conjunction with multigene panels for prediction of breast cancer risk are now available in the United States, and there is ongoing work optimizing these tests to aid in risk stratification in *BRCA1*, *BRCA2*, *ATM*, and *CHEK2* carriers (Gallagher et al. 2021; Hughes et al. 2021). The role of PRS in the management of patients, contribution to decision making, and in clinical outcomes is uncertain.

Identifying Clonal Hematopoiesis

Aberrant clonal expansion of somatic variants can occur and is often secondary to clonal hematopoiesis due to increasing age (Jaiswal et al. 2014). This can confound the results of germline

genetic testing. This typically affects germline testing for PVs in the *TP53* gene (Batalini et al. 2019). It is important to distinguish clonal hematopoiesis from germline PVs in *TP53* to avoid unnecessary interventions. Mutant to wild-type allele ratios of the germline PV may provide some guidance in this regard, as clonal hematopoiesis frequently has lower (<25%) ratios (Weitzel et al. 2018). If there is any doubt, ancillary testing should be undertaken, which may include testing of DNA obtained from fibroblasts cultured from skin or eyebrow plucks to confirm the results.

MANAGEMENT OF BREAST AND/OR OVARIAN CANCER RISK IN PV CARRIERS

In PV carriers, accurate estimation of risk is essential to determine the type of intervention needed to manage the breast, ovarian, and other cancer risks. The options could be risk-reducing surgery, intensive screening, or usual population-recommended screening as detailed by multiple guidelines (National Comprehensive Cancer Network 2023; Sessa et al. 2023). The selection of these options is dependent on the level of risk of breast or ovarian cancer, benefit from the intervention, and patient preference. For *BRCA1/2* PV carriers, risk-reducing salpingo-oophorectomy has been associated with reduced breast and ovarian cancer and improved all-cause mortality (Domchek et al. 2010; Finch et al. 2014). Risk-reducing mastectomy significantly reduces the risk of breast cancer. Due to the significant benefits of screening (Hadar et al. 2020), the impact of this risk reduction on mortality is less clear, particularly for *BRCA2* PV carriers (Heemskerk-Gerritsen et al. 2019). The long-term outcomes of surgical or nonsurgical interventions have not been evaluated for carriers of PVs in other genes. Consensus guidelines (National Comprehensive Cancer Network 2023) are helpful in deciding on the appropriate management strategy for these PV carriers. For the majority of moderate-risk genes, intensive surveillance for breast cancer risk through MRI and mammograms is likely appropriate (Whitworth et al. 2022; National Comprehensive Cancer Network 2023). For increased ovarian cancer risk, however, there are limited options for screening and surgical intervention is usually warranted (National Comprehensive Cancer Network 2023); gene-dependent absolute risk of ovarian cancer and timing need to be carefully considered. In *BRCA1/2* PV carriers with breast cancer, adjuvant treatment with tamoxifen may decrease the risk of contralateral breast cancer (Metcalfe et al. 2004; Gronwald et al. 2006; Phillips et al. 2013). However, the role of tamoxifen in the primary prevention of breast cancer in PV carriers has not been studied. Finally, it is also important to recognize that risk assessment and management of the cancer risk is a dynamic process and periodic reassessment based on the changes in personal and family history and the update in guidelines is essential.

From a therapeutic standpoint, PARP inhibitors are standard of care for treatment of *BRCA1/2*-associated ovarian cancer (Mirza et al. 2016; Coleman et al. 2017; Moore et al. 2018), metastatic breast cancer (Robson et al. 2017; Litton et al. 2020), metastatic prostate cancer (Hussain et al. 2020), and metastatic pancreatic cancer (Golan et al. 2019). More recently, one year of olaparib in early-stage, high-risk germline *BRCA1/2*-associated breast cancer has been shown to improve disease-free, distanced disease-free, and overall survival (Tutt et al. 2021; Geyer et al. 2022). PARP inhibitors are also known to lead to tumor response in *PALB2* PV carriers with metastatic breast cancer (Tung et al. 2020). Several other targeted agents, either alone or in combination with PARP inhibitors, are currently in clinical trials for germline PV carriers.

CONCLUSIONS

In the past two decades, there has been a significant improvement in our understanding of hereditary aspects of the breast and ovarian cancer risk. In addition to *BRCA1/2*, several other DNA-repair genes have been identified and multigene panels are now routinely used to evaluate for breast or ovarian cancer risk. In women who test positive for a germline PV, the risk of breast and ovarian cancer is dependent on several factors, including the specific variant and family history, among others. In addition, PRS may fur-

ther help stratify the level of risk in carriers of PV. Ultimately, accurate risk estimation through the incorporation of all modifying factors is essential to provide personalized recommendations for risk management of breast and/or ovarian cancer risk in PV carriers.

REFERENCES

Abeliovich D, Kaduri L, Lerer I, Weinberg N, Amir G, Sagi M, Zlotogora J, Heching N, Peretz T. 1997. The founder mutations 185delAG and 5382insC in BRCA1 and 6174delT in BRCA2 appear in 60% of ovarian cancer and 30% of early-onset breast cancer patients among Ashkenazi women. *Am J Hum Genet* **60:** 505–514.

Al-Mulla F, Bland JM, Serratt D, Miller J, Chu C, Taylor GT. 2009. Age-dependent penetrance of different germline mutations in the BRCA1 gene. *J Clin Pathol* **62:** 350–356. doi:10.1136/jcp.2008.062646

Alsop K, Fereday S, Meldrum C, deFazio A, Emmanuel C, George J, Dobrovic A, Birrer MJ, Webb PM, Stewart C, et al. 2012. BRCA mutation frequency and patterns of treatment response in BRCA mutation-positive women with ovarian cancer: a report from the Australian Ovarian Cancer Study Group. *J Clin Oncol* **30:** 2654–2663. doi:10.1200/JCO.2011.39.8545

Antoniou AC, Casadei S, Heikkinen T, Barrowdale D, Pylkäs K, Roberts J, Lee A, Subramanian D, De Leeneer K, Fostira F, et al. 2014. Breast-cancer risk in families with mutations in PALB2. *N Engl J Med* **371:** 497–506. doi:10.1056/NEJMoa1400382

Axilbund JE. 2016. Panel testing is not a panacea. *J Clin Oncol* **34:** 1433–1435. doi:10.1200/JCO.2015.65.5522

Barnes DR, Silvestri V, Leslie G, McGuffog L, Dennis J, Yang X, Adlard J, Agnarsson BA, Ahmed M, Aittomäki K, et al. 2022. Breast and prostate cancer risks for male BRCA1 and BRCA2 pathogenic variant carriers using polygenic risk scores. *J Natl Cancer Inst* **114:** 109–122. doi:10.1093/jnci/djab147

Batalini F, Peacock EG, Stobie L, Robertson A, Garber J, Weitzel JN, Tung NM. 2019. Li–Fraumeni syndrome: not a straightforward diagnosis anymore—the interpretation of pathogenic variants of low allele frequency and the differences between germline PVs, mosaicism, and clonal hematopoiesis. *Breast Cancer Res* **21:** 107. doi:10.1186/s13058-019-1193-1

Beitsch PD, Whitworth PW, Hughes K, Patel R, Rosen B, Compagnoni G, Baron P, Simmons R, Smith LA, Grady I, et al. 2019. Underdiagnosis of hereditary breast cancer: are genetic testing guidelines a tool or an obstacle? *J Clin Oncol* **37:** 453–460. doi:10.1200/JCO.18.01631

Black MH, Li S, LaDuca H, Chen J, Hoiness R, Gutierrez S, Lu HM, Dolinsky JS, Xu J, Vachon C, et al. 2018. Polygenic risk score for breast cancer in high-risk women. *J Clin Oncol* **36:** 1508–1508. doi:10.1200/JCO.2018.36.15_suppl.1508

Blair VR, McLeod M, Carneiro F, Coit DG, D'Addario JL, van Dieren JM, Harris KL, Hoogerbrugge N, Oliveira C, van der Post RS, et al. 2020. Hereditary diffuse gastric cancer: updated clinical practice guidelines. *Lancet Oncol* **21:** e386–e397. doi:10.1016/S1470-2045(20)30219-9

Boddicker NJ, Hu C, Weitzel JN, Kraft P, Nathanson KL, Goldgar DE, Na J, Huang H, Gnanaolivu RD, Larson N, et al. 2021. Risk of late-onset breast cancer in genetically predisposed women. *J Clin Oncol* **39:** 3430–3440. doi:10.1200/JCO.21.00531

Boonen R, Rodrigue A, Stoepker C, Wiegant WW, Vroling B, Sharma M, Rother MB, Celosse N, Vreeswijk MPG, Couch F, et al. 2019. Functional analysis of genetic variants in the high-risk breast cancer susceptibility gene PALB2. *Nat Commun* **10:** 5296. doi:10.1038/s41467-019-13194-2

Caswell-Jin JL, Gupta T, Hall E, Petrovchich IM, Mills MA, Kingham KE, Koff R, Chun NM, Levonian P, Lebensohn AP, et al. 2018. Racial/ethnic differences in multiple-gene sequencing results for hereditary cancer risk. *Genet Med* **20:** 234–239. doi:10.1038/gim.2017.96

CHEK2 Breast Cancer Case-Control Consortium. 2004. CHEK2*1100delC and susceptibility to breast cancer: a collaborative analysis involving 10,860 breast cancer cases and 9,065 controls from 10 studies. *Am J Hum Genet* **74:** 1175–1182. doi:10.1086/421251

Chen J, Bae E, Zhang L, Hughes K, Parmigiani G, Braun D, Rebbeck TR. 2020. Penetrance of breast and ovarian cancer in women who carry a BRCA1/2 mutation and do not use risk-reducing salpingo-oophorectomy: an updated meta-analysis. *JNCI Cancer Spectr* **4:** pkaa029. doi:10.1093/jncics/pkaa029

Chen F, Park SL, Wilkens LR, Wan P, Hart SN, Hu C, Yadav S, Couch FJ, Conti DV, de Smith AJ, et al. 2022. Genetic risk of second primary cancer in breast cancer survivors: The Multiethnic Cohort Study. *Cancer Res* **82:** 3201–3208. doi:10.1158/0008-5472.CAN-21-4461

Chiba A, Hoskin TL, Hallberg EJ, Cogswell JA, Heins CN, Couch FJ, Boughey JC. 2016. Impact that timing of genetic mutation diagnosis has on surgical decision making and outcome for BRCA1/BRCA2 mutation carriers with breast cancer. *Ann Surg Oncol* **23:** 3232–3238. doi:10.1245/s10434-016-5328-7

Childers CP, Childers KK, Maggard-Gibbons M, Macinko J. 2017. National estimates of genetic testing in women with a history of breast or ovarian cancer. *J Clin Oncol* **35:** 3800–3806. doi:10.1200/JCO.2017.73.6314

Chompret A, Abel A, Stoppa-Lyonnet D, Brugieres L, Pages S, Feunteun J, Bonaiti-Pellie C. 2001. Sensitivity and predictive value of criteria for p53 germline mutation screening. *J Med Genet* **38:** 43–47. doi:10.1136/jmg.38.1.43

Coleman RL, Oza AM, Lorusso D, Aghajanian C, Oaknin A, Dean A, Colombo N, Weberpals JI, Clamp A, Scambia G, et al. 2017. Rucaparib maintenance treatment for recurrent ovarian carcinoma after response to platinum therapy (ARIEL3): a randomised, double-blind, placebo-controlled, phase 3 trial. *Lancet* **390:** 1949–1961. doi:10.1016/S0140-6736(17)32440-6

Couch FJ, Shimelis H, Hu C, Hart SN, Polley EC, Na J, Hallberg E, Moore R, Thomas A, Lilyquist J, et al. 2017. Associations between cancer predisposition testing panel genes and breast cancer. *JAMA Oncol* **3:** 1190–1196. doi:10.1001/jamaoncol.2017.0424

Cragun D, Weidner A, Lewis C, Bonner D, Kim J, Vadaparampil ST, Pal T. 2017. Racial disparities in BRCA testing

and cancer risk management across a population-based sample of young breast cancer survivors. *Cancer* **123:** 2497–2505. doi:10.1002/cncr.30621

Cybulski C, Wokołorczyk D, Kładny J, Kurzwaski G, Suchy J, Grabowska E, Gronwald J, Huzarski T, Byrski T, Górski B, et al. 2007. Germline CHEK2 mutations and colorectal cancer risk: different effects of a missense and truncating mutations? *Eur J Hum Genet* **15:** 237–241. doi:10.1038/sj.ejhg.5201734

Domchek SM, Robson ME. 2019. Update on genetic testing in gynecologic cancer. *J Clin Oncol* **37:** 2501–2509. doi:10.1200/JCO.19.00363

Domchek SM, Friebel TM, Singer CF, Evans DG, Lynch HT, Isaacs C, Garber JE, Neuhausen SL, Matloff E, Eeles R, et al. 2010. Association of risk-reducing surgery in BRCA1 or BRCA2 mutation carriers with cancer risk and mortality. *JAMA* **304:** 967–975. doi:10.1001/jama.2010.1237

Domchek SM, Yao S, Chen F, Hu C, Hart SN, Goldgar DE, Nathanson KL, Ambrosone CB, Haiman CA, Couch FJ, et al. 2021. Comparison of the prevalence of pathogenic variants in cancer susceptibility genes in Black women and non-Hispanic white women with breast cancer in the United States. *JAMA Oncol* **7:** 1045–1050. doi:10.1001/jamaoncol.2021.1492

Dorling L, Carvalho S, Allen J, González-Neira A, Luccarini C, Wahlström C, Pooley KA, Parsons MT, Fortuno C, Wang Q, et al. 2021. Breast cancer risk genes—association analysis in more than 113,000 women. *N Engl J Med* **384:** 428–439. doi:10.1056/NEJMoa1913948

Eccles DM, Mitchell G, Monteiro AN, Schmutzler R, Couch FJ, Spurdle AB, Gómez-García EB; ENIGMA Clinical Working Group. 2015. BRCA1 and BRCA2 genetic testing-pitfalls and recommendations for managing variants of uncertain clinical significance. *Ann Oncol* **26:** 2057–2065. doi:10.1093/annonc/mdv278

Eoh KJ, Kim JE, Park HS, Lee ST, Park JS, Han JW, Lee JY, Kim S, Kim SW, Kim JH, et al. 2018. Detection of germline mutations in patients with epithelial ovarian cancer using multi-gene panels: beyond BRCA1/2. *Cancer Res Treat* **50:** 917–925. doi:10.4143/crt.2017.220

Figlioli G, Bogliolo M, Catucci I, Caleca L, Lasheras SV, Pujol R, Kiiski JI, Muranen TA, Barnes DR, Dennis J, et al. 2019. The FANCM:p.Arg658* truncating variant is associated with risk of triple-negative breast cancer. *NPJ Breast Cancer* **5:** 38–38. doi:10.1038/s41523-019-0127-5

Finch AP, Lubinski J, Møller P, Singer CF, Karlan B, Senter L, Rosen B, Maehle L, Ghadirian P, Cybulski C, et al. 2014. Impact of oophorectomy on cancer incidence and mortality in women with a BRCA1 or BRCA2 mutation. *J Clin Oncol* **32:** 1547–1553. doi:10.1200/JCO.2013.53.2820

Findlay GM, Daza RM, Martin B, Zhang MD, Leith AP, Gasperini M, Janizek JD, Huang X, Starita LM, Shendure J. 2018. Accurate classification of BRCA1 variants with saturation genome editing. *Nature* **562:** 217–222. doi:10.1038/s41586-018-0461-z

Forbes C, Fayter D, de Kock S, Quek RG. 2019. A systematic review of international guidelines and recommendations for the genetic screening, diagnosis, genetic counseling, and treatment of BRCA-mutated breast cancer. *Cancer Manag Res* **11:** 2321–2337. doi:10.2147/CMAR.S189627

Gallagher S, Hughes E, Kurian AW, Domchek SM, Garber J, Probst B, Morris B, Tshiaba P, Meek S, Rosenthal E, et al.

2021. Comprehensive breast cancer risk assessment for *CHEK2* and *ATM* pathogenic variant carriers incorporating a polygenic risk score and the Tyrer-Cuzick model. *JCO Precis Oncol* **5**: 1073–1081. doi:10.1200/PO.20.00484

Gao C, Polley EC, Hu C, Hart SN, Gnanaolivu R, Lee KY, Na J, Boddicker NJ, Samara R, Auer P, et al. 2019. PD3-02. Polygenic risk scores provide clinically meaningful risk stratification among women carrying moderate penetrance pathogenic variants in breast cancer predisposition genes: results from the CARRIERS study. In *San Antonio Breast Cancer Symposium*, San Antonio, TX.

Geyer CE Jr, Garber JE, Gelber RD, Yothers G, Taboada M, Ross L, Rastogi P, Cui K, Arahmani A, Aktan G, et al. 2022. Overall survival in the OlympiA phase III trial of adjuvant olaparib in patients with germline pathogenic variants in BRCA1/2 and high-risk, early breast cancer. *Ann Oncol* **33**: 1250–1268. doi:10.1016/j.annonc.2022.09.159

Golan T, Hammel P, Reni M, Van Cutsem E, Macarulla T, Hall MJ, Park JO, Hochhauser D, Arnold D, Oh DY, et al. 2019. Maintenance olaparib for germline *BRCA*-mutated metastatic pancreatic cancer. *N Engl J Med* **381**: 317–327. doi:10.1056/NEJMoa1903387

Goldberg M, Bell K, Aronson M, Semotiuk K, Pond G, Gallinger S, Zbuk K. 2017. Association between the Lynch syndrome gene MSH2 and breast cancer susceptibility in a Canadian familial cancer registry. *J Med Genet* **54**: 742–746. doi:10.1136/jmedgenet-2017-104542

Gronwald J, Tung N, Foulkes WD, Offit K, Gershoni R, Daly M, Kim-Sing C, Olsson H, Ainsworth P, Eisen A, et al. 2006. Tamoxifen and contralateral breast cancer in BRCA1 and BRCA2 carriers: an update. *Int J Cancer* **118**: 2281–2284. doi:10.1002/ijc.21536

Hadar T, Mor P, Amit G, Lieberman S, Gekhtman D, Rabinovitch R, Levy-Lahad E. 2020. Presymptomatic awareness of germline pathogenic *BRCA* variants and associated outcomes in women with breast cancer. *JAMA Oncol* **6**: 1460–1463. doi:10.1001/jamaoncol.2020.2059

Hall JM, Lee MK, Newman B, Morrow JE, Anderson LA, Huey B, King MC. 1990. Linkage of early-onset familial breast cancer to chromosome 17q21. *Science* **50**: 1684–1689. doi:10.1126/science.2270482

Hall MJ, Forman AD, Pilarski R, Wiesner G, Giri VN. 2014. Gene panel testing for inherited cancer risk. *J Natl Compr Canc Netw* **12**: 1339–1346. doi:10.6004/jnccn.2014.0128

Hamilton JG, Lobel M, Moyer A. 2009. Emotional distress following genetic testing for hereditary breast and ovarian cancer: a meta-analytic review. *Health Psychol* **28**: 510–518. doi:10.1037/a0014778

Han FF, Guo CL, Liu LH. 2013. The effect of *CHEK2* variant I157T on cancer susceptibility: evidence from a meta-analysis. *DNA Cell Biol* **32**: 329–335. doi:10.1089/dna.2013.1970

Hansford S, Kaurah P, Li-Chang H, Woo M, Senz J, Pinheiro H, Schrader KA, Schaeffer DF, Shumansky K, Zogopoulos G, et al. 2015. Hereditary diffuse gastric cancer syndrome: CDH1 mutations and beyond. *JAMA Oncol* **1**: 23–32. doi:10.1001/jamaoncol.2014.168

Harkness EF, Barrow E, Newton K, Green K, Clancy T, Lalloo F, Hill J, Evans DG. 2015. Lynch syndrome caused by *MLH1* mutations is associated with an increased risk of breast cancer: a cohort study. *J Med Genet* **52**: 553–556. doi:10.1136/jmedgenet-2015-103216

Hartmann LC, Sellers TA, Schaid DJ, Frank TS, Soderberg CL, Sitta DL, Frost MH, Grant CS, Donohue JH, Woods JE, et al. 2001. Efficacy of bilateral prophylactic mastectomy in BRCA1 and BRCA2 gene mutation carriers. *J Natl Cancer Inst* **93**: 1633–1637. doi:10.1093/jnci/93.21.1633

Heemskerk-Gerritsen BAM, Jager A, Koppert LB, Obdeijn AI, Collée M, Meijers-Heijboer HEJ, Jenner DJ, Oldenburg HSA, van Engelen K, de Vries J, et al. 2019. Survival after bilateral risk-reducing mastectomy in healthy BRCA1 and BRCA2 mutation carriers. *Breast Cancer Res Treat* **177**: 723–733. doi:10.1007/s10549-019-05345-2

Hu C, Polley EC, Yadav S, Lilyquist J, Shimelis H, Na J, Hart SN, Goldgar DE, Shah S, Pesaran T, et al. 2020. The contribution of germline predisposition gene mutations to clinical subtypes of invasive breast cancer from a clinical genetic testing cohort. *J Natl Cancer Inst* **112**: 1231–1241. doi:10.1093/jnci/djaa023

Hu C, Hart SN, Gnanaolivu R, Huang H, Lee KY, Na J, Gao C, Lilyquist J, Yadav S, Boddicker NJ, et al. 2021. A population-based study of genes previously implicated in breast cancer. *N Engl J Med* **384**: 440–451. doi:10.1056/NEJMoa2005936

Hu C, Susswein LR, Roberts ME, Yang H, Marshall ML, Hiraki S, Berkofsky-Fessler W, Gupta S, Shen W, Dunn CA, et al. 2022. Classification of BRCA2 variants of uncertain significance (VUS) using an ACMG/AMP model incorporating a homology-directed repair (HDR) functional assay. *Clin Cancer Res* **28**: 3742–3751. doi:10.1158/1078-0432.CCR-22-0203

Hughes E, Tshiaba P, Wagner S, Judkins T, Rosenthal E, Roa B, Gallagher S, Meek S, Dalton K, Hedegard W, et al. 2021. Integrating clinical and polygenic factors to predict breast cancer risk in women undergoing genetic testing. *JCO Precis Oncol* **5**: PO.20.00246. doi:10.1200/PO.20.00246

Hussain M, Mateo J, Fizazi K, Saad F, Shore N, Sandhu S, Chi KN, Sartor O, Agarwal N, Olmos D, et al. 2020. Survival with olaparib in metastatic castration-resistant prostate cancer. *N Engl J Med* **383**: 2345–2357. doi:10.1056/NEJMoa2022485

Jaiswal S, Fontanillas P, Flannick J, Manning A, Grauman PV, Mar BG, Lindsley RC, Mermel CH, Burtt N, Chavez A, et al. 2014. Age-related clonal hematopoiesis associated with adverse outcomes. *N Engl J Med* **371**: 2488–2498. doi:10.1056/NEJMoa1408617

Kaas R, Verhoef S, Wesseling J, Rookus MA, Oldenburg HS, Peeters MJ, Rutgers EJ. 2010. Prophylactic mastectomy in BRCA1 and BRCA2 mutation carriers: very low risk for subsequent breast cancer. *Ann Surg* **251**: 488–492. doi:10.1097/SLA.0b013e3181c3c36d

Katona BW, Clark DF, Domchek SM. 2020. CDH1 on multigene panel testing: look before you leap. *J Natl Cancer Inst* **112**: 330–334. doi:10.1093/jnci/djz229

Kaurah P, MacMillan A, Boyd N, Senz J, De Luca A, Chun N, Suriano G, Zaor S, Van Manen L, Gilpin C, et al. 2007. Founder and recurrent CDH1 mutations in families with hereditary diffuse gastric cancer. *JAMA* **297**: 2360–2372. doi:10.1001/jama.297.21.2360

Kilpivaara O, Vahteristo P, Falck J, Syrjäkoski K, Eerola H, Easton D, Bartkova J, Lukas J, Heikkilä P, Aittomäki K, et al. 2004. CHEK2 variant I157T may be associated with increased breast cancer risk. *Int J Cancer* **111**: 543–547. doi:10.1002/ijc.20299

King MC, Levy-Lahad E, Lahad A. 2014. Population-based screening for *BRCA1* and *BRCA2*: 2014 Lasker Award. *JAMA* **312**: 1091–1092. doi:10.1001/jama.2014.12483

Kratz CP, Freycon C, Maxwell KN, Nichols KE, Schiffman JD, Evans DG, Achatz MI, Savage SA, Weitzel JN, Garber JE, et al. 2021. Analysis of the Li–Fraumeni spectrum based on an international germline TP53 variant data set: an international agency for research on cancer *TP53* database analysis. *JAMA Oncol* **7**: 1800–1805. doi:10.1001/jamaoncol.2021.4398

Kuchenbaecker KB, Hopper JL, Barnes DR, Phillips KA, Mooij TM, Roos-Blom MJ, Jervis S, van Leeuwen FE, Milne RL, Andrieu N, et al. 2017a. Risks of breast, ovarian, and contralateral breast cancer for *BRCA1* and *BRCA2* mutation carriers. *JAMA* **317**: 2402–2416. doi:10.1001/jama.2017.7112

Kuchenbaecker KB, McGuffog L, Barrowdale D, Lee A, Soucy P, Dennis J, Domchek SM, Robson M, Spurdle AB, Ramus SJ, et al. 2017b. Evaluation of polygenic risk scores for breast and ovarian cancer risk prediction in BRCA1 and BRCA2 mutation carriers. *J Natl Cancer Inst* **109**: djw302. doi:10.1093/jnci/djw302

Kurian AW, Hare EE, Mills MA, Kingham KE, McPherson L, Whittemore AS, McGuire V, Ladabaum U, Kobayashi Y, Lincoln SE, et al. 2014. Clinical evaluation of a multiple-gene sequencing panel for hereditary cancer risk assessment. *J Clin Oncol* **32**: 2001–2009. doi:10.1200/JCO.2013.53.6607

Kurian AW, Hughes E, Handorf EA, Gutin A, Allen B, Hartman AR, Hall MJ. 2017. Breast and ovarian cancer penetrance estimates derived from germline multiple-gene sequencing results in women. *JCO Precis Oncol* doi:10.1200/PO.16.00066

Kurian AW, Ward KC, Hamilton AS, Deapen DM, Abrahamse P, Bondarenko I, Li Y, Hawley ST, Morrow M, Jagsi R, et al. 2018. Uptake, results, and outcomes of germline multiple-gene sequencing after diagnosis of breast cancer. *JAMA Oncol* **4**: 1066–1072. doi:10.1001/jamaoncol.2018.0644

Kurian AW, Ward KC, Howlader N, Deapen D, Hamilton AS, Mariotto A, Miller D, Penberthy LS, Katz SJ. 2019. Genetic testing and results in a population-based cohort of breast cancer patients and ovarian cancer patients. *J Clin Oncol* **37**: 1305–1315. doi:10.1200/JCO.18.01854

LaDuca H, Polley EC, Yussuf A, Hoang L, Gutierrez S, Hart SN, Yadav S, Hu C, Na J, Goldgar DE, et al. 2019. A clinical guide to hereditary cancer panel testing: evaluation of gene-specific cancer associations and sensitivity of genetic testing criteria in a cohort of 165,000 high-risk patients. *Genet Med* **22**: 407–415. doi:10.1038/s41436-019-0633-8

Laitman Y, Nielsen SM, Hatchell KE, Truty R, Bernstein-Molho R, Esplin ED, Friedman E. 2022. Re-evaluating cancer risks associated with the CHEK2 p.Ser428Phe Ashkenazi Jewish founder pathogenic variant. *Fam Cancer* **21**: 305–308. doi:10.1007/s10689-021-00278-6

Lakeman IMM, van den Broek AJ, Vos JAM, Barnes DR, Adlard J, Andrulis IL, Arason A, Arnold N, Arun BK, Balmaña J, et al. 2021. The predictive ability of the 313 variant-based polygenic risk score for contralateral breast cancer risk prediction in women of European ancestry with a heterozygous BRCA1 or BRCA2 pathogenic variant. *Genet Med* **23**: 1726–1737. doi:10.1038/s41436-021-01198-7

Levy-Lahad E, Catane R, Eisenberg S, Kaufman B, Hornreich G, Lishinsky E, Shohat M, Weber BL, Beller U, Lahad A, et al. 1997. Founder BRCA1 and BRCA2 mutations in Ashkenazi Jews in Israel: frequency and differential penetrance in ovarian cancer and in breast-ovarian cancer families. *Am J Hum Genet* **60**: 1059–1067.

Li FP, Fraumeni JF Jr, Mulvihill JJ, Blattner WA, Dreyfus MG, Tucker MA, Miller RW. 1988. A cancer family syndrome in twenty-four kindreds. *Cancer Res* **48**: 5358–5362.

Lilyquist J, LaDuca H, Polley E, Davis BT, Shimelis H, Hu C, Hart SN, Dolinsky JS, Couch FJ, Goldgar DE. 2017. Frequency of mutations in a large series of clinically ascertained ovarian cancer cases tested on multi-gene panels compared to reference controls. *Gynecol Oncol* **147**: 375–380. doi:10.1016/j.ygyno.2017.08.030

Litton JK, Hurvitz SA, Mina LA, Rugo HS, Lee KH, Gonçalves A, Diab S, Woodward N, Goodwin A, Yerushalmi R, et al. 2020. Talazoparib versus chemotherapy in patients with germline BRCA1/2-mutated HER2-negative advanced breast cancer: final overall survival results from the EMBRACA trial. *Ann Oncol* **31**: 1526–1535. doi:10.1016/j.annonc.2020.08.2098

Lokich E, Stuckey A, Raker C, Wilbur JS, Laprise J, Gass J. 2014. Preoperative genetic testing affects surgical decision making in breast cancer patients. *Gynecol Oncol* **134**: 326–330. doi:10.1016/j.ygyno.2014.05.028

Lynch HT, Shaw TG, Lynch JF. 2004. Inherited predisposition to cancer: a historical overview. *Am J Med Genet C Semin Med Genet* **129c**: 5–22. doi:10.1002/ajmg.c.30026

Manahan ER, Kuerer HM, Sebastian M, Hughes KS, Boughey JC, Euhus DM, Boolbol SK, Taylor WA. 2019. Consensus guidelines on genetic testing for hereditary breast cancer from the American Society of Breast Surgeons. *Ann Surg Oncol* **26**: 3025–3031. doi:10.1245/s10434-019-07549-8

Manchanda R, Legood R, Burnell M, McGuire A, Raikou M, Loggenberg K, Wardle J, Sanderson S, Gessler S, Side L, et al. 2015. Cost-effectiveness of population screening for BRCA mutations in Ashkenazi Jewish women compared with family history-based testing. *J Natl Cancer Inst* **107**: 380. doi:10.1093/jnci/dju380

Manchanda R, Patel S, Gordeev VS, Antoniou AC, Smith S, Lee A, Hopper JL, MacInnis RJ, Turnbull C, Ramus SJ, et al. 2018. Cost-effectiveness of population-based BRCA1, BRCA2, RAD51C, RAD51D, BRIP1, PALB2 mutation testing in unselected general population women. *J Natl Cancer Inst* **110**: 714–725. doi:10.1093/jnci/djx265

Marmolejo DH, Wong MYZ, Bajalica-Lagercrantz S, Tischkowitz M, Balmaña J; extended ERN-GENTURIS Thematic Group 3. 2021. Overview of hereditary breast and ovarian cancer (HBOC) guidelines across Europe. *Eur J Med Genet* **64**: 104350. doi:10.1016/j.ejmg.2021.104350

Mavaddat N, Pharoah PD, Michailidou K, Tyrer J, Brook MN, Bolla MK, Wang Q, Dennis J, Dunning AM, Shah M, et al. 2015. Prediction of breast cancer risk based on profiling with common genetic variants. *J Natl Cancer Inst* **107**: djv036. doi:10.1093/jnci/djv036

Mavaddat N, Michailidou K, Dennis J, Lush M, Fachal L, Lee A, Tyrer JP, Chen TH, Wang Q, Bolla MK, et al. 2019. Polygenic risk scores for prediction of breast cancer and breast cancer subtypes. *Am J Hum Genet* **104:** 21–34. doi:10.1016/j.ajhg.2018.11.002

Mavaddat N, Dorling L, Carvalho S, Allen J, González-Neira A, Keeman R, Bolla MK, Dennis J, Wang Q, Ahearn TU, et al. 2022. Pathology of tumors associated with pathogenic germline variants in 9 breast cancer susceptibility genes. *JAMA Oncol* **8:** e216744. doi:10.1001/jamaoncol.2021.6744

Maxwell KN, Domchek SM, Nathanson KL, Robson ME. 2016. Population frequency of germline *BRCA1/2* mutations. *J Clin Oncol* **34:** 4183–4185. doi:10.1200/JCO.2016.67.0554

Mersch J, Jackson MA, Park M, Nebgen D, Peterson SK, Singletary C, Arun BK, Litton JK. 2015. Cancers associated with *BRCA1* and *BRCA2* mutations other than breast and ovarian. *Cancer* **121:** 269–275. doi:10.1002/cncr.29041

Metcalfe K, Lynch HT, Ghadirian P, Tung N, Olivotto I, Warner E, Olopade OI, Eisen A, Weber B, McLennan J, et al. 2004. Contralateral breast cancer in BRCA1 and BRCA2 mutation carriers. *J Clin Oncol* **22:** 2328–2335. doi:10.1200/JCO.2004.04.033

Metcalfe K, Gershman S, Ghadirian P, Lynch HT, Snyder C, Tung N, Kim-Sing C, Eisen A, Foulkes WD, Rosen B, et al. 2014. Contralateral mastectomy and survival after breast cancer in carriers of BRCA1 and BRCA2 mutations: retrospective analysis. *BMJ* **348:** g226. doi:10.1136/bmj.g226

Miki Y, Swensen J, Shattuck-Eidens D, Futreal PA, Harshman K, Tavtigian S, Liu Q, Cochran C, Bennett LM, Ding W, et al. 1994. A strong candidate for the breast and ovarian cancer susceptibility gene *BRCA1*. *Science* **266:** 66–71. doi:10.1126/science.7545954

Mirza MR, Monk BJ, Herrstedt J, Oza AM, Mahner S, Redondo A, Fabbro M, Ledermann JA, Lorusso D, Vergote I, et al. 2016. Niraparib maintenance therapy in platinum-sensitive, recurrent ovarian cancer. *N Engl J Med* **375:** 2154–2164. doi:10.1056/NEJMoa1611310

Moore K, Colombo N, Scambia G, Kim BG, Oaknin A, Friedlander M, Lisyanskaya A, Floquet A, Leary A, Sonke GS, et al. 2018. Maintenance olaparib in patients with newly diagnosed advanced ovarian cancer. *N Engl J Med* **379:** 2495–2505. doi:10.1056/NEJMoa1810858

Muranen TA, Greco D, Blomqvist C, Aittomäki K, Khan S, Hogervorst F, Verhoef S, Pharoah PDP, Dunning AM, Shah M, et al. 2017. Genetic modifiers of CHEK2*1100del C-associated breast cancer risk. *Genet Med* **19:** 599–603. doi:10.1038/gim.2016.147

Narod SA, Feunteun J, Lynch HT, Watson P, Conway T, Lynch J, Lenoir GM. 1991. Familial breast-ovarian cancer locus on chromosome 17q12-q23. *Lancet* **338:** 82–83. doi:10.1097/00006254-199203000-00017

National Collaborating Centre for Cancer. 2013. National Institute for Health and Clinical Excellence: Guidance. In *Familial breast cancer: classification and care of people at risk of familial breast cancer and management of breast cancer and related risks in people with a family history of breast cancer.* National Collaborating Centre for Cancer Cardiff, Wales.

National Comprehensive Cancer Network. 2022. Genetic/familial high-risk assessment: colorectal (Version 2.2022).

National Comprehensive Cancer Network. 2023. Genetic/familial high-risk assessment: breast, ovarian, and pancreatic (Version 2.2023).

Norquist BM, Harrell MI, Brady MF, Walsh T, Lee MK, Gulsuner S, Bernards SS, Casadei S, Yi Q, Burger RA, et al. 2016. Inherited mutations in women with ovarian carcinoma. *JAMA Oncol* **2:** 482–490. doi:10.1001/jamaoncol.2015.5495

O'Neill SC, Rini C, Goldsmith RE, Valdimarsdottir H, Cohen LH, Schwartz MD. 2009. Distress among women receiving uninformative *BRCA1/2* results: 12-month outcomes. *Psychooncology* **18:** 1088–1096. doi:10.1002/pon.1467

Owens DK, Davidson KW, Krist AH, Barry MJ, Cabana M, Caughey AB, Doubeni CA, Epling JW, Kubik M, Landefeld CS, et al. 2019. Risk assessment, genetic counseling, and genetic testing for *BRCA*-related cancer: US Preventive Services Task Force Recommendation Statement. *JAMA* **322:** 652–665. doi:10.1001/jama.2019.10987

Pal T, Akbari MR, Sun P, Lee JH, Fulp J, Thompson Z, Coppola D, Nicosia S, Sellers TA, McLaughlin J, et al. 2012. Frequency of mutations in mismatch repair genes in a population-based study of women with ovarian cancer. *Br J Cancer* **107:** 1783–1790. doi:10.1038/bjc.2012.452

Pal T, Agnese D, Daly M, La Spada A, Litton J, Wick M, Klugman S, Esplin ED, Jarvik GP. 2019. Points to consider: is there evidence to support BRCA1/2 and other inherited breast cancer genetic testing for all breast cancer patients? A statement of the American College of Medical Genetics and Genomics (ACMG). *Genet Med* **22:** 681–685. doi:10.1038/s41436-019-0712-x

Palmer JR, Polley EC, Hu C, John EM, Haiman C, Hart SN, Gaudet M, Pal T, Anton-Culver H, Trentham-Dietz A, et al. 2020. Contribution of germline predisposition gene mutations to breast cancer risk in African American women. *J Natl Cancer Inst* **112:** 1213–1221. doi:10.1093/jnci/djaa040

Park JG, Yang HK, Kim WH, Caldas C, Yokota J, Guilford PJ. 2000. Report on the first meeting of the International Collaborative Group on Hereditary Gastric Cancer. *J Natl Cancer Inst* **92:** 1781–1782. doi:10.1093/jnci/92.21.1781

Phillips KA, Milne RL, Rookus MA, Daly MB, Antoniou AC, Peock S, Frost D, Easton DF, Ellis S, Friedlander ML, et al. 2013. Tamoxifen and risk of contralateral breast cancer for *BRCA1* and *BRCA2* mutation carriers. *J Clin Oncol* **31:** 3091–3099. doi:10.1200/JCO.2012.47.8313

Pritzlaff M, Summerour P, McFarland R, Li S, Reineke P, Dolinsky JS, Goldgar DE, Shimelis H, Couch FJ, Chao EC, et al. 2017. Male breast cancer in a multi-gene panel testing cohort: insights and unexpected results. *Breast Cancer Res Treat* **161:** 575–586. doi:10.1007/s10549-016-4085-4

Pujol P, Barberis M, Beer P, Friedman E, Piulats JM, Capoluongo ED, Garcia Foncillas J, Ray-Coquard I, Penault-Llorca F, Foulkes WD, et al. 2021. Clinical practice guidelines for BRCA1 and BRCA2 genetic testing. *Eur J Cancer* **146:** 30–47. doi:10.1016/j.ejca.2020.12.023

Ramus SJ, Harrington PA, Pye C, DiCioccio RA, Cox MJ, Garlinghouse-Jones K, Oakley-Girvan I, Jacobs IJ, Hardy RM, Whittemore AS, et al. 2007. Contribution of *BRCA1* and *BRCA2* mutations to inherited ovarian cancer. *Hum Mutat* **28:** 1207–1215. doi:10.1002/humu.20599

Ramus SJ, Song H, Dicks E, Tyrer JP, Rosenthal AN, Intermaggio MP, Fraser L, Gentry-Maharaj A, Hayward J, Philpott S, et al. 2015. Germline mutations in the BRIP1, BARD1, PALB2, and NBN genes in women with ovarian cancer. *J Natl Cancer Inst* **107:** djv214. doi:10.1093/jnci/djv214

Rebbeck TR, Mitra N, Wan F, Sinilnikova OM, Healey S, McGuffog L, Mazoyer S, Chenevix-Trench G, Easton DF, Antoniou AC, et al. 2015. Association of type and location of *BRCA1* and *BRCA2* mutations with risk of breast and ovarian cancer. *JAMA* **313:** 1347–1361. doi:10.1001/jama.2014.5985

Richardson ME, Hu C, Lee KY, LaDuca H, Fulk K, Durda KM, Deckman AM, Goldgar DE, Monteiro ANA, Gnanaolivu R, et al. 2021. Strong functional data for pathogenicity or neutrality classify BRCA2 DNA-binding-domain variants of uncertain significance. *Am J Hum Genet* **108:** 458–468. doi:10.1016/j.ajhg.2021.02.005

Roberts ME, Ranola JMO, Marshall ML, Susswein LR, Graceffo S, Bohnert K, Tsai G, Klein RT, Hruska KS, Shirts BH. 2019. Comparison of *CDH1* penetrance estimates in clinically ascertained families vs families ascertained for multiple gastric cancers. *JAMA Oncol* **5:** 1325–1331. doi:10.1001/jamaoncol.2019.1208

Robson M. 2014. Multigene panel testing: planning the next generation of research studies in clinical cancer genetics. *J Clin Oncol* **32:** 1987–1989. doi:10.1200/JCO.2014.56.0474

Robson M, Domchek S. 2019. Broad application of multigene panel testing for breast cancer susceptibility—Pandora's box is opening wider. *JAMA Oncol* doi:10.1001/jamaoncol.2019.4004

Robson M, Im SA, Senkus E, Xu B, Domchek SM, Masuda N, Delaloge S, Li W, Tung N, Armstrong A, et al. 2017. Olaparib for metastatic breast cancer in patients with a germline *BRCA* mutation. *N Engl J Med* **377:** 523–533. doi:10.1056/NEJMoa1706450

Rubinstein WS. 2004. Hereditary breast cancer in Jews. *Fam Cancer* **3:** 249–257. doi:10.1007/s10689-004-9550-2

Rudolph A, Song M, Brook MN, Milne RL, Mavaddat N, Michailidou K, Bolla MK, Wang Q, Dennis J, Wilcox AN, et al. 2018. Joint associations of a polygenic risk score and environmental risk factors for breast cancer in the Breast Cancer Association Consortium. *Int J Epidemiol* **47:** 526–536. doi:10.1093/ije/dyx242

Seemanová E, Jarolim P, Seeman P, Varon R, Digweed M, Swift M, Sperling K. 2007. Cancer risk of heterozygotes with the NBN founder mutation. *J Natl Cancer Inst* **99:** 1875–1880. doi:10.1093/jnci/djm251

Sessa C, Balmaña J, Bober SL, Cardoso MJ, Colombo N, Curigliano G, Domchek SM, Evans DG, Fischerova D, Harbeck N, et al. 2023. Risk reduction and screening of cancer in hereditary breast-ovarian cancer syndromes: ESMO Clinical Practice Guideline. *Ann Oncol* **34:** 33–47. doi:10.1016/j.annonc.2022.10.004

Shimelis H, LaDuca H, Hu C, Hart SN, Na J, Thomas A, Akinhanmi M, Moore RM, Brauch H, Cox A, et al. 2018.

Triple-negative breast cancer risk genes identified by multigene hereditary cancer panel testing. *J Natl Cancer Inst* **110:** 855–862. doi:10.1093/jnci/djy106

Slavin TP, Van Tongeren LR, Behrendt CE, Solomon I, Rybak C, Nehoray B, Kuzmich L, Niell-Swiller M, Blazer KR, Tao S, et al. 2018. Prospective study of cancer genetic variants: variation in rate of reclassification by ancestry. *J Natl Cancer Inst* **110:** 1059–1066. doi:10.1093/jnci/djy027

Song H, Cicek MS, Dicks E, Harrington P, Ramus SJ, Cunningham JM, Fridley BL, Tyrer JP, Alsop J, Jimenez-Linan M, et al. 2014. The contribution of deleterious germline mutations in BRCA1, BRCA2 and the mismatch repair genes to ovarian cancer in the population. *Hum Mol Genet* **23:** 4703–4709. doi:10.1093/hmg/ddu172

Song H, Dicks E, Ramus SJ, Tyrer JP, Intermaggio MP, Hayward J, Edlund CK, Conti D, Harrington P, Fraser L, et al. 2015. Contribution of germline mutations in the *RAD51B, RAD51C,* and *RAD51D* genes to ovarian cancer in the population. *J Clin Oncol* **33:** 2901–2907. doi:10.1200/JCO.2015.61.2408

Song H, Dicks EM, Tyrer J, Intermaggio M, Chenevix-Trench G, Bowtell DD, Traficante N, Group A, Brenton J, Goranova T, et al. 2021. Population-based targeted sequencing of 54 candidate genes identifies PALB2 as a susceptibility gene for high-grade serous ovarian cancer. *J Med Genet* **58:** 305–313. doi:10.1136/jmedgenet-2019-106739

Steffen J, Nowakowska D, Niwińska A, Czapczak D, Kluska A, Piatkowska M, Wiśniewska A, Paszko Z. 2006. Germline mutations 657del5 of the NBS1 gene contribute significantly to the incidence of breast cancer in Central Poland. *Int J Cancer* **119:** 472–475. doi:10.1002/ijc.21853

Suszynska M, Ratajska M, Kozlowski P. 2020. BRIP1, RAD51C, and RAD51D mutations are associated with high susceptibility to ovarian cancer: mutation prevalence and precise risk estimates based on a pooled analysis of ~30,000 cases. *J Ovarian Res* **13:** 50. doi:10.1186/s13048-020-00654-3

Tan MH, Mester JL, Ngeow J, Rybicki LA, Orloff MS, Eng C. 2012. Lifetime cancer risks in individuals with germline PTEN mutations. *Clin Cancer Res* **18:** 400–407. doi:10.1158/1078-0432.CCR-11-2283

Tavtigian SV, Simard J, Rommens J, Couch F, Shattuck-Eidens D, Neuhausen S, Merajver S, Thorlacius S, Offit K, Stoppa-Lyonnet D, et al. 1996. The complete BRCA2 gene and mutations in chromosome 13q-linked kindreds. *Nat Genet* **12:** 333–337. doi:10.1038/ng0396-333

Thompson ER, Rowley SM, Li N, McInerny S, Devereux L, Wong-Brown MW, Trainer AH, Mitchell G, Scott RJ, James PA, et al. 2016. Panel testing for familial breast cancer: calibrating the tension between research and clinical care. *J Clin Oncol* **34:** 1455–1459. doi:10.1200/JCO.2015.63.7454

Tung N, Battelli C, Allen B, Kaldate R, Bhatnagar S, Bowles K, Timms K, Garber JE, Herold C, Ellisen L, et al. 2015. Frequency of mutations in individuals with breast cancer referred for *BRCA1* and *BRCA2* testing using next-generation sequencing with a 25-gene panel. *Cancer* **121:** 25–33. doi:10.1002/cncr.29010

Tung NM, Robson ME, Ventz S, Santa-Maria CA, Nanda R, Marcom PK, Shah PD, Ballinger TJ, Yang ES, Vinayak S, et al. 2020. TBCRC 048: Phase II Study of Olaparib for Metastatic Breast Cancer and Mutations in Homologous Recombination-Related Genes. *J Clin Oncol* **38:** 4274–4282. doi:10.1200/JCO.20.02151

Tutt ANJ, Garber JE, Kaufman B, Viale G, Fumagalli D, Rastogi P, Gelber RD, de Azambuja E, Fielding A, Balmaña J, et al. 2021. Adjuvant olaparib for patients with *BRCA1-* or *BRCA2*-mutated breast cancer. *N Engl J Med* **384:** 2394–2405. doi:10.1056/NEJMoa2105215

van Os NJ, Roeleveld N, Weemaes CM, Jongmans MC, Janssens GO, Taylor AM, Hoogerbrugge N, Willemsen MA. 2016. Health risks for ataxia-telangiectasia mutated heterozygotes: a systematic review, meta-analysis and evidence-based guideline. *Clin Genet* **90:** 105–117. doi:10.1111/cge.12710

van Sprundel TC, Schmidt MK, Rookus MA, Brohet R, van Asperen CJ, Rutgers EJ, Van't Veer LJ, Tollenaar RA. 2005. Risk reduction of contralateral breast cancer and survival after contralateral prophylactic mastectomy in BRCA1 or BRCA2 mutation carriers. *Br J Cancer* **93:** 287–292. doi:10.1038/sj.bjc.6602703

Walsh T, Casadei S, Lee MK, Pennil CC, Nord AS, Thornton AM, Roeb W, Agnew KJ, Stray SM, Wickramanayake A, et al. 2011. Mutations in 12 genes for inherited ovarian, fallopian tube, and peritoneal carcinoma identified by massively parallel sequencing. *Proc Natl Acad Sci* **108:** 18032–18037. doi:10.1073/pnas.1115052108

Weinmann S, Phillips S, Sweet K, Cosgrove CM, Senter L. 2022. Hospital-based ovarian cancer patient traceback program results in minimal genetic testing uptake. *Gynecol Oncol* **164:** 615–621. doi:10.1016/j.ygyno.2021.12.027

Weitzel JN, Chao EC, Nehoray B, Van Tongeren LR, LaDuca H, Blazer KR, Slavin T, Facmg DABMD, Pesaran T, Rybak C, et al. 2018. Somatic TP53 variants frequently confound germ-line testing results. *Genet Med* **20:** 809–816. doi:10.1038/gim.2017.196

Weitzel JN, Kidd J, Bernhisel R, Shehayeb S, Frankel P, Blazer KR, Turco D, Nehoray B, McGreevy K, Svirsky K, et al. 2021. Multigene assessment of genetic risk for women for two or more breast cancers. *Breast Cancer Res Treat* **188:** 759–768. doi:10.1007/s10549-021-06201-y

Welsh JL, Hoskin TL, Day CN, Thomas AS, Cogswell JA, Couch FJ, Boughey JC. 2017. Clinical decision-making in patients with variant of uncertain significance in BRCA1 or BRCA2 genes. *Ann Surg Oncol* **24:** 3067–3072. doi:10.1245/s10434-017-5959-3

Whitworth PW, Beitsch PD, Patel R, Rosen B, Compagnoni G, Baron PL, Simmons R, Brown EA, Gold L, Holmes D, et al. 2022. Clinical utility of universal germline genetic testing for patients with breast cancer. *JAMA Netwk Open* **5:** e2232787. doi:10.1001/jamanetworkopen.2022.32787

Wooster R, Bignell G, Lancaster J, Swift S, Seal S, Mangion J, Collins N, Gregory S, Gumbs C, Micklem G. 1995. Identification of the breast cancer susceptibility gene BRCA2. *Nature* **378:** 789–792. doi:10.1038/378789a0

Yadav S, Couch FJ. 2019. Germline genetic testing for breast cancer risk: the past, present, and future. *Am Soc Clin Oncol Educ Book* **39:** 61–74. doi:10.1200/EDBK_238987

Yadav S, Reeves A, Campian S, Sufka A, Zakalik D. 2017. Preoperative genetic testing impacts surgical decision making in BRCA mutation carriers with breast cancer: a retrospective cohort analysis. *Hered Cancer Clin Pract* **15:** 11. doi:10.1186/s13053-017-0071-z

Yadav S, Jinna S, Pereira-Rodrigues O, Reeves A, Campian S, Sufka A, Zakalik D. 2018. Impact of preoperative BRCA1/2 testing on surgical decision making in patients with newly diagnosed breast cancer. *Breast J* **24:** 541–548. doi:10.1111/tbj.13007

Yadav S, Hart SN, Hu C, Hillman D, Lee KY, Gnanaolivu R, Na J, Polley EC, Couch FJ, Kohli M. 2019. Contribution of inherited DNA-repair gene mutations to hormone-sensitive and castrate-resistant metastatic prostate cancer and implications for clinical outcome. *JCO Precis Oncol* 1–12. doi:10.1200/PO.19.00067

Yadav S, Hu C, Hart SN, Boddicker N, Polley EC, Na J, Gnanaolivu R, Lee KY, Lindstrom T, Armasu S, et al. 2020a. Evaluation of germline genetic testing criteria in a hospital-based series of women with breast cancer. *J Clin Oncol* **38:** 1409–1418. doi:10.1200/JCO.19.02190

Yadav S, LaDuca H, Polley EC, Hu C, Niguidula N, Shimelis H, Lilyquist J, Na J, Lee KY, Gutierrez S, et al. 2020b. Racial and ethnic differences in multigene hereditary cancer panel test results for women with breast cancer. *J Natl Cancer Inst* **113:** 1429–1433. doi:10.1093/jnci/djaa167

Yadav S, Hu C, Nathanson KL, Weitzel JN, Goldgar DE, Kraft P, Gnanaolivu RD, Na J, Huang H, Boddicker NJ, et al. 2021. Germline pathogenic variants in cancer predisposition genes among women with invasive lobular carcinoma of the breast. *J Clin Oncol* **39:** 3918–3926. doi:10.1200/JCO.21.00640

Yadav S, Boddicker NJ, Na J, Polley EC, Hu C, Hart SN, Gnanaolivu RD, Larson N, Holtegaard S, Huang H, et al. 2023. Contralateral breast cancer risk among carriers of germline pathogenic variants in *ATM, BRCA1, BRCA2, CHEK2,* and *PALB2. J Clin Oncol* 1703–1713. doi:10.1200/JCO.22.01239

Yang S, Axilbund JE, O'Leary E, Michalski ST, Evans R, Lincoln SE, Esplin ED, Nussbaum RL. 2018a. Underdiagnosis of hereditary breast and ovarian cancer in Medicare patients: genetic testing criteria miss the mark. *Ann Surg Oncol* **25:** 2925–2931. doi:10.1245/s10434-018-6621-4

Yang X, Leslie G, Gentry-Maharaj A, Ryan A, Intermaggio M, Lee A, Kalsi JK, Tyrer J, Gaba F, Manchanda R, et al. 2018b. Evaluation of polygenic risk scores for ovarian cancer risk prediction in a prospective cohort study. *J Med Genet* **55:** 546–554. doi:10.1136/jmedgenet-2018-105313

Yang X, Leslie G, Doroszuk A, Schneider S, Allen J, Decker B, Dunning AM, Redman J, Scarth J, Plaskocinska I, et al. 2020. Cancer risks associated with germline PALB2 pathogenic variants: an international study of 524 families. *J Clin Oncol* **38:** 674–685. doi:10.1200/JCO.19.01907

Zhang S, Royer R, Li S, McLaughlin JR, Rosen B, Risch HA, Fan I, Bradley L, Shaw PA, Narod SA. 2011. Frequencies of BRCA1 and BRCA2 mutations among 1,342 unselected patients with invasive ovarian cancer. *Gynecol Oncol* **121:** 353–357. doi:10.1016/j.ygyno.2011.01.020

Zhang G, Zeng Y, Liu Z, Wei W. 2013. Significant association between Nijmegen breakage syndrome 1 657del5 polymorphism and breast cancer risk. *Tumour Biol* **34:** 2753–2757. doi:10.1007/s13277-013-0830-z

Human Ductal Carcinoma In Situ: Advances and Future Perspectives

Fariba Behbod,[1] Jennifer H. Chen,[2] and Alastair Thompson[3]

[1]Department of Pathology and Laboratory Medicine, MS 3045, The University of Kansas Medical Center, Kansas City, Kansas 66160, USA

[2]Michael E. Debakey Department of Surgery, Baylor College of Medicine, Houston, Texas 77030, USA

[3]Section of Breast Surgery, Baylor College of Medicine, Co-Director, Lester and Sue Smith Breast Center, Dan L Duncan Comprehensive Cancer Center, Houston, Texas 77030, USA

Correspondence: fbehbod@kumc.edu; alastair.thompson@bcm.edu

Due to widespread adoption of screening mammography, there has been a significant increase in new diagnoses of ductal carcinoma in situ (DCIS). However, DCIS outcomes remain unclear. A large fraction of human DCIS (>50%) may not need the multimodality treatment options currently offered to all DCIS patients. More importantly, while we may be overtreating many, we cannot identify those most at risk of invasion or metastasis following a DCIS diagnosis. This review summarizes the studies that have furthered our understanding of DCIS pathology and mechanisms of invasive progression by using advanced technologies including spatial genomics, transcriptomics, and multiplex proteomics. This review also highlights a need for rethinking DCIS with a more focused view on epithelial states and programs and their cross talk with the microenvironment.

Many currently believe that the multimodality treatment approach to ductal carcinoma in situ (DCIS), including surgery, radiation, and antihormonal therapy, are unnecessary and lead to overdiagnosis and overtreatment. Despite current therapies, the 20-yr mortality rate following a DCIS diagnosis is ~3.3%. Studies investigating the natural progression of human DCIS have reported untreated DCIS to progress to invasive breast cancer at a rate of 14% to 75% (Page et al. 1982; Leonard and Swain 2004; Collins et al. 2005; Sanders et al. 2005; Maxwell et al. 2018, 2022). These studies consist of case reports ranging from 4 to 311 women followed for as short as 1 yr up to 30 yr (Page et al. 1982; Leonard and Swain 2004; Collins et al. 2005; Sanders et al.

2005; Maxwell et al. 2018). Untreated DCIS was defined as those who were originally misdiagnosed with benign breast diseases but had subsequent examination with DCIS or those with biopsy-proven DCIS who underwent nonoperative management (Page et al. 1982; Collins et al. 2005; Sanders et al. 2005). Maxwell and colleagues reported the first systematic and retrospective longitudinal study of 89 untreated DCIS cases. Overall, 33% developed invasive breast cancer after a median follow-up time of 3.75 yr (range: 1–12 yr). Among the invasive breast cancers that developed, 48% of tumors were high grade, 32% were intermediate grade, and 18% were low grade. The risk factors significantly associated with the development of invasive disease were high grade, calcifications, younger age (<60 yr), and the absence of endocrine therapy (Maxwell et al. 2018). A follow-up study of 311 women also in the United Kingdom showed a 10-yr cumulative risk of 20% for development of invasive breast cancer in unresected DCIS. Like the previous study, intermediate to high-grade DCIS was associated with an increased risk for development of a future invasive breast cancer. Additionally, median size of DCIS was a risk factor associated with the development of future invasive breast cancer (Maxwell et al. 2022). Furthermore, autopsies on middle-aged women (40–70 yr old) with no known breast disease revealed that 8.9% had undiagnosed DCIS (Welch and Black 1997). These results have provided the basis for current clinical trials evaluating the safety of "active monitoring" for low-grade DCIS (Ozanne et al. 2011; Elshof et al. 2015; Francis et al. 2015; Sagara et al. 2015; Han and Khan 2018). A DCIS score has been marketed as a tool for DCIS risk stratification. However, this score has not demonstrated predictive value for the benefit of radiotherapy (Marshall 2014). More recently, a biological signature combined with clinical data aimed to identify women who will benefit from radiation (DCISionRT) (Bremer et al. 2018; Wärnberg et al. 2021). These tools have not yet been widely adopted and their clinical use is unclear.

These results collectively support the notion that a fraction of human DCIS (>50%) may not need the multimodality treatment options that are currently implemented. Moreover, identifying early biomarkers of long-term risk, in particular predictors of mortality following a DCIS diagnosis, is urgently needed. The following sections summarize findings from selected research that have advanced our knowledge of DCIS pathobiology. Despite these studies, there remains much uncertainty.

GREATER CHROMOSOMAL COPY NUMBER ABERRATIONS MAY PREDICT HIGH-RISK DCIS

Genomic studies of DCIS date back to the 1990s. These studies have shown a higher frequency of copy number losses and gains in intermediate to high-grade compared to low-grade DCIS. The higher rate of chromosomal aberrations may account for a higher risk of invasive disease in untreated intermediate to high-grade DCIS previously reported by Maxwell and colleagues (Maxwell et al. 2018, 2022). These chromosomal aberrations include higher frequency of losses of 1p, 1q, 6q, 8p, 9p, 11p, 11q, 13q, 16q, and 17q as well as gains in chromosomes 1q, 1p13, 8q24, 11p13, 11q13, 12p13, 16p13, 17q12, 17q21-q23, and 20q13.

One early study performed polymerase chain reaction (PCR)-amplified microsatellite markers on ten chromosomal arms in 41 cases of pure DCIS. They reported that the number of allelic losses was significantly greater in lesions of intermediate or high grade compared to low grade. Allelic losses of 16q and 17p were commonly found in all nuclear grades of DCIS. However, allelic losses of other chromosomal arms examined (1p, 1q, 6q, 9p, 11p, 11q, 13q, and 17q) were found at higher frequencies in intermediate- and high-grade DCIS. In 10 of these cases (24%), researchers identified patterns of allelic loss heterogeneity suggestive of intralesional diversity. For tumors with allelic loss heterogeneity, they hypothesized that chromosomal losses common to all tumor foci most likely preceded the chromosomal losses observed only in tumor foci of a more advanced genetic stage. These results suggest that early chromosomal aberrations indicative of low-grade lesions when found concurrently with chromosomal aberrations of higher-grade lesions may signal the development of future overall higher-risk lesions (Fujii et al. 1996). These studies were later validated by Allred and colleagues who demonstrated that ~48% of DCIS exhibited intratumoral heterogeneity, as evidenced by the presence of multiple grades of DCIS within the same lesion including 30.0% with grades 1 and 2, 6.6% with grades 2 and 3, and 9.2% with grades 1, 2, and 3. Furthermore, the level of intratumoral heterogeneity correlated with *TP53* mutation (Allred et al. 2008).

Two decades ago, chromosomal alterations of 32 cases of DCIS and six cases of DCIS with associated invasive breast cancer were evaluated by comparative genomic hybridization (CGH)

(Buerger et al. 1999). Similar to the previous study (Fujii et al. 1996), loss of 16q was seen only in well and intermediately differentiated DCIS. A higher frequency of gains of 1q and losses of 11q was seen in intermediately differentiated compared to well differentiated DCIS. Poorly differentiated DCIS displayed a higher frequency of amplification of 17q12 and 11q13 and a higher average rate of genomic imbalances. A later study in 2003 performed CGH on 34 cases of DCIS and 12 cases of DCIS with associated invasive breast cancer confirmed that DCIS showed the same degree of chromosomal instability as the associated invasive breast cancer. When compared with low-grade disease, high-grade DCIS exhibited more frequent gain of 17q and higher frequency of loss of 8p. These results led to the conclusion that low intermediate-grade and high-grade DCIS may exhibit distinct genomic aberrations. In 2006, Yao and colleagues performed cDNA array CGH and serial analysis of gene expression (SAGE) on 10 DCIS, 11 invasive breast cancer, and two lymph node metastases (Yao et al. 2006). This group identified 49 commonly amplified regions that included known chromosomal regions of 1q, 8q24, 11q13, 17q21-q23, and 20q13, as well as copy number gains on previously unidentified regions such as 12p13 and 16p13. When comparing DCIS to invasive breast cancer, there was a trend toward an increase in the number and amplitude of gains and losses associated with invasive breast cancer. The group evaluated gene expression changes simultaneously and found a number of previously known oncogenes associated with common amplifications, including 17q21 with ERBB2, PNMT (phenylethanolamine N-methyltransferase; enzyme involved in catecholamine biosynthesis), and PERLD (per1-like domain containing; involved in lipid remodeling steps of GPI-anchor maturation) overexpression. Interestingly, the 8q24 amplicon did not identify the presumed target, MYC. Therefore, a correlation between amplification and gene expression may not exist for all aberrant genomic regions. These studies overall have provided several interesting findings. For one, high-grade DCIS and low-grade DCIS may exhibit distinct mechanisms of DCIS to invasive breast cancer transition. This may be the reason that DCIS and related invasive breast cancers demonstrate similar genetic and gene expression patterns. The finding that there was a trend toward an increase in the number and amplitude of gains and losses associated with invasive breast cancer suggests that invasion is associated with the selection of tumor cells with higher chromosomal aberrations. Therefore, low-grade and high-grade DCIS could be treated as distinct conditions with the exception, perhaps, of those with intratumoral heterogeneity. Furthermore, simultaneous genomic and gene expression studies may help identify potential drivers of DCIS malignancy and targets for prevention in a subtype- and grade-specific manner.

As mentioned previously, untreated high-grade DCIS showed the highest rate of progression to invasive breast cancer (Maxwell et al. 2018, 2022). Supporting these data, Abba et al. (2015) performed comprehensive exome, transcriptome, and methylome analysis of 30 cases of pure high-grade DCIS (HG-DCIS) and their matched normal breast epithelial cells. HG-DCIS lesions displayed molecular profiles very similar to invasive breast cancers. Furthermore, copy number variation (CNV) analysis showed that 83% of high-grade DCIS lesions displayed large chromosomal alterations. Interestingly, most DCIS with mutations in cancer driver genes also showed significant chromosomal copy number aberrations (CNAs). These data support the hypothesis that large chromosomal CNAs may result in the selection of alleles with mutations in driver genes. More importantly, at the transcriptional level, pathway-based analysis pointed to *TP53* pathway inactivation as extremely common in high-grade DCIS. Interestingly, hypermethylation of *HOXA5* was more frequently associated with basal-like/HER2[+] DCIS. Since HOXA5 regulates the expression of *TP53*, HOXA5 hypermethylation may be an early event preceding TP53 inactivation and the development of basal invasive breast cancers. Integrated pathway analysis of RNA-seq data showed the existence of two DCIS subgroups (DCIS-C1 and DCIS-C2). The more aggressive DCIS-C1 (highly proliferative, basal-like, or ERBB2[+]) displayed signatures characteristic of activated Treg cells (CD4$^+$/CD25$^+$/FOXP3$^+$) and CTLA4$^+$/CD86$^+$ complexes, indicative of a tumor-associated immunosuppressive pheno type. In summary, high-grade DCIS, in particular those with *TP53* inactivation, may be at a higher risk for a future invasive breast cancer.

Afghahi et al. (2015) studied three common chromosomal CNVs in invasive breast cancer and designed a fluorescence in situ hybridization-based assay to measure copy number at these loci in DCIS. They examined the presence of copy number gains of 1q, 8q24, and 11q13 in 271 patients with DCIS (120 that did not develop invasive breast cancer and 151 with concurrent invasive disease). Compared to DCIS only, patients with concurrent invasive breast cancer had higher frequencies of CNAs 1q, 8q24, and 11q13 in their DCIS samples. On multivariable analysis with conventional clinicopathologic features, the copy number gains were significantly associated with concurrent invasion. The state of two of the three copy number gains in DCIS was associated with a ninefold increased risk of invasive breast cancer compared to no copy number gains, and the presence of gains at all three genomic loci in DCIS was associated with a more than 17-fold increased risk ($P = 0.0013$). While this is a novel approach in identifying DCIS with potential future invasive potential, these findings should be further validated in untreated DCIS with long-term follow-up.

Newburger and colleagues performed comparative whole genome sequencing on early lesions, matched normal, DCIS, and invasive breast cancer from six patients. They used somatic mutations as lineage markers to build a tree to relate tissue samples within each patient (Newburger et al. 2013). In four of the six patients, early lesions and cancer shared a mutated common ancestor with similar chromosomal aberration patterns. Their studies support the hypothesis that the accumulation of somatic mutations is a result of increased cell division rather than specific mutations or DNA repair defects. On average, invasive breast cancer accumulated a higher frequency of private mutations compared to DCIS and early lesions. The common ancestors of early lesions carried extensive aneuploidy and nuclear atypia, which predisposed the cells to

further genomic changes leading to invasion. All early lesions with aneuploidies shared common cellular ancestors with DCIS and invasive breast cancer, while none of the early lesions that were devoid of aneuploidy were related to DCIS or invasive breast cancer. Among the aneuploidies, gain of chromosome 1q was most prevalent among invasive breast cancers. While genomic aberrations (i.e., chromosome 1q amplification) may not be sufficient to predispose to invasion, additional aneuploidies may accumulate with increased cell divisions and, at some point, epigenetic or stromal changes ultimately induce development of invasive carcinoma. Therefore, aneuploidy in early lesions may be a risk factor for future advancement to DCIS and/or invasive breast cancer.

In summary, several important findings resulted from genomic studies of DCIS with or without associated invasive disease. First, low-grade and high-grade DCIS may need to be treated as conditions with differing potential for invasion, except perhaps those with intratumoral heterogeneity. Second, high-grade DCIS, in particular with *TP53* inactivation, may be at a higher risk for future invasion. Third, aneuploidy in early lesions may be a risk factor for future advancement to DCIS and/or invasive disease. Fourth, higher frequency of CNA and chromosomal imbalances may signal a high-risk DCIS with a future invasive potential.

CLONAL TRACKING IN DCIS TO INVASIVE DUCTAL CARCINOMA (IDC) TRANSITION

Casasent and colleagues (Casasent et al. 2018) developed topographic single-cell sequencing (TSCS) to measure genomic copy number changes of single tumor cells (1293 cells from 10 DCIS with associated invasive breast cancer) while preserving their spatial topography on tissue sections. They found that the mutations and copy number changes were initiated inside ducts prior to progression to invasion. Their results supported a multiclonal invasion model in which one or multiple clones leave the ducts and invade into adjacent tissues to establish the invasive cancer. This group also evaluated point mutations in cancer-related genes in in situ and invasive breast cancers. They found that most nonsynonymous mutations including those in known breast cancer genes, such as *TP53*, *PIK3CA*, and *RUNX1*, were similar in the ducts and invasive regions. At a higher sequencing depth, researchers found in situ–specific mutations to be present in the invasive areas, while the majority of invasive-specific mutations were exclusive to the invasive regions. Altogether, these suggest that some private mutations were acquired after the tumor cells left the ducts. However, only the mutations that were present in the duct cells could have enabled cells to escape to the invasive regions. In some cases, the frequency of some mutations (i.e., MMP8) increased during invasion while in other cases there were minor changes in frequency of mutations from in situ to invasive lesions. There were many shared truncal mutations and CNAs suggesting that all subclones in the ducts arose from a single initiating cell. Their results are consistent with a punctuated model of copy number evolution, in which short bursts of genome instability give rise to multiple clones that stably expand to form the tumor mass. The comigration of multiple clones into the invasive regions suggests that invasion occurs either through the complete breakdown of the basement membrane and random escape of tumor clones into the adjacent tissues or the cooperation of tumor clones that collectively break down the basement membrane. Others have investigated the frequency of nonsynonymous mutations and CNAs in pure DCIS that did not progress to invasion during a median follow-up of 72 mo compared with concurrent DCIS and invasive breast cancer. Their studies showed no significant differences in frequency of nonsynonymous mutations or CNAs. However, there was a numerically higher, but statistically nonsignificant, number of mutations in key cancer genes such as *TP53* and *PIK3CA* in DCIS that progressed to invasion. Additionally, the repertoire of somatic mutations was significantly different according to estrogen receptor (ER) and HER2 status (Pareja et al. 2020). These studies have demonstrated that clonal heterogeneity exists for a large fraction of DCIS. Furthermore, invasive clone(s) arise early in DCIS and promote invasion. At the DCIS stage, there may be a clonal selection of one or

multiple invasive clones that continue to expand and give rise to invasive breast cancer.

COMMONLY DYSREGULATED PATHWAYS RATHER THAN GENE EXPRESSION SIGNATURE(S) MAY BETTER PREDICT DCIS WITH A FUTURE INVASIVE POTENTIAL

Significant efforts have been made to find a gene expression signature that defines DCIS with a future invasive potential. Nevertheless, there is still no generally accepted gene expression signature by which to stratify DCIS patients into risk categories. However, some common pathways/themes have been identified. Table 1 lists studies that have implicated certain genes and signaling pathways in DCIS pathogenesis and progression to invasive breast cancer. Findings from a few selected studies are summarized below.

Ma et al. generated gene expression profiles of breast cancer using DNA microarray of laser capture microdissection (LCM) tissues from atypical ductal hyperplasia, DCIS, and invasive breast cancers (Ma et al. 2003). They found that DCIS and invasive breast cancers did not form distinct clusters. Indeed, the DCIS and their associated invasive lesions from the same patient clustered more closely together. These findings supported the previous studies that the genes conferring invasiveness were already active in the preinvasive stages. Interestingly though, different tumor grades did express distinct transcriptional signatures. The expression of a subset of these differentially expressed and grade-specific genes increased during the transition from DCIS to invasive disease. These differentially expressed genes included genes involved in the cell cycle (e.g., MCM6, TOP2A, CKS2, CDC25C, and UBE2C), centrosome function (TACC3, CENPA, and STK15), and DNA repair (RAD51 and RRM2). RRM2 was also found as one of the seven genes by Kretschmer et al. (2011b) to be up-regulated in DCIS compared to control. Thus, a significant subset of genes that are expressed at higher levels in grade III DCIS relative to grade I DCIS may further increase during the transition from DCIS to invasive breast cancer. Another conclusion may be that cells expressing cancer driver genes are

Table 1. Summary of genes/pathways implicated in ductal carcinoma in situ (DCIS) pathogenesis and progression to invasive breast cancer

Function	Gene symbol	Gene name	Key functions	References
Cell signaling	TP53	Tumor protein p53	Tumor suppressor protein; TP53 down-regulation has been associated with high-grade DCIS	Casasent et al. 2018
	HOXA5	Homeobox A5	Encodes transcription factor, regulates gene expression, development, and differentiation; hypermethylation of HOXA5 promoter has been associated with high-grade DCIS	Abba et al. 2015
	PIK3CA	Phosphatidylinositol-4,5-bisphosphate 3-kinase catalytic subunit α	Signal transduction molecule, protooncogenic	Casasent et al. 2018
	RUNX1	RUNX family transcription factor 1	Transcription factor, binds to enhancers and promoters	Casasent et al. 2018
	ERBB2	Erb-b2 receptor tyrosine kinase 2	Epidermal growth factor receptor tyrosine kinase	Yao et al. 2006
	APC2	APC regulator of WNT signaling pathway 2	Wnt pathway signaling, microtubule formation and assembly	Hannemann et al. 2006
	DAPK3	Death-associated protein kinase 3	Induction of apoptosis	
	ADM	Adrenomedullin	Signaling peptide	
	ARF1	ADP ribosylation factor 1	Vesicular trafficking, intra-Golgi transport	
	IQGAP1	IQ motif containing GTPase-activating protein 1	Signaling molecules involved in cell morphology and motility	
	TNK2	Tyrosine kinase nonreceptor 2	Tyrosine phosphorylation signal transduction pathway	
	CELSR2	Cadherin EGF LAG seven-pass G-type receptor 2	Contact-mediated communication, cell adhesion, receptor–ligand interactions	
	CCL19	C-C motif chemokine ligand 19	Lymphocyte trafficking and migration, binds chemokine receptor CCR7	
	SHH	Sonic hedgehog signaling molecule	Early embryo patterning, regulate myoepithelial cell function (loss of function in the myoepithelium)	Hu et al. 2008
	TP63	Tumor protein p63	Member of p53 transcription factors, regulate epithelial stem cell function and differentiation, myoepithelial cell differentiation (loss of function in the myoepithelium)	
	SOX11	SRY-box transcription factor 11	Transcription factor involved in embryonic development, role in tumorigenesis	Oliemuller et al. 2017, 2020

Continued

Table 1. *Continued*

Function	Gene symbol	Gene name	Key functions	References
Migration	LRRC15	Leucine rich repeat containing 15	Enables collagen, fibronectin, and laminin binding activity	Schuetz et al. 2006
	CXCL10	C-X-C motif chemokine ligand 10	Stimulates migration of immune cells, modifies adhesion molecule expression	Kim et al. 2021
Proliferation	RRM2	Ribonucleotide reductase regulatory subunit M2	DNA replication	Kretschmer et al. 2011a
	TGFb2	Transforming growth factor β2	Recruit and activate SMAD transcription factors	
	PTMS	Parathymosin	DNA replication	
	PSAP	Prosaposin	Lipid metabolism	
	TUBB2A	Tubulin β2A class IIa	Encode β-tubulin	
	MAP7	Microtubule-associated protein 7	Microtubule assembly and stabilization	
	CXCL12	C-X-C motif chemokine ligand 12	Tumor growth and metastasis, immune surveillance, inflammatory response	Allinen et al. 2004
	CXCL14	C-X-C motif chemokine ligand 14	Epithelial cell proliferation, migration, and invasion	
	FOXM1	Forkhead box M1	Transcriptional activator in cell proliferation	Kretschmer et al. 2011a
Cell-cycle regulation	EXO1	Exonuclease 1	DNA repair	
	NUSAP1	Nucleolar and spindle-associated protein 1	Spindle and microtubule organization	
	MCM6	Minichromosome maintenance complex component 6	Initiation of cell replication	Ma et al. 2003
	TOP2A	DNA topoisomerase IIα	Regulates DNA topology during transcription	
	CKS2	CDC28 protein kinase regulatory subunit 2	Binds cyclin-dependent kinases	
	CDC25C	Cell division cycle 25C	Regulates mitosis initiation, suppresses p53-induced growth arrest	
	UBE2C	Ubiquitin-conjugating enzyme E2 C	Degradation of cyclin proteins necessary for cell cycle progression	
	TACC3	Transforming acidic coiled-coil-containing protein 3	Stabilize mitotic spindle	Ma et al. 2003
	CENPA	Centromere protein A	Centromere function	
	AURKA (STK15)	Aurora kinase A	Microtubule stabilization and formation	

Continued

Table 1. *Continued*

Function	Gene symbol	Gene name	Key functions	References
	BCL9	BCL9 (B-cell lymphoma 9) transcription coactivator	Associated with B-cell acute lymphoblastic leukemia; may play a role in invasive breast cancer proliferation, EMT, angiogenesis, and immunotolerance	Elsarraj et al. 2015, 2020; Feng et al. 2019
Cell-to-cell interaction	*FAT1*	FAT atypical cadherin 1	Adhesion molecule (suppression associated with DCIS invasive progression)	Lee et al. 2012
	TMEM45A	Transmembrane protein 45A	Cell membrane component (suppression associated with DCIS invasive progression)	Schuetz et al. 2006
	DST (BPAG1)	Dystonin	Adhesion junction protein (suppression associated with DCIS invasive progression)	
	MUC1	Mucin 1, cell surface associated	Membrane-bound protein, intracellular signaling	Kretschmer et al. 2011a
	SPP1	Secreted phosphoprotein 1	Bone turnover, associated with breast cancer progression, invasion, metastasis	Kretschmer et al. 2011a
	SPARC	Secreted protein acidic and cysteine rich, osteonectin	Stromal glycoprotein, also involved in extracellular matrix (ECM) synthesis	Witkiewicz et al. 2009
	CAV1	Caveolin-1	Membrane protein involved in endocytosis, ECM remodeling, and cell migration (loss in fibroblasts associated with invasive progression)	Witkiewicz et al. 2009; Martinez-Outschocrn et al. 2010; Martins et al. 2013
	THBS2	Thrombospondin 2	Glycoprotein, mediates cell–cell and cell–matrix interaction, also inhibits tumor growth and angiogenesis	Allinen et al. 2004
	PERLD (PGAP3)	Post-GPI attachment to proteins phospholipase 3	Lipid remodeling in GPI-anchor maturation	Yao et al. 2006
ECM synthesis	*DCN*	Decorin	Proteoglycan involved in matrix assembly (high stromal expression associated with DCIS with invasive recurrence)	Van Bockstal et al. 2013
ECM degradation	*MMP2*	Matrix metallopeptidase 2	Breakdown of ECM	Allinen et al. 2004
	MMP8	Matrix metallopeptidase 8	Breakdown of ECM	Casasent et al. 2018
	MMP11	Matrix metallopeptidase 11	Breakdown of ECM	Schuetz et al. 2006
	PLAU	Plasminogen activator, urokinase	Protease, involved in cell migration and tissue remodeling	Schuetz et al. 2006
	PLAT	Plasminogen activator, tissue type	Protease, involved in cell migration and tissue remodeling	Hannemann et al. 2006

Continued

Table 1. *Continued*

Function	Gene symbol	Gene name	Key functions	References
	CTSF	Cathepsin F	Lysosomal proteinase, bone remodeling and resorption, contribute to tumor invasion in breast cancer	Allinen et al. 2004
	CTSK	Cathepsin K		
	CTSL	Cathepsin L	Intracellular protein catabolism	
	HTRA1 (PRSS11)	HtrA serine peptidase 1	Serine protease, regulator of cell growth	
Protease inhibitors	*SERPINB5 (Maspin)*	Serpin family B member 5	Inhibit endopeptidase activity (myoepithelial expression prevents invasion)	Sternlicht et al. 1997
	TIMP1	TIMP metallopeptidase inhibitor 1	Inhibit matrix metalloproteinases, may have antiapoptosis effects (myoepithelial expression prevents invasion)	
Cell metabolism	*BTD*	Biotinidase	Recycles biotin	Hannemann et al. 2006
	GMFG	Glia maturation factor γ	Actin binding, protein phosphorylation	
	PNMT	Phenylethanolamine N-methyltransferase	Catecholamine synthesis	Yao et al. 2006
DNA repair	*RAD51*	RAD51 recombinase	Homologous recombination and repair of DNA, interact with BRCA1 and BRCA2	Ma et al. 2003
Cellular stress	*COX-2 (MT-CO2)*	Mitochondrially encoded cytochrome *c* oxidase II	Mitochondrial electron transport, oxidative stress response	Fordyce et al. 2012
	SIAH2	Siah E3 ubiquitin protein ligase 2	Regulates cellular response to hypoxia, up-regulates HIF-a	Chan et al. 2011; Sun and Denko 2014

selected and expanded during the transition from DCIS to invasion.

Lee and colleagues used gene expression profiling of human DCIS ($n = 53$) and invasive breast cancer ($n = 51$) to discover 470 differentially expressed genes (Lee et al. 2012). They compared the 470 genes to nine similar studies, which showed that 74 (42 increased and 32 decreased) of the 470 genes had an overlap with two or more studies. Using hierarchical clustering, the 74-gene signature profile correctly identified more than 85% of samples in this study and two independent studies demonstrating the reliability of the 74-gene signature in correctly identifying DCIS from invasive breast cancer. This group further validated the role of four genes, CSTA, FAT1, DST, and TMEM45A in suppression of DCIS to invasive progression using the MIND models. Gene ontologies and pathways associated with differentially expressed genes revealed significant associations with migration, proliferation, cell–cell adhesion, cell–substrate adhesion, angiogenesis, response to wounding, and synthesis of extracellular matrix (ECM). In general, ECM-related pathways are restricted to epithelial cells and are elevated in invasive versus in situ disease.

Muggerud and colleagues (2010) studied gene expression patterns of 31 pure DCIS, 36 pure invasive cancers, and 42 cases of mixed samples (invasive cancer with an in situ component) using Agilent Whole Human Genome Oligo Microarrays 44K. Similar to previously reported studies of DCIS, all DCIS and invasive samples were classified into "intrinsic" molecular subtypes. Hierarchical clustering established that samples should be grouped by intrinsic subtypes and not by diagnosis. They identified a distinct subgroup of high-grade DCIS with gene expression signatures similar to invasive breast cancer and DCIS with associated invasion. The gene signature that differentiated this group of DCIS was the epithelial mesenchymal transition (EMT) and NF-κB-induced gene sets. The association of EMT gene signature to DCIS with invasive potential has also been reported by others (Ma et al. 2003; Lee et al. 2012).

Lesurf et al. (2016) performed mRNA, miRNA, and DNA copy number profiles of 59 DCIS and 85 invasive cancer samples. Like previous studies, this study showed that intrinsic subtypes, as opposed to disease stage (invasive vs. noninvasive), were the main sources of variability in gene expression. For example, basal DCIS coclustered with basal invasive disease and luminal DCIS with luminal invasive breast cancer. Neither subtype was enriched for DCIS or invasive breast cancer, stage, or lymph node–positive status. However, within each of the two clusters, subclusters consisted of DCIS or invasive breast cancer. This indicated that invasiveness is not the primary source of variability but gene expression changes could be observed between invasive and noninvasive lesions within each subtype. One hundred eighty-eight genes were found to be differentially expressed between DCIS versus invasive breast cancer of the same subtype. This set of genes was enriched in collagen fibril organization, ECM interactions, and focal adhesions, which consisted of a significant degree of overlap with signatures previously published. A genomic grade index (CGI) was used to evaluate CNAs. This showed that higher grade samples had higher frequency of CNAs compared to low-grade samples. Grade III had higher CGI scores than both grades I and II. Normal-like samples had the lowest CGI score while basal subtypes had the highest CGI scores. This study also collected sequencing data on TP53 and PIK3CA, two genes most frequently mutated in breast cancer. The mutation rates were similar in DCIS and invasive breast cancer, supporting the notion that TP53 and PIK3CA may be early drivers in DCIS.

In summary, study of gene expression changes associated with the transition from DCIS to invasive breast cancer have demonstrated commonly dysregulated pathways rather than a specific gene signature. The overlapping dysregulated pathways include cellular migration, cell adhesion, cell-cycle progression, cellular proliferation and response to wounding, synthesis of ECM, organization of ECM, EMT, NF-κB-regulated genes, metabolism, and DNA repair. Since most of the studies did not validate their gene signatures, it is difficult to interpret whether these pathways/signatures are simply associated with progression or serve as drivers of DCIS invasive progression. Another important finding was that the molecular mechanisms of progression are subtype specific. This is because most DCIS and invasive breast cancer lesions at the gene expression level cluster according to their subtypes and not by stage of disease (i.e., DCIS, invasion, lymph node status, or grades). Furthermore, apart from two studies (Muggerud et al. 2010; Lee et al. 2012), which found some overlap with other data sets, there was minimal agreement in the list of differentially gene expression genes when comparing DCIS to invasive breast cancer. In the future, perhaps more attention needs to be given to DCIS subtypes and their associated gene expression changes for invasive progression, similar to the studies by Lesurf et al. (2016). Furthermore, efforts should be directed toward validation of specific genes and pathways using relevant models of DCIS (i.e., mouse-intraductal models) (Hong et al. 2021).

THE ROLE OF BIDIRECTIONAL INTERACTIONS BETWEEN THE EPITHELIUM AND STROMA IN DCIS PROGRESSION

There is ample evidence that the cross talk between the noncancer cells in the tumor microenvironment, such as fibroblasts, adipocytes, endothelial cells, immune cells, the ECM, and premalignant epithelial cells, influence the transition from DCIS to invasion (Tlsty 2008; Nelson et al. 2018). There is even some evidence that the tumor microenvironment, rather than epithelial cells, may have a dominant role in DCIS progression. For example, invasive breast cancer cells behave like nonmalignant cells when placed in a nonpermissive environment (Weaver et al. 1997). Furthermore, primary epithelial cells without mutations progress to malignancy by coculture with stroma from a cancer lesion (Olumi et al. 1999; Barcellos-Hoff and Ravani 2000; Tlsty and Hein 2001). The finding of DCIS at autopsy further supports that the microenvironment may suppress progression of DCIS to invasive cancer (Welch and Black 1997). One study linked dysregulation of stroma-specific factors to risk of DCIS progression. For example, high stromal expression of decorin (a proteoglycan involved in matrix assembly) (Van Bockstal

et al. 2013) and osteonectin (a stromal glycoprotein) (Witkiewicz et al. 2010) have been linked to DCIS with a risk for invasive recurrence. Additionally, loss of caveolin-1 (CAV1) (a membrane protein involved in endocytosis, ECM remodeling, and cell migration) in fibroblasts was associated with poor outcome in DCIS (Witkiewicz et al. 2009; Martins et al. 2013; Martinez-Outschoorn et al. 2015). Cotransplantation of CAV1-null fibroblasts increased growth and angiogenesis in xenografted breast cancer cells and induced epithelial aerobic glycolysis (the Warburg effect) (Bonuccelli et al. 2010) and glycolytic inhibitors inhibited protumorigenic effects of CAV1-null fibroblasts. Additionally, carcinoma-associated fibroblasts (CAFs), through the secretion of interleukin (IL)-6, stimulate DCIS epithelial cell proliferation and migration (Osuala et al. 2015).

Del Alcazar and colleagues analyzed the composition of leukocytes by RNA-sequencing of CD45/CD3 T cells in normal breast tissues, DCIS, and invasive breast cancer (Del Alcazar et al. 2017). They reported a decrease in CD8[+] cells in invasive breast cancer compared to DCIS. Immunofluorescence analysis showed fewer activated GZMB[+] CD8[+] T cells in invasive breast cancer compared to DCIS, including in matched DCIS and invasive breast cancer (Wärnberg et al. 2021). T-cell receptor (TCR) clonotype diversity was significantly higher in DCIS than for invasive disease. Immune checkpoint protein TIGIT expressing T cells were more frequent in DCIS, whereas higher PD-L1 expression and CD274 amplification (encoding PD-L1) was only found in triple-negative invasive breast cancer. A follow-up study by the same group performed integrated transcriptomic, genomic, and whole slide image analysis to investigate how genomic profiles of tumors can shape the immune microenvironment and how coevolution of tumor and immune system influence DCIS to invasive breast cancer transition. They showed that DCIS and invasive breast cancer were similar in their CNA profiles including 1q, 8q, 17q, and 20q amplification and 8p, 11q, 16q, and 17p losses (Trinh et al. 2021). Furthermore, immune-hot and -cold tumors arose early during DCIS development and continued with invasive disease.

Importantly, the immune-hot triple-negative DCIS became more immunosuppressive during the transition from DCIS to invasive breast cancer due to recruitment of Treg cells. However, the immune-cold ER⁺ DCIS showed high expression of macrophages and Tregs that maintained similar expression patterns during DCIS to invasive breast cancer transition. Tumor-specific aberrations in MHC-1 presentation and losses in 3p, 4q, and 5p were associated with differences in immune signaling in ER-negative invasive breast cancer. Common mutations in genes including *TP53* and *PIk3CA* were predicated to be associated with differences in immune signaling. Therefore, genetic differences in tumor cells may promote DCIS to invasive breast cancer transition by altering the immune microenvironment (Trinh et al. 2021).

In further support of the role of bidirectional interactions between epithelial and stroma in DCIS, a recent study found CXCL10 to be the most highly expressed immune-regulated gene in DCIS with associated invasive disease compared to pure DCIS (Kim et al. 2021). CXCL10 induced epithelial cell migration, invasion, down-regulation of E-cadherin, and EMT in breast cancer cell lines. Additionally, CXCL10-positive tumors showed higher infiltrations of CD4⁺, CD8⁺, FOXP3⁺ TILs, and PDL1⁺ immune cells. Therefore, CXCL10 may play a key role in the progression of DCIS to invasive breast cancer through epithelial/immune cell interactions.

Biomarkers of altered cellular states, cellular stress responses, and stemness may drive aggressive DCIS (Fordyce et al. 2012). Fordyce and colleagues have proposed that a cellular program represented by epithelial-specific DNA damage and stress responses may drive DCIS invasiveness (Fordyce et al. 2012). This study showed that epithelial cells that undergo DNA damage and/or exhibit stress responses (via oncogenic activation) secrete factors (i.e., Activin A, IL6, IL8), which in turn induce a protumorigenic stromal microenvironment. In response to epithelial stress, a reprogrammed and remodeled tumor microenvironment together with desmoplasia provide positive feedback to epithelial cells and induce the up-regulation of epithelial stress response markers, including cyclo-oxygenase2 (COX-2) (Fordyce et al. 2012). Consequently, the expression of epithelial stress responses together with a compromised cellular proliferation pathway (marked by p16⁺/Ki67⁺) induce uncontrolled proliferation. In multivariate analyses, it was demonstrated that DCIS detected by palpation and those with p16/COX-2/Ki67-triple-positive (p16⁺COX-2⁺Ki67⁺) lesions were at increased risk of developing a subsequent invasive breast cancer, with a 10-year risk of ~22% (Molinaro et al. 2016). Among women with a subsequent invasive breast cancer, 63% were either detected by palpation or were p16⁺COX-2⁺Ki67⁺. This validated epithelial signature is the basis for the DCISionRT test, which identifies a low-risk and a high-risk group (63% with a 4% 10-year risk for subsequent invasive breast cancer vs. 37% with a 20% 10-year risk for subsequent invasive breast cancer, respectively), with the latter group benefiting from radiotherapy with a risk reduction from 20% to 7% (Bremer et al. 2018; Wärnberg et al. 2021). This epithelial stress response program is also referred to as an abrogated response to cellular stress (ARCS) signature (Gauthier et al. 2007; Kerlikowske et al. 2010). Most significantly, ARCS⁺ DCIS that were ER-HER2⁺ were at a significantly higher risk of metastasis (Molinaro et al. 2016). Another marker of cellular stress is ubiquitin ligase *seven in absentia homolog 2 (SIAH2)*. *SIAH2* up-regulates *HIF1-α* under hypoxia stress. *SIAH2* expression levels increase from the progression of normal breast epithelial cells to DCIS and to invasive breast cancers. *SIAH2* expression was positively correlated with tumor grade, HER2, p53, and intrinsic basal-like subtype in DCIS. Furthermore, *SIAH2*-positive tumors were associated with a shorter relapse-free survival (Chan et al. 2011; Sun and Denko 2014).

Both B-cell lymphoma-9 (BCL9) protein and mRNA expression have been associated with DCIS progression to invasive disease through epigenomic reprogramming with subsequent induction of epithelial cell proliferation, invasiveness, EMT, angiogenesis, and immunotolerance (Elsarraj et al. 2015, 2020; Feng et al. 2019). Additionally, an increase in matrix stiffness was shown to engage the epithelial BCL9/Wnt/β-catenin pathway

(Tao et al. 2021). BCL9 is therefore a prime candidate biomarker for stratification of high-risk DCIS and may also represent a therapeutic target for disease prevention (Elsarraj et al. 2020). Considering the observation that DCIS lesions with high BCL9 expression are more often ER⁻ HER2⁺ lesions, addition of BCL9 to the ARCS signature may further improve its predictive value.

Cellular stemness may be another cellular program/state driving DCIS. In fact, gain of cellular stemness has been demonstrated to predict tumour aggressiveness and clinical outcome (Ginestier et al. 2007). In support of this hypothesis, SOX11, an embryonic mammary epithelial marker, is significantly associated with distant metastasis and poor outcomes in breast cancer. Studies have demonstrated that SOX11⁺ DCIS tumor cells metastasize to brain and bone at greater frequency compared to cells with lower SOX11 levels using xenograft models of DCIS. High levels of SOX11 were also associated with mesenchymal and stemness phenotypes (Oliemuller et al. 2017, 2020).

In summary, the above studies have demonstrated that epithelial and stromal-derived secreted factors may promote cellular stress and stemness programs that drive progression of DCIS to invasive cancer. Therefore, the expression of biomarkers of cellular stress and stemness (i.e., COX2, ARCS⁺, ALDH1A), epithelial/stromal secreted factors (i.e., IL-6, IL-8, Activin A), and their cognate receptors, when combined with existing biomarkers, may provide more powerful predictive tools in stratifying risk for DCIS patients.

MYOEPITHELIAL CELLS: FRIEND OR FOE?

The classic model of DCIS transition to invasive breast cancer involves the escape of DCIS epithelial cells into the surrounding breast stroma by invasion through the myoepithelial cell layer and the basement membrane. This process involves the gradual loss of myoepithelial function marked by the loss of p63 followed by calponin and finally smooth muscle actin (SMA) expression (Russell et al. 2015). The loss of myoepithelial cell layer during DCIS to invasive progression has led many to believe that the myoepithelial cell layer serves as a gate keeper for DCIS progression

and may have tumor-suppressive properties. In fact, many previous studies have proposed that the suppressive role of myoepithelial cells due to secretion of protease inhibitors (maspin, and TIMP-1), down-regulation of metalloproteinases, and laminin (Sternlicht et al. 1997).

Hu and colleagues (2008) performed immunohistochemical analyses and gene expression profiling of luminal and myoepithelial cells from DCIS xenografts and human breast tissue, which led to the identification of transforming growth factor β (TGF-β), HedgeHog (HH), and p63 signaling pathways as potential regulators of myoepithelial cell function. They found extensive cross talk among these pathways in MCFDCIS.COM cells and demonstrated that decreasing TGF-β and HH pathway activity via down-regulating TGFBR2/SMAD4 and Gli2 expression resulted in loss of myoepithelial cells and accelerated progression to invasion. For example, both p63 and TGF-β1 up-regulate ITGB6 expression, which in turn can activate latent TGF-β1 and generate a positive feedback loop to activate the suppressive functions of myoepithelial cells. Based on these results, the authors concluded that epithelial/stromal factors that disrupt the myoepithelial regulatory mechanisms may induce DCIS invasive progression.

Allinen et al. (2004) generated SAGE libraries from various stromal cells including fibroblasts, myofibroblasts, endothelial cells, infiltrating lymphocytes, and epithelial cells from one normal reduction mammoplasty, two DCIS, three invasive breast cancers, one fibroadenoma, and one phyllodes tumor sample. They found that the most extensive gene expression changes occurred in the myoepithelial cell layer of DCIS, including genes encoding secreted factors or cell-surface proteins, suggesting there are extensive interactions between myoepithelial cells/stroma and epithelial cells. Several proteases (cathepsin F, K, and L, MMP2, and PRSS11), protease inhibitors (thrombospondin 2, SERPING1, cystatin C, and TIMP3), and various collagens were highly up-regulated in DCIS myoepithelial cells, indicating a role for these cells in ECM remodeling. Additionally, CXCL14 and CXCL12 chemokines were overexpressed in tumor myoepithelial cells and myofibroblasts, respectively. The authors demonstrated that receptors for CXCL14 and CXCL12 were expressed on DCIS (epithelial) cells adjacent to myoepithelial cells and validated the role of these chemokines and their cognate receptors in epithelial cell proliferation, migration, and invasion by using in vitro and in vivo studies. These studies support the hypothesis that the expression of myoepithelial CXCL14/CXCL12 may serve as myoepithelial biomarkers of DCIS invasive progression.

One interesting study recently published by Risom et al. (2022) used multiplexed ion beam imaging by time of light (MIBI-TOF) and a 37-plex antibody staining to evaluate stromal changes associated with invasive recurrence in DCIS. Contrary to previous studies, this study showed that myoepithelial loss and a reactive stroma was more pronounced in DCIS patients that did not recur compared to those that showed a subsequent recurrence. To explain how this loss of myoepithelial layer may be protective, they evaluated ontologies that correlated with high or low myoepithelial characters. Low scores typical of non-progressors were enriched in hypoxia, glycolysis, stromal immune density, and desmoplasia/remodeling of ECM. In contrast, high myoepithelial scores typical of progressors were enriched in immunomodulatory markers such as PDL1, IDO1, COX2, and PD1 in tumor and immune cells. These data suggest that loss of myoepithelial layer has a protective tumor-sensing function that favors immune cell activation in the surrounding stroma, thus preventing DCIS recurrence.

In summary, the role of myoepithelial cells as protective and tumor suppressive versus tumor promoting is currently under debate. Substantial evidence exists in the literature in support of, as well as against, their protective role in DCIS. As such, further investigation of myoepithelial cells are needed in order to better delineate their role in DCIS progression.

DCIS RECURRENCE AND CLONAL RELATEDNESS

A small fraction of DCIS (5%–10%) will show a subsequent recurrence following treatment (Shaaban et al. 2021). A study of 7934 patients treated with or without radiotherapy following lumpectomy revealed that 5.3% showed a subsequent invasive recurrence, while 3.8% showed recurrence as DCIS when followed for a median of 9.4 yr. A key question is whether DCIS that subsequently recur in the same breast show a common genetic lineage and are thus clonally related to the original DCIS, or they emerge from different initiating cells. To address this, Lips and colleagues (Lips et al. 2022) performed comprehensive genomic analysis on the original DCIS and paired invasive ($n = 95$) or paired DCIS ($n = 35$) recurrences that occurred 5–17 yr later. Their results showed that among the samples, 75% of recurrences were clonally related to the initial DCIS, while 18% were unrelated and 7% were ambiguous. Figure 1 shows the differences in CNA profiles observed in two representative, clonally unrelated and two representative, clonally related samples. As shown, the clonally unrelated pairs showed no common CNAs, while the clonally related pairs showed many shared CNAs. The authors also evaluated the frequency of specific mutations and/or CNAs that predicated future invasive recurrences. Interestingly, as shown previously by others, the frequency of mutations and CNAs were similar in DCIS and recurrent invasive disease. The only exception was gains in 1q and 8p11 that were significantly more common in recurrent invasive disease compared to DCIS, while 3p21 loss was more frequent in DCIS. While the authors did not find a genomic predisposition to explain the occurrence of a clonally unrelated breast cancer, field cancerization was proposed as a potential mechanism (Lips et al. 2022).

DCIS INVASION: DOES IT REALLY MATTER?

There is accumulating evidence that the focus on prevention of DCIS invasiveness may not necessarily provide survival benefit. A study of nearly 145,000 women diagnosed with primary DCIS using surveillance, epidemiology, and end results (SEER) registry data showed that the cumulative 20-yr risk of breast cancer–specific mortality following DCIS was ∼3%, with an increased risk to 8% for Black women (Narod et al. 2015). Indeed, women who are diagnosed with

DCIS are three times more likely to die from breast cancer relative to women in the general population without a breast cancer diagnosis (Giannakeas et al. 2020). Furthermore, while the risk of invasive recurrence was the highest among women who received lumpectomy alone compared to lumpectomy with radiation or mastectomy, the risk of mortality was similar (Narod et al. 2015). Other groups have reported similar findings (Wadsten et al. 2017; Elshof et al. 2018). Furthermore, disseminated epithelial (tumor) cells have been found in the peripheral blood of patients with DCIS at a rate similar to invasive breast cancers (13%–25%) (Sänger et al. 2011, 2014; Franken et al. 2012). In keeping with this, about 3% of patients diagnosed with pure DCIS already have evidence of lymph node metastasis (El Hage Chehade et al. 2017). These findings highlight the need for early detection of DCIS among women with the highest mortality risk. Although there remains a concern for overtreatment in DCIS, we may also be undertreating a subset of women at higher risk of developing metastatic breast cancer.

DIAGNOSTIC TOOLS FOR RISK STRATIFICATION IN DCIS

The role of radiation therapy (RT) following breast-conserving surgery (BCS) for DCIS remains controversial. The trials that evaluated the role of RT have all demonstrated a significant reduction in breast events with RT, but with no survival benefit (Wärnberg et al. 2001; Wapnir et al. 2011; Donker et al. 2013; Thorat et al. 2021). In an attempt to find those DCIS that do not benefit from and can forego RT, two diagnostic tests have been developed, DCIS-Score and DCISionRT.

DCIS-Score was the first of the two diagnostic tests that provide information on the risk of local recurrence based on DCIS biology independent of clinicopathologic features. DCIS-Score reports the expression levels of seven cancer-related genes (Ki67, STK15, Survivin, CCNB1, MYBL2, PgR, and GSTM1) plus five reference genes (β-actin, GAPDH, RPLPO, GUS, and TFRC). The score ranges from 0 to 100, and the results can be used as a continuous or categorical

variable. The risk categories are defined as DCIS-Score <39, low risk; DCIS-Score 39 to 54, intermediate risk; DCIS-Score 55 to 100, high risk. The 10-yr rate of ipsilateral events in the low-, intermediate-, and high-risk groups were 10.6%, 26.7%, and 25.9%, respectively ($P = 0.006$), and the 10-yr rates of invasive breast cancer were 3.7%, 12.3%, and 19.2%, respectively ($P = 0.003$). Although DCIS score has not been widely adopted, it may serve as a guide for therapeutic decision-making regarding surgery (mastectomy vs. BCS) and/or the use of adjuvant RT following BCS (Wood et al. 2014). However, the benefit from RT was not formally tested in high- versus low-risk groups. Furthermore, validation studies included a small number of HER2[+] or triple-negative DCIS; thus, the utility of DCIS-Score for risk stratification in these subtypes are limited.

DCISionRT is the second tool that is based on the immunohistochemical expression of seven cancer-related genes (COX-2, FOXA1, HER2, Ki-67, p16/INK4A, PgR, and SIAH2) in addition to four clinicopathologic features (age, size, margin status, and palpability). The test is based on research at the University of California San Francisco in an attempt to evaluate DCIS biology in combination with clinicopathologic features (Gauthier et al. 2007; Kerlikowske et al. 2010). DCISionRT uses an algorithm to generate a continuous decision score (DS) divided into DS low ≤3 and DS high >3. An advantage of DCISionRT is that it provides information regarding radiotherapy benefit. Prospective studies have shown that the use of DCISionRT resulted in a significant change in the recommendation for RT. The use of DS resulted in a 20% reduction in the use of RT. Additionally, the use of DS also resulted in a recommendation to use RT in 35% of women not recommended to receive RT prior to the test. Patients with elevated scores showed an absolute reduction of 9%–15% in 10-yr risk of subsequent invasive cancer with RT while those with low DS scores showed a small reduction (1%–2%).

The above two tests improve the prediction of local recurrence after BCS for DCIS but do not predict breast cancer mortality. A recent study showed that the use of 21-gene recurrence score

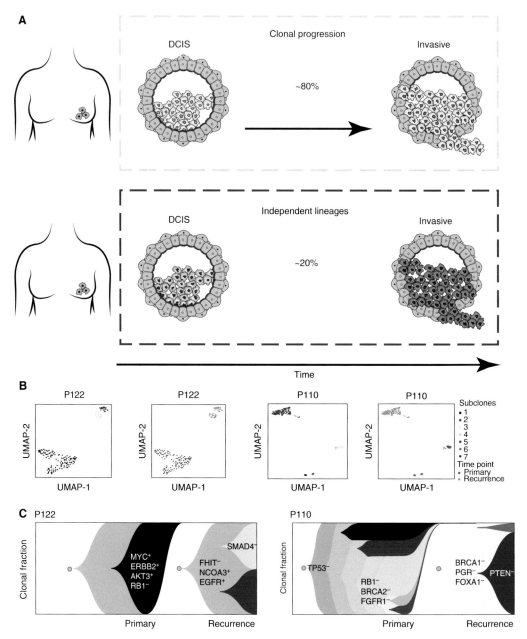

Figure 1. Not all ductal carcinoma in situ (DCIS) and recurrent breast cancers are clonally related. Eighty percent of breast cancer recurrences following a DCIS diagnosis are clonally related, while about 20% are unrelated (*A*). UMAP plots showing clonal distribution in two clonally unrelated breast cancer recurrences; P122 recurred as invasive while P110 recurred as pure DCIS. As shown, the clones were specific to either the primary (purple) or recurrent breast cancers (orange) (*B*). Muller plots confirming independent clonal lineages in the primary and recurrent breast cancers (*C*). UMAP plots showing clonal distribution in two clonally related breast cancer recurrences; both DCIS recurred as invasive lesions. As shown, many copy number alteration (CNA) events were shared between the primary and recurrent breast cancers (*D*). Muller plots showing subclones that harbored common CNAs in primary and recurrent breast cancers. As shown, some subclones expanded in the transition from DCIS to invasive breast cancer (*E*). (Figure reprinted from Lips et al. 2022 under the Creative Commons Attribution 4.0 International License.)

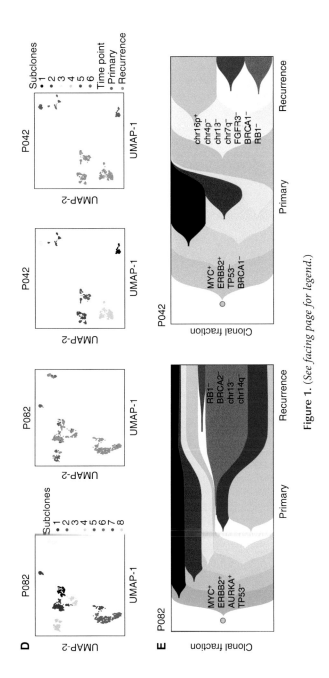

Figure 1. (*See facing page for legend.*)

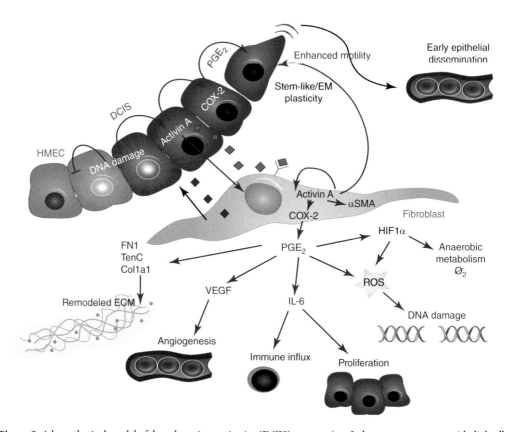

Figure 2. A hypothetical model of ductal carcinoma in situ (DCIS) progression. In human mammary epithelial cells with an intact cell-cycle pathway (human mammary epithelial cell [HMEC], green cells), cellular stress due to DNA damage/aneuploidy causes cell-cycle arrest. In contrast, the same type of insult in human mammary epithelial cells with a compromised cell-cycle pathway (i.e., DCIS; brown cells) induces uncontrolled proliferation, secretion of growth factors, and cytokines (i.e., Activin A, IL-6, IL-8, vascular endothelial growth factor [VEGF]) and up-regulation of biomarkers of cellular stress and stemness (i.e., ARCS[+], ALDH1A, SIAH2, SOX11, BCL9). Fibroblasts adjacent to stressed DCIS epithelial cells and exposed to epithelial-derived secreted factors up-regulate a protumorigenic microenvironment, typified by altered extracellular matrix (ECM), angiogenesis, immune cell influx, and altered metabolism, among others. In turn, protumorigenic stroma secretes factors that further fuel cellular stress and stemness programs via epigenetic changes that induce a hybrid epithelial/mesenchymal (E/M) state. Stem-like (E/M) cells characterized by their plasticity and invasive capacity will ultimately promote DCIS cell invasion and potentially early epithelial cell dissemination, leading to metastasis. (Figure modified Fordyce et al. 2012 under the terms of the Creative Commons Attribution License http://creativecommons.org/licenses/by/2.0.)

assays, which in addition to the DCIS-Score includes additional genes implicated in invasion, cellular adhesion, and metastasis, can predict risk of mortality in DCIS patients (Rakovitch et al. 2021). For women younger than 50 yr treated with BCS alone, the presence of a high RS was associated with an 11-fold increase in risk of breast cancer mortality or a 20-yr risk of 9.4% vs. 0.8% for low RS. For those treated with BCS plus RT, there was no statistically significant association between high RS and breast cancer mortality. For women older than 50, RS was not associated with mortality regardless of treatment (BCS with or without RT). Additionally, those with a high RS (>25), treatment with RT was associated with a 71% reduction in 20-yr cancer mortality and an absolute reduction of 5% regardless of age at diagnosis. In women with low RS, the risk of mortality was low regardless of RT therapy (Rakovitch et al. 2021).

CONCLUDING REMARKS

Our viewpoint of molecular mechanisms underlying DCIS progression and metastasis is summarized in the hypothetical model shown in Figure 2. Although much progress has been made toward the understanding of DCIS pathogenesis and mechanisms of progression, we still do not have clinically feasible, actionable, and validated tools to distinguish those with greatest mortality risk following a DCIS diagnosis.

Current studies have focused heavily on the role of epithelial cells, stroma, or myoepithelial cells in disease progression. Future diagnostic tests that focus on epithelial and stromal interactions as well as cellular states may further improve our prediction power. Additionally, emphasis on high-risk lesions and patient populations with increased risk of metastasis at the DCIS stage, rather than prevention of DCIS recurrence as invasive breast cancer, may provide further insights to improve patient survival.

ACKNOWLEDGMENTS

This work was supported by Cancer Research UK and KWF Kankerbestrijding (ref. C38317/A24043) (https://www.dcisprecision.org) (F.B., A.T.); KU Cancer Center Support Grant (P30 CA168524) and The Kansas Institute for Precision Medicine – COBRE (P20 GM130423); and Pilot Grant: NCATS Frontiers-CTSA grant from NCATS awarded to the University of Kansas for Frontiers, ref# UL1TR002366 (F.B.).

REFERENCES

Abba MC, Gong T, Lu Y, Lee J, Zhong Y, Lacunza E, Butti M, Takata Y, Gaddis S, Shen J, et al. 2015. A molecular portrait of high-grade ductal carcinoma in situ. *Cancer Res* **75:** 3980–3990. doi:10.1158/0008-5472.CAN-15-0506

Afghahi A, Forgo E, Mitani AA, Desai M, Varma S, Seto T, Rigdon J, Jensen KC, Troxell ML, Gomez SL, et al. 2015. Chromosomal copy number alterations for associations of ductal carcinoma in situ with invasive breast cancer. *Breast Cancer Res* **17:** 108. doi:10.1186/s13058-015-0623-y

Allinen M, Beroukhim R, Cai L, Brennan C, Lahti-Domenici J, Huang H, Porter D, Hu M, Chin L, Richardson A, et al. 2004. Molecular characterization of the tumor microenvironment in breast cancer. *Cancer Cell* **6:** 17–32. doi:10.1016/j.ccr.2004.06.010

Allred DC, Wu Y, Mao S, Nagtegaal ID, Lee S, Perou CM, Mohsin SK, O'Connell P, Tsimelzon A, Medina D. 2008. Ductal carcinoma in situ and the emergence of diversity during breast cancer evolution. *Clin Cancer Res* **14:** 370–378. doi:10.1158/1078-0432.CCR-07-1127

Barcellos-Hoff MH, Ravani SA. 2000. Irradiated mammary gland stroma promotes the expression of tumorigenic potential by unirradiated epithelial cells. *Cancer Res* **60:** 1254–1260.

Bonuccelli G, Whitaker-Menezes D, Castello-Cros R, Pavlides S, Pestell RG, Fatatis A, Witkiewicz AK, Vander Heiden MG, Migneco G, Chiavarina B, et al. 2010. The reverse Warburg effect: glycolysis inhibitors prevent the tumor promoting effects of caveolin-1 deficient cancer associated fibroblasts. *Cell Cycle* **9:** 1960–1971. doi:10.4161/cc.9.10.11601

Bremer T, Whitworth PW, Patel R, Savala J, Barry T, Lyle S, Leesman G, Linke SP, Jirström K, Zhou W, et al. 2018. A biological signature for breast ductal carcinoma in situ to predict radiotherapy benefit and assess recurrence risk. *Clin Cancer Res* **24:** 5895–5901. doi:10.1158/1078-0432.CCR-18-0842

Buerger H, Otterbach F, Simon R, Poremba C, Diallo R, Decker T, Riethdorf L, Brinkschmidt C, Dockhorn-Dworniczak B, Boecker W. 1999. Comparative genomic hybridization of ductal carcinoma in situ of the breast-evidence of multiple genetic pathways. *J Pathol* **187:** 396–402. doi:10.1002/(SICI)1096-9896(199903)187:4<396::AID-PATH286>3.0.CO;2-L

Casasent AK, Schalck A, Gao R, Sei E, Long A, Pangburn W, Casasent T, Meric-Bernstam F, Edgerton ME, Navin NE. 2018. Multiclonal invasion in breast tumors identified by topographic single cell sequencing. *Cell* **172:** 205–217.e212. doi:10.1016/j.cell.2017.12.007

Chan P, Möller A, Liu MC, Sceneay JE, Wong CS, Waddell N, Huang KT, Dobrovic A, Millar EK, O'Toole SA, et al. 2011. The expression of the ubiquitin ligase SIAH2 (seven in absentia homolog 2) is mediated through gene copy number in breast cancer and is associated with a basal-like phenotype and p53 expression. *Breast Cancer Res* **13:** R19. doi:10.1186/bcr2828

Collins LC, Tamimi RM, Baer HJ, Connolly JL, Colditz GA, Schnitt SJ. 2005. Outcome of patients with ductal carcinoma in situ untreated after diagnostic biopsy: results from the Nurses' Health Study. *Cancer* **103:** 1778–1784. doi:10.1002/cncr.20979

Del Alcazar CRG, Huh SJ, Ekram MB, Trinh A, Liu LL, Beca F, Zi X, Kwak M, Bergholtz H, Su Y, et al. 2017. Immune escape in breast cancer during in situ to invasive carcinoma transition. *Cancer Discov* **7:** 1098–1115. doi:10.1158/2159-8290.CD-17-0222

Donker M, Litière S, Werutsky G, Julien JP, Fentiman IS, Agresti R, Rouanet P, de Lara CT, Bartelink H, Duez N, et al. 2013. Breast-conserving treatment with or without radiotherapy in ductal carcinoma in situ: 15-year recurrence rates and outcome after a recurrence, from the

EORTC 10853 randomized phase III trial. *J Clin Oncol* **31:** 4054–4059. doi:10.1200/JCO.2013.49.5077

El Hage Chehade H, Headon H, Wazir U, Abtar H, Kasem A, Mokbel K. 2017. Is sentinel lymph node biopsy indicated in patients with a diagnosis of ductal carcinoma in situ? A systematic literature review and meta-analysis. *Am J Surg* **213:** 171–180. doi:10.1016/j.amjsurg.2016.04.019

Elsarraj HS, Hong Y, Valdez KE, Michaels W, Hook M, Smith WP, Chien J, Herschkowitz JI, Troester MA, Beck M, et al. 2015. Expression profiling of in vivo ductal carcinoma in situ progression models identified B cell lymphoma-9 as a molecular driver of breast cancer invasion. *Breast Cancer Res* **17:** 128. doi:10.1186/s13058-015-0630-z

Elsarraj HS, Hong Y, Limback D, Zhao R, Berger J, Bishop SC, Sabbagh A, Oppenheimer L, Harper HE, Tsimelzon A, et al. 2020. BCL9/STAT3 regulation of transcriptional enhancer networks promote DCIS progression. *NPJ Breast Cancer* **6:** 12. doi:10.1038/s41523-020-0157-z

Elshof LE, Tryfonidis K, Slaets L, van Leeuwen-Stok AE, Skinner VP, Dif N, Pijnappel RM, Bijker N, Rutgers EJ, Wesseling J. 2015. Feasibility of a prospective, randomised, open-label, international multicentre, phase III, non-inferiority trial to assess the safety of active surveillance for low risk ductal carcinoma in situ—the LORD study. *Eur J Cancer* **51:** 1497–1510. doi:10.1016/j.ejca.2015.05.008

Elshof LE, Schmidt MK, Rutgers EJT, van Leeuwen FE, Wesseling J, Schaapveld M. 2018. Cause-specific mortality in a population-based cohort of 9799 women treated for ductal carcinoma in situ. *Ann Surg* **267:** 952–958. doi:10.1097/SLA.0000000000002239

Feng M, Jin JQ, Xia L, Xiao T, Mei S, Wang X, Huang X, Chen J, Liu M, Chen C, et al. 2019. Pharmacological inhibition of β-catenin/BCL9 interaction overcomes resistance to immune checkpoint blockades by modulating Treg cells. *Sci Adv* **5:** eaau5240. doi:10.1126/sciadv.aau5240

Fordyce CA, Patten KT, Fessenden TB, Defilippis R, Hwang ES, Zhao J, Tlsty TD. 2012. Cell-extrinsic consequences of epithelial stress: activation of protumorigenic tissue phenotypes. *Breast Cancer Res* **14:** R155. doi:10.1186/bcr3368

Francis A, Thomas J, Fallowfield L, Wallis M, Bartlett JM, Brookes C, Roberts T, Pirrie S, Gaunt C, Young J, et al. 2015. Addressing overtreatment of screen detected DCIS; the LORIS trial. *Eur J Cancer* **51:** 2296–2303. doi:10.1016/j.ejca.2015.07.017

Franken B, de Groot MR, Mastboom WJ, Vermes I, van der Palen J, Tibbe AG, Terstappen LW. 2012. Circulating tumor cells, disease recurrence and survival in newly diagnosed breast cancer. *Breast Cancer Res* **14:** R133. doi:10.1186/bcr3333

Fujii H, Szumel R, Marsh C, Zhou W, Gabrielson E. 1996. Genetic progression, histological grade, and allelic loss in ductal carcinoma in situ of the breast. *Cancer Res* **56:** 5260–5265.

Gauthier I, Ding K, Winton T, Shepherd FA, Livingston R, Johnson DH, Rigas JR, Whitehead M, Graham B, Seymour L. 2007. Impact of hemoglobin levels on outcomes of adjuvant chemotherapy in resected non-small cell lung cancer: the JBR.10 trial experience. *Lung Cancer* **55:** 357–363. doi:10.1016/j.lungcan.2006.10.021

Giannakeas V, Sopik V, Narod SA. 2020. Association of a diagnosis of ductal carcinoma in situ with death from breast cancer. *JAMA Netw Open* **3:** e2017124. doi:10.1001/jamanetworkopen.2020.17124

Ginestier C, Hur MH, Charafe-Jauffret E, Monville F, Dutcher J, Brown M, Jacquemier J, Viens P, Kleer CG, Liu S, et al. 2007. ALDH1 is a marker of normal and malignant human mammary stem cells and a predictor of poor clinical outcome. *Cell Stem Cell* **1:** 555–567. doi:10.1016/j.stem.2007.08.014

Han MS, Khan SA. 2018. Clinical trials for ductal carcinoma in situ of the breast. *J Mammary Gland Biol Neoplasia* **23:** 293–301. doi:10.1007/s10911-018-9413-3

Hannemann J, Velds A, Halfwerk JBG, Kreike B, Peterse JL, van de Vijver MJ. 2006. Classification of ductal carcinoma in situ by gene expression profiling. *Breast Cancer Res* **8:** R61. doi:10.1186/bcr1613

Hong Y, Limback D, Elsarraj HS, Harper H, Haines H, Hansford H, Ricci M, Kaufman C, Wedlock E, Xu M, et al. 2021. Mouse-INtraDuctal (MIND): An in vivo model for studying the underlying mechanisms of DCIS malignancy. *J Pathol* **256:** 186–201. doi:10.1002/path.5820

Hu M, Yao J, Carroll DK, Weremowicz S, Chen H, Carrasco D, Richardson A, Violette S, Nikolskaya T, Nikolsky Y, et al. 2008. Regulation of in situ to invasive breast carcinoma transition. *Cancer Cell* **13:** 394–406. doi:10.1016/j.ccr.2008.03.007

Kerlikowske K, Molinaro AM, Gauthier ML, Berman HK, Waldman F, Bennington J, Sanchez H, Jimenez C, Stewart K, Chew K, et al. 2010. Biomarker expression and risk of subsequent tumors after initial ductal carcinoma in situ diagnosis. *J Natl Cancer Inst* **102:** 627–637. doi:10.1093/jnci/djq101

Kim M, Choi HY, Woo JW, Chung YR, Park SY. 2021. Role of CXCL10 in the progression of in situ to invasive carcinoma of the breast. *Sci Rep* **11:** 18007. doi:10.1038/s41598-021-97390-5

Kretschmer C, Sterner-Kock A, Siedentopf F, Schoenegg W, Schlag PM, Kemmner W. 2011a. Identification of early molecular markers for breast cancer. *Mol Cancer* **10:** 15. doi:10.1186/1476-4598-10-15

Kretschmer C, Sterner-Kock A, Siedentopf F, Schoenegg W, Schlag PM, Kemmner W. 2011b. Identification of early molecular markers for breast cancer. *Mol Cancer* **10:** 15. doi:10.1186/1476-4598-10-15

Lee S, Stewart S, Nagtegaal I, Luo J, Wu Y, Colditz G, Medina D, Allred DC. 2012. Differentially expressed genes regulating the progression of ductal carcinoma in situ to invasive breast cancer. *Cancer Res* **72:** 4574–4586. doi:10.1158/0008-5472.CAN-12-0636

Leonard GD, Swain SM. 2004. Ductal carcinoma in situ, complexities and challenges. *J Natl Cancer Inst* **96:** 906–920. doi:10.1093/jnci/djh164

Lesurf R, Aure MR, Mork HH, Vitelli V, Oslo Breast Cancer Research C, Lundgren S, Børresen-Dale AL, Kristensen V, Wärnberg F, Hallett M, et al. 2016. Molecular features of subtype-specific progression from ductal carcinoma in situ to invasive breast cancer. *Cell Rep* **16:** 1166–1179. doi:10.1016/j.celrep.2016.06.051

Lips EH, Kumar T, Megalios A, Visser LL, Sheinman M, Fortunato A, Shah V, Hoogstraat M, Sei E, Mallo D, et al. 2022. Genomic analysis defines clonal relationships of ductal carcinoma in situ and recurrent invasive breast cancer. *Nat Genet* **54:** 850–860. doi:10.1038/s41588-022-01082-3

Ma XJ, Salunga R, Tuggle JT, Gaudet J, Enright E, McQuary P, Payette T, Pistone M, Stecker K, Zhang BM, et al. 2003. Gene expression profiles of human breast cancer progression. *P Natl Acad Sci* **100:** 5974–5979. doi:10.1073/pnas.0931261100

Marshall E. 2014. Breast cancer. Dare to do less. *Science* **343:** 1454–1456. doi:10.1126/science.343.6178.1454

Martinez-Outschoorn UE, Pavlides S, Whitaker-Menezes D, Daumer KM, Milliman JN, Chiavarina B, Migneco G, Witkiewicz AK, Martinez-Cantarin MP, Flomenberg N, et al. 2010. Tumor cells induce the cancer associated fibroblast phenotype via caveolin-1 degradation: implications for breast cancer and DCIS therapy with autophagy inhibitors. *Cell Cycle* **9:** 2423–2433. doi:10.4161/cc.9.12.12048

Martinez-Outschoorn UE, Sotgia F, Lisanti MP. 2015. Caveolae and signalling in cancer. *Nat Rev Cancer* **15:** 225–237. doi:10.1038/nrc3915

Martins D, Beça FF, Sousa B, Baltazar F, Paredes J, Schmitt F. 2013. Loss of caveolin-1 and gain of MCT4 expression in the tumor stroma: key events in the progression from an in situ to an invasive breast carcinoma. *Cell Cycle* **12:** 2684–2690. doi:10.4161/cc.25794

Maxwell AJ, Clements K, Hilton B, Dodwell DJ, Evans A, Kearins O, Pinder SE, Thomas J, Wallis MG, Thompson AM, et al. 2018. Risk factors for the development of invasive cancer in unresected ductal carcinoma in situ. *Eur J Surg Oncol* **44:** 429–435. doi:10.1016/j.ejso.2017.12.007

Maxwell AJ, Hilton B, Clements K, Dodwell D, Dulson-Cox J, Kearins O, Kirwan C, Litherland J, Mylvaganam S, Provenzano E, et al. 2022. Unresected screen-detected ductal carcinoma in situ: outcomes of 311 women in the Forget-Me-Not 2 study. *Breast* **61:** 145–155. doi:10.1016/j.breast.2022.01.001

Molinaro AM, Sison JD, Ljung BM, Tlsty TD, Kerlikowske K. 2016. Risk prediction for local versus regional/metastatic tumors after initial ductal carcinoma in situ diagnosis treated by lumpectomy. *Breast Cancer Res Treat* **157:** 351–361. doi:10.1007/s10549-016-3814-z

Muggerud AA, Hallett M, Johnsen H, Kleivi K, Zhou WJ, Tahmasebpoor S, Amini RM, Botling J, Borresen-Dale AL, Sørlie T, et al. 2010. Molecular diversity in ductal carcinoma in situ (DCIS) and early invasive breast cancer. *Mol Oncol* **4:** 357–368. doi:10.1016/j.molonc.2010.06.007

Narod SA, Iqbal J, Giannakeas V, Sopik V, Sun P. 2015. Breast cancer mortality after a diagnosis of ductal carcinoma in situ. *JAMA Oncol* **1:** 888–896. doi:10.1001/jamaoncol.2015.2510

Nelson AC, Machado HL, Schwertfeger KL. 2018. Breaking through to the other side: microenvironment contributions to DCIS initiation and progression. *J Mammary Gland Biol Neoplasia* **23:** 207–221. doi:10.1007/s10911-018-9409-z

Newburger DE, Kashef-Haghighi D, Weng Z, Salari R, Sweeney RT, Brunner AL, Zhu SX, Guo X, Varma S, Troxell ML, et al. 2013. Genome evolution during progression to breast cancer. *Genome Res* **23:** 1097–1108. doi:10.1101/gr.151670.112

Oliemuller E, Kogata N, Bland P, Kriplani D, Daley F, Haider S, Shah V, Sawyer EJ, Howard BA. 2017. SOX11 promotes invasive growth and ductal carcinoma in situ progression. *J Pathol* **243:** 193–207. doi:10.1002/path.4939

Oliemuller E, Newman R, Tsang SM, Foo S, Muirhead G, Noor F, Haider S, Aurrekoetxea-Rodriguez I, Vivanco MD, Howard BA. 2020. SOX11 promotes epithelial/mesenchymal hybrid state and alters tropism of invasive breast cancer cells. *eLife* **9:** e58374. doi:10.7554/eLife.58374

Olumi AF, Grossfeld GD, Hayward SW, Carroll PR, Tlsty TD, Cunha GR. 1999. Carcinoma-associated fibroblasts direct tumor progression of initiated human prostatic epithelium. *Cancer Res* **59:** 5002–5011.

Osuala KO, Sameni M, Shah S, Aggarwal N, Simonait ML, Franco OE, Hong Y, Hayward SW, Behbod F, Mattingly RR, et al. 2015. IL-6 signaling between ductal carcinoma in situ cells and carcinoma-associated fibroblasts mediates tumor cell growth and migration. *BMC Cancer* **15:** 584. doi:10.1186/s12885-015-1576-3

Ozanne EM, Shieh Y, Barnes J, Bouzan C, Hwang ES, Esserman LJ. 2011. Characterizing the impact of 25 years of DCIS treatment. *Breast Cancer Res Treat* **129:** 165–173. doi:10.1007/s10549-011-1430-5

Page DL, Dupont WD, Rogers LW, Landenberger M. 1982. Intraductal carcinoma of the breast: follow-up after biopsy only. *Cancer* **49:** 751–758. doi:10.1002/1097-0142(19820215)49:4<751::AID-CNCR2820490426>3.0.CO;2-Y

Pareja F, Brown DN, Lee JY, Da Cruz Paula A, Selenica P, Bi R, Geyer FC, Gazzo A, da Silva EM, Vahdatinia M, et al. 2020. Whole-exome sequencing analysis of the progression from non-low grade ductal carcinoma in situ to invasive ductal carcinoma. *Clin Cancer Res* **26:** 3682–3693. doi:10.1158/1078-0432.CCR-19-2563

Rakovitch E, Sutradhar R, Nofech-Mozes S, Gu S, Fong C, Hanna W, Paszat L. 2021. 21-gene assay and breast cancer mortality in ductal carcinoma in situ. *J Natl Cancer Inst* **113:** 572–579. doi:10.1093/jnci/djaa179

Risom T, Glass DR, Averbukh I, Liu CC, Baranski A, Kagel A, McCaffrey EF, Greenwald NF, Rivero-Gutiérrez B, Strand SH, et al. 2022. Transition to invasive breast cancer is associated with progressive changes in the structure and composition of tumor stroma. *Cell* **185:** 299–310 e218. doi:10.1016/j.cell.2021.12.023

Russell TD, Jindal S, Agunbiade S, Gao D, Troxell M, Borges VF, Schedin P. 2015. Myoepithelial cell differentiation markers in ductal carcinoma in situ progression. *Am J Pathol* **185:** 3076–3089. doi:10.1016/j.ajpath.2015.07.004

Sagara Y, Mallory MA, Wong S, Aydogan F, DeSantis S, Barry WT, Golshan M. 2015. Survival benefit of breast surgery for low-grade ductal carcinoma in situ: a population-based cohort study. *JAMA Surg* **150:** 739–745. doi:10.1001/jamasurg.2015.0876

Sanders ME, Schuyler PA, Dupont WD, Page DL. 2005. The natural history of low-grade ductal carcinoma in situ of the breast in women treated by biopsy only revealed over 30 years of long-term follow-up. *Cancer* **103:** 2481–2484. doi:10.1002/cncr.21069

Sänger N, Effenberger KE, Riethdorf S, Van Haasteren V, Gauwerky J, Wiegratz I, Strebhardt K, Kaufmann M, Pantel K. 2011. Disseminated tumor cells in the bone marrow of patients with ductal carcinoma in situ. *Int J Cancer* **129:** 2522–2526. doi:10.1002/ijc.25895

Sänger N, Engels K, Graf A, Ruckhäberle E, Effenberger KE, Fehm T, Holtrich U, Becker S, Karn T. 2014. Molecular markers as prognostic factors in DCIS and small invasive breast cancers. *Geburtshilfe Frauenheilkd* **74:** 1016–1022. doi:10.1055/s-0034-1383033

Schuetz CS, Bonin M, Clare SE, Nieselt K, Sotlar K, Walter M, Fehm T, Solomayer E, Riess O, Wallwiener D, et al. 2006. Progression-specific genes identified by expression profiling of matched ductal carcinomas in situ and invasive breast tumors, combining laser capture microdissection and oligonucleotide microarray analysis. *Cancer Res* **66:** 5278–5286. doi:10.1158/0008-5472.CAN-05-4610

Shaaban AM, Hilton B, Clements K, Provenzano E, Cheung S, Wallis MG, Sawyer E, Thomas JS, Hanby AM, Pinder SE, et al. 2021. Pathological features of 11,337 patients with primary ductal carcinoma in situ (DCIS) and subsequent events: results from the UK Sloane project. *Br J Cancer* **124:** 1009–1017. doi:10.1038/s41416-020-01152-5

Sternlicht MD, Kedeshian P, Shao ZM, Safarians S, Barsky SH. 1997. The human myoepithelial cell is a natural tumor suppressor. *Clin Cancer Res* **3:** 1949–1958.

Sun RC, Denko NC. 2014. Hypoxic regulation of glutamine metabolism through HIF1 and SIAH2 supports lipid synthesis that is necessary for tumor growth. *Cell Metab* **19:** 285–292. doi:10.1016/j.cmet.2013.11.022

Tao B, Song Y, Wu Y, Yang X, Peng T, Peng L, Xia K, Xia X, Chen L, Zhong C. 2021. Matrix stiffness promotes glioma cell stemness by activating BCL9L/Wnt/β-catenin signaling. *Aging (Albany NY)* **13:** 5284–5296. doi:10.18632/aging.202449

Thorat MA, Levey PM, Jones JL, Pinder SE, Bundred NJ, Fentiman IS, Cuzick J. 2021. Prognostic and predictive value of HER2 expression in ductal carcinoma in situ: results from the UK/ANZ DCIS randomized trial. *Clin Cancer Res* **27:** 5317–5324. doi:10.1158/1078-0432.CCR-21-1239

Tlsty T. 2008. Cancer: whispering sweet somethings. *Nature* **453:** 604–605. doi:10.1038/453604a

Tlsty TD, Hein PW. 2001. Know thy neighbor: stromal cells can contribute oncogenic signals. *Curr Opin Genet Dev* **11:** 54–59. doi:10.1016/S0959-437X(00)00156-8

Trinh A, Gil Del Alcazar CR, Shukla SA, Chin K, Chang YH, Thibault G, Eng J, Jovanovic B, Aldaz CM, Park SY, et al. 2021. Genomic alterations during the in situ to invasive ductal breast carcinoma transition shaped by the immune system. *Mol Cancer Res* **19:** 623–635. doi:10.1158/1541-7786.MCR-20-0949

Van Bockstal M, Lambein K, Gevaert O, De Wever O, Praet M, Cocquyt V, Van den Broecke R, Braems G, Denys H, Libbrecht L. 2013. Stromal architecture and periductal decorin are potential prognostic markers for ipsilateral locoregional recurrence in ductal carcinoma in situ of the breast. *Histopathology* **63:** 520–533.

Wadsten C, Garmo H, Fredriksson I, Sund M, Wärnberg F. 2017. Risk of death from breast cancer after treatment for ductal carcinoma in situ. *Br J Surg* **104:** 1506–1513. doi:10.1002/bjs.10589

Wapnir IL, Dignam JJ, Fisher B, Mamounas EP, Anderson SJ, Julian TB, Land SR, Margolese RG, Swain SM, Costantino JP, et al. 2011. Long-term outcomes of invasive ipsilateral breast tumor recurrences after lumpectomy in NSABP B-17 and B-24 randomized clinical trials for DCIS. *J Natl Cancer Inst* **103:** 478–488. doi:10.1093/jnci/djr027

Wärnberg F, Bergh J, Zack M, Holmberg L. 2001. Risk factors for subsequent invasive breast cancer and breast cancer death after ductal carcinoma in situ: a population-based case-control study in Sweden. *Cancer Epidemiol Biomarkers Prev* **10:** 495–499.

Wärnberg F, Karlsson P, Holmberg E, Sandelin K, Whitworth PW, Savala J, Barry T, Leesman G, Linke SP, Shivers SC, et al. 2021. Prognostic risk assessment and prediction of radiotherapy benefit for women with ductal carcinoma in situ (DCIS) of the breast, in a randomized clinical trial (SweDCIS). *Cancers (Basel)* **13:** 6103. doi:10.3390/cancers13236103

Weaver VM, Petersen OW, Wang F, Larabell CA, Briand P, Damsky C, Bissell MJ. 1997. Reversion of the malignant phenotype of human breast cells in three-dimensional culture and in vivo by integrin blocking antibodies. *J Cell Biol* **137:** 231–245. doi:10.1083/jcb.137.1.231

Welch HG, Black WC. 1997. Using autopsy series to estimate the disease "reservoir" for ductal carcinoma in situ of the breast: how much more breast cancer can we find? *Ann Intern Med* **127:** 1023–1028. doi:10.7326/0003-4819-127-11-199712010-00014

Witkiewicz AK, Dasgupta A, Nguyen KH, Liu C, Kovatich AJ, Schwartz GF, Pestell RG, Sotgia F, Rui H, Lisanti MP. 2009. Stromal caveolin-1 levels predict early DCIS progression to invasive breast cancer. *Cancer Biol Ther* **8:** 1071–1079. doi:10.4161/cbt.8.11.8874

Witkiewicz AK, Freydin B, Chervoneva I, Potoczek M, Rizzo W, Rui H, Brody JR, Schwartz GF, Lisanti MP. 2010. Stromal CD10 and SPARC expression in ductal carcinoma in situ (DCIS) patients predicts disease recurrence. *Cancer Biol Ther* **10:** 391–396. doi:10.4161/cbt.10.4.12449

Wood WC, Alvarado M, Buchholz DJ, Hyams D, Hwang S, Manders J, Park C, Solin LJ, White J, Willey S. 2014. The current clinical value of the DCIS score. *Oncology (Williston Park)* **28** (Suppl 2): C2, 1-8, C3.

Yao J, Weremowicz S, Feng B, Gentleman RC, Marks JR, Gelman R, Brennan C, Polyak K. 2006. Combined cDNA array comparative genomic hybridization and serial analysis of gene expression analysis of breast tumor progression. *Cancer Res* **66:** 4065–4078. doi:10.1158/0008-5472.CAN-05-4083

Gene-Expression Profiling to Decipher Breast Cancer Inter- and Intratumor Heterogeneity

Alexander Swarbrick,[1,2] Aranzazu Fernandez-Martinez,[3,4] and Charles M. Perou[3,4]

[1]Cancer Ecosystems Program, Garvan Institute of Medical Research, Darlinghurst, New South Wales 2010, Australia

[2]St Vincent's Clinical School, Faculty of Medicine, University of New South Wales, Sydney, New South Wales 2052, Australia

[3]Lineberger Comprehensive Center, University of North Carolina, Chapel Hill, North Carolina 27599, USA

[4]Department of Genetics, University of North Carolina, Chapel Hill, North Carolina 27514, USA

Correspondence: a.swarbrick@garvan.org.au

Breast cancer is heterogeneous and differs substantially across different tumors (intertumor heterogeneity) and even within an individual tumor (intratumor heterogeneity). Gene-expression profiling has considerably impacted our understanding of breast cancer biology. Four main "intrinsic subtypes" of breast cancer (i.e., luminal A, luminal B, HER2-enriched, and basal-like) have been consistently identified by gene expression, showing significant prognostic and predictive value in multiple clinical scenarios. Thanks to the molecular profiling of breast tumors, breast cancer is a paradigm of treatment personalization. Several standardized prognostic gene-expression assays are presently being used in the clinic to guide treatment decisions. Moreover, the development of single-cell-level resolution molecular profiling has allowed us to appreciate that breast cancer is also heterogeneous within a single tumor. There is an evident functional heterogeneity within the neoplastic and tumor microenvironment cells. Finally, emerging insights from these studies suggest a substantial cellular organization of neoplastic and tumor microenvironment cells, thus defining breast cancer ecosystems and highlighting the importance of spatial localizations.

Breast cancer has long been recognized as a heterogeneous disease. For many decades and to this day, breast cancers have been pathologically diagnosed as estrogen receptor (ER)-positive/-negative, human epidermal receptor 2 (HER2)-positive/-negative, and progesterone receptor (PR)-positive/-negative. These three therapeutic biomarkers then, in essence, define three clinical groups that are ER/PR$^+$/HER2$^-$, HER2$^+$, and triple-negative breast cancer (TNBC) (i.e., negative for ER, PR, and HER2). This classification provides not only prognostic information but also treatment directions, which is the main driving reason for ER, PR, and HER2 testing. With the development of DNA microarray technology for profiling transcriptomes, breast cancers have been additionally classified into four main genomic "intrinsic" subtypes: luminal A, luminal B,

HER2-enriched, and basal-like (Perou et al. 2000; Sørlie et al. 2001). Molecular profiling of hormone receptor (HR)-positive (i.e., ER- and/or PR-positive), HER2-negative breast cancers have provided several standardized prognostic gene-expression signatures that are being used in clinical practice, all of which provide additional information for prognostication and prediction of treatment benefit. With the development of single-cell-level resolution molecular profiling approaches, it is now recognized that breast cancers are heterogeneous not only among different tumors but also within a single tumor. There is an evident functional heterogeneity within the neoplastic cell population in breast cancers. In addition, there is growing awareness of the importance of the tumor microenvironment (TME), which includes host stromal cells and immune cells, beyond the neoplastic cell-intrinsic heterogeneity. In this review, we will discuss breast cancer heterogeneity, both intertumoral and intratumoral, based on gene-expression and related "omics" approaches, focusing on insights for human disease. We will first discuss studies analyzing tissue bulk-based approaches and then those using single-cell and spatially resolved techniques. The emerging insights from these studies suggest a substantial cellular organization of neoplastic and TME cells, thus defining breast cancer ecosystems.

DNA MICROARRAYS TO NEXT-GENERATION SEQUENCING FOR GENE EXPRESSION

Gene-expression profiling with DNA microarrays and high-throughput sequencing technologies like RNA sequencing (RNA-seq) allows for the identification of tens of thousands of transcripts in a single experiment. Thus, we can now identify gene-expression signatures, defined as groups of genes with a unique and characteristic expression pattern, representing specific cell types and biological pathways (Massagué 2007; Sotiriou and Pusztai 2009; Fan et al. 2011) instead of focusing on single gene candidates. These signatures usually have a biologic, prognostic, or predictive value, and some of them are currently commercially available and commonly used for breast cancer management.

The first signature identification approaches in breast cancer were based on clustering methods. These experiments allowed for the classification of breast tumors into luminal A, luminal B, HER2-enriched, and basal-like as four main molecular entities, or intrinsic subtypes, with different tumor biology and prognostic impact (Perou et al. 2000; Sørlie et al. 2001). The first generation of clinically available gene-expression prognostic signatures (i.e., MammaPrint and Oncotype DX) was created by genomically comparing low- versus high-risk breast tumors. Over time, multiple gene-expression signatures have been developed and validated to track different molecular pathways and specific biological processes (i.e., proliferation, hypoxia, cell differentiation, immune-cell states), and some of these signatures also predict sensitivity to treatments (Gatza et al. 2010; Fan et al. 2011; Iglesia et al. 2014; Newman et al. 2015; Hollern et al. 2019; Garcia-Recio et al. 2020). Moreover, the integration of gene-expression signatures and clinicopathologic factors has proven to significantly increase the prognostic ability compared to either one alone (Fan et al. 2011); and the integration of gene signatures, clinicopathologic features, DNA copy number alterations (CNAs), and DNA somatic mutations can predict response to treatment with improved accuracy over clinical factors alone (Tanioka et al. 2018).

For the first 15 yr of gene-expression profiling, hybridization-based platforms like DNA microarrays were the gold standard technology for gene-expression analysis. However, with the advent of RNA-seq based on next-generation sequencing technologies, a new gold standard has been set that has clear advantages when compared to other gene-expression profiling platforms (Table 1). In recent years, RNA-seq has largely replaced microarrays for the discovery of breast cancer prognostic and predictive biomarkers.

Hybridization platforms are high throughput and relatively inexpensive. However, they have several limitations (Wang et al. 2009):

- They can analyze only pre-defined sequences,

- Their dynamic range is limited,

Table 1. Main characters of DNA microarrays and RNA sequencing (RNA-seq)

	Microarrays	RNA-seq
Number of genes	High throughput	High throughput
	>20,000	Whole transcriptome
Sample preparation	RNA extraction	RNA extraction
	Reverse transcribe sample	Reverse transcribe sample
		Fragmentation
		Library preparation
Processing steps	Label cDNAs	Sequencing
	Hybridization to array	
Instrument	Microarray scanner	Sequencer
Time on sample processing	17–18 h (overnight)	24 h
Time on data processing	Short	Long
Cost	Moderate	Expensive

- The hybridization can be nonspecific (i.e., cross-hybridization) (Okoniewski and Miller 2006), and

- They usually show highly variable results for lowly expressed genes.

RNA-seq is technically a significant improvement beyond hybridization-based approaches in many ways. Compared to DNA microarrays, RNA-seq has several advantages (Wang et al. 2009):

- No prior sequence information is required,

- Allows for a precise and more quantitative measure of the gene expression,

- High sensitivity for the detection of low-abundance transcripts,

- Wider dynamic range,

- Can reveal sequence variations (i.e., RNA-seq), thus facilitating the discovery of alternative splicing variants, structural variants, and single-nucleotide variations (SNVs), and

- Greater reproducibility.

Due to all these advantages and the fact that RNA-seq provides the entire transcriptome to be surveyed in a high-throughput and quantitative manner, it has become the current gold standard laboratory tool for gene-expression profiling. However, this technology also comes with several challenges, including:

- RNA-seq is computationally challenging and requires the constant development of new methods to store, retrieve, process, and analyze a large amount of data,

- The lack of standardization among sequencing laboratories and read depths can compromise reproducibility, and

- Although the cost of RNA-seq has decreased in recent years, it is relatively high.

PROGNOSTIC SIGNATURES IN BREAST CANCER

With these robust technologies in hand, be they DNA microarrays or RNA-seq, breast cancer biologists used these tools to identify gene-expression patterns that might be of clinical usefulness. During the last two decades, gene-expression platforms have significantly increased our ability to estimate distant recurrence of early-stage HR-positive/HER2 negative breast cancers. Comparing gene-expression profiles of high-risk and low-risk early-stage HR-positive/HER2-negative breast tumors yielded different prognostic gene-expression signatures. Some of these are currently commercially available, endorsed by clinical guidelines, and routinely used in clinical practice. These signatures include Oncotype DX Recurrence Score (RS), MammaPrint 70-gene signature, Prosigna risk of recurrence (ROR) score, EndoPredict, and Breast Cancer Index (Table 2; Cardoso et al. 2019; Andre et al. 2022).

Oncotype DX (Genomic Health, Redwood, CA, USA) assesses the expression of 21 genes (16 breast cancer–related genes and five reference genes) using reverse transcription polymerase chain reaction (RT-PCR), providing a risk score (RS) that classifies the patient in low-, intermediate-, and high-risk prognostic groups (Paik et al. 2004). The RS prognostic value was first retrospectively evaluated in two clinical trials, NSABP (Paik et al. 2004) and TransATAC (Dowsett et al. 2010). The RS predictive value for chemotherapy benefit was also studied in two retrospective studies: NSABP B-20 (Paik et al. 2006) and SWOG-8814 (Albain et al. 2010). According to the ASCO guidelines, NSABP B-20 data were confounded by the data set initially used to generate the assay, and the SWOG-8814 data was hypothesis-generating due to the small sample set and no additional prediction beyond 5 years. The prognostic ability of Oncotype DX RS has been analyzed in the prospective clinical trial TAILORx (Sparano et al. 2018), where the RS cutoffs between different prognostic groups were changed for this purpose. Patients with RS ≤ 10 were considered low risk and were assigned not to receive chemotherapy, patients with RS ≥ 26 were assigned to receive chemotherapy, and patients with RS between 11 and 25 were randomly assigned to not receive or receive adjuvant chemotherapy. This clinical trial demonstrated the noninferiority of endocrine therapy alone compared to a chemotherapy and endocrine therapy combination in patients with an RS score between 16 and 25, with a small benefit from chemotherapy in patients 50 years old and younger. Data from the RxPONDER trial suggests that the benefit of adjuvant chemotherapy is small in patients with RS < 25 and 1–3 positive nodes (Kalinsky et al. 2021). However, RS failed to predict chemotherapy benefit (i.e., statistically nonsignificant interaction test). In addition, premenopausal patients younger than 50 again showed a large absolute benefit (i.e., an absolute difference of 4.9 percentage points) from adjuvant chemotherapy. Whether the observed chemotherapy benefits in these studies is directly or indirectly from chemotherapy-induced ovarian suppression remains unknown (Sparano et al. 2018; Kalinsky et al. 2021) and thus remains a pressing question to

be addressed. Oncotype DX RS, therefore, provides relevant prognostic information and is currently recommended in clinical guidelines with the strongest level of evidence (Cardoso et al. 2019; Andre et al. 2022).

MammaPrint (Agendia, Amsterdam, The Netherlands) is a 70-gene signature that uses the microarray technology that required initially sending fresh frozen breast tumor samples to a central laboratory, and then switched to formalin-fixed paraffin-embedded (FFPE). The prognostic value of this assay has been validated in multiple retrospective (Van't Veer et al. 2002; Buyse et al. 2006) and prospective (Drukker et al. 2013; Cardoso et al. 2016) studies. Similar to Oncotype DX, MammaPrint has not proven to have predictive value for chemotherapy benefit. MammaPrint is also included in most breast cancer guidelines with a high level of evidence (Cardoso et al. 2019; Andre et al. 2022).

Prosigna (Veracyte, San Francisco, CA, USA) is a prognostic gene-expression assay that uses 58 genes (PAM50 signature) along with clinicopathological features to calculate the ROR score, a prognostic index that predicts the probability of cancer recurrence. Prosigna is a decentralized test that can be run using FFPE breast tumor samples. Apart from the ROR score (and outside the United States), Prosigna also provides the breast cancer–intrinsic subtype information. Retrospectively validated in two different clinical trials (Dowsett et al. 2010; Gnant et al. 2014), Prosigna can be used as a prognostic tool to guide the use of adjuvant therapy (Cardoso et al. 2019; Andre et al. 2022) and, unlike Oncotype DX and MammaPrint, the ROR score also has a prognostic value for late recurrences that might occur between 5 and 10 years post-diagnosis.

The Breast Cancer Index (Biotheranostics, San Diego, CA, USA) is a gene-expression signature that includes the *HOXB13/IL17BR* ratio and the molecular grade index. The prognostic score combines both pieces of information and provides an individual risk assessment for overall and late distant recurrence (Sgroi et al. 2013a, b). The predictive component is based on the *HOXB13/IL17BR* ratio and provides a classification (i.e., high vs. low) for the benefit of extended

Cite this article as *Cold Spring Harb Perspect Med* doi: 10.1101/cshperspect.a041320

Table 2. Characteristics of the prognostic gene-expression assays in early-stage HR-positive, HER2-negative breast cancer

Signature	Oncotype DX	MammaPrint	Prosigna	Breast Cancer Index	EndoPredict
Provider	Genomic Health	Agendia	Veracyte	Biotheranostics	Myriad Genetics
Tissue for analysis[a]	FFPE	FFPE	FFPE	FFPE	FFPE
Platform[b]	RT-PCR	Microarray	nCounter	RT-PCR	RT-PCR
Number of genes	21 genes	70 genes	58 genes	11 genes	12 genes
Central testing	Yes	Yes	No	Yes	No
Risk stratification	Three groups	Two groups	Three groups	Two groups	Two groups
Indications	Pre- and postmenopausal N0 Postmenopausal 1–3 positive nodes	Pre- and postmenopausal N0 Postmenopausal 1–3 positive nodes	Postmenopausal N0	Postmenopausal N0 Postmenopausal 1–3 positive nodes Prediction of benefit from extended (>5 yr) endocrine therapy	Postmenopausal N0 Postmenopausal 1–3 positive nodes

[a](FFPE) Formalin-fixed paraffin-embedded.
[b](RT-PCR) Reverse transcription polymerase chain reaction.

endocrine therapy (Bartlett et al. 2016). Unlike other assays, the Breast Cancer Index can be used to guide an extension of the endocrine treatment in patients with node-negative or node-positive (one to three nodes only) disease beyond 5 years of treatment (Andre et al. 2022).

EndoPredict (Myriad Genetics, Salt Lake City, UT, USA) is a prognostic gene-expression signature of 12 genes (eight cancer-related genes, three reference genes, and one DNA control gene). The commercialized test (EPclin) also considers clinical factors (tumor size and node status) to inform about the risk of recurrence and indication of adjuvant chemotherapy. This assay has been retrospectively tested in the ABCSG6, the ABCSG8, and the TransATAC clinical trials (Dubsky et al. 2013; Buus et al. 2016), and the European and American guidelines recommend it as a prognostic tool for treatment decisions (Cardoso et al. 2019; Andre et al. 2022).

These genomic assays help clinicians and patients to decide the need for adjuvant chemotherapy and the extension of hormone therapy to treat early-stage HR-positive/HER2-negative breast cancer. Despite multiple efforts in the molecular characterization of HER2-positive breast cancer and TNBC, gene-expression assays are not yet recommended by clinical guidelines to guide treatment decisions within these two clinical subtypes of breast cancer.

MOLECULAR CHARACTERIZATION OF BREAST CANCER BY TCGA

The Cancer Genome Atlas (TCGA) revolutionized the study of human cancers in two ways: first, it applied a set of at least six different "omic" technologies onto a common set of tumors (i.e., whole-exome sequencing, DNA–single-nucleotide polymorphism (SNP) microarrays, DNA methylation microarrays, gene expression by DNA microarrays and/or RNA-seq, miRNA sequencing, and reverse-phase protein arrays [RPPAs]), thus providing an unprecedented "multiplatform" data set. Second, all these data were made immediately public, and have literally been used in thousands of publications across tens of countries. Breast cancers were highly rep-

resented among the TCGA tumor types, which in the end represented 1100/10,000 cancers profiled (Hoadley et al. 2018). Multiple TCGA Breast Cancer Studies were published, including the most highly cited TCGA-sponsored paper from 2012, one of the first multiplatform analyses of a human tumor type (Cancer Genome Atlas Network 2012). A highly novel computational approach was employed in this paper where a single classification of breast cancers was made via the integrated analysis of all six data types at once, as opposed to most previous methods like those described above, where classifications are made based upon a single technology. This method, called "clustering of cluster assignments" (COCA), identified four main groups of breast tumors, which were highly similar to the four major breast cancer intrinsic subtypes based upon gene-expression patterns alone (Fig. 1); note that in the COCA method, each of the six omics data types was equally weighted, thus demonstrating that the gene-expression data well represents all these data and that these gene-expression-defined intrinsic subtypes are also highly correlated with DNA and protein features, which also define similar subtypes.

Luminal A is the most common intrinsic subtype among all breast cancers and within HR-positive/HER2-negative breast cancer. These tumors have heterogeneous CNAs and somatic mutations (Ciriello et al. 2013). Luminal A tumors have the highest number of recurrently mutated genes but the smallest number of mutations per tumor. Among the most frequently mutated genes are *PIK3CA* (45%), followed by *MAP3K1*, *GATA3*, *TP53*, *CDH1*, and *MAP2K4*. This subtype is characterized by typically high expression of ER and PR (both mRNA and protein levels) and an increased expression of the luminal signature, which contains genes that are also highly expressed in the mammary ducts' luminal epithelium (*ESR1*, *GATA3*, *FOXA1*, *XBP1*, and *MYB*). Compared to luminal B, luminal A tumors usually present lower proliferation rates (Cancer Genome Atlas Network 2012).

Luminal B breast cancer is similar and related to luminal A but is also unique in the profile of DNA CNAs and somatic mutations, and often shows a DNA hypermethylated phenotype.

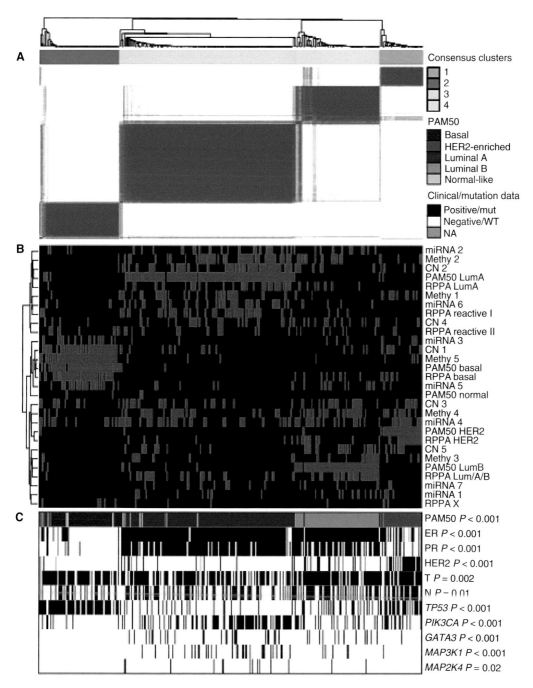

Figure 1. Coordinated analysis of breast cancer subtypes defined from five different genomic/proteomic platforms. (*A*) Consensus clustering analysis of the subtypes identifies four major groups (samples, *n* = 348). The blue and white heat map displays sample consensus. (*B*) Heat-map display of the subtypes defined independently by miRNAs, DNA methylation, copy number (CN), PAM50 mRNA expression, and reverse-phase protein array (RPPA) expression. The red bar indicates membership of a cluster type. (*C*) Associations with molecular and clinical features. *P* values were calculated using a chi-squared test. (Figure is reprinted from the Cancer Genome Atlas Network (2012) with permission from the author, C.M. Perou.)

Compared to luminal A tumors, luminal B tumors have a lower frequency of *PIK3CA* mutations and a higher frequency of *TP53* mutations. Luminal B tumors tend to have a lower expression of luminal signature, lower expression of PR (protein and mRNA), and a higher proliferation rate than luminal A tumors (Cancer Genome Atlas Network 2012).

HER2-Enriched is the predominant genomic subtype within pathological HER2-positive tumors, but not all HER2-positive tumors are HER2-enriched, and not all clinically defined HER2$^+$ tumors are of the HER2-enriched gene-expression phenotype. Clinically defined HER2-positive tumors are probably the most heterogeneous breast cancer clinical type among those analyzed by TCGA, where ~50% of HER2-positive tumors are HER2-enriched. An HER2-enriched subtype is mainly characterized by DNA amplification of the HER2 gene (*ERBB2*) and a high frequency of *TP53* mutation (~75%). Tumors that are both HER2-enriched and HER2-positive show a significantly higher expression of several tyrosine kinase receptors, including FGFR4, EGFR, HER2 itself, and genes within the *ERBB2* amplicon, including GRB7 and sometimes TOP2A. HER2-enriched tumors usually have a high frequency of *PIK3CA* mutations (39%), a lower frequency of *PTEN* mutations, high aneuploidy, and the highest somatic mutation rate (Cancer Genome Atlas Network 2012).

Basal-like is the predominant genomic subtype within TNBC, with ~75% of all TNBCs being basal-like. Basal-like breast tumors harbor a high frequency of *TP53* mutations (80%), and most of these tumors have lost the TP53 function based on the combination of the *TP53* mutation frequency and the inferred TP53 pathway activity. Although *PIK3CA* is the second-most-mutated gene in basal-like breast tumors, its frequency is relatively low (~9%) compared to *TP53* and much lower than that in luminal tumors (~40%); however, despite this low *PIK3CA* mutation rate, the inferred PI3K pathway activity based on RPPA proteomics is the highest in this subtype (Cancer Genome Atlas Network 2012). At the gene-expression level, basal-like tumors are characterized by high expressions of keratins 5, 6, and 17, low expression of *RB1* (which also shows a 5% mutation rate), and increased expression of proliferation genes. Hyperactivation of MYC and H1F1 pathways are also frequent. Basal-like tumors usually have the lowest level of global DNA methylation and are confirmed to have a strong association with germline *BRCA1* mutations (Cancer Genome Atlas Network 2012).

Last, several other TCGA Breast Cancer publications should be highlighted, including a multiplatform molecular description of lobular cancers (Ciriello et al. 2015), a comparison of omics features between breast cancer patients of African versus European ancestry (Huo et al. 2017), and a thorough comparison of multiple pathology features including grade versus all the genomic data (Heng et al. 2017). Very recently, a complete analysis of all 1100 TCGA breast cancers and all histologies, including a focus on six rare subtypes (cribriform, micropapillary, mucinous, papillary, metaplastic, and invasive ductal carcinoma [IDC] with medullary pattern), was completed, and, importantly, where histology and genomics were linked together to arrive at 12 consensus groups (Thennavan et al. 2021); these two consensus groups, highlighted in Figure 2, link together histology and genomics into a single framework that should improve our ability to understand the complex biology of each breast cancer clinical study.

In addition to TCGA, multiple international efforts have helped decipher breast cancer biology using a multi-omics approach. Thus, Metabric (Molecular Taxonomy of Breast Cancer International Consortium) used an integrated clustering approach of genome and transcriptome data from 2000 breast tumors describing 10 integrative clusters with distinct clinical outcomes (Curtis et al. 2012). Other international efforts, like the International Cancer Genome Consortium (ICGC/TCGA Pan-Cancer Analysis of Whole Genomes Consortium 2020), the Broad Institute (Banerji et al. 2012), and the National Cancer Institute's Clinical Proteomic Tumor Analysis Consortium (Krug et al. 2020), have also contributed to a better understanding of the molecular mechanisms of breast tumors. More recently, the Sweden Cancerome Analysis Network—Breast (SCAN-B) consortium was created as a multicen-

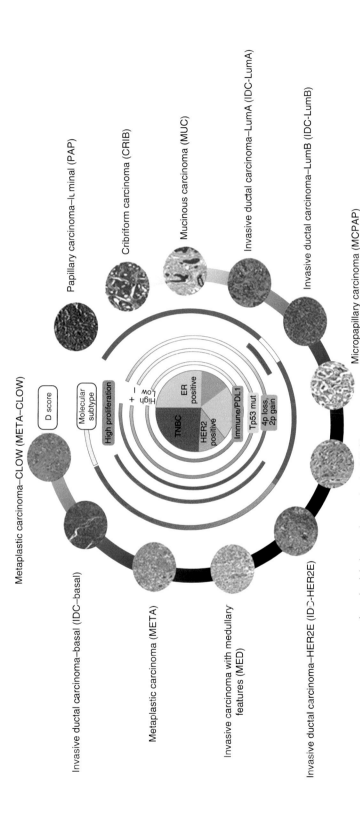

Figure 2. The Cancer Genome Atlas (TCGA) breast cancer classification based upon molecular and histologic features combined. Schematic representation of 12 consensus groups defined by histology and gene-expression analyses of the TCGA-BRCA data set and organized by differentiation score. These groups are connected by an outer ring based on differentiation score (D score: lowest differentiation to highest differentiation arranged in an anticlockwise direction). From the D score ring inward: the second ring exhibits PAM50 subtype association with D score (red: basal, pink: HER2E, blue: LumA and LumB, yellow: claudin low). The next ring highlights the proliferation gene signature, which is high in all biological groups with a low D score but also in one group with a high D score, namely, IDC-LumB. The next two rings represent the descending abundance of copy number alteration (CNA) events (4p loss, 2p gain) and mutation events (Tp53 mutation) associated with ascending D score. The final ring exhibits decreasing immunological gene signatures in relation to ascending D score. The innermost pie chart exhibits the clinical immunohistochemistry (IHC) status found in these 12 breast cancer consensus groups.

ter prospective study to analyze breast tumors with next-generation technologies to develop and validate new biomarker assays to better personalize cancer treatment (Saal et al. 2015). Thanks to this initiative, gene expression and clinical data from a cohort of 7743 patients with breast cancer have been analyzed (Staaf et al. 2022). Genomic and clinical data from these international consortiums is usually shared with the scientific community, providing the opportunity to accelerate breast cancer research.

SINGLE-CELL AND SPATIALLY RESOLVED GENE-EXPRESSION ANALYSIS

Breast cancers typically exhibit extensive immune cells and stromal cells infiltration, along with functional heterogeneity among neoplastic cells. There is a growing awareness of the importance of the TME, beyond neoplastic cell-intrinsic heterogeneity, to the role of the TME in breast cancer etiology, response to therapy, and patient survival outcomes. Gene-expression analyses of bulk tissues have led to paradigm shifts in our understanding of breast cancer biology. However, a limitation of such approaches is that they only measure tissue-average gene expression, and thus it is often challenging to know what gene(s) are being expressed in what cell types. Therefore, traditional analysis of bulk gene-expression data might result in a loss of signal from less abundant cell types, reducing our ability to resolve unique features of cellular states and limiting the conclusions of an experiment. Traditional methods for studying cell heterogeneity, such as immunohistochemistry and flow cytometry, rely on a small number of phenotypic markers and are therefore limited in their utility for studying cancer heterogeneity.

The field of single-cell genomics has rapidly grown in recent years. Early methods for single-cell RNA sequencing (scRNA-seq) (e.g., SMART-seq, Fluidigm C1) relied on physically separating cells into microfluidics or high-density microtiter plates by flow cytometric cell sorting, which greatly limited the throughput. The emergence of methods based on high-throughput encapsulation of cells and barcoded capture beads into aqueous droplets within an oil emulsion (Macosko et al. 2015) has revolutionized the field, permitting whole transcriptome analysis of hundreds of thousands of cells. These methods generally use poly-A-based capture of mRNA and so are sensitive to RNA integrity. mRNA capture is followed by reverse transcription, fragmentation of cDNA, and short-read sequencing of the ends of fragments. To address the limitations of short-read sequencing, single-molecule sequencing on the PacBio or Oxford Nanopore systems can be used to study splicing and structural variants or to identify SNPs not contained within the sequenced fragments. For instance, single-molecule long-read sequencing of full-length cDNA libraries was used to analyze alternative splicing and somatic hypermutation events in T- and B-cell receptors in breast cancers (Singh et al. 2019). These methods can be scaled to whole transcriptomes, opening up the possibility of analyzing the association of splice variants, mRNA editing, or genomically encoded cancer variants at single-cell resolution.

scRNA-seq is challenging to implement on clinical specimens, as it typically requires intact, alive cells and thus is not compatible with flash-frozen or paraffin-embedded material. Although sequencing of nuclei from flash-frozen tissue can be applied to avoid dissociation methods (Habib et al. 2017) and better represent the composition of complex cellular mixtures (Slyper et al. 2020), these methods have several distinct drawbacks in their application to oncology. These drawbacks include that the data generated is typical of substantially reduced sensitivity compared to using whole cells, that samples cannot be processed to remove dead or low-quality cells, which are common in solid cancers, that subsets of interest cannot be prospectively selected, and that the approach is not compatible with cell-surface immunophenotyping methods such as CITE-scq (Stoeckius et al. 2017). The consequence is that most clinical studies have used freshly processed samples, which is logistically difficult, inefficient, and may exacerbate variability introduced by batch effects within the resulting data set. One approach to solve these challenges is to cryopreserve or fix samples, which allows later sample batching. Studies (Guillaumet-Adkins et al. 2017; Wu et al. 2021c) have shown that cryopreservation effectively conserves the transcriptome and cell abundance seen in freshly processed breast cancer samples while maintaining maximum compatibility with downstream applications, including CITE-seq (Wu et al. 2021c), spatial transcriptomics (Wu et al. 2021b), and cell culture or xenotransplantation. The recent development of single-nuclei RNA sequencing (snRNA-seq) and scRNA-seq methods based on probe-based detection in fixed cells has markedly reduced the technical sensitivity to RNA integrity and will dramatically increase the application of these technologies to the analysis of clinical breast cancer cohorts (Vallejo et al. 2022).

It is increasingly apparent that cellular context plays a significant role in determining cellular phenotypes in breast cancer, such as heterogeneity in drug sensitivity or T-cell exhaustion. Understanding the organization of breast cancers requires complex high-dimensional methods to resolve the many cell states within a tumor and their expression of molecules. What was once an abstract dream has become a reality with the emergence of methods to perform spatially resolved transcriptomics. The ideal spatial transcriptomics platform would combine single-cell-level resolution, high-target multiplexing, high-sample throughput, and sensitivity with flexibility in sample preparation. Current technologies do not meet all these goals, often compromising on throughput, resolution, or sensitivity.

Until recently, commercially available platforms have used a variety of strategies to map transcripts to a spatial region, typically >50 μm in diameter and comprising numerous cells. These include the Visium platform from 10X Genomics and GeoMX DSP from Nanostring Technologies. Academic methods for spatial transcriptomics that use optical imaging of hybridized gene probes, such as Seq-FISH (Lubeck et al. 2014), have existed for approximately a decade and provide subcellular resolution for a panel of gene targets. However, these methods have generally been difficult for most laboratories to implement and often only work well on carefully prepared fresh tissues from model systems. However, numerous new commercial platforms converge on methods for spatial transcriptomics via probe hybridization that markedly improves the sensitivity

and robustness of analyzing archival FFPE tissues. These platforms provide the capability to spatially resolve hundreds to thousands of genes and proteins at the subcellular resolution level in routine clinical specimens. The new capabilities offered by these platforms make it possible for the next generation of spatial methods to become the dominant cellular genomics technology, relegating scRNA-seq to specific applications.

A Model for Cellular Heterogeneity in Breast Cancer

The emergence of standardized methods for cellular genomics, particularly single-cell, and spatial transcriptomics, is revolutionizing our understanding of breast cancer heterogeneity. To understand breast cancer heterogeneity, it is helpful to take a view of breast cancers as cellular ecosystems, whereby the etiology of disease is a function of interactions among neoplastic, stromal, and immune cells together with extracellular molecules such as matrix proteins, cytokines, and growth factors. This problem may be viewed at three levels (Fig. 3). The first level is one of taxonomy, to define the cell types and states that occupy tumors, particularly those of clinical relevance, and determine molecular programs that define their behavior and which of them we may aim to manipulate for a therapeutic purpose. The second is to understand the communities or cellular hubs that cells form in tumors, the mechanisms driving heterotypic cellular coalescence, and the impact those interactions have on tumor phenotype. Finally, there are widespread differences in cellular content in breast cancers and recurrent combinations of cellular composition, or ecotypes, across tumors that have recently been reported (Luca et al. 2021; Wu et al. 2021b). However, the drivers and clinical significance of ecotypes are mostly unknown.

Breast Cancer Cellular Taxonomies

Single-cell and spatial transcriptomics have diverse applications in the study of breast cancer heterogeneity (Fig. 4). A major focus in this area has been on defining cell types and cellular states that comprise breast cancers using scRNA-

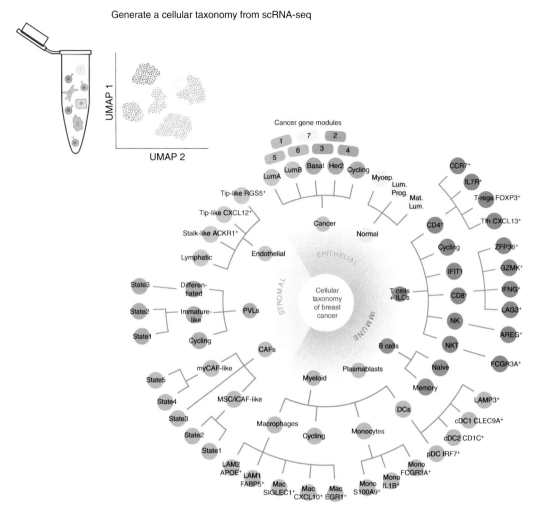

Figure 3. Breast cancer ecosystems can be viewed at the level of cellular taxonomies, cellular communities, and recurrent patterns of cellular composition, or ecotypes.

seq. The first generation of clinical scRNA-seq studies have typically produced unbiased descriptive atlases from early breast cancers (Chung et al. 2017; Singh et al. 2019; Qian et al. 2020; Wu et al. 2020, 2021; Gao et al. 2021; Pal et al. 2021), broadly sampling the cellular diversity of disease across neoplastic, immune, and stromal cells.

Neoplastic Cells

Heterogeneity and plasticity among neoplastic cells provide a source of diversity that drives drug resistance and are major challenges in the diagnosis and treatment of breast cancer. Howev-

er, identifying the recurrent features of tumor heterogeneity by scRNA-seq is complicated by the dominance of patient-specific features when using unbiased clustering approaches (Pal et al. 2021; Wu et al. 2021b). The result is that cells from different patients typically cluster separately unless a stringent batch correction is applied, which can skew gene expression. Several approaches have been used to address this problem. The first is to apply the PAM50 intrinsic gene set to annotate individual tumor cell phenotypes. However, the majority of PAM50 genes return zero values at single-cell resolution in scRNA-seq data sets. An approach based on the concepts of intrinsic sub-

Deconvolute bulk RNA-seq into cellular
compositions (ecotypes)

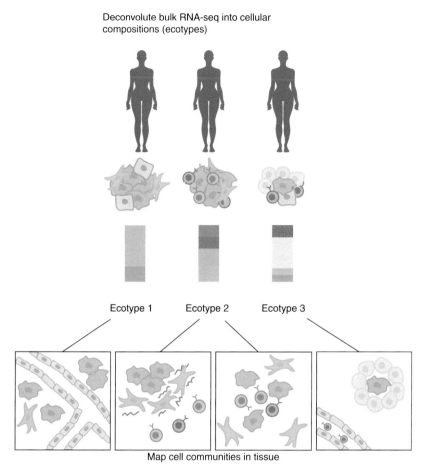

Ecotype 1   Ecotype 2   Ecotype 3

Map cell communities in tissue

Figure 3. (*Continued.*)

types but trained on scRNA-seq data rather than bulk RNA-seq data, "scSubtyper," has been developed to address this issue, and the results reveal that intrinsic subtypes can be applied to individual cells. Notably, while the majority of cells within a tumor have subtypes concordant with the "bulk" annotation, a minor subset of cells with an scSubtype "discordant" to the bulk subtype is seen in a majority of cancers (Wu et al. 2021). Further work is needed to develop this method and test whether quantifying intratumoral heterogeneity with scSubtype predicts drug response or relapse (Wu et al. 2021b), where one would hypothesize that high intratumor heterogeneity subtype diversity might predict worse outcomes.

A second approach to define neoplastic heterogeneity is identifying modules of genes associated with intratumoral heterogeneity and then applying these modules across cancers (Wu et al. 2021b). Using modules, or gene sets, instead of individual genes was an idea developed in the bulk world that has been used widely, and it is likely this approach may be even more important in single-cell data where biological and technological noise is even greater. The result is a series of modules representing major biological features, such as proliferation, epithelial–mesenchymal transition (EMT), and ER signaling. Some of these clusters of gene modules are spatially segregated (Wu et al. 2021), suggesting either that the local microenvironment controls their expression or that they associate with epi/genomic subclones. Methods to infer DNA copy number states from scRNA-seq data have been developed (Patel et al. 2014; Gao et al. 2021) and will be important tools in linking genotype to

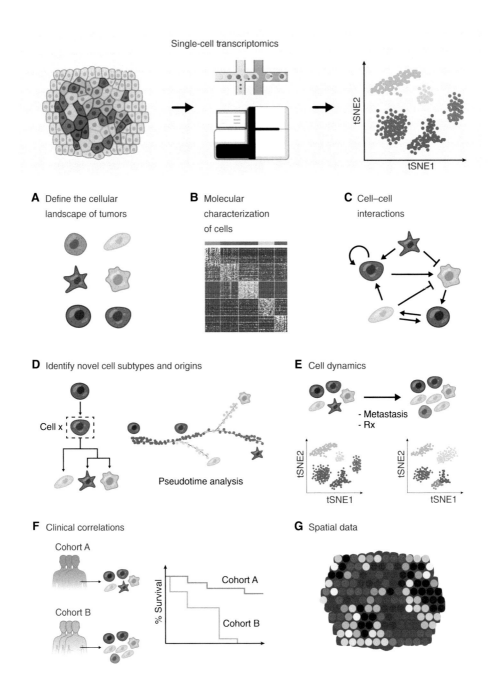

Figure 4. Single-cell transcriptomics can address a diversity of research questions. (*A*) Through integrated RNA and barcoded antibody measurements, cell types and states can be defined. (*B*) Gene-expression analysis can identify unique RNA and protein markers of populations. (*C*) Using algorithms that encompass known ligand–receptor pairings (Ramilowski et al. 2015; Efremova et al. 2020), putative cell–cell interactions can be identified. (*D*) Using dynamic changes in gene expression, branching trajectories of differentiation can be predicted (Trapnell et al. 2014). (*E*) By analyzing samples taken at different times or tissues, cellular dynamics during these processes can be discovered. (*F*) Cellular features of different clinical cohorts may reveal genes or signatures that are prognostic or predictive of response. (*G*) Patient- or disease-matched single-cell data allows accurate "deconvolution" of spatial data sampled from >1 cell into pseudo single-cell data. (tSNE) t-distributed stochastic neighbor embedding.

phenotype in single-cell and spatial transcriptomic data sets going forward.

Immune Cells

Many breast cancers contain an extensive immune infiltrate, and numerous immune cell populations are prognostic or predictive of response to chemotherapies and immunotherapies, typically T-cell and/or B-cell features (Hollern et al. 2019). Immune checkpoint inhibitors have made an impact on breast cancer, with these now being approved for use in metastatic and early-stage TNBC and being tested in all other breast clinical subtypes as well (see Bahl et al. 2022). Immune cell populations in breast cancer demonstrate substantially expanded phenotypic diversity compared to those in normal breast or blood (Azizi et al. 2018), suggesting that residence within malignant tissues provides an environment for diverse immune activation states. There are many similarities between the cell states of immune cells in breast cancer and those in other solid cancers, with the major variation being in relative abundance rather than unique cellular states. Analysis of resting and naive immune cell populations using scRNA-seq alone is challenging, as these cells are relatively transcriptionally inactive. It is here that multi-omic methods, such as CITE-seq (Stoeckius et al. 2017) that measure cell-surface epitopes using barcoded antibodies in parallel to transcriptomes, are particularly valuable in immunophenotyping (Al-Eryani et al. 2022).

T cells are the targets of currently approved immune checkpoint inhibitors. Several groups have reported a diverse repertoire of T cells along continuous spectra of activation, differentiation, and exhaustion within breast cancers (Azizi et al. 2018). Indeed, immune cell plasticity appears very high in breast cancer TME, resulting in novel populations not found outside of cancers (Al-Eryani et al. 2022), such as exhausted CXCL13[+] CD4 T follicular helper cells. T-cell states vary by breast cancer intrinsic subtype, with more proliferating and exhausted T cells found in TNBC (Pal et al. 2021; Wu et al. 2021b), mirroring the greater responses to

immune checkpoint inhibitors in this clinical subgroup than other subgroups. Interestingly, individual expanded clones of T cells, identified by parallel V(D)J sequencing, occupy restricted spaces in this continuum, suggesting that signaling through the T-cell receptor is vital to defining gene-expression states. Tissue-resident memory CD8 T cells (Trm) are found in breast cancers, particularly ER-negative disease (Pal et al. 2021), and the signatures derived from CD103-positive CD8 T cells correlate with good prognosis, suggesting that Trm may mediate antitumor immune responses (Savas et al. 2018). However, caution must be used when interpreting the markers for Trm, which are not proven to mark tissue-resident cells in human cancer and may instead mark activation and/or exhaustion (Al-Eryani et al. 2022).

B cells and plasma cells are frequently present in breast cancers, particularly in HER2-positive and TNBC (Pal et al. 2021; Wu et al. 2021b), and B-cell abundance and immunoglobulin transcript levels are strongly associated with good outcomes (Iglesia et al. 2014; Hollern et al. 2019; Hu et al. 2021). The function of these cells in breast cancer is poorly known and subject to intense research activity. Most B cells in TNBC are memory B cells that have typically undergone affinity maturation (Singh et al. 2019; Hu et al. 2021). Clonally expanded B cells can be found in paired tumors and draining lymph nodes (Singh et al. 2019), suggesting antigen presentation, trafficking to tumors, and proliferation in response to antigens of these cells. The function of B cells in breast cancers and antigens to which they respond are not yet known and are important areas for future research.

Breast cancers contain a substantial number of myeloid cells across diverse states. Granulocytes such as neutrophils are poorly represented in scRNA-seq data sets, which may result from their low abundance, low transcript levels, and their sensitivity to the dissociation and capture protocols employed for scRNA-seq. Methods such as snRNA-seq are more likely to capture granulocytes. Macrophages in breast cancer form a number of clusters, including two substantial populations of TREM2-positive cells that resemble lipid-associated macrophages initially identified in obese mice and humans (Jaitin et al. 2019). These populations, now found in many types of solid cancers (Mulder et al. 2021), are notable for their expression of lipid metabolism molecules and immunosuppressive molecules, such as CCL18, PDL1, and PDL2, and their association with poor patient survival (Wu et al. 2021b). TREM2-positive macrophages may be essential to understanding breast cancer immunity, as their abundance is associated with response to PDL1 blockade in TNBC (Zheng et al. 2021) and their depletion can sensitize mouse models of breast cancer to immune checkpoint inhibition (Molgora et al. 2020).

Stromal Cells

Breast cancers exhibit a diverse stromal milieu, and breast cancer specimens and models have been at the forefront of research into the role of stroma in cancer for decades. More recently, scRNA-seq has revealed surprising new insights into stromal biology. It is now clear that cancer-associated fibroblasts (CAFs) adopt diverse phenotypes, ranging from myofibroblasts with features of contractility and extracellular matrix remodeling through to less-differentiated cells resembling mesenchymal stromal cells. Trajectory analysis (Fig. 4D) suggests a continuum of differentiation states between these two extreme states (Wu et al. 2021b). These two states are commonly called myofibroblastic CAF (myCAF) and inflammatory CAF (iCAF), adopting the nomenclatures for similar cells reported in pancreatic cancer (Sahai et al. 2020) with which they share many similarities. iCAFs express many immunoregulatory molecules, including immune checkpoint ligands PDL1 and PDL2 and chemokines such as CXCL12, and colocalize with T and B lymphocytes, supporting the suggestion that they may be essential players in immunoregulation within tumors. Although iCAF may be the progenitor of myCAF, they occupy different regions of breast cancers, with myCAF typically closely associated with cancer cells (Wu et al. 2021b). The distinct localization of these cells suggests a role for unique microenvironments in either differentially recruiting CAF subsets or regulating CAF differentiation.

scRNA-seq has also revealed diversities within other stromal cell populations in the TME. Endothelial cells lie across a spectrum of differentiation, from cells resembling stalk-like cells to tip-like cells and a subset characterized by migration and angiogenic signatures (Wu et al. 2021b). Cells with features of smooth muscle cells and pericytes, termed perivascular-like cells (PVLs), are as frequent as CAFs in early breast cancers (Wu et al. 2021a). PVLs are also found across a range of differentiation states. Surprisingly, PVLs can be frequently found distant from blood vessels in breast cancers (Wu et al. 2021a). Given the overlapping morphology and markers with CAFs (e.g., smooth muscle actin), PVLs have almost certainly been misclassified as CAFs in the past. The function of PVL, particularly those disseminated throughout the tumor parenchyma, is a major knowledge gap.

A recurring observation among these studies is that rather than occupying discrete states, or cell types, cells of all three major lineages differ across a continuous spectrum of phenotypes or states, representing multiple axes comprising differentiation, activation, and others. Cells are presumably transitioning dynamically between states, so a significant challenge will be to understand the drivers of this plasticity, be it intrinsic such as genomic, or extrinsic such as extracellular interactions. These insights will drive a new generation of therapeutics based on manipulating the cellular composition or signaling within the TME.

Cellular Communities

While the field has long recognized the existence of breast cancer cell heterogeneity, few studies have illuminated its topological associations. A recent pan-cancer analysis examining associations between cancer gene-expression modules and cellular contexts revealed the regulation of interferon response signatures in cancer cells by adjacent macrophages and T cells (Barkley et al. 2022), which was also observed in HER2-positive breast cancers (Andersson et al. 2021). Recently, iCAFs were shown to associate with T and B lymphocytes, and spatial-resolved ligand-receptor mapping revealed potential signaling intermediates between these populations (Wu et al. 2020).

Tertiary lymphoid structures (TLS) are complex cellular structures comprised of lymphocytes, stromal cells, and myeloid cells that resemble secondary lymphoid organs such as lymph nodes, and they are observed in some breast tumors. Most cancers with TLS are associated with superior prognosis and response to immunotherapy (Schumacher and Thommen 2022). It is likely that many important stromal and immune cell types identified recently through single-cell methods, including memory B cells, iCAFs, CD4 T follicular helper cells, and dendritic cell subsets, interact within TLS to alter the antitumor immunity. The likely importance of these structures in the breast cancer immune context demonstrates the potential for spatial transcriptomics in mapping breast cancer ecosystems (Andersson et al. 2021). By placing cells within an actionable distance, spatial transcriptomics may provide an opportunity for new drug target discovery.

Ecotypes

Single-cell transcriptomics has revealed highly variable cell-type compositions among breast cancers (Pal et al. 2021; Wu et al. 2021b), which raises the question of whether these variabilities in cell-type composition are the result of a generally random process or instead reflects recurring patterns of cellular frequencies. Because single-cell atlases to date contain no more than 34 cases of breast cancer (Pal et al. 2021), it is not possible to address this question robustly yet. However, single-cell reference profiles of every cell type can be used to infer the proportion of cell types in large, published bulk RNA-seq data sets (Fig. 5).

Several computational methods have been developed to infer the abundance of different

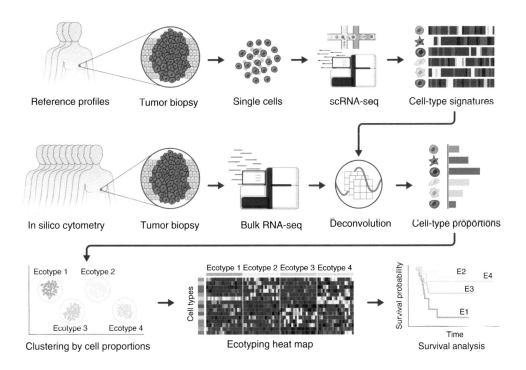

Figure 5. Disease-specific single-cell transcriptomics references can be used to predict cellular frequencies within bulk RNA sequencing (RNA-seq) data sets via in silico cytometry. This strategy can be used to study cellular proportions across large clinical cohorts.

cell types in bulk tissue. One example is CIBER-SORT, a deconvolution method based on support vector regression that allows the characterization of the immune cell composition of bulk tissues (Newman et al. 2015). CIBERSORT has been demonstrated to have prognostic value in breast cancer (Craven et al. 2021). Methods such as CIBERSORT use canonical gene signatures, often derived from blood, for deconvolution. Because of the marked influence of the TME on cellular phenotypes, the accuracy of predictions is markedly increased by using high-resolution and disease-specific reference profiles. Using scRNA-seq data from 26 breast cancer cases as reference, ~2000 cases of breast cancer in the Metabric cohort (Curtis et al. 2012) have been deconvoluted to reveal a number of recurrent compositional clusters, which are called ecotypes (Wu et al. 2021b). Ecotypes are composed of immune, stromal, and neoplastic cells in recurring proportions, suggesting that all these three cell lineages contribute to building these recurrent tumor neighborhoods. Ecotypes, in part, recapitulate the PAM50 and IntClusters derived from the Metabric and build upon these by integrating the TME and in finding additional biological groups that are associated with prognosis. Ecotypes appear to be conserved across diverse epithelial tumor types (Luca et al. 2021), but the mechanisms organizing ecotypes are mostly unknown. Although it remains to be tested, ecotypes may predict response to therapy, particularly immunotherapy.

FUTURE DIRECTIONS

In many ways, the age of cellular- and spatial-resolved "omics" is only beginning. A challenge in the single-cell space is to systematically define cell states and a common ontology while avoiding duplication of nomenclature. This will require integrative studies that find the shared features across studies and disease subsets and rely on methods that enable less biased capture of cell types and states, particularly for complex cell types such as granulocytes, neurons, and adipocytes, while retaining sensitivity. Methods compatible with fixed cells, particularly those embedded in paraffin, will open up large clini-

cally valuable cohorts to cellular genomics to definitively test the value of these new technologies to translational oncology.

Current methods of single-cell analysis typically sample ~1000–10,000 cells per tumor, and the majority of studies have captured several cell types, so any one cell type is shallowly sampled. Future work in this area will need to capture far more cells per tumor or focus on specific cellular populations to more comprehensively resolve cellular states and plasticity. In breast cancer, this is particularly relevant for understanding neoplastic cell heterogeneity. For instance, ER, PR, and androgen receptor (AR) are critically important to breast cancer biology and treatment. Intriguingly, histochemical studies reveal that ER, PR, and AR are generally heterogeneously expressed in breast cancers (Lindström et al. 2018), yet cancers with only 1% ER-positive cells may respond to endocrine therapies. Single-cell transcriptomics, in particular, when combined with newer methods such as single-cell ATAC-seq, may provide a path to better understand endocrine receptor function and endocrine therapy responsiveness.

Cellular genomics has the potential to unveil insights into the cellular basis for drug response and resistance. Although most studies have focused on treatment-naive disease, Kim and colleagues used scRNA-seq to monitor the cellular dynamics of TNBC in patients undergoing neoadjuvant chemotherapy treatment (Kim et al. 2018). Although this study cannot resolve the dynamics of stromal and immune cells through therapy, it reveals dynamic changes in neoplastic cell gene expression in samples from patients who had an incomplete response, including activation of pathways relating to extracellular matrix remodeling, AKT signaling, and EMT. Similarly, Griffiths et al. (2021) applied scRNA-seq to ER-positive breast cancers from a cohort of patients receiving the endocrine therapy letrozole with or without the CDK4/6 inhibitor ribociclib. This analysis revealed the emergence of conserved resistance mechanisms, including ER activation downstream of ERBB4 or JNK activation downstream of FGFR2 activation. This dynamic plasticity observed across cancer types and drug classes is a significant challenge for cancer therapy and is an area of investigation where cellular genomics will be a powerful tool. These insights may inform future trials of rational combination therapy and establish the groundwork to formally evaluate the benefit of cellular genomics to translational science through the analysis of substantial clinical trial cohorts.

REFERENCES

*Reference is also in this subject collection.

Albain KS, Barlow WE, Shak S, Hortobagyi GN, Livingston RB, Yeh IT, Ravdin P, Bugarini R, Baehner FL, Davidson NE, et al. 2010. Prognostic and predictive value of the 21-gene recurrence score assay in postmenopausal women with node-positive, oestrogen-receptor-positive breast cancer on chemotherapy: a retrospective analysis of a randomised trial. *Lancet Oncol* **11:** 55–65. doi:10.1016/S1470-2045(09)70314-6

Al-Eryani G, Bartonicek N, Chan CL, Anderson A, Harvey K, Wu SZ, Roden D, Wang T, Reeves J, Yeung BZ, et al. 2022. Integration of single-cell RNA and protein data identifies novel clinically-relevant lymphocyte phenotypes in breast cancers. bioRxiv doi:10.1101/2022.05.31.494081

Andersson A, Larsson L, Stenbeck L, Salmén F, Ehinger A, Wu SZ, Al-Eryani G, Roden D, Swarbrick A, Borg A, et al. 2021. Spatial deconvolution of HER2-positive breast cancer delineates tumor-associated cell type interactions. *Nat Commun* **12:** 6012. doi:10.1038/s41467-021-26271-2

Andre F, Ismaila N, Allison KH, Barlow WE, Collyar DE, Damodaran S, Henry NL, Jhaveri K, Kalinsky K, Kuderer NM, et al. 2022. Biomarkers for adjuvant endocrine and chemotherapy in early-stage breast cancer: ASCO guideline update. *J Clin Oncol* **40:** 1816–1837. doi:10.1200/JCO.22.00069

Azizi E, Carr AJ, Plitas G, Cornish AE, Konopacki C, Prabhakaran S, Nainys J, Wu K, Kiseliovas V, Setty M, et al. 2018. Single-cell map of diverse immune phenotypes in the breast tumor microenvironment. *Cell* **174:** 1293–1308.e36. doi:10.1016/j.cell.2018.05.060

* Bahl S, Carroll JS, Lupien M. 2022. Chromatin variants reveal the genetic determinants of oncogenesis in breast cancer. *Cold Spring Harb Perspect Med* **12:** a041322. doi:10.1101/cshperspect.a041322

Banerji S, Cibulskis K, Rangel-Escareno C, Brown KK, Carter SL, Frederick AM, Lawrence MS, Sivachenko AY, Sougnez C, Zou L, et al. 2012. Sequence analysis of mutations and translocations across breast cancer subtypes. *Nature* **486:** 405–409. doi:10.1038/nature11154

Barkley D, Moncada R, Pour M, Liberman DA, Dryg I, Werba G, Wang W, Baron M, Rao A, Xia B, et al. 2022. Cancer cell states recur across tumor types and form specific interactions with the tumor microenvironment. *Nat Genet* **54:** 1192–1201. doi:10.1038/s41588-022-01141-9

Bartlett J, Bayani J, Marshall A, Dunn JA, Campbell A, Cunningham C, Sobol MS, Hall PS, Poole CJ, Cameron DA, et al. 2016. Comparing breast cancer multiparameter tests in the OPTIMA prelim trial: no test is more equal than the others. *J Natl Cancer Inst* **108:** djw050. doi:10.1093/jnci/djw050

Buus R, Sestak I, Kronenwett R, Denkert C, Dubsky P, Krappmann K, Scheer M, Petry C, Cuzick J, Dowsett M. 2016. Comparison of EndoPredict and EPclin with Oncotype DX recurrence score for prediction of risk of distant recurrence after endocrine therapy. *J Natl Cancer Inst* **108:** djw149. doi:10.1093/jnci/djw149

Buyse M, Loi S, Van't Veer L, Viale G, Delorenzi M, Glas AM, Saghatchian d'Assignies M, Bergh J, Lidereau R, Ellis P, et al. 2006. Validation and clinical utility of a 70-gene prognostic signature for women with node-negative breast cancer. *J Natl Cancer Inst* **98:** 1183–1192. doi:10.1093/jnci/djj329

Cancer Genome Atlas Network. 2012. Comprehensive molecular portraits of human breast tumours. *Nature* **490:** 61–70. doi:10.1038/nature11412

Cardoso F, van't Veer LJ, Bogaerts J, Slaets L, Viale G, Delaloge S, Pierga JY, Brain E, Causeret S, DeLorenzi M, et al. 2016. 70-gene signature as an aid to treatment decisions in early-stage breast cancer. *N Engl J Med* **375:** 717–729. doi:10.1056/NEJMoa1602253

Cardoso F, Kyriakides S, Ohno S, Penault-Llorca F, Poortmans P, Rubio I, Zackrisson S, Senkus E. 2019. Early breast cancer: ESMO clinical practice guidelines for diagnosis, treatment and follow-up. *Ann Oncol* **30:** 1194–1220. doi:10.1093/annonc/mdz173

Chung W, Eum HH, Lee HO, Lee KM, Lee HB, Kim KT, Ryu HS, Kim S, Lee JE, Park YH, et al. 2017. Single-cell RNA-seq enables comprehensive tumour and immune cell profiling in primary breast cancer. *Nat Commun* **8:** 1–12. doi:10.1038/ncomms15081

Ciriello G, Sinha R, Hoadley KA, Jacobsen AS, Reva B, Perou CM, Sander C, Schultz N. 2013. The molecular diversity of luminal A breast tumors. *Breast Cancer Res Treat* **141:** 409–420. doi:10.1007/s10549-013-2699-3

Ciriello G, Gatza ML, Beck AH, Wilkerson MD, Rhie SK, Pastore A, Zhang H, McLellan M, Yau C, Kandoth C, et al. 2015. Comprehensive molecular portraits of invasive lobular breast cancer. *Cell* **163:** 506–519. doi:10.1016/j.cell.2015.09.033

Craven KE, Gökmen-Polar Y, Badve SS. 2021. CIBERSORT analysis of TCGA and METABRIC identifies subgroups with better outcomes in triple negative breast cancer. *Sci Rep* **11:** 1–19. doi:10.1038/s41598-021-83913-7

Curtis C, Shah SP, Chin SF, Turashvili G, Rueda OM, Dunning MJ, Speed D, Lynch AG, Samarajiwa S, Yuan Y, et al. 2012. The genomic and transcriptomic architecture of 2,000 breast tumours reveals novel subgroups. *Nature* **486:** 346–352. doi:10.1038/nature10983

Dowsett M, Cuzick J, Wale C, Forbes J, Mallon EA, Salter J, Quinn E, Dunbier A, Baum M, Buzdar A, et al. 2010. Prediction of risk of distant recurrence using the 21-gene recurrence score in node-negative and node-positive postmenopausal patients with breast cancer treated with anastrozole or tamoxifen: a TransATAC study. *J Clin Oncol* **28:** 1829–1834. doi:10.1200/JCO.2009.24.4798

Drukker CA, Bueno-de-Mesquita J, Retèl VP, van Harten WH, van Tinteren H, Wesseling J, Roumen RM, Knauer M, van't Veer LJ, Sonke GS, et al. 2013. A prospective evaluation of a breast cancer prognosis signature in the observational RASTER study. *Int J Cancer* **133:** 929–936. doi:10.1002/ijc.28082

Dubsky P, Brase J, Jakesz R, Rudas M, Singer C, Greil R, Dietze O, Luisser I, Klug E, Sedivy R, et al. 2013. The EndoPredict score provides prognostic information on late distant metastases in ER$^+$/HER2$^-$ breast cancer patients. *Br J Cancer* **109:** 2959–2964. doi:10.1038/bjc.2013.671

Efremova M, Vento-Tormo M, Teichmann SA, Vento-Tormo R. 2020. CellPhoneDB: inferring cell–cell communication from combined expression of multi-subunit ligand–receptor complexes. *Nat Protoc* **15:** 1484–1506. doi:10.1038/s41596-020-0292-x

Fan C, Prat A, Parker JS, Liu Y, Carey LA, Troester MA, Perou CM. 2011. Building prognostic models for breast cancer patients using clinical variables and hundreds of gene expression signatures. *BMC Med Genomics* **4:** 3. doi:10.1186/1755-8794-4-3

Gao R, Bai S, Henderson YC, Lin Y, Schalck A, Yan Y, Kumar T, Hu M, Sei E, Davis A, et al. 2021. Delineating copy number and clonal substructure in human tumors from single-cell transcriptomes. *Nat Biotechnol* **39:** 599–608. doi:10.1038/s41587-020-00795-2

Garcia-Recio S, Thennavan A, East MP, Parker JS, Cejalvo JM, Garay JP, Hollern DP, He X, Mott KR, Galván P, et al. 2020. FGFR4 regulates tumor subtype differentiation in luminal breast cancer and metastatic disease. *J Clin Invest* **130:** 4871–4887. doi:10.1172/JCI130323

Gatza ML, Lucas JE, Barry WT, Kim JW, Wang Q, Crawford MD, Datto MB, Kelley M, Mathey-Prevot B, Potti A, et al. 2010. A pathway-based classification of human breast cancer. *Proc Natl Acad Sci* **107:** 6994–6999. doi:10.1073/pnas.0912708107

Gnant M, Filipits M, Greil R, Stoeger H, Rudas M, Bago-Horvath Z, Mlineritsch B, Kwasny W, Knauer M, Singer C, et al. 2014. Predicting distant recurrence in receptor-positive breast cancer patients with limited clinicopathological risk: using the PAM50 risk of recurrence score in 1478 postmenopausal patients of the ABCSG-8 trial treated with adjuvant endocrine therapy alone. *Ann Oncol* **25:** 339–345. doi:10.1093/annonc/mdt494

Griffiths JI, Chen J, Cosgrove PA, O'Dea A, Sharma P, Ma C, Trivedi M, Kalinsky K, Wisinski KB, O'Regan R, et al. 2021. Serial single-cell genomics reveals convergent subclonal evolution of resistance as patients with early-stage breast cancer progress on endocrine plus CDK4/6 therapy. *Nat Cancer* **2:** 658–671. doi:10.1038/s43018-021-00215-7

Guillaumet-Adkins A, Rodríguez-Esteban G, Mereu E, Mendez-Lago M, Jaitin DA, Villanueva A, Vidal A, Martinez-Marti A, Felip E, Vivancos A, et al. 2017. Single-cell transcriptome conservation in cryopreserved cells and tissues. *Genome Biol* **18:** 1–15. doi:10.1186/s13059-017-1171-9

Habib N, Avraham-Davidi I, Basu A, Burks T, Shekhar K, Hofree M, Choudhury SR, Aguet F, Gelfand E, Ardlie K, et al. 2017. Massively parallel single-nucleus RNA-seq with DroNc-seq. *Nat Methods* **14:** 955–958. doi:10.1038/nmeth.4407

Heng YJ, Lester SC, Tse GM, Factor RE, Allison KH, Collins LC, Chen YY, Jensen KC, Johnson NB, Jeong JC, et al. 2017. The molecular basis of breast cancer pathological phenotypes. *J Pathol* **241:** 375–391. doi:10.1002/path.4847

Hoadley KA, Yau C, Hinoue T, Wolf DM, Lazar AJ, Drill E, Shen R, Taylor AM, Cherniack AD, Thorsson V, et al. 2018. Cell-of-origin patterns dominate the molecular classification of 10,000 tumors from 33 types of cancer. *Cell* **173:** 291–304.e6. doi:10.1016/j.cell.2018.03.022

Hollern DP, Xu N, Thennavan A, Glodowski C, Garcia-Recio S, Mott KR, He X, Garay JP, Carey-Ewend K, Marron D, et al. 2019. B cells and T follicular helper cells mediate response to checkpoint inhibitors in high mutation burden mouse models of breast cancer. *Cell* **179:** 1191–1206.e21. doi:10.1016/j.cell.2019.10.028

Hu Q, Hong Y, Qi P, Lu G, Mai X, Xu S, He X, Guo Y, Gao L, Jing Z, et al. 2021. Atlas of breast cancer infiltrated B-lymphocytes revealed by paired single-cell RNA-sequencing and antigen receptor profiling. *Nat Commun* **12:** 1–13. doi:10.1038/s41467-021-22300-2

Huo D, Hu H, Rhie SK, Gamazon ER, Cherniack AD, Liu J, Yoshimatsu TF, Pitt JJ, Hoadley KA, Troester M, et al. 2017. Comparison of breast cancer molecular features and survival by African and European ancestry in The Cancer Genome Atlas. *JAMA Oncol* **3:** 1654–1662. doi:10.1001/jamaoncol.2017.0595

ICGC/TCGA Pan-Cancer Analysis of Whole Genomes Consortium. 2020. Pan-cancer analysis of whole genomes. *Nature* **578:** 82–93. doi:10.1038/s41586-020-1969-6

Iglesia MD, Vincent BG, Parker JS, Hoadley KA, Carey LA, Perou CM, Serody JS. 2014. Prognostic B-cell signatures using mRNA-seq in patients with subtype-specific breast and ovarian cancer. *Clin Cancer Res* **20:** 3818–3829. doi:10.1158/1078-0432.CCR-13-3368

Jaitin DA, Adlung L, Thaiss CA, Weiner A, Li B, Descamps H, Lundgren P, Bleriot C, Liu Z, Deczkowska A, et al. 2019. Lipid-associated macrophages control metabolic homeostasis in a Trem2-dependent manner. *Cell* **178:** 686–698.e14. doi:10.1016/j.cell.2019.05.054

Kalinsky K, Barlow WE, Gralow JR, Meric-Bernstam F, Albain KS, Hayes DF, Lin NU, Perez EA, Goldstein LJ, Chia SK, et al. 2021. 21-gene assay to inform chemotherapy benefit in node-positive breast cancer. *N Engl J Med* **385:** 2336–2347. doi:10.1056/NEJMoa2108873

Kim C, Gao R, Sei E, Brandt R, Hartman J, Hatschek T, Crosetto N, Foukakis T, Navin NE. 2018. Chemoresistance evolution in triple-negative breast cancer delineated by single-cell sequencing. *Cell* **173:** 879–893.e13. doi:10.1016/j.cell.2018.03.041

Krug K, Jaehnig EJ, Satpathy S, Blumenberg L, Karpova A, Anurag M, Miles G, Mertins P, Geffen Y, Tang LC, et al. 2020. Proteogenomic landscape of breast cancer tumorigenesis and targeted therapy. *Cell* **183:** 1436–1456.e31. doi:10.1016/j.cell.2020.10.036

Lindström LS, Yau C, Czene K, Thompson CK, Hoadley KA, Van't Veer LJ, Balassanian R, Bishop JW, Carpenter PM, Chen YY, et al. 2018. Intratumor heterogeneity of the estrogen receptor and the long-term risk of fatal breast cancer. *J Natl Cancer Inst* **110:** 726–733. doi:10.1093/jnci/djx270

Lubeck E, Coskun AF, Zhiyentayev T, Ahmad M, Cai L. 2014. Single-cell in situ RNA profiling by sequential hybridization. *Nat Methods* **11**: 360–361. doi:10.1038/nmeth.2892

Luca BA, Steen CB, Matusiak M, Azizi A, Varma S, Zhu C, Przybyl J, Espín-Pérez A, Diehn M, Alizadeh AA, et al. 2021. Atlas of clinically distinct cell states and ecosystems across human solid tumors. *Cell* **184**: 5482–5496.e28. doi:10.1016/j.cell.2021.09.014

Macosko EZ, Basu A, Satija R, Nemesh J, Shekhar K, Goldman M, Tirosh I, Bialas AR, Kamitaki N, Martersteck EM, et al. 2015. Highly parallel genome-wide expression profiling of individual cells using nanoliter droplets. *Cell* **161**: 1202–1214. doi:10.1016/j.cell.2015.05.002

Massagué J. 2007. Sorting out breast-cancer gene signatures. *N Engl J Med* **356**: 294–297. doi:10.1056/NEJMe068292

Molgora M, Esaulova E, Vermi W, Hou J, Chen Y, Luo J, Brioschi S, Bugatti M, Omodei AS, Ricci B, et al. 2020. TREM2 modulation remodels the tumor myeloid landscape enhancing anti-PD-1 immunotherapy. *Cell* **182**: 886–900.e17. doi:10.1016/j.cell.2020.07.013

Mulder K, Patel AA, Kong WT, Piot C, Halitzki E, Dunsmore G, Khalilnezhad S, Irac SE, Dubuisson A, Chevrier M, et al. 2021. Cross-tissue single-cell landscape of human monocytes and macrophages in health and disease. *Immunity* **54**: 1883–1900.e5. doi:10.1016/j.immuni.2021.07.007

Newman AM, Liu CL, Green MR, Gentles AJ, Feng W, Xu Y, Hoang CD, Diehn M, Alizadeh AA. 2015. Robust enumeration of cell subsets from tissue expression profiles. *Nat Methods* **12**: 453–457. doi:10.1038/nmeth.3337

Okoniewski MJ, Miller CJ. 2006. Hybridization interactions between probesets in short oligo microarrays lead to spurious correlations. *BMC Bioinformatics* **7**: 1–14. doi:10.1186/1471-2105-7-276

Paik S, Shak S, Tang G, Kim C, Baker J, Cronin M, Baehner FL, Walker MG, Watson D, Park T, et al. 2004. A multigene assay to predict recurrence of tamoxifen-treated, node-negative breast cancer. *N Engl J Med* **351**: 2817–2826. doi:10.1056/NEJMoa041588

Paik S, Tang G, Shak S, Kim C, Baker J, Kim W, Cronin M, Baehner FL, Watson D, Bryant J, et al. 2006. Gene expression and benefit of chemotherapy in women with node-negative, estrogen receptor–positive breast cancer. *J Clin Oncol* **24**: 3726–3734. doi:10.1200/JCO.2005.04.7985

Pal B, Chen Y, Vaillant F, Capaldo BD, Joyce R, Song X, Bryant VL, Penington JS, Di Stefano L, Tubau Ribera N, et al. 2021. A single-cell RNA expression atlas of normal, preneoplastic and tumorigenic states in the human breast. *EMBO J* **40**: e107333. doi:10.15252/embj.2020107333

Patel AP, Tirosh I, Trombetta JJ, Shalek AK, Gillespie SM, Wakimoto H, Cahill DP, Nahed BV, Curry WT, Martuza RL, et al. 2014. Single-cell RNA-seq highlights intratumoral heterogeneity in primary glioblastoma. *Science* **344**: 1396–1401. doi:10.1126/science.1254257

Perou CM, Sørlie T, Eisen MB, van de Rijn M, Jeffrey SS, Rees CA, Pollack JR, Ross DT, Johnsen H, Akslen LA, et al. 2000. Molecular portraits of human breast tumours. *Nature* **406**: 747–752. doi:10.1038/35021093

Qian J, Olbrecht S, Boeckx B, Vos H, Laoui D, Etlioglu E, Wauters E, Pomella V, Verbandt S, Busschaert P, et al. 2020. A pan-cancer blueprint of the heterogeneous tumor microenvironment revealed by single-cell profiling. *Cell Res* **30**: 745–762. doi:10.1038/s41422-020-0355-0

Ramilowski JA, Goldberg T, Harshbarger J, Kloppmann E, Lizio M, Satagopam VP, Itoh M, Kawaji H, Carninci P, Rost B, et al. 2015. A draft network of ligand-receptor-mediated multicellular signalling in human. *Nat Commun* **6**: 1–12. doi:10.1038/ncomms8866

Saal LH, Vallon-Christersson J, Häkkinen J, Hegardt C, Grabau D, Winter C, Brueffer C, Tang MH, Reutersward C, Schulz R, et al. 2015. The Sweden Cancerome Analysis Network—Breast (SCAN-B) Initiative: a large-scale multicenter infrastructure towards implementation of breast cancer genomic analyses in the clinical routine. *Genome Med* **7**: 20. doi:10.1186/s13073-015-0131-9

Sahai E, Astsaturov I, Cukierman E, DeNardo DG, Egeblad M, Evans RM, Fearon D, Greten FR, Hingorani SR, Hunter T, et al. 2020. A framework for advancing our understanding of cancer-associated fibroblasts. *Nat Rev Cancer* **20**: 174–186. doi:10.1038/s41568-019-0238-1

Savas P, Virassamy B, Ye C, Salim A, Mintoff CP, Caramia F, Salgado R, Byrne DJ, Teo ZL, Dushyanthen S, et al. 2018. Single-cell profiling of breast cancer T cells reveals a tissue-resident memory subset associated with improved prognosis. *Nat Med* **24**: 986 993. doi:10.1038/s41591-018-0078-7

Schumacher TN, Thommen DS. 2022. Tertiary lymphoid structures in cancer. *Science* **375**: eabf9419. doi:10.1126/science.abf9419

Sgroi DC, Carney E, Zarrella E, Steffel L, Binns SN, Finkelstein DM, Szymonifka J, Bhan AK, Shepherd LE, Zhang Y, et al. 2013a. Prediction of late disease recurrence and extended adjuvant letrozole benefit by the HOXB13/IL17BR biomarker. *J Natl Cancer Inst* **105**: 1036–1042. doi:10.1093/jnci/djt146

Sgroi DC, Sestak I, Cuzick J, Zhang Y, Schnabel CA, Schroeder B, Erlander MG, Dunbier A, Sidhu K, Lopez-Knowles E, et al. 2013b. Prediction of late distant recurrence in patients with oestrogen-receptor-positive breast cancer: a prospective comparison of the breast-cancer index (BCI) assay, 21-gene recurrence score, and IHC4 in the TransATAC study population. *Lancet Oncol* **14**: 1067–1076. doi:10.1016/S1470-2045(13)70387-5

Singh M, Al-Eryani G, Carswell S, Ferguson JM, Blackburn J, Barton K, Roden D, Luciani F, Giang Phan T, Junankar S, et al. 2019. High-throughput targeted long-read single cell sequencing reveals the clonal and transcriptional landscape of lymphocytes. *Nat Commun* **10**: 1–13. doi:10.1038/s41467-019-11049-4

Slyper M, Porter C, Ashenberg O, Waldman J, Drokhlyansky E, Wakiro I, Smillie C, Smith-Rosario G, Wu J, Dionne D, et al. 2020. A single-cell and single-nucleus RNA-seq toolbox for fresh and frozen human tumors. *Nat Med* **26**: 792–802. doi:10.1038/s41591-020-0844-1

Sørlie T, Perou CM, Tibshirani R, Aas T, Geisler S, Johnsen H, Hastie T, Eisen MB, Van De Rijn M, Jeffrey SS, et al. 2001. Gene expression patterns of breast carcinomas distinguish tumor subclasses with clinical implications. *Proc Natl Acad Sci* **98**: 10869–10874. doi:10.1073/pnas.191367098

Sotiriou C, Pusztai L. 2009. Gene-expression signatures in breast cancer. *N Engl J Med* **360**: 790–800. doi:10.1056/NEJMra0801289

Sparano JA, Gray RJ, Makower DF, Pritchard KI, Albain KS, Hayes DF, Geyer CE Jr, Dees EC, Goetz MP, Olson JA Jr, et al. 2018. Adjuvant chemotherapy guided by a 21-gene expression assay in breast cancer. *N Engl J Med* **379:** 111–121. doi:10.1056/NEJMoa1804710

Staaf J, Häkkinen J, Hegardt C, Saal LH, Kimbung S, Hedenfalk I, Lien T, Sørlie T, Naume B, Russnes H, et al. 2022. RNA sequencing-based single sample predictors of molecular subtype and risk of recurrence for clinical assessment of early-stage breast cancer. *NPJ Breast Cancer* **8:** 94. doi:10.1038/s41523-022-00465-3

Stoeckius M, Hafemeister C, Stephenson W, Houck-Loomis B, Chattopadhyay PK, Swerdlow H, Satija R, Smibert P. 2017. Simultaneous epitope and transcriptome measurement in single cells. *Nat Methods* **14:** 865–868. doi:10.1038/nmeth.4380

Tanioka M, Fan C, Parker JS, Hoadley KA, Hu Z, Li Y, Hyslop TM, Pitcher BN, Soloway MG, Spears PA, et al. 2018. Integrated analysis of RNA and DNA from the phase III trial CALGB 40601 identifies predictors of response to trastuzumab-based neoadjuvant chemotherapy in HER2-positive breast cancer. *Clin Cancer Res* **24:** 5292–5304. doi:10.1158/1078-0432.CCR-17-3431

Thennavan A, Beca F, Xia Y, Recio SG, Allison K, Collins LC, Tse GM, Chen YY, Schnitt SJ, Hoadley KA, et al. 2021. Molecular analysis of TCGA breast cancer histologic types. *Cell Genom* **1:** 100067. doi:10.1016/j.xgen.2021.100067

Trapnell C, Cacchiarelli D, Grimsby J, Pokharel P, Li S, Morse M, Lennon NJ, Livak KJ, Mikkelsen TS, Rinn JL. 2014. The dynamics and regulators of cell fate decisions are revealed by pseudotemporal ordering of single cells. *Nat Biotechnol* **32:** 381–386. doi:10.1038/nbt.2859

Vallejo AF, Harvey K, Wang T, Wise K, Butler LM, Polo J, Plummer J, Swarbrick A, Martelotto LG. 2022. snPATHO-seq: unlocking the FFPE archives for single nucleus RNA profiling. bioRxiv doi:10.1101/2022.08.23.505054

Van't Veer LJ, Dai H, Van De Vijver MJ, He YD, Hart AA, Mao M, Peterse HL, Van Der Kooy K, Marton MJ, Witteveen AT, et al. 2002. Gene expression profiling predicts clinical outcome of breast cancer. *Nature* **415:** 530–536. doi:10.1038/415530a

Wang Z, Gerstein M, Snyder M. 2009. RNA-Seq: a revolutionary tool for transcriptomics. *Nat Rev Genet* **10:** 57–63. doi:10.1038/nrg2484

Wu SZ, Roden DL, Wang C, Holliday H, Harvey K, Cazet AS, Murphy KJ, Pereira B, Al-Eryani G, Bartonicek N, et al. 2020. Stromal cell diversity associated with immune evasion in human triple-negative breast cancer. *EMBO J* **39:** e104063.

Wu F, Yang J, Liu J, Wang Y, Mu J, Zeng Q, Deng S, Zhou H. 2021a. Signaling pathways in cancer-associated fibroblasts and targeted therapy for cancer. *Signal Transduct Target Ther* **6:** 218. doi:10.1038/s41392-021-00641-0

Wu SZ, Al-Eryani G, Roden DL, Junankar S, Harvey K, Andersson A, Thennavan A, Wang C, Torpy JR, Bartonicek N, et al. 2021b. A single-cell and spatially resolved atlas of human breast cancers. *Nat Genet* **53:** 1334–1347. doi:10.1038/s41588-021-00911-1

Wu SZ, Roden DL, Al-Eryani G, Bartonicek N, Harvey K, Cazet AS, Chan CL, Junankar S, Hui MN, Millar EA, et al. 2021c. Cryopreservation of human cancers conserves tumour heterogeneity for single-cell multi-omics analysis. *Genome Med* **13:** 1–17. doi:10.1186/s13073-021-00885-z

Zheng L, Qin S, Si W, Wang A, Xing B, Gao R, Ren X, Wang L, Wu X, Zhang J, et al. 2021. Pan-cancer single-cell landscape of tumor-infiltrating T cells. *Science* **374:** abe6474. doi:10.1126/science.abe6474

 Cite this article as *Cold Spring Harb Perspect Med* doi: 10.1101/cshperspect.a041320

Chromatin Variants Reveal the Genetic Determinants of Oncogenesis in Breast Cancer

Shalini Bahl,[1,2] Jason S. Carroll,[3] and Mathieu Lupien[1,2,4]

[1]Princess Margaret Cancer Centre, Toronto, Ontario M5G 1L7, Canada

[2]Department of Medical Biophysics, University of Toronto, Toronto, Ontario M5G 1L7, Canada

[3]Cancer Research UK Cambridge Institute, University of Cambridge, Cambridge CB2 0RE, United Kingdom

[4]Ontario Institute for Cancer Research, Toronto, Ontario M5G 0A3, Canada

Correspondence: Jason.Carroll@cruk.cam.ac.uk; Mathieu.Lupien@uhnresearch.ca

Breast cancer presents as multiple distinct disease entities. Each tumor harbors diverse cell populations defining a phenotypic heterogeneity that impinges on our ability to treat patients. To date, efforts mainly focused on genetic variants to find drivers of inter- and intratumor phenotypic heterogeneity. However, these efforts have failed to fully capture the genetic basis of breast cancer. Through recent technological and analytical approaches, the genetic basis of phenotypes can now be decoded by characterizing chromatin variants. These variants correspond to polymorphisms in chromatin states at DNA sequences that serve a distinct role across cell populations. Here, we review the function and causes of chromatin variants as they relate to breast cancer inter- and intratumor heterogeneity and how they can guide the development of treatment alternatives to fulfill the goal of precision cancer medicine.

BREAST CANCER IS A COMPLEX DISEASE IN NEED OF PRECISION MEDICINE

Globally, breast cancer is the most commonly diagnosed cancer and the fifth-leading cause of cancer-related death. According to the Global Cancer Statistics 2020 (GLOBOCAN), there were an estimated 2.3 million breast cancer cases and 685,000 deaths worldwide in 2020 (Sung et al. 2021). Among women, breast cancer accounted for ~24.5% of all cancer cases and 15.5% of cancer deaths ranking first for incidence and mortality in the majority of countries in 2020 (Lei et al. 2021; Sung et al. 2021). Despite the increase in incidence, mortality is decreasing in developed countries related to early detection from screening and access to therapy (Hashim et al. 2016). However, the effectiveness of current treatment regimens poses a challenge for some patients who go on to develop recurrent disease (Hilsenbeck et al. 1998; Dent et al. 2007; Gasparini 2012). Whereas recurrence for other cancers tend to occur months to years after remission, breast cancer has a long window of relapse spanning months to decades (Riggio et al. 2021) However, not all patients relapse; rather, between 8% and 10% of patients diagnosed with breast cancer show locoregional recurrence and 15%–30% develop metastatic disease (Lafourcade et al. 2018). Hence, the underlying causes of breast tumor progression and the differential clinical response remains elusive. The goal to eradicate breast cancer requires the adoption of precision medicine to tailor treatment for each patient.

Comparing breast tumors across patients reveals that breast cancer is a molecularly and clinically heterogeneous disease (Polyak 2011). For instance, gene expression signatures, such as the PAM50 (Parker et al. 2009) and SCMOD signatures (Sotiriou et al. 2003), classify breast tumors into five molecular subtypes, referred to as luminal A, luminal B, HER2-enriched, basal, and normal-like. These encompass histopathological subtypes defined according to the quantification of protein levels for the estrogen receptor α (ERα), progesterone receptor (PGR), and the human epidermal growth factor receptor 2 (HER2). Whereas luminal A consists of tumors expressing ERα, PGR, and HER2 (ERα$^+$/PGR$^+$/HER2$^+$), luminal B tumors can either be ERα$^+$/PGR$^+$/HER2$^+$ or ERα$^+$/PGR$^+$/HER2$^-$ and highly proliferative (KI67$^+$). HER2-enriched breast cancers lack expression for ERα and PGR (ERα$^-$/PGR$^-$/HER2$^+$). Basals lack expression of all three receptors and are referred to as triple-negative breast cancer (TNBC) (ERα$^-$/PGR$^-$/HER2$^-$) (Rivenbark et al. 2013). Normal-like tumors are similar to luminal A based on ERα, PGR, and HER2 protein levels (ERα$^+$/PGR$^+$/HER2$^-$). Combining additional biological insights further exemplify the variability between breast cancer tumors. For instance, the inclusion of copy number events can stratify the breast cancer molecular subtypes into 11 subgroups (Curtis et al. 2012). Subtyping of breast tumors according to copy number variants (CNVs) can also be informative of patients' response to therapy and clinical outcome (Pereira et al. 2016). Genome instability captured by CNVs is also linked to the presence of extrachromosomal DNAs (ecDNAs) that can further contribute to breast cancer heterogeneity by promoting oncogene overexpression (Kim et al. 2020; Keshavarzian and Lupien 2022). Considering the diversity, delivering on the promise of precision medicine relies on understanding and expanding our exploration into the mechanisms that account for the unique expression profiles that define each breast cancer subtype.

BREAST CANCER HETEROGENEITY ARISES FROM MECHANISMS REGULATING GENE EXPRESSION

Although breast cancer subtypes can be defined based on the differential expression of genes, the determinants of subtype-specific expression patterns lie in DNA elements found outside of genes, including *cis*-regulatory elements (CREs) (Zhou et al. 2016). CREs can be promoters located immediately upstream of the genes they regulate and enhancers that lie distal (up to hundreds of thousands of base pairs) from their target promoter(s). Promoters commonly recruit RNA polymerase II and the preinitiation complex that ensures the transcription of genes being activated upon. Enhancers are bound by transcription factors and associated coactivators to modulate RNA polymerase II activation. These transcription factors bind to CREs through their DNA-binding domains recognizing unique short (~9 mer) DNA sequences known as DNA recognition motifs (Castro-Mondragon et al. 2022). Together, the expression of each gene is under the regulation of one promoter and one or more distal enhancers. The sum of all CREs controlling the expression of a gene is known as a regulatory plexus (Bailey et al. 2016; Sallari et al. 2017). To achieve differential gene expression, each breast cancer subtype relies on cells engaging a unique set of CREs bound by transcription factors, RNA polymerase II, and associated proteins. (See Box 1 for definitions of terms used throughout.)

Studies that identified CREs differentially engaged across breast cancer subtypes have found the DNA elements and associated transcriptional machinery driving subtype-specific expression programs. For instance, CREs bound by the transcription factors ERα, FOXA1, and GATA3 help establish the luminal A and B breast cancer subtypes (Carroll et al. 2005; Chaudhary et al. 2017; Corces et al. 2018). The luminal A breast cancer subtype can be further characterized by CREs permissive for the binding of other transcription factors (Zhu et al. 2020) including ZNF217, that promotes cell proliferation by acting as a transcriptional repressor of proapoptotic gene expression (Nguyen et al. 2014). The HER2$^+$ breast cancer subtype engages CREs per-

BOX 1. TERMS AND DEFINITIONS USED THROUGHOUT

Term	Definition
DNA element	Segment of DNA in a genome with a distinctive chromatin state. DNA elements can be composed of a few hundred DNA base pairs or span kilobases to megabases. Types of DNA elements include but are not limited to coding (gene) and noncoding transcribed DNA, promoter, enhancer, insulator, and anchor of chromatin interaction.
Chromatin state	Defined according to the level of chromatin accessibility, the pattern of chromatin modifications (DNA methylation and/or histone post-translational modifications), and/or nucleosome composition (core vs. histone variants). Chromatin states differ between each type of DNA element.
Chromatin variant	Segments of DNA in a genome found in a different chromatin state between distinct cell states or across cells in the same state but from different individuals.
Cis-regulatory element	Type of DNA element that regulates the transcription of a coding (gene) or non-coding transcript in *cis*, such as promoters and enhancers.
Regulatory plexus	The sum of the regulatory elements collectively modulating the expression of a coding (gene) or noncoding transcript. A regulatory plexus includes the promoter and all enhancers of a transcribed DNA in a given cell state.
Cistrome	The collective sum of the regulatory elements bound by one transcription factor across the genome of one cell state.

missive for the NRF1 transcription factor to regulate the expression of genes linked to a variety of hallmark breast cancer processes including epithelial–mesenchymal transcription, cell apoptosis, and cell-cycle regulation (Ramos et al. 2018). In contrast, basal-like breast cancers engage CREs permissive for the transcription factors BCL11a and GRHL to mediate cell proliferation and tumorigenicity-promoting activity (Khaled et al. 2015; Corces et al. 2018; Wang et al. 2020). Furthermore, a subset of transcription factors is active across diverse breast cancer subtypes but engage with different sets of CREs in each subtype. This is sufficient for the same transcription factor to contribute to distinct regulatory plexuses and therefore lead to differential gene expression across subtypes. For example, the c-MYC transcription factor binds to CREs of the regulatory plexuses for the cell-cycle-associated genes *ATM*, *Cyclin B1*, *PIK3CA*, and *MKi67* in luminal and HER2 breast tumors facilitating tumor growth (Green et al. 2016). In basal-like breast tumors, c-MYC instead regulates the expression of the *Cyclin E* and *p16* genes involved

in glucose metabolism (Green et al. 2016). These results highlight the need to understand how the same transcription factor can access distinct CREs to mediate context-specific gene expression events across breast cancer subtypes. Collectively, these discoveries showcase the importance of distinguishing active CREs across breast cancer subtypes to identify the DNA elements contributing to subtype-specific phenotypes.

CHROMATIN STATES CLASSIFY THE BREAST CANCER GENOME INTO DNA ELEMENTS RELEVANT TO REGULATING GENE EXPRESSION

Differences in the activation state of a DNA element, such as a CRE, is possible because the DNA exists as a macromolecular complex with proteins, commonly histones, to form chromatin. This allows the DNA to either be wrapped around an octamer of histones to form a nucleosome or to bridge nucleosomes together, by establishing what is known as linker DNA (Luger et al. 2012). While nucleosomal DNA has a fixed length of 147 bp (Luger et al. 1997), linker DNA can vary in length between two nucleosomes. Short linker DNA is typical of compacted chromatin. In contrast, long linker DNA (hundreds of base pairs) characterizes accessible chromatin where DNA is readily recognized by proteins with DNA-binding domains, such as transcription factors (Buenrostro et al. 2015). A DNA sequence can be part of a nucleosome in certain cells but will serve as a long linker DNA in another cell, thereby presenting itself in one of two different chromatin states that discriminate the activation state of DNA elements (Thurman et al. 2012; Stergachis et al. 2013; Meuleman et al. 2020).

Whereas chromatin accessibility allows to stratify DNA elements into either accessible or inaccessible chromatin states, the presence of chemical modifications to the chromatin or histone variants provides additional ways of defining chromatin states at a higher resolution. Chemical modifications to chromatin include DNA methylation and post-translational modifications to the histones (Bowman and Poirier 2015). A multitude of different chromatin states can be observed based on distinct combinations of chromatin modifications and histone variants (Ernst and Kellis 2010; Ernst et al. 2011; Hoffman et al. 2012; Kundaje et al. 2015). These chromatin states can segment the genome of any cell into small (base pair scale) and large (kilobases to megabases scale) DNA elements. Chromatin states can be associated with transcriptional potential depending on the modifications present. For instance, the chromatin state composed of histone 3 lysine 4 trimethylation (H3K4me3) and histone 3 lysine 27 acetylation (H3K27ac) defines DNA elements active as promoters (Barski et al. 2007; Henikoff et al. 2020). Similarly, the chromatin state composed of H3K4me1/me2 and histone acetylation captures DNA elements active as enhancers (Heintzman et al. 2007, 2009; Lupien et al. 2008). In contrast, a distinct chromatin state composed of H3K27me3 with or without H3K4me3 reveals DNA elements serving as poised and silent promoters, respectively (Bernstein et al. 2006). A separate but well-established mechanism for silencing is via DNA

methylation of promoters and repetitive DNA sequences (Baylin 2005; Chiappinelli et al. 2015; Roulois et al. 2015). Distinct chromatin states based on H3K36me3 also distinguish actively transcribed for silent coding (gene) and noncoding DNA (Barski et al. 2007; Ernst and Kellis 2010; Hoffman et al. 2012; Kundaje et al. 2015). Therefore, it is possible to classify the human genome of any cell into DNA elements, including CREs and genes, based on the nature of their chromatin state. Dynamic changes in the chromatin state of DNA elements can therefore relate to alterations in gene expression events such as those differentiating breast cancer subtypes (Fig. 1).

CHROMATIN VARIANTS CAPTURE THE GENETIC DETERMINANTS SPECIFIC TO EACH BREAST CANCER SUBTYPE

Akin to genetic polymorphism giving rise to distinct phenotypes across individuals, chromatin state polymorphisms observed at a DNA sequence common between cell states define chromatin variants. In other words, chromatin variants correspond to a segment of DNA that adopts a different chromatin state across cell states. Chromatin variants are identified independently of the presence or absence of genetic variants and reflect changes in the contribution of DNA sequences toward a phenotype (Lupien et al. 2008; Akhtar-Zaidi et al. 2012; Magnani et al. 2013; Grosselin et al. 2019; Deblois et al. 2020). Hence, chromatin variants associated with specific cell states harbor the DNA sequences that define the genetic make-up of a unique phenotype. With respect to breast cancers, large-scale initiatives such as The Cancer Genome Atlas (TCGA) and The Lonestar Oncology Network for EpigeneticS Therapy And Research (LONESTAR) have found that unique genome-wide sets of chromatin variants can accurately segregate each subtype (Chen et al. 2018; Corces et al. 2018; Xi et al. 2018; Ruff et al. 2021). Analyzing the DNA sequence of subtype-specific chromatin variants classified as CREs identified transcription factors driving breast cancer subtype-specific gene expression (Chen et al. 2018; Corces et al. 2018). Chromatin variants can also capture differences in the contribution of DNA elements to intra-subtype heterogeneity. Tumors of the TNBC subtype can be further classified into one of four subgroups based on gene expression, namely, basal-like 1 (BL1), basal-like 2 (BL2), mesenchymal, or luminal androgen receptor (LAR) subgroups (Lehmann et al. 2011, 2021; Burstein et al. 2015; Jiang et al. 2019). Each of these TNBC subgroups can also be accurately defined based on subgroup-specific chromatin variants driven by subgroup-specific transcription factor activity patterns (Lehmann et al. 2021). Between TNBC subtypes, androgen receptor (AR) transcriptional activity distinguishes the LAR subtype from others, while BL1 and BL2 subtypes are enriched with transcription factors such as those from the SOX and STAT families, respectively (Burstein et al. 2015; Gerratana et al. 2018). Chromatin variants define DNA elements resulting from their interactions with subtype-specific transcription factors. These chromatin variants, such as CREs of the immune-related genes *IFNG, NLRC5,* and *CIITA,* are active in the BL1, BL2, and LAR subgroups but silent in the mesenchymal subgroup (Lehmann et al. 2021). Altogether, results to date demonstrate how chromatin variants can resolve inter- as well as intra-subtype heterogeneity in breast cancer.

Beyond the inter- and intra-subtype heterogeneity, individual breast tumors are composed of diverse cell populations, inclusive of normal and cancer cells, which account for "intratumoral heterogeneity" (ITH). ITH is associated with poor prognosis as it is thought to be one of the leading drivers of therapeutic resistance and metastasis (Swanton 2012; Koren and Bentires-Alj 2015). Pathological and molecular assessments can document ITH of breast tumors. For instance, ITH is reflected in common histological cell populations found in varying proportions within a breast tumor (Kreike et al. 2007; Gerashchenko et al. 2017; do Nascimento and Otoni 2020; Tan et al. 2020). Since the advent of next-genome sequencing, multiregional biopsy whole-genome sequencing identified temporal and spatially distinct genetic clones within breast tumors (Shah et al. 2009; Navin et al. 2011; Gerlinger et al. 2012). More recently, chromatin variants were used to characterize ITH in luminal breast cancer and

find populations related to tumor progression (Fig. 2; Patten et al. 2018). This has been made possible through recent technological advances in microfluidics and combinatorial indexing strategies, and reduced high-throughput sequencing costs have spurred the development of assays that capture chromatin states across the genome of single cells (Smallwood et al. 2014; Kaya-Okur et al. 2019; Bartosovic et al. 2021; Carter and Zhao 2021). In breast cancer, single-cell genetic or transcriptomic profiling showed that breast tumors are composed of various clonal populations (Dentro et al. 2021) as well as cell states including immune, epithelial, mesenchymal, and cycling cells in varying proportions across tumors (Wu et al. 2021b). In parallel, assays measuring single-cell chromatin states led to the identification of population-specific chromatin variants in breast tumors (Zhang et al. 2021a). For instance, single-cell chromatin state assays performed on patient-derived breast cancer xenografts distinguished populations that enable the progression of luminal breast cancer cells under endocrine therapy (Grosselin et al. 2019). Determining chromatin variants of these subpopulations have allowed for the investigation of CREs that govern processes of tumor progression including therapeutic resistance and cancer metabolism (Zhang et al. 2021b; Kusi et al. 2022). Therefore, we now have the ability to address the clinical problem of distinguishing intratumor populations by reporting chromatin variants reflecting the genetic basis of distinct populations within tumors.

CHROMATIN VARIANTS REVEAL THE ROLE OF GENETIC VARIANTS IN BREAST CANCER

Finding chromatin variants in breast cancer has allowed us to understand the functional contribution of genetic variants, both inherited risk variants and acquired somatic mutations, to breast cancer development. This stems from the fact that chromatin variants annotate DNA elements in the coding as well as noncoding genome of breast tumors, where most inherited and acquired genetic variants (risk-associated single nucleotide polymorphisms (raSNPs) and somatic mutations) for this disease are found (Beroukhim et al. 2010; Nik-Zainal et al.

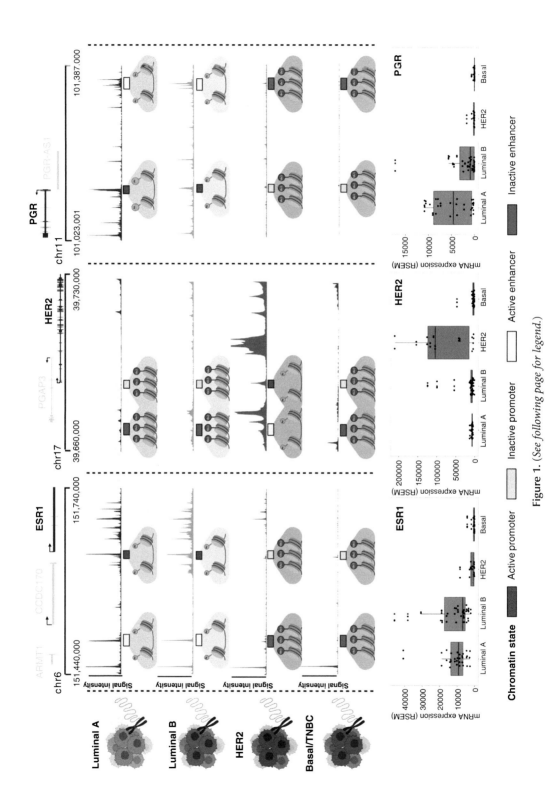

Figure 1. (*See following page for legend.*)

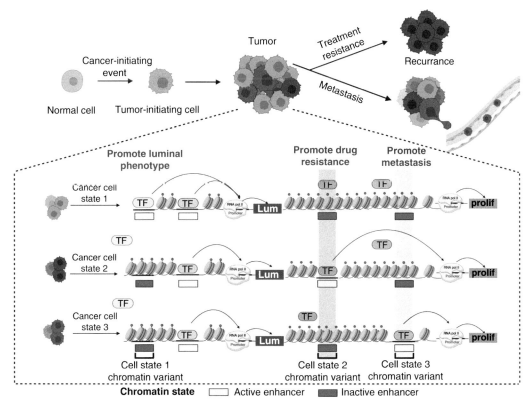

Figure 2. Chromatin variants capture cell-state population shifts across oncogenesis. Oncogenesis involves normal cells acquiring tumor-initiating properties from accumulating cancer-initiating genetic and chromatin variants. Finding regions of the genome differing in chromatin states across cancer cells in different states, such as primary versus drug-resistant or metastatic state, identifies cell-state-associated chromatin variants. Cancer cell-state-associated chromatin variants commonly define *cis*-regulatory elements permissive to transcription factor (TF)-binding guiding cell-state-specific transcriptional programs. Figure created with BioRender.com.

2016; Zhang et al. 2020). For instance, comprehensive characterization of breast cancer raSNPs reveal that they primarily target breast cancer–specific chromatin variants that define CREs for

transcription factors, including FOXA1, ERα, and GATA3 (Cowper-Sal lari et al. 2012; Fletcher et al. 2013; Ghoussaini et al. 2014; Zhang et al. 2014; Michailidou et al. 2017; Beesley et

---

Figure 1. Differences in chromatin states differentially delineate breast cancer subtype-specific gene regulatory landscapes. *Cis*-regulatory elements such as promoters and enhancers can harbor different combinations of chromatin modifications, known as chromatin states. These can denote active enhancers (e.g., H3K27ac, H3K4me1), active promoters (e.g., H3K27ac, H3K4me3), and inactive enhancers or promoters (e.g., H3K27me3, H3K9me3). Additionally, active *cis*-regulatory elements typically reside in accessible chromatin, whereas inactive ones are hidden in closed chromatin. Differences in chromatin state over enhancers and promoters of *ESR1*, *HER2 (ERBB2)*, and *PGR* (*top*) can be associated with differences in gene expression of the luminal A, luminal B, HER2, and basal/triple-negative breast cancer (TNBC) ($n = 15$) subtypes. High *ESR1* and *PGR* gene expression were typical of luminal A ($n = 18$) and B ($n = 25$) and high *ERBB2* for HER2 ($n = 11$) (*bottom*). Breast cancer samples were obtained from The Cancer Genome Atlas (TCGA) cohort with matched PAM50 subtyping and processed RSEM normalized gene expression levels were downloaded from cBioPortal (Cerami et al. 2012; Corces et al. 2018). Figure created with BioRender.com.

al. 2020). In fact, chromatin variants defining breast cancer-subtype-specific CREs are preferentially enriched for subtype-specific risk-associated genetic variants (Zhang et al. 2020). This is exemplified by non-overlapping sets of 1238 and 646 risk-associated genetic variants of ERα-positive and ERα-negative breast cancer, respectively (Fachal et al. 2020), linked to subtype-specific CREs relevant for subtype-specific gene expression. Similarly, an analysis of chromatin variants led to the identification of cancer driver CREs (Rheinbay et al. 2017) and cancer driver regulatory plexuses (Bailey et al. 2016) in breast cancer (Fig. 3A). The former were reported based on the excess of somatic mutations in individual CREs, such as the promoter of the *FOXA1* gene (Rheinbay et al. 2017). The latter were reported from the excess of inherited and acquired genetic variants across the sum of CREs regulating the expression of the *ESR1* gene (Fig. 3A; Bailey et al. 2016), an approach adopted in other cancer types (Kim et al. 2016; Sallari et al. 2017; Zhou et al. 2020). More recently, chromatin variants were used to partition the luminal breast cancer genome into transcription factor cistromes (Fig. 3B), using a method defined in prostate cancer to combine and analyze the collection of individual DNA elements bound by one transcription factor (Mazrooei et al. 2019). Integrating transcription factor cistromes with somatic mutations from luminal breast tumors showed enrichment of mutations within a subset of cistromes (Fig. 3B), thereby defining them as cancer driver cistromes (Ghamrasni et al. 2022). Altogether, identifying chromatin variants has proven use-

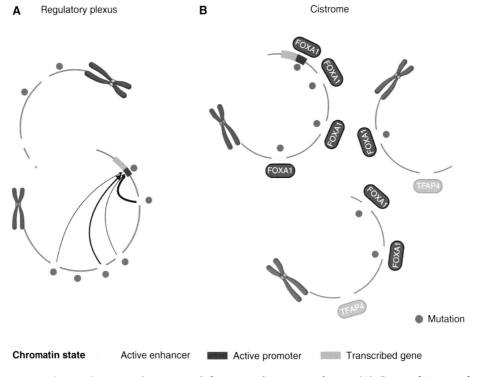

Figure 3. Regulatory plexuses and cistromes define noncoding cancer drivers. (*A*) Cancer driver regulatory plexuses are defined based on the excess of mutations (blue dots) across the sum of all *cis*-regulatory elements of a gene, inclusive of all enhancers (yellow boxes) and the promoter (red box) regulating the expression from one gene (green box). The excess of mutations relies on comparing the frequency of mutations across regulatory plexuses in a cell state. (*B*) Cancer driver cistromes are defined based on the excess of mutations across the sum of all *cis*-regulatory elements bound by the same transcription factor, such as FOXA1, in a cell state. Not all cistromes are excessively burdened by mutations (e.g., TFAP4 cistrome). Figure created with BioRender.com.

ful to decipher the role of inherited and acquired genetic variants, allowing the identification of cancer drivers across both the coding and non-coding breast cancer genome.

CHROMATIN VARIANTS CAN ARISE FROM DISRUPTIONS TO METABOLISM

Chromatin variants are maintained in a relatively stable chromatin state in terminally differentiated cells (Klemm et al. 2019). Changes in chromatin state giving rise to chromatin variants normally occur during cellular differentiation, under response to environmental stimuli or across disease development (Stergachis et al. 2013; Chronis et al. 2017; Hanahan 2022). Understanding how chromatin variants arise offers an avenue to identify means to prevent those associated with the development of diseases such as breast cancer.

In cancer, chromatin variants can arise following environmental changes. For instance, the appearance of chromatin variants in cancer cells acquiring resistance to standard-of-care therapy is well documented in breast cancer. Specifically, endocrine therapy resistance in luminal breast cancer is related to chromatin variants that reflect the enrichment of Notch signaling as a treatment-resistance mechanism, resulting in increased sensitivity to γ-secretase inhibitors (Magnani et al. 2013). Similarly, doxorubicin-resistant luminal breast cancers harbor chromatin variants that preferentially drive the expression of genes essential in regulating cell proliferation, metabolism, and inflammatory responses (Wang et al. 2021). In TNBC, standard-of-care consists of chemotherapy, such as anti-microtubule taxanes (e.g., paclitaxel), DNA intercalating anthracyclines, alkylating agents (e.g., cyclophosphamide, carboplatin), or the anti-metabolite fluorouracil (5-FU). Resistance to 5-FU is characterized by chromatin variants associated with loss of H3K27me3 at gene promoter regions (Marsolier et al. 2021). In TNBC, chronic exposure to paclitaxel also leads to chromatin variants captured from H3K27me3 and DNA methylation chromatin states. These chromatin variants are enabled through changes in

the one-carbon metabolic pathway (Deblois et al. 2020). Specifically, Deblois et al. (2020) showed that paclitaxel treatment leads to reduction in the levels of S-adenosylmethionine (SAM), the universal methyl donor, which impacts the levels and distribution of DNA methylation and H3K27me3 over repetitive DNA sequences in the genome of TNBC cells to drive paclitaxel resistance (Fig. 4A; Deblois et al. 2020). Collectively, these studies show how environmental insults, such as those disrupting metabolic pathways, can favor oncogenic chromatin variants.

PIONEER FACTORS AS DRIVERS OF CHANGE IN CHROMATIN STATES LEADING TO CHROMATIN VARIANTS

Beyond metabolic disruptions, classes of chromatin-bound proteins, including pioneer factors, hold the ability to shape chromatin states and give rise to chromatin variants. Pioneer factors are well-known DNA-binding proteins involved in establishing cell-state-specific transcriptional programs (Magnani et al. 2011b; Zaret and Carroll 2011; Iwafuchi-Doi et al. 2016; Zaret 2020). While condensed chromatin impedes access to the majority of transcription factors, pioneer factors can engage with condensed chromatin by recognizing their DNA recognition motif to initiate loosening of the chromatin (Magnani et al. 2011b; Zaret and Carroll 2011; Iwafuchi-Doi et al. 2016; Zaret 2020). This renders chromatin accessible to serve as CREs through the recruitment of transcription factors. For example, FOXA1 is a pioneer factor in breast tissue that renders chromatin accessible for the ERα transcription factor (Fig. 4B; Carroll et al. 2005; Lupien et al. 2008; Eeckhoute et al. 2009). This allows for ERα-dependent activity during mammary gland development as well as ERα-dependent activity in pathological states, when cancer has formed (Bernardo et al. 2010; Hurtado et al. 2011a; Jozwik and Carroll 2012). While capable of interacting with compacted chromatin, FOXA1 binding benefits from permissive chromatin states, characterized by the absence of DNA methylation (Sérandour et al. 2011) and the presence of methylated nucleo-

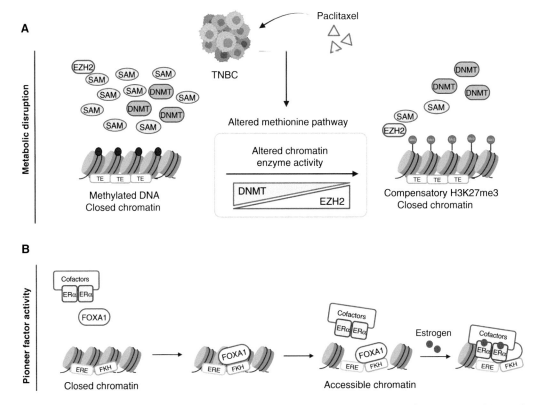

Figure 4. Chromatin variants can arise from disruption to metabolism and pioneer factor activity. (*A*) S-Adenosylmethionine (SAM) is the universal methyl donor. It is used by DNA methyltransferase (DNMT) enzymes to methylate DNA, which defines a repressive chromatin state over repetitive DNA sequences, such as transposable elements (TEs). The chemotherapeutic agent paclitaxel, used for the treatment of TNBCs, alters the methionine pathway responsible for SAM production. This leads to decreased SAM levels limiting DNMT activity and DNA hypomethylation giving rise to chromatin variants. Paclitaxel-resistant cancer cells emerge by compensating DNA hypomethylation over TEs with an alternative chromatin state relying on EZH2-dependent H3K27me3, thereby establishing chemotherapy-associated chromatin variants. (*B*) Pioneer factors, such as FOXA1, can access their binding sites (forkhead motif [FKH] for FOXA1) within closed chromatin and initiate nucleosome displacement to create chromatin variants by rendering the chromatin accessible to other transcription factors. In luminal breast cancer subtypes, this pioneering activity of FOXA1 exposes DNA-recognition motifs for the estrogen receptor α (ERα), known as EREs. Under estrogen stimulation (red circle), ERα can then bind those sequences and impact target gene expression. Figure created with BioRender.com.

somes (H3K4me1/me2) (Lupien et al. 2008; Wang et al. 2009), with evidence that FOXA1 can also recruit the COMPASS complex, which is responsible for depositing H3K4 methylation marks (Jozwik et al. 2016). The ability of pioneer factors to promote chromatin variants warrants a careful assessment of their role in breast cancer development.

The role of pioneer factors in breast cancer has drawn a lot of interest due to their ability to change chromatin states to impact transcription factor binding at CREs. Genetic variants within genes encoding pioneer factors can be disruptive to their activity and contribute to breast cancer. For example, somatic mutations within the promoter of the *FOXA1* gene can influence expression of *FOXA1* and these genetic variants are enriched in luminal breast cancer (Rheinbay et al. 2017). Somatic mutations within the *FOXA1* gene can also give rise to a FOXA1 protein with an altered affinity for its consensus DNA recognition motif, resulting in FOXA1 binding to a new set of CREs

(Cowper-Sal lari et al. 2012; Robinson et al. 2013; Arruabarrena-Aristorena et al. 2020). This in turn can impact where ERα binds across the genome of breast cancer cells, thereby favoring an altered transcriptional response upon estrogen stimulation (Arruabarrena-Aristorena et al. 2020). Additionally, overexpression of pioneer factors can increase their activity on the chromatin in breast cancer. For instance, high FOXA1 levels trigger genome-wide alterations in accessible chromatin that substantially increase the number of sites within the FOXA1 cistrome and enhance the activity of ERα in luminal breast cancers (Fu et al. 2016). Elevated levels of FOXA1 can also enable breast tumor progression, metastasis, and endocrine therapy resistance (Hurtado et al. 2011b; Fu et al. 2016; He et al. 2021). While FOXA1 is the prototypical pioneer factor in breast cancer, other factors including PBX1 (pre-B-cell leukemia homeobox 1) (Magnani et al. 2011a) and GATA3 (Takaku et al. 2020) have pioneering capacity and are also essential for ERα-associated transcriptional activity in breast cancer. Taken together, alterations in the activity of pioneer factors are sufficient to generate cancer-associated chromatin variants that can shift regulatory dependencies in breast cancer.

CHROMATIN VARIANTS OFFER AN ALTERNATIVE TO GENETIC VARIANTS FOR FINDING ACTIONABLE TARGETS

The future of cancer care lies in achieving the goal of precision medicine for treating patients. This requires a thorough investigation of the unique tumoral phenotype of each patient that identifies therapeutic options to guide treatment. Current approaches to guide clinical decisions of hard-to-treat tumors, such as high-risk breast tumors, primarily use genetic profiling for identifying genetic variants including raSNPs or somatic mutations. However, the success of large-scale genome sequencing efforts to yield actionable targets is limited (Rheinbay et al. 2020). In breast cancers, only 46 associations between drugs and genetic variants across 23 actionable genes have been identified to date (www.oncokb.org/actionableGenes). This current state of precision-based medicine reflects

a crucial need to consider complementary approaches to find the genetic basis of breast cancer beyond genetic variants. Expanding investigations for actionable targets to DNA elements revealed by chromatin variants specific to breast cancer can help fill the gap and improve our ability to match cancer patients to curative treatment options.

Targeting chromatin modifier enzymes that deposit or remove chemical modifications on chromatin is a promising avenue of breast cancer treatment, often referred to as epigenetic therapy, that takes advantage of chromatin variants (Arrowsmith et al. 2012; Wu et al. 2019; Bates 2020). Chromatin modifier enzymes, such as DNA methyltransferases (DNMTs), histone lysine methyltransferases (KMTs), histone lysine acetylases (KATs), histone lysine deacetylases (KDACs), or bromodomain and extraterminal motif proteins (BETs) involved in the establishment and maintenance of DNA and histone post-translational modifications, are primary targets for altering chromatin variants (Fig. 5A; Arrowsmith et al. 2012; Schapira and Arrowsmith 2016; Wu et al. 2019, 2021a; Jin et al. 2021). For example, since the promoters of tumor suppressor genes in breast cancer tend to be hypermethylated, many studies have focused on determining how DNMT inhibition impacts chromatin state and resulting breast cancer cell state (Chik and Szyf 2011). Oftentimes, multiple chromatin modifier enzymes are involved in maintaining the chromatin state over a DNA element. DNA hypomethylation resulting from DNMT inhibition can be accompanied by KMT-dependent compensation reflected in the deposition of repressive histone modifications such as H3K27me3 and H3K9me3 over DNA hypomethylated regions (Hon et al. 2012). The resulting chromatin variants define a vulnerability for additional intervention based on KMT inhibition (Song et al. 2016; Deblois et al. 2020). Until now, the use of chromatin modifier-based therapeutics alone has not successfully translated from the preclinical to clinical setting (Brown et al. 2022). However, with the growing interest in chromatin-based therapies and improving technologies to interrogate chromatin variants, there is promise in its clinical utility. This may especially apply to combinatorial

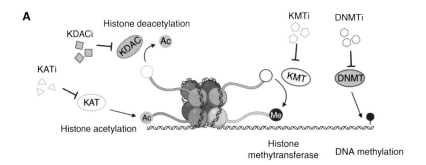

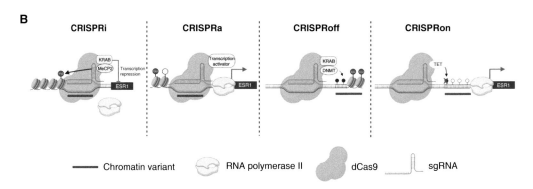

Figure 5. Perturbing chromatin variants can provide a means for delivering precision medicine. (*A*) Chromatin modifier enzymes deposit modifications on histones or DNA. DNA methyltransferases (DNMTs) methylate DNA. Histone lysine methyltransferases (KMTs) transfer one, two, or three methyl groups to lysine or arginine residues on histones. Histone lysine deacetylases (KDACs) remove acetyl groups from lysine residues of histones, whereas histone lysine acetylases (KATs) add acetyl groups. Chromatin modifiers can be targets of chemical inhibitors to alter chromatin states, defining epigenetic therapy. (*B*) Chromatin variants can be directly manipulated by using genome hacking approaches, such as CRISPRi, CRISPRa, CRISPRoff, or CRISPRon. Catalytically dead Cas9 nuclease (dCas9) can be used to modify a target site such as a putative chromatin variant on the genome specified by the sgRNA. CRISPRi involves the fusion of dCas9 to a repressor domain such as MeCP2 and KRAB. This changes the active chromatin state to an inactive chromatin state at target regions, which can serve to repress gene expression. CRISPRa consists of dCas9 fused to a transcriptional activator that can activate *cis*-regulatory elements of choice to increase target gene expression. CRISPRoff and CRISPRon introduce heritable changes to the chromatin state at target regions. The former consists of a dCas9-KRAB-DNMT construct that methylates DNA while the latter uses dCas9-TET to remove methyl groups on DNA. (CRISPRi) CRISPR interference, (KRAB) Krüppel-associated box, (sgRNA) short-guide RNA, (MeCP2) methyl-CpG-binding protein 2. Figure created with BioRender.com.

regimens. For instance, initial findings show that metastatic TNBC patients respond to combined BET inhibition and poly-ADP-ribose polymerase (PARP) targeting therapy, although phase 2 of the clinical trial is ongoing (Aftimos et al. 2021). Altogether, inhibiting the activity of chromatin-modifying enzymes presents as a valid therapeutic approach to guide precision medicine from chromatin variants in breast cancer.

Inhibition of pioneer factors stands as an attractive approach to breast cancer through their ability to alter chromatin states. However, the development of small molecule inhibitors that target pioneer factors within breast cancer are limited. The major obstacle for designing small molecule inhibitors is that transcription factors including pioneer factors such as FOXA1 lack ligand-binding pockets (Berg 2008) and the lack of structural information hinders the development of rationally designed inhibitors. Despite this limitation, Shen et al. (2021) developed a potent lead compound, known as T417, that can interfere with the interactions between the pioneer factor PBX1 and its target DNA sequence (Shen et al. 2021). The T417 compound reduces PBX1 activity and shows low toxicity within mouse models of ovarian cancer, which is of importance for clinical utility (Shen et al. 2021). Since PBX1 is reported to enable chromatin variants that favor ERα binding to the chromatin in breast cancer cells (Magnani et al. 2013, 2015), the T417 compound may reveal benefits in treating luminal breast cancer subtypes. Although approaches targeting pioneer factors are preliminary, successes such as that with PBX1 set the stage for future investigations in breast cancer. Future work may require the therapeutic inhibition of upstream regulatory enzymes that stabilize key pioneer factors in cancer as an indirect way to inhibit pioneer factor function.

GENOME HACKING AND ENGINEERING WITH CHROMATIN MODIFIER ENZYMES TO TARGET CHROMATIN VARIANTS

Advances in genome engineering defines another promising option to target chromatin variants. Specifically, RNA-guided nucleases derived from clustered regularly interspersed palindromic repeat (CRISPR) microbial systems (e.g., Cas9) has served to test the functional contribution of CREs bound by FOXA1 in breast cancer cells by editing their DNA sequence (Fei et al. 2019). In parallel, Lopes et al. (2021) assessed the functional contribution of approximately 15,000 CREs bound by ERα breast cancer cells using the CRISPR interference (CRISPRi) tool. In contrast to CRISPR/Cas9 tools, CRISPRi does not edit the genome. Instead, CRISPRi is a genome hacking tool that alters the chromatin state (Larson et al. 2013). CRISPRi consists of a CRISPR/deadCas9 (dCas9) fused to a KRAB and a short-guide RNA (sgRNA), which recognizes the DNA sequence of the CRE of interest (Fig. 5B; Larson et al. 2013; Alerasool et al. 2020). Hence, CRISPRi interferes with ERα binding to DNA by changing the chromatin state at its CREs (Lopes et al. 2021). This work revealed the role of a subset of ERα CREs toward breast cancer cell proliferation (Lopes et al. 2021). Recently, Slaven et al. (2022), adopted CRISPRi tools to perturb CREs in breast cancer

cells and assess their role toward proliferation in treatment of naive cells and endocrine therapy conditions. They uncovered CREs associated with NF-κB signaling as critical to driving cell survival, endocrine resistance, and disease progression (Slaven et al. 2022). These studies show the capability of using CRISPRi to alter chromatin variants that promote cancer progression.

In addition to CRISPRi, genome hacking technologies include CRISPR activation (CRISPRa), CRISPRon, and CRISPRoff (Fig. 5B). Rather than a CRISPR/dCas9-KRAB construct, CRISPRa consists of a transcriptional activator for the CRE defined by the sgRNA (Dong et al. 2018). In contrast with CRISPRi and CRISPRa, CRISPRoff and CRISPRon introduce heritable changes that are reminiscent of chromatin modifiers by depositing chromatin chemical modifications at target DNA sites (Nuñez et al. 2021). CRISPRoff is composed of a CRISPR/dCas9 fusion with KRAB and DNMT3A modifier enzymes that allows for DNA methylation near the target site defined by sgRNA (Nuñez et al. 2021). In opposition to CRISPRoff, CRISPRon reverses DNA methylation at target sites due to a DNA demethylase (Nuñez et al. 2021). Nuñez et al. (2021) transiently expressed the CRISPRoff construct in a TNBC breast cancer cell line with sgRNAs targeting the promoter and four enhancers of the *PVT1* long noncoding RNA gene. They validated the ability for CRISPRoff to yield expected decreased expression of *PVT1* highlighting the relevance of CRISPRoff dissecting function of chromatin variants. The opportunity to target chromatin variants with genomic hacking techniques, such as those based on CRISPR/dCas9, paves the way for the development of genome engineering methods to guide precision medicine in breast cancers targeting cancer drivers regardless of their mutational status.

CONCLUDING REMARKS

Understanding the genetic basis of breast cancer has been of focus in the field as a way of devising effective treatment options for patients. Because traits can be assigned to unique genetic variants when compared across individuals, for decades, the approach of choice to understand the genetic basis of breast cancer has relied on finding genetic variants specific to each breast cancer subtype. However, efforts to date have failed to capture the complete picture. Because heterogeneity in traits can arise from a common genome, such as across tissues or cell states in an individual, and the DNA is packaged into chromatin in living cells, attention has recently been placed on chromatin variants to identify the genetic basis of traits. This new era of studying the genome from chromatin variants is meant to complement efforts to map genetic variants in cancer to find the genetic drivers of oncogenesis.

Whereas distinct chromatin variants can readily be assigned to unique phenotypes, there is still a debate as to whether chromatin variants are causal or a consequence of phenotype. Under one hypothesis, chromatin variants could directly dictate the cell-state-specific transcriptional patterns. Alternatively, chromatin variants may simply reflect the transcriptional mechanisms accounting for cell-state identity. A recent study by Wang et al. (2022) reported a dependency for some chromatin states, namely, the H3K4me3- and H3K27ac-defined state typical of active promoters, toward transcription. However, this study also provided indications for chromatin accessibility and other chromatin modifications, namely, the enhancer-associated H3K4me1 state, being independent of transcription (Wang et al. 2022). Future work should be mindful of these contrasting working models and consider genome hacking-based assessments to help clarify the causal versus consequential role of chromatin variants on phenotype.

ACKNOWLEDGMENTS

The authors apologize to colleagues whose work was not cited in this document due to space restrictions. This work is supported by the Canadian Institute of Health Research (CIHR) (FRN-153234 and 168933 to M.L.), the Canadian Epigenetics, Environment, and Health Research Consortium (CEEHRC) (FRN-158225 to M.L.), the Ontario Institute for Cancer Research (OICR) through funding provided by the Government of Ontario (Investigator Award to M.L.), the Canadian Cancer Society (706373 to M.L.), and the Princess Margaret Cancer Foundation. J.S.C. acknowledges support from the University of Cambridge, Cancer Research UK core funding (grants A20411, A31344, A29580, and DRCPGM\100088) and Hutchison Whampoa Ltd.

REFERENCES

Aftimos P, Oliveira M, Punie K, Boni V, Hamilton E, Gucalp A, Shah PD, Mina L, Sharma P, Bauman L, et al. 2021. PS11-10: a phase 1b/2 study of the BET inhibitor ZEN003694 in combination with talazoparib for treatment of patients with TNBC without gBRCA1/2 mutations. *Age (Omaha)* **56:** 28–74.

Akhtar-Zaidi B, Cowper-Sal Lari R, Corradin O, Saiakhova A, Bartels CF, Balasubramanian D, Myeroff L, Lutterbaugh J, Jarrar A, Kalady MF, et al. 2012. Epigenomic enhancer profiling defines a signature of colon cancer. *Science* **336:** 736–739. doi:10.1126/science.1217277

Alerasool N, Segal D, Lee H, Taipale M. 2020. An efficient KRAB domain for CRISPRi applications in human cells. *Nat Methods* **17:** 1093–1096. doi:10.1038/s41592-020-0966-x

Arrowsmith CH, Bountra C, Fish PV, Lee K, Schapira M. 2012. Epigenetic protein families: a new frontier for drug discovery. *Nat Rev Drug Discov* **11:** 384–400. doi:10.1038/nrd3674

Arruabarrena-Aristorena A, Maag JLV, Kittane S, Cai Y, Karthaus WR, Ladewig E, Park J, Kannan S, Ferrando L, Cocco E, et al. 2020. FOXA1 mutations reveal distinct chromatin profiles and influence therapeutic response in breast cancer. *Cancer Cell* **38:** 534–550.e9. doi:10.1016/j.ccell.2020.08.003

Bailey SD, Desai K, Kron KJ, Mazrooei P, Sinnott-Armstrong NA, Treloar AE, Dowar M, Thu KL, Cescon DW, Silvester J, et al. 2016. Noncoding somatic and inherited single-nucleotide variants converge to promote ESR1 expression in breast cancer. *Nat Genet* **48:** 1260–1266. doi:10.1038/ng.3650

Barski A, Cuddapah S, Cui K, Roh T-Y, Schones DE, Wang Z, Wei G, Chepelev I, Zhao K. 2007. High-resolution profiling of histone methylations in the human genome. *Cell* **129:** 823–837. doi:10.1016/j.cell.2007.05.009

Bartosovic M, Kabbe M, Castelo-Branco G. 2021. Single-cell CUT&Tag profiles histone modifications and transcription factors in complex tissues. *Nat Biotechnol* **39:** 825–835. doi:10.1038/s41587-021-00869-9

Bates SE. 2020. Epigenetic therapies for cancer. *N Engl J Med* **383:** 650–663. doi:10.1056/NEJMra1805035

Baylin SB. 2005. DNA methylation and gene silencing in cancer. *Nat Clin Pract Oncol* **2:** S4–S11. doi:10.1038/ncponc0354

Beesley J, Sivakumaran H, Marjaneh MM, Lima LG, Hillman KM, Kaufmann S, Tuano N, Hussein N, Ham S, Mukhopadhyay P, et al. 2020. Chromatin interactome mapping at 139 independent breast cancer risk signals. *Genome Biol* **21:** 8. doi:10.1186/s13059-019-1877-y

Berg T. 2008. Inhibition of transcription factors with small organic molecules. *Curr Opin Chem Biol* **12:** 464–471. doi:10.1016/j.cbpa.2008.07.023

Bernardo GM, Lozada KL, Miedler JD, Harburg G, Hewitt SC, Mosley JD, Godwin AK, Korach KS, Visvader JE, Kaestner KH, et al. 2010. FOXA1 is an essential determinant of ERα expression and mammary ductal morphogenesis. *Development* **137:** 2045–2054. doi:10.1242/dev.043299

Bernstein BE, Mikkelsen TS, Xie X, Kamal M, Huebert DJ, Cuff J, Fry B, Meissner A, Wernig M, Plath K, et al. 2006. A bivalent chromatin structure marks key developmental genes in embryonic stem cells. *Cell* **125:** 315–326. doi:10.1016/j.cell.2006.02.041

Beroukhim R, Mermel CH, Porter D, Wei G, Raychaudhuri S, Donovan J, Barretina J, Boehm JS, Dobson J, Urashima M, et al. 2010. The landscape of somatic copy-number alteration across human cancers. *Nature* **463:** 899–905. doi:10.1038/nature08822

Bowman GD, Poirier MG. 2015. Post-translational modifications of histones that influence nucleosome dynamics. *Chem Rev* **115:** 2274–2295. doi:10.1021/cr500350x

Brown LJ, Achinger-Kawecka J, Portman N, Clark S, Stirzaker C, Lim E. 2022. Epigenetic therapies and biomarkers in breast cancer. *Cancers (Basel)* **14:** 47.

Buenrostro JD, Wu B, Chang HY, Greenleaf WJ. 2015. ATAC-seq: A method for assaying chromatin accessibility genome-wide. *Curr Protoc Mol Biol* **109:** 21.29.1–21.29.9. doi:10.1002/0471142727.mb2129s109

Burstein MD, Tsimelzon A, Poage GM, Covington KR, Contreras A, Fuqua SAW, Savage MI, Osborne CK, Hilsenbeck SG, Chang JC, et al. 2015. Comprehensive genomic analysis identifies novel subtypes and targets of triple-negative breast cancer. *Clin Cancer Res* **21:** 1688–1698. doi:10.1158/1078-0432.CCR-14-0432

Carroll JS, Liu XS, Brodsky AS, Li W, Meyer CA, Szary AJ, Eeckhoute J, Shao W, Hestermann EV, Geistlinger TR, et al. 2005. Chromosome-wide mapping of estrogen receptor binding reveals long-range regulation requiring the forkhead protein FoxA1. *Cell* **122:** 33–43. doi:10.1016/j.cell.2005.05.008

Carter B, Zhao K. 2021. The epigenetic basis of cellular heterogeneity. *Nat Rev Genet* **22:** 235–250. doi:10.1038/s41576-020-00300-0

Castro-Mondragon JA, Riudavets-Puig R, Rauluseviciute I, Berhanu Lemma R, Turchi L, Blanc-Mathieu R, Lucas J, Boddie P, Khan A, Manosalva Pérez N, et al. 2022. JASPAR 2022: the 9th release of the open-access database of transcription factor binding profiles. *Nucleic Acids Res* **50:** D165–D173. doi:10.1093/nar/gkab1113

Cerami E, Gao J, Dogrusoz U, Gross BE, Sumer SO, Aksoy BA, Jacobsen A, Byrne CJ, Heuer ML, Larsson E, et al. 2012. The cBio cancer genomics portal: an open platform for exploring multidimensional cancer genomics data. *Cancer Discov* **2:** 401–404. doi:10.1158/2159-8290.CD-12-0095

Chaudhary S, Madhu Krishna B, Mishra SK. 2017. A novel FOXA1/ESR1 interacting pathway: A study of Oncomine breast cancer microarrays. *Oncol Lett* **14:** 1247–1264. doi:10.3892/ol.2017.6329

Chen H, Li C, Peng X, Zhou Z, Weinstein JN, The Cancer Genome Atlas Research Network, Caesar-Johnson SJ,

Demchok JA, Felau I, Kasapi M, et al. 2018. A pan-cancer analysis of enhancer expression in nearly 9000 patient samples. *Cell* **173:** 386–399.e12. doi:10.1016/j.cell.2018.03.027

Chiappinelli KB, Strissel PL, Desrichard A, Li H, Henke C, Akman B, Hein A, Rote NS, Cope LM, Snyder A, et al. 2015. Inhibiting DNA methylation causes an interferon response in cancer via dsRNA including endogenous retroviruses. *Cell* **162:** 974–986. doi:10.1016/j.cell.2015.07.011

Chik F, Szyf M. 2011. Effects of specific DNMT gene depletion on cancer cell transformation and breast cancer cell invasion; toward selective DNMT inhibitors. *Carcinogenesis* **32:** 224–232. doi:10.1093/carcin/bgq221

Chronis C, Fiziev P, Papp B, Butz S, Bonora G, Sabri S, Ernst J, Plath K. 2017. Cooperative binding of transcription factors orchestrates reprogramming. *Cell* **168:** 442–459. e20. doi:10.1016/j.cell.2016.12.016

Corces MR, Granja JM, Shams S, Louie BH, Seoane JA, Zhou W, Silva TC, Groeneveld C, Wong CK, Cho SW, et al. 2018. The chromatin accessibility landscape of primary human cancers. *Science* **362:** eaav1898. doi:10.1126/science.aav1898

Cowper-Sal lari R, Zhang X, Wright JB, Bailey SD, Cole MD, Eeckhoute J, Moore JH, Lupien M. 2012. Breast cancer risk-associated SNPs modulate the affinity of chromatin for FOXA1 and alter gene expression. *Nat Genet* **44:** 1191–1198. doi:10.1038/ng.2416

Curtis C, Shah SP, Chin SF, Turashvili G, Rueda OM, Dunning MJ, Speed D, Lynch AG, Samarajiwa S, Yuan Y, et al. 2012. The genomic and transcriptomic architecture of 2,000 breast tumours reveals novel subgroups. *Nature* **486:** 346–352. doi:10.1038/nature10983

Deblois G, Madani Tonekaboni SA, Grillo G, Martinez C, Kao YI, Tai F, Ettayebi I, Fortier AM, Savage P, Fedor AN, et al. 2020. Epigenetic switch–induced viral mimicry evasion in chemotherapy resistant breast cancer. *Cancer Discov* **10:** 1312–1329. doi:10.1158/2159-8290.CD-19-1493

Dent R, Trudeau M, Pritchard KI, Hanna WM, Kahn HK, Sawka CA, Lickley LA, Rawlinson E, Sun P, Narod SA. 2007. Triple-negative breast cancer: clinical features and patterns of recurrence. *Clin Cancer Res* **13:** 4429–4434. doi:10.1158/1078-0432.CCR-06-3045

Dentro SC, Leshchiner I, Haase K, Tarabichi M, Wintersinger J, Deshwar AG, Yu K, Rubanova Y, Macintyre G, Demeulemeester J, et al. 2021. Characterizing genetic intra-tumor heterogeneity across 2,658 human cancer genomes. *Cell* **184:** 2239–2254.e39. doi:10.1016/j.cell.2021.03.009

do Nascimento RG, Otoni KM. 2020. Histological and molecular classification of breast cancer: what do we know? *Mastology* **30:** e20200024. doi:10.29289/25945394202020200024. https://www.mastology.org/wp-content/uploads/2020/09/MAS_2020024_AOP.pdf

Dong C, Fontana J, Patel A, Carothers JM, Zalatan JG. 2018. Synthetic CRISPR-Cas gene activators for transcriptional reprogramming in bacteria. *Nat Commun* **9:** 2489. doi:10.1038/s41467-018-04901-6

Eeckhoute J, Lupien M, Meyer CA, Verzi MP, Shivdasani RA, Liu XS, Brown M. 2009. Cell-type selective chromatin remodeling defines the active subset of FOXA1-bound

enhancers. *Genome Res* **19**: 372–380. doi:10.1101/gr
.084582.108

Ernst J, Kellis M. 2010. Discovery and characterization of
chromatin states for systematic annotation of the human
genome. *Nat Biotechnol* **28**: 817–825. doi:10.1038/nbt
.1662

Ernst J, Kheradpour P, Mikkelsen TS, Shoresh N, Ward LD,
Epstein CB, Zhang X, Wang L, Issner R, Coyne M, et al.
2011. Mapping and analysis of chromatin state dynamics
in nine human cell types. *Nature* **473**: 43–49. doi:10.1038/
nature09906

Fachal L, Aschard H, Beesley J, Barnes DR, Allen J, Kar S,
Pooley KA, Dennis J, Michailidou K, Turman C, et al.
2020. Fine-mapping of 150 breast cancer risk regions
identifies 191 likely target genes. *Nat Genet* **52**: 56–73.
doi:10.1038/s41588-019-0537-1

Fei T, Li W, Peng J, Xiao T, Chen CH, Wu A, Huang J, Zang
C, Liu XS, Brown M. 2019. Deciphering essential cistromes using genome-wide CRISPR screens. *Proc Natl
Acad Sci* **116**: 25186–25195. doi:10.1073/pnas.1908
155116

Fletcher MNC, Castro MAA, Wang X, de Santiago I, O'Reilly
M, Chin SF, Rueda OM, Caldas C, Ponder BAJ, Markowetz F, et al. 2013. Master regulators of FGFR2 signalling
and breast cancer risk. *Nat Commun* **4**: 2464. doi:10.1038/
ncomms3464

Fu X, Jeselsohn R, Pereira R, Hollingsworth EF, Creighton
CJ, Li F, Shea M, Nardone A, Angelis CD, Heiser LM, et al.
2016. FOXA1 overexpression mediates endocrine resistance by altering the ER transcriptome and IL-8 expression in ER-positive breast cancer. *Proc Natl Acad Sci* **113**:
E6600–E6609.

Gasparini G. 2012. *Prognostic variables in node-negative and
node-positive breast cancer.* Springer Science+Business
Media, Berlin.

Gerashchenko TS, Zavyalova MV, Denisov EV, Krakhmal
NV, Pautova DN, Litviakov NV, Vtorushin SV, Cher-
dyntseva NV, Perelmuter VM. 2017. Intratumoral morphological heterogeneity of breast cancer as an indicator
of the metastatic potential and tumor chemosensitivity.
Acta Naturae **9**: 56–67. doi:10.32607/20758251-2017-9-
1-56-67

Gerlinger M, Rowan AJ, Horswell S, Math M, Larkin J, Endesfelder D, Gronroos E, Martinez P, Matthews N, Stewart A, et al. 2012. Intratumor heterogeneity and branched
evolution revealed by multiregion sequencing. *N Engl J
Med* **366**: 883–892. doi:10.1056/NEJMoa1113205

Gerratana L, Basile D, Buono G, De Placido S, Giuliano M,
Minichillo S, Coinu A, Martorana F, De Santo I, Del
Mastro L, et al. 2018. Androgen receptor in triple negative
breast cancer: a potential target for the targetless subtype.
Cancer Treat Rev **68**: 102–110. doi:10.1016/j.ctrv.2018.06
.005

Ghamrasni SE, Quevedo R, Hawley J, Mazrooei P, Hanna Y,
Cirlan I, Zhu H, Bruce J, Oldfield LE, Yang SYC, et al.
2022. Mutations in non-coding *cis*-regulatory elements
reveal cancer driver cistromes in luminal breast cancer.
Mol Cancer Res **20**: 102–113. doi:10.1158/1541-7786
.MCR-21-0471

Ghoussaini M, Edwards SL, Michailidou K, Nord S, Cowper-
Sal Lari R, Desai K, Kar S, Hillman KM, Kaufmann S,
Glubb DM, et al. 2014. Evidence that breast cancer risk at

the 2q35 locus is mediated through IGFBP5 regulation.
Nat Commun **5**: 4999. doi:10.1038/ncomms5999

Green AR, Aleskandarany MA, Agarwal D, Elsheikh S, Nolan CC, Diez-Rodriguez M, Macmillan RD, Ball GR, Caldas C, Madhusudan S, et al. 2016. MYC functions are
specific in biological subtypes of breast cancer and confers
resistance to endocrine therapy in luminal tumours. *Br J
Cancer* **114**: 917–928. doi:10.1038/bjc.2016.46

Grosselin K, Durand A, Marsolier J, Poitou A, Marangoni E,
Nemati F, Dahmani A, Lameiras S, Reyal F, Frenoy O, et
al. 2019. High-throughput single-cell ChIP-seq identifies
heterogeneity of chromatin states in breast cancer. *Nat
Genet* **51**: 1060–1066. doi:10.1038/s41588-019-0424-9

Hanahan D. 2022. Hallmarks of cancer: new dimensions.
Cancer Discov **12**: 31–46. doi:10.1158/2159-8290.CD-
21-1059

Hashim D, Boffetta P, La Vecchia C, Rota M, Bertuccio P,
Malvezzi M, Negri E. 2016. The global decrease in cancer
mortality: trends and disparities. *Ann Oncol* **27**: 926–933.
doi:10.1093/annonc/mdw027

He Y, Wang L, Wei T, Xiao YT, Sheng H, Su H, Hollern DP,
Zhang X, Ma J, Wen S, et al. 2021. FOXA1 overexpression
suppresses interferon signaling and immune response in
cancer. *J Clin Invest* **131**: e147025. doi:10.1172/JCI147025

Heintzman ND, Stuart RK, Hon G, Fu Y, Ching CW, Hawkins RD, Barrera LO, Van Calcar S, Qu C, Ching KA, et al.
2007. Distinct and predictive chromatin signatures of
transcriptional promoters and enhancers in the human
genome. *Nat Genet* **39**: 311–318. doi:10.1038/ng1966

Heintzman ND, Hon GC, Hawkins RD, Kheradpour P, Stark
A, Harp LF, Ye Z, Lee LK, Stuart RK, Ching CW, et al.
2009. Histone modifications at human enhancers reflect
global cell-type-specific gene expression. *Nature* **459**:
108–112. doi:10.1038/nature07829

Henikoff S, Henikoff JG, Kaya-Okur HS, Ahmad K. 2020.
Efficient chromatin accessibility mapping in situ by nu-
cleosome-tethered tagmentation. *eLife* **9**: e63274. doi:10
.7554/eLife.63274

Hilsenbeck SG, Ravdin PM, de Moor CA, Chamness GC,
Osborne CK, Clark GM. 1998. Time-dependence of hazard ratios for prognostic factors in primary breast cancer.
Breast Cancer Res Treat **52**: 227–237. doi:10.1023/A
:1006133418245

Hoffman MM, Buske OJ, Wang J, Weng Z, Bilmes JA, Noble
WS. 2012. Unsupervised pattern discovery in human
chromatin structure through genomic segmentation.
Nat Methods **9**: 473–476. doi:10.1038/nmeth.1937

Hon GC, Hawkins RD, Caballero OL, Lo C, Lister R, Peliz-
zola M, Valsesia A, Ye Z, Kuan S, Edsall LE, et al. 2012.
Global DNA hypomethylation coupled to repressive
chromatin domain formation and gene silencing in breast
cancer. *Genome Res* **22**: 246–258. doi:10.1101/gr.125872
.111

Hurtado A, Holmes KA, Ross-Innes CS, Schmidt D, Carroll
JS. 2011a. FOXA1 is a key determinant of estrogen receptor function and endocrine response. *Nat Genet* **43**: 27–
33. doi:10.1038/ng.730

Hurtado A, Holmes KA, Ross-Innes CS, Schmidt D, Carroll
JS. 2011b. FOXA1 is a critical determinant of estrogen
receptor function and endocrine response. *Nat Genet*
43: 27–33. doi:10.1038/ng.730

Iwafuchi-Doi M, Donahue G, Kakumanu A, Watts JA, Mahony S, Pugh BF, Lee D, Kaestner KH, Zaret KS. 2016. The pioneer transcription factor FoxA maintains an accessible nucleosome configuration at enhancers for tissue-specific gene activation. *Mol Cell* **62:** 79–91. doi:10.1016/j.molcel.2016.03.001

Jiang YZ, Ma D, Suo C, Shi J, Xue M, Hu X, Xiao Y, Yu KD, Liu YR, Yu Y, et al. 2019. Genomic and transcriptomic landscape of triple-negative breast cancers: subtypes and treatment strategies. *Cancer Cell* **35:** 428–440.e5. doi:10.1016/j.ccell.2019.02.001

Jin N, George TL, Otterson GA, Verschraegen C, Wen H, Carbone D, Herman J, Bertino EM, He K. 2021. Advances in epigenetic therapeutics with focus on solid tumors. *Clin Epigenetics* **13:** 83. doi:10.1186/s13148-021-01069-7

Jozwik KM, Carroll JS. 2012. Pioneer factors in hormone-dependent cancers. *Nat Rev Cancer* **12:** 381–385. doi:10.1038/nrc3263

Jozwik KM, Chernukhin I, Serandour AA, Nagarajan S, Carroll JS. 2016. FOXA1 directs H3K4 monomethylation at enhancers via recruitment of the methyltransferase MLL3. *Cell Rep* **17:** 2715–2723. doi:10.1016/j.celrep.2016.11.028

Kaya-Okur HS, Wu SJ, Codomo CA, Pledger ES, Bryson TD, Henikoff JG, Ahmad K, Henikoff S. 2019. CUT&tag for efficient epigenomic profiling of small samples and single cells. *Nat Commun* **10:** 1930. doi:10.1038/s41467-019-09982-5

Keshavarzian T, Lupien M. 2022. ecDNAs personify cancer gangsters. *Mol Cell* **82:** 500–502. doi:10.1016/j.molcel.2022.01.003

Khaled WT, Choon Lee S, Stingl J, Chen X, Raza Ali H, Rueda OM, Hadi F, Wang J, Yu Y, Chin SF, et al. 2015. BCL11A is a triple-negative breast cancer gene with critical functions in stem and progenitor cells. *Nat Commun* **6:** 5987. doi:10.1038/ncomms6987

Kim K, Jang K, Yang W, Choi EY, Park SM, Bae M, Kim YJ, Choi JK. 2016. Chromatin structure–based prediction of recurrent noncoding mutations in cancer. *Nat Genet* **48:** 1321–1326. doi:10.1038/ng.3682

Kim H, Nguyen NP, Turner K, Wu S, Gujar AD, Luebeck J, Liu J, Deshpande V, Rajkumar U, Namburi S, et al. 2020. Extrachromosomal DNA is associated with oncogene amplification and poor outcome across multiple cancers. *Nat Genet* **52:** 891–897. doi:10.1038/s41588-020-0678-2

Klemm SL, Shipony Z, Greenleaf WJ. 2019. Chromatin accessibility and the regulatory epigenome. *Nat Rev Genet* **20:** 207–220. doi:10.1038/s41576-018-0089-8

Koren S, Bentires-Alj M. 2015. Breast tumor heterogeneity: source of fitness, hurdle for therapy. *Mol Cell* **60:** 537–546. doi:10.1016/j.molcel.2015.10.031

Kreike B, van Kouwenhove M, Horlings H, Weigelt B, Peterse H, Bartelink H, van de Vijver MJ. 2007. Gene expression profiling and histopathological characterization of triple-negative/basal-like breast carcinomas. *Breast Cancer Res* **9:** R65. doi:10.1186/bcr1771

Kundaje A, Meuleman W, Ernst J, Bilenky M, Yen A, Heravi-Moussavi A, Kheradpour P, Zhang Z, Wang J, Ziller MJ, et al. 2015. Integrative analysis of 111 reference human epigenomes. *Nature* **518:** 317–330. doi:10.1038/nature14248

Kusi M, Zand M, Lin L-L, Chen M, Lopez A, Lin C-L, Wang C-M, Lucio ND, Kirma NB, Ruan J. 2022. 2-Hydroxyglu-

tarate destabilizes chromatin regulatory landscape and lineage fidelity to promote cellular heterogeneity. *Cell Rep* **38:** 110220. doi:10.1016/j.celrep.2021.110220

Lafourcade A, His M, Baglietto L, Boutron-Ruault MC, Dossus L, Rondeau V. 2018. Factors associated with breast cancer recurrences or mortality and dynamic prediction of death using history of cancer recurrences: the French E3N cohort. *BMC Cancer* **18:** 171. doi:10.1186/s12885-018-4076-4

Larson MH, Gilbert LA, Wang X, Lim WA, Weissman JS, Qi LS. 2013. CRISPR interference (CRISPRi) for sequence-specific control of gene expression. *Nat Protoc* **8:** 2180–2196. doi:10.1038/nprot.2013.132

Lehmann BD, Bauer JA, Chen X, Sanders ME, Chakravarthy AB, Shyr Y, Pietenpol JA. 2011. Identification of human triple-negative breast cancer subtypes and preclinical models for selection of targeted therapies. *J Clin Invest* **121:** 2750–2767. doi:10.1172/JCI45014

Lehmann BD, Colaprico A, Silva TC, Chen J, An H, Ban Y, Huang H, Wang L, James JL, Balko JM, et al. 2021. Multi-omics analysis identifies therapeutic vulnerabilities in triple-negative breast cancer subtypes. *Nat Commun* **12:** 6276. doi:10.1038/s41467-021-26502-6

Lei S, Zheng R, Zhang S, Wang S, Chen R, Sun K, Zeng H, Zhou J, Wei W. 2021. Global patterns of breast cancer incidence and mortality: a population-based cancer registry data analysis from 2000 to 2020. *Cancer Commun* **41:** 1183–1194. doi:10.1002/cac2.12207

Lopes R, Sprouffske K, Sheng C, Uijttewaal ECH, Wesdorp AE, Dahinden J, Wengert S, Diaz-Miyar J, Yildiz U, Bleu M, et al. 2021. Systematic dissection of transcriptional regulatory networks by genome-scale and single-cell CRISPR screens. *Sci Adv* **7:** eabf5733. doi:10.1126/sciadv.abf5733

Luger K, Mäder AW, Richmond RK, Sargent DF, Richmond TJ. 1997. Crystal structure of the nucleosome core particle at 2.8 Å resolution. *Nature* **389:** 251–260. doi:10.1038/38444

Luger K, Dechassa ML, Tremethick DJ. 2012. New insights into nucleosome and chromatin structure: an ordered state or a disordered affair? *Nat Rev Mol Cell Biol* **13:** 436–447. doi:10.1038/nrm3382

Lupien M, Eeckhoute J, Meyer CA, Wang Q, Zhang Y, Li W, Carroll JS, Liu XS, Brown M. 2008. Foxa1 translates epigenetic signatures into enhancer-driven lineage-specific transcription. *Cell* **132:** 958–970. doi:10.1016/j.cell.2008.01.018

Magnani L, Ballantyne EB, Zhang X, Lupien M. 2011a. PBX1 genomic pioneer function drives ERα signaling underlying progression in breast cancer. *PLoS Genet* **7:** e1002368. doi:10.1371/journal.pgen.1002368

Magnani L, Eeckhoute J, Lupien M. 2011b. Pioneer factors: directing transcriptional regulators within the chromatin environment. *Trends Genet* **27:** 465–474. doi:10.1016/j.tig.2011.07.002

Magnani L, Stoeck A, Zhang X, Lánczky A, Mirabella AC, Wang TL, Győrffy B, Lupien M. 2013. Genome-wide reprogramming of the chromatin landscape underlies endocrine therapy resistance in breast cancer. *Proc Natl Acad Sci* **110:** E1490–E1499. doi:10.1073/pnas.1219992110

Magnani L, Patten DK, Nguyen VTM, Hong SP, Steel JH, Patel N, Lombardo Y, Faronato M, Gomes AR, Woodley L, et al. 2015. The pioneer factor PBX1 is a novel driver of metastatic progression in ERα-positive breast cancer. *Oncotarget* **6:** 21878–21891. doi:10.18632/oncotarget.4243

Marsolier J, Prompsy P, Durand A, Lyne AM. 2021. H3k27me3 is a determinant of chemotolerance in triple-negative breast cancer. bioRxiv doi:10.1101/2021.01.04.423386

Mazrooei P, Kron KJ, Zhu Y, Zhou S, Grillo G, Mehdi T, Ahmed M, Severson TM, Guilhamon P, Armstrong NS, et al. 2019. Cistrome partitioning reveals convergence of somatic mutations and risk variants on master transcription regulators in primary prostate tumors. *Cancer Cell* **36:** 674–689.e6. doi:10.1016/j.ccell.2019.10.005

Meuleman W, Muratov A, Rynes E, Halow J, Lee K, Bates D, Diegel M, Dunn D, Neri F, Teodosiadis A, et al. 2020. Index and biological spectrum of human DNase I hypersensitive sites. *Nature* **584:** 244–251. doi:10.1038/s41586-020-2559-3

Michailidou K, Lindström S, Dennis J, Beesley J, Hui S, Kar S, Lemaçon A, Soucy P, Glubb D, Rostamianfar A, et al. 2017. Association analysis identifies 65 new breast cancer risk loci. *Nature* **551:** 92–94. doi:10.1038/nature24284

Navin N, Kendall J, Troge J, Andrews P, Rodgers L, McIndoo J, Cook K, Stepansky A, Levy D, Esposito D, et al. 2011. Tumour evolution inferred by single-cell sequencing. *Nature* **472:** 90–94. doi:10.1038/nature09807

Nguyen NT, Vendrell JA, Poulard C, Győrffy B, Goddard-Léon S, Bièche I, Corbo L, Le Romancer M, Bachelot T, Treilleux I, et al. 2014. A functional interplay between ZNF217 and estrogen receptor α exists in luminal breast cancers. *Mol Oncol* **8:** 1441–1457. doi:10.1016/j.molonc.2014.05.013

Nik-Zainal S, Davies H, Staaf J, Ramakrishna M, Glodzik D, Zou X, Martincorena I, Alexandrov LB, Martin S, Wedge DC, et al. 2016. Landscape of somatic mutations in 560 breast cancer whole-genome sequences. *Nature* **534:** 47–54. doi:10.1038/nature17676

Nuñez JK, Chen J, Pommier GC, Cogan JZ, Replogle JM, Adriaens C, Ramadoss GN, Shi Q, Hung KL, Samelson AJ, et al. 2021. Genome-wide programmable transcriptional memory by CRISPR-based epigenome editing. *Cell* **184:** 2503–2519.e17. doi:10.1016/j.cell.2021.03.025

Parker JS, Mullins M, Cheang MCU, Leung S, Voduc D, Vickery T, Davies S, Fauron C, He X, Hu Z, et al. 2009. Supervised risk predictor of breast cancer based on intrinsic subtypes. *J Clin Oncol* **27:** 1160–1167. doi:10.1200/JCO.2008.18.1370

Patten DK, Corleone G, Győrffy B, Perone Y, Slaven N, Barozzi I, Erdős E, Saiakhova A, Goddard K, Vingiani A, et al. 2018. Enhancer mapping uncovers phenotypic heterogeneity and evolution in patients with luminal breast cancer. *Nat Med* **24:** 1469–1480. doi:10.1038/s41591-018-0091-x

Pereira B, Chin SF, Rueda OM, Vollan HKM, Provenzano E, Bardwell HA, Pugh M, Jones L, Russell R, Sammut SJ, et al. 2016. The somatic mutation profiles of 2,433 breast cancers refine their genomic and transcriptomic landscapes. *Nat Commun* **7:** 11479. doi:10.1038/ncomms11479

Polyak K. 2011. Heterogeneity in breast cancer. *J Clin Invest* **121:** 3786–3788. doi:10.1172/JCI60534

Ramos J, Das J, Felty Q, Yoo C, Poppiti R, Murrell D, Foster PJ, Roy D. 2018. NRF1 motif sequence-enriched genes involved in ER/PR −ve HER2 +ve breast cancer signaling pathways. *Breast Cancer Res Treat* **172:** 469–485. doi:10.1007/s10549-018-4905-9

Rheinbay E, Parasuraman P, Grimsby J, Tiao G, Engreitz JM, Kim J, Lawrence MS, Taylor-Weiner A, Rodriguez-Cuevas S, Rosenberg M, et al. 2017. Recurrent and functional regulatory mutations in breast cancer. *Nature* **547:** 55–60. doi:10.1038/nature22992

Rheinbay E, Nielsen MM, Abascal F, Wala JA, Shapira O, Tiao G, Hornshøj H, Hess JM, Juul RI, Lin Z, et al. 2020. Analyses of non-coding somatic drivers in 2,658 cancer whole genomes. *Nature* **578:** 102–111. doi:10.1038/s41586-020-1965-x

Riggio AI, Varley KE, Welm AL. 2021. The lingering mysteries of metastatic recurrence in breast cancer. *Br J Cancer* **124:** 13–26. doi:10.1038/s41416-020-01161-4

Rivenbark AG, O'Connor SM, Coleman WB. 2013. Molecular and cellular heterogeneity in breast cancer challenges for personalized medicine. *Am J Pathol* **183:** 1113–1124. doi:10.1016/j.ajpath.2013.08.002

Robinson JLL, Holmes KA, Carroll JS. 2013. FOXA1 mutations in hormone-dependent cancers. *Front Oncol* **3:** 20.

Roulois D, Loo Yau H, Singhania R, Wang Y, Danesh A, Shen SY, Han H, Liang G, Jones PA, Pugh TJ, et al. 2015. DNA-demethylating agents target colorectal cancer cells by inducing viral mimicry by endogenous transcripts. *Cell* **162:** 961–973. doi:10.1016/j.cell.2015.07.056

Ruff GL, Murphy KE, Smith ZR, Vertino PM, Murphy PJ. 2021. Subtype-independent ANP32E reduction during breast cancer progression in accordance with chromatin relaxation. *BMC Cancer* **21:** 1342. doi:10.1186/s12885-021-09077-9

Sallari RC, Sinnott-Armstrong NA, French JD, Kron KJ, Ho J, Moore JH, Stambolic V, Edwards SL, Lupien M, Kellis M. 2017. Convergence of dispersed regulatory mutations predicts driver genes in prostate cancer. bioRxiv doi:10.1101/097451v2

Schapira M, Arrowsmith CH. 2016. Methyltransferase inhibitors for modulation of the epigenome and beyond. *Curr Opin Chem Biol* **33:** 81–87. doi:10.1016/j.cbpa.2016.05.030

Sérandour AA, Avner S, Percevault F, Demay F, Bizot M, Lucchetti-Miganeh C, Barloy-Hubler F, Brown M, Lupien M, Métivier R, et al. 2011. Epigenetic switch involved in activation of pioneer factor FOXA1-dependent enhancers. *Genome Res* **21:** 555–565. doi:10.1101/gr.111534.110

Shah SP, Morin RD, Khattra J, Prentice L, Pugh T, Burleigh A, Delaney A, Gelmon K, Guliany R, Senz J, et al. 2009. Mutational evolution in a lobular breast tumour profiled at single nucleotide resolution. *Nature* **461:** 809–813. doi:10.1038/nature08489

Shen YA, Jung J, Shimberg GD, Hsu FC, Rahmanto YS, Gaillard SL, Hong J, Bosch J, Shih IM, Chuang CM, et al. 2021. Development of small molecule inhibitors targeting PBX1 transcription signaling as a novel cancer therapeutic strategy. *iScience* **24:** 103297. doi:10.1016/j.isci.2021.103297

Slaven N, Lopes R, Canale E, Ivanoiu D, Pacini C, Monteiro Barbosa IA, Bleu M, Bravaccini S, Ravaioli S, Dieci MV, et al. 2022. Genetic and epigenetic driven variation in regulatory regions activity contribute to adaptation and evolution under endocrine treatment. bioRxiv doi:10.1101/2022.02.15.480537

Smallwood SA, Lee HJ, Angermueller C, Krueger F, Saadeh H, Peat J, Andrews SR, Stegle O, Reik W, Kelsey G. 2014. Single-cell genome-wide bisulfite sequencing for assessing epigenetic heterogeneity. *Nat Methods* **11:** 817–820. doi:10.1038/nmeth.3035

Song X, Gao T, Wang N, Feng Q, You X, Ye T, Lei Q, Zhu Y, Xiong M, Xia Y, et al. 2016. Selective inhibition of EZH2 by ZLD1039 blocks H3K27methylation and leads to potent anti-tumor activity in breast cancer. *Sci Re* **6:** 2489.

Sotiriou C, Neo S-Y, McShane LM, Korn EL, Long PM, Jazaeri A, Martiat P, Fox SB, Harris AL, Liu ET. 2003. Breast cancer classification and prognosis based on gene expression profiles from a population-based study. *Proc Natl Acad Sci* **100:** 10393–10398. doi:10.1073/pnas.1732912100

Stergachis AB, Neph S, Reynolds A, Humbert R, Miller B, Paige SL, Vernot B, Cheng JB, Thurman RE, Sandstrom R, et al. 2013. Developmental fate and cellular maturity encoded in human regulatory DNA landscapes. *Cell* **154:** 888–903. doi:10.1016/j.cell.2013.07.020

Sung H, Ferlay J, Siegel RL, Laversanne M, Soerjomataram I, Jemal A, Bray F. 2021. Global cancer statistics 2020: GLOBOCAN estimates of incidence and mortality worldwide for 36 cancers in 185 countries. *CA Cancer J Clin* **71:** 209–249. doi:10.3322/caac.21660

Swanton C. 2012. Intratumor heterogeneity: evolution through space and time. *Cancer Res* **72:** 4875–4882. doi:10.1158/0008-5472.CAN-12-2217

Takaku M, Grimm SA, De Kumar B, Bennett BD, Wade PA. 2020. Cancer-specific mutation of GATA3 disrupts the transcriptional regulatory network governed by estrogen receptor α, FOXA1 and GATA3. *Nucleic Acids Res* **48:** 4756–4768. doi:10.1093/nar/gkaa179

Tan PH, Ellis I, Allison K, Brogi E, Fox SB, Lakhani S, Lazar AJ, Morris EA, Sahin A, Salgado R, et al. 2020. The 2019 world health organization classification of tumours of the breast. *Histopathology* **77:** 181–185. doi:10.1111/his.14091

Thurman RE, Rynes E, Humbert R, Vierstra J, Maurano MT, Haugen E, Sheffield NC, Stergachis AB, Wang H, Vernot B, et al. 2012. The accessible chromatin landscape of the human genome. *Nature* **489:** 75–82. doi:10.1038/nature11232

Wang Q, Li W, Zhang Y, Yuan X, Xu K, Yu J, Chen Z, Beroukhim R, Wang H, Lupien M, et al. 2009. Androgen receptor regulates a distinct transcription program in androgen-independent prostate cancer. *Cell* **138:** 245–256. doi:10.1016/j.cell.2009.04.056

Wang Z, Wu H, Daxinger L, Danen EHJ. 2020. Genome-wide identification of binding sites of GRHL2 in luminal-like and basal A subtypes of breast cancer. bioRxiv doi:10.1101/2020.02.13.946947

Wang X, Yan J, Shen B, Wei G. 2021. Integrated chromatin accessibility and transcriptome landscapes of doxorubicin-resistant breast cancer cells. *Front Cell Dev Biol* **9:** 708066. doi:10.3389/fcell.2021.708066

Wang Z, Chivu AG, Choate LA, Rice EJ, Miller DC, Chu T, Chou S-P, Kingsley NB, Petersen JL, Finno CJ, et al. 2022. Prediction of histone post-translational modification patterns based on nascent transcription data. *Nat Genet* **54:** 295–305. doi:10.1038/s41588-022-01026-x

Wu Q, Heidenreich D, Zhou S, Ackloo S, Krämer A, Nakka K, Lima-Fernandes E, Deblois G, Duan S, Vellanki RN, et al. 2019. A chemical toolbox for the study of bromodomains and epigenetic signaling. *Nat Commun* **10:** 1915. doi:10.1038/s41467-019-09672-2

Wu Q, Schapira M, Arrowsmith CH, Barsyte-Lovejoy D. 2021a. Protein arginine methylation: from enigmatic functions to therapeutic targeting. *Nat Rev Drug Discov* **20:** 509–530. doi:10.1038/s41573-021-00159-8

Wu SZ, Al-Eryani G, Roden DL, Junankar S, Harvey K, Andersson A, Thennavan A, Wang C, Torpy JR, Bartonicek N, et al. 2021b. A single-cell and spatially resolved atlas of human breast cancers. *Nat Genet* **53:** 1334–1347. doi:10.1038/s41588-021-00911-1

Xi Y, Shi J, Li W, Tanaka K, Allton KL, Richardson D, Li J, Franco HL, Nagari A, Malladi VS, et al. 2018. Histone modification profiling in breast cancer cell lines highlights commonalities and differences among subtypes. *BMC Genomics* **19:** 150. doi:10.1186/s12864-018-4533-0

Zaret KS. 2020. Pioneer transcription factors initiating gene network changes. *Annu Rev Genet* **54:** 367–385. doi:10.1146/annurev-genet-030220-015007

Zaret KS, Carroll JS. 2011. Pioneer transcription factors: establishing competence for gene expression. *Genes Dev* **25:** 2227–2241. doi:10.1101/gad.176826.111

Zhang X, Bailey SD, Lupien M. 2014. Laying a solid foundation for Manhattan—"setting the functional basis for the post-GWAS era." *Trends Genet* **30:** 140–149. doi:10.1016/j.tig.2014.02.006

Zhang H, Ahearn TU, Lecarpentier J, Barnes D, Beesley J, Qi G, Jiang X, O'Mara TA, Zhao N, Bolla MK, et al. 2020. Genome-wide association study identifies 32 novel breast cancer susceptibility loci from overall and subtype-specific analyses. *Nat Genet* **52:** 572–581. doi:10.1038/s41588-020-0609-2

Zhang K, Hocker JD, Miller M, Hou X, Chiou J, Poirion OB, Qiu Y, Li YE, Gaulton KJ, Wang A, et al. 2021a. A single-cell atlas of chromatin accessibility in the human genome. *Cell* **184:** 5985–6001.e19. doi:10.1016/j.cell.2021.10.024

Zhang Y, Chen H, Mo H, Hu X, Gao R, Zhao Y, Liu B, Niu L, Sun X, Yu X, et al. 2021b. Single-cell analyses reveal key immune cell subsets associated with response to PD-L1 blockade in triple-negative breast cancer. *Cancer Cell* **39:** 1578–1593.e8. doi:10.1016/j.ccell.2021.09.010

Zhou S, Treloar AE, Lupien M. 2016. Emergence of the noncoding cancer genome: a target of genetic and epigenetic alterations. *Cancer Discov* **6:** 1215–1229. doi:10.1158/2159-8290.CD-16-0745

Zhou S, Hawley JR, Soares F, Grillo G, Teng M, Tonekaboni SAM, Hua JT, Kron KJ, Mazrooei P, Ahmed M, et al. 2020. Noncoding mutations target cis-regulatory elements of the FOXA1 plexus in prostate cancer. *Nat Commun* **11:** 441. doi:10.1038/s41467-020-14318-9

Zhu Q, Tekpli X, Troyanskaya OG, Kristensen VN. 2020. Subtype-specific transcriptional regulators in breast tumors subjected to genetic and epigenetic alterations. *Bioinformatics* **36:** 994–999. doi:10.1093/bioinformatics/btz709

The Breast Cancer Proteome and Precision Oncology

Jonathan T. Lei,[1] Eric J. Jaehnig,[1] Hannah Smith,[2] Matthew V. Holt,[1] Xi Li,[2] Meenakshi Anurag,[1] Matthew J. Ellis,[1,3] Gordon B. Mills,[2] Bing Zhang,[1] and Marilyne Labrie[2,4,5]

[1]Lester and Sue Smith Breast Center and Dan L. Duncan Comprehensive Cancer Center, Baylor College of Medicine, Houston, Texas 77030, USA

[2]Knight Cancer Institute, Oregon Health & Science University, Portland, Oregon 97239, USA

Correspondence: millsg@ohsu.edu; bing.zhang@bcm.edu; marilyne.labrie@usherbrooke.ca

The goal of precision oncology is to translate the molecular features of cancer into predictive and prognostic tests that can be used to individualize treatment leading to improved outcomes and decreased toxicity. Success for this strategy in breast cancer is exemplified by efficacy of trastuzumab in tumors overexpressing ERBB2 and endocrine therapy for tumors that are estrogen receptor positive. However, other effective treatments, including chemotherapy, immune checkpoint inhibitors, and CDK4/6 inhibitors are not associated with strong predictive biomarkers. Proteomics promises another tier of information that, when added to genomic and transcriptomic features (proteogenomics), may create new opportunities to improve both treatment precision and therapeutic hypotheses. Here, we review both mass spectrometry-based and antibody-dependent proteomics as complementary approaches. We highlight how these methods have contributed toward a more complete understanding of breast cancer and describe the potential to guide diagnosis and treatment more accurately.

Genomic and transcriptomic profiling of >10,000 tumors and normal tissues across 33 cancer types by The Cancer Genome Atlas (TCGA) consortium has revealed enormous genomic complexity and transcriptomic subtypes that may stratify tumors for specific treatments (Blum et al. 2018; Rodon et al. 2019). However, understanding how genomic and transcriptomic aberrations drive clinically relevant cancer phenotypes to inform precision oncology remains a work in progress. As the functional readout of genomic and transcriptomic aberrations, proteins, with their posttranslational modifications (PTMs) and the accompanied formation of complexes and molecular machines, provide the critical connection between genotype and phenotype. Due to prevalent translational and posttranslational regulation,

---

[3]Present address: Early Oncology, Oncology Research and Development, AstraZeneca, Gaithersburg, MD 20878.

[4]Present address: Department of Immunology and Cellular Biology, Université de Sherbrooke, Sherbrooke, QC J1H 5N4, Canada.

[5]Present address: Department of Obstetrics and Gynecology, Université de Sherbrooke, Sherbrooke, QC J1H 5N4, Canada.

Cite this article as *Cold Spring Harb Perspect Med* doi: 10.1101/cshperspect.a041323

the functional impact of cancer mutations is therefore incompletely predicted from genomic and transcriptomic measurements alone. In deed, a single mutation can have markedly different effects in different tumors and in different tumor lineages due to intrinsic gene expression patterns. To provide direct measurements of proteins and PTMs, the antibody-based reverse-phase protein assay (RPPA) has quantified over 250 protein markers (total and phosphorylated proteins, now increased to 486 protein markers) in ~8,000 tumors across 32 cancer types from the TCGA cohort (Li et al. 2017; Chen et al. 2019). Subsequent efforts led by the National Cancer Institute's Clinical Proteomic Tumor Analysis Consortium (CPTAC) have combined genomic and transcriptomic profiling with mass spectrometry (MS)-based profiling of proteins, phosphorylation sites, and other PTMs for >1,000 tumors across 10 cancer types. The retrospectively collected TCGA breast cancer cohort was characterized in one of the early CPTAC studies (Mertins et al. 2016) followed by the analysis of an independent prospectively collected breast cancer cohort (Krug et al. 2020). Other proteogenomic analyses have recapitulated and extended observations from CPTAC (Johansson et al. 2019). In both breast cancer and other cancer types, integrated analyses have consistently shown moderate correlation (with an average r value of approximately 0.5) between mRNA and protein levels for many genes and frequent lack of *cis*-impact for genomic alterations on corresponding protein levels (Wang et al. 2017). This emphasizes the need to directly measure protein levels and function rather than to impute these from RNA measurements and importantly to implement approaches to confirm that genomic aberrations manifest at the protein level.

Both antibody- and MS-based proteomic approaches have unique strengths and requirements for data generation and analysis. In this perspective, we first describe MS-based proteomic methods, covering both discovery and targeted and single-cell proteomics, and then highlight important findings and translational efforts from breast cancer studies with implications for accurate diagnosis and treatment of breast cancer. Afterward, we present antibody-based methods with a focus on spatially resolved techniques. Finally, we give concluding remarks regarding the strengths and weaknesses of these two complementary approaches.

MASS SPECTROMETRY–BASED PROTEOMICS AND APPLICATIONS TO BREAST CANCER

Discovery Proteomics

MS-based shotgun proteomics represents the primary technology for unbiased, genome-wide profiling of proteins and their PTMs (Wu and MacCoss 2002). In a shotgun proteomics experiment, proteins extracted from tissue or cell samples are digested by proteases, typically trypsin, into peptides, which can be fractionated to reduce sample complexity, and then analyzed by liquid chromatography/tandem MS (LC–MS/MS) (Fig. 1). For an unbiased analysis of PTM peptides, because of their overall low abundance compared with unmodified peptides, PTM-specific peptide enrichment is required to improve their detection by LC-MS/MS. This can be achieved using PTM-directed affinity chromatography or immunoprecipitation with PTM-specific antibodies (Zhao and Jensen 2009). For example, phosphorylated peptides can be enriched with immobilized metal affinity chromatography (IMAC) or titanium dioxide, whereas acetylated peptides can be enriched with antibodies directed against the acetylated epsilon amino group of lysine residues. In addition to enriching PTM peptides, other subgroups of proteins or peptides of interest can also be for enrichment. For instance, kinase inhibitor–conjugated beads have been used for the enrichment of protein kinases (Oppermann et al. 2009), and major histocompatibility complex (MHC)-bound peptides can be purified by immunoprecipitation using an antibody specific to the desired MHC species (MHC-IP) (Purcell et al. 2019). Moreover, serial enrichment can be applied to obtain multiple types of proteomic measurements from the same sample. As an example, the multi-omic networked tissue enrichment (MONTE) workflow enables serial immu-

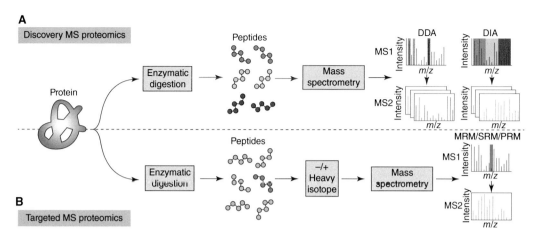

Figure 1. Overview of mass-spectrometry (MS)-based approaches. (*A*) For discovery MS-based methods, extracted proteins are subjected to enzymatic digestion to generate peptides, which are subsequently analyzed by liquid chromatography/tandem MS. Downstream data-dependent acquisition (DDA) selects a predefined number of MS1 precursor ions for fragmentation. In contrast, data-independent acquisition (DIA) scans all MS1 precursor ions and selects a window containing ions for fragmentation in MS2. (*B*) For targeted MS-based methods, predetermined peptides are quantified and heavy-isotope-labeled peptides can be spiked-in as internal standards for precise quantification. Multiple reaction monitoring (MRM), single-reaction monitoring (SRM), and parallel reaction monitoring (PRM) acquisition methods can be used to capture and measure specific peptides of interest and allows for quantification of low-abundance peptides.

nopeptidome, ubiquitylome, proteome, phosphoproteome, and acetylome analyses of sample-limited tissues (Abelin et al. 2022). MS-based shotgun proteomics is also routinely used in the discovery of protein–protein interactions (PPIs). Affinity purification–MS (AP-MS) can use protein-specific antibodies or alternatively epitope tags on target "bait" proteins of interest as affinity capture probes to capture interacting "prey" proteins, which are then identified by LC-MS/MS (Bauer and Kuster 2003). By using epitope tags on "bait" proteins as affinity capture probes, AP-MS may bypass the requirement of specific antibodies for each bait protein. Other orthogonal MS-based methods for PPI discovery are continuously being introduced, such as proximity-dependent biotinylation coupled to MS (PDB-MS), cross-linking MS (XL-MS), cofractionation MS (coFrac-MS), and thermal proximity coaggregation (TPCA) (Low et al. 2021).

There are two methods to acquire data on a mass spectrometer for discovery proteomics, data-dependent acquisition (DDA) and data-independent acquisition (DIA) (Doerr 2015). The DDA approach scans all precursor peptide ions during the survey scan (MS1), followed by selecting a predefined number of precursor ions for subsequent fragmentation (MS2) to gain detailed information for the selected precursor ions (Fig. 1A). Generally, the most abundant peptides are selected for fragmentation. This method is likely to miss low-abundance peptides, leading to many missing values across experiments in a study. The DIA approach sequentially analyzes all incoming ions and thus has high reproducibility in identification of the selected peptide across experiments (Fig. 1B). However, DIA generates complex results for a mixture of peptides; many methods require an experimentally defined library to match peptides to, and deconvolution is challenging computationally.

Shotgun proteomics experiments can be performed for one sample at a time, referred to as label-free proteomics. Several labeling approaches are also available to enable multiplexed proteome analysis, including stable isotope labeling with amino acids in cell culture (SILAC) (Ong et al. 2002), which is a metabolic labeling approach, and post-sample collection chemical-labeling approaches using an isotope-coded affinity tag (ICAT) (Chan et al. 2015), isobaric tags for relative and absolute quantitation (iTRAQ) (Wiese et al. 2007), or tandem mass tag (TMT) (Thompson et al. 2003). Currently, up to 18 samples can be combined into one TMT experiment increasing the amount of information that can be obtained from a single run (Li et al. 2021b). The label-free approach does not require extensive sample preparation and allows flexible experimental design because each sample is analyzed independently. However, highly reproducible LC separation across experiments is critical to enable meaningful comparisons, even for studies with small sample size that require only a single TMT. Labeled approaches typically take less machine time and facilitate a direct comparison of samples analyzed in the same experiment. However, they require additional sample preparation and usually a prespecified study design.

A typical LC-MS/MS experiment can produce hundreds of thousands of MS/MS spectra. Computational tools are required for raw data processing, data quality control, peptide and protein identification and quantification, PTM site localization, and downstream analyses. Peptide identification from MS/MS spectra is the most characteristic data analysis process of shotgun proteomics. For DDA data, peptide identification can be achieved using either database searching or de novo peptide sequencing. Database searching methods compare experimentally collected MS/MS spectra against a preselected reference protein database, such as UniProt, RefSeq, or Ensembl (Nesvizhskii 2007). In PTM experiments, additional algorithms, such as Ascore (Beausoleil et al. 2006) or PhosphoRS (Taus et al. 2011), are required to determine PTM site localization within a peptide sequence. To identify sample-specific peptides derived from somatic mutations, fusions, and abnormal transcription, which are missing from the reference databases, novel protein sequences predicted from DNA or RNA sequencing data can be incorporated into a customized database for MS/MS data searching (Wang and Zhang 2013; Nesvizhskii 2014). False discovery is a major concern in novel peptide identification, and methods have been developed to address this concern

(Yi et al. 2018; Zhu et al. 2018a; Wen et al. 2019); however, this remains a challenge. De novo peptide sequencing does not rely on a reference protein database and is particularly attractive for the identification of novel protein sequences, but it is more challenging computationally (Ma and Johnson 2012). Many peptide identification algorithms and tools have been developed, including database search algorithms such as SEQUEST (Eng et al. 1994), X!Tandem (Craig and Beavis 2003, 2004), X!!Tandem (Bjornson et al. 2008), Mascot (Perkins et al. 1999), Andromeda (Cox et al. 2011), MS-GF+ (Kim and Pevzner 2014), MSFragger (Kong et al. 2017), and de novo peptide-sequencing algorithms such as PepNovo (Frank and Pevzner 2005), PEAKS (Ma et al. 2003), and pNovo (Chi et al. 2010). Methods have also been developed to combine the two approaches, such as DirecTag (Tabb et al. 2008) and JUMP (Wang et al. 2014). More recently, deep learning has been leveraged to improve peptide identification, as exemplified by DeepNovo (Tran et al. 2017) and DeepRescore (Li et al. 2020). Visualization tools such as PDV (Li et al. 2019) can be used for manual examination of the peptide-spectrum match results, which is particularly important for validation of novel peptide identifications.

Peptide identification from DIA data is complicated by the fact that all precursor ions within an isolation window are fragmented (Fig. 1A), leading to a highly complex fragment ion mass spectrum that cannot be easily interpreted. There are two major classes of methods for DIA data analysis, library-based methods, and library-free methods. Library-based methods rely on the construction of study-specific spectrum libraries by multiple fractionated DDA analysis of the same or similar samples in the DIA study and then searching DIA spectra against the spectra in the constructed library. Library-based analysis tools include Open-SWATH (Röst et al. 2014), Spectronaut (Bruderer et al. 2015), Skyline (Pino et al. 2020), Peak-View (www.sciex.com), DIA-NN (Demichev et al. 2020), and EncyclopeDIA (Searle et al. 2019), among others. Library-free methods analyze the DIA spectra without the need of a spectral library, and tools in this class include PEAKS

Cite this article as *Cold Spring Harb Perspect Med* doi: 10.1101/cshperspect.a041323

(Ma et al. 2003), directDIA (Spectronaut) (Bruderer et al. 2015), DIA-Umpire (Tsou et al. 2015), and Group-DIA (Li et al. 2015), among others. Despite obvious advantages compared to library-based methods, library-free methods bear some unique challenges such as the effective control of false discovery rate.

For label-free data generated by both DDA and DIA workflows, quantitation is typically based on precursor signal intensity in MS1 spectra. For labeled data generated by the SILAC workflow, quantification is achieved by comparing the intensities of the light- and heavy-labeled precursor ions in MS1 spectra. For the iTRAQ and TMT workflows, quantification is achieved by comparing the intensity ratios of the iTRAQ or TMT reporter ions in MS2 spectra. In MS2-based quantification, contaminating peptide ions may be co-isolated with the target peptide ion, and the resulting interference may affect the accuracy and precision of peptide quantification. Some studies collect an additional isolation and fragmentation event (MS3 scan) to overcome the interference effect, and MS3 data is used for peptide quantification in these studies (Ting et al. 2011). After appropriate data quality checking, normalization, and missing value imputation, quantitative data can be used for downstream analyses, which may include differential protein identification, unsupervised clustering, predictive modeling, and pathway and network analysis, etc.

Targeted Proteomics

In contrast to unbiased, genome-wide profiling approaches described above, targeted proteomic approaches only measure specific, predetermined peptides (Doerr 2013; van Bentum and Selbach 2021). Instead of systematically analyzing incoming signals, the mass spectrometer is set to continuously analyze a narrow band of incoming peptides. Single-reaction monitoring (SRM) is a technique used in triple-quadrupole instruments (Yost and Enke 1979) where a narrow isolation band is used for precursor selection and fragmentation, and a second narrow isolation band is used to measure a single fragment ion from the peptide (Lange et al. 2008). The result is highly sensitive, quantitative, and easy to interpret. Instead of analyzing a single fragment, multiple-reaction monitoring (MRM) analyzes multiple fragments sequentially. Parallel-reaction monitoring (PRM) allows simultaneous measurement of all fragment ions possible on orbitrap and time-of-flight (ToF) instruments (Peterson et al. 2012). PRM is designed to track and quantify preprogrammed "target" peptides of interest, offering better sensitivity and linearity in quantification than regular data-dependent acquisition. There are promising improvements in proteomics technology that may enable precise measurement of hundreds of targets. For example, PRM/MRM assays can be run with stable isotope-labeled peptides as spike-in standards (Fig. 1B). Spiked-in isotope-labeled peptides of known concentration are co-eluted with endogenous peptides in each sample. The MS recognizes them as a matched pair, thus allowing the endogenous peptides to be quantified accurately with respect to the standard. Immuno-MRM assays improve the ability to quantify limited amounts of peptide targets using immunoaffinity enrichment of those targets (Whiteaker and Paulovich 2011; Whiteaker et al. 2011, 2012).

Identification of target peptides is achieved by matching predetermined fragment ion peaks. The expected peaks can be selected through untargeted experimental approaches, machine learning–based solutions such as Prosit (Gessulat et al. 2019), database-mining methods such as SRMAtlas (Kusebauch et al. 2016) and PeptideAtlas (Deutsch 2010), and synthetic libraries available in ProteomeTools (Zolg et al. 2017). Data visualization and matching tools such as MRMer (Martin et al. 2008) and Skyline (Pino et al. 2020) are commonly used to compare identified peaks with theoretical peaks. There is no universally accepted standard for peak matching as different instruments have weaknesses and strengths with regard to mass accuracy, resolution, sensitivity, and signal-to-noise (van Bentum and Selbach 2021). These parameters are neither consistent nor linear within a single spectrum. Fourier transform (FT) instruments have greater resolution at lower m/z values (it is best to consider resolution a function rather than a single value) while ToF has slightly higher resolution at higher *m/z* values (Makarov et al. 2009). Despite these challenges, most targeted proteomics analyses yield extremely high confidence identifications by matching multiple ions at predicted retention times with complimentary stable isotope-labeled control peptides if they are used (Gallien et al. 2015). Although still at the early stage, targeted MS-based assays have great potential for clinical implementation (Zhang et al. 2019).

Spatial and Single-Cell Proteomics

To better understand the highly heterogeneous nature of tumors, MS-based approaches can also be used to investigate proteins in specific regions of interest. Laser-capture microdissection (LCM) of histopathologically defined tumor areas coupled with MS-based proteomic profiling has been used to reveal spatially associated tumor features with potential clinical implications (Buczak et al. 2018; Herrera et al. 2020). Recent advancements in instrumentation, analytical tools, and sample preparation protocols have enabled scientists to probe single-cell protein content using MS. For example, the nanoPOTS (nanodroplet processing in one pot for trace samples) platform downscales sample processing volumes to <200 nL to minimize surface losses (Zhu et al. 2018b). When combined with ultrasensitive LC-MS and the Match Between Runs algorithm of MaxQuant, nanoPOTS allows consistent identification of >3,000 proteins from as few as 10 cells. The nanoPOTS protocol also enabled quantification of ~2,400 proteins from a single human pancreatic islet thin section from type 1 diabetic and control donors, demonstrating its potential application for spatially resolved proteome measurements from clinical tissues. Using a multiplexed experimental design in which proteins from single cells and from a protein carrier (total cell lysate of a small group of cells) are barcoded and then combined to increase the amount of material available for MS analysis. Single-Cell ProtEomics by MS (SCoPE-MS) and its second-generation implementation, SCoPE2, allow the quantification of thousands of proteins across hundreds of single-cell samples

(Slavov 2020). In deep visual proteomics (DVP), advances in LCM microscopy in combination with machine-learning recognition of cell phenotypes have made image-guided isolation of single cells possible from tissue sections that can then be subsequently processed by ultralow sample input methods followed by MS (Mund et al. 2022). This method retains precise spatial proteomic information in the tissue context and thus can connect the physiological characteristics of cells with protein measurements, which is critical for understanding tumor microenvironment and heterogeneity.

Breast Cancer Applications

In its initial effort to provide deep proteomic characterization of breast cancer, the CPTAC was able to use the iTRAQ platform to quantify >10,000 proteins and >30,000 phosphorylation sites from 77 TCGA breast cancer tumors (Mertins et al. 2016). More recently, CPTAC integrated genomic and transcriptomic profiles from 122 prospectively collected breast cancer tumors with TMT-based profiles for >10,000 proteins, ~39,000 phosphorylation sites, and ~10,000 acetylation sites (Krug et al. 2020). MS-based proteomics profiling studies spanning all breast cancer clinical subtypes have been also described elsewhere. Most of these studies focused solely on quantifying proteins (Panis et al. 2014; Tyanova et al. 2016; Bouchal et al. 2019; Asleh et al. 2022), but some also included gene expression profiling (Tang et al. 2018; Johansson et al. 2019).

Proteomics data from the CPTAC breast cancer studies and their integration with genomic and transcriptomic data have generated new biological and clinical insights that may help improve breast cancer classification and treatment (Krug et al. 2020). Integrative multi-omics classification of breast cancer revealed four clusters of tumors, designated as luminal A-inclusive (LumA-I), luminal B-inclusive (LumB-I), basal-inclusive (Basal-I), and HER2-inclusive (HER2-I), based on their alignment with the PAM50 classification. Despite the overall concordance between the multi-omics and PAM50 classifications, the LumB-I cluster comprised all but one PAM50 luminal B case but also included a subset of luminal A samples. Importantly, applying an mRNA-based classifier trained to distinguish PAM50 luminal A samples assigned to the LumB-I cluster from those assigned to the LumA-I cluster separated PAM50 luminal A samples in the METABRIC study into two classes with distinct survival outcomes. Phosphoproteomic analysis identified kinases abnormally hyperphosphorylated in each multi-omics cluster, including PRKDC, MAP4K4, and SPEG for the Basal-I subtype, ERBB2 and CDK12 for the HER2-I subtype, DCLK1 for the LumA-I subtype, and SIK3 for the LumB-I subtype, which may potentially help stratify patients for treatment with specific kinase inhibitors. While immune checkpoint inhibitors (ICIs) are currently only approved for triple-negative breast cancer (TNBC), integrated analysis of mutational signatures, DNA repair processes, immune and tumor microenvironment environment profiles, and expression of ICI targets identified subsets of PAM50 luminal tumors, where elevation of immune microenvironment- and checkpoint-related features was associated with APOBEC-mediated mutagenesis or single-strand break repair defects leading to an important question whether these patients could benefit from ICI treatment. Moreover, although CDK4/6 inhibitors are currently only approved for hormone-receptor-positive, HER2-negative, advanced breast cancer, data from the CPTAC study suggested that accurate proteogenomic assessment of Rb could prove useful as a predictive marker for applying CDK4/6 inhibitors in a subset of TNBC.

The standardized CPTAC protocol required at least 100 mg of tumor tissue as input for proteogenomic profiling (Mertins et al. 2018). This sample input requirement exceeds many sources of clinically important material, hindering the clinical translation of the proteogenomic technology. To address this limitation, a microscaled proteogenomic method has been developed that can generate high-quality, deep-scale proteogenomic profiles from a single snap-frozen, tumor-rich core needle biopsy (<20 mg) (Satpathy et al. 2020). This technology was applied to profile core biopsies from breast cancer patients enrolled in the DP1 clinical trial (Satpathy et al.

2020). Patients in this trial were classified as ERBB2-positive (ERBB2$^+$) based on clinical diagnostic assays that directly quantify ERBB2 protein by immunohistochemistry (IHC) and/or copy number status by fluorescence in situ hybridization (FISH) and then treated with trastuzumab (or trastuzumab plus pertuzumab) plus chemotherapy. Samples were collected at baseline prior to and 48 h to 72 h following treatment. Five out of the 14 patients (~35%) with tumors available for molecular profiling did not show pathologic complete response (pCR) to the treatment, consistent with the typical response rate for ERBB2-targeted therapies.

Proteogenomic profiling of the core needle breast tumor biopsies from the DP1 trial quantified between 10,000 to 20,000 genes at the copy number, mRNA, and protein level (TMT-based proteomics) and >17,000 phosphosites. Integrative analysis of these data allowed us to better understand the molecular features that may contribute to resistance to anti-ERBB2 treatment (Satpathy et al. 2020). All of the tumors from patients showed that pCR had ERBB2 gene amplification and concordantly high levels of ERBB2 mRNA, protein, and phosphorylation, and they also had evidence of impairment of ERBB2 signaling at the protein (down-regulation of ERBB2) and phosphorylation (down-regulation of ERBB2 as well as MTOR target phosphorylation) level after treatment. Detailed proteogenomic examination of the ERBB2 locus in baseline tumors refined ERBB2$^+$ classification and suggested possibilities for treatment resistance for individual cases. In one of the resistant tumors, ERBB2 was copy number neutral with low levels of ERBB2 mRNA, protein, and phosphoprotein, thereby reclassifying this tumor as "false ERBB2$^+$" and suggesting misdiagnosis of ERBB2$^+$ disease that may explain resistance due to lack of the therapeutic target. Two other resistant samples were reclassified as "pseudo-ERBB2$^+$" tumors where ERBB2 amplification does not lead to up-regulation of downstream analytes (ERBB2 mRNA, protein, and phosphosite levels). Similar to the "false ERBB2$^+$" specimen, these "pseudo-ERBB2$^+$" samples lacked protein expression of the therapeutic target. These results reveal limitations of the copy number and IHC-based clinical assays and highlight a critical need for more accurate protein quantification to improve patient diagnosis. A recent study showed that immuno-MRM-MS (Kennedy et al. 2021) can provide highly accurate HER2 protein quantification in FFPE biopsies and frozen breast cancer bulk tumors. Samples where HER2 was low and challenging to accurately measure by standard IHC assays were present in both studies. Although MS-based proteomic quantification of ERBB2 was clearly higher in ERBB2$^+$ samples (primarily IHC 3+) than in IHC 0, 1+, and 2+ samples, there was an overlapping, wide range of ERBB2 abundance for these ERBB2$^-$ samples. MS may provide an alternative and potentially more quantitative assay to help stratify patients with low levels of HER2 for treatment with HER2-targeted therapy, which is particularly important given the demonstration that the trastuzumab deruxtecan antibody drug conjugate is active in HER2 low tumors (Modi et al. 2022).

In one of the "pseudo-ERBB2$^+$" cases, the amplification event was centered on a region adjacent to ERBB2 rather than ERBB2 itself. The neighboring gene TOP2A had higher copy number levels than ERBB2, and this shift in amplification was supported by elevated protein measurements, suggesting TOP2A as a driver and putative target for this case. The false ERBB2$^+$ sample and one pseudo-ERBB2$^+$ sample had high levels of immune infiltration, raising the possibility for ICI treatment. Proteogenomic analysis of the ERBB2 locus in the CPTAC breast cancer data sets described above also revealed similar frequencies of "psuedo-ERBB2$^+$" cases (~15% of ERBB2 amplified tumors) in larger cohorts (Krug et al. 2020). One of the "true ERBB2$^+$" samples that showed amplification for ERBB2 with concordantly high mRNA, protein, and phosphoprotein and that was resistant to trastuzumab showed elevated PI3K/MTOR signaling pathways at multiple omic levels along with up-regulation of multiple mucin proteins. PI3K/MTOR signaling has been proposed to be a mechanism of anti-HER2 targeted therapy (Berns et al. 2007). Mucins were shown to block ERBB2 antibodies from recognizing their epitopes in experimental models (Vu and Claret 2012; Farahmand et al. 2018;

Namba et al. 2019). To test the hypothesis that targeting the PI3K/MTOR signaling pathway with a small molecule inhibitor (everolimus) can overcome trastuzumab resistance when mucin is also elevated, two ERBB2$^+$ patient-derived xenograft (PDX) models, one expressing low levels and another expressing high levels of mucin, were treated with trastuzumab and everolimus as single agents and in combination (Satpathy et al. 2020). Tumor regression was most effective in the mucin low PDX after trastuzumab monotherapy while tumor regression was only observed in the mucin high PDX after combination treatment. These results suggest the potential use of ERBB2 antibody therapy in combination with everolimus to overcome resistance and demonstrate the application of proteogenomics to identify mechanisms of resistance and therapeutic targets.

MS-based assays have also been used to better understand cancer driver gene function in breast cancer through examining their interaction partners. Recent work has investigated the physical PPI network of frequently altered genes in breast cancer by pulling down the protein products of these genes in breast cancer cell lines by AP using FLAG-tagged proteins followed by MS (AP-MS) (Kim et al. 2021). Results of the approach revealed previously unknown interacting proteins with well-established oncogenes and also identified interactors that could be used as novel markers to indicate treatment response. For example, Kim et al. (2021) identified UBE2N protein as a functional interactor of BRCA1, and they demonstrated that its expression negatively correlates with patient response to DNA repair–targeted therapies such as poly (ADP-ribose) polymerase inhibitor.

More recently, a study has investigated spatial heterogeneity at the proteomic level by using LCM to sample multiple regions from treatment-naive breast cancer biopsies followed by MS analysis (Mardamshina et al. 2021). After integrating spatial information with proteomic profiling, subtype-specific associations were observed. For example, levels of immune-related and metabolic genes varied given the degree of hormone receptor positivity (HR$^+$) heterogeneity in a specific tumor region. The authors also observed that luminal markers typically expressed by HR$^+$ tumors were found to be enriched in heterogeneous triple-negative tumors. Collectively, this study highlights the potential of spatially resolved proteomics to refine our knowledge on the complexity underlying breast cancer subtypes.

Together, these results clearly demonstrate the utility of MS proteomics in generating diagnostic and treatment insights from clinical biopsies.

BIOMARKER DISCOVERY AND VALIDATION WITH ANTIBODY-BASED APPROACHES

Antibody-based approaches have been extensively used in breast cancer studies for bulk analysis of protein lysate, as well as for single-cell and spatially resolved analysis (Fig. 2).

Bulk Proteomics

Multiplex proteomics assays can be performed in either forward (antibody or aptamer printed) or reverse (sample printed)-phase formats. Forward-phase arrays have been used to characterize multiple sample sets based on printing of antibodies or aptamers in an array format and analyzing a single sample at a time. Recently, a large-scale aptamer array has demonstrated the ability to assess up to 7,000 proteins in tissue lysates and in circulating material (somalogic .com). The forward-phase format is limited to analyzing all antibodies or aptamers in a single hybridization condition and the one sample at a time format. RPPA represents a robust high-throughput method that has been applied to large numbers of breast cancer samples and models. In RPPA, samples are "printed" onto an array, usually a glass slide, and then each array is probed with a single antibody. Based on the availability of validated high-quality antibodies with both high sensitivity and specificity, RPPA can simultaneously measure up to 500 total and posttranslationally modified proteins in up to 1,000 samples (Akbani et al. 2014a; Lu et al. 2016). RPPA is particularly efficient for the analysis of proteins with PTMs and, when high-affinity antibodies are available, can be more sensitive than MS-based approaches, and can identify phosphosites that are not readily detect-

ed by other proteomic approaches such as MS (Mertins et al. 2014). In breast cancer research, RPPA has been especially useful in translational studies for breast tumor classification and uncovering actionable signaling pathways (Gromova et al. 2021), and to define signatures of response and resistance to therapy (Xing et al. 2019; Westin et al. 2021; Shi et al. 2022). It has also been used extensively in preclinical studies for the characterization of breast cancer cell lines, xenograft, and PDX models of response to anticancer therapies (Lindholm et al. 2014; Evans et al. 2017; Hsu et al. 2018).

RPPA has been integrated into several large studies with publicly available data sets, including TCGA, the Cancer Cell Line Encyclopedia (CCLE), and the Cancer-Perturbed Proteomics Atlas (CPPA), that have large breast cancer sample sets. To facilitate access and exploration of RPPA data generated from TCGA samples, the data and additional RPPA studies, including detailed characterization of cell lines, have been centralized into the publicly available Cancer Proteome Atlas (TCPA) portal (TCPAportal .org). TCGA RPPA data can also be found on the NIH GDC data portal, with a total of 7,539 samples analyzed with a total of 487 antibodies. RPPA characterization of cell lines is also available through Broad's CCLE (sites.broadinstitute .org/ccle) and the TCGA RPPA data through cBioPortal.org. The CPPA, another RPPA data set, containing a large data set of cell lines treated with a panel of drugs is also available at TCPA-portal.org. The availability of a perturbation data set from untreated and treated cells is a valuable tool for studying cancer biology and oncogenic pathways. Because of the large content of RPPA data sets, protein expression can be compared across samples using different approaches, including clustering analysis, pathway analysis, and single-protein expression comparisons (Akbani et al. 2014b; Labrie et al. 2019).

Although forward- and reverse-phase arrays can be used to assess tumor heterogeneity across patients or across tumors from a same patient (Labrie et al. 2019, 2021; Westin et al. 2021), they do not provide information on intratumoral heterogeneity and give little input about interactions between the tumor and the immune system

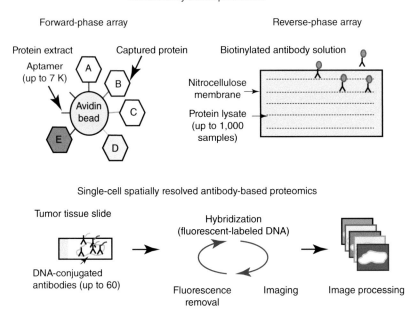

Figure 2. Antibody-based proteomic approaches. In forward-phased arrays, antibodies or aptamers are bound to a solid-phase platform, usually a glass slide or beads. The array is incubated with a protein extract or with plasma/serum to measure up to 7,000 proteins simultaneously in a single sample. In reverse-phase arrays, protein lysates from up to 1,000 samples are printed onto a solid phase such as glass slides or nitrocellulose membranes. The arrays are then incubated with a single antibody in optimized conditions. Up to 500 antibodies can be measured in a single experiment by staining replicate arrays. As an example of spatially resolved single-cell proteomics in the CODEX platform, slides with the tissue of interest are incubated with a DNA-conjugated antibody solution, which can contain up to 60 different antibodies. Antibody binding is then revealed through multiple cycles of hybridization with fluorescent-labeled complementary DNA, imaging, and removal of the fluorescent DNA. Image processing then allows alignment of images on the tissue slide with segmentation of individual cells facilitating spatial single-cell analysis. Other spatially oriented approaches such as cyclic-immunofluorescence (cyc-IF) and multiplex IHC (mIHC) use a variation of the approach encompassing the concept of staining, imaging, and erasing signals through multiple cycles.

or tumor microenvironment as they are basically a "grind and bind" approach that homogenizes the tissue and its cellular content. For that reason, multiplexed antibody-based tissue imaging approaches have become increasingly popular, and, as they are becoming more accessible, the field of oncology is achieving major advances.

Single-Cell Proteomics

There is growing evidence that phenotypic heterogeneity is an important mediator of breast cancer progression and response to therapy (Georgopoulou et al. 2021). Importantly, phenotypic heterogeneity is measured at the protein level and is genotype-independent (Luo et al. 2009; Labrie et al. 2022). Tumor architecture and composition can be assessed through spatially resolved antibody-based, single-cell proteomic approaches, including codetection by indexing (CODEX) (Phillips et al. 2021), multiplex cyclic-immunofluorescence (cyc-IF) (Lin et al. 2015), and multiplex IHC (mIHC) (Banik et al. 2020). Antibody-based MS approaches for spatial resolution also exist and are described below. Importantly, each of those approaches has their own strengths and limitations and can be used for different purposes. For example, CODEX is a spatially resolved single-cell proteomic approach that has been used extensively in the field of immuno-oncology and has helped delineate the immune and tumor microenvironment contexture in many cancer models. This approach relies on multiplexing of DNA-conjugated antibodies and allows the spatial analysis of up to 60 proteins at the single-cell level (Phillips et al. 2021). Briefly, after a single incubation of a tissue slide with a panel of conjugated antibodies, several rounds of hybridization with fluorescent-labeled DNA, imaging and stripping of the probes are performed. Images are then incorporated into a bioinformatic pipeline for registration, segmentation, and single-cell analysis. Other approaches such as Cyc-IF and mIHC are comparable to CODEX and might represent cheaper and more flexible alternatives. Cyc-IF consists of sequential staining, imaging, and quenching of a tissue slide and relies on fluorescently labeled primary antibodies. mIHC consists of rounds of IHC staining and imaging, followed by antibody stripping on tissue slides between each round. Both approaches are flexible in terms of antibody choices and incubation method, but Cyc-IF is limited by the lack of signal amplification, which can affect the intensity of the staining. mIHC is limited by the number of antibodies that can be used per round and the number of rounds of antibody stripping is limited by the tissue remaining after each round. Recently, Cyc-IF and mIHC have been used by our group to study the adaptive responses and resistance to targeted therapies in breast tumors (Labrie et al. 2021; Li et al. 2021a), helping us to understand how therapy alters tumor composition and immune contexture and informing possible personalized treatments for those patients. Taken together, these new technologies add an additional dimension to our understanding of the breast tumor microenvironment by helping delineate the tumor-immune microenvironment, precisely identifying cell subpopulations, and allowing detection of rare cell populations and interactions between different populations in the tumor.

Combining Antibody-Based Approaches with Mass Spectrometry

Antibody-based assays and MS are not mutually exclusive and can be combined into very sensitive and robust proteomic approaches. Immunoprecipitation-MS (IP-MS) is routinely used for sensitive and accurate characterization of protein complexes. The immuno-MRM assay mentioned above couples antibody-based enrichment of targets with highly specific information on the peptide sequence from MS, providing high sensitivity and specificity in targeted immunopeptide quantification.

Spatially oriented single-cell analysis can also benefit from the combination of antibody and MS. For example, imaging mass cytometry (Jackson et al. 2020; Georgopoulou et al. 2021), which relies on metal conjugated primary antibodies, has been used to study the phenotypic landscape of breast tumors (Ali et al. 2020; Jackson et al. 2020; Kuett et al. 2022). These studies have helped characterize patient-specific intratumor phenotypic architecture, composition, and heterogeneity, while associating those features to distinct clinical outcomes. Multiplexed ion beam imaging (MIBI), also relies on metal-labeled primary antibodies but uses a tunable ion beam instead of a laser for sample acquisition (Baharlou et al. 2019). MIBI has been shown to be highly sensitive, with a quantitative dynamic range for up to 36 antibodies measured simultaneously (Keren et al. 2019). MIBI has also been used to investigate composition of breast tumors and has revealed the relationship between immune populations, tumor stroma composition, and breast cancer progression (Keren et al. 2019; Risom et al. 2022).

Combining Antibody-Based Approaches with Other Omic Platforms

In addition to being limited by the availability of high-quality commercial antibodies, antibody-based approaches are also limited by the number of proteins that can be measured in each assay. Compared to transcriptomics, the field of proteomics has lagged behind in terms of capacity. Nonetheless, a combined transcriptomic and proteomic profile of a sample in a spatially resolved and single-cell manner adds valuable information to analysis of breast cancer samples and models. A few approaches can be used to combine proteomics and transcriptomics at the single-cell level. Some of those approaches are the Cellular Indexing of Transcriptomes and Epitopes by Sequencing (CITE-seq) and the Nanostring CosMx spatial molecular imager (SMI). Additional companies such as 10X and Akoya are in the process of implementing new technologies that combine spatially resolved analysis of protein and RNA.

CITE-seq combines single-cell transcriptome information with epitope detection of thousands of single cells by unbiased transcriptome profiling and detection of antibody-based protein markers in parallel (Stoeckius et al. 2017). CITE-seq extends invaluable single-cell RNA sequencing (scRNA-seq) data to the analysis of the immune landscape to further characterize the interaction between tumor cells and the immune milieu. Recently, BioLegend released its universal antibody cocktail, validated by flow cytometry, for both human and mouse tissue containing over 100 pretitrated antibodies for comprehensive immune cell profiling that can enable characterizing immune profiles in concert with transcriptomics (Stoeckius et al. 2017). Combining CITE-seq data to spatial transcriptomic architecture along with pathology has identified nine tumor ecotypes that had high power for predicting different outcomes in breast cancer (Wu et al. 2021). Combined CITE-seq and single-cell T-cell receptor sequencing (scTCR-seq) has identified multiple immunotypes and associated gene sets that were correlated with T-cell expansion after anti-PD1 treatment (Bassez et al. 2021). CITE-seq has been used to extend predictive capacity of RNA velocity analysis by adding protein information, which increases estimation of future cell states of individual cells (Gorin et al. 2020). Although CITE-seq has great potential in combining RNA and protein information, it is still being optimized and is limited by several factors, including the inability to detect cytoplasmic proteins and the lack of spatial information. Also, as CITE-seq requires single-cell preparation, the effects of isolation on cell viability might influence the results.

SMI consists of cyclic in situ hybridization of tissue sections (frozen or FFPE) and is able to capture hundreds to thousands of RNAs and up to 100 proteins from single cells with subcellular resolution (Rad et al. 2021). Its capacity for subcellular detection of 1,000 plex RNA as well as 100-plex protein markers with 3D spatial mapping information makes Nanostring SMI a useful tool for cell typing and characterizing cell states and cell–cell interactions. Using MER-FISH to assess subcellular RNA distribution, with the assumption that the first-order time derivation of mRNA abundance in the cytoplasm can be determined by the balance between RNA export from nuclei and degradation in the cytoplasm, has extended the utility of RNA velocity analysis (Xia et al. 2019). The integration of subcellular localization of RNA, protein distribution, and PTM, and spatial milieu of individual cell types suggests that simultaneous high plex analysis of protein and RNA with subcellular resolution on the Nanostring, 10X, and other emerging platforms promises to revolutionize studies in intratumoral heterogeneity. However, current platforms are significantly limited in terms of the numbers of assessable RNAs and proteins compared to current scRNA-seq proteomic technologies, indicating the need for continued assay optimization and increasing the number of RNA and protein probes included on panels.

Translation to the Clinic

While clinical standard of care for breast cancer diagnosis includes IHC for a panel of standard markers, RPPA has been applied in preclinical studies that have subsequently inspired the opening of clinical trials, as well as analysis of clinical trial specimens. In 2010, I-SPY (Investigation of Serial Studies to Predict Your Therapeutic Response with Imaging and Molecular Analysis) became the first clinical study to incorporate RPPA into the analysis of frozen core needle biopsies and standardized the approach (Mueller et al. 2010). This spurred the creation of I-SPY2, which has been used to classify stage II/III breast cancer into 10 subtypes as well as establish an infrastructure where imaging, FISH, IHC, and RPPA could be integrated into a data-sharing system to find noninvasive biomarkers and determine drug therapy responses (Wang and Yee 2019). The individualized approach has expedited the selection of drug combination therapies, development, and market approval (Wang and Yee 2019).

Similarly, Side Out 2 has employed multiomic molecular profiling (MoMP) to create personalized therapies in treatment-resistant metastatic breast cancer (Mueller et al. 2010). Samples are analyzed using IHC, transcript arrays, and RPPA, which then informs treatment selection. The outcomes are measured via a growth modulation index (GMI), which is calculated as a ratio of progression-free survival (PFS) to the time to progression (TTP) (Pierobon et al. 2022). The trial displayed an improved response to irinotecan-based regimens in patients with high TOPO1 expression. After multiple unsuccessful attempts with standard treatments for treatment-refractory metastatic breast cancer, most of the study participants achieved a GMI ≥ 1.3 following MoMP-informed therapy, indicating a positive therapeutic response (Pierobon et al. 2022).

Additionally, antibody-based technologies like Cyc-IF and mIHC allow researchers to study the tumor microenvironment in 3D, enabling the identification of information critical for understanding growth and metastasis, like tumor cell invasion and tissue architecture, that was previously lost in 2D models (Kuett et al. 2022). Combining this methodology with DNA, RNA, and proteomic profiles and 2D and 3D SEM, researchers evaluated an individual with HR-positive, HER2-normal, metastatic breast cancer over 3 and a half years, taking multiple biopsies throughout the duration of treatment (Johnson et al. 2022). This study, through the SMMART (Serial Measurements of Molecular and Architectural Responses to Therapy) program and tumor board, revealed the feasibility of obtaining and analyzing longitudinal metadata in a clinical setting to further the understanding of mechanisms of drug resistance (Johnson et al. 2022). We have translated the Nanostring Digital Spatial Profiling (DSP) platform to the CLIA environment allowing the simultaneous measurement of 90 protein and phosphoprotein targets on a single slide with spatial resolution for the tumor and adjacent stroma. This DSP platform is routinely used in our tumor board approaches in SMMART to identify approaches for patient management and to select combination therapy for patients in clinical trials.

Antibody-based technology was also recently used to investigate proteomic changes following PARP and MEK inhibitor combination therapy in RAS-mutant cancers, a resistant tumor lineage with limited treatment options. The therapy was effective regardless of BRCA1/2 and p53 status and produced cytotoxic effects in RAS-mutant ovarian, pancreatic, and melanoma tumor models (Sun et al. 2017). In another study, RPPA analysis was able to show increased androgen receptor (AR), BCL2, FASN, and phospho-ACC protein levels in TNBC with AR positivity, opening the door for potential therapeutic vulnerabilities and further demonstrating the need for antibody-based proteomic research in the understanding of cancer (Blucher et al. 2020).

To advance the study of adaptive responses in breast cancers, researchers established the combinatorial adaptive response therapy (CART) platform. Following the application of RPPA, a subsequent rational combination therapy was designed through investigation of individual proteins and geared toward the inhibition of key signaling pathways (Sun et al. 2020). By blocking "rewired" signaling pathways that a cancer cell relies on to resist therapy, the combination is then expected to produce synthetic lethality. This technique thus offers an opportunity to identify, prevent, and overcome therapeutic resistance. The CART strategy can uncover biomarkers of sensitivity for single and combination therapies including chemotherapy, radiation, and immunotherapy. The platform has already been used to examine possible posttreatment adaptive responses with targets such as PARP, PI3K/AKT/mTOR, FAK, MEK/ERK inhibitors, EGFR antibodies, HDAC, BRD4, CHK1, WEE1, and ATR inhibitors (Sun et al. 2020).

In summary, antibody-based platforms have been critical for establishing drug targets at the posttranslational pathway level, profiling intracellular and extracellular oncogenic signaling pathways, providing more accurate EC50 values, engaging in predictive pharmacodynamics, and exploring rational drug combinations (Akbani et al. 2014a). These advances show how proteomic mapping of phosphorylation, acetylation, and protein cleavage end points to critical signaling pathways is a robust approach for informing combinatorial therapy and can be used to understand drug resistance mechanisms and identify biomarkers (Sheehan et al. 2005).

CONCLUDING REMARKS ON STRENGTHS/ WEAKNESSES OF MASS SPECTROMETRY- AND ANTIBODY-BASED PROTEOMIC TECHNOLOGIES

To facilitate the selection of MS- and antibody-based proteomic technologies for breast cancer research, we provide a high-level summary of the strengths and weaknesses of the two technol-

ogies (Table 1). Both approaches are able to analyze multiple samples at the same time. Using the TMT system where peptides from different sam ples are labeled with a distinct isobaric tag, up to 18 samples can be analyzed in a single MS experiment while with RPPA, it is possible to examine up to 1,000 samples simultaneously. The number of proteins quantified in antibody arrays like RPPA is more limited compared to MS-based methods and also critically dependent on the availability of high-quality antibodies against the protein of interest. MS is more appropriate for unbiased profiling where there are no or limited assumptions on proteins of interest and also well-suited toward novel protein biomarker discovery for which no antibody exists. In contrast, panels can be designed for antibody arrays targeting cancer-specific proteins with well-described roles. Validated antibodies must pass rigorous quantity

Table 1. Strengths and weaknesses of mass spectrometry (MS)- and antibody-based proteomic technologies

Feature	MS	Antibody
Typical input sample amount	20 mg–100 mg tumor tissue	40 µg protein lysate (reverse-phase protein assay [RPPA]) 55 µL plasma/serum (aptamer array)
Sample throughput	Up to 18 samples per tandem mass tag (TMT) plex, integration of data from TMT plexes, or individual label-free experiments requires appropriate normalization	Up to 1,000 samples per slide (RPPA)
Number of analytes	10–15 K proteins 30–50 K phosphopeptides 5–10 K acetylated peptides 10–20 K ubiquitinated peptides 10–15 K glycosylated peptides 1–10 K MHC-bound peptides	500 proteins and posttranslationally modified proteins (RPPA) 7 K proteins and posttranslationally modified proteins (aptamer array)
Protein detection	Limited by sample complexity	Limited by antibody availability
Low abundance protein quantification	Requires enrichment	Yes
Absolute quantification	Yes, with spiked-in, heavy-isotope-labeled peptides	Yes, with standard curve from purified proteins
Protein–protein interaction detection	Affinity purification–MS (AP-MS)	Immunoprecipitation
Single and spatial resolution	Yes	Yes
Data processing	Computationally intensive	Relatively less computationally demanding
Studies on clinical utility	Yes	Yes

Cite this article as *Cold Spring Harb Perspect Med* doi: 10.1101/cshperspect.a041323

control metrics and are able to quantify low-abundance proteins such as PTMs. Enrichment of PTM peptides with metal columns or antibodies against peptides of interest is required to detect low-abundance proteins prior to MS, emphasizing the complementarity of these approaches. Once detected, both approaches are able to provide quantitative data using spiked-in, heavy-isotope-labeled peptides for MS and standard curves constructed from purified proteins from antibody arrays to precisely quantify proteins of interest. Both approaches are also able to provide spatially resolved and single-cell-level proteomic characterization. Although data processing can be more computationally intensive for discovery-based MS proteomics compared to antibody arrays, spatial and single-cell analyses are more complex and require additional informatics skills to analyze for both approaches.

In conclusion, both MS- and antibody-based methods have unique advantages and disadvantages. These two methods are not replacements for one another, but complementary approaches that have defined and characterized the breast cancer proteome, leading toward a better understanding of underlying disease mechanisms and identifying markers that may better guide treatment decisions for patients.

REFERENCES

Abelin JG, Bergstrom EJ, Taylor HB, Rivera KD, Klaeger S, Xu C, White CJ, Olive ME, Maynard M, Kane MH, et al. 2022. MONTE enables serial immunopeptidome, ubiquitylome, proteome, phosphoproteome, acetylome analyses of sample-limited tissues. bioRxiv doi:10.1101/2021 06 22 449417

Akbani R, Becker KF, Carragher N, Goldstein T, de Koning L, Korf U, Liotta L, Mills GB, Nishizuka SS, Pawlak M, et al. 2014a. Realizing the promise of reverse phase protein arrays for clinical, translational, and basic research: a workshop report: the RPPA (reverse-phase protein array) society. Mol Cell Proteomics 13: 1625–1643. doi:10.1074/mcp.O113.034918

Akbani R, Ng PK, Werner HM, Shahmoradgoli M, Zhang F, Ju Z, Liu W, Yang JY, Yoshihara K, Li J, et al. 2014b. A pan-cancer proteomic perspective on The Cancer Genome Atlas. Nat Commun 5: 3887. doi:10.1038/ncomms4887

Ali HR, Jackson HW, Zanotelli VRT, Danenberg E, Fischer JR, Bardwell H, Provenzano E; CRUK IMAXT Grand Challenge Team; Rueda OM, Chin SF, et al. 2020. Imaging mass cytometry and multiplatform genomics define the phenogenomic landscape of breast cancer. Nat Cancer 1: 163–175. doi:10.1038/s43018-020-0026-6

Asleh K, Negri GL, Spencer Miko SE, Colborne S, Hughes CS, Wang XQ, Gao D, Gilks CB, Chia SKL, Nielsen TO, et al. 2022. Proteomic analysis of archival breast cancer clinical specimens identifies biological subtypes with distinct survival outcomes. Nat Commun 13: 896. doi:10.1038/s41467-022-28524-0

Baharlou H, Canete NP, Cunningham AL, Harman AN, Patrick E. 2019. Mass cytometry imaging for the study of human diseases—applications and data analysis strategies. Front Immunol 10: 2657. doi:10.3389/fimmu.2019.02657

Banik G, Betts CB, Liudahl SM, Sivagnanam S, Kawashima R, Cotechini T, Larson W, Goecks J, Pai SI, Clayburgh DR, et al. 2020. High-dimensional multiplexed immunohistochemical characterization of immune contexture in human cancers. Methods Enzymol 635: 1–20. doi:10.1016/bs.mie.2019.05.039

Bassez A, Vos H, Van Dyck L, Floris G, Arijs I, Desmedt C, Boeckx B, Vanden Bempt M, Nevelsteen I, Lambein K, et al. 2021. A single-cell map of intratumoral changes during anti-PD1 treatment of patients with breast cancer. Nat Med 27: 820–832. doi:10.1038/s41591-021-01323-8

Bauer A, Kuster B. 2003. Affinity purification-mass spectrometry. Powerful tools for the characterization of protein complexes. Eur J Biochem 270: 570–578. doi:10.1046/j.1432-1033.2003.03428.x

Beausoleil SA, Villén J, Gerber SA, Rush J, Gygi SP. 2006. A probability-based approach for high-throughput protein phosphorylation analysis and site localization. Nat Biotechnol 24: 1285–1292. doi:10.1038/nbt1240

Berns K, Horlings HM, Hennessy BT, Madiredjo M, Hijmans EM, Beelen K, Linn SC, Gonzalez-Angulo AM, Stemke-Hale K, Hauptmann M, et al. 2007. A functional genetic approach identifies the PI3K pathway as a major determinant of trastuzumab resistance in breast cancer. Cancer Cell 12: 395–402. doi:10.1016/j.ccr.2007.08.030

Bjornson RD, Carriero NJ, Colangelo C, Shifman M, Cheung KH, Miller PL, Williams K. 2008. X!!tandem, an improved method for running X!tandem in parallel on collections of commodity computers. J Proteome Res 7: 293–299. doi:10.1021/pr0701198

Blucher AS, Mills GB, Tsang YH. 2020. Which path to follow? Utilizing proteomics to improve therapy choices for breast cancer patients. Expert Rev Proteomics 17: 187–190. doi:10.1080/14789450.2020.1757442

Blum A, Wang P, Zenklusen JC. 2018. Snapshot: TCGA-analyzed tumors. Cell 173: 530. doi:10.1016/j.cell.2018.03.059

Bouchal P, Schubert OT, Faktor J, Capkova L, Imrichova H, Zoufalova K, Paralova V, Hrstka R, Liu Y, Ebhardt HA, et al. 2019. Breast cancer classification based on proteotypes obtained by SWATH mass spectrometry. Cell Rep 28: 832–843.e7. doi:10.1016/j.celrep.2019.06.046

Bruderer R, Bernhardt OM, Gandhi T, Miladinović SM, Cheng LY, Messner S, Ehrenberger T, Zanotelli V, Butscheid Y, Escher C, et al. 2015. Extending the limits of quantitative proteome profiling with data-independent acquisition and application to acetaminophen-treated three-dimensional liver microtissues. Mol Cell Proteomics 14: 1400–1410. doi:10.1074/mcp.M114.044305

Buczak K, Ori A, Kirkpatrick JM, Holzer K, Dauch D, Roessler S, Endris V, Lasitschka F, Parca L, Schmidt A, et al. 2018. Spatial tissue proteomics quantifies inter- and intratumor heterogeneity in hepatocellular carcinoma (HCC). *Mol Cell Proteomics* **17**: 810–825. doi:10.1074/mcp.RA117.000189

Chan JCY, Zhou L, Chan ECY. 2015. The isotope-coded affinity tag method for quantitative protein profile comparison and relative quantitation of cysteine redox modifications. *Curr Protoc Protein Sci* **82**: 23.2.1–23.2.19.

Chen MM, Li J, Wang Y, Akbani R, Lu Y, Mills GB, Liang H. 2019. TCPA v3.0: an integrative platform to explore the pan-cancer analysis of functional proteomic data. *Mol Cell Proteomics* **18**: S15–S25. doi:10.1074/mcp.RA118.001260

Chi H, Sun RX, Yang B, Song CQ, Wang LH, Liu C, Fu Y, Yuan ZF, Wang HP, He SM, et al. 2010. Pnovo: de novo peptide sequencing and identification using HCD spectra. *J Proteome Res* **9**: 2713–2724. doi:10.1021/pr100182k

Cox J, Neuhauser N, Michalski A, Scheltema RA, Olsen JV, Mann M. 2011. Andromeda: a peptide search engine integrated into the MaxQuant environment. *J Proteome Res* **10**: 1794–1805. doi:10.1021/pr101065j

Craig R, Beavis RC. 2003. A method for reducing the time required to match protein sequences with tandem mass spectra. *Rapid Commun Mass Spectrom* **17**: 2310–2316. doi:10.1002/rcm.1198

Craig R, Beavis RC. 2004. TANDEM: matching proteins with tandem mass spectra. *Bioinformatics* **20**: 1466–1467. doi:10.1093/bioinformatics/bth092

Demichev V, Messner CB, Vernardis SI, Lilley KS, Ralser M. 2020. DIA-NN: neural networks and interference correction enable deep proteome coverage in high throughput. *Nat Methods* **17**: 41–44. doi:10.1038/s41592-019-0638-x

Deutsch EW. 2010. The PeptideAtlas Project. *Methods Mol Biol* **604**: 285–296. doi:10.1007/978-1-60761-444-9_19

Doerr A. 2013. Mass spectrometry-based targeted proteomics. *Nat Methods* **10**: 23. doi:10.1038/nmeth.2286

Doerr A. 2015. DIA mass spectrometry. *Nat Methods* **12**: 35–35. doi:10.1038/nmeth.3234

Eng JK, McCormack AL, Yates JR. 1994. An approach to correlate tandem mass spectral data of peptides with amino acid sequences in a protein database. *J Am Soc Mass Spectrom* **5**: 976–989. doi:10.1016/1044-0305(94)80016-2

Evans KW, Yuca E, Akcakanat A, Scott SM, Arango NP, Zheng X, Chen K, Tapia C, Tarco E, Eterovic AK, et al. 2017. A population of heterogeneous breast cancer patient-derived xenografts demonstrate broad activity of PARP inhibitor in BRCA1/2 wild-type tumors. *Clin Cancer Res* **23**: 6468–6477. doi:10.1158/1078-0432.CCR-17-0615

Farahmand L, Merikhian P, Jalili N, Darvishi B, Majidzadeh AK. 2018. Significant role of MUC1 in development of resistance to currently existing anti-cancer therapeutic agents. *Curr Cancer Drug Targets* **18**: 737–748. doi:10.2174/1568009617666170623113520

Frank A, Pevzner P. 2005. Pepnovo: de novo peptide sequencing via probabilistic network modeling. *Anal Chem* **77**: 964–973. doi:10.1021/ac048788h

Gallien S, Kim SY, Domon B. 2015. Large-scale targeted proteomics using internal standard triggered-parallel reaction monitoring (IS-PRM). *Mol Cell Proteomics* **14**: 1630–1644. doi:10.1074/mcp.O114.043968

Georgopoulou D, Callari M, Rueda OM, Shea A, Martin A, Giovannetti A, Qosaj F, Dariush A, Chin SF, Carnevalli LS, et al. 2021. Landscapes of cellular phenotypic diversity in breast cancer xenografts and their impact on drug response. *Nat Commun* **12**: 1998. doi:10.1038/s41467-021-22303-z

Gessulat S, Schmidt T, Zolg DP, Samaras P, Schnatbaum K, Zerweck J, Knaute T, Rechenberger J, Delanghe B, Huhmer A, et al. 2019. Prosit: proteome-wide prediction of peptide tandem mass spectra by deep learning. *Nat Methods* **16**: 509–518. doi:10.1038/s41592-019-0426-7

Gorin G, Svensson V, Pachter L. 2020. Protein velocity and acceleration from single-cell multiomics experiments. *Genome Biol* **21**: 39. doi:10.1186/s13059-020-1945-3

Gromova I, Espinoza JA, Grauslund M, Santoni-Rugiu E, Møller Talman ML, van Oostrum J, Moreira JMA. 2021. Functional proteomic profiling of triple-negative breast cancer. *Cells* **10**: 2768. doi:10.3390/cells10102768

Herrera JA, Mallikarjun V, Rosini S, Montero MA, Lawless C, Warwood S, O'Cualain R, Knight D, Schwartz MA, Swift J. 2020. Laser capture microdissection coupled mass spectrometry (LCM-MS) for spatially resolved analysis of formalin-fixed and stained human lung tissues. *Clin Proteomics* **17**: 24. doi:10.1186/s12014-020-09287-6

Hsu PY, Wu VS, Kanaya N, Petrossian K, Hsu HK, Nguyen D, Schmolze D, Kai M, Liu CY, Lu H, et al. 2018. Dual mTOR kinase inhibitor MLN0128 sensitizes HR$^+$/HER2$^+$ breast cancer patient-derived xenografts to trastuzumab or fulvestrant. *Clin Cancer Res* **24**: 395–406. doi:10.1158/1078-0432.CCR-17-1983

Jackson HW, Fischer JR, Zanotelli VRT, Ali HR, Mechera R, Soysal SD, Moch H, Muenst S, Varga Z, Weber WP, et al. 2020. The single-cell pathology landscape of breast cancer. *Nature* **578**: 615–620. doi:10.1038/s41586-019-1876-x

Johansson HJ, Socciarelli F, Vacanti NM, Haugen MH, Zhu Y, Siavelis I, Fernandez-Woodbridge A, Aure MR, Sennblad B, Vesterlund M, et al. 2019. Breast cancer quantitative proteome and proteogenomic landscape. *Nat Commun* **10**: 1600. doi:10.1038/s41467-019-09018-y

Johnson BE, Creason AL, Stommel JM, Keck JM, Parmar S, Betts CB, Blucher A, Boniface C, Bucher E, Burlingame E, et al. 2022. An omic and multidimensional spatial atlas from serial biopsies of an evolving metastatic breast cancer. *Cell Rep Med* **3**: 100525. doi:10.1016/j.xcrm.2022.100525

Kennedy JJ, Whiteaker JR, Kennedy LC, Bosch DE, Lerch ML, Schoenherr RM, Zhao L, Lin C, Chowdhury S, Kilgore MR, et al. 2021. Quantification of human epidermal growth factor receptor 2 by immunopeptide enrichment and targeted mass spectrometry in formalin-fixed paraffin-embedded and frozen breast cancer tissues. *Clin Chem* **67**: 1008–1018. doi:10.1093/clinchem/hvab047

Keren L, Bosse M, Thompson S, Risom T, Vijayaragavan K, McCaffrey E, Marquez D, Angoshtari R, Greenwald NF, Fienberg H, et al. 2019. MIBI-TOF: a multiplexed imaging platform relates cellular phenotypes and tissue structure. *Sci Adv* **5**: eaax5851. doi:10.1126/sciadv.aax5851

Cite this article as *Cold Spring Harb Perspect Med* doi: 10.1101/cshperspect.a041323

Kim S, Pevzner PA. 2014. MS-GF⁺ makes progress towards a universal database search tool for proteomics. *Nat Commun* **5**: 5277. doi:10.1038/ncomms6277

Kim M, Park J, Bouhaddou M, Kim K, Rojc A, Modak M, Soucheray M, McGregor MJ, O'Leary P, Wolf D, et al. 2021. A protein interaction landscape of breast cancer. *Science* **374**: eabf3066. doi:10.1126/science.abf3066

Kong AT, Leprevost FV, Avtonomov DM, Mellacheruvu D, Nesvizhskii AI. 2017. MSFragger: ultrafast and comprehensive peptide identification in mass spectrometry-based proteomics. *Nat Methods* **14**: 513–520. doi:10.1038/nmeth.4256

Krug K, Jaehnig EJ, Satpathy S, Blumenberg L, Karpova A, Anurag M, Miles G, Mertins P, Geffen Y, Tang LC, et al 2020. Proteogenomic landscape of breast cancer tumorigenesis and targeted therapy. *Cell* **183**: 1436–1456.e31. doi:10.1016/j.cell.2020.10.036

Kuett L, Catena R, Özcan A, Plüss A; Cancer Grand Challenges IMAXT Consortium; Schraml P, Moch H, de Souza N, Bodenmiller B. 2022. Three-dimensional imaging mass cytometry for highly multiplexed molecular and cellular mapping of tissues and the tumor microenvironment. *Nat Cancer* **3**: 122–133. doi:10.1038/s43018-021-00301-w

Kusebauch U, Campbell DS, Deutsch EW, Chu CS, Spicer DA, Brusniak MY, Slagel J, Sun Z, Stevens J, Grimes B, et al. 2016. Human SRMAtlas: a resource of targeted assays to quantify the complete human proteome. *Cell* **166**: 766–778. doi:10.1016/j.cell.2016.06.041

Labrie M, Kim TB, Ju Z, Lee S, Zhao W, Fang Y, Lu Y, Chen K, Ramirez P, Frumovitz M, et al. 2019. Adaptive responses in a PARP inhibitor window of opportunity trial illustrate limited functional interlesional heterogeneity and potential combination therapy options. *Oncotarget* **10**: 3533–3546. doi:10.18632/oncotarget.26947

Labrie M, Li A, Creason A, Betts C, Keck J, Johnson B, Sivagnanam S, Boniface C, Ma H, Blucher A, et al. 2021. Multiomics analysis of serial PARP inhibitor treated metastatic TNBC inform on rational combination therapies. *NPJ Precis Oncol* **5**: 92. doi:10.1038/s41698-021-00232-w

Labrie M, Brugge JS, Mills GB, Zervantonakis IK. 2022. Therapy resistance: opportunities created by adaptive responses to targeted therapies in cancer. *Nat Rev Cancer* **22**: 323–339. doi:10.1038/s41568-022-00454-5

Lange V, Picotti P, Domon B, Aebersold R. 2008. Selected reaction monitoring for quantitative proteomics: a tutorial. *Mol Syst Biol* **4**: 222. doi:10.1038/msb.2008.61

Li Y, Zhong CQ, Xu X, Cai S, Wu X, Zhang Y, Chen J, Shi J, Lin S, Han J. 2015. Group-DIA: analyzing multiple data-independent acquisition mass spectrometry data files. *Nat Methods* **12**: 1105–1106. doi:10.1038/nmeth.3593

Li J, Akbani R, Zhao W, Lu Y, Weinstein JN, Mills GB, Liang H. 2017. Explore, visualize, and analyze functional cancer proteomic data using the Cancer Proteome Atlas. *Cancer Res* **77**: e51–e54. doi:10.1158/0008-5472.CAN-17-0369

Li K, Vaudel M, Zhang B, Ren Y, Wen B. 2019. PDV: an integrative proteomics data viewer. *Bioinformatics* **35**: 1249–1251. doi:10.1093/bioinformatics/bty770

Li K, Jain A, Malovannaya A, Wen B, Zhang B. 2020. Deep-rescore: leveraging deep learning to improve peptide iden-

tification in immunopeptidomics. *Proteomics* **20**: 1900334. doi:10.1002/pmic.201900334

Li A, Keck JM, Parmar S, Patterson J, Labrie M, Creason AL, Johnson BE, Downey M, Thomas G, Beadling C, et al. 2021a. Characterizing advanced breast cancer heterogeneity and treatment resistance through serial biopsies and comprehensive analytics. *NPJ Precis Oncol* **5**: 28. doi:10.1038/s41698-021-00165-4

Li J, Cai Z, Bomgarden RD, Pike I, Kuhn K, Rogers JC, Roberts TM, Gygi SP, Paulo JA. 2021b. TMTpro-18plex: the expanded and complete set of TMTpro reagents for sample multiplexing. *J Proteome Res* **20**: 2964–2972. doi:10.1021/acs.jproteome.1c00168

Lin JR, Fallahi-Sichani M, Sorger PK. 2015. Highly multiplexed imaging of single cells using a high-throughput cyclic immunofluorescence method. *Nat Commun* **6**: 8390. doi:10.1038/ncomms9390

Lindholm EM, Krohn M, Iadevaia S, Kristian A, Mills GB, Mælandsmo GM, Engebraaten O. 2014. Proteomic characterization of breast cancer xenografts identifies early and late bevacizumab-induced responses and predicts effective drug combinations. *Clin Cancer Res* **20**: 404–412. doi:10.1158/1078-0432.CCR-13-1865

Low TY, Syafruddin SE, Mohtar MA, Vellaichamy A, A Rahman NS, Pung YF, Tan CSH. 2021. Recent progress in mass spectrometry-based strategies for elucidating protein–protein interactions. *Cell Mol Life Sci* **78**: 5325–5339. doi:10.1007/s00018-021-03856-0

Lu Y, Ling S, Hegde AM, Byers LA, Coombes K, Mills GB, Akbani R. 2016. Using reverse-phase protein arrays as pharmacodynamic assays for functional proteomics, biomarker discovery, and drug development in cancer. *Semin Oncol* **43**: 476–483. doi:10.1053/j.seminoncol.2016.06.005

Luo J, Solimini NL, Elledge SJ. 2009. Principles of cancer therapy: oncogene and non-oncogene addiction. *Cell* **136**: 823–837. doi:10.1016/j.cell.2009.02.024

Ma B, Johnson R. 2012. De novo sequencing and homology searching. *Mol Cell Proteomics* **11**: O111.014902. doi:10.1074/mcp.O111.014902

Ma B, Zhang K, Hendrie C, Liang C, Li M, Doherty-Kirby A, Lajoie G. 2003. PEAKS: powerful software for peptide de novo sequencing by tandem mass spectrometry. *Rapid Commun Mass Spectrom* **17**: 2337–2342. doi:10.1002/rcm.1196

Makarov A, Denisov E, Lange O. 2009. Performance evaluation of a high-field Orbitrap mass analyzer. *J Am Soc Mass Spectrom* **20**: 1391–1396. doi:10.1016/j.jasms.2009.01.005

Mardamshina M, Shenoy A, Necula D, Krol K, Pirak D, Itzhacky N, Marin I, Shalmon B, Sharan R, Gal-Yam E, et al. 2021. Proteomic landscape of multi-layered breast cancer internal tumor heterogeneity. bioRxiv doi:10.1101/2021.08.05.455361

Martin DB, Holzman T, May D, Peterson A, Eastham A, Eng J, McIntosh M. 2008. MRMer, an interactive open source and cross-platform system for data extraction and visualization of multiple reaction monitoring experiments. *Mol Cell Proteomics* **7**: 2270–2278. doi:10.1074/mcp.M700504-MCP200

Mertins P, Yang F, Liu T, Mani DR, Petyuk VA, Gillette MA, Clauser KR, Qiao JW, Gritsenko MA, Moore RJ, et al.

2014. Ischemia in tumors induces early and sustained phosphorylation changes in stress kinase pathways but does not affect global protein levels. *Mol Cell Proteomics* **13:** 1690–1704. doi:10.1074/mcp.M113.036392

Mertins P, Mani DR, Ruggles KV, Gillette MA, Clauser KR, Wang P, Wang X, Qiao JW, Cao S, Petralia F, et al. 2016. Proteogenomics connects somatic mutations to signalling in breast cancer. *Nature* **534:** 55–62. doi:10.1038/nature18003

Mertins P, Tang LC, Krug K, Clark DJ, Gritsenko MA, Chen L, Clauser KR, Clauss TR, Shah P, Gillette MA, et al. 2018. Reproducible workflow for multiplexed deep-scale proteome and phosphoproteome analysis of tumor tissues by liquid chromatography-mass spectrometry. *Nat Protoc* **13:** 1632–1661. doi:10.1038/s41596-018-0006-9

Modi S, Jacot W, Yamashita T, Sohn J, Vidal M, Tokunaga E, Tsurutani J, Ueno NT, Prat A, Chae YS, et al. 2022. Trastuzumab deruxtecan in previously treated HER2-low advanced breast cancer. *N Engl J Med* **387:** 9–20. doi:10.1056/NEJMoa2203690

Mueller C, Liotta LA, Espina V. 2010. Reverse phase protein microarrays advance to use in clinical trials. *Mol Oncol* **4:** 461–481. doi:10.1016/j.molonc.2010.09.003

Mund A, Coscia F, Kriston A, Hollandi R, Kovács F, Brunner AD, Migh E, Schweizer L, Santos A, Bzorek M, et al. 2022. Deep visual proteomics defines single-cell identity and heterogeneity. *Nat Biotechnol* **40:** 1231–1240. doi:10.1038/s41587-022-01302-5

Namba M, Hattori N, Hamada H, Yamaguchi K, Okamoto Y, Nakashima T, Masuda T, Sakamoto S, Horimasu Y, Miyamoto S, et al. 2019. Anti-KL-6/MUC1 monoclonal antibody reverses resistance to trastuzumab-mediated antibody-dependent cell-mediated cytotoxicity by capping MUC1. *Cancer Lett* **442:** 31–39. doi:10.1016/j.canlet.2018.10.037

Nesvizhskii AI. 2007. Protein identification by tandem mass spectrometry and sequence database searching. *Methods Mol Biol* **367:** 87–119.

Nesvizhskii AI. 2014. Proteogenomics: concepts, applications and computational strategies. *Nat Methods* **11:** 1114–1125. doi:10.1038/nmeth.3144

Ong SE, Blagoev B, Kratchmarova I, Kristensen DB, Steen H, Pandey A, Mann M. 2002. Stable isotope labeling by amino acids in cell culture, SILAC, as a simple and accurate approach to expression proteomics. *Mol Cell Proteomics* **1:** 376–386. doi:10.1074/mcp.M200025-MCP200

Oppermann FS, Gnad F, Olsen JV, Hornberger R, Greff Z, Kéri G, Mann M, Daub H. 2009. Large-scale proteomics analysis of the human kinome. *Mol Cell Proteomics* **8:** 1751–1764. doi:10.1074/mcp.M800588-MCP200

Panis C, Pizzatti L, Herrera AC, Corrêa S, Binato R, Abdelhay E. 2014. Label-free proteomic analysis of breast cancer molecular subtypes. *J Proteome Res* **13:** 4752–4772. doi:10.1021/pr500676x

Perkins DN, Pappin DJ, Creasy DM, Cottrell JS. 1999. Probability-based protein identification by searching sequence databases using mass spectrometry data. *Electrophoresis* **20:** 3551–3567. doi:10.1002/(SICI)1522-2683(19991201)20:18<3551::AID-ELPS3551>3.0.CO;2-2

Peterson AC, Russell JD, Bailey DJ, Westphall MS, Coon JJ. 2012. Parallel reaction monitoring for high resolution and high mass accuracy quantitative, targeted proteomics.

Mol Cell Proteomics **11:** 1475–1488. doi:10.1074/mcp.O112.020131

Phillips D, Schürch CM, Khodadoust MS, Kim YH, Nolan GP, Jiang S. 2021. Highly multiplexed phenotyping of immunoregulatory proteins in the tumor microenvironment by CODEX tissue imaging. *Front Immunol* **12:** 687673. doi:10.3389/fimmu.2021.687673

Pierobon M, Robert NJ, Northfelt DW, Jahanzeb M, Wong S, Hodge KA, Baldelli E, Aldrich J, Craig DW, Liotta LA, et al. 2022. Multi-omic molecular profiling guide's efficacious treatment selection in refractory metastatic breast cancer: a prospective phase II clinical trial. *Mol Oncol* **16:** 104–115. doi:10.1002/1878-0261.13091

Pino LK, Searle BC, Bollinger JG, Nunn B, MacLean B, MacCoss MJ. 2020. The Skyline ecosystem: informatics for quantitative mass spectrometry proteomics. *Mass Spectrom Rev* **39:** 229–244. doi:10.1002/mas.21540

Purcell AW, Ramarathinam SH, Ternette N. 2019. Mass spectrometry-based identification of MHC-bound peptides for immunopeptidomics. *Nat Protoc* **14:** 1687–1707. doi:10.1038/s41596-019-0133-y

Rad HS, Rad HS, Shiravand Y, Radfar P, Arpon D, Warkiani ME, O'Byrne K, Kulasinghe A. 2021. The Pandora's box of novel technologies that may revolutionize lung cancer. *Lung Cancer* **159:** 34–41. doi:10.1016/j.lungcan.2021.06.022

Risom T, Glass DR, Averbukh I, Liu CC, Baranski A, Kagel A, McCaffrey EF, Greenwald NF, Rivero-Gutiérrez B, Strand SH, et al. 2022. Transition to invasive breast cancer is associated with progressive changes in the structure and composition of tumor stroma. *Cell* **185:** 299–310.e18. doi:10.1016/j.cell.2021.12.023

Rodon J, Soria JC, Berger R, Miller WH, Rubin E, Kugel A, Tsimberidou A, Saintigny P, Ackerstein A, Braña I, et al. 2019. Genomic and transcriptomic profiling expands precision cancer medicine: the WINTHER trial. *Nat Med* **25:** 751–758. doi:10.1038/s41591-019-0424-4

Röst HL, Rosenberger G, Navarro P, Gillet L, Miladinović SM, Schubert OT, Wolski W, Collins BC, Malmström J, Malmstrom L, et al. 2014. OpenSWATH enables automated, targeted analysis of data-independent acquisition MS data. *Nat Biotechnol* **32:** 219–223. doi:10.1038/nbt.2841

Satpathy S, Jaehnig EJ, Krug K, Kim BJ, Saltzman AB, Chan DW, Holloway KR, Anurag M, Huang C, Singh P, et al. 2020. Microscaled proteogenomic methods for precision oncology. *Nat Commun* **11:** 532. doi:10.1038/s41467-020-14381-2

Searle BC, Lawrence RT, MacCoss MJ, Villén J. 2019. Thesaurus: quantifying phosphopeptide positional isomers. *Nat Methods* **16:** 703–706. doi:10.1038/s41592-019-0498-4

Sheehan KM, Calvert VS, Kay EW, Lu Y, Fishman D, Espina V, Aquino J, Speer R, Araujo R, Mills GB, et al. 2005. Use of reverse phase protein microarrays and reference standard development for molecular network analysis of metastatic ovarian carcinoma. *Mol Cell Proteomics* **4:** 346–355. doi:10.1074/mcp.T500003-MCP200

Shi Z, Wulfkuhle J, Nowicka M, Gallagher RI, Saura C, Nuciforo PG, Calvo I, Andersen J, Passos-Coelho JL, Gil-Gil MJ, et al. 2022. Functional mapping of AKT signaling and biomarkers of response from the FAIRLANE trial of neoadjuvant ipatasertib plus paclitaxel for triple-negative breast cancer. *Clin Cancer Res* **28:** 993–1003. doi:10.1158/1078-0432.CCR-21-2498

Slavov N. 2020. Unpicking the proteome in single cells. *Science* **367:** 512–513. doi:10.1126/science.aaz6695

Stoeckius M, Hafemeister C, Stephenson W, Houck-Loomis B, Chattopadhyay PK, Swerdlow H, Satija R, Smibert P. 2017. Simultaneous epitope and transcriptome measurement in single cells. *Nat Methods* **14:** 865–868. doi:10.1038/nmeth.4380

Sun C, Fang Y, Yin J, Chen J, Ju Z, Zhang D, Chen X, Vellano CP, Jeong KJ, Ng PK, et al. 2017. Rational combination therapy with PARP and MEK inhibitors capitalizes on therapeutic liabilities in RAS mutant cancers. *Sci Transl Med* **9:** eaal5148. doi:10.1126/scitranslmed.aal5148

Sun C, Fang Y, Labrie M, Li X, Mills GB. 2020. Systems approach to rational combination therapy: PARP inhibitors. *Biochem Soc Trans* **48:** 1101–1108. doi:10.1042/BST20191092

Tabb DL, Ma ZQ, Martin DB, Ham AJ, Chambers MC. 2008. Directag: accurate sequence tags from peptide MS/MS through statistical scoring. *J Proteome Res* **7:** 3838–3846. doi:10.1021/pr800154p

Tang W, Zhou M, Dorsey TH, Prieto DA, Wang XW, Ruppin E, Veenstra TD, Ambs S. 2018. Integrated proteotranscriptomics of breast cancer reveals globally increased protein-mRNA concordance associated with subtypes and survival. *Genome Med* **10:** 94. doi:10.1186/s13073-018-0602-x

Taus T, Köcher T, Pichler P, Paschke C, Schmidt A, Henrich C, Mechtler K. 2011. Universal and confident phosphorylation site localization using phosphoRS. *J Proteome Res* **10:** 5354–5362. doi:10.1021/pr200611n

Thompson A, Schäfer J, Kuhn K, Kienle S, Schwarz J, Schmidt G, Neumann T, Johnstone R, Mohammed AK, Hamon C. 2003. Tandem mass tags: a novel quantification strategy for comparative analysis of complex protein mixtures by MS/MS. *Anal Chem* **75:** 1895–1904. doi:10.1021/ac0262560

Ting L, Rad R, Gygi SP, Haas W. 2011. MS3 eliminates ratio distortion in isobaric multiplexed quantitative proteomics. *Nat Methods* **8:** 937–940. doi:10.1038/nmeth.1714

Tran NH, Zhang X, Xin L, Shan B, Li M. 2017. De novo peptide sequencing by deep learning. *Proc Natl Acad Sci* **114:** 8247–8252. doi:10.1073/pnas.1705691114

Tsou CC, Avtonomov D, Larsen B, Tucholska M, Choi H, Gingras AC, Nesvizhskii AI. 2015. DIA-umpire: comprehensive computational framework for data-independent acquisition proteomics. *Nat Methods* **12:** 258–264, 257 p following 264. doi:10.1038/nmeth.3255

Tyanova S, Albrechtsen R, Kronqvist P, Cox J, Mann M, Geiger T. 2016. Proteomic maps of breast cancer subtypes. *Nat Commun* **7:** 10259. doi:10.1038/ncomms10259

van Bentum M, Selbach M. 2021. An introduction to advanced targeted acquisition methods. *Mol Cell Proteomics* **20:** 100165. doi:10.1016/j.mcpro.2021.100165

Vu T, Claret FX. 2012. Trastuzumab: updated mechanisms of action and resistance in breast cancer. *Front Oncol* **2:** 62.

Wang H, Yee D. 2019. I-SPY 2: a neoadjuvant adaptive clinical trial designed to improve outcomes in high-risk breast cancer. *Curr Breast Cancer Rep* **11:** 303–310. doi:10.1007/s12609-019-00334-2

Wang X, Zhang B. 2013. *customProDB*: an R package to generate customized protein databases from RNA-seq data for proteomics search. *Bioinformatics* **29:** 3235–3237. doi:10.1093/bioinformatics/btt543

Wang X, Li Y, Wu Z, Wang H, Tan H, Peng J. 2014. JUMP: a tag-based database search tool for peptide identification with high sensitivity and accuracy. *Mol Cell Proteomics* **13:** 3663–3673. doi:10.1074/mcp.O114.039586

Wang J, Ma Z, Carr SA, Mertins P, Zhang H, Zhang Z, Chan DW, Ellis MJ, Townsend RR, Smith RD, et al. 2017. Proteome profiling outperforms transcriptome profiling for coexpression based gene function prediction. *Mol Cell Proteomics* **16:** 121–134. doi:10.1074/mcp.M116.060301

Wen B, Wang X, Zhang B. 2019. Pepquery enables fast, accurate, and convenient proteomic validation of novel genomic alterations. *Genome Res* **29:** 485–493. doi:10.1101/gr.235028.118

Westin SN, Labrie M, Litton JK, Blucher A, Fang Y, Vellano CP, Marszalek JR, Feng N, Ma X, Creason A, et al. 2021. Phase Ib dose expansion and translational analyses of olaparib in combination with capivasertib in recurrent endometrial, triple-negative breast, and ovarian cancer. *Clin Cancer Res* **27:** 6354–6365. doi:10.1158/1078-0432.CCR-21-1656

Whiteaker JR, Paulovich AG. 2011. Peptide immunoaffinity enrichment coupled with mass spectrometry for peptide and protein quantification. *Clin Lab Med* **31:** 385–396. doi:10.1016/j.cll.2011.07.004

Whiteaker JR, Zhao L, Abbatiello SE, Burgess M, Kuhn E, Lin C, Pope ME, Razavi M, Anderson NL, Pearson TW, et al. 2011. Evaluation of large scale quantitative proteomic assay development using peptide affinity-based mass spectrometry. *Mol Cell Proteomics* **10:** M110.005645. doi:10.1074/mcp.M110.005645

Whiteaker JR, Zhao L, Lin C, Yan P, Wang P, Paulovich AG. 2012. Sequential multiplexed analyte quantification using peptide immunoaffinity enrichment coupled to mass spectrometry. *Mol Cell Proteomics* **11:** M111.015347. doi:10.1074/mcp.M111.015347

Wiese S, Reidegeld KA, Meyer HE, Warscheid B. 2007. Protein labeling by iTRAQ: a new tool for quantitative mass spectrometry in proteome research. *Proteomics* **7:** 340–350. doi:10.1002/pmic.200600422

Wu CC, MacCoss MJ. 2002. Shotgun proteomics: tools for the analysis of complex biological systems. *Curr Opin Mol Ther* **4:** 242–250.

Wu SZ, Al-Eryani G, Roden DL, Junankar S, Harvey K, Andersson A, Thennavan A, Wang C, Torpy JR, Bartonicek N, et al. 2021. A single-cell and spatially resolved atlas of human breast cancers. *Nat Genet* **53:** 1334–1347. doi:10.1038/s41588-021-00911-1

Xia C, Fan J, Emanuel G, Hao J, Zhuang X. 2019. Spatial transcriptome profiling by MERFISH reveals subcellular RNA compartmentalization and cell cycle-dependent gene expression. *Proc Natl Acad Sci* **116:** 19490–19499. doi:10.1073/pnas.1912459116

Xing Y, Lin NU, Maurer MA, Chen H, Mahvash A, Sahin A, Akcakanat A, Li Y, Abramson V, Litton J, et al. 2019.

Phase II trial of AKT inhibitor MK-2206 in patients with advanced breast cancer who have tumors with PIK3CA or AKT mutations, and/or PTEN loss/PTEN mutation. *Breast Cancer Res* 21: 78. doi:10.1186/s13058 019-1154-8

Yi X, Wang B, An Z, Gong F, Li J, Fu Y. 2018. Quality control of single amino acid variations detected by tandem mass spectrometry. *J Proteomics* 187: 144–151. doi:10.1016/j .jprot.2018.07.004

Yost RA, Enke CG. 1979. Triple quadrupole mass spectrometry for direct mixture analysis and structure elucidation. *Anal Chem* 51: 1251–1264. doi:10.1021/ac50048a002

Zhang B, Whiteaker JR, Hoofnagle AN, Baird GS, Rodland KD, Paulovich AG. 2019. Clinical potential of mass spectrometry-based proteogenomics. *Nat Rev Clin Oncol* 16: 256–268. doi:10.1038/s41571-018-0135-7

Zhao Y, Jensen ON. 2009. Modification-specific proteomics: strategies for characterization of post-translational mod-

ifications using enrichment techniques. *Proteomics* 9: 4632–4641. doi:10.1002/pmic.200900398

Zhu Y, Orre LM, Johansson HJ, Huss M, Boekel J, Vesterlund M, Fernandez-Woodbridge A, Branca RMM, Lehtiö J. 2018a. Discovery of coding regions in the human genome by integrated proteogenomics analysis workflow. *Nat Commun* 9: 903. doi:10.1038/s41467-018-03311-y

Zhu Y, Piehowski PD, Zhao R, Chen J, Shen Y, Moore RJ, Shukla AK, Petyuk VA, Campbell-Thompson M, Mathews CE, et al. 2018b. Nanodroplet processing platform for deep and quantitative proteome profiling of 10-100 mammalian cells. *Nat Commun* 9: 882. doi:10.1038/s41467-018-03367-w

Zolg DP, Wilhelm M, Schnatbaum K, Zerweck J, Knaute T, Delanghe B, Bailey DJ, Gessulat S, Ehrlich HC, Weininger M, et al. 2017. Building ProteomeTools based on a complete synthetic human proteome. *Nat Methods* 14: 259–262. doi:10.1038/nmeth.4153

Mechanisms of Organ-Specific Metastasis of Breast Cancer

Emma Nolan,[1,2] Yibin Kang,[3,4] and Ilaria Malanchi[1]

[1]Tumour Host Interaction laboratory, The Francis Crick Institute, NW1 1AT London, United Kingdom

[2]Auckland Cancer Society Research Centre, University of Auckland, Auckland 1023, New Zealand

[3]Department of Molecular Biology, Princeton University, Princeton, New Jersey 08544, USA

[4]Ludwig Institute for Cancer Research Princeton Branch, Princeton, New Jersey 08544, USA

Correspondence: emma.nolan@auckland.ac.nz; ykang@princeton.edu; Ilaria.Malanchi@crick.ac.uk

Cancer metastasis, or the development of secondary tumors in distant tissues, accounts for the vast majority of fatalities in patients with breast cancer. Breast cancer cells show a striking proclivity to metastasize to distinct organs, specifically the lung, liver, bone, and brain, where they face unique environmental pressures and a wide variety of tissue-resident cells that together create a strong barrier for tumor survival and growth. As a consequence, successful metastatic colonization is critically dependent on reciprocal cross talk between cancer cells and host cells within the target organ, a relationship that shapes the formation of a tumor-supportive microenvironment. Here, we discuss the mechanisms governing organ-specific metastasis in breast cancer, focusing on the intricate interactions between metastatic cells and specific niche cells within a secondary organ, and the remarkable adaptations of both compartments that cooperatively support cancer growth. More broadly, we aim to provide a framework for the microenvironmental prerequisites within each distinct metastatic site for successful breast cancer metastatic seeding and outgrowth.

Distant metastasis is the leading cause of cancer-related mortality in patients with breast cancer. Metastatic breast cancer carries a poor prognosis, with an overall 5-yr survival rate of ~27% (Yousefi et al. 2018). Breast cancer exhibits a distinct metastatic pattern, with patients most frequently presenting with metastasis in the bone, liver, lung, and brain. Different breast cancer subtypes also show preferences for distinct secondary sites, with luminal breast tumors commonly metastasizing to bone. Triple-negative breast cancer (TNBC) displays lung tropism while HER2[+] breast cancers exhibit a high propensity for brain and liver metastasis (Wu et al. 2017). Metastasis is a complex multistep process that begins with local invasion of tumor cells through the basement membrane, intravasation, and then dissemination via the circulation and/or lymphatic systems. After extravasation at the target distant organ, disseminated tumor cell (DTC) seeding and outgrowth leads to the formation of micrometastatic lesions, and eventually life-threatening macrometastases.

In this review, we will focus on the major rate-limiting step of the metastatic cascade, the successful colonization of distant organs (Vanharanta and Massagué 2013). DTCs arriving in the secondary organ are vulnerable to immune surveillance, host-tissue defense, as well as the distinct nutrient and environmental constraints in the foreign environment. Thus, the generation of a favorable metastatic niche is fundamental for successful metastatic seeding and the expansion from micrometastases to overt lesions. Cancer cells have the remarkable ability to modify metastatic sites, both at the systemic level in preparation for metastatic colonization (premetastatic niche), and after their arrival at the secondary organ to create a specialized environment more suitable for outgrowth. Cancer cells themselves also undergo profound adaptations, such as metabolic reprogramming, to adapt to the specific requirements of the distant tissue. This review will explore the extraordinary coevolution of metastatic breast cancer cells and their target organ, with a special focus on the intricate tumor–niche interactions that nurture the survival and outgrowth of DTCs and govern metastatic success. Increasing our understanding of the drivers of cancer-niche evolution, as well as the unique microenvironmental prerequisites for successful colonization, may shed new light on effective anti-metastasis therapies.

LUNG NICHE

TNBC, which accounts for ∼10%–15% of all breast cancers, are especially prone to metastasize to the lung, with an incidence of pulmonary metastasis in up to 40% of patients (Foulkes et al. 2010). Lung metastases are a major cause of breast cancer–related mortality, with patients typically facing a median survival of 22 mo after treatment (Smid et al. 2008). The lung metastatic niche represents a remarkable example of the complex and constantly evolving symbiosis between tumor cells and host cells during metastatic colonization. Breast DTCs evoke profound phenotypic changes in lung-resident and recruited cells, shaping the formation of a tumor-supportive niche, which in turn fuels metastatic progression (Fig. 1). In this section,

we unravel this ultimate partner-in-crime relationship, identifying the key players and mechanisms that determine metastatic success in the lung environment.

Tumor–Immune Cell Interplay in the Lung Niche

The cooperation between newly arriving breast cancer cells and the immune compartment of the lung showcases the ability of tumor cells to fine-tune their environment and, in doing so, license their growth. Macrophages, an abundant component of the lung niche, play a critical role in supporting the extravasation, seeding, and outgrowth of malignant cells (Pollard 2004). A distinct subset of tumor-promoting macrophages has been identified in the metastatic lung, which are preferentially recruited by extravasating breast cancer cells (Qian et al. 2009). These originate from CCR2-expressing inflammatory monocytes that infiltrate the lung in response to cancer cell–derived CCL2, where they differentiate into macrophages and directly support metastatic seeding (Qian et al. 2011). In addition to promoting macrophage infiltration, lung-trophic breast cancer cells are intrinsically primed to exploit macrophage-derived protumorigenic signals (Song et al. 2016). By up-regulating their expression of GALNT14, arriving breast cancer cells can maximize their responsiveness to macrophage-derived fibroblast growth factor (FGF) signaling, increasing their chance of successful colonization. Interestingly, GALNT14 up-regulation is also an adaptive mechanism to overcome the inhibitory effect of anti-metastatic bone morphogenetic protein (BMP) signaling in the lung (Song et al. 2016).

Perhaps the most well-established driver of pulmonary metastasis are neutrophils, a major component of the lung immune microenvironment. Both pro- and antitumoral functions of neutrophils have been described, with recent reviews extensively covering this context-dependent dichotomy (Jaillon et al. 2020; Hedrick and Malanchi 2022). In metastasis, neutrophils typically behave as tumor accomplices, secreting an array of cytokines and chemokines that support cancer cell seeding, survival, immune-eva-

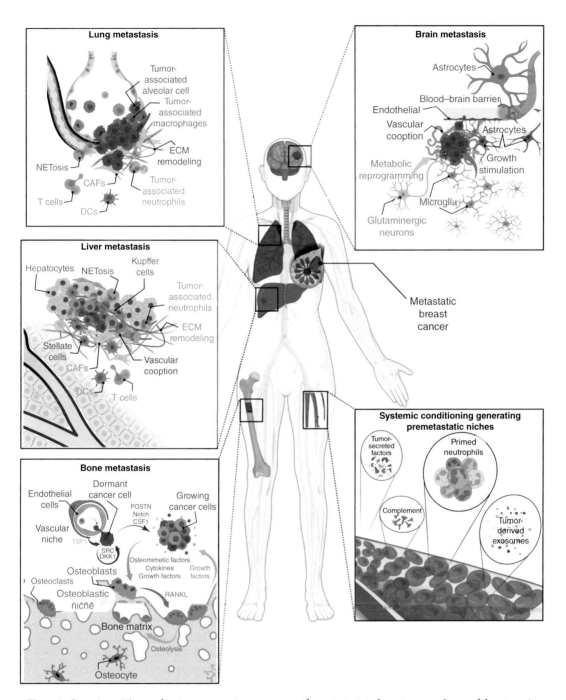

Figure 1. Overview of the mechanisms governing organ-specific metastasis in breast cancer. Successful metastatic seeding in the lung, liver, bone, and brain environment is highly dependent on intricate cross talk between cancer cells and tissue-specific niche cells. Depicted here are the main players, biological processes, and adaptations that contribute to the evolution of a protumorigenic niche within each distinct metastatic site. Cancer cells are capable of modifying their intended secondary organ at both the systemic level in preparation for metastatic colonization (premetastatic niche; *lower right* panel) and after their arrival in the distant tissue. (ECM) Extracellular matrix, (CAFs) cancer-associated fibroblasts, (DCs) dendritic cells.

sion, and outgrowth. Neutrophils also promote metastasis via neutrophil extracellular traps (NETs), the web-like lattice of DNA coated in proteolytic enzymes that trap and kill pathogens during an infection. Arriving metastatic breast cancer cells can induce NETosis in lung neutrophils in the absence of pathogens, fostering invasion and the expansion of disseminated cells (Park et al. 2016). Thus, cancer cells are able to hijack this intrinsic neutrophil activity to support their growth in noninfectious conditions. NET-driven metastasis is profoundly exacerbated by sustained lung inflammation induced by tobacco smoke exposure or lipopolysaccharide (LPS), in a mechanism linked to extracellular matrix (ECM) remodeling by NET-DNA-associated proteases (Albrengues et al. 2018). More recently, NET-fueled metastasis was linked to cancer-cell-secreted protease cathepsin C, which promotes breast cancer metastatic outgrowth by enhancing IL-6/CCL3-mediated neutrophil recruitment and NETosis (Xiao et al. 2021). Importantly, targeting cathepsin C effectively suppressed lung metastasis in mice, arguing for the use of niche-modifying agents for the treatment of metastasis.

Recent studies have highlighted the dynamics and complexity of the tumor–immune cell interplay in the lung niche, suggesting multiple layers of cell cross talk that create a "domino effect." For example, the prometastatic effect of lung neutrophils is tunable, dictated by the status of host natural killer (NK) cells (Li et al. 2020a). In an elegant study using mice with varying levels of immune system integrity, the authors showed that while neutrophils themselves display ROS-mediated tumoricidal activity, they simultaneously suppress the more potent tumoricidal ability of NK cells. Thus, in the lung tissue of an NK-competent host, the net contribution of neutrophils is prometastatic. Interestingly, the NK cell status of the host has also recently been found to impact clonal evolution in metastatic breast cancer (Lo et al. 2020). Circulating tumor cell (CTC) clusters were shown to be less sensitive to NK-mediated immunosurveillance compared to single CTCs; thus, NK cells confer a selective advantage for polyclonal seeding in the lung. Another example of coordinated tu-

mor–immune cell interplay is T-cell immunoregulation, in which cancer and immune cells conspire to create an immunologically permissive environment for metastatic colonization. T-cell immunosuppression can be driven by γδ-T-cell-induced neutrophil recruitment and polarization (Coffelt et al. 2015), myeloid-derived osteopontin production (Sangaletti et al. 2014), or via a T-cell-intrinsic mechanism of oxygen sensing specific to the lung environment (Clever et al. 2016). In all cases, the induction of immunotolerance licensed breast tumor colonization in the lung, highlighting the power of immunoregulation to either restrain or unleash metastatic outgrowth.

Importantly, some of the mechanisms enabling immune cells to support metastatic breast cancer cell growth within the lung are influenced by systemic perturbations orchestrated by primary tumor-derived signals. Those systemic changes influencing distant organs drive the formation of a "premetastatic niche," and they will be discussed later in this article.

Tumor-Mediated Education of Lung Stromal Cells

Breast DTCs arriving in the lung elicit a profound rewiring of the host stromal cells such as fibroblasts, an obligatory event that nurtures their expansion from micrometastases to overt macrometastases. The activation and perturbation of lung fibroblasts and other mesenchymal cells lead to the generation of carcinoma-associated fibroblasts (CAFs), a general term now used to define a highly heterogenous collection of cells with distinct features and activities within the metastatic environment (Sahai et al. 2020; Chen et al. 2021b). At the very early stages of metastatic colonization, lung-infiltrating breast cancer cells trigger the induction of stromal periostin (POSTN) expression, supporting cancer stem cell (CSC) maintenance and metastatic colonization by augmenting Wnt signaling (Malanchi et al. 2012). Indeed, human breast metastatic cells with enhanced stem cell features have been linked with a higher metastatic-initiating capacity, with their early arrival into distant tissues preceding advanced metastatic disease

(Lawson et al. 2015). Another matricellular protein, tenascin C (TNC), is highly implicated in promoting cancer cell stemness and metastatic lung colonization (Oskarsson et al. 2011). While initially secreted by tumor cells themselves to support their survival in the new host tissue, lung stromal cells eventually take over as the source of TNC and foster metastatic outgrowth. More recently, transcriptomic analysis has painted a picture of a dynamic global reprogramming of fibroblasts in the lung metastatic niche. Interestingly, these studies showed that fibroblast evolution is influenced by both the stage of metastatic progression and by the metastatic potential of the arriving breast cancer cells (Pein et al. 2020; Shani et al. 2021). Breast DTC seeding in the lung evoke an IL-1α/IL-1β-mediated proinflammatory phenotype in lung fibroblasts, reminiscent of fibroblast activation in wound healing and primary tumors. Cancer cells then exploit this reactive fibroblast niche to fuel their colonization (Pein et al. 2020). Using transgenic mice at defined stages of breast-to-lung metastases, Erez and colleagues suggested that fibroblasts functionally adapt to the evolving lung niche via distinct *Myc*-regulated transcriptional programs (Shani et al. 2021). ECM remodeling and cellular stress programs were up-regulated in fibroblasts in early micrometastases, while inflammatory signaling was instigated in advanced metastases to perpetuate growth. This demonstration of coevolution of lung stromal and breast cancer cells is in line with previous work showing a reciprocal cross talk between cancer cells and the lung stromal niche, modulating the epithelial-to-mesenchymal (EMT) state of cancer cells, and dictating the outgrowth potential of DTCs (Martin et al. 2015). Overall, these studies reveal the fascinating ability of lung-infiltrating breast cancer cells to trigger the rewiring of their local stromal niche, which, in turn, plays a crucial role in shaping their progression into overt metastases.

New Players in the Lung Niche

More recently, in addition to the more abundant and well-characterized tumor niche cell types, the engagement of other organ-specific cellular components in the lung metastatic niche has emerged. Metastasis in the lung initiates from cancer cells arriving via the microvasculature, and outgrowth mainly occurs within the alveolar environment. During metastatic expansion, malignant cells can reenter the alveolar wall and migrate between endothelial and epithelial layers, whereby they coopt the existing pulmonary microvasculature to enable tumor vascularization (Szabo et al. 2015). Recently, an active participation of lung alveolar cells was discovered in the breast cancer metastatic niche (Ombrato et al. 2019; Montagner et al. 2020). Using a novel niche-labeling system, the arrival of breast DTCs in the lung was found to trigger the regenerative-like activation of neighboring alveolar type II cells (Ombrato et al. 2019). This epithelial response, characterized by the acquisition of stem cell–like features, directly supported the growth of cancer cells ex vivo, suggesting it may contribute to the formation of a favorable metastatic niche. Importantly, cross talk between DTCs and the lung epithelium, in particular alveolar type I cells, may also mediate breast cancer dormancy and the survival of indolent cancer cells (Montagner et al. 2020). Tumor cell dormancy is highly problematic for the clinical management of cancer, a topic covered elsewhere in the literature (see, for example, Dalla et al. 2022); thus, exposing niche factors that mediate tumor cell latency could have major implications for cancer treatment.

In addition to the epithelium, Zheng et al. (2019) recently identified a novel population of atypical cytokine-producing large platelets specifically enriched in lung metastases but not in primary mammary tumors. These are locally produced in response to lung-infiltrating DTCs, likely from lung-resident megakaryocytes that egress from extrapulmonary sites such as bone marrow (Lefrançais et al. 2017). While the contribution of these megakaryocytes to metastatic growth remains to be determined, their identification further shows the complexity of the tumor microenvironment in metastases. A recent study revealed a method for in situ transcriptomic profiling of rare cell populations ("Flura-seq") to examine organ-specific metabolic reprogramming during the early stage of breast cancer metastatic colonization (Basnet et al. 2019). The authors showed that the pulmonary microenvironment induced specific cancer cell adaptations, in particular an enrichment of mitochondrial electron transport, oxidative stress, and antioxidant programs, when compared to primary tumors or brain micrometastases. The high-sensitivity of Flura-seq could reveal the existence of rare cell populations that, despite representing a small fraction of the metastatic niche, may still be of clinical significance.

BONE NICHE

Bone marrow is a preferred metastatic site for multiple solid tumors including breast, with ~70% of advanced breast cancer patients harboring bone metastasis (Coleman 2001). The vast majority of breast cancer bone metastases are classified as bone-destructive osteolytic lesions, which are associated with multiple bone complications that can profoundly impact the patient's quality of life, such as bone pain, fracture, hypercalcemia, and paralysis. The success of breast DTC colonization of the bone is dictated by the intrinsic features of tumor cells, favorable tumor–niche interactions, and by the cancer cell–directed protumorigenic rewiring of the local bone environment. There are several recent reviews that extensively dissect the bone marrow metastatic niche, providing a wealth of knowledge on the role of the specialized bone environment in supporting tumor colonization (Esposito et al. 2018; Chen et al. 2021a; Satcher and Zhang 2022). In this section, we provide an update on the intricate tumor–stromal interactions governing the progression from initial seeding to overt macrometastasis, focusing on the evolving symbiosis between tumor cells and the bone microenvironment during colonization (Fig. 1).

Tumor–Niche Interactions

After extravasation into the bone marrow compartment, cancer cells preferentially occupy one of two specialized bone microenvironments or niches—the vascular niche and the osteogenic (or endosteal) niche. The vascular niche harbors bone marrow stromal cells, pericytes, and endothelial cells that closely line the sinusoids, while the osteogenic niche primarily includes all osteoblast and osteoclast lineage cells and is the central hub of bone remodeling. Both niches play a key role in shaping the metastatic process, and are capable of providing both a safe haven for indolent cancer cells or supplying a fertile soil to unleash metastatic growth.

In the osteogenic niche, early-stage colonization of breast DTCs is dependent on the formation of heterotypic adherens junctions between cancer-derived E-cadherin and osteogenic N-cadherin, triggering mTOR activation in cancer cells (Wang et al. 2015). Direct calcium influx from the osteogenic niche to cancer cells is also essential for the proliferation of early-stage micrometastatic lesions (Wang et al. 2018). Once cancer cells have established a foothold within the bone marrow, metastatic progression is driven by the osteolytic vicious cycle, a formidable positive-feedback cycle composed of tumor cells, osteoclasts, osteoblasts, and the bone matrix. Differentiation and maturation of osteoclasts from their monocytic precursors critically depends on the cytokine RANKL, which is produced by osteoblasts. Signaling of RANKL is antagonized by its decoy receptor osteoprotegerin (OPG). Therefore, imbalance in the level of RANKL and OPG may result in disruption of bone homeostasis, leading to osteoporosis or osteopetrosis conditions. In osteolytic bone metastasis of breast cancer, elevated osteoclast activity can be induced by cancer-derived secreted factors such as PTHrP and prostaglandin E2, which stimulate dysregulation of osteoblastic RANKL/OPG expression (for reviews, see Esposito et al. 2018; Wu et al. 2020). Breast cancer cells themselves can also directly secrete RANKL in response to enhanced BMP signaling (Liu et al. 2018) or CCL20 secreted by both cancer cells and osteoblasts (Lee et al. 2017). Alternatively, tumor-derived matrix metalloproteinases can suppress OPG expression by proteolytically releasing membrane-bound EGF family growth factors and subsequent downstream EGFR signaling in osteoblasts (Lu et al. 2009). Activation of osteoclastogenesis can occur independently of RANKL via IL-11-mediated JAK1/STAT3 signaling (Liang et al. 2019) or VCAM1-induced NF-κB signaling (Lu et al. 2011). Bone destruction triggers the release of growth factors deposited in the bone matrix such as IGF1, PDGF, and TGF-β, which in turn stimulates tumor growth and further accelerates bone loss. TGF-β plays a crucial role in perpetuating the osteolytic vicious cycle, and an enrichment of TGF-β signaling is a well-established feature of highly bone-metastatic breast cancer cells (Kang et al. 2003; Korpal et al. 2011; Sethi et al. 2011; Esposito et al. 2021a). TGF-β stimulates the expression of PTHrP, Jagged1, CTGF, and IL-11, all of which are strong activators of osteoclastogenesis (Yin et al. 1999; Kang et al. 2005; Shimo et al. 2006; Sethi et al. 2011). TGF-β-mediated induction of DACT1 is critical for bone metastatic progression via suppression of Wnt signaling, which is important for initial cancer cell seeding but is constrained during macrometastasis outgrowth (Esposito et al. 2021a). The temporal modulation of Wnt signaling in both cancer cells and the bone niche is indeed one of the key aspects in bone metastasis. During the early stage of metastasis, Wnt activity in breast cancer cells induced by IL-1β derived from the bone microenvironment can promote breast cancer stemness and metastatic outgrowth (Eyre et al. 2019). Moreover, breast cancer cells were also reported to secrete DKK1, which boosts osteoclastogenesis by suppressing canonical Wnt signaling–mediated expression of OPG, increasing osteolytic bone metastatic progression (Zhuang et al. 2017).

In another demonstration of bidirectional cross talk involving TGF-β, its release by bone destruction was shown to trigger cancer cell secretion of Jagged1, which in turn engages Notch signaling in tumor-associated osteoblasts and preosteoclasts. This confers both a proliferative advantage to tumor cells by enhancing osteoblast-derived protumorigenic growth factors IL-6 and CTGF and stimulates osteoclast maturation, together fostering osteolytic metastasis (Sethi et al. 2011). Interestingly, chemotherapy appears to induce the inverse cross talk, in which treatment-induced Jagged1 secretion by osteoblasts creates a prosurvival niche for breast DTCs via Notch activation in cancer cells. Anti-Jag1 therapy sensitizes bone metastasis to

chemotherapy, with a nearly 100-fold reduction in bone metastatic burden demonstrated following combination therapy in vivo (Zheng et al. 2017). In line with this study, cancer treatment–induced alterations in the bone niche were recently linked to augmented bone loss and breast-to-bone metastasis (Zuo et al. 2020). The dual PARP1/2 inhibitor olaparib was found to promote osteoclast differentiation and bone resorption and foster the recruitment of immune-suppressive immature myeloid cells to the bone niche. The overall effect was exacerbated bone metastasis, warranting a careful examination of the use of current PARP inhibitors in the clinic.

The perivascular niche has increasingly been recognized to play a key role in the development of bone metastasis. Stable microvasculature has been reported to promote the dormancy of DTCs via endothelial-derived thrombospondin-1, TGF-β1, and POSTN as active tumor-promoting factors (Ghajar et al. 2013). The perivascular niche also protects DTCs from chemotherapy through integrin-mediated interactions. Furthermore, vascular niche–derived E-selectin promotes bone metastasis by directly engaging with cancer cells and inducing mesenchymal-to-epithelial transition and Wnt-dependent stemness (Esposito et al. 2019). Using quantitative 3D imaging of mouse bone marrow, Yip and colleagues recently showed that mouse mammary DTCs preferentially home to a distinct vascular niche comprised of type H capillaries during bone colonization. Metastatic progression was critically dependent on cancer cell–directed remodeling of the local vasculature, creating a self-sustaining microenvironment (Yip et al. 2021).

Other important players fueling metastatic growth in the bone environment are emerging. Sympathetic nerves have been shown to play an active role in pathogenesis of bone metastasis (Elefteriou 2016). Ma et al. recently identified a specific subset of protumorigenic macrophages derived from CCL2-recruited inflammatory monocytes, which promote breast cancer bone metastasis in an IL-4R-dependent manner (Ma et al. 2020). Bone marrow adipocytes, the most abundant component of the bone marrow environment, can provide a supportive niche for metastatic cancer cells via the secretion of a plethora of adipokines such as leptin and IL-6. Adipokines play an assertive role in metastatic colonization and progression in the bone marrow through the regulation of tumor cell recruitment, invasion, survival, proliferation, angiogenesis, and immune modulation (for reviews, see Liu et al. 2020; Soni et al. 2021).

Cancer Cell Adaptations in the Bone Environment

Metastatic breast cancer cells are capable of expressing genes normally restricted to bone cells to support their growth in the bone niche. This is known as osteomimicry, in which breast tumor cells acquire an osteocyte-like phenotype and secrete osteomimetic factors that enhance their survival, proliferation, and colonization of the bone environment. These factors include cadherin-11 (Tamura et al. 2008), osteoactivin (Rose et al. 2007), BSP (Wang et al. 2013), Runx2 (Tan et al. 2016), Src (Zhang et al. 2009), cathepsin K (Gall et al. 2007), and osteopontin (Anborgh et al. 2010), among others (Brook et al. 2018). The master switch for this organ-specific cancer cell reprogramming has been accredited to forkhead box F2 (FOXF2), a transcription factor that drives pleiotropic transactivation of the BMP4/SMAD1 signaling pathway (Wang et al. 2019). Epigenetic rewiring of ER$^+$ breast cancer cells in the osteogenic niche also influences clonal evolution and fosters endocrine therapy resistance (Bado et al. 2021). Osteoblast–cancer cell interactions during early metastatic colonization drive a transient EZH2-dependent loss in ER expression, allowing cancer cells to escape endocrine therapy–mediated destruction. This partially recovers as the lesion progresses beyond the osteogenic niche, contributing to tumor heterogeneity in late-stage bone metastasis (Bado et al. 2021). Interestingly, EZH2-mediated epigenetic reprogramming in the bone environment also plays a role in metastasis-to-metastasis multiorgan seeding, by enhancing cancer cell stemness and invigorating their dissemination (Zhang et al. 2021).

Cancer cells are also able to adapt to the bone microenvironment by adjusting their metabolic activities. Bone metastatic breast cancer cells are found to overexpress three enzymes required for de novo serine synthesis: phosphoglycerate dehydrogenase (PHGDH), phosphoserine aminotransferase 1 (PSAT1), and phosphoserine phosphatase (PSPH), as well as the serine transporter SLC1A4 (Pollari et al. 2011). These enzymes promote the production of L-serine, an essential amino acid for differentiation of osteoclasts. Bone metastatic tumor cells also release large amounts of lactate to fuel osteoclastogenesis and osteolytic progression (Lemma et al. 2017).

BRAIN NICHE

Brain metastasis is associated with poor survival outcomes and poses unique clinical challenges such as poor drug delivery across the blood–brain barrier (BBB). The distinct microenvironment of the brain, characterized by highly specialized resident cells and metabolic constraints imposes intense selective pressures on breast DTCs. Thus, successful metastatic colonization is dictated by the ability of cancer cells to adapt to the brain microenvironment and orchestrate coevolution of brain-resident cells. In addition, the anatomical location of the brain in which the metastatic cells colonize profoundly influences the metastatic process. The specific microenvironmental requirements associated with each location results in distinct types of metastatic lesions in the brain: parenchymal metastases, leptomeningeal metastasis (in which cancer cells colonize the cerebral spinal fluid), and the rare form, choroid plexus metastasis. The molecular mechanisms underlying breast cancer brain metastases have been recently reviewed in detail (Wang et al. 2021b). In this section, we will focus on the interplay between arriving breast DTCs and the brain microenvironment postextravasation, a relationship that profoundly shapes the metastatic process and empowers tumor colonization (Fig. 1). While we will focus on parenchymal metastasis due to the body of literature in this area, we will also touch on leptomeningeal metastases, a less common site but which is associated with an alarmingly poor prognosis (Franzoi and Hortobagyi 2019).

Tumor–Niche Cross Talk in the Brain Microenvironment

The survival and outgrowth of breast parenchymal metastases critically depends on favorable interactions between cancer cells and the unique cellular compartments within the neural niche. Astrocytes are the most abundant glial cells in the brain and become activated by direct contact with breast DTCs following extravasation through the BBB. Reactive GFAP⁺ astrocytes secrete a plethora of soluble proteins including IL-1, IL-6, IGF-1, TNF-α, and TGF-β that directly supports the invasion, growth, and survival of metastatic cancer cells (Sierra et al. 1997; Wasilewski et al. 2017). Recent studies have highlighted an array of additional metastasis-promoting functions of reactive astrocytes. A subpopulation of pSTAT3⁺ astrocytes found at the periphery of breast, lung, and melanoma brain metastatic lesions form an immunosuppression barrier for metastatic-initiating cells, via the regulation of innate and adaptive immune cells (Priego et al. 2018). Astrocyte–cancer cell cross talk can shape metastatic progression through the regulation of tumor cell gene expression. Astrocyte-derived miR-19a-containing exosomes induce PTEN loss in breast DTCs, priming metastatic outgrowth by boosting CCL2-mediated recruitment of protumorigenic myeloid cells (Zhang et al. 2015). Astrocyte-induced epigenetic regulation of the glycoprotein Reelin specifically in HER2⁺ metastatic breast cancer cells leads to increased proliferation and supports their prometastatic phenotype (Jandial et al. 2017). This may also help to explain the high incidence of brain metastasis in patients with recurrent HER2⁺ breast cancer (Heitz et al. 2011; Martin et al. 2017). In an elegant demonstration of reciprocal cross talk, research from the Massagué laboratory showed that transfer of cGAMP from brain metastatic breast cancer cells to astrocytes triggers the secretion of proinflammatory cytokines, which in turn activates STAT1/NF-κB signaling in cancer cells. The overall effect is augmented metastatic outgrowth and increased chemoresistance (Chen et al. 2016). Bidirectional tumor–astrocyte interactions were also shown to support the growth of metastatic breast cancer cells in the brain in a study by Watabe and colleagues. Cancer cell–derived COX2 and prostaglandins activate astrocytes to secrete chemokine CCL7, which in turn augments the self-renewal activity of metastatic-initiating cells in the brain (Wu et al. 2015). Importantly, astrocytes may also foster metastatic progression by allowing cancer cells to take advantage of extrinsic factors and nutrients in the brain microenvironment. For example, astrocytes can serve as a source of polyunsaturated fatty acids for arriving breast DTCs, promoting cancer cell proliferation via the activation of peroxisome proliferator–activated receptor γ (PPARγ) (Zou et al. 2019). Astrocytes also act as paracrine mediators of estrogen signaling in TNBC brain metastasis, transmitting mitogenic signals to otherwise unresponsive ER⁻ breast cancer cells and supporting prometastatic behaviors (Sartorius et al. 2016). These studies suggest that cancer cell–astrocyte cross talk allows metastatic cancer cells to exploit the lipid and hormone-rich milieu of the brain microenvironment, fueling their transition to overt metastasis.

Microglia are the unique resident macrophages of the brain parenchyma, and become activated in response to physical contact from arriving DTCs. Activated microglia are capable of carrying out cytotoxic functions, but this can be cleverly thwarted by the secretion of neurotrophin-3 (Louie et al. 2013) or MYC-induced production of the antioxidant glutathione peroxidase 1 by brain metastatic cancer cells (Klotz et al. 2020). On the other hand, tumor-educated microglia can favor metastatic outgrowth by promoting mesenchymal-to-epithelial transition via E-cadherin up-regulation (Louie et al. 2013), or supporting proinvasive behavior in a Wnt-dependent manner (Pukrop et al. 2010). Macrophage-derived cathepsin C supports tumor cell survival and outgrowth in the brain (Sevenich et al. 2014). Interestingly, breast cancer cells themselves are the initial source of cathepsin C to promote BBB transmigration, while macrophages take over as the predominant supplier after extravasation and foster metastatic expansion. More recently, the loss of X-inactive-specific-transcript (XIST) in brain metastatic breast cancer cells was shown to trigger prometastatic microglia reprogramming via cancer cell–derived exosomal miRNA-503 (Xing et al. 2018). Augmented secretion of immunomodulating cytokines by the rewired microglia led to suppression of local T-cell immunity, thereby enhancing metastatic growth (Xing et al. 2018). Interestingly, brain-resident microglia may also be responsible for the more aggressive disease commonly observed in younger breast cancer patients and in mouse models of both TNBC and luminal B breast cancers (Evans et al. 2004; Hung et al. 2014; Wu et al. 2021a). The increase in brain metastatic propensity in younger versus older mice, an effect not observed in liver or lung metastasis, was recently attributed to age-associated loss of protumoral-resident microglia, and myeloid cell depletion preferentially reduced brain metastatic burden in younger mice (Wu et al. 2021a). Importantly, the complexity and plasticity of the immune landscape in the neural niche was recently unveiled through single-cell analyses combined with transgenic mouse models (Guldner et al. 2020). Central nervous system (CNS)-native myeloid cells, primarily microglia, were found to cultivate an immunosuppressive niche via Cxcl10 signaling, promoting the outgrowth of breast cancer brain metastasis (Guldner et al. 2020). More recently, Gonzalez and colleagues generated an extensive single-cell transcriptomic data set for both malignant and nonmalignant niche cells from 15 human brain metastases, including three breast metastatic lesions (Gonzalez et al. 2022). Interestingly, comparable metastatic niches were observed across cancers of different primary tumor origin, which included two functionally distinct subsets of immune-modulating macrophages.

A critical step in successful brain metastatic colonization is vascular cooption, in which interactions between DTCs and endothelial cells allow newly arriving cancer cells to exploit the preexisting blood supply, supporting their outgrowth (for review, see García-Gómez and Valiente 2020). This is unlike leptomeningeal metastasis, in which de novo angiogenesis is a key feature (Lorger and Felding-Habermann 2010). In vivo experimental metastasis models using

both human and mouse breast cancer cells showed vascular cooption in over 95% of early brain parenchymal micrometastases, rather than neoangiogenesis. This process is dependent on integrin β1–mediated breast tumor cell adhesion to the vascular basement membrane (Carbonell et al. 2009), while others have also implicated integrin β4 (Fan et al. 2011) and adhesion molecule L1CAM (Valiente et al. 2014). Importantly, niche cells can foster cancer cell–endothelial interactions, elegantly demonstrated by the JAK2/STAT3-dependent promotion of cooption by tumor-recruited macrophages (Wang et al. 2017).

Cancer Cell Adaptation to the Neural Niche

Successful metastatic colonization of the brain requires cancer cells to undergo metabolic reprogramming to adapt to nutrient deprivation in the brain environment. Patient-derived brain metastatic breast cancer cells exhibit neural characteristics and coopt γ-aminobutyric acid (GABA) as an oncometabolite, providing a proliferative advantage (Neman et al. 2014). More recently, brain-colonizing breast DTCs were shown to coopt glutamate-mediated neuronal signaling to support metastatic growth (Zeng et al. 2019). The formation of pseudosynapses between tumor cells and glutamatergic neurons enabled seeding cancer cells to access a supply of glutamate, enhancing their proliferation and supporting their outgrowth into overt metastasis.

Similarly, increased expression of the fatty acid–binding protein 7 in brain-trophic HER2$^+$ breast cancer cells enhances fatty-acid utilization and supports the adoption of a glycolytic phenotype in the brain environment (Cordero et al. 2019). In response to glucose deprivation in the neural niche, GRP9A-expressing breast cancer cells can induce a prosurvival autophagy response, relieving metabolic stress and enhancing survival (Santana-Codina et al. 2020). Brain metastatic breast cancer cells also display enhanced gluconeogenesis and enhanced oxidation of glutamine and branched chain amino acids, facilitating glucose-independent growth and enhancing brain metastasis in vivo (Chen et al. 2015). Finally, as well as supporting the nutrient requirements of metastatic cells, metabolic reprogramming in the neural niche can increase cancer cell resistance to ROS-mediated oxidative stress. Increased LEF1 expression by brain-colonizing breast cancer cells boosts levels of glutathione, improving the antioxidant capability of metastatic cells (Blazquez et al. 2020).

Brain-colonizing breast DTCs also adapt protective mechanisms to shield themselves from brain defenses. By producing high levels of the antiplasminogen activator inhibitory Serpins, brain metastatic cancer cells can subvert the lethal action of plasmin produced by the reactive brain stroma, in particular by activated astrocytes (Valiente et al. 2014). Brain metastatic cells can dampen the adaptive immune response in the neural niche through the recruitment of PD-L1$^+$ immunosuppressive neutrophils into the brain TME (Zhang et al. 2020). This is instigated by Src-dependent phosphorylation of EZH2 in brain metastatic cells, activating the c-Jun/G-CSF/neutrophil axis to deter T-cell functions and foster metastatic colonization. Interestingly, in leptomeningeal metastases, breast and lung metastatic cancer cells can also orchestrate protumorigenic modifications of their local environment (Boire et al. 2017). Through the secretion of complement component 3, cancer cells disrupt the blood–CSF barrier and facilitate the entry of tumor-supportive mitogens into the nutrient-poor CSF. Thus, by adapting to the hostile environment of the CSF, cancer cells are able to overcome microenvironmental challenges and thrive against the odds (Boire et al. 2017). Interestingly, studies analyzing the CSF metabolome before and during progression of brain metastasis have shown potential for CSF sampling for the early diagnosis of cancer patients with leptomeningeal metastasis (Dekker et al. 2005; Yoo et al. 2017).

HEPATIC NICHE

The liver is a frequent site of metastasis for the majority of solid tumors including breast. This metastatic propensity can be partly attributed to its unique biological structure: the liver is a highly vascularized organ, characterized by a fenes-

trated endothelium that lacks a subendothelial basement membrane. These features, combined with an exceptionally low blood-flow rate, make the organ intrinsically susceptible to DTC extravasation. After extravasation, successful metastatic outgrowth is then dependent on reciprocal cross talk between tumor cells and the unique network of highly specialized tissue-resident cells, including parenchymal hepatocytes, liver sinusoidal endothelial cells, hepatic stellate cells, and Kupffer cells. Recruited and resident immune cells such as neutrophils can also influence the metastatic potential of DTCs in the liver. In this section, we summarize the requirements for the formation of a hospitable hepatic niche in metastatic breast cancer (Fig. 1).

Cancer Cell Adaptations to the Liver Niche

Metastatic breast cancer cells can achieve angiogenesis-independent growth in the liver by replacing hepatocytes and coopting sinusoidal blood vessels (Stessels et al. 2004). This hijacking may be crucial for overt liver metastasis, since lesions that use the existing blood supply were shown to thrive in the liver microenvironment (Martin et al. 2010; Frentzas et al. 2016). Postextravasation, liver-metastatic breast cancer cells undergo metabolic reprogramming to acclimatize to the hepatic microenvironment. In a mechanism dependent on the HIF-1α target pyruvate dehydrogenase kinase-1 (PDK1), breast cancer cells adapt a specific glycolytic phenotype in the liver in response to nutrient limitations and hypoxia (Dupuy et al. 2015). This distinct metabolic profile is crucial for efficient hepatic metastasis, since PDK1-knockdown resulted in an 80% reduction in liver metastasis in vivo but had no effect on outgrowth in the lung (Dupuy et al. 2015). Niche-induced metabolic plasticity may also drive endocrine resistance in liver metastatic ER$^+$ breast cancer (Zuo et al. 2022). In response to fulvestrant treatment, ER$^+$ metastatic tumors from the liver but not lung demonstrated increased glucose metabolism in vivo, differential ERα activity, and reduced treatment response. Interestingly, this metabolic vulnerability could be exploited by giving mice a fasting-mimicking diet to block glycogen accumulation in the liver, resulting in an improved fulvestrant response (Zuo et al. 2022).

Cancer Cell–Niche Interactions

Interactions between breast cancer cells and the liver ECM is a crucial step in the formation of a favorable hepatic metastatic niche. This is mediated by the tight-junction protein claudin-2, which is up-regulated in liver-trophic breast cancer cells and liver metastases from breast cancer patients (Tabariès et al. 2011). By enhancing cell surface integrin expression, claudin-2 drives cancer cell adhesion to the liver ECM and promotes metastatic colonization. Expression of the adhesion molecule CD44 is also highest in liver-trophic breast cancer cells, and its induction fosters breast cancer liver metastasis in vivo (Ouhtit et al. 2007; Erin et al. 2013). In addition, ECM remodeling in response to liver fibrosis creates a growth-permissive microenvironment capable of enhancing metastatic colonization (Cox et al. 2013). This is mediated by enhanced lysyl oxidase (LOX) expression in activated hepatic stellate cells, which triggers collagen cross-linking and favors the outgrowth of breast DTCs.

Cross talk between liver-infiltrating breast cancer cells and tissue-resident hepatocytes has also been linked to successful metastatic colonization. As with matrix adhesion, hepatocyte–breast cancer cell interactions are facilitated by claudin-2, although this heterotypic interaction occurs independently of integrin complexes (Tabariès et al. 2012). Reduction of claudin-2 levels in liver-aggressive breast cancer cells significantly decreases metastasis in vivo, which can be rescued by specifically restoring breast–hepatocyte interactions using chimeric claudin-2 constructs. This study supports the historic observation that breast cancer cells, upon seeding the liver, extend cellular projections through the fenestrated endothelium and make direct contact with hepatocytes (Roos et al. 1978). Mechanistically, breast DTCs and hepatocyte cross talk may support metastatic outgrowth by triggering mesenchymal-to-epithelial transition. Hepatocytes have been shown to directly induce E-cad-

herin reexpression in neighboring breast cancer cells, driving reversion to an epithelial phenotype and boosting postextravasation survival and chemoresistance (Chao et al. 2012).

The liver-metastatic potential of breast DTCs is highly reliant on interactions with infiltrating prometastatic neutrophils within the liver microenvironment (Tabariès et al. 2015). A distinct subset of immature low-density neutrophils (iLDNs) is preferentially mobilized in mice bearing liver metastases and is required for efficient hepatic colonization (Hsu et al. 2019). This metabolically distinct subset exhibits an enhanced bioenergetic capacity, enabling the execution of prometastatic functions such as NETosis under conditions of nutrient deprivation. Both this group and others (Yang et al. 2020) have demonstrated the functional importance of NETosis during breast cancer liver metastasis, with a significant reduction in metastatic colonization observed following the treatment of mice with the nuclease DNase1. In the latter study, the authors demonstrated a profound abundance of NETs in liver metastases of patients with breast and colon cancers, and identified a specific NET-DNA receptor, CCDC25, that mediates NET-dependent liver metastasis (Yang et al. 2020).

Hepatic Niche Perturbations

Recent evidence suggests the hepatic metastatic niche can be greatly influenced by systemic perturbations in the host. For example, a high-fat diet was shown to enhance spontaneous breast cancer metastasis to the liver, but not the lungs, by enhancing the accumulation of myeloid cells with immune-suppressive functions that dampen antitumor immunity (Clements et al. 2018). In a mouse model of obesity, metastatic breast cancer cells triggered triglyceride lipolysis in adjacent hepatocytes, the products of which were then transferred to tumor cells via fatty acid transporter protein 1 and used as a source of energy for tumor growth. This reciprocal relationship led to a dramatic increase in liver metastasis (Li et al. 2020b).

Interestingly, it was recently shown that weaning-induced involution in the postpartum liver conveys a metastatic advantage to seeding breast cancer cells via the induction of immune tolerance (Bartlett et al. 2021). Mechanistically, the massive influx of immune cells, hepatocyte cell death, and substantial ECM remodeling that accompanies liver involution resulted in the formation of a prometastatic hepatic niche, which favored the transition of micrometastases to overt metastatic lesions (Bartlett et al. 2021). This finding is in line with a study that found an increased propensity for liver metastasis in young postpartum breast cancer patients compared to age-, tumor-, and subtype-matched nulliparous women (Goddard et al. 2019). Thus, in addition to tumor-induced niche evolution, normal physiological remodeling of secondary sites may also support the formation of a prometastatic niche.

PREMETASTATIC NICHE

In the previous sections, we have outlined the symbiosis of metastatic cancer cells and their host organ, which is a prerequisite for cancer cells to thrive in a secondary site. However, it is well established that distant tissues are not idle, and niche evolution does not just begin upon the arrival of DTCs. Instead, the engraftment of cancer cells is preceded by active modifications in the target tissue to create a permissive microenvironment, the premetastatic niche, for the subsequent establishment of metastatic foci (for review, see Peinado et al. 2017). Premetastatic niches are generated through a complex interplay of soluble factors and exosomes released from the primary tumor, the mobilization and recruitment of immune cells to the intended site, and tumor-elicited reprogramming of tissue-resident cells (Fig. 1). This specialized microenvironment sets the stage for subsequent metastasis, allowing newly arriving DTCs to gain a foothold in the foreign tissue by supporting their survival and growth during seeding.

Tumor-Driven Education of a Distant Site

Cross talk between cancer cells and secondary organs is a requirement for the formation of a tumor-supportive premetastatic niche and is augmented by soluble factors released from cancer cells. An excellent example of tumor-mediated niche preparation is the systemic mobilization and accumulation of myeloid cells in intended metastatic sites such as the lung and brain, prior to cancer cell infiltration (Liu et al. 2013; Wculek and Malanchi 2015). In the premetastatic lung, CD11b$^+$Ly6G$^+$ neutrophils support metastatic initiation by selectively expanding the subpool of arriving breast DTCs with high metastatic potential (Wculek and Malanchi 2015). Similarly, neutrophil accumulation in the premetastatic liver, triggered by primary tumor-derived tissue inhibitor of metalloproteinases (TIMP)-1, increases its susceptibility to breast cancer metastasis (Seubert et al. 2015).

The preparation of secondary tissues for metastatic growth can also be augmented by hypoxia in the primary tumor, triggering the release of protumorigenic factors. Hypoxic breast tumors secrete LOX, which accumulates in premetastatic lungs and catalyzes collagen remodeling, facilitating the recruitment of metastasis-promoting myeloid cells (Erler et al. 2009). The secretion of LOX and other ECM-modifying family members is critically regulated by hypoxia-inducible factor (HIF-1), and the suppression of HIF activity abrogates premetastatic niche formation and metastatic colonization (Wong et al. 2011). More recently, HIF-dependent expression of ADAM12 was found to promote breast-to-lung metastasis through the activation of EGFR signaling, endowing cancer cells with an increased capability for migration, invasion, and outgrowth in distant tissues (Wang et al. 2021a). In the premetastatic bone, hypoxia-induced LOX secretion by primary tumor cells can also modulate bone homeostasis, triggering RANKL-independent osteoclastogenesis and the formation of bone lesions (Cox et al. 2015). The requirement for premetastatic osteolytic lesions for successful bone metastasis was also demonstrated by an earlier study, in this instance triggered by RANKL production by tumor-educated T cells (Monteiro et al. 2013). More recently, breast cancer–derived RSPO2 and RANKL was shown to recruit osteoclast progenitors and promote osteoclastogenesis in the premetastatic niche via Wnt signaling modulation (Yue et al. 2022). Regardless of their origin, focal osteolytic bone lesions serve as a highly supportive niche to foster the colonization of newly arrived breast DTCs.

Nonmalignant cells within the primary tumor microenvironment can also prime cancer cells for organ-specific metastasis. CXCL12 and IGF-1 secreted by CAFs in the primary site can select for cancer cells fit to thrive in the bone environment, leading to gain in the predisposition of cancer cells for bone metastasis (Zhang et al. 2013). Similarly, microenvironment-derived TGF-β primes departing cancer cells for lung metastasis by inducing the expression of ANGPTL4, facilitating lung seeding via destabilization of the pulmonary vasculature (Padua et al. 2008).

Exosome-Mediated Niche Preparation

Tumor-derived exosomes are small membrane-encapsulated vesicles (30–100 nm) that carry a cargo of functional biomolecules including DNA, RNA, proteins, and lipids. By facilitating intercellular communication between cancer cells and their intended niche, they play a crucial role in shaping distant microenvironments for subsequent metastatic colonization. Seminal work by the Lyden group demonstrated the influence of exosomal priming on metastatic organotropism. Exosomal integrins $a_6\beta_4$ and $a_6\beta_1$ from lung-tropic breast cancer cells was found to fuse preferentially with lung fibroblasts and epithelial cells, promoting lung metastasis; whereas, integrin $a_v\beta_5$ specifically binds to Kupffer cells, mediating liver tropism (Hoshino et al. 2015). Thus, the preferential transfer of exosomal integrins to resident cells in intended metastatic sites elicits organ-specific niche preparation. Breast tumor–derived exosomes can also foster immune suppression in the premetastatic microenvironment. NK cell attenuation and reduced adaptive immune surveillance favored metastatic growth in the lung and liver following the uptake of breast tumor–derived exosomes (Wen et al. 2016), while exosome uptake has also been linked to proinflammatory activation of distant macrophages (Chow et al. 2014). In brain metastasis, exosomal cell migra-

tion–inducing and hyaluronan-binding protein (CEMIP) secreted by brain metastatic cancer cells induces inflammation and vascular remodeling in the perivascular niche, supporting brain metastatic colonization (Rodrigues et al. 2019). In addition, metastatic outgrowth in the brain and lung niche can be fostered by exosomal annexin II, which promotes tPA-dependent angiogenesis and macrophage-mediated premetastatic niche formation (Maji et al. 2017).

Tumor-driven education of distant tissues can also be achieved by exosome-mediated transfer of cancer cell–derived microRNAs (miRNAs). In the bone, the uptake of miR-21-containing exosomes by osteoclasts was recently shown to promote premetastatic niche formation, priming metastasis through increased osteoclastogenesis and osteolysis (Yuan et al. 2021). Osteoclastogenesis can be induced by exosomal miR-19a from bone-trophic ER$^+$ breast cancer cells, suggesting a potential mechanism for the increased incidence of bone metastasis in ER$^+$ patients (Wu et al. 2021b). In an elegant demonstration of the diversity of biological processes influenced by miRNAs, exosomal uptake of miR-122 was found to orchestrate metabolic reprogramming in lung and brain-resident premetastatic niche cells. This fueled metastatic outgrowth by increasing the glucose availability for arriving cancer cells (Fong et al. 2015). In a more direct example of miRNA-induced metastasis, the uptake of miR-105 (Zhou et al. 2014) and miR-181c (Tominaga et al. 2015) can promote metastatic seeding through the destruction of endothelial tight junctions and loss of barrier function in distant tissues, including the lung and brain.

Importantly, chemotherapy was recently shown to stimulate the release of tumor-derived prometastatic extracellular vesicles (EVs), including exosomes. Enriched in annexin-6, chemotherapy-elicited EVs induced endothelial cell activation, Ccl-2 induction, and Ly6C$^+$CCR2$^+$ monocyte expansion in the premetastatic lung niche, resulting in augmented breast-to-lung metastasis (Keklikoglou et al. 2019). The monitoring of cancer-associated EVs has shown promise as biomarkers for the management of metastatic breast cancer. Protein profiling of EVs from the plasma of cancer patients could accurately monitor and predict treatment responses in metastatic breast cancer patients, as well as serve as an independent prognostic factor for progression-free survival (Tian et al. 2021).

Distant Tissues as Active Players

Host cells in the premetastatic niche are themselves capable of influencing the metastatic cascade and encouraging the recruitment and outgrowth of breast DTCs. Thus, the remodeled secondary organ serves as a tumor accomplice, coordinating a second line of attack to facilitate successful metastatic colonization of that tissue. For example, the accumulation of NETs in the premetastatic liver and lung acts as a chemotactic factor to attract breast cancer cells and enhance distant metastasis (Yang et al. 2020). This is orchestrated via the specific NET-DNA receptor CCDC25 present on cancer cells, which senses extracellular DNA and responds by activating cancer cell motility. In the premetastatic lung, mesenchymal stromal cells display potent metastatic-promoting activity via the production of complement C3 (Zheng et al. 2021). C3 promotes neutrophil recruitment and NET formation, which fosters subsequent breast-to-lung metastasis. The role of premetastatic niche cells in aiding metastasis has also been demonstrated in the bone microenvironment. Bone-derived CXCL12 preferentially recruits tumor cells expressing CXCR4 (Müller et al. 2001; Devignes et al. 2018). In the latter study, the boost in CXCL12 in the circulation was attributed to HIF activation in osteoprogenitors residing in hypoxic bone marrow niches. This led to a CXCR4-dependent increase in breast cancer cell proliferation and dissemination (Devignes et al. 2018). Bone metastasis is also promoted by bone-derived RANKL, which favors the recruitment and colonization of RANK$^+$ breast DTCs (Jones et al. 2006).

Recent studies have shown that premetastatic conditioning of secondary organs can also be elicited by tumor-independent mechanisms, via local or systemic perturbations in the host. Obesity, a condition associated with chronic, low-grade inflammation, enhances neutrophil accumulation in the premetastatic lung niche and supports subsequent lung met-

astatic seeding via elevated IL-5 and GM-CSF (Quail et al. 2017). In addition to neutrophil recruitment, obesity can drive neutrophil reprogramming in the premetastatic lung, enhancing ROS production and NETosis (McDowell et al. 2021). This results in a loss of vascular integrity, and boosts extravasation of breast DTCs. Even in obese tumor-free mice, their lungs demonstrate key features reminiscent of the premetastatic niche formed by primary tumors, including myeloid cell accumulation, increased expression of inflammatory cytokines, and elevated collagen deposition (Hillers-Ziemer et al. 2021). This response, orchestrated by obesity-activated lung stromal cells, suggests obesity can create a premetastatic niche that is conducive for metastatic growth in the absence of a primary tumor. Local tissue insults can also elicit a tumor-supportive environment in a secondary organ that fosters subsequent metastasis. Chronic nicotine exposure promotes the influx of prometastatic neutrophils that secrete lipocalin 2 (LCN2), promoting breast tumor cell colonization and metastatic outgrowth (Tyagi et al. 2021). Similarly, activated neutrophils recruited to the lungs following radiation-induced injury elicit tissue perturbations that strongly fuel metastatic colonization (Nolan et al. 2022). This tumor-supportive preconditioning of the lung environment was governed by enhanced regenerative Notch signaling in the lung epithelium, a process dependent on neutrophil activation (Nolan et al. 2022). Together, these findings suggest the status of the host tissue, independent of tumor-elicited changes, significantly influences its capacity to support metastatic growth.

CONCLUDING REMARKS

Metastatic disease is a devastating and usually incurable complication of cancer that remains the underlying cause of death for the majority of breast cancer patients. In this work, we have provided an overview of organ-specific breast cancer metastasis, with a focus on the site-dependent interactions between tissue-resident cells and arriving cancer cells that govern successful colonization of the lung, bone, brain, and liver environments (summarized in Table 1).

Table 1. Comparison of metastatic niches in breast cancer, highlighting the key processes and cellular subsets that govern successful metastatic seeding and outgrowth

Metastatic niches	Lung	Bone	Brain	Liver
Immune recruitment and activation	TAMs (CCL2-recruited) Neutrophils (pro- and antitumor) NK cells, T cells	TAMs (CCL2-recruited) Myeloid-derived suppressor cells	Myeloid cells (CCL2-recruited) Protumor neutrophils Distinct TAM subsets	Protumor neutrophils (iLDNs)
NETosis	Enhanced by inflammation			NET-DNA receptor CCDC25
T-cell suppression	γδ-T-cell/neutrophil interplay Myeloid-derived osteopontin Oxygen sensing	Immune-suppressive myeloid cells (chemotherapy) Adipokines	Astrocyte immunomodulation Arg1/PD-L1$^+$ neutrophils	Immune-suppressive myeloid cells (high-fat diet, involution)
Vascular engagement	Distinct growth pattern: angiogenic or vascular cooption		Distinct growth pattern: angiogenic or vascular cooption	Distinct growth pattern angiogenic or vascular cooption
Metabolic reprogramming	↑ Electron transport, oxidative stress, antioxidant programs	↑ Serine synthesis, lactate production	↑ FA utilization, gluconeogenesis, antioxidants, autophagy Neuron cooption (GABA, glutamate)	↑ Glycolysis (PDK1-dependent) Endocrine resistance ↑ FA utilization (obesity)
Resident cell interactions	Lung alveolar cells Stromal cells Endothelial cells	Osteogenic niche Perivascular niche Adipocytes	Reactive astrocytes Microglia reprogramming Neurons, endothelial cells	Hepatocytes Hepatic stellate cells Endothelial cells
ECM remodeling	↑ ECM proteins (TNC, POSTN)	Bone resorption (vicious cycle) RANKL/OPG dysregulation		Matrix adhesion LOX-induced collagen cross-linking, fibrosis
Fibroblast activation	CAF-induced stemness Dynamic reprogramming EMT modulation			Stellate cell activation, ↑ LOX expression
Premetastatic niche	Myeloid cell infiltration LOX-induced ECM remodeling Exome-mediated priming NET accumulation	LOX-induced ECM remodeling Exome-mediated priming Bone-derived factors (CXCL12/RANKL)	Myeloid cell infiltration Exome-mediated priming	Myeloid cell infiltration Exome-mediated priming NET accumulation

(TAM) Tumor-associated macrophage, (iLDNs) immature low-density neutrophils, (FA) fatty acid, (GABA) γ-aminobutyric acid, (PDK1) pyruvate dehydrogenase kinase-1, (TNC) tenascin C, (POSTN) periostin, (CAF) cancer-associated fibroblast, (EMT) epithelial-

The ability of both metastatic cancer cells and their target organ to coevolve and cooperatively support outgrowth is perhaps the most remarkable and intriguing step in the metastatic cascade. This partner-in-crime relationship is complex, dynamic, and context-dependent, influenced by tumor intrinsic factors, the unique players, and environmental constraints within each organ, as well as by local and systemic perturbations of the host. Undoubtedly, a deeper understanding of the tissue-specific prerequisites for metastatic growth will lay the foundation for the development of more effective therapies for metastatic breast cancer.

There is still much to be learned about the molecular mechanisms of breast cancer metastasis. Recent studies have begun unraveling the influence of tumor extrinsic factors such as tissue injury, metabolic diseases, physical exercise, and normal physiological processes (such as postpartum involution) on the development of tumor-supportive niches, a list certain to grow in the future. More work is needed to understand the complexity and evolution of metastatic niches, for example, the existence of spatially and temporally dynamic niches within the same organ, and whether an established metastatic niche in one organ can support niche development in other tissues, facilitating multiorgan metastasis. The influence of cancer-niche evolution on treatment response and the development of resistance is a topic of great interest to the field. Indeed, a recent study using optical barcoding revealed that the metastatic niche can profoundly influence the degree of intrametastatic breast tumor heterogeneity, a known barrier for treatment efficacy (Berthelet et al. 2021). Finally, the concept of "personalized niches" needs exploring, in which the intrinsic genomic profile of cancer cells plus the extrinsic host environment together dictate the development of patient-specific personalized niches.

It remains a daunting challenge to develop TME-targeting agents for cancer therapy, since it requires identification of the fundamental drivers of tumor-promoting niches. Progress has been made in breast-to-bone metastasis, with bone resorption inhibitors such as bisphosphonates (such as zoledronic acid) and denosumab (an anti-RANKL antibody) now widely used for the management of metastatic breast cancer. Other bone-modifying agents are emerging that have demonstrated potential clinical utility, including cathepsin K and Src inhibitors (for review, see Brook et al. 2018). The complexity and diversity of organ-specific niches present as major challenges to develop therapies that can be broadly applicable to a large fraction of patients. One possible approach is to identify common stresses experienced by metastatic cancer cells at different organ sites, and the shared fitness-promoting pathways that promote their survival under stress (Esposito et al. 2021b). Although rapid improvement has been made in our understanding of organ-specific breast cancer metastasis, it is clear that we have only scratched the surface of this remarkably complicated process. Probably an even bigger challenge will be translating our growing knowledge of cancer–niche interactions into better preventative and curative options for metastatic breast cancer patients.

ACKNOWLEDGMENTS

The work of E.N. and I.M. is supported by the Francis Crick Institute, which receives its core funding from Cancer Research UK (FC001112), the UK Medical Research Council (FC001112), the Wellcome Trust (FC001112), and a European Research Council grant (ERC CoG-H2020-725492). E.N. is supported by the Auckland Medical Research Foundation's Douglas Goodfellow Repatriation Fellowship (1421001). The work in the laboratory of Y.K. is supported by Ludwig Cancer Research, Brewster Foundation, and grants from the American Cancer Society, Susan G. Komen Foundation, Breast Cancer Research Foundation, METAvivor, Department of Defense, and the National Cancer Institute to Y.K.

REFERENCES

*Reference is also in this subject collection.

Albrengues J, Shields MA, Ng D, Park CG, Ambrico A, Poindexter ME, Upadhyay P, Uyeminami DL, Pommier A, Küttner V, et al. 2018. Neutrophil extracellular traps produced during inflammation awaken dormant cancer cells in mice. *Science* **361**: eaao4227. doi:10.1126/science.aao4227

Anborgh PH, Mutrie JC, Tuck AB, Chambers AF. 2010. Role of the metastasis-promoting protein osteopontin in the tumour microenvironment. *J Cell Mol Med* **14**: 2037–2044. doi:10.1111/j.1582-4934.2010.01115.x

Bado IL, Zhang W, Hu J, Xu Z, Wang H, Sarkar P, Li L, Wan YW, Liu J, Wu W, et al. 2021. The bone microenvironment increases phenotypic plasticity of ER⁺ breast cancer cells. *Dev Cell* **56**: 1100–1117.e9. doi:10.1016/j.devcel.2021.03.008

Bartlett AQ, Pennock ND, Klug A, Schedin P. 2021. Immune milieu established by postpartum liver involution promotes breast cancer liver metastasis. *Cancers (Basel)* **13**: 1698. doi:10.3390/cancers13071698

Basnet H, Tian L, Ganesh K, Huang YH, Macalinao DG, Brogi E, Finley LW, Massagué J. 2019. Flura-seq identifies organ-specific metabolic adaptations during early metastatic colonization. *eLife* **8**: e43627. doi:10.7554/eLife.43627

Berthelet J, Wimmer VC, Whitfield HJ, Serrano A, Boudier T, Mangiola S, Merdas M, El-Saafin F, Baloyan D, Wilcox J, et al. 2021. The site of breast cancer metastases dictates their clonal composition and reversible transcriptomic profile. *Sci Adv* **7**: eabf4408. doi:10.1126/sciadv.abf4408

Blazquez R, Rietkötter E, Wenske B, Wlochowitz D, Sparrer D, Vollmer E, Müller G, Seegerer J, Sun X, Dettmer K, et al. 2020. LEF1 supports metastatic brain colonization by regulating glutathione metabolism and increasing ROS resistance in breast cancer. *Int J Cancer* **146**: 3170–3183. doi:10.1002/ijc.32742

Boire A, Zou Y, Shieh J, Macalinao DG, Pentsova E, Massagué J. 2017. Complement component 3 adapts the cerebrospinal fluid for leptomeningeal metastasis. *Cell* **168**: 1101–1113.e13. doi:10.1016/j.cell.2017.02.025

Brook N, Brook E, Dharmarajan A, Dass CR, Chan A. 2018. Breast cancer bone metastases: pathogenesis and therapeutic targets. *Int J Biochem Cell Biol* **96**: 63–78. doi:10.1016/j.biocel.2018.01.003

Carbonell WS, Ansorge O, Sibson N, Muschel R. 2009. The vascular basement membrane as "soil" in brain metastasis. *PLoS ONE* **4**: e5857. doi:10.1371/journal.pone.0005857

Chao Y, Wu Q, Shepard C, Wells A. 2012. Hepatocyte induced re-expression of E-cadherin in breast and prostate cancer cells increases chemoresistance. *Clin Exp Metastas* **29**: 39–50. doi:10.1007/s10585-011-9427-3

Chen J, Lee HJ, Wu X, Huo L, Kim SJ, Xu L, Wang Y, He J, Bollu LR, Gao G, et al. 2015. Gain of glucose-independent growth upon metastasis of breast cancer cells to the brain. *Cancer Res* **75**: 554–565. doi:10.1158/0008-5472.CAN-14-2268

Chen Q, Boire A, Jin X, Valiente M, Er EE, Lopez-Soto A, Jacob LS, Patwa R, Shah H, Xu K, et al. 2016. Carcinoma–astrocyte gap junctions promote brain metastasis by cGAMP transfer. *Nature* **533**: 493–498. doi:10.1038/nature18268

Chen F, Han Y, Kang Y. 2021a. Bone marrow niches in the regulation of bone metastasis. *Brit J Cancer* **124**: 1912–1920. doi:10.1038/s41416-021-01329-6

Chen Y, McAndrews KM, Kalluri R. 2021b. Clinical and therapeutic relevance of cancer-associated fibroblasts. *Nat Rev Clin Oncol* **18**: 792–804. doi:10.1038/s41571-021-00546-5

Chow A, Zhou W, Liu L, Fong MY, Champer J, Haute DV, Chin AR, Ren X, Gugiu BG, Meng Z, et al. 2014. Macrophage immunomodulation by breast cancer–derived exosomes requires toll-like receptor 2-mediated activation of NF-κB. *Sci Rep* **4**: 5750. doi:10.1038/srep05750

Clements VK, Long T, Long R, Figley C, Smith DMC, Ostrand-Rosenberg S. 2018. Frontline science: high fat diet and leptin promote tumor progression by inducing myeloid-derived suppressor cells. *J Leukocyte Biol* **103**: 395–407. doi:10.1002/JLB.4HI0517-210R

Clever D, Roychoudhuri R, Constantinides MG, Askenase MH, Sukumar M, Klebanoff CA, Eil RL, Hickman HD, Yu Z, Pan JH, et al. 2016. Oxygen sensing by T cells establishes an immunologically tolerant metastatic niche. *Cell* **166**: 1117–1131.e14. doi:10.1016/j.cell.2016.07.032

Coffelt SB, Kersten K, Doornebal CW, Weiden J, Vrijland K, Hau C-S, Verstegen NJM, Ciampricotti M, Hawinkels LJAC, Jonkers J, et al. 2015. IL17-producing γδ T cells and neutrophils conspire to promote breast cancer metastasis. *Nature* **522**: 345–348. doi:10.1038/nature14282

Coleman RE. 2001. Metastatic bone disease: clinical features, pathophysiology and treatment strategies. *Cancer Treat Rev* **27**: 165–176. doi:10.1053/ctrv.2000.0210

Cordero A, Kanojia D, Miska J, Panek WK, Xiao A, Han Y, Bonamici N, Zhou W, Xiao T, Wu M, et al. 2019. FABP7 is a key metabolic regulator in HER2⁺ breast cancer brain metastasis. *Oncogene* **38**: 6445–6460. doi:10.1038/s41388-019-0893-4

Cox TR, Bird D, Baker AM, Barker HE, Ho MWY, Lang G, Erler JT. 2013. LOX-mediated collagen crosslinking is responsible for fibrosis-enhanced metastasis. *Cancer Res* **73**: 1721–1732. doi:10.1158/0008-5472.CAN-12-2233

Cox TR, Rumney RMH, Schoof EM, Perryman L, Høye AM, Agrawal A, Bird D, Latif NA, Forrest H, Evans HR, et al. 2015. The hypoxic cancer secretome induces pre-metastatic bone lesions through lysyl oxidase. *Nature* **522**: 106–110. doi:10.1038/nature14492

* Dalla E, Sreekumar A, Aguirre-Ghiso JA, Chodosh LA. 2022. Dormancy in breast cancer. *Cold Spring Harb Perspect Med* doi:10.1101/cshperspect.a041331

Dekker LJ, Boogerd W, Stockhammer G, Dalebout JC, Siccama I, Zheng P, Bonfrer JM, Verschuuren JJ, Jenster G, Verbeek MM, et al. 2005. MALDI-TOF mass spectrometry analysis of cerebrospinal fluid tryptic peptide profiles to diagnose leptomeningeal metastases in patients with breast cancer. *Mol Cell Proteomics* **4**: 1341–1349. doi:10.1074/mcp.M500081-MCP200

Devignes CS, Aslan Y, Brenot A, Devillers A, Schepers K, Fabre S, Chou J, Casbon AJ, Werb Z, Provot S. 2018. HIF signaling in osteoblast-lineage cells promotes systemic breast cancer growth and metastasis in mice. *Proc Natl Acad Sci* **115**: E992–E1001. doi:10.1073/pnas.1718009115

Dupuy F, Tabariès S, Andrzejewski S, Dong Z, Blagih J, Annis MG, Omeroglu A, Gao D, Leung S, Amir E, et al. 2015. PDK1-dependent metabolic reprogramming dictates metastatic potential in breast cancer. *Cell Metab* **22**: 577–589. doi:10.1016/j.cmet.2015.08.007

Elefteriou F. 2016. Role of sympathetic nerves in the establishment of metastatic breast cancer cells in bone. *J Bone Oncol* **5:** 132–134. doi:10.1016/j.jbo.2016.03.003

Erin N, Kale Ş, Tanrıöver G, Köksoy S, Duymuş Ö, Korcum AF. 2013. Differential characteristics of heart, liver, and brain metastatic subsets of murine breast carcinoma. *Breast Cancer Res Treat* **139:** 677–689. doi:10.1007/s10549-013-2584-0

Erler JT, Bennewith KL, Cox TR, Lang G, Bird D, Koong A, Le QT, Giaccia AJ. 2009. Hypoxia-induced lysyl oxidase is a critical mediator of bone marrow cell recruitment to form the premetastatic niche. *Cancer Cell* **15:** 35–44. doi:10.1016/j.ccr.2008.11.012

Esposito M, Guise T, Kang Y. 2018. The biology of bone metastasis. *Cold Spring Harb Perspect Med* **8:** a031252. doi:10.1101/cshperspect.a031252

Esposito M, Mondal N, Greco TM, Wei Y, Spadazzi C, Lin S-C, Zheng H, Cheung C, Magnani JL, Lin SH, et al. 2019. Bone vascular niche e-selectin induces mesenchymal-epithelial transition and Wnt activation in cancer cells to promote bone metastasis. *Nat Cell Biol* **21:** 627–639. doi:10.1038/s41556-019-0309-2

Esposito M, Fang C, Cook KC, Park N, Wei Y, Spadazzi C, Bracha D, Gunaratna RT, Laevsky G, DeCoste CJ, et al. 2021a. TGF-β-induced DACT1 biomolecular condensates repress Wnt signalling to promote bone metastasis. *Nat Cell Biol* **23:** 257–267. doi:10.1038/s41556-021-00641-w

Esposito M, Ganesan S, Kang Y. 2021b. Emerging strategies for treating metastasis. *Nat Cancer* **2:** 258–270. doi:10.1038/s43018-021-00181-0

Evans AJ, James JJ, Cornford EJ, Chan SY, Burrell HC, Pinder SE, Gutteridge E, Robertson JFR, Hornbuckle J, Cheung KL. 2004. Brain metastases from breast cancer: identification of a high-risk group. *Clin Oncol* **16:** 345–349. doi:10.1016/j.clon.2004.03.012

Eyre R, Alférez DG, Santiago-Gómez A, Spence K, McConnell JC, Hart C, Simões BM, Lefley D, Tulotta C, Storer J, et al. 2019. Microenvironmental IL1β promotes breast cancer metastatic colonisation in the bone via activation of Wnt signalling. *Nat Commun* **10:** 5016. doi:10.1038/s41467-019-12807-0

Fan J, Cai B, Zeng M, Hao Y, Giancotti FG, Fu BM. 2011. Integrin β4 signaling promotes mammary tumor cell adhesion to brain microvascular endothelium by inducing erbb2-mediated secretion of VEGF. *Ann Biomed Eng* **39:** 2223–2241. doi:10.1007/s10439-011-0321-6

Fong MY, Zhou W, Liu L, Alontaga AY, Chandra M, Ashby J, Chow A, O'Connor STF, Li S, Chin AR, et al. 2015. Breast-cancer-secreted miR-122 reprograms glucose metabolism in premetastatic niche to promote metastasis. *Nat Cell Biol* **17:** 183–194. doi:10.1038/ncb3094

Foulkes WD, Smith IE, Reis-Filho JS. 2010. Triple-negative breast cancer. *N Engl J Med* **363:** 1938–1948. doi:10.1056/NEJMra1001389

Franzoi MA, Hortobagyi GN. 2019. Leptomeningeal carcinomatosis in patients with breast cancer. *Crit Rev Oncol Hemat* **135:** 85–94. doi:10.1016/j.critrevonc.2019.01.020

Frentzas S, Simoneau E, Bridgeman VL, Vermeulen PB, Foo S, Kostaras E, Nathan MR, Wotherspoon A, Gao Z, Shi Y, et al. 2016. Vessel co-option mediates resistance to anti-angiogenic therapy in liver metastases. *Nat Med* **22:** 1294–1302. doi:10.1038/nm.4197

Gall CL, Bellahcène A, Bonnelye E, Gasser JA, Castronovo V, Green J, Zimmermann J, Clézardin P. 2007. A cathepsin K inhibitor reduces breast cancer–induced osteolysis and skeletal tumor burden. *Cancer Res* **67:** 9894–9902. doi:10.1158/0008-5472.CAN-06-3940

García-Gómez P, Valiente M. 2020. Vascular co-option in brain metastasis. *Angiogenesis* **23:** 3–8. doi:10.1007/s10456-019-09693-x

Ghajar CM, Peinado H, Mori H, Matei IR, Evason KJ, Brazier H, Almeida D, Koller A, Hajjar KA, Stainier DYR, et al. 2013. The perivascular niche regulates breast tumour dormancy. *Nat Cell Biol* **15:** 807–817. doi:10.1038/ncb2767

Goddard ET, Bassale S, Schedin T, Jindal S, Johnston J, Cabral E, Latour E, Lyons TR, Mori M, Schedin PJ, et al. 2019. Association between postpartum breast cancer diagnosis and metastasis and the clinical features underlying risk. *JAMA Netw Open* **2:** e186997. doi:10.1001/jamanetworkopen.2018.6997

Gonzalez H, Mei W, Robles I, Hagerling C, Allen BM, Okholm TLH, Nanjaraj A, Verbeek T, Kalavacherla S, van Gogh M, et al. 2022. Cellular architecture of human brain metastases. *Cell* **185:** 729–745.e20. doi:10.1016/j.cell.2021.12.043

Guldner IH, Wang Q, Yang L, Golomb SM, Zhao Z, Lopez JA, Brunory A, Howe EN, Zhang Y, Palakurthi B, et al. 2020. CNS-native myeloid cells drive immune suppression in the brain metastatic niche through Cxcl10. *Cell* **183:** 1234–1248.e25. doi:10.1016/j.cell.2020.09.064

Hedrick CC, Malanchi I. 2022. Neutrophils in cancer: heterogeneous and multifaceted. *Nat Rev Immunol* **22:** 173–187. doi:10.1038/s41577-021-00571-6

Heitz F, Rochon J, Harter P, Lueck HJ, Fisseler-Eckhoff A, Barinoff J, Traut A, Lorenz-Salehi F, du Bois A. 2011. Cerebral metastases in metastatic breast cancer: disease-specific risk factors and survival. *Ann Oncol* **22:** 1571–1581. doi:10.1093/annonc/mdq625

Hillers-Ziemer LE, Williams AE, Janquart A, Grogan C, Thompson V, Sanchez A, Arendt LM. 2021. Obesity-activated lung stromal cells promote myeloid lineage cell accumulation and breast cancer metastasis. *Cancers (Basel)* **13:** 1005. doi:10.3390/cancers13051005

Hoshino A, Costa-Silva B, Shen TL, Rodrigues G, Hashimoto A, Mark MT, Molina H, Kohsaka S, Giannatale AD, Ceder S, et al. 2015. Tumour exosome integrins determine organotropic metastasis. *Nature* **527:** 329–335. doi:10.1038/nature15756

Hsu BE, Tabariès S, Johnson RM, Andrzejewski S, Senecal J, Lehuédé C, Annis MG, Ma EH, Völs S, Ramsay L, et al. 2019. Immature low-density neutrophils exhibit metabolic flexibility that facilitates breast cancer liver metastasis. *Cell Rep* **27:** 3902–3915.e6. doi:10.1016/j.celrep.2019.05.091

Hung MH, Liu CY, Shiau CY, Hsu CY, Tsai YF, Wang YL, Tai LC, King KL, Chao TC, Chiu JH, et al. 2014. Effect of age and biological subtype on the risk and timing of brain metastasis in breast cancer patients. *PLoS ONE* **9:** e89389. doi:10.1371/journal.pone.0089389

Jaillon S, Ponzetta A, Mitri DD, Santoni A, Bonecchi R, Mantovani A. 2020. Neutrophil diversity and plasticity in tumour progression and therapy. *Nat Rev Cancer* **20**: 485–503. doi:10.1038/s41568-020-0281-y

Jandial R, Choy C, Levy DM, Chen MY, Ansari KI. 2017. Astrocyte-induced Reelin expression drives proliferation of Her2⁺ breast cancer metastases. *Clin Exp Metastasis* **34**: 185–196. doi:10.1007/s10585-017-9839-9

Jones DH, Nakashima T, Sanchez OH, Kozieradzki I, Komarova SV, Sarosi I, Morony S, Rubin E, Sarao R, Hojilla CV, et al. 2006. Regulation of cancer cell migration and bone metastasis by RANKL. *Nature* **440**: 692–696. doi:10.1038/nature04524

Kang Y, Siegel PM, Shu W, Drobnjak M, Kakonen SM, Cordón-Cardo C, Guise TA, Massagué J. 2003. A multigenic program mediating breast cancer metastasis to bone. *Cancer Cell* **3**: 537–549. doi:10.1016/S1535-6108(03)00132-6

Kang Y, He W, Tulley S, Gupta GP, Serganova I, Chen C-R, Manova-Todorova K, Blasberg R, Gerald WL, Massagué J. 2005. Breast cancer bone metastasis mediated by the Smad tumor suppressor pathway. *Proc Natl Acad Sci* **102**: 13909–13914. doi:10.1073/pnas.0506517102

Keklikoglou I, Cianciaruso C, Güç E, Squadrito ML, Spring LM, Tazzyman S, Lambein L, Poissonnier A, Ferraro GB, Baer C, et al. 2019. Chemotherapy elicits pro-metastatic extracellular vesicles in breast cancer models. *Nat Cell Biol* **21**: 190–202. doi:10.1038/s41556-018-0256-3

Klotz R, Thomas A, Teng T, Han SM, Iriondo O, Li L, Restrepo-Vassalli S, Wang A, Izadian N, MacKay M, et al. 2020. Circulating tumor cells exhibit metastatic tropism and reveal brain metastasis drivers. *Cancer Discov* **10**: 86–103. doi:10.1158/2159-8290.CD-19-0384

Korpal M, Ell BJ, Buffa FM, Ibrahim T, Blanco MA, Celià-Terrassa T, Mercatali L, Khan Z, Goodarzi H, Hua Y, et al. 2011. Direct targeting of Sec23a by miR-200s influences cancer cell secretome and promotes metastatic colonization. *Nat Med* **17**: 1101–1108. doi:10.1038/nm.2401

Lawson DA, Bhakta NR, Kessenbrock K, Prummel KD, Yu Y, Takai K, Zhou A, Eyob H, Balakrishnan S, Wang CY, et al. 2015. Single-cell analysis reveals a stem-cell program in human metastatic breast cancer cells. *Nature* **526**: 131–135. doi:10.1038/nature15260

Lee SK, Park KK, Kim H-J, Park J, Son SH, Kim KR, Chung WY. 2017. Human antigen R-regulated CCL20 contributes to osteolytic breast cancer bone metastasis. *Sci Rep* **7**: 9610. doi:10.1038/s41598-017-09040-4

Lefrançais E, Ortiz-Muñoz G, Caudrillier A, Mallavia B, Liu F, Sayah DM, Thornton EE, Headley MB, David T, Coughlin SR, et al. 2017. The lung is a site of platelet biogenesis and a reservoir for haematopoietic progenitors. *Nature* **544**: 105–109. doi:10.1038/nature21706

Lemma S, Pompo GD, Porporato PE, Sboarina M, Russell S, Gillies RJ, Baldini N, Sonveaux P, Avnet S. 2017. MDA-MB-231 breast cancer cells fuel osteoclast metabolism and activity: a new rationale for the pathogenesis of osteolytic bone metastases. *Biochim Biophys Acta* **1863**: 3254–3264.

Li P, Lu M, Shi J, Hua L, Gong Z, Li Q, Shultz LD, Ren G. 2020a. Dual roles of neutrophils in metastatic colonization are governed by the host NK cell status. *Nat Commun* **11**: 4387. doi:10.1038/s41467-020-18125-0

Li Y, Su X, Rohatgi N, Zhang Y, Brestoff JR, Shoghi KI, Xu Y, Semenkovich CF, Harris CA, Peterson LL, et al. 2020b.

Hepatic lipids promote liver metastasis. *JCI Insight* **5**: e136215. doi:10.1172/jci.insight.136215

Liang M, Ma Q, Ding N, Luo F, Bai Y, Kang F, Gong X, Dong R, Dai J, Dai Q, et al. 2019. IL-11 is essential in promoting osteolysis in breast cancer bone metastasis via RANKL-independent activation of osteoclastogenesis. *Cell Death Dis* **10**: 353. doi:10.1038/s41419-019-1594-1

Liu Y, Kosaka A, Ikeura M, Kohanbash G, Fellows-Mayle W, Snyder LA, Okada H. 2013. Premetastatic soil and prevention of breast cancer brain metastasis. *Neuro Oncol* **15**: 891–903. doi:10.1093/neuonc/not031

Liu C, Wang H, Wang W, Wang L, Liu W, Wang J, Geng Q, Lu Y. 2018. ENO2 promotes cell proliferation, glycolysis, and glucocorticoid-resistance in acute lymphoblastic leukemia. *Cell Physiol Biochem* **46**: 1525–1535. doi:10.1159/000489196

Liu C, Zhao Q, Yu X. 2020. Bone marrow adipocytes, adipocytokines, and breast cancer cells: novel implications in bone metastasis of breast cancer. *Front Oncol* **10**: 561595. doi:10.3389/fonc.2020.561595

Lo HC, Xu Z, Kim IS, Pingel B, Aguirre S, Kodali S, Liu J, Zhang W, Muscarella AM, Hein SM, et al. 2020. Resistance to natural killer cell immunosurveillance confers a selective advantage to polyclonal metastasis. *Nat Cancer* **1**: 709–722. doi:10.1038/s43018-020-0068-9

Lorger M, Felding-Habermann B. 2010. Capturing changes in the brain microenvironment during initial steps of breast cancer brain metastasis. *Am J Pathol* **176**: 2958–2971. doi:10.2353/ajpath.2010.090838

Louie E, Chen XF, Coomes A, Ji K, Tsirka S, Chen EI. 2013. Neurotrophin-3 modulates breast cancer cells and the microenvironment to promote the growth of breast cancer brain metastasis. *Oncogene* **32**: 4064–4077. doi:10.1038/onc.2012.417

Lu X, Wang Q, Hu G, Poznak CV, Fleisher M, Reiss M, Massagué J, Kang Y. 2009. ADAMTS1 and MMP1 proteolytically engage EGF-like ligands in an osteolytic signaling cascade for bone metastasis. *Gene Dev* **23**: 1882–1894. doi:10.1101/gad.1824809

Lu X, Mu E, Wei Y, Riethdorf S, Yang Q, Yuan M, Yan J, Hua Y, Tiede BJ, Lu X, et al. 2011. VCAM-1 promotes osteolytic expansion of indolent bone micrometastasis of breast cancer by engaging α4β1-positive osteoclast progenitors. *Cancer Cell* **20**: 701–714. doi:10.1016/j.ccr.2011.11.002

Ma R-Y, Zhang H, Li XF, Zhang C-B, Selli C, Tagliavini G, Lam AD, Prost S, Sims AH, Hu HY, et al. 2020. Monocyte derived macrophages promote breast cancer bone metastasis outgrowth. *J Exp Med* **217**: e20191820. doi:10.1084/jem.20191820

Maji S, Chaudhary P, Akopova I, Nguyen PM, Hare RJ, Gryczynski I, Vishwanatha JK. 2017. Exosomal annexin II promotes angiogenesis and breast cancer metastasis. *Mol Cancer Res* **15**: 93–105. doi:10.1158/1541-7786.MCR-16-0163

Malanchi I, Santamaria-Martínez A, Susanto E, Peng H, Lehr HA, Delaloye JF, Huelsken J. 2012. Interactions between cancer stem cells and their niche govern metastatic colonization. *Nature* **481**: 85–89. doi:10.1038/nature10694

Martin MD, Kremers GJ, Short KW, Rocheleau JV, Xu L, Piston DW, Matrisian LM, Gorden DL. 2010. Rapid extravasation and establishment of breast cancer micrometastases in the liver microenvironment. *Mol Cancer Res* **8:** 1319–1327. doi:10.1158/1541-7786.MCR-09-0551

Martin YDP, Park D, Ramachandran A, Ombrato L, Calvo F, Chakravarty P, Spencer-Dene B, Derzsi S, Hill CS, Sahai E, et al. 2015. Mesenchymal cancer cell-stroma crosstalk promotes niche activation, epithelial reversion, and metastatic colonization. *Cell Rep* **13:** 2456–2469. doi:10.1016/j.celrep.2015.11.025

Martin AM, Cagney DN, Catalano PJ, Warren LE, Bellon JR, Punglia RS, Claus EB, Lee EQ, Wen PY, Haas-Kogan DA, et al. 2017. Brain metastases in newly diagnosed breast cancer: a population-based study. *JAMA Oncol* **3:** 1069. doi:10.1001/jamaoncol.2017.0001

McDowell SAC, Luo RBE, Arabzadeh A, Doré S, Bennett NC, Breton V, Karimi E, Rezanejad M, Yang RR, Lach KD, et al. 2021. Neutrophil oxidative stress mediates obesity-associated vascular dysfunction and metastatic transmigration. *Nat Cancer* **2:** 545–562. doi:10.1038/s43018-021-00194-9

Montagner M, Bhome R, Hooper S, Chakravarty P, Qin X, Sufi J, Bhargava A, Ratcliffe CDH, Naito Y, Pocaterra A, et al. 2020. Crosstalk with lung epithelial cells regulates Sfrp2-mediated latency in breast cancer dissemination. *Nat Cell Biol* **22:** 289–296. doi:10.1038/s41556-020-0474-3

Monteiro AC, Leal AC, Gonçalves-Silva T, Mercadante ACT, Kestelman F, Chaves SB, Azevedo RB, Monteiro JP, Bonomo A. 2013. T cells induce pre-metastatic osteolytic disease and help bone metastases establishment in a mouse model of metastatic breast cancer. *PLoS ONE* **8:** e68171. doi:10.1371/journal.pone.0068171

Müller A, Homey B, Soto H, Ge N, Catron D, Buchanan ME, McClanahan T, Murphy E, Yuan W, Wagner SN, et al. 2001. Involvement of chemokine receptors in breast cancer metastasis. *Nature* **410:** 50–56. doi:10.1038/35065016

Neman J, Termini J, Wilczynski S, Vaidehi N, Choy C, Kowolik CM, Li H, Hambrecht AC, Roberts E, Jandial R. 2014. Human breast cancer metastases to the brain display GABAergic properties in the neural niche. *Proc Natl Acad Sci* **111:** 984–989. doi:10.1073/pnas.1322098111

Nolan E, Bridgeman VL, Ombrato L, Karoutas A, Rabas N, Sewnath CAN, Vasquez M, Rodrigues FS, Horswell S, Faull P, et al. 2022. Radiation exposure elicits a neutrophil-driven response in healthy lung tissue that enhances metastatic colonization. *Nat Cancer* **3:** 173–187. doi:10.1038/s43018-022-00336-7

Ombrato L, Nolan E, Kurelac I, Mavousian A, Bridgeman VL, Heinze I, Chakravarty P, Horswell S, Gonzalez-Gualda E, Matacchione G, et al. 2019. Metastatic-niche labelling reveals parenchymal cells with stem features. *Nature* **572:** 603–608. doi:10.1038/s41586-019-1487-6

Oskarsson T, Acharyya S, Zhang XHF, Vanharanta S, Tavazoie SF, Morris PG, Downey RJ, Manova-Todorova K, Brogi E, Massagué J. 2011. Breast cancer cells produce tenascin C as a metastatic niche component to colonize the lungs. *Nat Med* **17:** 867–874. doi:10.1038/nm.2379

Ouhtit A, Elmageed ZYA, Abdraboh ME, Lioe TF, Raj MHG. 2007. In vivo evidence for the role of CD44s in promoting breast cancer metastasis to the liver. *Am J Pathol* **171:** 2033–2039. doi:10.2353/ajpath.2007.070535

Padua D, Zhang XHF, Wang Q, Nadal C, Gerald WL, Gomis RR, Massagué J. 2008. TGFβ primes breast tumors for lung metastasis seeding through angiopoietin-like 4. *Cell* **133:** 66–77. doi:10.1016/j.cell.2008.01.046

Park J, Wysocki RW, Amoozgar Z, Maiorino L, Fein MR, Jorns J, Schott AF, Kinugasa-Katayama Y, Lee Y, Won NH, et al. 2016. Cancer cells induce metastasis-supporting neutrophil extracellular DNA traps. *Sci Transl Med* **8:** 361ra138. doi:10.1126/scitranslmed.aag1711

Pein M, Insua-Rodríguez J, Hongu T, Riedel A, Meier J, Wiedmann L, Decker K, Essers MAG, Sinn H-P, Spaich S, et al. 2020. Metastasis-initiating cells induce and exploit a fibroblast niche to fuel malignant colonization of the lungs. *Nat Commun* **11:** 1494. doi:10.1038/s41467-020-15188-x

Peinado H, Zhang H, Matei IR, Costa-Silva B, Hoshino A, Rodrigues G, Psaila B, Kaplan RN, Bromberg JF, Kang Y, et al. 2017. Pre-metastatic niches: organ-specific homes for metastases. *Nat Rev Cancer* **17:** 302–317. doi:10.1038/nrc.2017.6

Pollard JW. 2004. Tumour-educated macrophages promote tumour progression and metastasis. *Nat Rev Cancer* **4:** 71–78. doi:10.1038/nrc1256

Pollari S, Käkönen S-M, Edgren H, Wolf M, Kohonen P, Sara H, Guise T, Nees M, Kallioniemi O. 2011. Enhanced serine production by bone metastatic breast cancer cells stimulates osteoclastogenesis. *Breast Cancer Res Treat* **125:** 421–430. doi:10.1007/s10549-010-0848-5

Priego N, Zhu L, Monteiro C, Mulders M, Wasilewski D, Bindeman W, Doglio L, Martínez L, Martínez-Saez E, Cajal SRY, et al. 2018. STAT3 labels a subpopulation of reactive astrocytes required for brain metastasis. *Nat Med* **24:** 1024–1035. doi:10.1038/s41591-018-0044-4

Pukrop T, Dehghani F, Chuang H, Lohaus R, Bayanga K, Heermann S, Regen T, Rossum DV, Klemm F, Schulz M, et al. 2010. Microglia promote colonization of brain tissue by breast cancer cells in a Wnt-dependent way. *Glia* **58:** 1477–1489. doi:10.1002/glia.21022

Qian B, Deng Y, Im JH, Muschel RJ, Zou Y, Li J, Lang RA, Pollard JW. 2009. A distinct macrophage population mediates metastatic breast cancer cell extravasation, establishment and growth. *PLoS ONE* **4:** e6562. doi:10.1371/journal.pone.0006562

Qian BZ, Li J, Zhang H, Kitamura T, Zhang J, Campion LR, Kaiser EA, Snyder LA, Pollard JW. 2011. CCL2 recruits inflammatory monocytes to facilitate breast-tumor metastasis. *Nature* **475:** 222–225. doi:10.1038/nature10138

Quail DF, Olson OC, Bhardwaj P, Walsh LA, Akkari L, Quick ML, Chen I-C, Wendel N, Ben-Chetrit N, Walker J, et al. 2017. Obesity alters the lung myeloid cell landscape to enhance breast cancer metastasis through IL5 and GM-CSF. *Nat Cell Biol* **19:** 974–987. doi:10.1038/ncb3578

Rodrigues G, Hoshino A, Kenific CM, Matei IR, Steiner L, Freitas D, Kim HS, Oxley PR, Scandariato I, Casanova-Salas I, et al. 2019. Tumour exosomal CEMIP protein promotes cancer cell colonization in brain metastasis. *Nat Cell Biol* **21:** 1403–1412. doi:10.1038/s41556-019-0404-4

Roos E, Dingemans KP, Van de Pavert IV, Van den Bergh-Weerman MA. 1978. Mammary-carcinoma cells in mouse liver: infiltration of liver tissue and interaction with Kupffer cells. *Brit J Cancer* **38:** 88–99. doi:10.1038/bjc.1978.167

Rose AAN, Pepin F, Russo C, Khalil JEA, Hallett M, Siegel PM. 2007. Osteoactivin promotes breast cancer metastasis to bone. *Mol Cancer Res* **5**: 1001–1014. doi:10.1158/1541-7786.MCR-07-0119

Sahai E, Astsaturov I, Cukierman E, DeNardo DG, Egeblad M, Evans RM, Fearon D, Greten FR, Hingorani SR, Hunter T, et al. 2020. A framework for advancing our understanding of cancer-associated fibroblasts. *Nat Rev Cancer* **20**: 174–186. doi:10.1038/s41568-019-0238-1

Sangaletti S, Tripodo C, Sandri S, Torselli I, Vitali C, Ratti C, Botti L, Burocchi A, Porcasi R, Tomirotti A, et al. 2014. Osteopontin shapes immunosuppression in the metastatic niche. *Cancer Res* **74**: 4706–4719. doi:10.1158/0008-5472.CAN-13-3334

Santana-Codina N, Muixí L, Foj R, Sanz-Pamplona R, Badia-Villanueva M, Abramowicz A, Marcé-Grau A, Cosialls AM, Gil J, Archila I, et al. 2020. GRP94 promotes brain metastasis by engaging pro-survival autophagy. *Neuro Oncol* **22**: 652–664. doi:10.1093/neuonc/noz198

Sartorius CA, Hanna CT, Gril B, Cruz H, Serkova NJ, Huber KM, Kabos P, Schedin TB, Borges VF, Steeg PS, et al. 2016. Estrogen promotes the brain metastatic colonization of triple negative breast cancer cells via an astrocyte-mediated paracrine mechanism. *Oncogene* **35**: 2881–2892. doi:10.1038/onc.2015.353

Satcher RL, Zhang XH-F. 2022. Evolving cancer–niche interactions and therapeutic targets during bone metastasis. *Nat Rev Cancer* **22**: 85–101. doi:10.1038/s41568-021-00406-5

Sethi N, Dai X, Winter CG, Kang Y. 2011. Tumor-derived Jagged1 promotes osteolytic bone metastasis of breast cancer by engaging notch signaling in bone cells. *Cancer Cell* **19**: 192–205. doi:10.1016/j.ccr.2010.12.022

Seubert B, Grünwald B, Kobuch J, Cui H, Schelter F, Schaten S, Siveke JT, Lim NH, Nagase H, Simonavicius N, et al. 2015. Tissue inhibitor of metalloproteinases (TIMP)-1 creates a premetastatic niche in the liver through SDF-1/CXCR4-dependent neutrophil recruitment in mice. *Hepatology* **61**: 238–248. doi:10.1002/hep.27378

Sevenich L, Bowman RL, Mason SD, Quail DF, Rapaport F, Elie BT, Brogi E, Brastianos PK, Hahn WC, Holsinger LJ, et al. 2014. Analysis of tumor- and stroma-supplied proteolytic networks reveals a brain metastasis-promoting role for cathepsin S. *Nat Cell Biol* **16**: 876–888. doi:10.1038/ncb3011

Shani O, Raz Y, Monteran L, Scharff Y, Levi-Galibov O, Megides O, Shacham H, Cohen N, Silverbush D, Avivi C, et al. 2021. Evolution of fibroblasts in the lung metastatic microenvironment is driven by stage-specific transcriptional plasticity. *eLife* **10**: e60745. doi:10.7554/eLife.60745

Shimo T, Kubota S, Yoshioka N, Ibaragi S, Isowa S, Eguchi T, Sasaki A, Takigawa M. 2006. Pathogenic role of connective tissue growth factor (CTGF/CCN2) in osteolytic metastasis of breast cancer. *J Bone Miner Res* **21**: 1045–1059. doi:10.1359/jbmr.060416

Sierra A, Price J, García-Ramirez M, Méndez O, López L, Fabra A. 1997. Astrocyte-derived cytokines contribute to the metastatic brain specificity of breast cancer cells. *Lab Invest* **77**: 357–368.

Smid M, Wang Y, Zhang Y, Sieuwerts AM, Yu J, Klijn JGM, Foekens JA, Martens JWM. 2008. Subtypes of breast cancer show preferential site of relapse. *Cancer Res* **68**: 3108–3114. doi:10.1158/0008-5472.CAN-07-5644

Song KH, Park MS, Nandu TS, Gadad S, Kim SC, Kim MY. 2016. GALNT14 promotes lung-specific breast cancer metastasis by modulating self-renewal and interaction with the lung microenvironment. *Nat Commun* **7**: 13796. doi:10.1038/ncomms13796

Soni S, Torvund M, Mandal CC. 2021. Molecular insights into the interplay between adiposity, breast cancer and bone metastasis. *Clin Exp Metast* **38**: 119–138.

Stessels F, den Eynden GV, der Auwera IV, Salgado R, den Heuvel EV, Harris AL, Jackson DG, Colpaert CG, Marck EAV, Dirix LY, et al. 2004. Breast adenocarcinoma liver metastases, in contrast to colorectal cancer liver metastases, display a non-angiogenic growth pattern that preserves the stroma and lacks hypoxia. *Brit J Cancer* **90**: 1429–1436. doi:10.1038/sj.bjc.6601727

Szabo V, Bugyik E, Dezso K, Ecker N, Nagy P, Timar J, Tovari J, Laszlo V, Bridgeman VL, Wan E, et al. 2015. Mechanism of tumour vascularization in experimental lung metastases. *J Pathol* **235**: 384–396. doi:10.1002/path.4464

Tabariès S, Dong Z, Annis MG, Omeroglu A, Pepin F, Ouellet V, Russo C, Hassanain M, Metrakos P, Diaz Z, et al. 2011. Claudin-2 is selectively enriched in and promotes the formation of breast cancer liver metastases through engagement of integrin complexes. *Oncogene* **30**: 1318–1328. doi:10.1038/onc.2010.518

Tabariès S, Dupuy F, Dong Z, Monast A, Annis MG, Spicer J, Ferri LE, Omeroglu A, Basik M, Amir E, et al. 2012. Claudin-2 promotes breast cancer liver metastasis by facilitating tumor cell interactions with hepatocytes. *Mol Cell Biol* **32**: 2979–2991. doi:10.1128/MCB.00299-12

Tabariès S, Ouellet V, Hsu BE, Annis MG, Rose AA, Meunier L, Carmona E, Tam CE, Mes-Masson AM, Siegel PM. 2015. Granulocytic immune infiltrates are essential for the efficient formation of breast cancer liver metastases. *Breast Cancer Res* **17**: 45. doi:10.1186/s13058-015-0558-3

Tamura D, Hiraga T, Myoui A, Yoshikawa H, Yoneda T. 2008. Cadherin-11-mediated interactions with bone marrow stromal/osteoblastic cells support selective colonization of breast cancer cells in bone. *Int J Oncol* **33**: 17–24.

Tan CC, Li GX, Tan LD, Du X, Li XQ, He R, Wang QS, Feng YM. 2016. Breast cancer cells obtain an osteomimetic feature *via* epithelial-mesenchymal transition that have undergone BMP2/RUNX2 signaling pathway induction. *Oncotarget* **7**: 79688 79705. doi:10.18632/oncotarget.12939

Tian F, Zhang S, Liu C, Han Z, Liu Y, Deng J, Li Y, Wu X, Cai L, Qin L, et al. 2021. Protein analysis of extracellular vesicles to monitor and predict therapeutic response in metastatic breast cancer. *Nat Commun* **12**: 2536. doi:10.1038/s41467-021-22913-7

Tominaga N, Kosaka N, Ono M, Katsuda T, Yoshioka Y, Tamura K, Lötvall J, Nakagama H, Ochiya T. 2015. Brain metastatic cancer cells release microRNA-181c-containing extracellular vesicles capable of destroying blood–brain barrier. *Nat Commun* **6**: 6716. doi:10.1038/ncomms7716

Tyagi A, Sharma S, Wu K, Wu SY, Xing F, Liu Y, Zhao D, Deshpande RP, D'Agostino RB, Watabe K. 2021. Nicotine promotes breast cancer metastasis by stimulating N2 neutrophils and generating pre-metastatic niche in lung. *Nat Commun* 12: 474. doi:10.1038/s41467-020-20733-9

Valiente M, Obenauf AC, Jin X, Chen Q, Zhang XH-F, Lee DJ, Chaft JE, Kris MG, Huse JT, Brogi E, et al. 2014. Serpins promote cancer cell survival and vascular co-option in brain metastasis. *Cell* 156: 1002–1016. doi:10.1016/j.cell.2014.01.040

Vanharanta S, Massagué J. 2013. Origins of metastatic traits. *Cancer Cell* 24: 410–421. doi:10.1016/j.ccr.2013.09.007

Wang J, Wang L, Xia B, Yang C, Lai H, Chen X. 2013. BSP gene silencing inhibits migration, invasion, and bone metastasis of MDA-MB-231BO human breast cancer cells. *PLoS ONE* 8: e62936. doi:10.1371/journal.pone.0062936

Wang H, Yu C, Gao X, Welte T, Muscarella AM, Tian L, Zhao H, Zhao Z, Du S, Tao J, et al. 2015. The osteogenic niche promotes early-stage bone colonization of disseminated breast cancer cells. *Cancer Cell* 27: 193–210. doi:10.1016/j.ccell.2014.11.017

Wang S, Liang K, Hu Q, Li P, Song J, Yang Y, Yao J, Mangala LS, Li C, Yang W, et al. 2017. JAK2-binding long noncoding RNA promotes breast cancer brain metastasis. *J Clin Invest* 127: 4498–4515. doi:10.1172/JCI91553

Wang H, Tian L, Liu J, Goldstein A, Bado I, Zhang W, Arenkiel BR, Li Z, Yang M, Du S, et al. 2018. The osteogenic niche is a calcium reservoir of bone micrometastases and confers unexpected therapeutic vulnerability. *Cancer Cell* 34: 823–839.e7. doi:10.1016/j.ccell.2018.10.002

Wang S, Li GX, Tan CC, He R, Kang LJ, Lu JT, Li XQ, Wang QS, Liu PF, Zhai QL, et al. 2019. FOXF2 reprograms breast cancer cells into bone metastasis seeds. *Nat Commun* 10: 2707. doi:10.1038/s41467-019-10379-7

Wang R, Godet I, Yang Y, Salman S, Lu H, Lyu Y, Zuo Q, Wang Y, Zhu Y, Chen C, et al. 2021a. Hypoxia-inducible factor-dependent ADAM12 expression mediates breast cancer invasion and metastasis. *Proc Natl Acad Sci* 118: e2020490118. doi:10.1073/pnas.2020490118

Wang Y, Ye F, Liang Y, Yang Q. 2021b. Breast cancer brain metastasis: insight into molecular mechanisms and therapeutic strategies. *Br J Cancer* 125: 1056–1067. doi:10.1038/s41416-021-01424-8

Wasilewski D, Priego N, Fustero-Torre C, Valiente M. 2017. Reactive astrocytes in brain metastasis. *Front Oncol* 7: 298. doi:10.3389/fonc.2017.00298

Wculek SK, Malanchi I. 2015. Neutrophils support lung colonization of metastasis-initiating breast cancer cells. *Nature* 528: 413–417. doi:10.1038/nature16140

Wen SW, Sceneay J, Lima LG, Wong CSF, Becker M, Krumeich S, Lobb RJ, Castillo V, Wong KN, Ellis S, et al. 2016. The biodistribution and immune suppressive effects of breast cancer–derived exosomes. *Cancer Res* 76: 6816–6827. doi:10.1158/0008-5472.CAN-16-0868

Wong CCL, Gilkes DM, Zhang H, Chen J, Wei H, Chaturvedi P, Fraley SI, Wong CM, Khoo US, Ng IOL, et al. 2011. Hypoxia-inducible factor 1 is a master regulator of breast cancer metastatic niche formation. *Proc Natl Acad Sci* 108: 16369–16374. doi:10.1073/pnas.1113483108

Wu K, Fukuda K, Xing F, Zhang Y, Sharma S, Liu Y, Chan MD, Zhou X, Qasem SA, Pochampally R, et al. 2015. Roles of the cyclooxygenase 2 matrix metalloproteinase 1 pathway in brain metastasis of breast cancer. *J Biol Chem* 290: 9842–9854. doi:10.1074/jbc.M114.602185

Wu Q, Li J, Zhu S, Wu J, Chen C, Liu Q, Wei W, Zhang Y, Sun S. 2017. Breast cancer subtypes predict the preferential site of distant metastases: a SEER based study. *Oncotarget* 8: 27990–27996. doi:10.18632/oncotarget.15856

Wu X, Li F, Dang L, Liang C, Lu A, Zhang G. 2020. RANKL/RANK system-based mechanism for breast cancer bone metastasis and related therapeutic strategies. *Front Cell Dev Biol* 8: 76.

Wu AML, Gossa S, Samala R, Chung MA, Gril B, Yang HH, Thorsheim HR, Tran AD, Wei D, Taner E, et al. 2021a. Aging and CNS myeloid cell depletion attenuate breast cancer brain metastasis. *Clin Cancer Res* 27: 4422–4434. doi:10.1158/1078-0432.CCR-21-1549

Wu K, Feng J, Lyu F, Xing F, Sharma S, Liu Y, Wu S-Y, Zhao D, Tyagi A, Deshpande RP, et al. 2021b. Exosomal miR-19a and IBSP cooperate to induce osteolytic bone metastasis of estrogen receptor-positive breast cancer. *Nat Commun* 12: 5196. doi:10.1038/s41467-021-25473-y

Xiao Y, Cong M, Li J, He D, Wu Q, Tian P, Wang Y, Yang S, Liang C, Liang Y, et al. 2021. Cathepsin C promotes breast cancer lung metastasis by modulating neutrophil infiltration and neutrophil extracellular trap formation. *Cancer Cell* 39: 423–437.e7. doi:10.1016/j.ccell.2020.12.012

Xing F, Liu Y, Wu SY, Wu K, Sharma S, Mo YY, Feng J, Sanders S, Jin G, Singh R, et al. 2018. Loss of XIST in breast cancer activates MSN-c-Met and reprograms microglia via exosomal miRNA to promote brain metastasis. *Cancer Res* 78: 4316–4330. doi:10.1158/0008-5472.CAN-18-1102

Yang L, Liu Q, Zhang X, Liu X, Zhou B, Chen J, Huang D, Li J, Li H, Chen F, et al. 2020. DNA of neutrophil extracellular traps promotes cancer metastasis via CCDC25. *Nature* 583: 133–138. doi:10.1038/s41586-020-2394-6

Yin JJ, Selander K, Chirgwin JM, Dallas M, Grubbs BG, Wieser R, Massagué J, Mundy GR, Guise TA. 1999. TGF-β signaling blockade inhibits PTHrP secretion by breast cancer cells and bone metastases development. *J Clin Invest* 103: 197–206. doi:10.1172/JCI3523

Yip RKH, Rimes JS, Capaldo BD, Vaillant F, Mouchemore KA, Pal B, Chen Y, Surgenor E, Murphy AJ, Anderson RL, et al. 2021. Mammary tumour cells remodel the bone marrow vascular microenvironment to support metastasis. *Nat Commun* 12: 6920. doi:10.1038/s41467-021-26556-6

Yoo BC, Lee JH, Kim KH, Lin W, Kim JH, Park JB, Park HJ, Shin SH, Yoo H, Kwon JW, et al. 2017. Cerebrospinal fluid metabolomic profiles can discriminate patients with leptomeningeal carcinomatosis from patients at high risk for leptomeningeal metastasis. *Oncotarget* 8: 101203–101214. doi:10.18632/oncotarget.20983

Yousefi M, Nosrati R, Salmaninejad A, Dehghani S, Shahryari A, Saberi A. 2018. Organ-specific metastasis of breast cancer: molecular and cellular mechanisms underlying lung metastasis. *Cell Oncol* 41: 123–140. doi:10.1007/s13402-018-0376-6

Yuan X, Qian N, Ling S, Li Y, Sun W, Li J, Du R, Zhong G, Liu C, Yu G, et al. 2021. Breast cancer exosomes contribute to pre-metastatic niche formation and promote bone metastasis of tumor cells. *Theranostics* 11: 1429–1445. doi:10.7150/thno.45351

Yue Z, Niu X, Yuan Z, Qin Q, Jiang W, He L, Gao J, Ding Y, Liu Y, Xu Z, et al. 2022. RSPO2 and RANKL signal through LGR4 to regulate osteoclastic premetastatic niche formation and bone metastasis. *J Clin Investigation* **132:** e144579. doi:10.1172/JCI144579

Zeng Q, Michael IP, Zhang P, Saghafinia S, Knott G, Jiao W, McCabe BD, Galván JA, Robinson HPC, Zlobec I, et al. 2019. Synaptic proximity enables NMDAR signalling to promote brain metastasis. *Nature* **573:** 526–531. doi:10.1038/s41586-019-1576-6

Zhang XHF, Wang Q, Gerald W, Hudis CA, Norton L, Smid M, Foekens JA, Massagué J. 2009. Latent bone metastasis in breast cancer tied to Src-dependent survival signals. *Cancer Cell* **16:** 67–78. doi:10.1016/j.ccr.2009.05.017

Zhang XHF, Jin X, Malladi S, Zou Y, Wen YH, Brogi E, Smid M, Foekens JA, Massagué J. 2013. Selection of bone metastasis seeds by mesenchymal signals in the primary tumor stroma. *Cell* **154:** 1060–1073. doi:10.1016/j.cell.2013.07.036

Zhang L, Zhang S, Yao J, Lowery FJ, Zhang Q, Huang W-C, Li P, Li M, Wang X, Zhang C, et al. 2015. Microenvironment-induced PTEN loss by exosomal microRNA primes brain metastasis outgrowth. *Nature* **527:** 100–104. doi:10.1038/nature15376

Zhang L, Yao J, Wei Y, Zhou Z, Li P, Qu J, Badu-Nkansah A, Yuan X, Huang YW, Fukumura K, et al. 2020. Blocking immunosuppressive neutrophils deters pY696-EZH2–driven brain metastases. *Sci Transl Med* **12:** eaaz5387. doi:10.1126/scitranslmed.aaz5387

Zhang W, Bado IL, Hu J, Wan YW, Wu L, Wang H, Gao Y, Jeong HH, Xu Z, Hao X, et al. 2021. The bone microenvironment invigorates metastatic seeds for further dissemination. *Cell* **184:** 2471–2486.e20. doi:10.1016/j.cell.2021.03.011

Zheng H, Bae Y, Kasimir-Bauer S, Tang R, Chen J, Ren G, Yuan M, Esposito M, Li W, Wei Y, et al. 2017. Therapeutic antibody targeting tumor- and osteoblastic niche-derived Jagged1 sensitizes bone metastasis to chemotherapy. *Cancer Cell* **32:** 731–747.e6. doi:10.1016/j.ccell.2017.11.002

Zheng W, Zhang H, Zhao D, Zhang J, Pollard JW. 2019. Lung mammary metastases but not primary tumors induce accumulation of atypical large platelets and their chemokine expression. *Cell Rep* **29:** 1747–1755.e4. doi:10.1016/j.celrep.2019.10.016

Zheng Z, Li Y, Jia S, Zhu M, Cao L, Tao M, Jiang J, Zhan S, Chen Y, Gao PJ, et al. 2021. Lung mesenchymal stromal cells influenced by Th2 cytokines mobilize neutrophils and facilitate metastasis by producing complement C3. *Nat Commun* **12:** 6202. doi:10.1038/s41467-021-26460-z

Zhou W, Fong MY, Min Y, Somlo G, Liu L, Palomares MR, Yu Y, Chow A, O'Connor STF, Chin AR, et al. 2014. Cancer-secreted miR-105 destroys vascular endothelial barriers to promote metastasis. *Cancer Cell* **25:** 501–515. doi:10.1016/j.ccr.2014.03.007

Zhuang X, Zhang H, Li X, Li X, Cong M, Peng F, Yu J, Zhang X, Yang Q, Hu G. 2017. Differential effects on lung and bone metastasis of breast cancer by Wnt signalling inhibitor DKK1. *Nat Cell Biol* **19:** 1274–1285. doi:10.1038/ncb3613

Zou Y, Watters A, Cheng N, Perry CE, Xu K, Alicea GM, Parris JLD, Baraban E, Ray P, Nayak A, et al. 2019. Polyunsaturated fatty acids from astrocytes activate PPARγ signaling in cancer cells to promote brain metastasis. *Cancer Discov* **9:** 1720–1735. doi:10.1158/2159-8290.CD-19-0270

Zuo H, Yang D, Yang Q, Tang H, Fu YX, Wan Y. 2020. Differential regulation of breast cancer bone metastasis by PARP1 and PARP2. *Nat Commun* **11:** 1578. doi:10.1038/s41467-020-15429-z

Zuo Q, Mogol AN, Liu YJ, Casiano AS, Chien C, Drnevich J, Imir OB, Kulkoyluoglu-Cotul E, Park NH, Shapiro DJ, et al. 2022. Targeting metabolic adaptations in the breast cancer–liver metastatic niche using dietary approaches to improve endocrine therapy efficacy. *Mol Cancer Res* **20:** 923–937. doi:10.1158/1541-7786.MCR-21-0781

Dormancy in Breast Cancer

Erica Dalla,[1,5] Amulya Sreekumar,[2,5] Julio A. Aguirre-Ghiso,[3] and Lewis A. Chodosh[2,4]

[1]Division of Hematology and Oncology, Department of Medicine and Department of Otolaryngology, Department of Oncological Sciences, Black Family Stem Cell Institute, Tisch Cancer Institute, Icahn School of Medicine at Mount Sinai, New York, New York 10029, USA

[2]Department of Cancer Biology and Abramson Family Cancer Research Institute, Perelman School of Medicine, University of Pennsylvania, Philadelphia, Pennsylvania 19104, USA

[3]Department of Cell Biology, Department of Oncology, Cancer Dormancy and Tumor Microenvironment Institute, Montefiore Einstein Cancer Center, Gruss Lipper Biophotonics Center, Ruth L. and David S. Gottesman Institute for Stem Cell Research and Regenerative Medicine, Institute for Aging Research, Albert Einstein College of Medicine, Bronx, New York 10461, USA

[4]Department of Medicine, Abramson Cancer Center, and 2-PREVENT Translational Center of Excellence, Perelman School of Medicine, University of Pennsylvania, Philadelphia, Pennsylvania 19104, USA

Correspondence: julio.aguirre-ghiso@einsteinmed.edu; chodosh@pennmedicine.upenn.edu

The pattern of delayed recurrence in a subset of breast cancer patients has long been explained by a model that incorporates a variable period of cellular or tumor mass dormancy prior to disease relapse. In this review, we critically evaluate existing data to develop a framework for inferring the existence of dormancy in clinical contexts of breast cancer. We integrate these clinical data with rapidly evolving mechanistic insights into breast cancer dormancy derived from a broad array of genetically engineered mouse models as well as experimental models of metastasis. Finally, we propose actionable interventions and discuss ongoing clinical trials that translate the wealth of knowledge gained in the laboratory to the long-term clinical management of patients at a high risk of developing recurrence.

CLINICAL EVIDENCE FOR BREAST CANCER DORMANCY

In 1934, Rupert Willis postulated that "tumor dormancy" might explain the long-standing observation that patients whose primary tumors had been successfully treated, and who had no evidence of local recurrence, could nevertheless experience the delayed occurrence of metastases many years later. He proposed that "neoplastic cells must have lain dormant in the tissues in which they were arrested, and their resumption of growth must be attributed to some alteration in the qualities of these tissues, or to some release of growth restraints exercised by them on tumor cells" (Willis 1934).

Two decades later, Geoffrey Hadfield further advanced this insight by speculating that a "temporary mitotic arrest" of tumor cells could explain the protracted latency window between

---

[5]These authors contributed equally to this work.

Cite this article as *Cold Spring Harb Perspect Med* doi: 10.1101/cshperspect.a041331

the definitive treatment of a primary tumor and subsequent reemergence of a secondary tumor that had been observed in a variety of human cancers, including breast cancer (Hadfield 1954). Historically, a latency window of >5 yr between primary tumor resection and recurrence was considered unusually long (Hadfield 1954), and this threshold continues to be used as a clinical reference for distinguishing "early" from "late" tumor recurrence in dormancy-related research (Klein 2020).

An implicit assumption underlying the above concepts is that minimal residual disease (MRD) constitutes the source from which recurrent tumors arise. MRD is defined as the microscopic tumor cell burden that persists at primary or metastatic sites following definitive surgery, radiation therapy, and/or systemic (neo)adjuvant therapy. MRD is challenging to observe directly in patients, primarily due to the ultra-rare frequency of these cells coupled with clinical limitations associated with sampling tumor cells from particular anatomical sites. Nevertheless, several clinical observations support the possibility that MRD can exist in a latent state. The most striking example of this phenomenon derives from reported cases of cancer transmission from organ donors thought to be cancer-free at the time of donation to recipients (Au et al. 2018; Matser et al. 2018; Greenhall et al. 2022). Cancer transmission via the transplantation of organs that are themselves uncommon sites of metastases, such as the heart or kidney, suggests a model in which disease recurrence from otherwise undetectable residual tumor cells may occur when the growth restraints that they experience, including those that may be imposed by an intact immune system, are compromised following transplantation into the recipient.

Multiple additional lines of evidence support a model in which recurrent tumors are seeded by MRD. For example, the presence of cytokeratin-positive disseminated tumor cells (DTCs) in the bone marrow of patients is strongly associated with shorter time-to-recurrence at both local and at distant sites, as well as poorer disease-specific survival and overall survival (Wiedswang et al. 2003; Braun et al. 2005; Bidard et al. 2008; Hartkopf et al. 2014, 2021). In

particular, a pooled analysis of over 10,000 early-stage breast cancer patients detected DTCs in ~30% of patients, and the presence of detectable DTCs was confirmed as a prognostic marker that is independent of breast cancer subtype (Hartkopf et al. 2021). Further, a study designed to deplete the DTC pool that persists following adjuvant chemotherapy with additional chemotherapeutic agents reported improved metastasis-free survival in responders who converted to a DTC-negative status, compared to nonresponders who remained DTC-positive (Naume et al. 2014). Although a control group was not included in this study, these provocative data suggest a link between DTC presence and the development of recurrent metastases. The persistence of DTCs following neoadjuvant chemotherapy is also strongly associated with poor prognosis in breast cancer patients (Mathiesen et al. 2012).

Disease latency preceding the clinical detection of recurrent tumors could formally be explained by models in which residual tumor cells (1) are reversibly arrested in a dormant, G_0-like state, which could occur as a consequence of cell-intrinsic or extrinsic (i.e., microenvironmental) signals; (2) proliferate at an extremely low rate; or (3) maintain a constant tumor cell mass through balanced proliferation and cell death, such as might occur due to ongoing immune-mediated elimination of tumor cells or a restrictive microenvironment lacking a vascular supply sufficient to support sustained proliferative outgrowth (Aguirre-Ghiso 2007; Klein 2011). Consistent with the notion of a form of cellular dormancy or immune-mediated control of residual cancer cells, DTCs exist most commonly as single cells or small clusters of cells (Woelfle et al. 2005), are nonproliferative as evidenced by the absence of staining for the cell-cycle marker Ki67 (Pantel et al. 1993), and may express the dormancy marker NR2F1 (Sosa et al. 2015; Borgen et al. 2018; Khalil et al. 2022). Each of these observations supports a model in which residual tumor cells can persist in a resting, G_0-like state. However, functionally demonstrating in patients that even a majority of DTCs exist in a state of reversible quiescence is difficult with existing methods, thus making it challenging to definitively rule out that tumor cells proliferate very slowly and/ or are otherwise restrained by their vascular and/ or immune microenvironment. Nevertheless, in light of proof-of-principle studies targeting angiogenesis or residual tumor cell survival in dormancy models (Carlson et al. 2019; Calvo et al. 2021), these data support the possibility that therapeutic targeting of MRD during this latent phase could constitute a tractable approach to prevent lethal recurrences in patients.

The latency interval between initial tumor diagnosis and recurrent disease is especially pronounced in breast cancer patients in whom excess mortality is observed more than 30 yr after primary tumor diagnosis (Pedersen et al. 2022). At first glance, this phenomenon appears to be breast cancer–subtype dependent, wherein early-stage estrogen receptor (ER)-negative breast cancer patients demonstrate the highest risk of recurrence and death within the first 5 yr following surgery, whereas ER^+ breast cancer patients demonstrate a relatively constant annual rate of recurrence of ∼1%–2% from 5 to 20 yr following initial therapy (Pan et al. 2017). Consequently, a crossover is observed whereby a greater fraction of recurrences after 5 yr are attributable to ER^+ tumors.

However, closer inspection reveals some evidence that the distinction between the risk of late recurrence for ER^+ and ER-negative disease is not absolute. For example, a recent study stratified breast cancer patient samples from the Molecular Taxonomy of Breast Cancer International Consortium (METABRIC) into 11 genomic subgroups to define populations of patients at high risk of late relapse and poor prognosis (Rueda et al. 2019). Beyond identifying highly variable probabilities of relapse from 5 to 20 yr among ER^+ genomic subsets, this study also identified ER-negative genomic subsets that displayed an increasing probability of late recurrence within this same time frame. Consistent with these findings, long-term follow-up of breast cancer patients who were disease-free 10 yr after diagnosis revealed that ∼8% of ER-negative patients developed recurrent tumors over the ensuing 15 yr, demonstrating unequivocally that late recurrence can occur in patients with ER-negative primary tumors (Pedersen

et al. 2022). These data suggest that the propensity of breast cancers to pass through a dormant phase prior to undergoing stochastic reactivation to give rise to late recurrences is unlikely to be explained by ER status and anti-estrogen therapies alone.

In this regard, while delayed recurrence for some ER^+ breast cancers may typify an extreme form of latency, it is important to note that its existence does not preclude the possibility of clinically relevant shorter periods of dormancy in patients harboring ER-negative breast cancers. For example, pancreatic cancer is widely considered to be a tumor type that does not display clinical dormancy, based in part upon the observation that ∼75% of patients develop metastases within 2 yr of primary tumor removal. Nevertheless, Ki67-negative DTCs have been identified in the livers of such patients (Pommier et al. 2018). While it is unknown whether such Ki67-negative cells are indeed reversibly quiescent and capable of giving rise to "recurrent" pancreatic tumors, this finding shows that the kinetics of clinical recurrence alone may not be sufficient to distinguish cancers that do or do not pass through a dormant state at the cellular level. This conclusion is further underscored by laboratory and clinical evidence for early dissemination of breast and pancreatic cancers insofar as even those metastases that become clinically manifest soon after primary tumor diagnosis may have been seeded years earlier during primary tumor development (Schardt et al. 2005; Rhim et al. 2012; Harper et al. 2016; Hosseini et al. 2016; Linde et al. 2018; Hu et al. 2020). Consequently, therapies targeting the window of dormancy in MRD may have clinical utility for both ER-negative and ER^+ disease, as well as different modes of disease progression.

In addition to the above considerations, it is important to note that from a biological perspective the designation of 5 yr postdiagnosis as the distinction between "early" and "late" recurrence is arbitrary, particularly given the lack of knowledge about the cellular state of DTCs in patients as a function of time. Moreover, the presumption that a clinically defined recurrence threshold of >5 yr implies a dormant phase, whereas a shorter period rules out such a phase, is even less tenable when considering changes in diagnostic and treatment modalities that have occurred over time (Klein 2011). Indeed, large-scale meta-analyses of breast cancer patient responses to targeted therapies clearly show historical changes in the kinetics of disease recurrence. For example, data from randomized trials reporting long-term outcomes in ER$^+$ breast cancer patients demonstrate that adjuvant treatment with 5 yr of tamoxifen suppressed recurrence more strongly at 5 yr, compared to 10 yr, following diagnosis (Early Breast Cancer Trialists' Collaborative Group [EBCTCG] et al. 2011). This resulted in an increased proportion of patients that recurred at 10 yr versus 5 yr in the tamoxifen-treated arm compared to controls (37.9% vs. 30.7%). That is, among patients who recur, adjuvant treatment with tamoxifen for 5 yr increases the relative fraction of patients who recur after 5 yr compared to those who recur within 5 yr of diagnosis. Further, continuing adjuvant endocrine therapy in ER$^+$ patients for 10 yr versus 5 yr further delays disease recurrence. Thus, endocrine therapy itself, as well as changes in its duration, alter the temporal distribution of recurrent tumors. These observations suggest that anti-estrogen therapy might facilitate disease latency, perhaps via some of the dormancy mechanisms proposed above, and clearly show how changes in therapy over time result in changes in the distribution of recurrent tumors relative to the 5-yr threshold that is used to classify "early" versus "late" recurrence (Davies et al. 2013; Pan et al. 2017).

Analogous to therapeutic targeting of ER, meta-analyses of adjuvant trials assessing the impact of adding trastuzumab to chemotherapy in early-stage HER2$^+$ breast cancers revealed that the annual rate of recurrence was suppressed by about a third within the first 5 yr following diagnosis. However, as observed for tamoxifen, although the addition of trastuzumab suppressed the overall percentage of patients who recurred, trastuzumab treatment increased the relative proportion of patients that recurred at 10 yr versus 5 yr compared to controls (25.8% vs. 19.8%) (Early Breast Cancer Trialists' Collaborative Group [EBCTCG] 2021). This provides further evidence that targeted therapy changes the tem-

poral distribution of recurrent tumors, thereby altering the balance between "early" and "late" recurrence Notably, these therapy induced changes in the fraction of tumors that recur "early" versus "late" may be independent of any intrinsic biological tendency of a breast cancer subtype to pass through a dormant state.

The observation that the annual hazard rate for ER-negative breast cancer patients peaks within 2 yr after diagnosis could reflect a scenario in which recurrent tumors detected within this time frame arise from a proliferating pool of residual tumor cells that had escaped treatment. However, it is also possible that shorter durations of therapy for ER-negative compared to ER$^+$ patients, as well as the dearth of targeted therapies for ER-negative, HER2-negative disease, may contribute to the rapid kinetics of recurrence. For these reasons, it is intriguing to speculate that improvements in targeted therapies for ER-negative disease might increase the relative proportion of tumors that recur beyond the "early" threshold of 5 yr. Thus, it is likely that both the intrinsic biological properties of breast cancer subtypes, as well as the therapies used to treat them, impact on the relative proportion of tumors that recur after 5 yr and the propensity of residual tumor cells to pass through a dormant state. Further, there may be drug–tumor interactions such that different breast cancer subtypes may respond differently to the same therapy with respect to either the frequency or phenotype of dormancy.

In sum, an accumulating body of clinical evidence suggests that uncoupling historical definitions of clinical tumor latency from the histopathologic observation of quiescent residual tumor cells would provide greater clarity and advance mechanistic discussions regarding tumor dormancy. While a clinical definition predicated solely on temporal patterns of disease recurrence has been interpreted by many as suggesting that breast cancer dormancy is restricted to ER$^+$ disease, both clinical and histopathologic evidence are consistent with the possibility that reversible cellular quiescence may be a stage of cancer progression through which both ER$^+$ and ER-negative breast cancer may pass, as well as other human cancers not generally considered to undergo dormancy. Accordingly, the ability to identify tumors with a propensity to undergo dormancy in a more nuanced manner that does not rely on ER status alone would constitute an important advance in the field. Furthermore, the molecular characterization of DTCs in concert with their matched primary tumor may also refine the identification of patients with MRD who are more or less likely to undergo intrinsic or therapy-related dormancy.

WHAT IS DORMANCY?

Cancer dormancy has been proposed to present in two distinct modes (Aguirre-Ghiso 2007): (1) tumor mass dormancy, characterized by an equilibrium state where a small tumor mass or masses are kept constant and asymptomatic by either a lack of angiogenic support or immune surveillance (Holmgren et al. 1995; Farrar et al. 1999; Mahnke et al. 2005; Naumov et al. 2006a; Koebel et al. 2007), and (2) cellular dormancy, in which single or small clusters of DTCs predominantly exist in a growth-arrested state (Aguirre-Ghiso 2007, 2018; Giancotti 2013; Sosa et al. 2014). This does not imply that residual DTCs never divide, but rather that their divisions are separated by long periods of growth arrest, which defines their phenotype. Since once cells commit to mitosis, this process is not paused and progresses actively unless damage is encountered, still the timing of cellular dormancy would mainly be controlled by G_0/G_1 phases. These possibilities for cellular dormancy can occur in residual cancer cells at the site of the primary tumor (also called "local recurrence") and at the distant metastatic site (e.g., lungs, bone, liver). Tumor mass dormancy is an intriguing phenomenon that has been difficult to characterize and track in patients as recently described (Wiecek et al. 2021). In this study, overt tumors (~9000) available via public databases were profiled for the presence of dormancy signatures derived from angiogenic dormancy or immune-equilibrium models. This study did not really profile dormant tumor masses; nevertheless, the overt tumor masses revealed variable enrichment for tumor mass dormancy gene-expression signatures. While in some tumor sub-

types the signatures were represented in 26%–30% of samples (including breast), the rest showed low enrichment of these dormancy signatures (2%–7%). These data argue that in overt tumors, tumor mass dormancy gene profiles can be found, likely due to varying levels of hypoxic niches that can also influence immune infiltration (Baldominos et al. 2022). Another space where angiogenic tumor mass dormancy may be proven to exist in patients is with the use of metronomic chemotherapy or anti-angiogenic therapy (Montagna et al. 2014; Natale and Bocci 2018). Metronomic chemotherapy in breast cancer may have some advantage as it may target the endothelium and angiogenesis and this may be improved by the use of anti-VEGF- or VEGFR-targeted therapies (Montagna et al. 2014). However, whether metronomic use of anti-angiogenic therapies alone induces tumor mass dormancy was not reported (Montagna et al. 2014; Natale and Bocci 2018). Further, whether the above approaches recapitulate gene signatures from experimental models of tumor mass dormancy has also, to our knowledge, not been reported (Montagna et al. 2014; Natale and Bocci 2018). While open questions on tumor mass dormancy persist in breast cancer metastasis, the data above suggest that these could be explored by further modeling and focused clinical trials with molecular readouts.

In describing cellular dormancy in cancer, the field has employed different definitions based on potential states for the growth arrest: (1) quiescence, (2) senescence, (3) differentiation, or (4) embryonic diapause. A recent review addressed in detail the overlap and differences between cancer cell dormancy and the four potential states indicated above (Risson et al. 2020). The common denominator is the obvious acquisition of a G_0-G_1 arrest. The literature tends to favor the notion that cancer cells adopt a quiescent state with programs carrying characteristics of quiescent adult stem cells, interactions with their niches, and embryonic pluripotency programs found during diapause (Aguirre-Ghiso and Sosa 2018; Dhimolea et al. 2021; Lim and Ghajar 2022). Embryonic diapause is a unique developmental state in mammals where the blastocyst as a whole enters a reversible growth arrest in response to reduced maternal hormonal input signals (Scognamiglio et al. 2016). Interestingly, blastocysts in diapause display reduced Myc activity but maintain pluripotency, which is also found in models of cancer cell dormancy (Shachaf and Felsher 2005). These data further support that cellular dormancy is therefore distinct from senescence, a cell-cycle arrest activated by oncogene and replicative stress, and a first barrier to transformation (Gorgoulis et al. 2019). However, some stress-adaptive pathways, such as the unfolded protein response, and specific secretory programs may be shared between quiescent and senescent states. Differentiation programs, in particular lineage commitment transcriptional programs, can be partially activated in dormant cells that are quiescent (Sosa et al. 2015; Laughney et al. 2020). However, the spontaneously reversible nature of dormancy does not conform to the specialized functional state of fully differentiated cells. Finally, therapy-associated dormancy has been explored in the context of "persister cancer cells," a population of drug-tolerant cells that evades cell death induced by therapy, including in genetically engineered mouse (GEM) models for breast cancer dormancy and recurrence following targeted therapy (Moody et al. 2002, 2005; Gunther et al. 2003; Abravanel et al. 2015; Ruth et al. 2021). These cells may represent epigenetic variants, rather than selected genetic clones, partially regulated by niche-driven cues and damage signals that can also trigger senescence (Shen et al. 2020). As the microenvironment surrounding dormant cancer cells significantly impacts cellular dormancy, understanding the niche signals regulating the DTC state is of utmost importance.

EXPERIMENTAL MODELS OF BREAST CANCER DORMANCY

Clinical observations in patients strongly suggest that recurrent breast cancers are seeded by MRD, which in turn may exist in a state of reversible quiescence or balanced proliferation and cell death. Considering the limitations of what can reliably be inferred from clinical data, experimental modeling of disease progression is critical for probing the biological under-

pinnings of dormancy initiation, maintenance, and exit to identify novel treatment strategies for preventing tumor recurrence (Fig. 1A).

How a state of reversible quiescence is acquired and maintained remains an area of active research. GEM models of breast cancer dormancy rely on simulating disease progression as it is observed in patients. Importantly, given that the great majority of early-stage breast cancer patients receive some form of systemic therapy, breast cancer dormancy in patients typically occurs in the setting of therapy, either ongoing or completed. Accordingly, mouse models for breast cancer dormancy and recurrence have been developed that incorporate this critical feature (Fig. 1B). These include models for onco-gene-dependent primary tumor formation, cell survival following targeted therapy, persistence of tumor cells in a dormant state, and subsequent spontaneous recurrence (e.g., *Myc* [D'Cruz et al. 2001], *Her2* [Moody et al. 2002, 2005; Abravanel et al. 2015; Ruth et al. 2021], *Wnt1* [Gunther et al. 2003; Ruth et al. 2021], and *Fgfr1* [Janghorban et al. 2021]). In such models, targeted therapy imposes a bottleneck that results in either the de novo acquisition, or adaptive selection, of tumor cells with the ability to undergo dormancy. Notably, employing quiescence-associated gene-expression signatures (Janghorban et al. 2021) derived from GEM models to gene-expression data sets of patients treated with neoadjuvant chemotherapy reveals the acquisition of a dormancy-associated transcriptional phenotype following treatment.

The application of dormancy-related gene-expression signatures derived from in vivo GEM models of breast cancer to patient data sets demonstrates their potential utility in identifying the subset of patients who are at greatest risk for tumor recurrence. For example, the interrogation of large patient data sets using gene-expression signatures derived from dormant MRD in two different GEM models that mimic targeted therapy of local disease found that patients whose primary tumors exhibited a dormancy signature displayed decreased rates of recurrence, potentially reflecting an increased frequency or ability of tumor cells to persist in a latent state (Ruth et al. 2021). Further, this

Cite this article as *Cold Spring Harb Perspect Med* doi: 10.1101/cshperspect.a041331

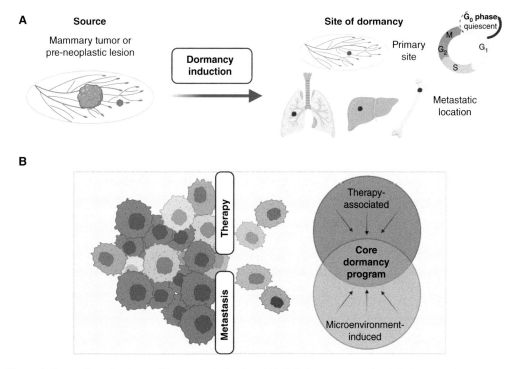

Figure 1. Sites and mechanisms of dormancy induction. (*A*) Cells from pre-neoplastic lesions or primary tumors can undergo dormancy at the primary site or at secondary metastatic locations. (*B*) Dormancy can be induced following therapy directed at the primary tumor and/or by the microenvironment in the process of dissemination. Several core regulators are likely to be conserved independently of the mechanism of dormancy induction.

study's demonstration that a gene-expression signature derived from local dormant residual disease in mouse models predicts recurrence-free survival in patient data sets composed predominantly of metastatic recurrences suggests that the properties of dormant tumor cells at local and distant sites may be related (Ruth et al. 2021). Defining the overlap between such cellular dormancy-associated gene-expression signatures and other models of persistent cancer cells that have not been confirmed to exhibit features of dormancy, such as local residual disease in patient samples following neoadjuvant chemotherapy (Creighton et al. 2009; Balko et al. 2014; Kim et al. 2018) or xenograft models (Echeverria et al. 2019), may help clarify the contribution of quiescent residual tumor cells to breast cancer recurrence.

A commonly cited limitation of existing GEM models that mimic human disease progression through a latent phase is the relative lack of models for estrogen-dependent breast cancers expressing ER (Bushnell et al. 2021). The development of such models would no doubt hold great value for teasing apart whether particular subtypes of breast cancer, such as ER$^+$ tumors, have a unique intrinsic propensity or ability to enter or persist in a dormant state, and whether this capability is amplified by the use of targeted therapies. Nevertheless, the recent demonstration that the gene-expression state of dormant tumor cells derived from ER-negative GEM models predicts the probability of recurrence in patient data sets composed primarily of ER$^+$ tumors indicates that ER-negative models can provide clinically relevant insights into dormancy mechanisms, while providing further evidence that cellular dormancy is unlikely to be restricted to ER$^+$ disease (Ruth et al. 2021).

Another paradigm for dormancy modeling involves mimicking the adaptation of tumor cells to foreign microenvironments encountered during metastasis, which is important given that breast cancer most commonly recurs at distant sites (Fig. 1B). Frequently used models include related sets of breast cancer cell lines derived from common origins that exhibit indolent (e.g., D2.OR, 4TO7) or aggressive growth (e.g., D2A1, 4T1) at metastatic sites including the lung, liver, and bone (Mahoney et al. 1985; Aslakson and Miller 1992; Morris et al. 1993, 1994). Additional strategies include the in vivo selection of stable clones that display dormant phenotypes (e.g., HCC1954-LCC1 [Malladi et al. 2016], MDA-MB-231-SCP6 [Lu et al. 2011]), or the use of established but unrelated human breast cancer cell lines with indolent (e.g., MCF7, T47D) or aggressive (e.g., MDA-MB-231) growth properties (Montagner and Sahai 2020). In aggregate, these models encompass a variety of breast cancer subtypes and enable the study of organ-specific barriers to macrometastatic outgrowth.

One recent study derived dormancy-associated gene-expression signatures from noncycling tumor cells in the lungs and bones of tumor-bearing *MMTV-PyMT* mice, as well as quiescent D2.OR tumor cells in the lung, and found the striking conservation of a core set of dormancy-related genes (Ren et al. 2022). Furthermore, dormancy-associated gene-expression signatures derived from indolent ER$^+$ D2.OR cells and from *MMTV-PyMT* noncycling tumor cells isolated from lung and bone were associated with better prognosis in breast cancer patients (Montagner et al. 2020; Ren et al. 2022).

Together, findings from gene-expression profiling of multiple clinically relevant models of dormancy support a model in which a conserved set of genes is enriched in dormant tumor cells, irrespective of the subtype of breast cancer or the site at which the tumor cells exist in a dormant state (Fig. 1B). Parsing these gene-expression signatures should help identify both novel markers of dormant MRD and functional regulators of disease progression through a dormant state. Additionally, developing dormancy models that mimic the high degree of heterogeneity and clonal evolution observed in patients as a consequence of therapy, metastatic dissemination, and cross talk with immune and stromal cell components in different microenvironments is likely to refine our understanding of mechanisms promoting tumor cell persistence in a dormant state.

MECHANISMS UNDERLYING DORMANCY ENTRY AND PERSISTENCE

Therapy-Associated Cancer Cell Dormancy

Cancer cell dormancy can be manifested following treatment directed toward the primary or secondary tumor. Therapies suppressing oncogene signaling or DNA-damaging, antiproliferative therapies invoke changes in signaling pathways and epigenetic programs, enabling persister cells, as one potential evasion mechanism, to enter dormancy and escape therapy (Sharma et al. 2010). Alternatively, preexisting spontaneously slow-cycling or deep dormant cells could survive treatment (Shaffer et al. 2017). While some cells may remain unable to reactivate posttreatment, a fraction of therapy-resistant or therapy-tolerant persister cells can eventually reactivate and proliferate, fueling cancer relapse.

In the context of breast cancer, chemotherapy has been shown to leave persister cancer cells behind. Echeverria and colleagues found that triple-negative breast cancer (TNBC) patient-derived xenograft tumors and biopsies treated with standard chemotherapy regrew from residual drug-tolerant cancer cells. Using clonal tracking and sequencing via barcodes, they showed that recurrent tumors maintained the architecture and heterogeneity of initial tumors and were thus not clonally selected upon chemotherapy (Echeverria et al. 2019). Further, a preclinical study using the *MMTV-PyMT* mouse model found that a small subset of primary tumor cells underwent an epithelial-to-mesenchymal transition (EMT), but these cells were not the predominant population in lung metastasis (Fischer et al. 2015). However, cells that underwent EMT resisted treatment with the chemotherapeutic agent cyclophosphamide and were thus responsible for metastatic recurrence.

These cells displayed reduced proliferation, increased resistance to apoptosis, and elevated expression of drug-metabolizing genes. Therefore, cells that underwent EMT that either preexisted in the primary tumor or were forced into this state by therapy may represent the persister cell population in the context of conventional chemotherapy.

Therapies targeting specific oncogene dependencies also fail to completely eradicate all cancer cells (Gu et al. 2016). Initial modeling of oncogene addiction and targeting was performed using doxycycline-inducible GEM models activating breast cancer–relevant pathways, including *Myc*, *Wnt1*, and *Her2/neu*. In these models, de-induction of oncogenes resulted in residual breast cancer cells that eventually reprogrammed and could fuel recurrences (Moody et al. 2002; Gunther et al. 2003; Boxer et al. 2004). These and other studies of residual disease and recurrence in *Her2/neu* and *Wnt1* doxycycline-inducible GEM models implicated dormant residual tumor cells as an intermediate in the stochastic spontaneous recurrence of mammary tumors following treatment, in some cases in the context of EMT, and revealed striking similarities between populations of dormant residual tumor cells isolated from *Her2/neu* and *Wnt1* GEM models (Moody et al. 2005; Ruth et al. 2021). In particular, Ruth et al. showed that, after oncogenic HER2 inhibition in mouse cells, or a combination of anti-HER2 and anti-ER therapy in human breast cancer cells, residual dormant cancer cells were found both at the primary site and in the lungs (Ruth et al. 2021). Furthermore, these cells were able to reenter the cell cycle and proliferate to form recurrent tumors after long intervals. Intriguingly, analysis of gene programs revealed that oncogene inhibition also differentially regulated genes associated with microenvironment-induced dormancy (see next section), such as TGF-β2 (Fluegen et al. 2017; Nobre et al. 2021b), supporting the idea that these persister cells might also generate niches that support their dormant state (Fig. 1B).

In previous studies of doxycycline-inducible GEM models, subsets of residual tumor cells and recurrences have been shown to activate Met or Notch signaling following HER2 inhibition

(Feng et al. 2014; Abravanel et al. 2015). Activation of both the Met and Notch pathways was demonstrated to be associated with tumor recurrence in mice as well as breast cancer patients (Feng et al. 2014; Abravanel et al. 2015). Fox et al. (2020) further found that HER2 inhibition promoted changes in cellular metabolism, leading to oxidative stress and up-regulation of the antioxidant transcription factor NRF2. Similar to the Notch study, activation of NRF2 accelerated tumor recurrence and up-regulated de novo nucleotide synthesis. These data reveal that residual cancer cells that enter dormancy in response to targeting of oncogenes use a multiplicity of adaptive pathways to enter a reversible growth arrest resembling quiescence and that they may also produce signals that mimic niches supporting dormancy (Fig. 1B).

More recent work further investigated the response of HER2-driven breast cancer cell lines and HER2[+] patients' tumors to lapatinib (a HER2 tyrosine kinase inhibitor). Using a barcode lentiviral tracing system, it was shown that drug-tolerant persister cells could be found as cycling or noncycling populations (Oren et al. 2021). Cycling persister cells maintained their proliferative capacity and up-regulated a reactive oxygen species (ROS) signature and antioxidant mechanisms, shifting their metabolic state.

Few studies in breast cancer focus on additional oncogenic pathways besides *Her2* and their correlation with cancer dormancy. This is mostly likely due to the limited number of driver-mutated kinases found in breast cancer. Use of a constitutive *Wnt1*-driven, inducible *Fgfr1* mouse mammary tumor model has been used to explore residual dormant tumor cell biology after treatment with a fibroblast growth factor receptor (FGFR) inhibitor (Holdman et al. 2015; Janghorban et al. 2021). These dormant cells gave rise to recurrent tumors that displayed activation of EGFR signaling and surrounding dense stroma (Holdman et al. 2015). Notably, single-cell RNA sequencing of residual lesions in this model uncovered significant remodeling of the tumor microenvironment, including the stromal and immune compartments, which is associated with the dormant state (Janghorban et al. 2021).

Focusing on ER$^+$ breast cancer and its response to hormonal therapy, a 2019 study by Hong et al. used single-cell RNA sequencing to characterize the phenotypic plasticity and adaptation of this subset of breast cancer after endocrine therapy (Hong et al. 2019). Here, they identified a population of cells (called preadapted) that expressed a dormancy transcriptional signature and preferentially survived treatment.

Another therapeutic intervention linked to dormancy in breast cancer is treatment with the histone deacetylase (HDAC) inhibitor entinostat. HDAC inhibitors were shown to epigenetically stimulate expression of leukemia inhibitory factor receptor (LIFR) in breast cancer cells, independent of ER status (Clements et al. 2021). They induced a pro-dormancy gene program and reduced tumor growth in mice. From clinical data, treatment of advanced breast cancer patients with entinostat and azacitidine (DNA methyltransferase inhibitor) showed some promise in a subset of patients (Connolly et al. 2017). The use of azacytidine combined with retinoic acid was also shown to induce a program of dormancy across various epithelial cancers, including a TNBC model, suggesting that such "reprogramming" approaches may prove fruitful in suppressing metastasis via dormancy induction (Sosa et al. 2015; Khalil et al. 2022).

In summary, the different types of therapy described herein leave behind cancer cells that survive treatment in a dormant state. Further work into understanding the mechanisms of therapy-induced DTC dormancy and subsequent reactivation is needed to optimize intervention strategies.

Niche-Derived Signals Instruct Cancer Cells to Activate Dormancy Programs

Signals from the microenvironmental niche, both at the primary site and at the site of metastatic colonization, appear to be integrated to induce or maintain DTC dormancy (Fig. 2).

Dormancy Signals from the Primary Site and Target Organ Niches

The earliest evolved lesions, such as ductal carcinoma in situ (DCIS), which are considered localized and noninvasive, were classically thought to be incapable of disseminating cancer cells as this would require invasion. However, early work revealed that DTCs can be detected in the bone marrow of patients with DCIS (Hüsemann et al. 2008), and that genetic alterations linked to malignancy in invasive cancer were already present at this DCIS stage (Schmidt-Kittler et al. 2003). These findings supported the idea that in situ lesions may be capable of disseminating cancer cells. However, this does not rule out that as lesions progress dissemination continues through late stages, albeit in some instances, but not all, with reduced impetus (Hosseini et al. 2016). A long-standing hypothesis was that, by virtue of being evolved early, these lesions would produce DTCs prone to enter dormancy. In three papers (Harper et al. 2016; Hosseini et al. 2016; Linde et al. 2018), the mechanisms driving early dissemination were identified in various mouse models (*MMTV-Her2* and *MMTV-PyMT*) and human samples across various breast cancer subtypes. In the two mouse models tested, a population of *Her2$^+$* or *PyMT$^+$* (a TNBC-like model) DTCs in *MMTV*-transgenic models of breast cancer activated a WNT-dependent EMT-like Twist1^{hi}E-cad^{lo} program to travel to and colonize distant sites, such as the lungs (Harper et al. 2016). Of note, during early stages of progression, these HER2 lesions express both ERα and progesterone receptor (Hosseini et al. 2016), suggesting that hormonal regulation could foster dissemination. In fact, Hosseini et al. (2016), elegantly showed how progesterone-induced signaling was key for the regulation of dissemination of the HER2$^+$ DTCs. These pioneering findings may also shed light into how pregnancy-associated breast cancer may foster dissemination seeding and metastasis (Goddard et al. 2019; Lefrère et al. 2021). However, further studies are needed to understand how ER$^+$/PR$^+$/HER2$^-$ lesions accomplish the process of early dissemination and how early DTCs are affected by anti-estrogen therapies.

Supporting that early DTCs may enter dormancy, upon arrival to the lung, the Twist1^{hi}E-cad^{lo} DTCs remained in a dormant, nonproliferative state for long periods, supporting that early DTCs can remain dormant and give rise to

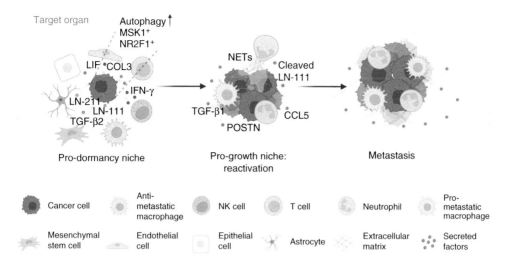

Figure 2. Cross talk between tumor cells and the tumor microenvironment regulates dormancy. In addition to epithelial cells present in the target organ, stromal and immune cells, as well as the extracellular matrix (ECM), can regulate whether the tumor cells remain dormant or reactivate to give rise to recurrences. (NK) Natural killer.

metastases. A recent expansion of this work has shown that a primed pluripotency regulator ZFP281 regulates early dissemination and subsequent dormancy of early DTCs in the MMTV-HER2 and -PyMT models also via regulation of TWIST1 and a class II cadherin, CDH11 (Nobre et al. 2022). It was also shown that *Her2*[+] and *PyMT*[+] DTCs were aided by CD206[+]/Tie2[+] macrophages in the mammary gland for efficient early dissemination (Linde et al. 2018). Impressively, eliminating these macrophages exclusively during the early lesion stage decreased metastasis by more than 50% months later, arguing for the strong contribution of early DTCs to metastasis after a protracted dormancy phase. The rest of the metastases were presumably contributed by later disseminating DTCs as they were not affected by the macrophage depletion (Linde et al. 2018). These data do not indicate that late DTCs are not fit to initiate metastasis, but rather reveal an unexpected and understudied role for early DTCs in the breast cancer metastatic process. Another unexplored aspect that may connect early and late DTCs is whether they cooperate in metastasis formation as proposed previously (Sosa et al. 2014; Aguirre-Ghiso and Sosa 2018).

The above results suggest that early DTCs, those that disseminate from early evolutionary stages of breast cancer, like DCIS (as defined by pathology), may be prone to enter dormancy. This could be due to signals from the microenvironment at the primary site that, while promoting dissemination, may also promote entry into dormancy in the target organ. When examining late-stage primary tumors (late stage means overt invasive breast cancer as defined by a pathologist) in patient-derived xenografts and transgenic mouse models, hypoxic microenvironments in breast primary tumors were found to induce the up-regulation of dormancy genes, such as NR2F1 and p27 (Fluegen et al. 2017). The DTCs leaving the tumor carried this signature and a TGF-$\beta 2^{hi}$ dormant phenotype to secondary sites.

Multiple studies have investigated the effects of signals coming from the target organ microenvironment surrounding DTCs on their phenotype and fate. Common sites of metastatic colonization from breast cancer cells derived for example from ductal carcinomas are bone, lungs, and brain (Kennecke et al. 2010).

While hypoxia in specific areas of the primary tumor initially imprints a dormancy program in disseminating cancer cells, high hypoxic conditions in the colonized organ have been shown to eventually reactivate DTCs. In breast cancer cells disseminated to the bone marrow, hypoxia down-regulated LIFR, leading to down-regulation of dormancy and cancer stem cell–associated genes (Johnson et al. 2016). Thus, LIFR is a positive regulator of dormancy in the bone and hypoxia can cause reactivation. In this same tissue, studying the function of bone-resident NG2$^+$/Nestin$^+$ mesenchymal stem cells (MSCs), MSCs in homeostasis were found to maintain DTC dormancy (Nobre et al. 2021b). In intracardiac injection models of metastasis to the bone, genetic depletion of MSCs awakened dormant DTCs, which formed metastases. MSC-mediated dormancy was driven by TGF-β2 and the induction of p27 in DTCs. Other studies on bone marrow components have found that the vascular endothelium and the perivascular niche protect DTCs from chemotherapy (Carlson et al. 2019). Targeting the niche, specifically the endothelial von Willebrand factor (vWF) and vascular cell adhesion molecule 1 (VCAM1), rendered dormant cancer cells sensitive to chemotherapy, without having them dangerously reenter the cell cycle.

The lung niche plays a relevant role in DTC fate. Dormant D2.OR breast cancer cells disseminated to the lung interact with alveolar epithelial type I (AT1) cells, favoring integrin-dependent survival pathways (Montagner et al. 2020). In vivo screening for genes required for cross talk identified the molecule SFRP2 on breast cancer cells. Further, targeting of SFRP2 by shRNA-mediated silencing was shown to eradicate dormant cells.

The microenvironment surrounding DTCs is not only composed of resident host cells, but also of the abundant ECM. Dormant cancer cells were found to produce a type III collagen–enriched ECM, which was essential to maintain DTC dormancy both at the primary site and in the lung (Di Martino et al. 2022). When type III collagen was not present, DDR1-mediated signaling was inactive and cells began proliferating.

A less studied organ that harbors dormant breast DTCs is the brain. A recent study identified signals from the brain niche that could drive DTC dormancy (Dai et al. 2022). Here, they showed that dormant DTCs in the brain resided close to astrocytes. Astrocytes produced laminin-211, inducing DTC dormancy via dystroglycan receptor signaling in cancer cells. It was also shown that a bone morphogenetic protein (BMP) signaling antagonist DAND5/Coco can awaken dormant breast cancer cells in the lung (Gao et al. 2012; Giancotti 2013), and BMP4-SMAD7 signaling is also a mediator of metastasis suppression in other breast cancer models (Eckhardt et al. 2020). The ECM protein thrombospondin-1 has also been implicated in both angiogenic dormancy (Naumov et al. 2006b; Aguirre-Ghiso 2007) and cellular dormancy (Ghajar et al. 2013), possibly by its ability to directly affect endothelial cell proliferation but also via direct effects on DTCs in the perivascular niche. These data support how a multiplicity of factors create a redundancy network to suppress metastasis initiation via dormancy induction and maintenance.

Immune-Mediated Tumor Cell Dormancy

Immune cells constitute a major component of the tissue microenvironment. Only recently has the dormancy community focused on examining the role of innate and adaptive immune cells in regulation of DTC dormancy (Fig. 2).

In a mouse model of chemotherapy-induced dormancy of triple-negative 4T1 cells, CD4$^+$ and CD8$^+$ T cells were required for dormancy (Lan et al. 2019). 4T1 cells treated with chemotherapeutic agents (methotrexate or doxorubicin) and dormant D2.OR cells were enriched for type 1 interferon (IFN) response genes. Specifically, they up-regulated IRF7, IRF9, STAT1, and STAT2. RNAi-mediated silencing of IRF7 allowed escape from dormancy, increasing local tumor growth and metastatic burden. Further, in clinical samples of serum of breast cancer patients collected during neoadjuvant chemotherapy, the presence of IFN-β was associated with longer metastasis-free survival. Another study found that 4TO7 primary tumor-derived CD8$^+$ T cells induced protective immunity and dormancy at distant sites (Tallón de Lara et al.

2021). A subpopulation of CD39$^+$ PD-1$^+$ CD8$^+$ T cells was proposed to be responsible for this phenotype, but functional validation was lacking.

Aside from T cells, natural killer (NK) cells have also been linked to breast cancer dormancy. Dormant HER2$^+$ DTCs evaded immune surveillance by NK cells via autocrine DKK1-mediated inhibition of WNT signaling (Malladi et al. 2016). After dormancy escape, DTCs could be detected and cleared by NK cells. In a more recent study, NK cells sustained triple-negative MDA-MB-231 dormancy in the liver through IFN-γ (Correia et al. 2021). Activation of hepatic stellate cells, as seen during liver injury, prevented the expansion of NK cells through CXCL12 and induced NK cell quiescence. Inhibited NK cells were unable to maintain IFN-γ DTC dormancy, leading to proliferation and metastatic growth.

Autophagy and Novel Pathways Regulating Dormant Cell Survival

Multiple studies have reported the role of autophagy in the survival and dormancy of DTCs (Fig. 2; Sosa et al. 2013). Autophagy is a process of degradation and recycling of cytoplasmic components that occurs intracellularly.

One of the models used to study the role of autophagy in spontaneously arising breast cancer dormancy is represented by the dormant D2.OR and proliferative D2A1 paired cell lines. D2.OR cells were found to activate autophagy, and autophagy inhibition with hydroxychloroquine (HCQ) or other drugs reduced viability of dormant cells injected intravenously to reach the lung (Vera-Ramirez et al. 2018). A genetic model of *Atg7* silencing corroborated the findings that autophagy is important for survival of dormant breast cancer cells. Autophagy may also be involved in the induction and or maintenance of the dormant state of DTCs. Autophagy inactivation by depletion of *Atg3*, *Atg7*, or *p62/sequestosome-1* led to exit from dormancy and metastatic outgrowth of D2.OR cells (La Belle Flynn et al. 2019). Metastatic recurrence was associated with increased 6-phospho-fructo-2-kinase/fructose-2,6-biphosphatase 3 (*Pfkfb3*) expression, which is an autophagy substrate normally degraded in dormant cells. Ac-

cordingly, autophagic degradation of NBR1 also restricts metastasis and its loss favored metastatic reactivation from DTCs (Marsh et al. 2020). However, whether the pro-dormancy effects of autophagy via NBR1 regulation were due to a regulation of the timing of DTC dormancy was not formally explored (Marsh et al. 2020).

An additional intracellular recycling-related pathway shown to be important for DTC survival is the TFEB-lysosomal biogenesis pathway. An in vivo shRNA screen identified genes required for breast DTC survival upon dissemination to the lungs, specifically the transmembrane protein EphB6, a regulator of the TFEB axis (Zangrossi et al. 2021). Interestingly, EphB6 expression was stimulated by the lung microenvironment and influenced the response of lung AT1 cells, pointing to the cross-regulation of DTCs and their surrounding milieu.

The discovery of novel pathways and genes involved in cancer dormancy has been accelerated by the use of unbiased genome-wide screens in recent years. A study by Gawrzak et al. reported the discovery of mitogen- and stress-activated kinase 1 (MSK1) as a regulator of bone micrometastatic dormancy in a model of ER$^+$ breast cancer (Gawrzak et al. 2018). MSK1 loss was associated with early relapse in ER$^+$ breast cancer patients and enhanced bone metastasis in mice.

Another novel regulator of breast cancer metastatic dormancy was found to be the long noncoding RNA NR2F1-AS1 (Liu et al. 2021). NR2F1 was previously implicated as a potent dormancy factor in breast and non-breast cancer cells (Sosa et al. 2015; Borgen et al. 2018). In a recent study, NR2F1-AS1 was up-regulated in latent mesenchymal cancer cells and promoted metastatic seeding. However, it inhibited their proliferation and forced them into dormancy via up-regulation of NR2F1.

MECHANISMS UNDERLYING DORMANCY EXIT AND RECURRENCE

After long periods of dormancy (in both spontaneous and therapy-associated conditions), DTCs can awaken and reenter the cell cycle by the combination of both intrinsic and extrinsic mechanisms. The former include accumulation of genetic mutations and epigenetic alterations, while the latter involve signals from the organ microenvironment related to inflammation (see below for breast cancer examples) and changes of target organs related to aging as shown in melanoma (Fane et al. 2022) and as seen for entry into dormancy (Fig. 2).

An example of genetic alteration-mediated escape from dormancy was described by Bui, Gu, and colleagues, who used a GEM model of *PyMT* oncogene activation with specific deletion of integrin β1 (ITGB1) in the epithelium upon doxycycline administration (Bui et al. 2022). Mammary epithelial deletion of ITGB1 impaired tumor initiation and lung metastasis. ITGB1-deficient tumors exhibited tumor mass dormancy, explained by a combination of proliferation, apoptosis, and senescence, and they were able to escape dormancy after p53 mutation and accumulation of a protumorigenic stromal compartment.

Genetic (loss of heterozygosity) loss of receptors linked to dormancy induction suggests how such changes could lead to the awakening of dormant cells in different cancers (Sharifi et al. 2007; Kobayashi et al. 2011; Bragado et al. 2013) but also support that DTCs are in constant communication with the surrounding environment and cannot be analyzed without taking into consideration physiological events happening around them. For example, loss of homeostatic niches for hematopoietic stem cell quiescence and coordinated by Nestin[+] MSCs led to rapid awakening of dormant breast cancer cells across luminal B and TNBC subtype models (Nobre et al. 2021b). Profibrotic environments in the lung have also been shown to cause awakening of dormant breast cancer DTCs by altering cytoskeletal and MAPK signaling (Barkan et al. 2010; Barkan and Green 2011; Weidenfeld et al. 2016). The ability of residual breast cancer cells to reactivate after blocking expression of the HER2 oncogene was linked to their ability to control redox homeostasis. Loss-of-function of the antioxidant transcription factor NRF2 blocked recurrence (Fox et al. 2020), suggesting that antioxidant responses are able to provide residual cancer cells with adaptive fitness during recurrence.

Inflammation, in both systemic and local contexts, has been shown to trigger escape from dormancy. Specifically, lipopolysaccharide treatment in the lung awakened dormant D2A1 cancer cells, resulting in proliferation and metastasis (De Cock et al. 2016). The effect appeared to be mediated by neutrophils, but this analysis lacked mechanistic insights, warranting further studies. Albrengues et al. found that during lung inflammation (as after exposure to cigarette smoke), lung neutrophils form neutrophil extracellular traps (NETs) (Albrengues et al. 2018). In mouse models of breast cancer metastasis (D2.0R/D2A1, 4TO7/4T1), the proteases neutrophil elastase (NE) and matrix metalloproteinase 9 (MMP9) on NETs cleaved laminin, generating a version that activated integrin-mediated signaling and proliferation of previous dormant cancer cells. Whether these mechanisms occur or constitute rate-limiting events in breast cancer recurrence in patients is unknown.

Another study investigated broader systemic inflammation and its effect on DTC dormancy (Krall et al. 2018). This study used mammary fat pad injection of D2A1 cells and subsequent sham surgery. After surgery, a systemic inflammatory response was induced, and specific T-cell-mediated tumor restriction was overrun. However, primary tumor removal in place of sham surgery would be a best-suited model to test the systemic response to surgery, as primary tumors are removed from patients.

Along with influencing DTC state at secondary sites, inflammation promotes breast cancer recurrence locally. HER2 down-regulation in a conditional mouse model of HER2-driven breast cancer caused an inflammatory program through TNF-α signaling (Walens et al. 2019). The cytokine CCL5 was found high in residual tumors and attracted CCR5[+] macrophages. Further, macrophage infiltration promoted collagen deposition and tumor regrowth.

OPPORTUNITIES FOR CLINICAL INTERVENTION

Advances in breast cancer diagnosis and treatment in recent decades have resulted in markedly improved outcome for patients. Nevertheless, in up to 30% of patients, residual tumor cells reawaken following extended periods of dormancy and resume growth, ultimately resulting in incurable recurrent cancers. At present, there are no clinically approved treatments that exploit the mechanisms underlying dormant MRD in breast cancer. New trails in other hormone receptor–driven cancers (Aguirre-Ghiso 2021) aimed at inducing dormancy in biochemically recurrent prostate cancer (NCT03572387) may offer new ideas on how to manage MRD in the breast cancer setting. However, it is possible that strategies in prostate cancer may not play out in breast cancer or may be limited to some subtypes. Insights into the mechanisms underlying breast cancer dormancy gained from patient-relevant experimental models have far-reaching consequences in the clinic. Careful application of the knowledge gained from such studies could revolutionize breast cancer management such that the remission period (i.e., no evidence of disease [NED]) following definitive treatment of the primary cancer becomes one of active surveillance and preemptive intervention, rather than passive monitoring as is currently the standard-of-care for breast cancer patients.

A framework for exploiting residual tumor cell dormancy to prevent tumor recurrence can be envisaged to (1) eliminate or deplete residual disease burden; (2) maintain residual tumor cells in a dormant state; or (3) induce a controlled exit from dormancy in residual tumor cells coupled with therapies that are highly effective in targeting proliferating cells (Fig. 3). Of these approaches, eradication of residual disease burden would ostensibly be associated with the least risk of recurrence and the greatest facility for monitoring treatment efficacy over time. The induction of dormancy as a maintenance therapy, along with proper biomarkers to monitor patients, may also constitute a safe approach. In contrast, intentionally reactivating dormant cells to then target them with antiproliferative therapies represents the most risky approach, as replication may be associated with clonal evolution and therapy resistance. Thus, even the most effective therapies for targeting proliferating cancer cells are unlikely to achieve complete eradication and may thereby leave a small, but potentially lethal, proliferating population of tumor cells.

The observation that breast cancer patients with detectable DTCs in their bone marrow are at a considerably higher risk of developing recurrences is the guiding principle driving the idea of rationally targeting MRD to prevent recurrence. Several benefits exist for such an approach. First, the massive debulking of primary tumors that accompanies (neo)adjuvant therapy, radiation, and/or surgery may render the more limited burden of residual disease a more tractable target for therapy. Second, some studies have detected a decrease in clonal heterogeneity following therapy in experimental models (Walens et al. 2020), as well as in a subset of patient samples (Almendro et al. 2014; Kim et al. 2018). Limited heterogeneity would theoretically limit the possibilities for developing resistance following residual disease-targeted therapy. Third, the presence of most residual tumor cells in a nonproliferative state may limit the rapid propagation of any recurrence-promoting inherited genotype that drives treatment refractory recurrent disease outgrowth. Fourth, dormancy-inducing therapies may provide additional means to mitigate acquired resistance while providing more time to employ therapies that eradicate MRD (Aguirre-Ghiso 2021; Khalil et al. 2022).

Conversely, some conceptual limitations exist for targeting MRD to prevent recurrences. An important consideration is that only about 40% of patients with detectable DTCs in their bone marrow develop recurrences over a 10-yr follow-up period (Braun et al. 2005). Thus, treating MRD should be approached with care to avoid overtreating patients who otherwise might not develop recurrent disease. Refining existing latency-associated genomic and transcriptomic signatures to more accurately identify those survivors at greatest risk for late recurrence would help mitigate this concern, although it is unlikely to eliminate it. Additionally, if the majority of residual tumor cells in patients exist in a nonproliferative state, it would be anticipated that these cells might be less conducive to elimination by chemotherapies whose efficacy is influenced by cell-cycle status. However, this idea has been challenged by a recent study that found that dormant DTCs can be sensitized to chemotherapy by disrupting their interactions with en-

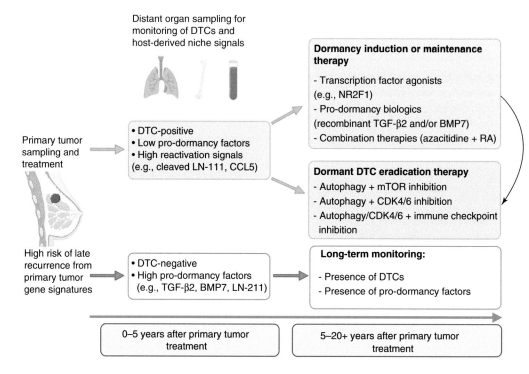

Figure 3. Proposed scheme for the application of dormancy-oriented therapies in the clinic. Breast cancer patients who undergo primary tumor-oriented therapy can be screened for their risk of developing recurrences by gene expression profiling of their primary tumor samples and/or monitoring for the presence of minimal residual disease (MRD) (e.g., disseminated tumor cells [DTCs]). High-risk patients who are DTC-positive can be enrolled into trials that either induce/maintain dormancy, or those that eradicate dormant DTCs. Patients classified as low risk based on their DTC status can be monitored long term for DTC presence as well as levels of niche-derived pro-dormancy cues. (RA) Retinoic acid.

dothelial cells in their niche without inducing cell-cycle reentry. However, the mechanism by which antiproliferative therapies targeting topoisomerase II (TOPOII) can kill quiescent cells was not elucidated (Carlson et al. 2019). Taken together, clinical trials designed to target MRD must thoughtfully maximize the benefits of an interventional approach while mitigating potential risks.

An ideal clinical trial in the residual disease surveillance period would combine the identification of patients at highest risk of recurrence with knowledge of signaling pathways that maintain residual tumor cell survival and dormancy to target MRD (Fig. 3; Cescon et al. 2022). Identifying breast cancer patients who are most likely to benefit from such an approach could involve considering pathological features

such as tumor grade and lymph node status (Pan et al. 2017), in combination with the application of predictive dormancy-associated gene-expression signatures to primary tumor samples harvested from patients. More refined risk classification might be achieved by direct sampling of bone marrow or blood of patients to quantify their residual tumor cell burden. The ability to isolate these ultrarare cells combined with improvements in single-cell technologies could additionally help characterize the dormancy status of residual tumor cells in patients. A complementary approach would be to determine whether host-derived niche signals known to maintain dormancy are altered, informing the potential risk of relapse. Such an approach revealed that breast cancer patients with low levels of the pro-dormancy factor BMP7 (Kobayashi et al. 2011) were at a higher risk of developing lethal bone metastasis (Nobre et al. 2021a). These types of information could be invaluable insofar as they might reveal the presence of potential targets and inform choices of therapeutic regimens. Moreover, MRD sampling could also be incorporated into the trial itself to measure the efficacy of the intervention at defined intervals as a potential surrogate end point for recurrence-free and overall survival in patients.

A proof-of-principle DTC-guided clinical trial (NCT00248703) enrolled patients with persistent DTCs in their bone marrow following standard-of-care chemotherapy and administered docetaxel in a postadjuvant therapy setting to deplete DTCs. Of the patients who remained DTC-positive after the secondary postadjuvant therapy, 46.7% experienced recurrence compared to 8.8% in the patients who were DTC-negative (Naume et al. 2014). Of note, this trial lacked a control arm consisting of DTC-positive patients who did not receive additional therapy. Thus, other biological differences may exist between patients who "clear" their bone marrow DTCs following additional chemotherapy as compared to those who do not.

Another trial (NCT00172068) enrolled patients who were DTC-positive and randomized them into a control arm that received adjuvant therapy alone or adjuvant therapy in addition to zoledronic acid (Banys et al. 2013). After 24 mo, 16% of patients in the control arm still had detectable DTCs, whereas none of the zoledronic acid–treated patients had detectable DTCs. A trending decrease in overall and disease-free survival was reported for patients in the control arm versus the treatment arm; however, the analysis was limited by the small sample size. These data support the important possibility that eliminating MRD might constitute a promising approach to prevent recurrence.

More recent trials have employed rational drug combinations designed to leverage dormancy biology to reduce or eliminate DTCs. Patients at elevated risk for recurrence enrolled in the Penn-SURMOUNT screening study (NCT02732171) undergo bone marrow sampling for detection of DTCs, and patients in whom DTCs are detected are offered enrollment in DTC-targeting phase II clinical trials (Bayne et al. 2018, 2021). The CLEVER trial (NCT030 32406), which was first among this cohort of trials, uses the autophagy inhibitor HCQ and the mTOR inhibitor everolimus, alone or in combination, in patients with detectable bone marrow DTCs who are at elevated risk of recurrence (Bayne et al. 2018). The ABBY trial (NCT04523857) tests the efficacy of a CDK4/6 inhibitor (abemaciclib) either alone or in combination with the autophagy inhibitor (HCQ) to eliminate DTCs. The PALAVY trial (NCT 04841148) assesses the efficacy of inhibiting DTC persistence in ER$^+$ breast cancer patients by inhibiting autophagy (HCQ) or CDK4/6 with a different inhibitor (palbociclib), in combination with a checkpoint inhibitor (avelumab) to circumvent immune evasion. Data from this cohort of controlled clinical trials in breast cancer patients harboring DTCs will provide critical insights into the feasibility and efficacy of a DTC-targeting approach to mitigate recurrences.

In conclusion, our evolving understanding of dormancy presents a unique therapeutic opportunity with the potential to radically redefine the clinical management of breast cancer by preventing recurrent disease and reducing its associated mortality.

REFERENCES

Abravanel DL, Belka GK, Pan T, Pant DK, Collins MA, Sterner CJ, Chodosh LA. 2015. Notch promotes recurrence of dormant tumor cells following HER2/neu-targeted therapy. *J Clin Invest* **125:** 2484–2496. doi:10.1172/JCI74883

Aguirre-Ghiso JA. 2007. Models, mechanisms and clinical evidence for cancer dormancy. *Nat Rev Cancer* **7:** 834–846. doi:10.1038/nrc2256

Aguirre-Ghiso JA. 2018. How dormant cancer persists and reawakens. *Science* **361:** 1314–1315. doi:10.1126/science.aav0191

Aguirre-Ghiso JA. 2021. Translating the science of cancer dormancy to the clinic. *Cancer Res* **81:** 4673–4675. doi:10.1158/0008-5472.CAN-21-1407

Aguirre-Ghiso JA, Sosa MS. 2018. Emerging topics on disseminated cancer cell dormancy and the paradigm of metastasis. *Annu Rev Cancer Biol* **2:** 377–393. doi:10.1146/annurev-cancerbio-030617-050446

Albrengues J, Shields MA, Ng D, Park CG, Ambrico A, Poindexter ME, Upadhyay P, Uyeminami DL, Pommier A, Küttner V, et al. 2018. Neutrophil extracellular traps produced during inflammation awaken dormant cancer

cells in mice. *Science* **361:** eaao4227. doi:10.1126/science
.aao4227

Almendro V, Cheng YK, Randles A, Itzkovitz S, Marusyk A, Ametller E, Gonzalez-Farre X, Muñoz M, Russnes HG, Helland A, et al. 2014. Inference of tumor evolution during chemotherapy by computational modeling and in situ analysis of genetic and phenotypic cellular diversity. *Cell Rep* **6:** 514–527. doi:10.1016/j.celrep.2013.12.041

Aslakson CJ, Miller FR. 1992. Selective events in the metastatic process defined by analysis of the sequential dissemination of subpopulations of a mouse mammary tumor. *Cancer Res* **52:** 1399–1405.

Au E, Wong G, Chapman JR. 2018. Cancer in kidney transplant recipients. *Nat Rev Nephrol* **14:** 508–520. doi:10.1038/s41581-018-0022-6

Baldominos P, Barbera-Mourelle A, Barreiro O, Huang Y, Wight A, Cho JW, Zhao X, Estivill G, Adam I, Sanchez X, et al. 2022. Quiescent cancer cells resist T cell attack by forming an immunosuppressive niche. *Cell* **185:** 1694–1708.e19. doi:10.1016/j.cell.2022.03.033

Balko JM, Giltnane JM, Wang K, Schwarz LJ, Young CD, Cook RS, Owens P, Sanders ME, Kuba MG, Sánchez V, et al. 2014. Molecular profiling of the residual disease of triple-negative breast cancers after neoadjuvant chemotherapy identifies actionable therapeutic targets. *Cancer Discov* **4:** 232–245. doi:10.1158/2159-8290.CD-13-0286

Banys M, Solomayer EF, Gebauer G, Janni W, Krawczyk N, Lueck HJ, Becker S, Huober J, Kraemer B, Wackwitz B, et al. 2013. Influence of zoledronic acid on disseminated tumor cells in bone marrow and survival: results of a prospective clinical trial. *BMC Cancer* **13:** 480. doi:10.1186/1471-2407-13-480

Barkan D, Green JE. 2011. An in vitro system to study tumor dormancy and the switch to metastatic growth. *J Vis Exp* **54:** 2914.

Barkan D, El Touny LH, Michalowski AM, Smith JA, Chu I, Davis AS, Webster JD, Hoover S, Simpson RM, Gauldie J, et al. 2010. Metastatic growth from dormant cells induced by a col-I-enriched fibrotic environment. *Cancer Res* **70:** 5706–5716. doi:10.1158/0008-5472.CAN-09-2356

Bayne LJ, Nivar I, Goodspeed B, Wileyto P, Savage J, Shih NNC, Feldman MD, Edwards J, Clark AS, Fox KR, et al. 2018. Abstract OT2-07-09: detection and targeting of minimal residual disease in breast cancer to reduce recurrence: the PENN-SURMOUNT and CLEVER trials. *Cancer Res* **78:** OT2-07-09. doi:10.1158/1538-7445.SABCS17-OT2-07-09

Bayne LJ, Nivar I, Goodspeed B, Deluca SE, Wileyto P, Shih NNC, Nayak A, Feldman MD, Edwards J, Fox K, et al. 2021. Abstract PD9-11: identifying breast cancer survivors with dormant disseminated tumor cells: the PENN-SURMOUNT screening study. *Cancer Res* **81:** PD9-11. doi:10.1158/1538-7445.SABCS20-PD9-11

Bidard FC, Vincent-Salomon A, Gomme S, Nos C, de Rycke Y, Thiery JP, Sigal-Zafrani B, Mignot L, Sastre-Garau X, Pierga JY, et al. 2008. Disseminated tumor cells of breast cancer patients: a strong prognostic factor for distant and local relapse. *Clin Cancer Res* **14:** 3306–3311. doi:10.1158/1078-0432.CCR-07-4749

Borgen E, Rypdal MC, Sosa MS, Renolen A, Schlichting E, Lønning PE, Synnestvedt M, Aguirre-Ghiso JA, Naume B. 2018. NR2F1 stratifies dormant disseminated tumor cells in breast cancer patients. *Breast Cancer Res* **20:** 120. doi:10.1186/s13058-018-1049-0

Boxer RB, Jang JW, Sintasath L, Chodosh LA. 2004. Lack of sustained regression of c-MYC-induced mammary adenocarcinomas following brief or prolonged MYC inactivation. *Cancer Cell* **6:** 577–586. doi:10.1016/j.ccr.2004.10.013

Bragado P, Estrada Y, Parikh F, Krause S, Capobianco C, Farina HG, Schewe DM, Aguirre-Ghiso JA. 2013. TGF-β2 dictates disseminated tumour cell fate in target organs through TGF-β-RIII and p38α/β signalling. *Nat Cell Biol* **15:** 1351–1361. doi:10.1038/ncb2861

Braun S, Vogl FD, Naume B, Janni W, Osborne MP, Coombes RC, Schlimok G, Diel IJ, Gerber B, Gebauer G, et al. 2005. A pooled analysis of bone marrow micrometastasis in breast cancer. *N Engl J Med* **353:** 793–802. doi:10.1056/NEJMoa050434

Bui T, Gu Y, Ancot F, Sanguin-Gendreau V, Zuo D, Muller WJ. 2022. Emergence of β1 integrin-deficient breast tumours from dormancy involves both inactivation of p53 and generation of a permissive tumour microenvironment. *Oncogene* **41:** 527–537. doi:10.1038/s41388-021-02107-7

Bushnell GG, Deshmukh AP, den Hollander P, Luo M, Soundararajan R, Jia D, Levine H, Mani SA, Wicha MS. 2021. Breast cancer dormancy: need for clinically relevant models to address current gaps in knowledge. *NPJ Breast Cancer* **7:** 66. doi:10.1038/s41523-021-00269-x

Calvo V, Zheng W, Staschke KA, Cheung J, Nobre AR, Farias EF, Nowacek A, Mulvihill M, Rigby AC, Aguirre-Ghiso JA. 2021. PERK inhibition blocks metastasis initiation by limiting UPR-dependent survival of dormant disseminated cancer cells. bioRxiv doi:10.1101/2021.07.30.454473

Carlson P, Dasgupta A, Grzelak CA, Kim J, Barrett A, Coleman IM, Shor RE, Goddard ET, Dai J, Schweitzer EM, et al. 2019. Targeting the perivascular niche sensitizes disseminated tumour cells to chemotherapy. *Nat Cell Biol* **21:** 238–250. doi:10.1038/s41556-018-0267-0

Cescon DW, Kalinsky K, Parsons HA, Smith KL, Spears PA, Thomas A, Zhao F, DeMichele A. 2022. Therapeutic targeting of minimal residual disease to prevent late recurrence in hormone-receptor positive breast cancer: challenges and new approaches. *Front Oncol* **11:** 667397. doi:10.3389/fonc.2021.667397

Clements ME, Holtslander L, Edwards C, Todd V, Dooyema SDR, Bullock K, Bergdorf K, Zahnow CA, Connolly RM, Johnson RW. 2021. HDAC inhibitors induce LIFR expression and promote a dormancy phenotype in breast cancer. *Oncogene* **40:** 5314–5326. doi:10.1038/s41388-021-01931-1

Connolly RM, Li H, Jankowitz RC, Zhang Z, Rudek MA, Jeter SC, Slater SA, Powers P, Wolff AC, Fetting JH, et al. 2017. Combination epigenetic therapy in advanced breast cancer with 5-azacitidine and entinostat: a phase II national cancer institute/stand up to cancer study. *Clin Cancer Res* **23:** 2691–2701. doi:10.1158/1078-0432.CCR-16-1729

Correia AL, Guimaraes JC, Auf der Maur P, De Silva D, Trefny MP, Okamoto R, Bruno S, Schmidt A, Mertz K, Volkmann K, et al. 2021. Hepatic stellate cells suppress NK cell-sustained breast cancer dormancy. *Nature* **594:** 566–571. doi:10.1038/s41586-021-03614-z

Creighton CJ, Li X, Landis M, Dixon JM, Neumeister VM, Sjolund A, Rimm DL, Wong H, Rodriguez A, Herschkowitz JI, et al. 2009. Residual breast cancers after conventional therapy display mesenchymal as well as tumor-initiating features. *Proc Natl Acad Sci* **106:** 13820–13825. doi:10.1073/pnas.0905718106

Dai J, Cimino PJ, Gouin KH, Grzelak CA, Barrett A, Lim AR, Long A, Weaver S, Saldin LT, Uzamere A, et al. 2022. Astrocytic laminin-211 drives disseminated breast tumor cell dormancy in brain. *Nat Cancer* **3:** 25–42. doi:10.1038/s43018-021-00297-3

Davies C, Pan H, Godwin J, Gray R, Arriagada R, Raina V, Abraham M, Medeiros Alencar VH, Badran A, Bonfill X, et al. 2013. Long-term effects of continuing adjuvant tamoxifen to 10 years versus stopping at 5 years after diagnosis of oestrogen receptor-positive breast cancer: ATLAS, a randomised trial. *Lancet* **381:** 805–816. doi:10.1016/S0140-6736(12)61963-1

D'Cruz CM, Gunther EJ, Boxer RB, Hartman JL, Sintasath L, Moody SE, Cox JD, Ha SI, Belka GK, Golant A, et al. 2001. c-MYC induces mammary tumorigenesis by means of a preferred pathway involving spontaneous Kras2 mutations. *Nat Med* **7:** 235–239. doi:10.1038/84691

De Cock JM, Shibue T, Dongre A, Keckesova Z, Reinhardt F, Weinberg RA. 2016. Inflammation triggers Zeb1-dependent escape from tumor latency. *Cancer Res* **76:** 6778–6784. doi:10.1158/0008-5472.CAN-16-0608

Dhimolea E, de Matos Simoes R, Kansara D, Al'Khafaji A, Bouyssou J, Weng X, Sharma S, Raja J, Awate P, Shirasaki R, et al. 2021. An embryonic diapause-like adaptation with suppressed Myc activity enables tumor treatment persistence. *Cancer Cell* **39:** 240–256.e11. doi:10.1016/j.ccell.2020.12.002

Di Martino JS, Nobre AR, Mondal C, Taha I, Farias EF, Fertig EJ, Naba A, Aguirre-Ghiso JA, Bravo-Cordero JJ. 2022. A tumor-derived type III collagen-rich ECM niche regulates tumor cell dormancy. *Nat Cancer* **3:** 90–107. doi:10.1038/s43018-021-00291-9

Early Breast Cancer Trialists' Collaborative Group (EBCTCG). 2021. Trastuzumab for early-stage, HER2-positive breast cancer: a meta-analysis of 13 864 women in seven randomised trials. *Lancet Oncol* **22:** 1139–1150. doi:10.1016/S1470-2045(21)00288-6

Early Breast Cancer Trialists' Collaborative Group (EBCTCG); Davies C, Godwin J, Gray R, Clarke M, Cutter D, Darby S, McGale P, Pan HC, Taylor C, et al. 2011. Relevance of breast cancer hormone receptors and other factors to the efficacy of adjuvant tamoxifen: patient-level meta-analysis of randomised trials. *Lancet* **378:** 771–784. doi:10.1016/S0140-6736(11)60993-8

Echeverria GV, Ge Z, Seth S, Zhang X, Jeter-Jones S, Zhou X, Cai S, Tu Y, McCoy A, Peoples M, et al. 2019. Resistance to neoadjuvant chemotherapy in triple-negative breast cancer mediated by a reversible drug-tolerant state. *Sci Transl Med* **11:** eaav0936. doi:10.1126/scitranslmed.aav0936

Eckhardt BL, Cao Y, Redfern AD, Chi LH, Burrows AD, Roslan S, Sloan EK, Parker BS, Loi S, Ueno NT, et al. 2020. Activation of canonical BMP4-SMAD7 signaling suppresses breast cancer metastasis. *Cancer Res* **80:** 1304–1315. doi:10.1158/0008-5472.CAN-19-0743

Fane ME, Chhabra Y, Alicea GM, Maranto DA, Douglass SM, Webster MR, Rebecca VW, Marino GE, Almeida F, Ecker BL, et al. 2022. Stromal changes in the aged lung induce an emergence from melanoma dormancy. *Nature* **606:** 396–405. doi:10.1038/s41586-022-04774-2

Farrar JD, Katz KH, Windsor J, Thrush G, Scheuermann RH, Uhr JW, Street NE. 1999. Cancer dormancy. VII: A regulatory role for CD8$^+$ T cells and IFN-γ in establishing and maintaining the tumor-dormant state. *J Immunol* **162:** 2842–2849. doi:10.4049/jimmunol.162.5.2842

Feng Y, Pan TC, Pant DK, Chakrabarti KR, Alvarez JV, Ruth JR, Chodosh LA. 2014. SPSB1 promotes breast cancer recurrence by potentiating c-MET signaling. *Cancer Discov* **4:** 790–803. doi:10.1158/2159-8290.CD-13-0548

Fischer KR, Durrans A, Lee S, Sheng J, Li F, Wong STC, Choi H, El Rayes T, Ryu S, Troeger J, et al. 2015. Epithelial-to-mesenchymal transition is not required for lung metastasis but contributes to chemoresistance. *Nature* **527:** 472–476. doi:10.1038/nature15748

Fluegen G, Avivar-Valderas A, Wang Y, Padgen MR, Williams JK, Nobre AR, Calvo V, Cheung JF, Bravo-Cordero JJ, Entenberg D, et al. 2017. Phenotypic heterogeneity of disseminated tumour cells is preset by primary tumour hypoxic microenvironments. *Nat Cell Biol* **19:** 120–132. doi:10.1038/ncb3465

Fox DB, Garcia NMG, McKinney BJ, Lupo R, Noteware LC, Newcomb R, Liu J, Locasale JW, Hirschey MD, Alvarez JV. 2020. NRF2 activation promotes the recurrence of dormant tumour cells through regulation of redox and nucleotide metabolism. *Nat Metab* **2:** 318–334. doi:10.1038/s42255-020-0191-z

Gao H, Chakraborty G, Lee-Lim AP, Mo Q, Decker M, Vonica A, Shen R, Brogi E, Brivanlou AH, Giancotti FG. 2012. The BMP inhibitor coco reactivates breast cancer cells at lung metastatic sites. *Cell* **150:** 764–779. doi:10.1016/j.cell.2012.06.035

Gawrzak S, Rinaldi L, Gregorio S, Arenas EJ, Salvador F, Urosevic J, Figueras-Puig C, Rojo F, Del Barco Barrantes I, Cejalvo JM, et al. 2018. MSK1 regulates luminal cell differentiation and metastatic dormancy in ER$^+$ breast cancer. *Nat Cell Biol* **20:** 211–221. doi:10.1038/s41556-017-0021-z

Ghajar CM, Peinado H, Mori H, Matei IR, Evason KJ, Brazier H, Almeida D, Koller A, Hajjar KA, Stainier DY, et al. 2013. The perivascular niche regulates breast tumour dormancy. *Nat Cell Biol* **15:** 807–817. doi:10.1038/ncb2767

Giancotti FG. 2013. Mechanisms governing metastatic dormancy and reactivation. *Cell* **155:** 750–764. doi:10.1016/j.cell.2013.10.029

Goddard ET, Bassale S, Schedin T, Jindal S, Johnston J, Cabral E, Latour E, Lyons TR, Mori M, Schedin PJ, et al. 2019. Association between postpartum breast cancer diagnosis and metastasis and the clinical features underlying risk. *JAMA Netw Open* **2:** e186997. doi:10.1001/jamanetworkopen.2018.6997

Gorgoulis V, Adams PD, Alimonti A, Bennett DC, Bischof O, Bishop C, Campisi J, Collado M, Evangelou K, Ferbeyre G, et al. 2019. Cellular senescence: defining a path forward. *Cell* **179:** 813–827. doi:10.1016/j.cell.2019.10.005

Greenhall GHB, Ibrahim M, Dutta U, Doree C, Brunskill SJ, Johnson RJ, Tomlinson LA, Callaghan CJ, Watson CJE.

2022. Donor-transmitted cancer in orthotopic solid organ transplant recipients: a systematic review. *Transpl Int* **35:** 10092. doi:10.3389/ti.2021.10092

Gu G, Dustin D, Fuqua SA. 2016. Targeted therapy for breast cancer and molecular mechanisms of resistance to treatment. *Curr Opin Pharmacol* **31:** 97–103. doi:10.1016/j.coph.2016.11.005

Gunther EJ, Moody SE, Belka GK, Hahn KT, Innocent N, Dugan KD, Cardiff RD, Chodosh LA. 2003. Impact of p53 loss on reversal and recurrence of conditional Wnt-induced tumorigenesis. *Genes Dev* **17:** 488–501. doi:10.1101/gad.1051603

Hadfield G. 1954. The dormant cancer cell. *Br Med J* **2:** 607–610. doi:10.1136/bmj.2.4888.607

Harper KL, Sosa MS, Entenberg D, Hosseini H, Cheung JF, Nobre R, Avivar-Valderas A, Nagi C, Girnius N, Davis RJ, et al. 2016. Mechanism of early dissemination and metastasis in Her2$^+$ mammary cancer. *Nature* **540:** 588–592. doi:10.1038/nature20609

Hartkopf AD, Taran FA, Wallwiener M, Hahn M, Becker S, Solomayer EF, Brucker SY, Fehm TN, Wallwiener D. 2014. Prognostic relevance of disseminated tumour cells from the bone marrow of early stage breast cancer patients—results from a large single-centre analysis. *Eur J Cancer* **50:** 2550–2559. doi:10.1016/j.ejca.2014.06.025

Hartkopf AD, Brucker SY, Taran FA, Harbeck N, von Au A, Naume B, Pierga JY, Hoffmann O, Beckmann MW, Rydén L, et al. 2021. Disseminated tumour cells from the bone marrow of early breast cancer patients: results from an international pooled analysis. *Eur J Cancer* **154:** 128–137. doi:10.1016/j.ejca.2021.06.028

Holdman XB, Welte T, Rajapakshe K, Pond A, Coarfa C, Mo Q, Huang S, Hilsenbeck SG, Edwards DP, Zhang X, et al. 2015. Upregulation of EGFR signaling is correlated with tumor stroma remodeling and tumor recurrence in FGFR1-driven breast cancer. *Breast Cancer Res* **17:** 141–141. doi:10.1186/s13058-015-0649-1

Holmgren L, O'Reilly MS, Folkman J. 1995. Dormancy of micrometastases: balanced proliferation and apoptosis in the presence of angiogenesis suppression. *Nat Med* **1:** 149–153. doi:10.1038/nm0295-149

Hong SP, Chan TE, Lombardo Y, Corleone G, Rotmensz N, Bravaccini S, Rocca A, Pruneri G, McEwen KR, Coombes RC, et al. 2019. Single-cell transcriptomics reveals multistep adaptations to endocrine therapy. *Nat Commun* **10:** 3840–3840. doi:10.1038/s41467-019-11721-9

Hosseini H, Obradović MMS, Hoffmann M, Harper KL, Sosa MS, Werner-Klein M, Nanduri LK, Werno C, Ehrl C, Maneck M, et al. 2016. Early dissemination seeds metastasis in breast cancer. *Nature* **540:** 552–558. doi:10.1038/nature20785

Hu Z, Li Z, Ma Z, Curtis C. 2020. Multi-cancer analysis of clonality and the timing of systemic spread in paired primary tumors and metastases. *Nat Genet* **52:** 701–708. doi:10.1038/s41588-020-0628-z

Hüsemann Y, Geigl JB, Schubert F, Musiani P, Meyer M, Burghart E, Forni G, Eils R, Fehm T, Riethmüller G, et al. 2008. Systemic spread is an early step in breast cancer. *Cancer Cell* **13:** 58–68. doi:10.1016/j.ccr.2007.12.003

Janghorban M, Yang Y, Zhao N, Hamor C, Nguyen TM, Zhang XH, Rosen JM. 2021. Single cell analysis unveils the role of the tumor immune microenvironment and notch signaling in dormant minimal residual disease. *Cancer Res* **82:** 855–899.

Johnson RW, Finger EC, Olcina MM, Vilalta M, Aguilera T, Miao Y, Merkel AR, Johnson JR, Sterling JA, Wu JY, et al. 2016. Induction of LIFR confers a dormancy phenotype in breast cancer cells disseminated to the bone marrow. *Nat Cell Biol* **18:** 1078–1089. doi:10.1038/ncb3408

Kennecke H, Yerushalmi R, Woods R, Cheang MCU, Voduc D, Speers CH, Nielsen TO, Gelmon K. 2010. Metastatic behavior of breast cancer subtypes. *J Clin Oncol* **28:** 3271–3277. doi:10.1200/JCO.2009.25.9820

Khalil BD, Sanchez R, Rahman T, Rodriguez-Tirado C, Moritsch S, Martinez AR, Miles B, Farias E, Mezei M, Nobre AR, et al. 2022. An NR2F1-specific agonist suppresses metastasis by inducing cancer cell dormancy. *J Exp Med* **219:** e20210836. doi:10.1084/jem.20210836

Kim C, Gao R, Sei E, Brandt R, Hartman J, Hatschek T, Crosetto N, Foukakis T, Navin NE. 2018. Chemoresistance evolution in triple-negative breast cancer delineated by single-cell sequencing. *Cell* **173:** 879–893.e13. doi:10.1016/j.cell.2018.03.041

Klein CA. 2011. Framework models of tumor dormancy from patient-derived observations. *Curr Opin Genet Dev* **21:** 42–49. doi:10.1016/j.gde.2010.10.011

Klein CA. 2020. Cancer progression and the invisible phase of metastatic colonization. *Nat Rev Cancer* **20:** 681–694. doi:10.1038/s41568-020-00300-6

Kobayashi A, Okuda H, Xing F, Pandey PR, Watabe M, Hirota S, Pai SK, Liu W, Fukuda K, Chambers C, et al. 2011. Bone morphogenetic protein 7 in dormancy and metastasis of prostate cancer stem-like cells in bone. *J Exp Med* **208:** 2641–2655. doi:10.1084/jem.20110840

Koebel CM, Vermi W, Swann JB, Zerafa N, Rodig SJ, Old LJ, Smyth MJ, Schreiber RD. 2007. Adaptive immunity maintains occult cancer in an equilibrium state. *Nature* **450:** 903–907. doi:10.1038/nature06309

Krall JA, Reinhardt F, Mercury OA, Pattabiraman DR, Brooks MW, Dougan M, Lambert AW, Bierie B, Ploegh HL, Dougan SK, et al. 2018. The systemic response to surgery triggers the outgrowth of distant immune-controlled tumors in mouse models of dormancy. *Sci Transl Med* **10:** eaan3464. doi:10.1126/scitranslmed.aan3464

La Belle Flynn A, Calhoun BC, Sharma A, Chang JC, Almasan A, Schiemann WP. 2019. Autophagy inhibition elicits emergence from metastatic dormancy by inducing and stabilizing Pfkfb3 expression. *Nat Commun* **10:** 3668–3668. doi:10.1038/s41467-019-11640-9

Lan Q, Peyvandi S, Duffey N, Huang YT, Barras D, Held W, Richard F, Delorenzi M, Sotiriou C, Desmedt C, et al. 2019. Type I interferon/IRF7 axis instigates chemotherapy-induced immunological dormancy in breast cancer. *Oncogene* **38:** 2814–2829. doi:10.1038/s41388-018-0624-2

Laughney AM, Hu J, Campbell NR, Bakhoum SF, Setty M, Lavallée V-P, Xie Y, Masilionis I, Carr AJ, Kottapalli S, et al. 2020. Regenerative lineages and immune-mediated pruning in lung cancer metastasis. *Nat Med* **26:** 259–269. doi:10.1038/s41591-019-0750-6

Lefrère H, Lenaerts L, Borges VF, Schedin P, Neven P, Amant F. 2021. Postpartum breast cancer: mechanisms underlying its worse prognosis, treatment implications, and fertility preservation. *Int J Gynecol Cancer* **31**: 412–422. doi:10.1136/ijgc-2020-002072

Lim AR, Ghajar CM. 2022. Thorny ground, rocky soil: tissue-specific mechanisms of tumor dormancy and relapse. *Semin Cancer Biol* **78**: 104–123. doi:10.1016/j.semcancer.2021.05.007

Linde N, Casanova-Acebes M, Sosa MS, Mortha A, Rahman A, Farias E, Harper K, Tardio E, Reyes Torres I, Jones J, et al. 2018. Macrophages orchestrate breast cancer early dissemination and metastasis. *Nat Commun* **9**: 21. doi:10.1038/s41467-017-02481-5

Liu Y, Zhang P, Wu Q, Fang H, Wang Y, Xiao Y, Cong M, Wang T, He Y, Ma C, et al. 2021. Long non-coding RNA NR2F1-AS1 induces breast cancer lung metastatic dormancy by regulating NR2F1 and ΔNp63. *Nat Commun* **12**: 5232–5232. doi:10.1038/s41467-021-25552-0

Lu X, Mu E, Wei Y, Riethdorf S, Yang Q, Yuan M, Yan J, Hua Y, Tiede BJ, Lu X, et al. 2011. VCAM-1 promotes osteolytic expansion of indolent bone micrometastasis of breast cancer by engaging α4β1-positive osteoclast progenitors. *Cancer Cell* **20**: 701–714. doi:10.1016/j.ccr.2011.11.002

Mahnke YD, Schwendemann J, Beckhove P, Schirrmacher V. 2005. Maintenance of long-term tumour-specific T-cell memory by residual dormant tumour cells. *Immunology* **115**: 325–336. doi:10.1111/j.1365-2567.2005.02163.x

Mahoney KH, Miller BE, Heppner GH. 1985. FACS quantitation of leucine aminopeptidase and acid phosphatase on tumor-associated macrophages from metastatic and nonmetastatic mouse mammary tumors. *J Leukoc Biol* **38**: 573–585. doi:10.1002/jlb.38.5.573

Malladi S, Macalinao DG, Jin X, He L, Basnet H, Zou Y, de Stanchina E, Massagué J. 2016. Metastatic latency and immune evasion through autocrine inhibition of WNT. *Cell* **165**: 45–60. doi:10.1016/j.cell.2016.02.025

Marsh T, Kenific CM, Suresh D, Gonzalez H, Shamir ER, Mei W, Tankka A, Leidal AM, Kalavacherla S, Woo K, et al. 2020. Autophagic degradation of NBR1 restricts metastatic outgrowth during mammary tumor progression. *Dev Cell* **52**: 591–604.e6. doi:10.1016/j.devcel.2020.01.025

Mathiesen RR, Borgen E, Renolen A, Løkkevik E, Nesland JM, Anker G, Østenstad B, Lundgren S, Risberg T, Mjaaland I, et al. 2012. Persistence of disseminated tumor cells after neoadjuvant treatment for locally advanced breast cancer predicts poor survival. *Breast Cancer Res* **14**: R117. doi:10.1186/bcr3242

Matser YAH, Terpstra ML, Nadalin S, Nossent GD, de Boer J, van Bemmel BC, van Eeden S, Budde K, Brakemeier S, Bemelman FJ. 2018. Transmission of breast cancer by a single multiorgan donor to 4 transplant recipients. *Am J Transplant* **18**: 1810–1814. doi:10.1111/ajt.14766

Montagna E, Cancello G, Dellapasqua S, Munzone E, Colleoni M. 2014. Metronomic therapy and breast cancer: a systematic review. *Cancer Treat Rev* **40**: 942–950. doi:10.1016/j.ctrv.2014.06.002

Montagner M, Sahai E. 2020. In vitro models of breast cancer metastatic dormancy. *Front Cell Dev Biol* **8**: 37. doi:10.3389/fcell.2020.00037

Montagner M, Bhome R, Hooper S, Chakravarty P, Qin X, Sufi J, Bhargava A, Ratcliffe CDH, Naito Y, Pocaterra A, et al. 2020. Crosstalk with lung epithelial cells regulates Sfrp2-mediated latency in breast cancer dissemination.

Nat Cell Biol **22**: 289–296. doi:10.1038/s41556-020-0474-3

Moody SE, Sarkisian CJ, Hahn KT, Gunther EJ, Pickup S, Dugan KD, Innocent N, Cardiff RD, Schnall MD, Chodosh LA. 2002. Conditional activation of Neu in the mammary epithelium of transgenic mice results in reversible pulmonary metastasis. *Cancer Cell* **2**: 451–461. doi:10.1016/S1535-6108(02)00212-X

Moody SE, Perez D, Pan TC, Sarkisian CJ, Portocarrero CP, Sterner CJ, Notorfrancesco KL, Cardiff RD, Chodosh LA. 2005. The transcriptional repressor snail promotes mammary tumor recurrence. *Cancer Cell* **8**: 197–209. doi:10.1016/j.ccr.2005.07.009

Morris VL, Tuck AB, Wilson SM, Percy D, Chambers AF. 1993. Tumor progression and metastasis in murine D2 hyperplastic alveolar nodule mammary tumor cell lines. *Clin Exp Metastasis* **11**: 103–112. doi:10.1007/BF00880071

Morris VL, Koop S, MacDonald IC, Schmidt EE, Grattan M, Percy D, Chambers AF, Groom AC. 1994. Mammary carcinoma cell lines of high and low metastatic potential differ not in extravasation but in subsequent migration and growth. *Clin Exp Metastasis* **12**: 357–367. doi:10.1007/BF01755879

Natale G, Bocci G. 2018. Does metronomic chemotherapy induce tumor angiogenic dormancy? A review of available preclinical and clinical data. *Cancer Lett* **432**: 28–37. doi:10.1016/j.canlet.2018.06.002

Naume B, Synnestvedt M, Falk RS, Wiedswang G, Weyde K, Risberg T, Kersten C, Mjaaland I, Vindi L, Sommer HH, et al. 2014. Clinical outcome with correlation to disseminated tumor cell (DTC) status after DTC-guided secondary adjuvant treatment with docetaxel in early breast cancer. *J Clin Oncol* **32**: 3848–3857. doi:10.1200/JCO.2014.56.9327

Naumov GN, Akslen LA, Folkman J. 2006a. Role of angiogenesis in human tumor dormancy: animal models of the angiogenic switch. *Cell Cycle* **5**: 1779–1787. doi:10.4161/cc.5.16.3018

Naumov GN, Bender E, Zurakowski D, Kang SY, Sampson D, Flynn E, Watnick RS, Straume O, Akslen LA, Folkman J, et al. 2006b. A model of human tumor dormancy: an angiogenic switch from the nonangiogenic phenotype. *J Natl Cancer Inst* **98**: 316–325. doi:10.1093/jnci/djj068

Nobre AR, Dalla E, Yang J, Huang X, Kenigsberg E, Wang J, Aguirre-Ghiso JA. 2021a. A mesenchymal-like program of dormancy controlled by ZFP281 serves as a barrier to metastatic progression of early disseminated cancer cells. *Research Square* doi:10.21203/rs.3.rs-145308/v1

Nobre AR, Risson E, Singh DK, Di Martino JS, Cheung JF, Wang J, Johnson J, Russnes HG, Bravo-Cordero JJ, Birbrair A, et al. 2021b. Bone marrow NG2$^+$/Nestin$^+$ mesenchymal stem cells drive DTC dormancy via TGFβ2. *Nat Cancer* **2**: 327–339. doi:10.1038/s43018-021-00179-8

Nobre AR, Dalla E, Yang J, Huang X, Wullkopf L, Risson E, Razghandi P, Anton ML, Zheng W, Seoane JA, et al. 2022. ZFP281 drives a mesenchymal-like dormancy program in early disseminated breast cancer cells that prevents metastatic outgrowth in the lung. *Nat Cancer* **3**: 1165–1180. doi:10.1038/s43018-022-00424-8

Oren Y, Tsabar M, Cuoco MS, Amir-Zilberstein L, Cabanos HF, Hütter J-C, Hu B, Thakore PI, Tabaka M, Fulco CP, et al. 2021. Cycling cancer persister cells arise from lineages with distinct programs. *Nature* **596:** 576–582. doi:10.1038/s41586-021-03796-6

Pan H, Gray R, Braybrooke J, Davies C, Taylor C, McGale P, Peto R, Pritchard KI, Bergh J, Dowsett M, et al. 2017. 20-year risks of breast-cancer recurrence after stopping endocrine therapy at 5 years. *N Engl J Med* **377:** 1836–1846. doi:10.1056/NEJMoa1701830

Pantel K, Schlimok G, Braun S, Kutter D, Lindemann F, Schaller G, Funke I, Izbicki JR, Riethmuller G. 1993. Differential expression of proliferation-associated molecules in individual micrometastatic carcinoma cells. *J Natl Cancer Inst* **85:** 1419–1424. doi:10.1093/jnci/85.17.1419

Pedersen RN, Esen BO, Mellemkjaer L, Christiansen P, Ejlertsen B, Lash TL, Nørgaard M, Cronin-Fenton D. 2022. The incidence of breast cancer recurrence 10–32 years after primary diagnosis. *J Natl Cancer Inst* **114:** 391–399. doi:10.1093/jnci/djab202

Pommier A, Anaparthy N, Memos N, Kelley ZL, Gouronnec A, Yan R, Auffray C, Albrengues J, Egeblad M, Iacobuzio-Donahue CA, et al. 2018. Unresolved endoplasmic reticulum stress engenders immune-resistant, latent pancreatic cancer metastases. *Science* **360:** eaao4908. doi:10.1126/science.aao4908

Ren Q, Khoo WH, Corr AP, Phan TG, Croucher PI, Stewart SA. 2022. Gene expression predicts dormant metastatic breast cancer cell phenotype. *Breast Cancer Res* **24:** 10. doi:10.1186/s13058-022-01503-5

Rhim AD, Mirek ET, Aiello NM, Maitra A, Bailey JM, McAllister F, Reichert M, Beatty GL, Rustgi AK, Vonderheide RH, et al. 2012. EMT and dissemination precede pancreatic tumor formation. *Cell* **148:** 349–361. doi:10.1016/j.cell.2011.11.025

Risson E, Nobre AR, Maguer-Satta V, Aguirre-Ghiso JA. 2020. The current paradigm and challenges ahead for the dormancy of disseminated tumor cells. *Nat Cancer* **1:** 672–680. doi:10.1038/s43018-020-0088-5

Rueda OM, Sammut SJ, Seoane JA, Chin SF, Caswell-Jin JL, Callari M, Batra R, Pereira B, Bruna A, Ali HR, et al. 2019. Dynamics of breast-cancer relapse reveal late-recurring ER-positive genomic subgroups. *Nature* **567:** 399–404. doi:10.1038/s41586-019-1007-8

Ruth JR, Pant DK, Pan TC, Seidel HE, Baksh SC, Keister BA, Singh R, Sterner CJ, Bakewell SJ, Moody SE, et al. 2021. Cellular dormancy in minimal residual disease following targeted therapy. *Breast Cancer Res* **23:** 63. doi:10.1186/s13058-021-01416-9

Schardt JA, Meyer M, Hartmann CH, Schubert F, Schmidt-Kittler O, Fuhrmann C, Polzer B, Petronio M, Eils R, Klein CA. 2005. Genomic analysis of single cytokeratin-positive cells from bone marrow reveals early mutational events in breast cancer. *Cancer Cell* **8:** 227–239. doi:10.1016/j.ccr.2005.08.003

Schmidt-Kittler O, Ragg T, Daskalakis A, Granzow M, Ahr A, Blankenstein TJF, Kaufmann M, Diebold J, Arnholdt H, Müller P, et al. 2003. From latent disseminated cells to overt metastasis: genetic analysis of systemic breast cancer progression. *Proc Natl Acad Sci* **100:** 7737–7742. doi:10.1073/pnas.1331931100

Scognamiglio R, Cabezas-Wallscheid N, Thier MC, Altamura S, Reyes A, Prendergast AM, Baumgärtner D, Carnevalli LS, Atzberger A, Haas S, et al. 2016. Myc depletion

induces a pluripotent dormant state mimicking diapause. *Cell* **164:** 668–680. doi:10.1016/j.cell.2015.12.033

Shachaf CM, Felsher DW. 2005. Tumor dormancy and MYC inactivation: pushing cancer to the brink of normalcy. *Cancer Res* **65:** 4471–4474. doi:10.1158/0008-5472.CAN-05-1172

Shaffer SM, Dunagin MC, Torborg SR, Torre EA, Emert B, Krepler C, Beqiri M, Sproesser K, Brafford PA, Xiao M, et al. 2017. Rare cell variability and drug-induced reprogramming as a mode of cancer drug resistance. *Nature* **546:** 431–435. doi:10.1038/nature22794

Sharifi N, Hurt EM, Kawasaki BT, Farrar WL. 2007. TGFBR3 loss and consequences in prostate cancer. *Prostate* **67:** 301–311. doi:10.1002/pros.20526

Sharma SV, Lee DY, Li B, Quinlan MP, Takahashi F, Maheswaran S, McDermott U, Azizian N, Zou L, Fischbach MA, et al. 2010. A chromatin-mediated reversible drug-tolerant state in cancer cell subpopulations. *Cell* **141:** 69–80. doi:10.1016/j.cell.2010.02.027

Shen S, Vagner S, Robert C. 2020. Persistent cancer cells: the deadly survivors. *Cell* **183:** 860–874. doi:10.1016/j.cell.2020.10.027

Sosa MS, Bragado P, Debnath J, Aguirre-Ghiso JA. 2013. Regulation of tumor cell dormancy by tissue microenvironments and autophagy. *Adv Exp Med Biol* **734:** 73–89. doi:10.1007/978-1-4614-1445-2_5

Sosa MS, Bragado P, Aguirre-Ghiso JA. 2014. Mechanisms of disseminated cancer cell dormancy: an awakening field. *Nat Rev Cancer* **14:** 611–622. doi:10.1038/nrc3793

Sosa MS, Parikh F, Maia AG, Estrada Y, Bosch A, Bragado P, Ekpin E, George A, Zheng Y, Lam HM, et al. 2015. NR2F1 controls tumour cell dormancy via SOX9- and RARβ-driven quiescence programmes. *Nat Commun* **6:** 6170. doi:10.1038/ncomms7170

Tallón de Lara P, Castañón H, Vermeer M, Núñez N, Silina K, Sobottka B, Urdinez J, Cecconi V, Yagita H, Movahedian Attar F, et al. 2021. CD39⁺PD-1⁺CD8⁺ T cells mediate metastatic dormancy in breast cancer. *Nat Commun* **12:** 769–769. doi:10.1038/s41467-021-21045-2

Vera-Ramirez L, Vodnala SK, Nini R, Hunter KW, Green JE. 2018. Autophagy promotes the survival of dormant breast cancer cells and metastatic tumour recurrence. *Nat Commun* **9:** 1944–1944. doi:10.1038/s41467-018-04070-6

Walens A, DiMarco AV, Lupo R, Kroger BR, Damrauer JS, Alvarez JV. 2019. CCL5 promotes breast cancer recurrence through macrophage recruitment in residual tumors. *eLife* **8:** e43653. doi:10.7554/eLife.43653

Walens A, Lin J, Damrauer JS, McKinney B, Lupo R, Newcomb R, Fox DB, Mabe NW, Gresham J, Sheng Z, et al. 2020. Adaptation and selection shape clonal evolution of tumors during residual disease and recurrence. *Nat Commun* **11:** 5017. doi:10.1038/s41467-020-18730-z

Weidenfeld K, Schif-Zuck S, Abu-Tayeh H, Kang K, Kessler O, Weissmann M, Neufeld G, Barkan D. 2016. Dormant tumor cells expressing LOXL2 acquire a stem-like phenotype mediating their transition to proliferative growth. *Oncotarget* **7:** 71362–71377. doi:10.18632/oncotarget.12109

Wiecek AJ, Jacobson DH, Lason W, Secrier M. 2021. Pan-cancer survey of tumor mass dormancy and underlying mutational processes. *Front Cell Dev Biol* **9:** 698659. doi:10.3389/fcell.2021.698659

Wiedswang G, Borgen E, Kåresen R, Kvalheim G, Nesland JM, Qvist H, Schlichting E, Sauer T, Janbu J, Harbitz T, et al. 2003. Detection of isolated tumor cells in bone marrow is an independent prognostic factor in breast cancer. *J Clin Oncol* **21:** 3469–3478. doi:10.1200/JCO.2003.02.009

Willis RA. 1934. *The spread of tumours in the human body.* J. & A. Chuchill, London.

Woelfle U, Breit E, Zafrakas K, Otte M, Schubert F, Müller V, Izbicki JR, Löning T, Pantel K. 2005. Bi-specific immu-nomagnetic enrichment of micrometastatic tumour cell clusters from bone marrow of cancer patients. *J Immunol Methods* **300:** 136–145. doi:10.1016/j.jim.2005.03.006

Zangrossi M, Romani P, Chakravarty P, Ratcliffe CDH, Hooper S, Dori M, Forcato M, Bicciato S, Dupont S, Sahai E, et al. 2021. Ephb6 regulates TFEB-Lysosomal pathway and survival of disseminated indolent breast cancer cells. *Cancers (Basel)* **13:** 1079. doi:10.3390/cancers 13051079

Organoid Cultures for the Study of Mammary Biology and Breast Cancer: The Promise and Challenges

Senthil K. Muthuswamy[1] and Joan S. Brugge[2,3]

[1]Laboratory of Cancer Biology and Genetics, Center for Cancer Research, National Cancer Institute, National Institutes of Health (NIH), Bethesda, Maryland 20894, USA

[2]Department of Cell Biology, Harvard Medical School, Boston, Massachusetts 02115, USA

[3]Ludwig Center at Harvard, Harvard Medical School Boston, Boston, Massachusetts 02115, USA

Correspondence: senthil.muthuswamy@nih.gov; joan_brugge@hms.harvard.edu

During the last decade, biomedical research has experienced a resurgence in the use of three-dimensional culture models for studies of normal and cancer biology. This resurgence has been driven by the development of models in which primary cells are grown in tissue-mimicking media and extracellular matrices to create organoid or organotypic cultures that more faithfully replicate the complex architecture and physiology of normal tissues and tumors. In addition, patient-derived tumor organoids preserve the three-dimensional organization and characteristics of the patient tumors ex vivo, becoming excellent preclinical models to supplement studies of tumor xenografts transplanted into immunocompromised mice. In this perspective, we provide an overview of how organoids are being used to investigate normal mammary biology and as preclinical models of breast cancer and discuss improvements that would enhance their utility and relevance to the field.

Organoid cultures offer an enormously powerful technology to investigate many different aspects of the development, physiology, pathology, and vulnerabilities of cells from a wide variety of tissues and tumors (Clevers 2016; Fatehullah et al. 2016; Simian and Bissell 2017; Tuveson and Clevers 2019; Corsini and Knoblich 2022). In general, organoid cultures are useful because they recapitulate many features of normal and pathological tissue composition and architecture, thus providing models that better resemble normal tissue states. Since 1957, investigators have cultured epithelial cells from mammary tissues to understand critical aspects of mammary development and function and to understand the properties of mammary cancer cells. More recently, variations in this methodology are being used to extend the lifetime of cultures, maintain the identity of epithelial subtypes, and replicate more complex ductal structures for understanding normal development and disease modeling. In particular, organoid cultures are proving to be an effective platform to investigate how interactions between different cell types within the normal human breast regulate normal biology and disease initiation.

Three-dimensional (3D)/organoid culture methods are also beginning to significantly impact breast cancer research and drug development efforts. In particular, the promise for the use of organoid cultures rests in their potential to serve as better preclinical models and reduce the cost associated with the failure of drugs in clinical trials. It is estimated that we spend about 2023 US $4.5 billion in research and development for every drug approved by the U.S. Food and Drug Administration (FDA) (DiMasi et al. 2016). The cost-driving factor is thought to be the >80% failure rate as drugs enter human trials, pointing in large part to the unreliable nature of preclinical models used to test the compounds (DiMasi et al. 2016).

Established human cell lines and patient-derived xenograft (PDX) models are the workhorse platforms for preclinical studies for drugs targeted at cancer patients. These models are effective in monitoring changes in cell proliferation, death, and, to some extent, tumor diversity; however, attributes related to tumor cell differentiation status and tissue structure disruption, among others, cannot be modeled in conventional cell-culture-based preclinical models. Tumor organoids tend to retain the overall tissue organization of the patient's tumors; hence, they can be used for discovering pathways and targets that impact changes in cellular organization and restore the differentiation status of tumors to a nonaggressive state (Zhao et al. 2021). Furthermore, most preclinical studies are designed to monitor the magnitude of response to a drug in a small set of preclinical models. The greater the magnitude of response in the majority of the models, the more likely the drug gets advanced in the development process. A recent meta-analysis of drug response in clinical trials has identified an unexpected role for patient-to-patient variation and individual drug response in a population as a significant driver of the clinical efficacy of drug action (Plana et al. 2022). For example, the clinical outcome of a drug combination is best revealed by variations in the patient-specific clinical response to each component of the drug combination, suggesting the efficacy of drug action in the patient cohort needs to be modeled as a

population response and not just as the magnitude of response in a few representative preclinical models. Although large cohorts of the traditional PDX models can address this need, it is time-consuming and costly, which will further increase the cost of drug development. Generation and use of patient tumor-derived organoid (PDO) models for translational research is both time- and cost-effective compared to PDX models, and we can now develop large banks of PDOs to represent population diversity for preclinical studies. In our opinion, the generation and use of large cohorts of PDO models are well within reach of every researcher and research institution. However, several factors contribute to the inertia for the implementation of such efforts. In this article, we will also discuss such challenges and offer our perspectives on their use in breast cancer research.

Many reviews have highlighted (1) the historical development of 3D mammary cultures, (2) the variety of conditions that have been developed for organoid cultures for normal and tumor (Roelofs et al. 2019; Srivastava et al. 2020; Wood and Ewald 2021; Lewis et al. 2022), (3) important findings from the studies (Vidi et al. 2013; Simian and Bissell 2017; Srivastava et al. 2020; Sumbal et al. 2020a), and (4) similar investigations of mammary cancer cell organoids (Fröhlich 2023; Ortiz and Andrechek 2023). We refer the readers to these review articles to get a detailed perspective on technical advances and discoveries made using organoid models and summaries outlining potential applications. In this perspective, we will provide an overview of the opportunities afforded by the different organoid culture conditions, examples of the nature of findings that can be derived from organoid cultures, improvements in the cultures that will benefit the field overall, and challenges going forward.

ORGANOIDS DERIVED FROM PRIMARY MAMMARY TISSUES

The term "organoid" has been used in many different contexts. For the purposes of this section, we predominantly discuss organoids generated from epithelial cell clusters or single cells dissociated from primary normal mammary mouse or human breast tissue. Generally, these structures retain all of the epithelial subtypes and the bilayer organization of ducts and lobules; however, over time in culture these properties are not always retained. We distinguish organoids from mammospheres, which are clonal structures formed in nonadherent conditions from stem-like normal or cancer cells. At the end of the section, we will discuss mixed "organotypic" cultures of epithelial cells with fibroblasts, immune cells, endothelial cells, or adipocytes.

Use of organoids derived from human breast tissue and mouse mammary glands makes it possible to examine mechanisms involved in the differentiation and function of these tissues as well as the influence of exogenous factors and mammary epithelial cell (MEC) intrinsic genes or pathways. Although some of these mechanisms can be explored in mice in vivo, MEC organoids can generate large quantities of cells, and genetic

manipulations, imaging, and molecular profiling are generally easier. In both human and mouse mammary organoids, it is also easier to control for the many variables associated with cells within tissues that are difficult to account for and control in vivo. Compared to monolayer cultures, organoid culture conditions allow cells to organize into structures that more faithfully replicate their structures in glands, and the cells are not subjected to the unnatural extreme stiffness of plastic tissue culture dishes. Furthermore, all three lineages of epithelial cells can be maintained in organoid cultures (basal/myoepithelial [BA] cells as well as the two major subtypes of luminal cells that are referred to as luminal progenitors [LP], alveolar [AV] cells, or secretory cells and mature luminal[(ML] or hormone-sensing [HS] cells) (Fig. 1A). Importantly, under some conditions, MECs fail to undergo replica-

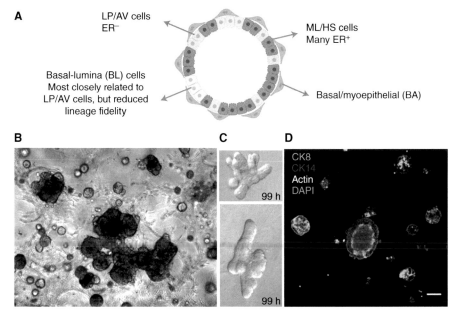

Figure 1. (*A*) Diagram of mammary epithelial cell types. (*B*) Spheroid and terminal duct lobular unit (TDLU)-like structures formed in reconstituted basement membrane (rBM) using Sachs/Clevers conditions. (Figure adapted from Rosenbluth et al. 2020 and reprinted under the Creative Commons Attribution 4.0 International License.) (*C*) Mouse organoids formed in COL I: rBM ratio of 3:7 (*upper*) versus 7:3 (*lower*). (Figure adapted from Nguyen-Ngoc and Ewald 2013 with permission from the authors, *Journal of Microscopy* © 2013; Royal Microscopical Society © 2013.) (*D*) Confocal image of human breast epithelial organoids stained as indicated for CK8 (green), CK14 (red), and actin (white), and counterstained with DAPI (blue). Scale bar, 100 µm. (LP) Luminal progenitor, (AV) alveolar, (ML) mature luminal, (HS) hormone-sensing. (Figure adapted from Rosenbluth et al. 2020 and reprinted under the Creative Commons Attribution 4.0 International License.)

tive senescence and thus can provide longer-term reproducible models.

Dissociation Conditions

Several excellent methods articles have been published recently describing optimized conditions for dissociation of human and rodent mammary glands for use in organoid cultures (mouse: Linnemann et al. 2017; Sheridan and Visvader 2019; human: Stingl et al. 2005; Shehata and Stingl 2017). From our experience and anecdotal discussions with others, the extent of mincing prior to the enzymatic digestion of the tissue as well as the enzymes utilized for dissociation and the timing of dissociation can affect the specific cell types and their viability; however, a careful comparison of the different methodologies has not been carried out. One report that evaluated the effects of different dissociation conditions using single-cell RNA sequencing (scRNA-seq) as a readout provided useful insights relating to the effects of different conditions on recovery of different cell types and effects on transcriptomes of the cells (Engelbrecht et al. 2021).

Conditions for Culturing Organoids

Many different conditions for culturing mouse and human epithelial mammary organoids have been reported and described in recent methods articles (human: Linnemann et al. 2017; Dekkers et al. 2021; mouse: Mroue and Bissell 2013; Jardé et al. 2016; Jamieson et al. 2017; Caruso et al. 2022). These methodologies vary not only in the use of different growth factors, small molecule modulators, and base media, but also the extracellular matrix (ECM) components or gels that the cells are embedded within. The two major ECM components that have been utilized are collagen I (COL I) and reconstituted basement membrane (rBM) (also referred to by commercial product names BME or Matrigel). rBM is a mixture of matrix proteins (laminin, collagen IV, heparan sulfate proteoglycans, and other proteins) secreted from the EHS sarcoma cell line (Swarm 1963; Orkin et al. 1977). COL I has also been mixed with rBM or with other ECM proteins in synthetic polymeric hydrogels. Hydrogels offer more uniform platforms than rBM and the option to include specific ECM proteins that mimic specific microenvironments (e.g., from virgin, parous, involuting glands) and avoid the batch-to-batch variability of the gel and growth factor components associated with Matrigel. ECM-derived from decellularized tissue has also been incorporated into hydrogels (Mollica et al. 2019).

Table 1 compares the properties of organoids cultured under these different conditions. Because these formulations were developed in different laboratories, usually with different media components, and were not directly compared in a rigorous fashion, the table is designed to describe reported features of the structures formed under the different conditions and not as a direct comparison (Table 1). Spherical, acinus-like structures or a mix of these with budding or duct-like structures were formed under most conditions; however, complex branched TDLU-like structures were generally generated much more efficiently when collagen was included in the matrix, and it is optimal to include collagen for studies of invasive behavior (see Fig. 1 for a comparison of different types of organoid structures formed under the different conditions).

Organoid culture conditions also vary significantly in the media components that are added to base media. We show some of the most commonly used components with notes regarding the rationale for inclusion and comments on the specific effects in organoid cultures (Table 2). Decisions regarding which media components to use depend to some extent on the intended use of the organoids (e.g., to study morphogenesis, branching, or lactation) and whether one or more subtypes of epithelial cells are required. There has only been one systematic analysis of the effects of elimination of individual components of one of the organoid culture media (Rosenbluth et al. 2020). Removal of EGF and B27 affected expansion of organoids overall. However, elimination of other components affected the proportion of different cell types. For example, removal of EGF caused an increase in HS/ML cells and a decrease in basal cells, and removal of either heregulin β1, p38 MAPK inhibitor, or

Cite this article as *Cold Spring Harb Perspect Med* doi: 10.1101/cshperspect.a041661

Table 1. Examples of different conditions for organoid culture

Species	Matrix	Growth factors	Notes and references
Human	Floating COL I	MECGM	Spherical and three types of branched structure (TDLU-like, thin, and star), branched structures lost after two passages (Linnemann et al. 2017)
			Produce casein after prolactin treatment
			Form-invasive polarized structures with purified LP cells, dependent on reduced ROCK activity (Linnemann et al. 2015)
	rBM domes	FGF7, 10 NRG1/HRG EGF RSPO1 Noggin	Mostly spherical structures, bilayered initially
			Can be passaged many times, proliferation ceases by P20 (Sachs et al. 2018; Dekkers et al. 2020)
			Maintain lineage fidelity based on CyTOF analysis (Rosenbluth et al. 2020)
	COL I + rBM	EGF FGF	BME:COL I (7:3 ratio); timed addition of FGF2 and EGF, cycles of branching and elongation, generates high-level branched 3D mammary organoids (Caruso et al. 2022)
	COL I, LN, FN in hydrogel	EGF Insulin hydrocortisone	Mixed acinar and TDLU-like, 60% bilayered
			Basal cell only cultures formed small spheroids
			Poor efficiency with single cells compared to clusters (0.16% vs. 33%), only 4.5% are TDLU-like
			Produce casein with prolactin (Sokol et al. 2016; Miller et al. 2017)
Mouse	rBM	NRG1 RSPO1 Noggin	Bilayered structures; optimal growth dependent on the addition of Nrg1 and low Rspo1
			Responsive to estradiol and progesterone
			Able to reconstitute mammary glands after 30 d
			Maintained for four months (Jardé et al. 2016)
		WNT3 RSPO2 FGF2 Heparin hydrocortisone	Purified basal cells or mixed epithelial cells form bilayered budding structures, produce casein with prolactin
			Proliferation enhanced by progesterone and β-estradiol
			Purified luminal cells form single layer structures, without basal cells
			Able to reconstitute ducts in vivo (Jamieson et al. 2017)
	rBM	EGF, Wnt3A, RSPO	Spherical structures with squamous centers with estradiol and progesterone (Zhang et al. 2017)
	rBM + COL I	FGF2	Provided evidence that mixed gels provide a more physiologically accurate model of mammary branching morphogenesis
			Strong correlation between the persistence of elongation and the fraction of COL I in the gel
			Collagen fibrils mimicked the patterned, axillary oriented interstitial fibrillar COL I and laminin-collagen IV basement membrane
			Myoepithelial coverage of buds depended on ratio of collagen: rBM (Nguyen Ngoc and Ewald 2013)

(MECGM) Commercial MEC Growth Medium (PromoCell, promocell.com/product/mammary-epithelial-cell-growth-medium), (LN) laminin, (FN) fibronectin, (RSPO) R-spondin, (NRG) neuregulin, (FGF) fibroblast growth factor, (EGF) epidermal growth factor, (TDLU) terminal duct lobular unit, (LP) luminal progenitor, (ROCK) Rho-associated protein kinase, (cyTOF) cytometry by time of flight, (rBM) reconstituted basement membrane.

Table 2. Components included in organoid culture media

Media component	Rationale and references
Fibroblast growth factors: FGF7 and FGF10	Produced by breast stromal fibroblasts (Palmieri et al. 2003)
	Induces proliferation and morphogenesis of both epithelial and myoepithelial cells (Sachs et al. 2018)
	Promotes alveologenesis and branching
	Removal decreased ML/HS cells, increased LP/AV cells (Rosenbluth et al. 2020)
FGF2	Promotes morphogenesis and invasion (Nguyen-Ngoc and Ewald 2013; Jamieson et al. 2017)
EGF	Critical in intestinal organoids, stimulates proliferation (Mizutani and Clevers 2020)
	Removal of EGF led to an increase in mature luminal cells but decrease in basal cells (Rosenbluth et al. 2020)
Neuregulin 1 (NRG1/ HRG)	Ligand of ERBB3/4
	Regulates proliferation and lobuloalveolar development in vivo (Yang et al. 1995; Li et al. 2002; Wansbury et al. 2008)
	Stimulated lobulo-alveolar budding and the production of milk proteins in vivo (Förster et al. 2014)
	Shown to mediate extended growth of for mouse organoids; basal to luminal signaling (Jardé et al. 2016)
RSPO1	Potentiates Wnt signaling
	Functions as a ligand for Lgr5, which forms a complex with Wnt coreceptors
	Critical in intestinal organoids (Mizutani and Clevers 2020)
	Promotes basal cell differentiation at high levels (Jardé et al. 2016)
	Removal led to decrease in LP/AV cells (Rosenbluth et al. 2020)
Noggin	Critical in intestinal organoids (Mizutani and Clevers 2020)
	Removal led to variable decrease in LP/AV cells (Rosenbluth et al. 2020)
Wnt3a	Promotes stem cell renewal and differentiation; increased proliferation of organoids observed (Jamieson et al. 2017)
B27 supplement	Mixture of vitamins, antioxidants, proteins, and other components (Mizutani and Clevers 2020)
	(Composition, www.thermofisher.com/us/en/home/technical-resources/media-formulation.250.html)
	Removal causes a decrease in the confluency and size of the organoids in all four cases (Rosenbluth et al. 2020)
Hydrocortisone	Reported to enhance maintenance of alveolar as well as multi-lobular branching morphogenesis and casein levels (Darcy et al. 1995)
Y-27632 ROCK inhibitor	Prevents anoikis during dissociation (Zhang et al. 2011)
	In early cultures, increases regenerative capacity, colony formation and thin branch formation (Linnemann et al. 2015; Zubeldia-Plazaola et al. 2015)
	Distinct effects of continuous treatment (Linnemann et al. 2015; Zubeldia-Plazaola et al. 2015; Jamieson et al. 2017; Ganz et al. 2021)
TGF-β inhibitor	To prevent suppression of proliferation of ER$^+$ cells
	Removal promotes accumulation of BL cells (Rosenbluth et al. 2020)
Forskolin	Agonist of adenyl cyclase, raises cAMP (Fradkin et al. 1982)
	Promote polarization and lumen formation (Nedvetsky et al. 2012)
	Critical in one report for generation of TDLU-like organoids (Linnemann et al. 2017)
Bovine pituitary extract (BPE)	Contain factors important for mammary development, including growth hormone, fibroblast growth factors, and follicle-stimulating hormone (Hammond et al. 1984)

(LN) Laminin, (RSPO) R-spondin, (FGF) fibroblast growth factor, (EGF) epidermal growth factor, (TDLU) terminal duct lobular unit, (LP) luminal progenitor, (AV) alveolar, (ML) mature luminal, (HS) hormone-sensing, (BL) B lymphocyte, (cAMP) cyclic adenosine monophosphate.

FGF7 and FGF10 resulted in a relative decrease in HS/ML cells with a concomitant increase in one of the other mammary lineages.

Another factor that contributes to differences in organoid morphology and physiology is the mechanical property of the embedding matrix. The concentration of collagen and anchorage of the collagen gel to the tissue culture dish affects the elastic modulus of the gels, and this in turn influences the size and shape of acinar structures as well as branching and casein production. For example, increasing collagen gel compliance by switching gels from an adherent, rigid state to free flotation suffices to promote alveologenesis and enhance casein stability and secretion, highlighting the importance of physical parameters in differentiation of MECs (Emerman et al. 1977; Emerman et al. 1979; Lee et al. 1984, 1985). In addition, the extent of fibril assembly, which was found to affect the frequency of subcellular protrusions into the ECM and the extent of epithelial growth, can be manipulated by varying the time at 4°C after pH neutralization of collagen (Nguyen-Ngoc and Ewald 2013). In turn, the epithelial cells can modify the structure of fibrillar collagen (Nguyen-Ngoc and Ewald 2013; Ganz et al. 2021). The collective back-and-forth motion of cells within the branches generates tension that is sufficiently strong to promote reorganization of the outlying collagen network resulting in the generation of mechanically stable collagen cages (Buchmann et al. 2021). This "matrix encasing" then creates enhanced tension, branch outgrowth, and deformation of the matrix.

The culture conditions originally developed by the Clevers laboratory involve embedding intestinal epithelial cells in domes of high rBM concentration (Mizutani and Clevers 2020). Under these conditions, breast epithelial structures have been reported to lose their myoepithelial/basal layer after a couple of passages (Rosenbluth et al. 2020). Although the basis for the loss of myoepithelial cells is not known, use of lower concentrations of rBM should be evaluated to examine whether this allows maintenance of basal/myoepithelial cells, which may be influenced by hyperactivation of integrin signaling in the dense matrix. Indeed, mouse MECs cultured in 2% rBM formed organoids that resembled those in domes (Sahu et al. 2022), suggesting that the high rBM concentration in domes is not necessary.

Maintenance of MEC Lineage Fidelity

An important question that has not been well addressed is how each of the different organoid culture conditions allows MECs to retain their lineage fidelity, and how each component of the medium contributes to the lineage fidelity, survival, and proliferative capacity of each epithelial cell type. With respect to growth factors, the growth and size of mouse mammary organoids over a time course of 10 days were significantly impaired when cultured with EGF, FGF2, or FGF10 alone as compared to with a cocktail of the growth factors in base media (Sahu et al. 2022). In addition, organoids cultured in single growth factors had reduced myoepithelial cells; however, there was no reduction in estrogen receptor-positive (ER$^+$) and progesterone receptor-positive (PR$^+$) cells except when cultured in EGF alone, where these cells were consistently reduced.

In another study of human primary MECs, cytometry by time of flight (CyTOF) analysis of 39 markers was used to define the lineage fidelity of individual cells cultured in the organoid medium developed by Sachs/Clevers medium (Fig. 2; Rosenbluth et al. 2020). The majority of the lineage markers were faithfully expressed in each epithelial cell type (note MUC1, GATA3, CD24, BRCA1 in the ML/HS cells; SMA, Laminin5, CD10, CD90, endothelial protein C receptor [EPCR] in basal cells; and CD133, alanyl aminopeptidase (ANPEP), medium levels of EPCAM and CD49f in the LP/AV cells). In addition, ~40% of the LP/AV cells also expressed the basal keratin CK14. Luminal cells that express basal keratins have been described before in association with embryonic mouse mammary glands (Ma et al. 2022) and were found to be enriched with age and in *BRCA1/BRCA2* mutation carriers (Hinz et al. 2021; Shalabi et al. 2021; Gray et al. 2022). scRNA-seq revealed that these "basal-luminal" (BL) ER$^-$ luminal cells have reduced lineage fidelity based on their expression of genes that are specifically enriched in BA, ML/HS MECs, or

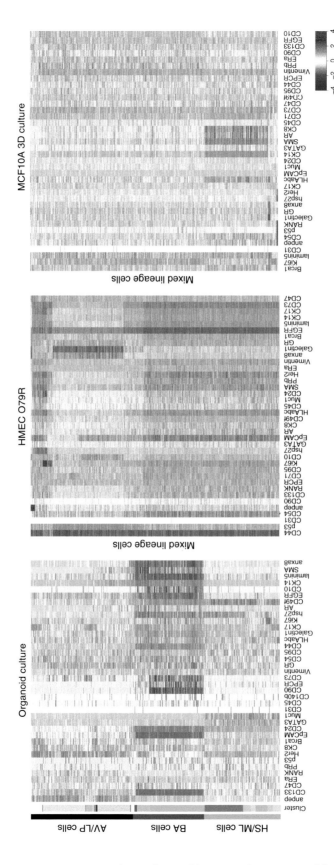

Figure 2. Heatmaps from cytometry by time of flight (CyTOF) analysis of a primary human breast organoid culture (*left*), primary breast epithelial cells cultured in human mammary epithelial cell (HMEC) medium in tissue culture plates (*center*), and the MCF10A cell line (*right*). (HS) Hormone-sending, (ML) mature luminal, (BA) basal/myoepithelial, (AV) alveolar, (LP) luminal progenitor. (Figure adapted from Rosenbluth et al. 2020 and reprinted under the Creative Commons Attribution 4.0 International License.)

expressed in non-MECs (Gray et al. 2022). It is currently unknown what mechanisms lead to BL cell accumulation or what epigenetic factors control their expression; however, cells expressing a BL signature are enriched in basal breast tumors (Gray et al. 2022) and treatment of organoids with TGF-β or elimination of TGF-β inhibitor from organoid cultures enriched for these cells (Rosenbluth et al. 2020). Overall, the high-fidelity lineage marker expression in these organoids contrasts dramatically with that observed for cells cultured in the standard HMEC medium in 2D (Fig. 2, center), which loses lineage fidelity. scRNA-seq would provide a much better assessment of the lineage fidelity of the organoids; it is also important to assess this under the different conditions used to culture organoids.

Single Epithelial Subtype Organoids

Another important question is whether any of the reported conditions will support continued passage of the three MEC populations cultured alone. There are many factors that have been implicated in cross talk between different MEC subtypes; however, the extent to which they are required for preservation of an individual lineage is not known. In mice, lineage tracing experiments in vivo have demonstrated that after birth, each lineage is maintained independently (Van Keymeulen et al. 2011, 2017). However, the hierarchical relationship of the MECs in adult human glands has not been established. Although adult mouse basal cells are unipotent in vivo, dissociated basal cells can reconstitute all mammary epithelial lineages after transplantation (Shackleton et al. 2006, Stingl et al. 2006), suggesting loss of lineage restriction following dissociation. Recently, it has been shown that elimination of the luminal cell layer in mouse mammary glands can induce expression of luminal genes in basal cells, and mixing experiments have provided evidence that luminal cells restrict the lineage plasticity of basal cells (Centonze et al. 2020). This finding is consistent with the ability of isolated single basal epithelial cells to form organoids containing all three epithelial lineages under some conditions (Eirew et al. 2008; Lim et al. 2009; Jamieson et al. 2017; Linnemann et al. 2017). In

contrast, two publications provided evidence that single basal cells were unable to form colonies (Jardé et al. 2016; Sokol et al. 2016). The basis for this discrepancy is not known. However, given the extent of differentiation drift in organoids, further studies are required to identify factors that allow maintenance of basal or luminal ER⁻ cells as pure monocultures. In mouse mammary cultures, high-level stimulation of the WNT pathway has been shown to promote basal cell differentiation leading to predominantly basal cultures (Jardé et al. 2016). Fridriksdottir et al. (2017) have reported that culture of human myoepithelial cells with interlobular fibroblasts prevents their differentiation and retains their multipotency in vitro and in vivo.

Culture of ER⁺ cells

Most of our understanding about the function and mechanisms of regulation of the estrogen receptor has been derived from ER⁺ breast tumor cell lines like MCF7 and T47D cells. In monolayer cultures, ER⁺ cells are not maintained and EpCAM⁺CD49f⁻ (ER⁺) primary mouse or breast epithelial cells are unable to form colonies in standard colony formation assays used to study lineage relationships. In addition, EPCAM⁺CD49⁻ luminal cells were unable to form structures in the collagen gels. This has severely limited studies of estrogen receptor function and regulation in normal cells, as well as the initiation of cancer in ER⁺ cells. A breakthrough in this area is the ability of isolated single ER⁺ cells to proliferate and form organoids under several conditions (Jardé et al. 2016; Jamieson et al. 2017; Rosenbluth et al. 2020). TGF-β inhibitor is critical for ER expression. In addition, fibroblast feeder cells have been shown to increase expression of ER and activity in 2D cultures (Haslam 1986; Fridriksdottir et al. 2015). Although these studies reported increased proliferation induced by estradiol treatment, inferring responsiveness, further studies are required to establish culture conditions for maintaining expression of estrogen receptor expression and hormone responsiveness over multiple passages. These conditions offer the opportunity to study events involved in ER⁺ breast tumor initiation.

Variability in Organoid Cultures

In multiple reports that included analysis of organoids cultured from many different donors, it was noted that there is significant heterogeneity in the variety of different structures as well as the proportion of different epithelial cell types (Linnemann et al. 2015; Rosenbluth et al. 2020). This reflects the expected variation in donor physiology as well as the extent of variation in the proportions of epithelial cell types that have been reported using phenotypic assays or messenger RNA (mRNA)-based approaches (Keller et al. 2010; Nakshatri et al. 2015; Henry et al. 2021; Pal et al. 2021; Gray et al. 2022). In one study, the proportion of different cell types was compared in primary tissue and organoids derived from five patients (Rosenbluth et al. 2020). Although the lineage distribution was similar in three cases, it was strikingly dissimilar in two others. The authors suggested that the differences in lineage distribution may be due to sampling artifact (e.g., due to the small amount of tissue that is used for the organoid culture samples) or differences in clinical variables such as age, parity, and inherited mutation status that could affect the fitness or proliferation capacity of the different populations; however, it is likely that variations in mammary density and the extent of tissue mincing and enzymatic digestions between investigators and tissue replicates could also contribute. Regardless of the basis for variability overall, this is a very important factor in considering use of organoids from different donor populations (e.g., those from donors that vary in genotype, age, parity, body mass index [BMI]) because the extent of variability dictates that organoids from very large numbers of donors will be required to derive statistically significant conclusions. Organoids derived from mouse tissues are much more homogeneous in morphology. Because mice are inbred and their diet and environment can be controlled, there are less variables than in human tissues.

INSIGHTS INTO MAMMARY BIOLOGY USING ORGANOID CULTURES

Mechanisms of Branching Morphogenesis

Organoids are especially well suited for studies of branching morphogenesis. This process can occur independent of stroma in cultured organoids and the use of green fluorescent protein (GFP) reporter and knockout mice, as well as genetic manipulation of epithelial cells, has facilitated mechanistic studies. Many of the mechanistic studies of tubulogenesis and branching morphogenesis have been carried out in mouse organoid cultures generated from dissociated clusters of epithelial cells incubated in either rBM or rBM and collagen (see Tables 1 and 3). Studies by Ewald and colleagues have shown that elongation of tubes in response to FGF treatment occurs without ECM-directed protrusions at the leading edge, being driven by collective migration of stratified epithelium that lack membrane polarity (Ewald et al. 2008; Nguyen-Ngoc and Ewald 2013). Cells within the stratified epithelial terminal end bud-like structures are able to migrate and rearrange during branching of the primary organoids. In addition, they were able to show that activation of the ERK pathway was enriched in the fast-migrating cells at the front of the elongating ducts, whereas the activated PI3K-AKT pathway was detected ubiquitously (Huebner et al. 2016). Inhibition of the ERK pathway blocked both cell migration and duct elongation, and inducible activation of MEK-ERK activity was sufficient to promote collective cell migration and ductal elongation. Matrix metalloproteinase (MMP) activity was also shown to be necessary for branching and is sufficient to drive branching in the absence of exogenous growth factors (Simian et al. 2001). Other genes that regulate branching have been characterized in mammary cultures through targeted approaches or screens (see Table 3).

The Scheel laboratory has investigated the behavior of purified human LP (ER⁻ luminal cells) in floating hydrogels containing COL I and found that the cells are able to form branched structures that collectively invade and realign collagen matrix in a fashion similar to invasive carcinoma (Ganz et al. 2021). This process was guided by leader cells that required reduction of the Rho/Rock/myosin II pathway (using the Rock inhibitor Y27632) for polarization, lumen formation, branching, and invasion. Because ROCK inhibitor is typically removed from organoid cultures after they are established, this and the absence of myoepithelial cells could at least in part explain the previous absence of such

Table 3. Examples of experimental findings from use of organoid cultures

Species	General morphogenesis	References
Murine	Effects of loss of E-cadherin on structure formation; Deletion of E-cadherin induced loss of simple epithelial organization	Ganz et al. 2021
Murine	Loss of PRC2 function in mammary organoids reduces proliferation and fitness of mammary organoids, affects chromatin compaction, and induces an expression signature similar to that of claudin-low tumor	Michalak et al. 2018
	Branching morphogenesis	
Murine	Interplay of TF MRTF-A and matrix stiffness controls acinar structure and protrusion	Melcher et al. 2022
Murine	The receptor protein tyrosine phosphatase PTPRB negatively regulates FGF2-dependent branching morphogenesis	Soady et al. 2017
Murine	DC-SCRIPT deficiency delays mouse mammary gland development and branching morphogenesis	Tang et al. 2019
Murine	Paxillin-dependent regulation of apical–basal polarity in mammary gland morphogenesis	Xu et al. 2019
Murine	Roles for Bcl-3 in mammary gland branching, stromal collagen invasion, involution and tumor pathology	Carr et al. 2022
Murine	Coordination of receptor tyrosine kinase signaling and interfacial tension dynamics drives radial intercalation and tube elongation	Neumann et al. 2018
Murine	Inhibition of the tyrosine phosphatase, PTPRB enhances branching through hyperactivation of FGFR	Soady et al. 2017
Murine	Chemical genetic screen reveals a role for desmosomal adhesion and AHR in mammary branching morphogenesis	Basham et al. 2013
Human	Surface-tension induced budding drives alveologenesis in human mammary gland organoids	Fernández et al. 2021
	Tumor initiation	
Human	Analysis of effects of KO of four tumor suppressors (P53, PTEN, RB1, NF1) on proliferation in organoids and tumor formation in mice	Dekkers et al. 2020
Mouse	Organoids used to validate genes identified in in vivo CRISPR-Cas9 screen. Focused on Axin1/Trp53 and Prkar1a/Trp53 double mutants	Heitink et al. 2022
Mouse	Effect of MYC expression in postpregnancy MECs to study how pregnancy can modify transcriptional output of MYC	Feigman et al. 2020
Mouse	TRPS1 acts as a context-dependent regulator of mammary epithelial cell growth/differentiation and breast cancer development	Cornelissen et al. 2020

(PRC2) Polycomb repressive complex 2, (TF) transcription factor, (MRTF-A) myocardin-related transcription factor A, (PTPRB) protein tyrosine phosphatase receptor type B, (FGF2) fibroblast growth factor 2, (Bcl3) B-cell leukemia 3, (FGFR) GFG receptor, (AHR) aryl hydrocarbon receptor, (KO) knockout, (TRPS1) transcriptional repressor GATA binding 1.

branched, invasive structures guided by leader cells.

Lactogenic Differentiation

Some of the earliest studies with floating collagen gels demonstrated that mammary cells stimulated with prolactin, which triggers alveoli maturation and lactogenic differentiation in vivo, can express the milk protein β-casein (Emerman et al. 1977, 1979; Lee et al. 1984, 1985). This process was found to be much more efficient in rBM, where 90% of the cells, compared to 30%, expressed β-casein (Li et al. 1987). Casein production has also been modeled in MECs grown on cell culture inserts where they feature lactation-specific tight junctions and membrane protein apical and basal associations (Kobayashi 2023). Recently, a more detailed analysis of lactogenic differentiation was carried out in mouse organoids cultured within rBM domes in Sachs/Clevers-like medium (Sumbal et al. 2020b). Prolactin and hydrocortisone stimulated casein (*Csn2*) expression and apical lipid droplets in cells. Addition of FGF2 to organoid cultures enhanced branching as well as casein and whey acidic protein (WAP) gene expression after prolactin stimulation suggesting that expansion of branched structures can enhance the lactogenic ability of mammary epithelium. Addition of oxytocin induced contraction of the organoids with a frequency similar to that in the mammary gland. Withdrawal of prolactin and hydrocortisone caused involution of the complex branched structures into small spheroids, expression of matrix-remodeling matrix metalloproteases Mmp2 and Mmp13, and reduced viability; however, the residual organoids were able to grow and undergo branching morphogenesis when replated, indicating reversibility of the regression process. In another report, the Bissel laboratory demonstrated that the expression of myoepithelial specific gap junction protein Cx43 is required for contraction, which is critical for lactation (Mroue et al. 2015). These studies highlight the feasibility of using organoid cultures to investigate mechanisms associated with lactogenic differentiation and involution as well as factors that affect milk production and myoepithelial contractability.

Tumorigenesis Studies

Organoids provide a useful context to investigate genetic and epigenetic alterations and tumor microenvironment (TME) factors that can influence tumorigenesis. Given the importance of the TME and epithelial cell context in suppressing or promoting tumorigenesis, organoids are being used extensively to study events involved in tumor initiation and progression. Table 3 includes examples of studies of tumor initiation using mammary organoids. In one study using human organoids engineered to express CAS9, lentiviral vectors targeting two tumor suppressors commonly altered in breast cancer (*PTEN* and *RB1*) were transduced into sorted primary basal and LP MECs to knock out these genes (Dekkers et al. 2020). Nutlin, which stabilizes p53 and kills cells expressing wild-type p53, was employed to select for cells lacking p53. *P53/PTEN/RB1*-mutated organoids displayed a higher proliferation rate relative to

controls; however, both mutant and control organoids ceased proliferating after 20 passages. One of the six triply mutated organoid lines formed $ER^+PR^+HER2^-$ luminal breast tumors when transplanted into mice. Deletion of a fourth tumor suppressor significantly increased the frequency of tumorigenesis. This study demonstrates the utility of CRISPR-Cas9 gene editing of human organoids for identifying drivers of tumorigenesis. It is of interest that the tumors generated in mice were ER^+ tumors given that basal and ER^- luminal cells were targeted. This suggests that loss of this combination of tumor suppressors may more specifically target an ER^+ cell that differentiated within the organoid cultures.

In another study, organoids were employed to obtain clues regarding the functional activity of the GATA-type zinc finger transcription factor TRPS1, which is mutated or amplified in breast cancer and has displayed both oncogenic and tumor suppressive activities. Genetic alteration of TRPS in different contexts in organoids revealed that TRPS is required for normal mammary organoid propagation; however, under conditions in which E-cadherin is also targeted, its loss leads to persistent proliferation of mammary organoids and accelerated mammary tumor formation in mice (Cornelissen et al. 2020). Thus, organoids facilitated an understanding of at least one aspect of the context-dependent effects of this gene implicated in breast cancer.

COCULTURES OF EPITHELIAL AND STROMAL CELLS

Stromal cells within mammary tissues play critical roles in regulating tissue structure and functions of the gland; thus, cocultures of epithelial and stromal cells can provide a good platform for elucidating the role of cross talk in gland physiology and the molecular components of stroma and epithelial cells that are involved. Given the complex nature of this cross talk, 3D cultures that retain the organization of epithelial cells and their interactions with stroma are essential. Human and mouse mammary tissue differ in the organization of epithelial and stromal cells,

Table 4. Examples of cocultures of mammary epithelial cells and other cell types

Type of cells cocultured with MEC	Matrix	Findings	References
Fibroblasts		Review article on mammary cell lines incubated with fibroblasts	Koledova 2017
Fibroblasts	rBM	Methods report: Primary mouse fibroblasts promote mouse primary MECs to undergo branching morphogenesis	Koledova and Lu 2017; Sumbal and Koledova 2019
		FGFs produced by mammary fibroblasts can drive mammary branching	
Fibroblasts and adipocytes	Silk protein scaffolds in 1:1 rBM and collagen	Differentiated human adipocytes enhanced human MEC ductal structure formation, polarity, casein production, and estrogen responses (proliferation, gene expression)	Wang and Kaplan 2012
Endothelial (EC) cells and fibroblasts	rBM	Mouse EC cells mediate activation of Wnt signaling in the fibroblasts, which in turn regulate epithelial cell branching	Wang et al. 2021
Adipocytes differentiated from 3T3-L1 cells	Microlithographic embedding in collagen	Adipocytes embedded in collagen gels induce branching of EpH4 mouse MECs through paracrine signals, including hepatocyte growth factor	Pavlovich et al. 2010; Wang et al. 2009
Primary pre- and mature human adipocytes	rBM and collagen in 3D porous silk scaffolds	Mixed population of preadipocytes and adipocytes differentiated from human adipose–derived stem cells induced branching of MCF10A breast epithelial cells, possibly through HGF	
Macrophages	rBM	Primary mouse M2 differentiated macrophages cocultured with mammary basal cells enhances the formation of branched functional mammary organoids through TNF-α activation of PI3K/Cdk1/Cyclin B1 signaling in mammary cells	Zhou et al. 2023

(MEC) Mammary epithelial cell, (rBM) reconstituted basemant membrane, (FGF) fibroblast growth factor, (HGF) hepatocyte growth factor.

with mouse stroma being dominated by adipose tissue and human with fibroblasts. With one exception (Pavlovich et al. 2010), our discussion is limited to studies involving primary mammary epithelial cells.

Table 4 provides representative examples of different types of organotypic cocultures. These methodologies are just being developed using cells derived from primary tissues; however, they offer significant promise to define factors involved in physiologically relevant cross talk.

For example, the Zeng laboratory used cocultures of fibroblasts (FBs) and endothelial cells (ECs) to address the role of ECs in regulation of WNT activation in FBs, which provide paracrine signals for epithelial branching (Wang et al. 2021). Unexpectedly, ECs were found to be required for activation of WNT in FBs, which sustained branching after just a pulse of FGF2. In another example, critical players involved in macrophage regulation of mammary morphogenesis were examined. Mammary resident macrophages were found to interact with mammary basal cells through the Notch ligand Dll1 expressed in basal cells, thereby activating Notch signaling in macrophages and promoting the secretion of WNT ligand, thus forming a feedback signal to maintain the mammary stem cell (MaSC) activity (Zhou et al. 2023).

ARE TUMOR ORGANOIDS PAR EXCELLENCE MODELS FOR BREAST CANCER?

The ability to maintain breast tumors in culture as tumor organoids has dramatically impacted our reliance on established cell lines as the go-to model for preclinical and mechanistic studies (Clevers 2016; Tuveson and Clevers 2019). As discussed above, there are a few different methods used to generate tumor organoids from either a surgical sample or a tumor biopsy. Similar to the conditions summarized for normal breast organoids, tumor organoid media show variance in the composition of the key ingredients (Table 2). For example, Sachs et al. (2018) used a media supplemented with WNT3A, R-Spondin, noggin, EGF, and FGFs, whereas the Welm laboratory used a simpler media containing 5% fetal bovine serum, EGF, and FGF2 or amphiregulin (Guillen et al. 2022). Another variation of the Sachs/Clevers medium used by the Deng group included hydrocortisone, estradiol, transferrin, and selenium (Chen et al. 2021). Whether these methods are equally competent for the generation and maintenance of organoids or whether they have their own advantages and challenges needs to be determined. Standardization of the methods to generate and use tumor organoids should enable better reproducibility of studies.

All breast tumor organoids are established from biopsy or surgical samples, raising questions about their ability to represent the genetic complexity of the patient's tumor. It is unlikely that a section or biopsy obtained from a patient's tumor will be representative of the entire genetic/ epigenetic diversity that is present in the patient's tumor because recent advances in cancer genomics have demonstrated that different regions of tumors differ in genetic, epigenetic, and sometimes histological features (Marusyk and Polyak

2010; Koren and Bentires-Alj 2015; Pasha and Turner 2021; Lewis et al. 2022; de Visser and Joyce 2023). By extension, no patient-derived model, be it PDX or PDO, generated from a section of the patient's tumor will be able to fully represent the heterogeneity of the original tumor. Nevertheless, PDX or PDO models retain tumor cell characteristics that are shared among most of the tumor, the so-called truncal changes, in addition to the clonal changes present within the sample; hence, developing a large and diverse collection of patient-derived models for breast cancer will provide the opportunity to have representation of most if not all genetic diversity observed in breast cancer. Although there are limitations of PDO models—the most significant being the lack of stromal cells and systemic factors—the lower cost and faster timeline for development compared to PDX models make them amenable to generating large biobanks that collectively represent the diversity of human breast cancer. Furthermore, the true value of patient-derived models should rely on their ability to predict clinical drug response because the phenotypic response to drug treatment is the outcome of a complex interaction between genetic, epigenetic, and posttranslational changes that occur within a cell.

The first large-scale effort to generate a living biobank of more than 100 breast tumor organoids was initiated and benchmarked to assess retention of the genetic and morphological traits comparable to the matched primary tumor (Sachs et al. 2018). Gene expression analysis showed a representation of most of the breast cancer subtypes; however, there were challenges in retaining ER expression in organoid cultures established from ER^+ or PR^+ breast cancers, suggesting that ER^-/PR^- cells are more efficient in establishing organoids than their ER^+ counterparts. Interestingly, functional drug sensitivity screens identified sensitivities that were not readily predictable by gene expression studies, highlighting the utility of breast tumor organoids for functional screens during the identification of personalized treatment options. Although the organoid response to therapeutic drugs, as measured by IC_{50} values, did not always match the expected trend in patients, the utility of these biobanks cannot be undervalued. Having access to a panel of breast tumor models that represent patient diversity will offer significant power for preclinical studies (Sachs et al. 2018). A recent study by Chen et al. (2021) reports a high (75%) success rate for generating short-term cultures of breast tumor organoids from patients. Although they observed a significant decrease in the rates of proliferation after two passages, they used 57 early passage lines to conduct pharmacotyping and demonstrated a high (71%) concordance with matched patients' clinical drug response for at least one of the drugs used to treat the patient. With a 2- to 5-week turnaround for generating PDO drug responses, the authors leveraged the PDO-drug response data to tailor a clinical treatment that led to improved outcomes for the six patients who had exhausted all other treatment options (Chen et al. 2021). In a similar vein, the Welm laboratory generated PDX models from patient tumors and generated PDX tumor–derived organoid (PDxO) models (Guillen et al. 2022). They performed extensive characterization of the PDxO models to establish whether they shared properties with the PDX tumor and the matched patient tumor. They went on to successfully relate the drug response in organoids to the patients' clinical responses. Using this approach, the authors were able to identify a new treatment and extend the life of a patient who had progressed on other treatments. The efficacious drugs identified by organoid testing varied between patients, identifying new treatment options such as bortezomib, vinorelbine, everolimus, or doxycycline (Chen et al. 2021; Guillen et al. 2022) are likely to be patient-specific. Although these studies represent a small number of patients, it is likely that the organoid-based phenotypic screening approach on a large cohort of patients, in conjunction with an integrated analysis of other phenotypic and genotypic traits of the patient, will identify treatment options that are applicable to a breast cancer patient that shares specific traits.

Together these studies demonstrate that despite the perceived inability of PDOs to retain the full heterogeneity of the patient's tumor, they can be used effectively to model a patient's clinical response. In our opinion, this should be the ultimate test for how representative a model is for the patient's tumor. These studies also offer hope for using tumor organoid models in preclinical studies as they may increase the chance of identifying clinically effective drugs or drug combinations. The generation of a large biobank of breast-tumor organoid models for preclinical studies should be supported by institutions and funding agencies with the goal of reducing the risk of failure of drugs in the clinic and thus reducing the cost of new drug development efforts.

TECHNICAL KNOW-HOW AND ADOPTION FOR RESEARCH

One of the major bottlenecks for adopting tumor organoid models in a research operation is familiarity with technical know-how. As discussed above for normal mammary organoids, this is a significant challenge as each study introduces its own protocol. For example, among the studies discussed above, the Deng laboratory modified the media conditions published by Sachs et al. (2018) to include hydrocortisone, estradiol, insulin, transferrin, and selenium (Chen et al. 2021), and the Welm laboratory used a simpler media containing 5% fetal bovine serum, hydrocortisone, and EGF (Guillen et al. 2022). The reproducibility and rigor of any given method is also an important consideration. For example, Petersen's group used the method reported by Sachs et al. (2018) to characterize the cellular composition of primary breast tumor organoids as a function of passage and reported that upon two to four passages, the tumor organoid cultures tended to drift toward predominantly normal like and basal cells (Goldhammer et al. 2019), suggesting care should be taken on the use of long-term tumor-organoid cultures under the conditions reported in Sachs et al. (2018). However, another study generated 32 organoids from advanced breast cancers and reported the conservation of histological and molecular features, albeit with some propensity for the enrichment of basal cell types in luminal tumor-derived organoids using conditions previously reported by Sachs et al. (2018) and Campaner et al. (2020). We have reported that the Sachs method is effective in supporting the growth of normal breast-derived organoids that retain basal, luminal, and ER$^+$ epithelial subtypes (Rosenbluth et al. 2020). We have also recently identified media conditions that support the establishment and long-term maintenance of ER expression and responsiveness in organoid cultures generated from the ER$^+$ subtype of breast cancer (Oliphant et al. 2023). Although it would benefit the field to have a consensus on media conditions for the establishment of breast cancer organoids, additional studies are needed before the criteria for the establishment and growth of different subtypes of breast cancer are better understood. In our experience, we have found that it is important to avoid the use of serum, to limit the proliferation of fibroblasts at the expense of tumor epithelia, and growth factors that promote the growth of undifferentiated tumor or normal epithelial or stem-like cells present in the tissue. It is important to include factors that are known to play homeostatic functions needed for maintaining the expression of differentiated epithelial traits, such as estrogen receptors or cytokeratin. The identification of the growth factors needed to support the growth of tumor epithelia that maintain their differentiation status will require empirical testing of different concentrations of specific growth factors and cytokines previously reported for tissue maintenance in developmental biology studies. Thus, standardization of culture conditions and definition of key benchmarking criteria will be necessary for the adoption of organoid-based research efforts by the broader breast cancer research community.

INNOVATION, VALIDATION, AND BENCHMARKING

There is no single method to generate and maintain breast tumor organoids to date because different conditions will select for different properties. It is likely that the previously published methods may need validation in the scientist's own hands to assess their ability to retain genomic and histological concordance with matched patient tumors and for rigor and reproducibility. Furthermore, any change to previously reported methods will require a significant amount of effort to validate its fidelity and utility. Journals must also be open to reporting studies that develop new methods and improvements on previously published methods supported with rigorous validation.

Most common benchmarking strategies involve mutation and gene expression analysis, where one expects a high concordance (>70%) with the genetic makeup of matched tumors. This is particularly important because long-term cultures can result in the expansion of normal epithelial at the expense of tumor cells (Tuveson and Clevers 2019). In addition, it would be essential to investigate the preservation of epithelial cell states and tissue organization between organoids and the matched tumor by immunostaining for expression of estrogen receptor, HER2, and epithelial differentiation state markers, including basal 5/6/14 and luminal 8/18/19 cytokeratins, and histopathology evaluated by a clinical pathologist. A rigorous benchmarking approach needs to be coupled with all methods and biobank development efforts to improve the rigor and reproducibility in using PDOs for preclinical studies.

Validation becomes more critical in efforts focused on finding alternatives to the basement membrane (BM) matrix. As powerful and biologically sophisticated as the EHS-derived BM matrix is a platform for growing organoids, the challenges associated with batch-to-batch variation and the lack of a complete understanding of the composition of this BM have motivated bioengineers to develop synthetic matrices. For example, a recent effort in breast cancer used alginate, a natural polymer from brown algae, to grow polyoma virus middle T antigen–induced mouse tumors as alginate microbeads (Fang et al. 2021). The resulting organoids tended to be cystic, so it is unclear if they mimic the complex structures that form when grown as rBM cultures. A nonfibrillar biomimetic hydrogel synthesized from gelatin and cellulose nanocrystals was able to support the initiation and growth of organoids from PDXs by promoting the maintenance of histopathological features, gene expression, and drug response while suppressing the growth of contaminating mouse cells (Prince et al. 2022). Most synthetic matrix platforms have advantages and disadvantages that need to be balanced depending on the experimental condition, but it is likely a matter of time before a synthetic gel that behaves similarly to basement membrane extract (BME) will be developed.

Bioengineers have also engaged in the generation of bioprinting technologies to enable large-scale efforts of using PDOs for genetic and chemical screens to find new drugs or drug targets. However, the ability of such platforms to retain the complex biology of tumor organoids needs rigorous validation. For example, Mollica et al. (2019) used decellularized matrix from a rat or human mammary gland and compared it to rat tail collagen or Geltrex. Although the system was amenable to 3D bioprinting applications, cells grown in different matrix substrata show different 3D histomorphology and biology, highlighting the challenges of restoring the biological complexity of tissue organization in combination with bioprinting platforms (Mollica et al. 2019). It is possible that one can introduce simple modifications to culture conditions to achieve scalability. Nonadherent culture conditions supplemented with dilute Matrigel (2%) can be sufficient to promote organoid-like growth that obviates the need for gelled substratum, which is ideal for large-scale expansion and genetic screens (Wrenn et al. 2020). Thus, every major technical innovation needs to be combined with rigorous validation with an eye for preserving the biological sophistication organoids are expected to have; if not, the new approaches will fail to read out on biological processes beyond the traditional readouts of cell proliferation and cell death.

CONCLUDING REMARKS

Although organoid cultures offer excellent models of normal mouse and human mammary tissues, additional studies are required to improve their value. Given the enormous variation in conditions used in the field, it would be useful to carry out systematic comparisons of different conditions to provide investigators with choices based on rigorous evidence. The variability in the composition and physiology of cells from different donors will necessitate analysis of organoids derived from multiple donors matched for age,

BMI, parity, menopause, and other factors. It is also important to assess the extent of lineage fidelity and its maintenance, optimally using scRNA-seq, to monitor the expression of different cell types over time of passage. Similarly, development of conditions that allow maintenance of cultures of individual epithelial cell types would allow investigations of research questions specifically related to ML/HS, LP/AV, and basal/myoepithelial cells. Last, given the underlying variability in the physiology of mammary cells due to the confounding factors mentioned above, along with unaccountable variables and difficulties in standardizing the isolation of similar epithelial content from breast tissues, it is critical to ensure that experimental studies are sufficiently powered to address this variability.

Another challenge in the field is to further optimize conditions for the coculture of normal epithelial or breast tumor cells with other cell types in the microenvironment, including T cells, adipocytes, endothelial cells, and stromal fibroblasts. Several studies of normal cells were highlighted in this review, and tumor cell cocultures with stromal fibroblasts or T cells have been developed using human, mouse, or rat mammary tumors where the organoid cultures are maintained for short periods in the presence of cytokines to support T-cell growth (Dijkstra et al. 2018; Tsai et al. 2018; Fiorini et al. 2020; Yuki et al. 2020; Meng et al. 2021; Prajapati et al. 2021; Gil Del Alcazar et al. 2022). For example, we recently reported the development of a platform to coculture tumor organoids and patients' peripheral blood-derived mononuclear cells and demonstrated a dramatic (1000–100,000-fold) clonal expansion of cytotoxic T cells (referred to as organoid-primed T, opT, cells) (Meng et al. 2021). The opT cells express T-cell receptors (TCRs) that, when cloned and expressed in heterologous T cells, confer a tumor-killing ability to the heterologous T cells, demonstrating the utility of opT culture as an empirical platform to identify tumor-targeting TCRs. However, this area of research is in its infancy. Organotypic cultures of cells derived from normal tissues promise to provide a better understanding of the specific factors and mechanisms involved in cross talk essential for tissue architecture and function. For tumor cell cocultures, the ability to model the complexities of TME interactions in culture will provide opportunities for identifying new ways to control cancer. Last, advances in imaging approaches to analyze the behavior of normal and cancer cells within organoid cultures will provide unprecedented insights. For example, the use of a nontoxic clearing agent for imaging organoids from 3D Matrigel can provide details of the 3D organization of nuclei and cell structure with in vivo relevance (Dekkers et al. 2019). In addition, long-term imaging using light sheet microscopy has shown how tumor-like structures can evolve after oncogene induction in a single epithelial cell within organoids (Alladin et al. 2020), highlighting the potential of combining the organoid platform with high-resolution imaging modalities. A combination of imaging and coculture methods will provide a powerful platform to understand how cancer cells respond to drug treatments by not only monitoring changes in cell division and death but also changes in cell state plasticity and the role they play in drug response. It is likely that advances in organoid platforms will create opportunities to conduct preclinical studies that have a high probability of clinical translation.

ACKNOWLEDGMENTS

We thank Angie Martinez-Gakidis for expert scientific editing as well as Drs. Andrew Ewald and Jennifer Rosenbluth for contributing figure panels.

REFERENCES

Alladin A, Chaible L, Garcia Del Valle L, Sabine R, Loeschinger M, Wachsmuth M, Heriche JK, Tischer C, Jechlinger M. 2020. Tracking cells in epithelial acini by light sheet microscopy reveals proximity effects in breast cancer initiation. *eLife* **9**: e54066. doi:10.7554/eLife.54066

Basham KJ, Kieffer C, Shelton DN, Leonard CJ, Bhonde VR, Vankayalapati H, Milash B, Bearss DJ, Looper RE, Welm BE. 2013. Chemical genetic screen reveals a role for desmosomal adhesion in mammary branching morphogenesis. *J Biol Chem* **288**: 2261–2270. doi:10.1074/jbc.M112.411033

Buchmann B, Engelbrecht LK, Fernandez P, Hutterer FP, Raich MK, Scheel CH, Bausch AR. 2021. Mechanical plasticity of collagen directs branch elongation in human mammary gland organoids. *Nat Commun* **12**: 2759. doi:10.1038/s41467-021-22988-2

Campaner E, Zannini A, Santorsola M, Bonazza D, Bottin C, Cancila V, Tripodo C, Bortul M, Zanconati F, Schoeftner S, et al. 2020. Breast cancer organoids model patient-specific response to drug treatment. *Cancers (Basel)* **12**: 3869. doi:10.3390/cancers12123869

Carr D, Zein A, Coulombe J, Jiang T, Cabrita MA, Ward G, Daneshmand M, Sau A, Pratt MAC. 2022. Multiple roles for Bcl-3 in mammary gland branching, stromal collagen invasion, involution and tumor pathology. *Breast Cancer Res* **24**: 40. doi:10.1186/s13058-022-01536-w

Caruso M, Huang S, Mourao L, Scheele C. 2022. A mammary organoid model to study branching morphogenesis. *Front Physiol* **13**: 826107. doi:10.3389/fphys.2022.826107

Centonze A, Lin S, Tika E, Sifrim A, Fioramonti M, Malfait M, Song Y, Wuidart A, Van Herck J, Dannau A, et al. 2020. Heterotypic cell–cell communication regulates glandular stem cell multipotency. *Nature* **584**: 608–613. doi:10.1038/s41586-020-2632-y

Chen P, Zhang X, Ding R, Yang L, Lyu X, Zeng J, Lei JH, Wang L, Bi J, Shao N, et al. 2021. Patient-derived organoids can guide personalized-therapies for patients with advanced breast cancer. *Adv Sci* **8**: e2101176. doi:10.1002/advs.202101176

Clevers H. 2016. Modeling development and disease with organoids. *Cell* **165**: 1586–1597. doi:10.1016/j.cell.2016.05.082

Cornelissen LM, Drenth AP, van der Burg E, de Bruijn R, Pritchard CEJ, Huijbers IJ, Zwart W, Jonkers J. 2020. TRPS1 acts as a context-dependent regulator of mammary epithelial cell growth/differentiation and breast cancer development. *Genes Dev* **34**: 179–193. doi:10.1101/gad.331371.119

Corsini NS, Knoblich JA. 2022. Human organoids: new strategies and methods for analyzing human development and disease. *Cell* **185**: 2756–2769. doi:10.1016/j.cell.2022.06.051

Darcy KM, Shoemaker SF, Lee PP, Ganis BA, Ip MM. 1995. Hydrocortisone and progesterone regulation of the proliferation, morphogenesis, and functional differentiation of normal rat mammary epithelial cells in three dimensional primary culture. *J Cell Physiol* **163**: 365–379. doi:10.1002/jcp.1041630217

Dekkers JF, Alieva M, Wellens LM, Ariese HCR, Jamieson PR, Vonk AM, Amatngalim GD, Hu H, Oost KC, Snippert HJG, et al. 2019. High-resolution 3D imaging of fixed and cleared organoids. *Nat Protoc* **14**: 1756–1771. doi:10.1038/s41596-019-0160-8

Dekkers JF, Whittle JR, Vaillant F, Chen HR, Dawson C, Liu K, Geurts MH, Herold MJ, Clevers H, Lindeman GJ, et al. 2020. Modeling breast cancer using CRISPR-Cas9-mediated engineering of human breast organoids. *J Natl Cancer Inst* **112**: 540–544. doi:10.1093/jnci/djz196

Dekkers JF, van Vliet EJ, Sachs N, Rosenbluth JM, Kopper O, Rebel HG, Wehrens EJ, Piani C, Visvader JE, Verissimo CS, et al. 2021. Long-term culture, genetic manipulation and xenotransplantation of human normal and breast cancer organoids. *Nat Protoc* **16**: 1936–1965. doi:10.1038/s41596-020-00474-1

de Visser KE, Joyce JA. 2023. The evolving tumor microenvironment: from cancer initiation to metastatic outgrowth. *Cancer Cell* **41**: 374–403. doi:10.1016/j.ccell.2023.02.016

Dijkstra KK, Cattaneo CM, Weeber F, Chalabi M, van de Haar J, Fanchi LF, Slagter M, van der Velden DL, Kaing S, Kelderman S, et al. 2018. Generation of tumor-reactive T cells by co-culture of peripheral blood lymphocytes and tumor organoids. *Cell* **174:** 1586–1598. doi:10.1016/j.cell.2018.07.009

DiMasi JA, Grabowski HG, Hansen RW. 2016. Innovation in the pharmaceutical industry: new estimates of R&D costs. *J Health Econ* **47:** 20–33. doi:10.1016/j.jhealeco.2016.01.012

Eirew P, Stingl J, Raouf A, Turashvili G, Aparicio S, Emerman JT, Eaves CJ. 2008. A method for quantifying normal human mammary epithelial stem cells with in vivo regenerative ability. *Nat Med* **11:** 1384–1389. doi:10.1038/nm.1791

Emerman JT, Enami J, Pitelka DR, Nandi S. 1977. Hormonal effects on intracellular and secreted casein in cultures of mouse mammary epithelial cells on floating collagen membranes. *Proc Natl Acad Sci* **74:** 4466–4470. doi:10.1073/pnas.74.10.4466

Emerman JT, Burwen SJ, Pitelka DR. 1979. Substrate properties influencing ultrastructural differentiation of mammary epithelial cells in culture. *Tissue Cell* **11:** 109–119. doi:10.1016/0040-8166(79)90011-9

Engelbrecht LK, Twigger AJ, Ganz HM, Gabka CJ, Bausch AR, Lickert H, Sterr M, Kunze I, Khaled WT, Scheel CH. 2021. A strategy to address dissociation-induced compositional and transcriptional bias for single-cell analysis of the human mammary gland. bioRxiv doi:10.1101/2021.02.11.430721

Ewald AJ, Brenot A, Duong M, Chan BS, Werb Z. 2008. Collective epithelial migration and cell rearrangements drive mammary branching morphogenesis. *Dev Cell* **14:** 570–581. doi:10.1016/j.devcel.2008.03.003

Fang G, Lu H, Rodriguez de la Fuente L, Law AMK, Lin G, Jin D, Gallego-Ortega D. 2021. Mammary tumor organoid culture in non-adhesive alginate for luminal mechanics and high-throughput drug screening. *Adv Sci* **8:** e2102418. doi:10.1002/advs.202102418

Fatehullah A, Tan SH, Barker N. 2016. Organoids as an in vitro model of human development and disease. *Nat Cell Biol* **18:** 246–254. doi:10.1038/ncb3312

Feigman MJ, Moss MA, Chen C, Cyrill SL, Ciccone MF, Trousdell MC, Yang ST, Frey WD, Wilkinson JE, Dos Santos CO. 2020. Pregnancy reprograms the epigenome of mammary epithelial cells and blocks the development of premalignant lesions. *Nat Commun* **11:** 2649. doi:10.1038/s41467-020-16479-z

Fernández PA, Buchmann B, Goychuk A, Engelbrecht LK, Raich MK, Scheel CH, Frey E, Bausch AR. 2021. Surface-tension-induced budding drives alveologenesis in human mammary gland organoids. *Nat Phys* **17:** 1130–1136. doi:10.1038/s41567-021-01336-7

Fiorini E, Veghini L, Corbo V. 2020. Modeling cell communication in cancer with organoids: making the complex simple. *Front Cell Dev Biol* **8:** 166. doi:10.3389/fcell.2020.00166

Fradkin JE, Cook GH, Kilhoffer MC, Wolff J. 1982. Forskolin stimulation of thyroid adenylate cyclase and cyclic 3′,5′-adenosine monophosphate accumulation. *J Endocrinol* **111:** 849–856. doi:10.1210/endo-111-3-849

Fridriksdottir AJ, Kim J, Villadsen R, Klitgaard MC, Hopkinson BM, Petersen OW, Rønnov-Jessen L. 2015. Propagation of oestrogen receptor–positive and oestrogen-responsive normal human breast cells in culture. *Nat Commun* **6:** 8786. doi:10.1038/ncomms9786

Fridriksdottir AJ, Villadsen R, Morsing M, Klitgaard MC, Kim J, Petersen OW, Rønnov-Jessen L. 2017. Proof of region-specific multipotent progenitors in human breast epithelia. *Proc Natl Acad Sci* **114:** E10102–E10111. doi:10.1073/pnas.1714063114

Fröhlich E. 2023. The variety of 3D breast cancer models for the study of tumor physiology and drug screening. *Int J Mol Sci* **24:** 7116. doi:10.3390/ijms24087116

Ganz HM, Buchmann B, Engelbrecht LK, Iesinghaus M, Eichelberger L, Gabka CJ, Schmidt GP, Muckenhuber A, Weichert W, Bausch AR, et al. 2021. Generation of ductal organoids from normal mammary luminal cells reveals invasive potential. *J Pathol* **255:** 451–463. doi:10.1002/path.5790

Gil Del Alcazar CR, Trinh A, Alečkovic M, Rojas Jimenez E, Harper NW, Oliphant MUJ, Xie S, Krop ED, Lulseged B, Murphy KC, et al. 2022. Insights into immune escape during tumor evolution and response to immunotherapy using a rat model of breast cancer. *Cancer Immunol Res* **10:** 680–697. doi:10.1158/2326-6066.CIR-21-0804

Goldhammer N, Kim J, Timmermans-Wielenga V, Petersen OW. 2019. Characterization of organoid cultured human breast cancer. *Breast Cancer Res* **21:** 141. doi:10.1186/s13058-019-1233-x

Gray GK, Li CM, Rosenbluth JM, Selfors LM, Girnius N, Lin JR, Schackmann RCJ, Goh WL, Moore K, Shapiro HK, et al. 2022. A human breast atlas integrating single-cell proteomics and transcriptomics. *Dev Cell* **57:** 1400–1420 e1407.e7. doi:10.1016/j.devcel.2022.05.003

Guillen KP, Fujita M, Butterfield AJ, Scherer SD, Bailey MH, Chu Z, DeRose YS, Zhao L, Cortes-Sanchez E, Yang CH, et al. 2022. A human breast cancer-derived xenograft and organoid platform for drug discovery and precision oncology. *Nat Cancer* **3:** 232–250. doi:10.1038/s43018-022-00337-6

Hammond SL, Ham RG, Stampfer MR. 1984. Serum-free growth of human mammary epithelial cells: rapid clonal growth in defined medium and extended serial passage with pituitary extract. *Proc Natl Acad Sci* **81:** 5435–5439. doi:10.1073/pnas.81.17.5435

Haslam SZ. 1986. Mammary fibroblast influence on normal mouse mammary epithelial cell responses to estrogen in vitro. *Cancer Res* **46:** 310–316.

Heitink L, Whittle JR, Vaillant F, Capaldo BD, Dekkers JF, Dawson CA, Milevskiy MJG, Surgenor E, Tsai M, Chen HR, et al. 2022. In vivo genome-editing screen identifies tumor suppressor genes that cooperate with *Trp53* loss during mammary tumorigenesis. *Mol Oncol* **16:** 1119–1131. doi:10.1002/1878-0261.13179

Henry S, Trousdell MC, Cyrill SL, Zhao Y, Feigman MJ, Bouhuis JM, Aylard DA, Siepel A, Dos Santos CO. 2021. Characterization of gene expression signatures for the identification of cellular heterogeneity in the developing mammary gland. *J Mammary Gland Biol Neoplasia* **26:** 43–66. doi:10.1007/s10911-021-09486-3

Hinz S, Manousopoulou A, Miyano M, Sayaman RW, Aguilera KY, Todhunter ME, Lopez JC, Sohn LL, Wang LD,

LaBarge MA. 2021. Deep proteome profiling of human mammary epithelia at lineage and age resolution. *iScience* **24:** 103026. doi:10.1016/j.isci.2021.103026

Huebner RJ, Neumann NM, Ewald AJ. 2016. Mammary epithelial tubes elongate through MAPK-dependent coordination of cell migration. *Development* **143:** 983–993. doi:10.1242/dev.127944

Jamieson PR, Dekkers JF, Rios AC, Fu NY, Lindeman GJ, Visvader JE. 2017. Derivation of a robust mouse mammary organoid system for studying tissue dynamics. *Development* **144:** 1065–1071. doi:10.1242/dev.145045

Jardé T, Lloyd-Lewis B, Thomas M, Kendrick H, Melchor L, Bougaret L, Watson PD, Ewan K, Smalley MJ, Dale TC. 2016. Wnt and Neuregulin1/ErbB signalling extends 3D culture of hormone responsive mammary organoids. *Nat Commun* **7:** 13207. doi:10.1038/ncomms13207

Keller PJ, Lin AF, Arendt LM, Klebba I, Jones AD, Rudnick JA, DiMeo TA, Gilmore H, Jefferson DM, Graham RA, et al. 2010. Mapping the cellular and molecular heterogeneity of normal and malignant breast tissues and cultured cell lines. *Breast Cancer Res* **12:** R87. doi:10.1186/bcr2755

Kobayashi K. 2023. Culture models to investigate mechanisms of milk production and blood–milk barrier in mammary epithelial cells: a review and a protocol. *J Mammary Gland Biol Neoplasia* **28:** 8. doi:10.1007/s10911-023-09536-y

Koledova Z. 2017. 3D coculture of mammary organoids with fibrospheres: a model for studying epithelial–stromal interactions during mammary branching morphogenesis. *Methods Mol Biol* **1612:** 107–124. doi:10.1007/978-1-4939-7021-6_8

Koledova Z, Lu P. 2017. A 3D fibroblast-epithelium co-culture model for understanding microenvironmental role in branching morphogenesis of the mammary gland. *Methods Mol Biol* **1501:** 217–231. doi:10.1007/978-1-4939-6475-8_10

Koren S, Bentires-Alj M. 2015. Breast tumor heterogeneity: source of fitness, hurdle for therapy. *Mol Cell* **60:** 537–546. doi:10.1016/j.molcel.2015.10.031

Lee EY, Parry G, Bissell MJ. 1984. Modulation of secreted proteins of mouse mammary epithelial cells by the collagenous substrata. *J Cell Biol* **98:** 146–155. doi:10.1083/jcb.98.1.146

Lee EY, Lee WH, Kaetzel CS, Parry G, Bissell MJ. 1985. Interaction of mouse mammary epithelial cells with collagen substrata: regulation of casein gene expression and secretion. *Proc Natl Acad Sci* **82:** 1419–1423. doi:10.1073/pnas.82.5.1419

Lewis SM, Callaway MK, Dos Santos CO. 2022. Clinical applications of 3D normal and breast cancer organoids: a review of concepts and methods. *Exp Biol Med* **247:** 2176–2183. doi:10.1177/15353702221131877

Li ML, Aggeler J, Farson DA, Hatier C, Hassell J, Bissell MJ. 1987. Influence of a reconstituted basement membrane and its components on casein gene expression and secretion in mouse mammary epithelial cells. *Proc Natl Acad Sci* **84:** 136–140. doi:10.1073/pnas.84.1.136

Li L, Cleary S, Mandarano MA, Long W, Birchmeier C, Jones FE. 2002. The breast proto-oncogene, HRGα regulates epithelial proliferation and lobuloalveolar development in the mouse mammary gland. *Oncogene* **21:** 4900–4907. doi:10.1038/sj.onc.1205634

Lim E, Vaillant F, Wu D, Forrest NC, Pal B, Hart AH, Asselin-Labat ML, Gyorki DE, Ward T, Partanen A, et al. 2009. Aberrant luminal progenitors as the candidate target population for basal tumor development in *BRCA1* mutation carriers. *Nat Med* **15:** 907–913. doi:10.1038/nm.2000

Linnemann JR, Miura H, Meixner LK, Irmler M, Kloos UJ, Hirschi B, Bartsch HS, Sass S, Beckers J, Theis FJ, et al. 2015. Quantification of regenerative potential in primary human mammary epithelial cells. *Development* **142:** 3239–3251. doi:10.1242/dev.123554

Linnemann JR, Meixner LK, Miura H, Scheel CH. 2017. An organotypic 3D assay for primary human mammary epithelial cells that recapitulates branching morphogenesis. *Methods Mol Biol* **1612:** 125–137. doi:10.1007/978-1-4939-7021-6_9

Ma Z, Lytle NK, Ramos C, Naeem RF, Wahl GM. 2022. Single-cell transcriptomic and epigenetic analyses of mouse mammary development starting with the embryo. *Methods Mol Biol* **2471:** 49–82. doi:10.1007/978-1-0716-2193-6_3

Marusyk A, Polyak K. 2010. Tumor heterogeneity: causes and consequences. *Biochim Biophys Acta* **1805:** 105–117. doi:10.1016/j.bbcan.2009.11.002

Melcher ML, Block I, Kropf K, Singh AK, Posern G. 2022. Interplay of the transcription factor MRTF-A and matrix stiffness controls mammary acinar structure and protrusion formation. *Cell Commun Signal* **20:** 158. doi:10.1186/s12964-022-00977-2

Meng Q, Xie S, Gray GK, Dezfulian MH, Li W, Huang L, Akshinthala D, Ferrer E, Conahan C, Perea Del Pino S, et al. 2021. Empirical identification and validation of tumor-targeting T cell receptors from circulation using autologous pancreatic tumor organoids. *J Immunother Cancer* **9:** e003213. doi:10.1136/jitc-2021-003213

Michalak EM, Milevskiy MJG, Joyce RM, Dekkers JF, Jamieson PR, Pal B, Dawson CA, Hu Y, Orkin SH, Alexander WS, et al. 2018. Canonical PRC2 function is essential for mammary gland development and affects chromatin compaction in mammary organoids. *PLoS Biol* **16:** e2004986. doi:10.1371/journal.pbio.2004986

Miller DH, Sokol ES, Gupta PB. 2017. 3D primary culture model to study human mammary development. *Methods Mol Biol* **1612:** 139–147. doi:10.1007/978-1-4939-7021-6_10

Mizutani T, Clevers H. 2020. Primary intestinal epithelial organoid culture. *Methods Mol Biol* **2171:** 185–200. doi:10.1007/978-1-0716-0747-3_11

Mollica PA, Booth-Creech EN, Reid JA, Zamponi M, Sullivan SM, Palmer XL, Sachs PC, Bruno RD. 2019. 3D bioprinted mammary organoids and tumoroids in human mammary derived ECM hydrogels. *Acta Biomater* **95:** 201–213. doi:10.1016/j.actbio.2019.06.017

Mroue R, Bissell MJ. 2013. Three-dimensional cultures of mouse mammary epithelial cells. *Methods Mol Biol* **945:** 221–250. doi:10.1007/978-1-62703-125-7_14

Mroue R, Inman J, Mott J, Budunova I, Bissell MJ. 2015. Asymmetric expression of connexins between luminal epithelial- and myoepithelial- cells is essential for contractile function of the mammary gland. *Dev Biol* **399:** 15–26. doi:10.1016/j.ydbio.2014.11.026

Nakshatri H, Anjanappa M, Bhat-Nakshatri P. 2015. Ethnicity-dependent and -independent heterogeneity in healthy normal breast hierarchy impacts tumor characterization. *Sci Rep* **5**: 13526. doi:10.1038/srep13526

Nedvetsky PI, Kwon SH, Debnath J, Mostov KE. 2012. Cyclic AMP regulates formation of mammary epithelial acini in vitro. *Mol Biol Cell* **15**: 2973–2981. doi:10.1091/mbc.E12-02-0078

Neumann NM, Perrone MC, Veldhuis JH, Huebner RJ, Zhan H, Devreotes PN, Brodland GW, Ewald AJ. 2018. Coordination of receptor tyrosine kinase signaling and interfacial tension dynamics drives radial intercalation and tube elongation. *Dev Cell* **45**: 67–82.e6. doi:10.1016/j.devcel.2018.03.011

Nguyen-Ngoc KV, Ewald AJ. 2013. Mammary ductal elongation and myoepithelial migration are regulated by the composition of the extracellular matrix. *J Microsc* **251**: 212–223. doi:10.1111/jmi.12017

Oliphant MUJ, Akshinthala D, Muthuswamy SK. 2023. Establishing conditions for the generation and maintenance of estrogen receptor-positive organoid models of breast cancer. bioRxiv doi: 10.1101/2023.08.09.552657

Orkin RW, Gehron P, McGoodwin EB, Martin GR, Valentine T, Swarm R. 1977. A murine tumor producing a matrix of basement membrane. *J Exp Med* **145**: 204–220. doi:10.1084/jem.145.1.204

Ortiz MMO, Andrechek ER. 2023. Molecular characterization and landscape of breast cancer models from a multiomics perspective. *J Mammary Gland Biol Neoplasia* **28**: 12. doi:10.1007/s10911-023-09540-2

Pal B, Chen Y, Vaillant F, Capaldo BD, Joyce R, Song X, Bryant VL, Penington JS, Di Stefano L, Tubau Ribera N, et al. 2021. A single-cell RNA expression atlas of normal, preneoplastic and tumorigenic states in the human breast. *EMBO J* **40**: e107333. doi:10.15252/embj.2020107333

Palmieri C, Roberts-Clark D, Assadi-Sabet A, Coope RC, O'Hare M, Sunters A, Hanby A, Slade MJ, Gomm JJ, Lam EW, et al. 2003. Fibroblast growth factor 7, secreted by breast fibroblasts, is an interleukin-1beta-induced paracrine growth factor for human breast cells. *J Endocrinol* **177**: 65–81. doi:10.1677/joe.0.1770065

Pasha N, Turner NC. 2021. Understanding and overcoming tumor heterogeneity in metastatic breast cancer treatment. *Nat Cancer* **2**: 680–692. doi:10.1038/s43018-021-00229-1

Pavlovich AL, Manivannan S, Nelson CM. 2010. Adipose stroma induces branching morphogenesis of engineered epithelial tubules. *Tissue Eng Part A* **16**: 3719–3726. doi:10.1089/ten.tea.2009.0836

Plana D, Palmer AC, Sorger PK. 2022. Independent drug action in combination therapy: implications for precision oncology. *Cancer Discov* **12**: 606–624. doi:10.1158/2159-8290.CD-21-0212

Prajapati E, Kumar S, Kumar S. 2021. Muscope: a miniature on-chip lensless microscope. *Lab Chip* **21**: 4357–4363. doi:10.1039/D1LC00792K

Prince E, Cruickshank J, Ba-Alawi W, Hodgson K, Haight J, Tobin C, Wakeman A, Avoulov A, Topolskaia V, Elliott MJ, et al. 2022. Biomimetic hydrogel supports initiation and growth of patient-derived breast tumor organoids. *Nat Commun* **13**: 1466. doi:10.1038/s41467-022-28788-6

Roelofs C, Hollande F, Redvers R, Anderson RL, Merino D. 2019. Breast tumour organoids: promising models for the genomic and functional characterisation of breast

cancer. *Biochem Soc Trans* **47**: 109–117. doi:10.1042/BST20180375

Rosenbluth JM, Schackmann RCJ, Gray GK, Selfors LM, Li CM, Boedicker M, Kuiken HJ, Richardson A, Brock J, Garber J, et al. 2020. Organoid cultures from normal and cancer-prone human breast tissues preserve complex epithelial lineages. *Nat Commun* **11**: 1711. doi:10.1038/s41467-020-15548-7

Sachs N, de Ligt J, Kopper O, Gogola E, Bounova G, Weeber F, Balgobind AV, Wind K, Gracanin A, Begthel H, et al. 2018. A living biobank of breast cancer organoids captures disease heterogeneity. *Cell* **172**: 373–386.e10. doi:10.1016/j.cell.2017.11.010

Sahu S, Albaugh ME, Martin BK, Patel NL, Riffle L, Mackem S, Kalen JD, Sharan SK. 2022. Growth factor dependency in mammary organoids regulates ductal morphogenesis during organ regeneration. *Sci Rep* **12**: 7200. doi:10.1038/s41598-022-11224-6

Shackleton M, Vaillant F, Simpson KJ, Stingl J, Smyth GK, Asselin-Labat ML, Wu L, Lindeman GJ, Visvader JE. 2006. Generation of a functional mammary gland from a single stem cell. *Nature* **439**: 84–88. doi:10.1038/nature04372

Shalabi SF, Miyano M, Sayaman RW, Lopez JC, Jokela TA, Todhunter ME, Hinz S, Garbe JC, Stampfer MR, Kessenbrock K, et al. 2021. Evidence for accelerated aging in mammary epithelia of women carrying germline *BRCA1* or *BRCA2* mutations. *Nat Aging* **1**: 838–849. doi:10.1038/s43587-021-00104-9

Shehata M, Stingl J. 2017. Purification of distinct subsets of epithelial cells from normal human breast tissue. *Methods Mol Biol* **1501**: 261–276. doi:10.1007/978-1-4939-6475-8_13

Sheridan JM, Visvader JE. 2019. Isolation and propagation of mammary epithelial stem and progenitor cells. *Methods Mol Biol* **1940**: 217–229. doi:10.1007/978-1-4939-9086-3_16

Simian M, Bissell MJ. 2017. Organoids: a historical perspective of thinking in three dimensions. *J Cell Biol* **216**: 31–40. doi:10.1083/jcb.201610056

Simian M, Hirai Y, Navre M, Werb Z, Lochter A, Bissell MJ. 2001. The interplay of matrix metalloproteinases, morphogens and growth factors is necessary for branching of mammary epithelia. *Development* **128**: 3117–3131. doi:10.1242/dev.128.16.3117

Soady KJ, Tornillo G, Kendrick H, Meniel V, Olijnyk Dallas D, Morris JS, Stein T, Gusterson BA, Isacke CM, Smalley MJ. 2017. The receptor protein tyrosine phosphatase PTPRB negatively regulates FGF2-dependent branching morphogenesis. *Development* **144**: 3777–3788. doi:10.1242/dev.149120

Sokol ES, Miller DH, Breggia A, Spencer KC, Arendt LM, Gupta PB. 2016. Growth of human breast tissues from patient cells in 3D hydrogel scaffolds. *Breast Cancer Res* **18**: 19. doi:10.1186/s13058-016-0677-5

Srivastava V, Huycke TR, Phong KT, Gartner ZJ. 2020. Organoid models for mammary gland dynamics and breast cancer. *Curr Opin Cell Biol* **66**: 51–58. doi:10.1016/j.ceb.2020.05.003

Stingl J, Emerman JT, Eaves CJ. 2005. Enzymatic dissociation and culture of normal human mammary tissue to detect progenitor activity. *Methods Mol Biol* **290**: 249–263. doi:10.1385/1-59259-838-2:249

Stingl J, Eirew P, Ricketson I, Shackleton M, Vaillant F, Choi D, Li HI, Eaves CJ. 2006. Purification and unique properties of mammary epithelial stem cells. *Nature* **439**: 993–997. doi:10.1038/nature04496

Sumbal J, Koledova Z. 2019. FGF signaling in mammary gland fibroblasts regulates multiple fibroblast functions and mammary epithelial morphogenesis. *Development* **146**.

Sumbal J, Budkova Z, Traustadóttir GA, Koledova Z. 2020a. Mammary organoids and 3D cell cultures: old dogs with new tricks. *J Mammary Gland Biol Neoplasia* **25**: 273–288. doi:10.1007/s10911-020-09468-x

Sumbal J, Chiche A, Charifou E, Koledova Z, Li H. 2020b. Primary mammary organoid model of lactation and involution. *Front Cell Dev Biol* **8**: 68. doi:10.3389/fcell.2020.00068

Swarm RL. 1963. Transplantation of a murine chondrosarcoma in mice of different inbred strains. *J Natl Cancer Inst* **31**: 953–975.

Tang C, van den Bijgaart RJE, Looman MWG, Tel-Karthaus N, de Graaf AMA, Gilfillan S, Colonna M, Ansems M, Adema GJ. 2019. DC-SCRIPT deficiency delays mouse mammary gland development and branching morphogenesis. *Dev Biol* **455**: 42–50. doi:10.1016/j.ydbio.2019.06.023

Tsai S, McOlash L, Palen K, Johnson B, Duris C, Yang Q, Dwinell MB, Hunt B, Evans DB, Gershan J, et al. 2018. Development of primary human pancreatic cancer organoids, matched stromal and immune cells and 3D tumor microenvironment models. *BMC Cancer* **18**: 335. doi:10.1186/s12885-018-4238-4

Tuveson D, Clevers H. 2019. Cancer modeling meets human organoid technology. *Science* **364**: 952–955. doi:10.1126/science.aaw6985

Van Keymeulen A, Rocha AS, Ousset M, Beck B, Bouvencourt G, Rock J, Sharma N, Dekoninck S, Blanpain C. 2011. Distinct stem cells contribute to mammary gland development and maintenance. *Nature* **479**: 189–193. doi:10.1038/nature10573

Van Keymeulen A, Fioramonti M, Centonze A, Bouvencourt G, Achouri Y, Blanpain C. 2017. Lineage-restricted mammary stem cells sustain the development, homeostasis, and regeneration of the estrogen receptor positive lineage. *Cell Rep* **20**: 1525–1532. doi:10.1016/j.celrep.2017.07.066

Vidi PA, Bissell MJ, Lelièvre SA. 2013. Three-dimensional culture of human breast epithelial cells: the how and the why. *Methods Mol Biol* **945**: 193–219. doi:10.1007/978-1-62703-125-7_13

Wang X, Kaplan DL. 2012. Hormone-responsive 3D multicellular culture model of human breast tissue. *Biomaterials* **33**: 3411–3420. doi:10.1016/j.biomaterials.2012.01.011

Wang X, Zhang X, Sun L, Subramanian B, Maffini MV, Soto A, Sonnenschein C, Kaplan DL. 2009. Preadipocytes stimulate ductal morphogenesis and functional differentiation of human mammary epithelial cells on 3D silk scaffolds. *Tissue Eng Part A* **15**: 3087–3098. doi:10.1089/ten.tea.2008.0670

Wang J, Song W, Yang R, Li C, Wu T, Dong XB, Zhou B, Guo X, Chen J, Liu Z, et al. 2021. Endothelial wnts control mammary epithelial patterning via fibroblast signaling. *Cell Rep* **34**: 108897. doi:10.1016/j.celrep.2021.108897

Wansbury O, Panchal H, James M, Parry S, Ashworth A, Howard B. 2008. Dynamic expression of Erbb pathway members during early mammary gland morphogenesis. *J Invest Dermatol* **128**: 1009–1021. doi:10.1038/sj.jid.5701118

Wood LD, Ewald AJ. 2021. Organoids in cancer research: a review for pathologist-scientists. *J Pathol* **254**: 395–404. doi:10.1002/path.5684

Wrenn ED, Moore BM, Greenwood E, McBirney M, Cheung KJ. 2020. Optimal, large-scale propagation of mouse mammary tumor organoids. *J Mammary Gland Biol Neoplasia* **25**: 337–350. doi:10.1007/s10911-020-09464-1

Xu W, Gulvady AC, Goreczny GJ, Olson EC, Turner CE. 2019. Paxillin-dependent regulation of apical-basal polarity in mammary gland morphogenesis. *Development* **146**. doi:10.1242/dev.174367

Yang Y, Spitzer E, Meyer D, Sachs M, Niemann C, Hartmann G, Weidner KM, Birchmeier C, Birchmeier W. 1995. Sequential requirement of hepatocyte growth factor and neuregulin in the morphogenesis and differentiation of the mammary gland. *J Cell Biol* **131**: 215–226. doi:10.1083/jcb.131.1.215

Yuki K, Cheng N, Nakano M, Kuo CJ. 2020. Organoid models of tumor immunology. *Trends Immunol* **41**: 652–664. doi:10.1016/j.it.2020.06.010

Zhang L, Valdez JM, Zhang B, Wei L, Chang J, Xin L. 2011. ROCK inhibitor Y-27632 suppresses dissociation-induced apoptosis of murine prostate stem/progenitor cells and increases their cloning efficiency. *PLoS ONE* **6**: e18271. doi:10.1371/journal.pone.0018271

Zhang L, Adileh M, Martin ML, Klingler S, White J, Ma X, Howe LR, Brown AM, Kolesnick R. 2017. Establishing estrogen-responsive mouse mammary organoids from single Lgr5+ cells. *Cell Signal* **29**: 41–51. doi:10.1016/j.cellsig.2016.08.001

Zhao N, Powell RT, Yuan X, Bae G, Roarty KP, Stossi F, Strempfl M, Toneff MJ, Johnson HL, Mani SA, et al. 2021. Morphological screening of mesenchymal mammary tumor organoids to identify drugs that reverse epithelial–mesenchymal transition. *Nat Commun* **12**: 4262. doi:10.1038/s41467-021-24545-3

Zhou Y, Ye Z, Wei W, Zhang M, Huang F, Li J, Cai C. 2023. Macrophages maintain mammary stem cell activity and mammary homeostasis via TNF-alpha-PI3K-Cdk1/Cyclin B1 axis. *NPJ Regen Med* **8**: 23. doi:10.1038/s41536-023-00296-1

Zubeldia-Plazaola A, Ametller E, Mancino M, Prats de Puig M, Lopez-Plana A, Guzman F, Vinyals L, Pastor-Arroyo EM, Almendro V, Fuster G, et al. 2015. Comparison of methods for the isolation of human breast epithelial and myoepithelial cells. *Front Cell Dev Biol* **3**: 32. doi:10.3389/fcell.2015.00032

Advances in Immunocompetent Mouse and Rat Models

Wen Bu and Yi Li

Lester and Sue Smith Breast Center, Baylor College of Medicine, Houston, Texas 77030, USA

Correspondence: liyi@bcm.edu

Rodent models of breast cancer have played critical roles in our understanding of breast cancer development and progression as well as preclinical testing of cancer prevention and therapeutics. In this article, we first review the values and challenges of conventional genetically engineered mouse (GEM) models and newer iterations of these models, especially those with inducible or conditional regulation of oncogenes and tumor suppressors. Then, we discuss nongermline (somatic) GEM models of breast cancer with temporospatial control, made possible by intraductal injection of viral vectors to deliver oncogenes or to manipulate the genome of mammary epithelial cells. Next, we introduce the latest development in precision editing of endogenous genes using in vivo CRISPR-Cas9 technology. We conclude with the recent development in generating somatic rat models for modeling estrogen receptor–positive breast cancer, something that has been difficult to accomplish in mice.

Rodent models have been highly valuable for understanding how breast cancer initiates and progresses. These models are also critical tools for preclinical testing of therapeutics for treatment and prevention. For most of the 20th century, rodent models relied on chemical carcinogens, physical agents (such as γ-irradiation), and retroviruses (e.g., mouse mammary tumor virus [MMTV]) for insertional mutagenesis, as well as oncogenic viruses (such as Rous-sarcoma virus). While valuable at that time and still used to date, the lack of defined oncogenic drivers as well as the lack of tissue and/or cell-type specificity has greatly limited the utility of these models.

Alongside these early models, cultured human breast tumor cell lines were also engrafted into immunodeficient mice (usually in the fat pad, but also subcutaneously and occasionally in the kidney capsule). These xenograft models offered the opportunity to investigate human breast tumor progression in vivo and to test therapeutic agents on human tumor cells in vivo. Since the cells can be genetically manipulated in various ways in culture and then xenografted in mice, these models have been used widely to elucidate genetic factors controlling breast cancer progression and spread (Wu et al. 2009). Some of these genetic factors influence signaling networks within human cancer cells, while others impact secreted molecules and receptors that interact with cells in the microenvironment. Along the same line, various stromal cells can be genetically manipulated and cotransplanted with human breast tumor cells to investigate molecular interactions between tumor and stromal cells (Kuperwasser et al. 2004). Alternatively, these human tumor cells can be engrafted into mice defective in different cellular components of the immune system and/or with altered stromal cells to investigate tumor cell and microenvironment interactions. These cell line xenograft models are reproducible and highly accessible to anyone with a vivarium, but cell lines often have drifted far from human tumors and cannot recapitulate the complexity and large intratumor heterogeneity in patient tumors. Some of the drawbacks of using cell line models can be overcome by using grafts from patient specimens (patient-derived xenografts [PDXs]), which is discussed in detail in an article in this collection by Drs. Lewis and Caldas (Lewis and Caldas 2023). However, a critical limitation of these xenograft models is the use of immunodeficient hosts, which largely prevents the study of the immune cell contribution. The immune system is known to have a fundamental contribution to tumor formation, progression, and therapeutic response as discussed in detail in the articles in this collection by Drs. Quail, Park, Welm, and Ekiz (Quail et al. 2023) and by Drs. Emens and Loi (Emens and Loi 2023). Many of cancer therapeutic drugs, antiestrogens included, target both tumor cells themselves and immune cells (Bottos et al. 2016; Chakraborty et al. 2021). Some of them may exert efficacy primarily via targeting noncancerous cells such as myeloid cells (Goswami et al. 2022) and cancer-associated fibroblasts (Chen et al. 2021) while immune checkpoint inhibitors (ICIs) target the cytotoxic interaction between T cells and tumor cells (Sharma and Allison 2020). Progress is being made in engrafting human hemopoietic stem cells into special mouse recipients to "humanize" the immune system of the mice, so that human tumor cells can survive in mice for an extended window of time, but the engrafted human immune cells can only restore part of the human immune system and inevitably attack the host (Jin et al. 2021).

Genetic engineering of mice can overcome the drawbacks of undefined genetic alterations and the lack of immune cell participation or incompatible interactions of mutated mammary cells and stromal cells. Hundreds of genetically engineered mouse (GEM) models have been created, and these preclinical rodent models of cancer play critical roles in investigating genes in cancer development and progression as well as preclinical testing of cancer prevention and therapeutics. The following sections will discuss these genetically engineered rodent models and the latest developments in this area.

VALUE AND CHALLENGES OF GERMLINE GENETICALLY ENGINEERED MOUSE MODELS AND THEIR NEWER ITERATIONS

GEM models have greatly advanced our understanding of breast cancer initiation and progression as well as therapeutic response and resistance. Individual models have also been bred together to generate compound mutant mouse lines to study oncogenic collaboration in driving tumor formation and progression as well as therapeutic resistance (Sinn et al. 1987; Donehower et al. 1995; Podsypanina et al. 2004). As these models are immunocompetent, they are especially valuable for studying the involvement of the immune system (Welte et al. 2016; Kim et al. 2019), a key advantage over xenograft models. Breeding tumor models into mouse lines whose immune and other stromal cell types have been genetically modified has led to an ever-expanding pool of models for studying cancer cell–immune cell interactions, epithelial–stromal interactions, and cancer cell–metastatic niche interactions. Commonly used and well-characterized GEM models often use MMTV long terminal repeat (LTR) to drive the expression of oncogenes. Examples include MMTV-*c-Myc* (the first mammary transgenic model) (Stewart et al. 1984), MMTV-*PyMT* (popular due to its speed to develop tumors and high metastatic burden) (Guy et al. 1992), MMTV-*ErbB2* (*Neu*) (for review, see Ursini-Siegel et al. 2007), and MMTV-*Wnt1* (for review, see Li et al. 2000). These models have been instrumental in understanding breast cancer formation and progression as well as cancer cell heterogeneity and therapeutic resistance.

However, there are some challenges with these transgenic models. The MMTV LTR is expressed in a subset of luminal epithelial cells. Depending on the insertion site of the MMTV

LTR, the expression intensity and distribution within the luminal cell layer can vary significantly. Therefore, multiple founder lines need to be screened to obtain useful lines to model breast cancer. In addition, beyond the mammary epithelia, the MMTV LTR also has activity in salivary gland, kidney, lung, and some hematopoietic cells (Stöcklin et al. 1993; Godley et al. 1996). Therefore, tumors sometimes appear in other sites besides the mammary glands. Systemic factors from these developing tumors may also affect the interpretation of mammary gland phenotypes. Furthermore, transgenic expression in nonmammary tissues may complicate the study of mammary tumor metastasis. For example, finding tumor cells in distant organs may not be the result of metastasis but rather the emergence of new primary tumors in these organs. This leakiness of transgene expression in other organs can potentially complicate the interpretation of studies to characterize the premetastatic niche. For example, microenvironment changes in lungs in these mice before the appearance of metastatic tumors may be the outcome of leaky transgene expression in lungs but not due to the lungs receiving signals from the primary site.

Besides these conventional transgenic promoters, other promoters have been used to express transgenes in selected cell subsets within the luminal or the basal/cap cell populations, since distinct cell subsets within the mammary ductal tree may contribute to different breast cancer subtypes. The whey acidic protein (WAP) (Andres et al. 1987; Pittius et al. 1988) and β-lactoglobulin (BLG) gene promoters (Whitelaw et al. 1992; Selbert et al. 1998), which can be stimulated by lactation, have been used to direct transgene expression to differentiated alveolar and secretory luminal cells. Not surprisingly, when different promoters are used to regulate the expression of the same oncogenic driver, tumors with different characteristics can develop (Herschkowitz et al. 2007), reflecting the impact of cell of origin on breast cancer phenotypes.

We have reported that cytokeratin 6 (KRT6) marks a luminal cell subpopulation with progenitor features, and targeting *PyMT* into this cell population leads to papillary tumors that resemble the normal-like breast cancer subtype in patients (Bu et al. 2011). On the other hand, multiple transgenic lines have been reported using the promoters of *Krt5* or *Krt14* to express oncogenes and other genes in the basal cell population (Kuraguchi et al. 2009; Bowman-Colin et al. 2013; Bao et al. 2019). One caveat is that these promoters are also not limited to the mammary epithelial population. These keratin promoters are active in epithelial cells in other nonmammary organs such as skin, salivary gland, and ovary (Vassar et al. 1989; Jonkers et al. 2001; Orsulic et al. 2002; Grimm et al. 2006). WAP is known to be expressed in brain, testes, and even muscle (Wagner et al. 1997), and the BLG promoter has activity in salivary gland (Whitelaw et al. 1992). Of note, a given oncogene may exhibit significant differences in potency in driving tumor formation in different cell types or organs (Leder et al. 1986).

Transgenic models have been used widely as preclinical models to test therapeutics. These immunocompetent models have certainly helped develop new and better therapeutics including immunotherapeutic agents, and they have helped uncover new mechanisms in drug resistance. However, the MMTV LTR is known to be responsive to steroid hormones including glucocorticoid, progesterone, and androgens (Cato et al. 1987). The *Wap* promoter is highly responsive to steroid and peptide hormones and is transcriptionally regulated by STAT5, as is MMTV LTR (Hobbs et al. 1982). Therefore, therapeutic agents that impact the production and signaling of these hormones may incidentally impact transgene expression levels and complicate the interpretation of the results from these preclinical studies. For example, ruxolitinib is an FDA approved small molecule inhibitor of Janus kinase (JAK), which phosphorylates and activates STAT5, but ruxolitinib treatment would shut down expression of transgenic oncogenes in WAP or MMTV transgenic models; therefore, these transgenic models should not be used for testing therapeutic efficacy of ruxolitinib or any drugs that target JAK-STAT signaling.

Transgenic and knockout (KO) models generally introduce genetic alterations into the germline. These genetic alterations can affect normal mammary gland development. For example, tumors begin to appear in MMTV-*PyMT* mice before puberty; widespread hyperplasia is present in MMTV-*Wnt1* mice even before puberty (Lin et al. 1992; Cunha and Hom 1996); and ductal development is impaired in MMTV-*ErbB2* mice (Mukherjee et al. 2000). On the other hand, germline KO leads to loss of gene function in all cells starting from embryogenesis, often causing embryogenic defects, impaired organ development, and tumor development in nonmammary tissue, impeding the study of gene functions in mammary tumorigenesis. For example, *Tp53* KO mice usually succumb to lymphomas and sarcomas before mammary tumors appear (Donehower et al. 1992). To overcome these drawbacks, mammary tissues from KO mice can be transplanted into wild-type mice so that the specific impact on mammary development and tumorigenesis can be studied. For example, transplantation of *Tp53* KO mammary gland pieces into the cleared fatpad of wild-type mice allowed the study of the impact of *Tp53* on mammary tumor development without the interference of other tumors in *Tp53*-null mice (Jerry et al. 2000). Interestingly, the mammary gland has potent reprograming capacity—when mixed with mammary epithelial cells and engrafted into a cleared fat pad, even neural stem cells (tagged by LacZ) have been reported to contribute to the outgrowth of the mammary ductal tree (Booth et al. 2008).

Another approach to study the impact of gene deletion selectively in the mammary gland is to use a conditional KO method. Cre is a bacterial recombinase that catalyzes recombination between a pair of DNA recognition sites termed LoxP sites (Sauer and Henderson 1988; Lewandoski 2001). Breeding a mouse line carrying Cre under the control of a mammary-specific promoter with a mouse line that has part or a whole tumor suppressor flanked by LoxP sites allows the deletion of the tumor suppressor in mammary epithelia. For example, MMTV-*Cre/BRCA1*$^{loxp/loxp}$ mice develop normally but exhibit blunted ductal morphogenesis and eventually tumors (Xu et al. 1999). Of note, the use of MMTV-*Cre* versus WAP-*Cre* can also lead to tumors of different properties (Lin et al. 2004; Herschkowitz et al. 2007; Jiang et al. 2010).

A *Cre* transgene may also be used to turn on the expression of a transgenic oncogene. For example, a loxP-flanked DNA fragment harboring transcriptional stop signals may be inserted between a transgenic promoter and the transgenic oncogene, preventing transgenic expression unless in cells expressing transgenic *Cre* (Andrechek et al. 2000). This hybrid system can be used to achieve more fine tuning in cell specificity of transgenic expression since the cell has to support both the promoter driving *Cre* and the promoter driving the STOP$^{loxp/loxp}$-oncogene to turn on the gene of interest. Alternatively, this method can be used to conditionally turn on a mutated version of a proto-oncogene engineered into the endogenous locus (i.e., knockin), so that the activated oncogene can be under the control of the endogenous promoter (Andrechek et al. 2000). Besides this Cre-LoxP system, the FLP-FRT recombination system is also commonly used and sometimes combined with the Cre-LoxP to achieve more sophisticated conditional deletion/activation of one or more genes (Dymecki and Tomasiewicz 1998; Branda and Dymecki 2004).

To overcome the drawback of premature oncogene expression before completion of organ development, inducible promoters have been used to drive oncogenic expression. These promoters are silent until induced by a drug, providing a temporal control to GEM models. While several methods can be used to achieve inducible expression (No et al. 1996; Cronin et al. 2001), the most common method is the tet-operon (tetO)/repressor bitransgenic system, in which TRE (tetO + a minimum promoter) is used to control the gene of interest while a mammary cell–selective promoter drives the expression of tTA, which is a transcriptional activator resulting from the fusion of tet repressor and the VP16 transcription activator and can bind and activate TRE under the tight control of tetracycline or doxycycline. The original tTA is active in the absence of tetracycline or doxycycline, while a later developed variant version (also called reverse tTA, or rtTA) requires tetracycline or doxycycline for transcriptional activation. Both Tet-Off and Tet-On systems are widely used to control oncogene and tumor suppressor gene expression in cancer modeling as well as other genes involved in development and diseases. A tetracycline-regulated vector can also be used to delete a tumor suppressor gene at a selected time. For example, Tet-On can be used to turn on Cre at a selected time to delete part or a whole tumor suppressor to investigate its impact on cancer development and progression.

Besides inducible gene promoters, another commonly used approach to temporal control of gene expression is the use of a transgene (Cre^{ERT} or Cre^{ERT2}) that fuses Cre to a mutated ligand-binding domain of estrogen receptor (ER), which does not bind to estrogen but binds to tamoxifen with high affinity (Littlewood et al. 1995; Feil et al. 1996; Indra et al. 1999). Tamoxifen is a selective ER modifier and is commonly used to suppress ER activity of breast cancer. In the absence of tamoxifen, this fusion protein is complexed with HSP90 and other intracellular proteins, masking the Cre recombinase activity, but tamoxifen binding releases Cre from this inhibitory complex (Whitfield et al. 2015). Therefore, tamoxifen administration can be used to unleash the Cre activity to delete a tumor suppressor gene. Likewise, this inducible Cre can also be used to conditionally activate a transgenic oncogene that is separated from its promoter by a floxed STOP nucleotide sequence. Besides controlling Cre, this ER gene fragment can also be fused with an oncogenic transcription factor (such as c-Myc) so that tamoxifen can be used to control its transcriptional activity (Littlewood et al. 1995).

Besides inducibly expressing an oncogene or deleting a tumor suppressor to initiate mammary tumors, inducible gene-expression tools have also been used to test whether an oncogene is required for maintenance, continued growth, progression, and metastasis of breast cancer. For example, following doxycycline administration to induce transgenic *c-Myc* to cause mammary tumors, withdrawal of doxycycline caused half of the tumors to rapidly regress, indicating that continued high levels of *c-Myc* are needed for tumor maintenance; therefore, these tumors are addicted to transgenic *c-Myc* (i.e., oncogene addiction) (D'Cruz et al. 2001). However, the remaining half of the tumors gained activating mutations in *Kras*, causing them to become independent of the initiating *c-Myc* oncogene (D'Cruz et al. 2001). These data have implications in cancer treatment, as they suggest that patients with *c-Myc*-overexpressed tumors may need to be treated with drugs targeting both Myc and Ras signaling for eradication.

While these GEM models have greatly advanced research in breast cancer initiation, progression, and treatment, conventional GEM models and even conditional and inducible models usually cause a tissue-wide impact, often leading to early stages of tumorigenesis throughout the whole organ. If the initiating oncogenic event is a potent oncogene, multifocal tumors rapidly emerge, as in the case of MMTV-*PyMT*. In most cases, stochastic tumors form after a much longer tumor latency because the mutated epithelia still need additional genetic (and epigenetic) changes to progress to cancer, and only a small subset of these precancerous cells gain sporadic secondary or tertiary genetic lesions, leading to clonally derived cancer. While these models are suitable for studying tumor progression, they are not highly relevant for the study of tumor initiation, especially the initiation of sporadic breast cancer, which arises from one or a few mutated cells in a normally developed ductal tree. These singly mutated cells then evolve into a cancer after additional rounds of mutations and clonal expansion. The next section will discuss models that are made to better mimic sporadic cancer formation.

INTRADUCTAL DELIVERY OF VIRAL VECTORS TO MANIPULATE SOMATIC MAMMARY CELLS TO MODEL SPORADIC BREAST CANCER

Most human breast tumors arise sporadically as a result of somatic mutations in a few cells that clonally expand into a cancer. To closely model the formation of these sporadic breast tumors in patients, we and others have used retrovirus and lentivirus to carry proto-oncogenes and their various mutants into a small number of mammary epithelial cells after the mammary gland is fully developed (Fig. 1; Du and Li 2007; Siwko et al. 2008). The mammary gland is uniquely amenable to this type of genetic manipulation since these viruses can be delivered by injection into

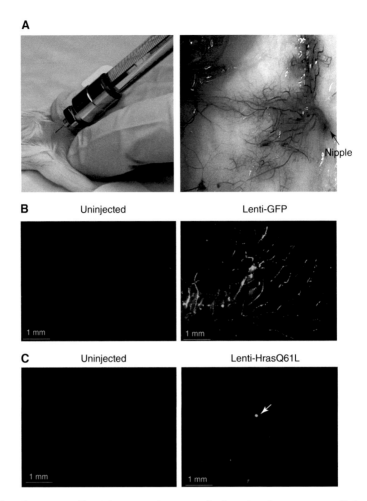

A

B Uninjected Lenti-GFP

C Uninjected Lenti-HrasQ61L

Figure 1. Intraductal injection of lentivirus to engineer genetic alterations in mammary cells in mice. (*A*) Photos showing a successful intraductal injection. Bromophenol blue spreads throughout the ductal tree. (*B*) Photos showing whole-mount GFP imaging of a control gland uninjected with virus and a gland injected by Lenti-*GFP* (3.6×10^6 IUs) 3 d earlier. (*C*) Photos showing whole-mount GFP imaging of a control gland uninjected with virus and a gland injected by Lenti-*HrasQ61L*-GFP (2.4×10^6 IUs) 8 d earlier. Several GFP$^+$ precancerous early lesions can be seen, and the largest one is noted by an arrow.

the nipple duct (i.e., up-the-teat injection) (Yang et al. 1995; Bu and Li 2023). While a human nipple harbors the openings of several lactiferous ducts, which branch into segmental ducts and then subsegmental ducts that further expand into ductile and acini (alveoli) forming lobules, a mouse teat is home to a single major duct that branches out to form the entire ductal tree. This duct is sealed off by keratinized skin cells in virgin mice, but can be easily exposed by transection using surgical scissors. Then a Hamilton syringe fitted with a blunt needle can be used to inject up to 30 µL of virus (>30 µL can lead to duct rupture; Fig. 1A). Viruses injected into the nipple duct can travel up the entire ductal tree, infecting any cell along the route (Fig. 1B). Depending on the titer of the virus injected, a few to thousands of cells can be infected. Since the luminal cells are "chained" tightly to each other by tight junctions, desmosomes, and gap junctions, and since this epithelial layer is also covered by the underneath myoepithelial layer and further sealed off from the stroma by basement membrane, viruses flushed into the lumen appear to infect luminal cells only, rarely infect myoepithelial cells, and generally do not infect cells beyond the basement membrane (Bu et al. 2009). Retroviruses generally require cell division for integration into the genome, but lentivirus can infect all cells in contact and therefore have a broader target population. Besides oncogenes, viruses can also be used to carry *Cre* into mammary cells to delete loxP-marked tumor suppressor genes to initiate cancer. For deleting a floxed gene, Cre is needed only for a brief time; therefore, adenovirus, which is a DNA virus and does not integrate into genome, can substitute for retrovirus and lentivirus for a transient expression of *Cre* (Visbal et al. 2011). On the other hand, for inducible expression, an *rtTA* or *Cre^{ERT}* needs to be delivered by a retroviral or lentiviral vector so that these genes are present when inducers are administered.

Lenti- and retrovirus-mediated expression can be highly valuable for studying the impact of reproduction on mammary tumorigenesis compared to conventional transgenic models. This is because mammary-selective promoters such as MMTV or WAP have the PR- and STAT5-binding sites (Chalepakis et al. 1988; Préfontaine et al. 1999) and are induced during pregnancy and lactation when high progesterone and prolactin levels activate PR and STAT5. As a result, they transcribe higher levels of MMTV or WAP-driven transgene during reproduction. These fluctuations of transgene expression can significantly complicate any study of reproduction on mammary tumorigenesis and progression. In contrast, ubiquitous promoters used in viral vectors do not become more transcriptionally active during reproduction; thus, viral vectors can generate more suitable models for studying the impact of reproduction on mammary tumor formation and progression. By using a retrovirus to deliver *ErbB2* and *Wnt1* and then impregnating the mice, we found that reproduction instigates precancerous lesions to advance to cancer (Haricharan et al. 2013). We further showed that this accelerated initiation is due to precancerous cells activating the prolactin-JAK2-STAT5 signaling in late pregnancy and lactation, leading to a weakened apoptosis anticancer barrier and thus accelerated progression to cancer (Haricharan et al. 2013). In addition, while involution

normally deactivates STAT5 and turns on STAT3 to activate apoptosis and clear excess alveolar cells gained during reproduction, it cannot distinguish in a timely manner the prolactin receptor-JAK2-STAT5 signaling in these precancerous cells; therefore, this involution-associated protection against cell expansion is also weakened in these precancerous lesions, leading to accelerated progression to cancer (Haricharan et al. 2013). Since precancerous lesions are more likely to be found in older than in younger women, this finding provides an explanation for why late-age childbirth increases breast cancer risk. Using a similar approach, we showed that hyperprolactinemia-inducing antipsychotics also stimulate precancerous early lesions to progress to cancer via activation of STAT5, exposing a potential breast cancer risk of using this group of antipsychotics in women who have developed atypia (Johnston et al. 2018).

Since tumor initiation in these somatic models closely mimics human breast cancer initiation, we have also used these models to study breast cancer chemoprevention, a field that has not received its deserved attention as discussed in the article in this collection by Dr. Brown (Brown 2023). Using a viral vector to initiate tumorigenesis and then administering therapeutic agents to test their efficacy on the progression of these precancerous lesions to cancer, we have found that even a short-term treatment with small molecular inhibitors against JAK1/2 (ruxolitinib) or the STAT protein family (C188-9) or the key anti-apoptosis protein Bcl2 (venetoclax) can restore the apoptosis anticancer barrier and prevent breast cancer (Haricharan et al. 2013; Young et al. 2022). Based on these preclinical data, we have already initiated a multicenter window-of-opportunity clinical trial to test the impact of ruxolitinib on apoptosis and biomarkers including pSTAT5 and its selected targets in premalignant lesions (Translational Breast Cancer Research Consortium TBCRC042, Clinical Trials.gov Identifier: NCT02928978).

However, this intraductal viral method of model generation has some drawbacks. (1) Each experimental animal needs to be intraductally injected, which is time-consuming and not trivial technically, while transgenic and KO models need only animal breeding to generate new experimental mice. (2) Another concern is that the use of foreign genes or reporters or epitope tags (e.g., HA, FLAG, Cre, rtTA, GFP, and luciferase) in these vectors may unnaturally provoke the immune system because they are neoantigens (Gambotto et al. 2000; Han et al. 2008; Limberis et al. 2009; Petkov et al. 2013). Luciferase is an especially potent neoantigen (Petkov et al. 2013). These neoantigens may both cause immune clearance of the infected cells and complicate the interpretation of the data. To avoid these complications, several tolerized models have been made (Day et al. 2014; Aoyama et al. 2018). For example, "glowing head" mice are transgenic for brain-specific expression of GFP and luciferase and they are reported to be peripherally tolerant to both luciferase and GFP (Day et al. 2014), although recent studies suggest that even the "glowing head" mice are not completely immune-tolerized (Grzelak et al. 2022). (3) This method also lacks specificity to infect a selected subset of the luminal layer, such as luminal stem cells, progenitor cells, ductal luminal cells, or alveolar cells. Cell-type-specific expression of an oncogene is important in understanding the cancer cell of origin and cancer heterogeneity, since it is becoming increasingly clear that the cell of origin may also contribute to the heterogeneity observed in human breast cancer (for review, see Visvader 2011).

INTRADUCTAL INJECTION OF RCAS VIRUS INTO *Tva*-TRANSGENIC MICE FOR INTRODUCING GENETIC MATERIALS INTO SPECIFIC SUBSETS OF MAMMARY EPITHELIAL CELLS

To achieve cell-type specificity, we have adapted the RCAS-TVA method (Fig. 2; Federspiel et al. 1994). RCAS is a vector derived from Rous sarcoma virus (Hughes 2004). This virus does not infect mammalian cells since the gene encoding its receptor TVA is present only in avian species. Ectopic expression of *Tva* sensitizes any cell to RCAS infection. Transgenic expression of *Tva* under the control of a cell-type-specific promoter thus introduces cell-type specificity to RCAS infection. The RCAS-TVA method was initially

used to introduce oncogenes into specific cells in the brain to model gliomas (Holland et al. 2000), and has subsequently been used to model cancer in many other tissues including the ovary (Orsulic et al. 2002), liver (Lewis et al. 2005), and pancreas (Lewis et al. 2003). We have created several *Tva*-transgenic lines driven off the MMTV LTR (Du et al. 2006), WAP promoter (Haricharan et al. 2014), *Krt6a* promoter (Bu et al. 2011), and TOP (a synthetic promoter that is active in cells with active Wnt signaling) (Bu et al. 2013) for RCAS infection of general luminal cells, alveolar cells, a luminal progenitor, and Wnt signaling-activated cells, respectively. Taking advantage of the cell-specific infection feature in these *Tva* transgenic lines, we have made a number of observations: (1) ErbB2 activation in Wnt-activated cells fails to cause tumors since the infected cells appear to undergo cell death rather than rapid tumorigenesis (Bu et al. 2013), potentially explaining why Wnt signaling is less often observed in HER2[+] breast cancer than in basal-like breast cancer (BLBC); (2) STAT5 activity in WAP[+] alveolar cells can stimulate tumor initiation (Haricharan et al. 2013); (3) KRT6A-positive mammary epithelial progenitors are not at increased vulnerability to tumorigenesis initiated by ErbB2 compared to the bulk luminal population (Holloway et al. 2015), suggesting that another cell subset in the luminal epithelia may be more susceptible to ErbB2-driven tumorigenesis.

We also used RCAS-TVA to study the cell subset in precancerous lesions that evolves into cancer, since precancerous lesions are heterogeneous in cell subsets just like the normal ductal epithelia. While the cell of origin in the normal epithelia has been shown by us and others to affect the characteristics of the resulting mammary tumors, it was previously unknown whether one or multiple cell subsets in precancerous lesions contribute to the eventual cancer. In MMTV-*Wnt1* transgenic mice, we identified a *Krt6a*-expressing precancerous stem cell (PcSC) subset and a more differentiated (WAP)-positive cell subset. RCAS-mediated introduction of constitutively active versions of either *Hras* or *Braf* into these two cell types (made possible through intraductal injection of RCAS into Krt6a-*Tva* and WAP-*Tva* bred with MMTV-*Wnt1*) led to rapid

Cite this article as *Cold Spring Harb Perspect Med* doi: 10.1101/cshperspect.a041328

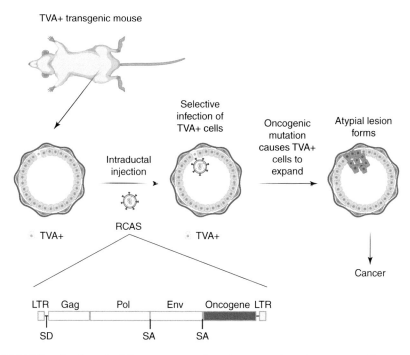

Figure 2. RCAS-TVA technology for delivering genetic alterations into selected mammary cells in mice. In a *Tva*-transgenic mouse line, intraductal injection delivers RCAS to mammary epithelial cells expressing TVA. The number of infected cells varies from a few (Bu et al. 2013, 2019) to tens of thousands (Du et al. 2006) depending upon both the TVA⁺ cells available for infection and the amount of virus injected. Stable expression of an oncogene from long terminal repeat (LTR) following RCAS integration causes infected cells to expand to form atypical lesions, which evolve into cancer.

tumorigenesis (Bu et al. 2019). However, the resulting tumors were dramatically different in protein profiles and histopathology (Bu et al. 2019). These observations indicate that different subsets of precancerous cells can contribute to the cancer etiology, but their differentiation status may impact the resulting cancer characteristics.

Besides using RCAS to deliver oncogenic mutations to selected cell subsets, we have also used this retroviral vector system more generally to study mammary tumor initiation and progression. For example, we found that RCAS-mediated expression of a constitutively activated ErbB2 in somatic mammary cells leads to the formation of an ATM/p53/ARF-dependent senescence barrier and an ATM-dependent but p53/ARF-independent apoptosis barrier (Reddy et al. 2010; Sinha et al. 2015). And our subsequent studies showed that among these ErbB2-activated cells, the loss of ER leads to the formation of a highly aggressive tumor subtype (Ding et al. 2021). In addition, by

injecting RCAS to express both *Cre* and *PyMT* into mice transgenic for *Tva* (K19-*Tva*) and floxed for *Fgfr1*, we deleted *Fgfr1* selectively in PyMT-activated mammary cells, and showed that *Fgfr1* was required for both primary tumor growth and metastasis (Wang et al. 2017).

We have also used this RCAS-TVA method to trace the infected cell population. For example, using RCAS to deliver an HA-tagged endogenous gene to trace the infected cell population, we found that the half-life of mammary epithelial cells as a group is ∼31 d (Dong et al. 2010). Using intraductal injection of a lentivirus to deliver both *ErbB2* and floxed *GFP* in mice that carried a *Cre* knocked into the *PR* locus, we were able to ask whether PR was ever expressed in the evolution to cancer. We found that 75% of tumors retained GFP in all tumor cells (Dong et al. 2016), suggesting that throughout the evolution of a majority of the tumors initiated by *ErbB2*, PR was never expressed.

There are some additional advantages in using RCAS as a viral vector. This virus is replication competent in avian cells; therefore, large amounts of high titer virus can be produced in an avian cell line such as the chicken fibroblast line DF1 by simply expanding the infected cell culture rather than using large-scale transfection of lentiviral DNA constructs mixed together with ancillary vectors, which are usually needed in making replication-deficient lentiviruses. However, in mammalian cells, this virus does not make any appreciable amount of viral proteins such as Gag, Pol, or Env for reasons that are not clear (Li et al. 2011); therefore, it does not elicit immune rejection of infected cells. RCAS is also much safer to handle than lentivirus—it does not infect mammals, so that concerns of accidental infection of the experimenter are minimal, while accidental human infection by lentivirus is a nonnegligible risk to the experimenter—even though replication-deficient vectors cannot spread, the oncogene carried by these vectors can nevertheless pose a significant cancer risk.

RCAS-TVA technology also has a few drawbacks. Shared with the lentiviral and other retroviral vectors are the time-consuming intraductal injection of each experimental mouse, and concern of immune response to any foreign antigen introduced by a viral vector. A technical drawback of RCAS is that this vector can accommodate only a gene <3.0 kb. Large inserts lead to diminished levels of expression and spontaneous truncation (Li et al. 2011). This weakness could be overcome using lentivirus that is pseudo-typed with the RSV *env* gene (Lewis et al. 2001; Siwko et al. 2008), so that it can accommodate large inserts and can integrate into the genomes of both dividing and nondividing cells.

Furthermore, all of these viral vector approaches to introducing oncogenes into somatic cells in mice suffer one common drawback. Following infection and integration into the genome, provirus expresses the gene of interest under the control of exogenous promoters (e.g., viral LTR or a housekeeping gene); therefore, these virus-delivered ectopic genes lack the physiological control of the oncogene in its native locus. Furthermore, retrovirus and lentivirus integrate largely randomly into the genome of the infected cells, as one or multiple copies (although usually preferentially in regions of active chromatin). Depending on the integration site and the number of integrated copies, the expression levels of the oncogene could vary greatly among the infected cells. Moreover, while the chance of a lentivirus/retrovirus landing in a transcribed region of the chromatin of a given cell and disrupting transcription of the critical gene is relatively small, when many cells are infected in multiple mice, a potential tumorigenic impact of virus-caused host gene disruption cannot be totally dismissed.

CRISPR-MEDIATED EDITING OF ENDOGENOUS GENES IN SOMATIC MAMMARY CELLS

To overcome the drawbacks of viral vector approaches including nonphysiologically relevant promoters, insertional effects, and potential immune responses to ectopic genes and tags, the CRISPR-Cas9 system has been adapted in animal models to edit endogenous genes in their native loci (Fig. 3; Anzalone et al. 2020), eliminating the drawbacks associated with the use of viral vector–mediated oncogene expression. CRISPR editing can introduce both indels and point mutations (Anzalone et al. 2020). gRNA-directed, Cas9-catalyzed double-strand breaks (DSBs) at specific genomic locations are repaired by nonhomologous end joining (NHEJ) repair, leading to insertions and deletions and subsequent loss of gene function (Fig. 3A). Multiple mouse models have been produced by using virus carrying gRNA to somatically indel-edit tumor suppressor genes in mice transgenic for Cas9 (Platt et al. 2014; Sánchez-Rivera et al. 2014; Xue et al. 2014; Annunziato et al. 2016, 2019, 2020; Oldrini et al. 2018; Teng et al. 2021).

Precise missense and activating mutations in proto-oncogenes can be achieved by a deaminase-mediated base-editing approach, which uses a nuclease-deficient Cas9 fused with a cytidine or adenine deaminase transition mutation (C:G to T:A base editing [CBE] or A:T to G:C base editing [ABE]) (Anzalone et al. 2020). The C:G to G:C transversion mutation has also been reported (Kurt et al. 2021; Yuan et al. 2021). But only the CBE technique has been used for tumor modeling in vivo (Annunziato et al. 2020), and this editing

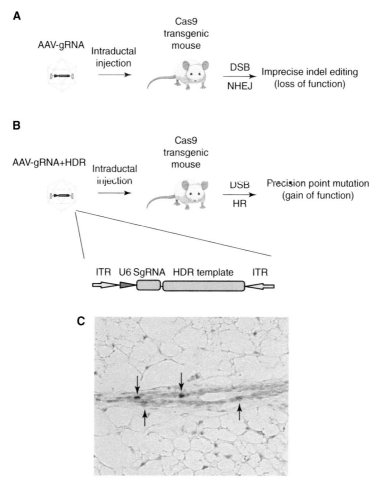

Figure 3. The CRISPR method for introducing indels and point mutations into endogenous genes in their native loci. (*A*) In a *Cas9* transgenic mouse line, intraductal injection of AAV carrying gRNA causes double-strand breaks (DSBs) at the targeted loci, which are repaired by nonhomologous end joining (NHEJ), leading to small insertions and deletions (indels). This method is most commonly used for loss-of-function studies, such as tumor suppressor genes. (*B*) We have found that a modified AAV vector carrying both gRNA and homology-directed repair (HDR) can lead to precise point mutations. This approach can be used to test gain-of function mutations of proto-oncogenes. Of note, indels still occur in this approach, but cells that suffered gain-of-function changes have competitive advantages and are thus selected for in tumor evolution. (*C*) Example of efficient infection of mammary epithelial cells by serotype 9 of AAV.

method is restricted by the availability of the protospacer-adjacent motif (PAM) sequence close to the editing candidate, and currently can perform only six out of 12 possible point mutations. Furthermore, as a specific deaminase cannot differentiate between the intended base and others in close proximity, this method often introduces unintended mutations proximal to the targeted nucleotide within the base-editing window (Annunziato et al. 2020). Prime editor is another approach

to introduce mutations (Anzalone et al. 2020). It is comprised of a Cas9 nickase fused with a reverse transcriptase and a pegRNA, which provide both gRNA and a template for reverse transcription. This method has been used to introduce an activating point mutation into *Ctnbb1* to initiate liver tumors in mice, but there are still technical challenges in vivo (Liu et al. 2021).

On the other hand, coupling homology-directed repair (HDR) to gRNA/Cas9 can introduce all 12 point mutations (Anzalone et al. 2020). The HDR donor sequence is a short DNA fragment matching the target sequence but with predesigned point mutations so that homologous recombination at the Cas9 cleavage site leads to precision mutations although indels also occur. This method is mature for in vitro experiments, but employing it to somatically edit proto-oncogenes in vivo for cancer modeling to date has not been very successful (Platt et al. 2014; Winters et al. 2017; Oldrini et al. 2018). For example, when Platt et al. (2014) constructed an AAV vector to both indel-edit *Tp53* and *Lkb1* and introduce an activating point mutation in the *Kras* proto-oncogene in respiratory epithelium to initiate lung cancer, they detected robust indel mutations of *Tp53* and *Lkb1* in the resulting tumors, but the *KrasG12D* signal was barely above the baseline even though cells with edited *KrasG12D* were predicted to be highly selected for in lung tumorigenesis (Ji et al. 2007). We confirmed that serotype 9 of AAV can efficiently infect mammary epithelial cells in vivo (Fig. 3C). And by modifying the original targeting vector (Platt et al. 2014) to remove reporter genes that may cause immune rejection and/or interfere with HDR function, we have achieved high efficiency precision editing of proto-oncogenes in mammary glands (Fig. 3B; Bu et al. 2023). Somatic editing of two common proto-oncogenes, *Kras* and *Pik3ca*, in either normal or hyperplastic mammary glands led to swift mammary tumor development (Bu et al. 2023). The resulting tumors overall showed less intertumor variation than tumors induced by lentivirus, likely due to the steadier expression levels of the edited genes across infected cells than with virus-expressed oncogenes. These CRISPR-edited tumor models therefore offer more consistent models for cancer biology studies and therapeutic development.

RAT MODELS OF BREAST CANCER: CAN WE FINALLY MODEL ALL ASPECTS OF ER$^+$ BREAST CANCER?

Rats are the first mammalian species domesticated for laboratory research (before the mid-nineteenth century) (Gill et al. 1989). They are not large mice—they are 5 million evolution years closer to humans than mice (Zhao et al. 2004). Rats are much more social than mice, can learn complex cognitive tasks, and are well suited for behavioral, psychological, and neuroscience studies (Jacob and Kwitek 2002). They are also closer pharmacologically to human than mice and are preferred over mice for toxicology studies (Kratchman et al. 2018; Jacob and Kwitek 2002). *Apc* mutant mice develop tumors in the small intestine, but *Apc* mutant rats develop colorectal cancer (Irving et al. 2014) just like human carriers of an *APC* mutant allele, suggesting that rats are more similar to humans with respect to tumor susceptibility than mice in some cases. Rats have six pairs of mammary glands, which form lobules as humans do while the mouse ductal tree largely lacks acini except during pregnancy and lactation (Cardiff et al. 2018). These lobules are also wrapped in modest amounts of connective tissue fibroblasts as lobules in humans are, although the involvement of connective tissue and fibroblasts in rat mammary glands is much less than in human breasts (Masso-Welch et al. 2000).

While hundreds of transgenic and KO mouse models of breast cancer exist and many more exist for development and disease in general, there are far fewer transgenic and KO rat lines. Although pronuclear microinjection can be used to generate transgenic mouse with high efficacy, the application in rats is met with low embryo survival. Likewise, while mouse embryonic stem (ES) cells can be maintained in vitro with relative ease and genetically manipulated before implantation into pseudo-pregnant hosts, rat ES cells lose multipotency fast. These technical hurdles plus the per-diem cost have contributed to the general shortage of genetically engineered rat lines. But technological improvements including improved rat ES cell culturing, genetic manipulation of the spermatogonial stem cells in vitro, and use of CRISPR have lessened the difficulty in generating transgenic and KO rats (Tong et al. 2011; Remy et al. 2014; Chapman et al. 2015; Bäck et al. 2019). On the other hand, intraductal injection of viruses for genetic manipulation was first established in rats, not mice (Wang et al. 1991), since the teat opening of rats is much larger and can be more easily injected. Besides retrovirus and lentivirus, we have found that intraductal injection of AAV can also be used to deliver gRNA to edit endogenous genes in situ in rats with similar ease as in mice (Bu et al., unpubl.).

While only a few GEM models develop ER⁺ mammary tumors and none appears to be estrogen-dependent (Lin et al. 2004; Zhang et al. 2005; Meyer et al. 2011; Tikoo et al. 2012; Chan et al. 2014; Dabydeen and Furth 2014; Van Keymeulen et al. 2015; Andò et al. 2017; Özdemir et al. 2018) (with the possible exception of *Stat1* KO [Chan et al. 2012]), rats readily develop estrogen-dependent mammary tumors following treatment with carcinogens such as 3-methylcholanthrene (MCA), *N*-methyl-*N*-nitrosourea (NMU), or 7,12-dimethybenz[a]anthracene (DMBA) (Huggins et al. 1959, 1961; McCormick et al. 1981), at the ages of ~45–60 d, which is a stage of sexual maturation and rapid mammary gland development (Russo et al. 1990). High proportions (50% or more) of these tumors harbor an activation mutation in *Hras* (Sukumar et al. 1983; Zarbl et al. 1985; Kumar et al. 1990), although mutations in *Pik3ca*, the most commonly mutated protooncogene in human ER⁺ breast cancer, appear to be modest in carcinogen-induced tumors (Showler et al. 2017; Gil Del Alcazar et al. 2022). Intraductal injection of retrovirus carrying *HrasG12E* and *KrasG12E* also led to ovarian hormone-dependent ER⁺ tumors (Wang et al. 1991; Thompson et al. 1998b), and further studies showed that the transforming potential of mutated RAS is mostly transmitted via the effector protein RAF (Showler et al. 2017). We found that intraductal injection of lentivirus carrying *Hras* mutated at codon 61 (Q61L) into rats also led to hormone-dependent ER⁺/PR⁺ adenocarcinoma (Bu and Li 2020), while the same virus in mice led to metaplastic carcinoma with heavy squamous differentiation (Fig. 4), a very rare histological subtype in women, further showing differences of rats versus mice in breast cancer

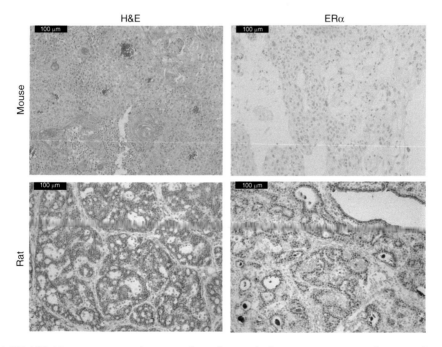

Figure 4. HRASQ61L causes ovarian hormone-dependent ER⁺ adenocarcinoma in rats but metaplastic carcinoma with heavy squamous differentiation in mice. Lenti-*HrasG61L* was intraductally injected into adult FVB/n mice and Sprague Dawley rats. Mammary tumors appear within a few weeks in both groups of animals. Tumors in mice are metaplastic carcinoma with heavy squamous differentiation (also ER-negative). However, tumors in rats are adenocarcinoma with no evidence of metaplasia, but with strong positivity for both ERα and PR (not shown) (Bu and Li 2020). These ER⁺/PR⁺ tumors are also ovarian hormone dependent (Bu and Li 2020).

modeling. Furthermore, intraductal injection into rats of retrovirus carrying a constitutively activated version of *ErbB2* also led to ER$^+$ tumors (Wang et al. 1992), and we have reported that intraductal injection of lentivirus carrying *PIK3-CAH1047R* likewise led to ER$^+$/PR$^+$ mammary tumors in rats (Bu and Li 2020). Moreover, Steensma and colleagues showed that germline *Nf1* KO, which derepresses both Ras signaling and ER transcription, also led to ER$^+$/PR$^+$ mammary tumors in rats (Dischinger et al. 2018). In addition, we found that intraductal injection of AAV carrying gRNA targeting *Nf1* also led to ER$^+$/PR$^+$ mammary tumors in rats (Fig. 5). Together, these data indicate that rats have a strong tendency to develop ER$^+$ tumors regardless of carcinogen-mediated gene mutations or virus-mediated oncogene delivery/editing.

It is unclear at this time why rats are highly prone to ER$^+$ mammary tumors. It is possible that compared to the ER$^-$ cells, the ER$^+$ subset of mammary epithelial population in rats is highly vulnerable to transformation by carcinogens and oncogenic mutations (at least in what we have tested so far). Alternatively, the stromal microenvironment or systemic factors in rats may also contribute to this difference, and it has been reported that the tumor-immune environment of rats appears to bear similarities to human breast cancers (Gil Del Alcazar et al. 2022), although rigorous comparisons among rats, mice, and humans are yet to be done. In addition, since even in mouse, ER is often expressed in precancerous lesions before it is down-regulated in progression to cancer, it is possible that ER expression is controlled differently in tumor evolution in rats versus mice.

Rat models may provide a valuable tool to investigate the progression from ductal carcinoma in situ (DCIS) to invasive cancer. DCIS is detected in high frequency in women—approximately 50,000 cases a year in the United States. It is considered to be the immediate precursor to invasive cancer. As detailed in the article in this collection by Drs. Behbod, Chen, and Thompson (Behbod et al. 2023), there is an intense interest in the breast cancer community to understand mechanisms controlling DCIS progression to invasive cancer, but DCIS is rarely reported in GEM models, hampering this area of research. However, DCIS appears to be a common occurrence in rats either treated with carcinogen (Middleton 1965; Russo and Russo 1978, 1987, 2000; Thompson et al. 1995, 1998a; Gil Del Alcazar et al. 2022) or intraductally injected with viruses to deliver oncogenic changes (Fig. 6; Wang et al. 1991).

Rat models may also offer a new tool to investigate ER$^+$ breast cancer bone metastasis. Human ER$^+$ breast cancer metastasizes primarily to bone, but none of the genetic models in mice to date have been reported to develop spontaneous bone metastasis, although a few mouse and rat tumor cell lines implanted into the fat pad can sometimes form bone metastasis (Simmons et al. 2015). In rats bearing tumors induced by carcinogens, two groups detected hypercalcemia (Gullino et al. 1975; Stoica et al. 1983, 1984), and one group even observed potential tumor cells in bone marrow, although adenocarcinomatous or papillary features were absent, raising concerns about their

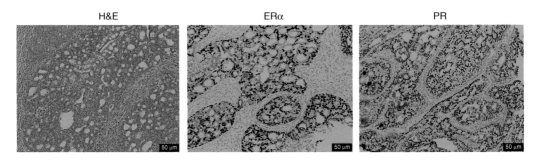

Figure 5. The *Nf1* CRISPR knockout (KO) rat model of ER$^+$/PR$^+$ breast cancer. AAV9 carrying gRNA targeting *Nf1* was intraductally injected into *Cas9* rats. Tumors were palpated within 2 wk to 3 mo. These tumors are adenocarcinoma-positive for both ERα and PR.

 Cite this article as *Cold Spring Harb Perspect Med* doi: 10.1101/cshperspect.a041328

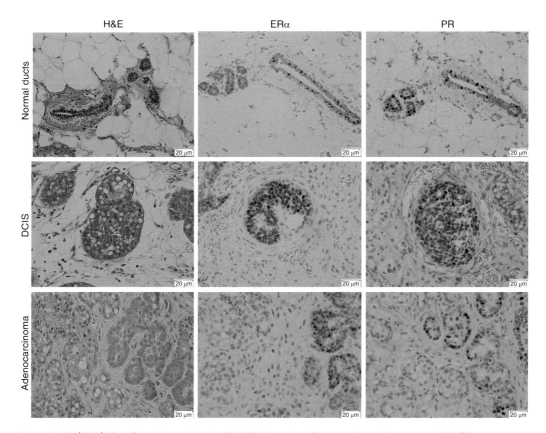

Figure 6. ER⁺/PR⁺ ductal carcinoma in situ (DCIS), invasive adenocarcinoma in a rat model of breast cancer. Lenti-*PIK3CAH1047R* was intraductally injected into adult Sprague Dawley rats. These rats develop ER⁺/PR⁺ DCIS, followed by invasive ER⁺/PR⁺ adenocarcinoma with a median latency of ~3.5 mo (Bu and Li 2020).

mammary origin (Gullino et al. 1975). Our preliminary characterization indicates that some of these rat models of ER⁺ breast cancer develop bone metastases that can be confirmed by X-ray and histology (Bu et al., unpubl.). These models may provide a much-needed tool to address multiple critical questions in ER⁺ breast cancer research. (1) What are the mechanisms controlling metastatic growth in bone versus visceral organs? (2) What are the mechanisms of endocrine resistance in bone? (3) Are immune cells involved in metastatic progression in bone? (4) Why do ER⁺ breast cancer bone metastases present as immune-cold when they are immersed in an immune-rich bone microenvironment? And (5) Could ER⁺ bone metastases be converted to immune-hot and then treated with anti-immune checkpoint therapy?

The rat's larger size also offers benefits for metastasis studies (1) since primary tumors can

be allowed to develop larger, seeding more metastatic cells in distant organs, and (2) since metastatic lesions may also be allowed to expand to a larger size than in mice. Furthermore, laboratory rats can live 3–5 yr or even longer depending on the strain, while mice typically live no more than 2 yr. Therefore, there is more time to observe metastatic development or residual disease development and recurrence following primary tumor resection or therapeutic interventions.

CONCLUDING REMARKS

Many rodent models have been developed in the past, but improved models are emerging due to technological advancement in genetic manipulation in rodents. Immunocompetence is a critical advantage of these models, but they also have some model-dependent drawbacks. Potential pitfalls and limitations should be thoroughly considered so that the most appropriate models are selected for the specific questions to be addressed. While the number of mouse models far exceeds that of rat models and mice cost much less to purchase and to house, rats do fill a critical gap—ER$^+$ breast cancer and bone metastasis.

ACKNOWLEDGMENTS

We thank Drs. Jeffrey Rosen and Gary Chamness for a critical review of this manuscript. This work was supported by DOD CDMRP BC191649 and BC191646 as well as NIH R01CA271498.

REFERENCES

*Reference is also in this subject collection.

Andò S, Malivindi R, Catalano S, Rizza P, Barone I, Panza S, Rovito D, Emprou C, Bornert JM, Laverny G, et al. 2017. Conditional expression of Ki-Ras^{G12V} in the mammary epithelium of transgenic mice induces estrogen receptor α (ERα)-positive adenocarcinoma. *Oncogene* **36:** 6420–6431. doi:10.1038/onc.2017.252

Andrechek ER, Hardy WR, Siegel PM, Rudnicki MA, Cardiff RD, Muller WJ. 2000. Amplification of the *neu/erbB-2* oncogene in a mouse model of mammary tumorigenesis. *Proc Natl Acad Sci* **97:** 3444–3449. doi:10.1073/pnas.97.7 .3444

Andres AC, Schönenberger CA, Groner B, Hennighausen L, LeMeur M, Gerlinger P. 1987. Ha-ras oncogene expression directed by a milk protein gene promoter: tissue specificity, hormonal regulation, and tumor induction in transgenic mice. *Proc Natl Acad Sci* **84:** 1299–1303. doi:10.1073/pnas.84.5.1299

Annunziato S, Kas SM, Nethe M, Yücel H, Del Bravo J, Pritchard C, Bin Ali R, van Gerwen B, Siteur B, Drenth AP, et al. 2016. Modeling invasive lobular breast carcinoma by CRISPR/Cas9-mediated somatic genome editing of the mammary gland. *Genes Dev* **30:** 1470–1480. doi:10 .1101/gad.279190.116

Annunziato S, de Ruiter JR, Henneman L, Brambillasca CS, Lutz C, Vaillant F, Ferrante F, Drenth AP, van der Burg E, Siteur B, et al. 2019. Comparative oncogenomics identifies combinations of driver genes and drug targets in BRCA1-mutated breast cancer. *Nat Commun* **10:** 397. doi:10.1038/s41467-019-08301-2

Annunziato S, Lutz C, Henneman L, Bhin J, Wong K, Siteur B, van Gerwen B, de Korte-Grimmerink R, Zafra MP, Schatoff EM, et al. 2020. In situ CRISPR-Cas9 base editing for the development of genetically engineered mouse models of breast cancer. *EMBO J* **39:** e102169. doi:10 .15252/embj.2019102169

Anzalone AV, Koblan LW, Liu DR. 2020. Genome editing with CRISPR-Cas nucleases, base editors, transposases and prime editors. *Nat Biotechnol* **38:** 824–844. doi:10 .1038/s41587-020-0561-9

Aoyama N, Miyoshi H, Miyachi H, Sonoshita M, Okabe M, Taketo MM. 2018. Transgenic mice that accept luciferase- or GFP-expressing syngeneic tumor cells at high efficiencies. *Genes Cells* **23:** 580–589. doi:10.1111/gtc.12592

Bäck S, Necarsulmer J, Whitaker LR, Coke LM, Koivula P, Heathward EJ, Fortuno LV, Zhang Y, Yeh CG, Baldwin HA, et al. 2019. Neuron-specific genome modification in the adult rat brain using CRISPR-Cas9 transgenic rats. *Neuron* **102:** 105–119.e8. doi:10.1016/j.neuron.2019.01 .035

Bao J, Di Lorenzo A, Lin K, Lu Y, Zhong Y, Sebastian MM, Muller WJ, Yang Y, Bedford MT. 2019. Mouse models of overexpression reveal distinct oncogenic roles for different type I protein arginine methyltransferases. *Cancer Res* **79:** 21–32. doi:10.1158/0008-5472.CAN-18-1995

* Behbod F, Chen JH, Thompson A. 2023. Human ductal carcinoma in situ: advances and future perspectives. *Cold Spring Harb Perspect Med* doi:10.1101/cshperspect .a041319

Booth BW, Mack DL, Androutsellis-Theotokis A, McKay RD, Boulanger CA, Smith GH. 2008. The mammary microenvironment alters the differentiation repertoire of neural stem cells. *Proc Natl Acad Sci* **105:** 14891–14896. doi:10.1073/pnas.0803214105

Bottos A, Gotthardt D, Gill JW, Gattelli A, Frei A, Tzankov A, Sexl V, Wodnar-Filipowicz A, Hynes NE. 2016. Decreased NK-cell tumour immunosurveillance consequent to JAK inhibition enhances metastasis in breast cancer models. *Nat Commun* **7:** 12258. doi:10.1038/ ncomms12258

Bowman-Colin C, Xia B, Bunting S, Klijn C, Drost R, Bowman P, Fineman L, Chen X, Culhane AC, Cai H, et al. 2013. *palb2* synergizes with *Trp53* to suppress mammary tumor formation in a model of inherited breast cancer. *Proc Natl Acad Sci* **110:** 8632–8637. doi:10.1073/pnas .1305362110

Branda CS, Dymecki SM. 2004. Talking about a revolution: the impact of site-specific recombinases on genetic analyses in mice. *Dev Cell* **6:** 7–28. doi:10.1016/S1534-5807 (03)00399-X

* Brown P. 2023. Prevention in breast cancer. *Cold Spring Harb Perspect Med* doi:10.1101/cshperspect.a041345

Bu W, Li Y. 2020. Intraductal injection of lentivirus vectors for stably introducing genes into rat mammary epithelial cells in vivo. *J Mammary Gland Biol Neoplasia* **25:** 389–396. doi:10.1007/s10911-020-09469-w

Bu W, Li Y. 2023. In vivo gene delivery into mouse mammary epithelial cells through mammary intraductal injection. *J Vis Exp* doi:10.3791/64718

Bu W, Xin L, Toneff M, Li L, Li Y. 2009. Lentivirus vectors for stably introducing genes into mammary epithelial cells in vivo. *J Mammary Gland Biol Neoplasia* **14:** 401–404. doi:10.1007/s10911-009-9154-4

Bu W, Chen J, Morrison GD, Huang S, Creighton CJ, Huang J, Chamness GC, Hilsenbeck SG, Roop DR, Leavitt AD, et al. 2011. Keratin 6a marks mammary bipotential progenitor cells that can give rise to a unique tumor model resembling human normal-like breast cancer. *Oncogene* **30**: 4399–4409. doi:10.1038/onc.2011.147

Bu W, Zhang X, Dai H, Huang S, Li Y. 2013. Mammary cells with active Wnt signaling resist ErbB2-induced tumorigenesis. *PLoS ONE* **8**: e78720. doi:10.1371/journal.pone .0078720

Bu W, Liu Z, Jiang W, Nagi C, Huang S, Edwards DP, Jo E, Mo Q, Creighton CJ, Hilsenbeck SG, et al. 2019. Mammary precancerous stem and non-stem cells evolve into cancers of distinct subtypes. *Cancer Res* **79**: 61–71. doi:10 .1158/0008-5472.CAN-18-1087

Bu W, Creighton CJ, Heavener KS, Gutierrez C, Dou Y, Ku AT, Zhang Y, Jiang W, Urrutia J, Jiang W, et al. 2023. Efficient cancer modeling through CRISPR/Cas9/HDR-based somatic precision gene editing in mice. *Sci Adv* (in press).

Cardiff RD, Jindal S, Treuting PM, Going JJ, Gusterson B, Thompson HJ. 2018. Mammary gland. In *Comparative anatomy and histology* (ed. Treuting P, Dintzis S, Montine KS), pp. 487–509. Academic, Cambridge, MA.

Cato AC, Henderson D, Ponta H. 1987. The hormone response element of the mouse mammary tumour virus DNA mediates the progestin and androgen induction of transcription in the proviral long terminal repeat region. *EMBO J* **6**: 363–368. doi:10.1002/j.1460-2075.1987.tb 04763.x

Chakraborty B, Byemerwa J, Shepherd J, Haines CN, Baldi R, Gong W, Liu W, Mukherjee D, Artham S, Lim F, et al. 2021. Inhibition of estrogen signaling in myeloid cells increases tumor immunity in melanoma. *J Clin Invest* **131**: e151347. doi:10.1172/JCI151347

Chalepakis G, Arnemann J, Slater E, Brüller HJ, Gross B, Beato M. 1988. Differential gene activation by glucocorticoids and progestins through the hormone regulatory element of mouse mammary tumor virus. *Cell* **53**: 371–382. doi:10.1016/0092-8674(88)90157-2

Chan SR, Vermi W, Luo J, Lucini L, Rickert C, Fowler AM, Lonardi S, Arthur C, Young LJ, Levy DE, et al. 2012. STAT1-deficient mice spontaneously develop estrogen receptor α-positive luminal mammary carcinomas. *Breast Cancer Res* **14**: R16. doi:10.1186/bcr3100

Chan SR, Rickert CG, Vermi W, Sheehan KC, Arthur C, Allen JA, White JM, Archambault J, Lonardi S, McDevitt TM, et al. 2014. Dysregulated STAT1-SOCS1 control of JAK2 promotes mammary luminal progenitor cell survival and drives ERα⁺ tumorigenesis. *Cell Death Differ* **21**: 234–246. doi:10.1038/cdd.2013.116

Chapman KM, Medrano GA, Jaichander P, Chaudhary J, Waits AE, Nobrega MA, Hotaling JM, Ober C, Hamra FK. 2015. Targeted germline modifications in rats using CRISPR/Cas9 and spermatogonial stem cells. *Cell Rep* **10**: 1828–1835. doi:10.1016/j.celrep.2015.02.040

Chen Y, McAndrews KM, Kalluri R. 2021. Clinical and therapeutic relevance of cancer-associated fibroblasts. *Nat Rev Clin Oncol* **18**: 792–804. doi:10.1038/s41571-021-00546-5

Cronin CA, Gluba W, Scrable H. 2001. The *lac* operator-repressor system is functional in the mouse. *Genes Dev* **15**: 1506–1517. doi:10.1101/gad.892001

Cunha GR, Hom YK. 1996. Role of mesenchymal-epithelial interactions in mammary gland development. *J Mam-*

mary Gland Biol Neoplasia **1**: 21–35. doi:10.1007/BF02096300

Dabydeen SA, Furth PA. 2014. Genetically engineered ERα-positive breast cancer mouse models. *Endocr Relat Cancer* **21**: R195–R208. doi:10.1530/ERC-13-0512

Day CP, Carter J, Weaver Ohler Z, Bonomi C, El Meskini R, Martin P, Graff-Cherry C, Feigenbaum L, Tüting T, Van Dyke T, et al. 2014. "Glowing head" mice: a genetic tool enabling reliable preclinical image-based evaluation of cancers in immunocompetent allografts. *PLoS ONE* **9**: e109956. doi:10.1371/journal.pone.0109956

D'Cruz CM, Gunther EJ, Boxer RB, Hartman JL, Sintasath L, Moody SE, Cox JD, Ha SI, Belka GK, Golant A, et al. 2001. c-MYC induces mammary tumorigenesis by means of a preferred pathway involving spontaneous Kras2 mutations. *Nat Med* **7**: 235–239. doi:10.1038/84691

Ding Y, Liu Y, Lee DK, Tong Z, Yu X, Li Y, Xu Y, Lanz RB, O'Malley BW, Xu J. 2021. Cell lineage tracing links ERα loss in Erbb2-positive breast cancers to the arising of a highly aggressive breast cancer subtype. *Proc Natl Acad Sci* **118**: e2100673118. doi:10.1073/pnas.2100673118

Dischinger PS, Tovar EA, Essenburg CJ, Madaj ZB, Gardner EE, Callaghan ME, Turner AN, Challa AK, Kempston T, Eagleson B, et al. 2018. NF1 deficiency correlates with estrogen receptor signaling and diminished survival in breast cancer. *NPJ Breast Cancer* **4**: 29. doi:10.1038/s41523-018-0080-8

Donehower LA, Harvey M, Slagle BL, McArthur MJ, Montgomery CA Jr, Butel JS, Bradley A. 1992. Mice deficient for p53 are developmentally normal but susceptible to spontaneous tumours. *Nature* **356**: 215–221. doi:10 .1038/356215a0

Donehower LA, Godley LA, Aldaz CM, Pyle R, Shi YP, Pinkel D, Gray J, Bradley A, Medina D, Varmus HE. 1995. Deficiency of p53 accelerates mammary tumorigenesis in Wnt-1 transgenic mice and promotes chromosomal instability. *Genes Dev* **9**: 882–895. doi:10.1101/gad.9.7.882

Dong J, Tong T, Reynado AM, Rosen JM, Huang S, Li Y. 2010. Genetic manipulation of individual somatic mammary cells in vivo reveals a master role of STAT5a in inducing alveolar fate commitment and lactogenesis even in the absence of ovarian hormones. *Dev Biol* **346**: 196–203. doi:10.1016/j.ydbio.2010.07.027

Dong J, Zhao W, Shi A, Toneff M, Lydon J, So D, Li Y. 2016. The PR status of the originating cell of ER/PR-negative mouse mammary tumors. *Oncogene* **35**: 4149–4154. doi:10.1038/onc.2015.465

Du Z, Li Y. 2007. RCAS-TVA in the mammary gland: an in vivo oncogene screen and a high fidelity model for breast transformation? *Cell Cycle* **6**: 823–826. doi:10.4161/cc.6.7 .4074

Du Z, Podsypanina K, Huang S, McGrath A, Toneff MJ, Bogoslovskaia E, Zhang X, Moraes RC, Fluck M, Allred DC, et al. 2006. Introduction of oncogenes into mammary glands in vivo with an avian retroviral vector initiates and promotes carcinogenesis in mouse models. *Proc Natl Acad Sci* **103**: 17396–17401. doi:10.1073/pnas.0608 607103

Dymecki SM, Tomasiewicz H. 1998. Using Flp-recombinase to characterize expansion of Wnt1-expressing neural progenitors in the mouse. *Dev Biol* **201**: 57–65. doi:10.1006/dbio.1998.8971

* Emens LA, Loi S. 2023. Immunotherapy approaches for breast cancer patients in 2023. *Cold Spring Harb Perspect Med* doi:10.1101/cshperspect.a041332

Federspiel MJ, Bates P, Young JA, Varmus HE, Hughes SH. 1994. A system for tissue-specific gene targeting: transgenic mice susceptible to subgroup A avian leukosis virus-based retroviral vectors. *Proc Natl Acad Sci* **91:** 11241–11245. doi:10.1073/pnas.91.23.11241

Feil R, Brocard J, Mascrez B, LeMeur M, Metzger D, Chambon P. 1996. Ligand-activated site-specific recombination in mice. *Proc Natl Acad Sci* **93:** 10887–10890. doi:10.1073/pnas.93.20.10887

Gambotto A, Dworacki G, Cicinnati V, Kenniston T, Steitz J, Tüting T, Robbins PD, DeLeo AB. 2000. Immunogenicity of enhanced green fluorescent protein (EGFP) in BALB/c mice: identification of an H2-Kd-restricted CTL epitope. *Gene Ther* **7:** 2036–2040. doi:10.1038/sj.gt.3301335

Gill TJ III, Smith GJ, Wissler RW, Kunz HW. 1989. The rat as an experimental animal. *Science* **245:** 269–276. doi:10.1126/science.2665079

Gil Del Alcazar CR, Trinh A, Alečković M, Rojas Jimenez E, Harper NW, Oliphant MUJ, Xie S, Krop ED, Lulseged B, Murphy KC, et al. 2022. Insights into immune escape during tumor evolution and response to immunotherapy using a rat model of breast cancer. *Cancer Immunol Res* **10:** 680–697. doi:10.1158/2326-6066.CIR-21-0804

Godley LA, Kopp JB, Eckhaus M, Paglino JJ, Owens J, Varmus HE. 1996. Wild-type p53 transgenic mice exhibit altered differentiation of the ureteric bud and possess small kidneys. *Genes Dev* **10:** 836–850. doi:10.1101/gad.10.7.836

Goswami S, Anandhan S, Raychaudhuri D, Sharma P. 2023. Myeloid cell-targeted therapies for solid tumours. *Nat Rev Immunol* **23:** 106–120. doi:10.1038/s41577-022-00737-w

Grimm SL, Bu W, Longley MA, Roop DR, Li Y, Rosen JM. 2006. Keratin 6 is not essential for mammary gland development. *Breast Cancer Res* **8:** R29. doi:10.1186/bcr1504

Grzelak CA, Goddard ET, Lederer EE, Rajaram K, Dai J, Shor RE, Lim AR, Kim J, Beronja S, Funnell APW, et al. 2022. Elimination of fluorescent protein immunogenicity permits modeling of metastasis in immune-competent settings. *Cancer Cell* **40:** 1–2. doi:10.1016/j.ccell.2021.11.004

Gullino PM, Pettigrew HM, Grantham FH. 1975. *N*-nitrosomethylurea as mammary gland carcinogen in rats. *J Natl Cancer Inst* **54:** 401–414.

Guy CT, Cardiff RD, Muller WJ. 1992. Induction of mammary tumors by expression of polyomavirus middle T oncogene: a transgenic mouse model for metastatic disease. *Mol Cell Biol* **12:** 954–961.

Han WGH, Unger WWJ, Wauben MHM. 2008. Identification of the immunodominant CTL epitope of EGFP in C57BL/6 mice. *Gene Ther* **15:** 700–701. doi:10.1038/sj.gt.3303104

Haricharan S, Dong J, Hein S, Reddy JP, Du Z, Toneff M, Holloway K, Hilsenbeck SG, Huang S, Atkinson R, et al. 2013. Mechanism and preclinical prevention of increased breast cancer risk caused by pregnancy. *eLife* **2:** e00996. doi:10.7554/eLife.00996

Haricharan S, Hein SM, Dong J, Toneff MJ, Aina OH, Rao PH, Cardiff RD, Li Y. 2014. Contribution of an alveolar cell of origin to the high-grade malignant phenotype of pregnancy-associated breast cancer. *Oncogene* **33:** 5729–5739. doi:10.1038/onc.2013.521

Herschkowitz JI, Simin K, Weigman VJ, Mikaelian I, Usary J, Hu Z, Rasmussen KE, Jones LP, Assefnia S, Chandrasekharan S, et al. 2007. Identification of conserved gene expression features between murine mammary carcinoma models and human breast tumors. *Genome Biol* **8:** R76. doi:10.1186/gb-2007-8-5-r76

Hobbs AA, Richards DA, Kessler DJ, Rosen JM. 1982. Complex hormonal regulation of rat casein gene expression. *J Biol Chem* **257:** 3598–3605. doi:10.1016/S0021-9258(18)34822-1

Holland EC, Li Y, Celestino J, Dai C, Schaefer L, Sawaya RA, Fuller GN. 2000. Astrocytes give rise to oligodendrogliomas and astrocytomas after gene transfer of polyoma virus middle T antigen in vivo. *Am J Pathol* **157:** 1031–1037. doi:10.1016/S0002-9440(10)64615-9

Holloway KR, Sinha VC, Toneff MJ, Bu W, Hilsenbeck SG, Li Y. 2015. Krt6a-positive mammary epithelial progenitors are not at increased vulnerability to tumorigenesis initiated by ErbB2. *PLoS ONE* **10:** e0117239. doi:10.1371/journal.pone.0117239

Huggins C, Briziarelli G, Sutton H Jr. 1959. Rapid induction of mammary carcinoma in the rat and the influence of hormones on the tumors. *J Exp Med* **109:** 25–42. doi:10.1084/jem.109.1.25

Huggins C, Grand LC, Brillantes FP. 1961. Mammary cancer induced by a single feeding of polynuclear hydrocarbons, and its suppression. *Nature* **189:** 204–207. doi:10.1038/189204a0

Hughes SH. 2004. The RCAS vector system. *Folia Biol (Praha)* **50:** 107–119.

Indra AK, Warot X, Brocard J, Bornert JM, Xiao JH, Chambon P, Metzger D. 1999. Temporally-controlled site-specific mutagenesis in the basal layer of the epidermis: comparison of the recombinase activity of the tamoxifen-inducible Cre-ER^T and Cre-ER^{T2} recombinases. *Nucleic Acids Res* **27:** 4324–4327. doi:10.1093/nar/27.22.4324

Irving AA, Yoshimi K, Hart ML, Parker T, Clipson L, Ford MR, Kuramoto T, Dove WF, Amos-Landgraf JM. 2014. The utility of Apc-mutant rats in modeling human colon cancer. *Dis Model Mech* **7:** 1215–1225.

Jacob HJ, Kwitek AE. 2002. Rat genetics: attaching physiology and pharmacology to the genome. *Nat Rev Genet* **3:** 33–42. doi:10.1038/nrg702

Jerry DJ, Kittrell FS, Kuperwasser C, Laucirica R, Dickinson ES, Bonilla PJ, Butel JS, Medina D. 2000. A mammary-specific model demonstrates the role of the p53 tumor suppressor gene in tumor development. *Oncogene* **19:** 1052–1058. doi:10.1038/sj.onc.1203270

Ji H, Ramsey MR, Hayes DN, Fan C, McNamara K, Kozlowski P, Torrice C, Wu MC, Shimamura T, Perera SA, et al. 2007. LKB1 modulates lung cancer differentiation and metastasis. *Nature* **448:** 807–810. doi:10.1038/nature06030

Jiang Z, Deng T, Jones R, Li H, Herschkowitz JI, Liu JC, Weigman VJ, Tsao MS, Lane TF, Perou CM, et al. 2010. Rb deletion in mouse mammary progenitors induces luminal-B or basal-like/EMT tumor subtypes depending on p53 status. *J Clin Invest* **120:** 3296–3309. doi:10.1172/JCI41490

Jin KT, Du WL, Lan HR, Liu YY, Mao CS, Du JL, Mou XZ. 2021. Development of humanized mouse with patient-derived xenografts for cancer immunotherapy studies: a comprehensive review. *Cancer Sci* **112:** 2592–2606. doi:10.1111/cas.14934

Johnston AN, Bu W, Hein S, Garcia S, Camacho L, Xue L, Qin L, Nagi C, Hilsenbeck SG, Kapali J, et al. 2018. Hyperprolactinemia-inducing antipsychotics increase breast cancer risk by activating JAK-STAT5 in precancerous lesions. *Breast Cancer Res* **20:** 42. doi:10.1186/s13058-018-0969-z

Jonkers J, Meuwissen R, van der Gulden H, Peterse H, van der Valk M, Berns A. 2001. Synergistic tumor suppressor activity of BRCA2 and p53 in a conditional mouse model for breast cancer. *Nat Genet* **29:** 418–425. doi:10.1038/ng747

Kim IS, Gao Y, Welte T, Wang H, Liu J, Janghorban M, Sheng K, Niu Y, Goldstein A, Zhao N, et al. 2019. Immuno-subtyping of breast cancer reveals distinct myeloid cell profiles and immunotherapy resistance mechanisms. *Nat Cell Biol* **21:** 1113–1126. doi:10.1038/s41556-019-0373-7

Kratchman J, Wang B, Gray G. 2018. Which is most sensitive? Assessing responses of mice and rats in toxicity bioassays. *J Toxicol Environ Health A* **81:** 173–183.

Kumar R, Sukumar S, Barbacid M. 1990. Activation of *ras* oncogenes preceding the onset of neoplasia. *Science* **248:** 1101–1104. doi:10.1126/science.2188364

Kuperwasser C, Chavarria T, Wu M, Magrane G, Gray JW, Carey L, Richardson A, Weinberg RA. 2004. Reconstruction of functionally normal and malignant human breast tissues in mice. *Proc Natl Acad Sci* **101:** 4966–4971. doi:10.1073/pnas.0401064101

Kuraguchi M, Ohene-Baah NY, Sonkin D, Bronson RT, Kucherlapati R. 2009. Genetic mechanisms in Apc-mediated mammary tumorigenesis. *PLoS Genet* **5:** e1000367. doi:10.1371/journal.pgen.1000367

Kurt IC, Zhou R, Iyer S, Garcia SP, Miller BR, Langner LM, Grünewald J, Joung JK. 2021. CRISPR C-to-G base editors for inducing targeted DNA transversions in human cells. *Nat Biotechnol* **39:** 41–46. doi:10.1038/s41587-020-0609-x

Leder A, Pattengale PK, Kuo A, Stewart TA, Leder P. 1986. Consequences of widespread deregulation of the c-*myc* gene in transgenic mice: multiple neoplasms and normal development. *Cell* **45:** 485–495. doi:10.1016/0092-8674(86)90280-1

Lewandoski M. 2001. Conditional control of gene expression in the mouse. *Nat Rev Genet* **2:** 743–755. doi:10.1038/35093537

* Lewis MT, Caldas C. 2023. The power and promise of patient-derived xenografts of human breast cancer. *Cold Spring Harb Perspect Med* doi:10.1101/cshperspect.a041329

Lewis BC, Chinnasamy N, Morgan RA, Varmus HE. 2001. Development of an avian leukosis-sarcoma virus subgroup A pseudotyped lentiviral vector. *J Virol* **75:** 9339–9344. doi:10.1128/JVI.75.19.9339-9344.2001

Lewis BC, Klimstra DS, Varmus HE. 2003. The c-*myc* and PyMT oncogenes induce different tumor types in a so-matic mouse model for pancreatic cancer. *Genes Dev* **17:** 3127–3138. doi:10.1101/gad.1140403

Lewis BC, Klimstra DS, Socci ND, Xu S, Koutcher JA, Varmus HE. 2005. The absence of *p53* promotes metastasis in a novel somatic mouse model for hepatocellular carcinoma. *Mol Cell Biol* **25:** 1228–1237. doi:10.1128/MCB.25.4.1228-1237.2005

Li Y, Hively WP, Varmus HE. 2000. Use of MMTV-Wnt-1 transgenic mice for studying the genetic basis of breast cancer. *Oncogene* **19:** 1002–1009. doi:10.1038/sj.onc.1203273

Li Y, Ferris A, Lewis BC, Orsulic S, Williams BO, Holland EC, Hughes SH. 2011. The RCAS/TVA somatic gene transfer method in modeling human cancer. In *Genetically engineered mice for cancer research: design, analysis, pathways, validation and pre-clinical testing* (ed. Green JE, Ried T), pp. 83–111. Springer, New York.

Limberis MP, Bell CL, Wilson JM. 2009. Identification of the murine firefly luciferase-specific CD8 T-cell epitopes. *Gene Ther* **16:** 441–447. doi:10.1038/gt.2008.177

Lin TP, Guzman RC, Osborn RC, Thordarson G, Nandi S. 1992. Role of endocrine, autocrine, and paracrine interactions in the development of mammary hyperplasia in Wnt-1 transgenic mice. *Cancer Res* **52:** 4413–4419.

Lin SC, Lee KF, Nikitin AY, Hilsenbeck SG, Cardiff RD, Li A, Kang KW, Frank SA, Lee WH, Lee EY. 2004. Somatic mutation of *p53* leads to estrogen receptor α-positive and -negative mouse mammary tumors with high frequency of metastasis. *Cancer Res* **64:** 3525–3532. doi:10.1158/0008-5472.CAN-03-3524

Littlewood TD, Hancock DC, Danielian PS, Parker MG, Evan GI. 1995. A modified oestrogen receptor ligand-binding domain as an improved switch for the regulation of heterologous proteins. *Nucleic Acids Res* **23:** 1686–1690. doi:10.1093/nar/23.10.1686

Liu P, Liang SQ, Zheng C, Mintzer E, Zhao YG, Ponnienselvan K, Mir A, Sontheimer EJ, Gao G, Flotte TR, et al. 2021. Improved prime editors enable pathogenic allele correction and cancer modelling in adult mice. *Nat Commun* **12:** 2121. doi:10.1038/s41467-021-22295-w

Masso-Welch PA, Darcy KM, Stangle-Castor NC, Ip MM. 2000. A developmental atlas of rat mammary gland histology. *J Mammary Gland Biol Neoplasia* **5:** 165–185. doi:10.1023/A:1026491221687

McCormick DL, Adamowski CB, Fiks A, Moon RC. 1981. Lifetime dose-response relationships for mammary tumor induction by a single administration of *N*-methyl-*N*-nitrosourea. *Cancer Res* **41:** 1690–1694.

Meyer DS, Brinkhaus H, Müller U, Müller M, Cardiff RD, Bentires-Alj M. 2011. Luminal expression of *PIK3CA* mutant H1047R in the mammary gland induces heterogeneous tumors. *Cancer Res* **71:** 4344–4351. doi:10.1158/0008-5472.CAN-10-3827

Middleton PJ. 1965. The histogenesis of mammary tumours induced in the rat by chemical carcinogens. *Br J Cancer* **19:** 830–839. doi:10.1038/bjc.1965.96

Mukherjee S, Louie SG, Campbell M, Esserman L, Shyamala G. 2000. Ductal growth is impeded in mammary glands of C-neu transgenic mice. *Oncogene* **19:** 5982–5987. doi:10.1038/sj.onc.1203964

No D, Yao TP, Evans RM. 1996. Ecdysone-inducible gene expression in mammalian cells and transgenic mice. *Proc Natl Acad Sci* **93**: 3346–3351. doi:10.1073/pnas.93.8.3346

Oldrini B, Curiel-García A, Marques C, Matia V, Uluçkan O, Graña-Castro O, Torres-Ruiz R, Rodriguez-Perales S, Huse JT, Squatrito M. 2018. Somatic genome editing with the RCAS-TVA-CRISPR-Cas9 system for precision tumor modeling. *Nat Commun* **9**: 1466. doi:10.1038/s41467-018-03731-w

Orsulic S, Li Y, Soslow RA, Vitale-Cross LA, Gutkind JS, Varmus HE. 2002. Induction of ovarian cancer by defined multiple genetic changes in a mouse model system. *Cancer Cell* **1**: 53–62. doi:10.1016/S1535-6108(01)00002-2

Özdemir BC, Sflomos G, Brisken C. 2018. The challenges of modeling hormone receptor-positive breast cancer in mice. *Endocr Relat Cancer* **25**: R319–R330. doi:10.1530/ERC-18-0063

Petkov SP, Heuts F, Krotova OA, Kilpelainen A, Engström G, Starodubova ES, Isagulients MG. 2013. Evaluation of immunogen delivery by DNA immunization using non-invasive bioluminescence imaging. *Hum Vaccin Immunother* **9**: 2228–2236. doi:10.4161/hv.25561

Pittius CW, Hennighausen L, Lee E, Westphal H, Nicols E, Vitale J, Gordon K. 1988. A milk protein gene promoter directs the expression of human tissue plasminogen activator cDNA to the mammary gland in transgenic mice. *Proc Natl Acad Sci* **85**: 5874–5878. doi:10.1073/pnas.85.16.5874

Platt RJ, Chen S, Zhou Y, Yim MJ, Swiech L, Kempton HR, Dahlman JE, Parnas O, Eisenhaure TM, Jovanovic M, et al. 2014. CRISPR-Cas9 knockin mice for genome editing and cancer modeling. *Cell* **159**: 440–455. doi:10.1016/j.cell.2014.09.014

Podsypanina K, Li Y, Varmus HE. 2004. Evolution of somatic mutations in mammary tumors in transgenic mice is influenced by the inherited genotype. *BMC Med* **2**: 24. doi:10.1186/1741-7015-2-24

Préfontaine GG, Walther R, Giffin W, Lemieux ME, Pope L, Haché RJ. 1999. Selective binding of steroid hormone receptors to octamer transcription factors determines transcriptional synergism at the mouse mammary tumor virus promoter. *J Biol Chem* **274**: 26713–26719. doi:10.1074/jbc.274.38.26713

* Quail DF, Park M, Welm AL, Ekiz HA. 2023. Breast cancer immunity: it is TIME for the next chapter. *Cold Spring Harb Perspect Med* doi:10.1101/cshperspect.a041324

Reddy JP, Peddibhotla S, Bu W, Zhao J, Haricharan S, Du YC, Podsypanina K, Rosen JM, Donehower LA, Li Y. 2010. Defining the ATM-mediated barrier to tumorigenesis in somatic mammary cells following ErbB2 activation. *Proc Natl Acad Sci* **107**: 3728–3733. doi:10.1073/pnas.0910665107

Remy S, Tesson L, Menoret S, Usal C, De Cian A, Thepenier V, Thinard R, Baron D, Charpentier M, Renaud JB, et al. 2014. Efficient gene targeting by homology-directed repair in rat zygotes using TALE nucleases. *Genome Res* **24**: 1371–1383. doi:10.1101/gr.171538.113

Russo IH, Russo J. 1978. Developmental stage of the rat mammary gland as determinant of its susceptibility to 7,12-dimethylbenz[a]anthracene. *J Natl Cancer Inst* **61**: 1439–1449.

Russo J, Russo IH. 1987. Biological and molecular bases of mammary carcinogenesis. *Lab Invest* **57**: 112–137.

Russo J, Russo IH. 2000. Atlas and histologic classification of tumors of the rat mammary gland. *J Mammary Gland Biol Neoplasia* **5**: 187–200. doi:10.1023/A:1026443305758

Russo J, Gusterson BA, Rogers AE, Russo IH, Wellings SR, van Zwieten MJ. 1990. Comparative study of human and rat mammary tumorigenesis. *Lab Invest* **62**: 244–278.

Sánchez-Rivera FJ, Papagiannakopoulos T, Romero R, Tammela T, Bauer MR, Bhutkar A, Joshi NS, Subbaraj L, Bronson RT, Xue W, et al. 2014. Rapid modelling of cooperating genetic events in cancer through somatic genome editing. *Nature* **516**: 428–431. doi:10.1038/nature13906

Sauer B, Henderson N. 1988. Site-specific DNA recombination in mammalian cells by the Cre recombinase of bacteriophage P1. *Proc Natl Acad Sci* **85**: 5166–5170. doi:10.1073/pnas.85.14.5166

Selbert S, Bentley DJ, Melton DW, Rannie D, Lourenço P, Watson CJ, Clarke AR. 1998. Efficient BLG-Cre mediated gene deletion in the mammary gland. *Transgenic Res* **7**: 387–398. doi:10.1023/A:1008848304391

Sharma P, Allison JP. 2020. Dissecting the mechanisms of immune checkpoint therapy. *Nat Rev Immunol* **20**: 75–76. doi:10.1038/s41577-020-0275-8

Showler K, Nishimura M, Daino K, Imaoka T, Nishimura Y, Morioka T, Blyth BJ, Kokubo T, Takabatake M, Fukuda M, et al. 2017. Analysis of genes involved in the PI3K/Akt pathway in radiation- and MNU-induced rat mammary carcinomas. *J Radiat Res* **58**: 183–194. doi:10.1093/jrr/rrw097

Simmons JK, Hildreth BE III, Supsavhad W, Elshafae SM, Hassan BB, Dirksen WP, Toribio RE, Rosol TJ. 2015. Animal models of bone metastasis. *Vet Pathol* **52**: 827–841. doi:10.1177/0300985815586223

Sinha VC, Qin L, Li Y. 2015. A p53/ARF-dependent anticancer barrier activates senescence and blocks tumorigenesis without impacting apoptosis. *Mol Cancer Res* **13**: 231–238. doi:10.1158/1541-7786.MCR-14-0481-T

Sinn E, Muller W, Pattengale P, Tepler I, Wallace R, Leder P. 1987. Coexpression of MMTV/v-Ha-ras and MMTV/c-myc genes in transgenic mice: synergistic action of oncogenes in vivo. *Cell* **49**: 465–475. doi:10.1016/0092-8674(87)90449-1

Siwko SK, Bu W, Gutierrez C, Lewis B, Jechlinger M, Schaffhausen B, Li Y. 2008. Lentivirus-mediated oncogene introduction into mammary cells in vivo induces tumors. *Neoplasia* **10**: 653–662. doi:10.1593/neo.08266

Stewart TA, Pattengale PK, Leder P. 1984. Spontaneous mammary adenocarcinomas in transgenic mice that carry and express MTV/*myc* fusion genes. *Cell* **38**: 627–637. doi:10.1016/0092-8674(84)90257-5

Stöcklin E, Botteri F, Groner B. 1993. An activated allele of the c-erbB-2 oncogene impairs kidney and lung function and causes early death of transgenic mice. *J Cell Biol* **122**: 199–208. doi:10.1083/jcb.122.1.199

Stoica G, Koestner A, Capen CC. 1983. Characterization of N-ethyl-N-nitrosourea–induced mammary tumors in the rat. *Am J Pathol* **110**: 161–169.

Stoica G, Koestner A, Capen CC. 1984. Neoplasms induced with high single doses of N-ethyl-N-nitrosourea in 30-day-old sprague-dawley rats, with special emphasis on mammary neoplasia. *Anticancer Res* **4:** 5–12.

Sukumar S, Notario V, Martin-Zanca D, Barbacid M. 1983. Induction of mammary carcinomas in rats by nitroso-methylurea involves malignant activation of H-ras-1 locus by single point mutations. *Nature* **306:** 658–661. doi:10.1038/306658a0

Teng K, Ford MJ, Harwalkar K, Li Y, Pacis AS, Farnell D, Yamanaka N, Wang YC, Badescu D, Ton Nu TN, et al. 2021. Modeling high-grade serous ovarian carcinoma using a combination of in vivo fallopian tube electroporation and CRISPR-Cas9–mediated genome editing. *Cancer Res* **81:** 5147–5160. doi:10.1158/0008-5472.CAN-20-1518

Thompson HJ, McGinley JN, Rothhammer K, Singh M. 1995. Rapid induction of mammary intraductal proliferations, ductal carcinoma in situ and carcinomas by the injection of sexually immature female rats with 1-methyl-1-nitrosourea. *Carcinogenesis* **16:** 2407–2412. doi:10.1093/carcin/16.10.2407

Thompson HJ, McGinley JN, Wolfe P, Singh M, Steele VE, Kelloff GJ. 1998a. Temporal sequence of mammary intraductal proliferations, ductal carcinomas in situ and adenocarcinomas induced by 1-methyl-1-nitrosourea in rats. *Carcinogenesis* **19:** 2181–2185. doi:10.1093/carcin/19.12.2181

Thompson TA, Kim K, Gould MN. 1998b. Harvey ras results in a higher frequency of mammary carcinomas than Kirsten ras after direct retroviral transfer into the rat mammary gland. *Cancer Res* **58:** 5097–5104.

Tikoo A, Roh V, Montgomery KG, Ivetac I, Waring P, Pelzer R, Hare L, Shackleton M, Humbert P, Phillips WA. 2012. Physiological levels of *Pik3ca*^H1047R mutation in the mouse mammary gland results in ductal hyperplasia and formation of ERα-positive tumors. *PLoS ONE* **7:** e36924. doi:10.1371/journal.pone.0036924

Tong C, Huang G, Ashton C, Li P, Ying QL. 2011. Generating gene knockout rats by homologous recombination in embryonic stem cells. *Nat Protoc* **6:** 827–844. doi:10.1038/nprot.2011.338

Ursini-Siegel J, Schade B, Cardiff RD, Muller WJ. 2007. Insights from transgenic mouse models of ERBB2-induced breast cancer. *Nat Rev Cancer* **7:** 389–397. doi:10.1038/nrc2127

Van Keymeulen A, Lee MY, Ousset M, Brohée S, Rorive S, Giraddi RR, Wuidart A, Bouvencourt G, Dubois C, Salmon I, et al. 2015. Reactivation of multipotency by oncogenic PIK3CA induces breast tumour heterogeneity. *Nature* **525:** 119–123. doi:10.1038/nature14665

Vassar R, Rosenberg M, Ross S, Tyner A, Fuchs E. 1989. Tissue-specific and differentiation-specific expression of a human K14 keratin gene in transgenic mice. *Proc Natl Acad Sci* **86:** 1563–1567. doi:10.1073/pnas.86.5.1563

Visbal AP, LaMarca HL, Villanueva H, Toneff MJ, Li Y, Rosen JM, Lewis MT. 2011. Altered differentiation and paracrine stimulation of mammary epithelial cell proliferation by conditionally activated smoothened. *Dev Biol* **352:** 116–127. doi:10.1016/j.ydbio.2011.01.025

Visvader JE. 2011. Cells of origin in cancer. *Nature* **469:** 314–322. doi:10.1038/nature09781

Wagner KU, Wall RJ, St-Onge L, Gruss P, Wynshaw-Boris A, Garrett L, Li M, Furth PA, Hennighausen L. 1997. Cre-mediated gene deletion in the mammary gland. *Nucleic Acids Res* **25:** 4323–4330. doi:10.1093/nar/25.21.4323

Wang BC, Kennan WS, Yasukawa-Barnes J, Lindstrom MJ, Gould MN. 1991. Carcinoma induction following direct in situ transfer of v-Ha-ras into rat mammary epithelial cells using replication-defective retrovirus vectors. *Cancer Res* **51:** 2642–2648.

Wang B, Kennan WS, Yasukawa-Barnes J, Lindstrom MJ, Gould MN. 1992. Difference in the response of neu and ras oncogene-induced rat mammary carcinomas to early and late ovariectomy. *Cancer Res* **52:** 4102–4105.

Wang W, Meng Y, Dong B, Dong J, Ittmann MM, Creighton CJ, Lu Y, Zhang H, Shen T, Wang J, et al. 2017. A versatile tumor gene deletion system reveals a crucial role for FGFR1 in breast cancer metastasis. *Neoplasia* **19:** 421–428. doi:10.1016/j.neo.2017.03.003

Welte T, Kim IS, Tian L, Gao X, Wang H, Li J, Holdman XB, Herschkowitz JI, Pond A, Xie G, et al. 2016. Oncogenic mTOR signalling recruits myeloid-derived suppressor cells to promote tumour initiation. *Nat Cell Biol* **18:** 632–644. doi:10.1038/ncb3355

Whitelaw CB, Harris S, McClenaghan M, Simons JP, Clark AJ. 1992. Position-independent expression of the ovine β-lactoglobulin gene in transgenic mice. *Biochem J* **286:** 31–39. doi:10.1042/bj2860031

Whitfield J, Littlewood T, Evan GI, Soucek L. 2015. The estrogen receptor fusion system in mouse models: a reversible switch. *Cold Spring Harb Protoc* **2015:** 227–234.

Winters IP, Chiou SH, Paulk NK, McFarland CD, Lalgudi PV, Ma RK, Lisowski L, Connolly AJ, Petrov DA, Kay MA, et al. 2017. Multiplexed in vivo homology-directed repair and tumor barcoding enables parallel quantification of Kras variant oncogenicity. *Nat Commun* **8:** 2053. doi:10.1038/s41467-017-01519-y

Wu M, Jung L, Cooper AB, Fleet C, Chen L, Breault L, Clark K, Cai Z, Vincent S, Bottega S, et al. 2009. Dissecting genetic requirements of human breast tumorigenesis in a tissue transgenic model of human breast cancer in mice. *Proc Natl Acad Sci* **106:** 7022–7027. doi:10.1073/pnas.0811785106

Xu X, Wagner KU, Larson D, Weaver Z, Li C, Ried T, Hennighausen L, Wynshaw-Boris A, Deng CX. 1999. Conditional mutation of Brca1 in mammary epithelial cells results in blunted ductal morphogenesis and tumour formation. *Nat Genet* **22:** 37–43. doi:10.1038/8743

Xue W, Chen S, Yin H, Tammela T, Papagiannakopoulos T, Joshi NS, Cai W, Yang G, Bronson R, Crowley DG, et al. 2014. CRISPR-mediated direct mutation of cancer genes in the mouse liver. *Nature* **514:** 380–384. doi:10.1038/nature13589

Yang J, Tsukamoto T, Popnikolov N, Guzman RC, Chen X, Yang JH, Nandi S. 1995. Adenoviral-mediated gene transfer into primary human and mouse mammary epithelial cells in vitro and in vivo. *Cancer Lett* **98:** 9–17. doi:10.1016/S0304-3835(06)80004-2

Young A, Bu W, Jiang W, Ku A, Kapali J, Dhamne S, Qin L, Hilsenbeck SG, Du YN, Li Y. 2022. Targeting the pro-survival protein BCL-2 to prevent breast cancer. *Cancer Prev Res (Phila)* **15:** 3–10. doi:10.1158/1940-6207.CAPR-21-0031

Yuan T, Yan N, Fei T, Zheng J, Meng J, Li N, Liu J, Zhang H, Xie L, Ying W, et al. 2021. Optimization of C-to-G base editors with sequence context preference predictable by machine learning methods. *Nat Commun* **12:** 4902. doi:10.1038/s41467-021-25217-y

Zarbl H, Sukumar S, Arthur AV, Martin-Zanca D, Barbacid M. 1985. Direct mutagenesis of Ha-ras-1 oncogenes by *N*-nitroso-*N*-methylurea during initiation of mammary carcinogenesis in rats. *Nature* **315:** 382–385. doi:10.1038/315382a0

Zhang X, Podsypanina K, Huang S, Mohsin SK, Chamness GC, Hatsell S, Cowin P, Schiff R, Li Y. 2005. Estrogen receptor positivity in mammary tumors of Wnt-1 transgenic mice is influenced by collaborating oncogenic mutations. *Oncogene* **24:** 4220–4231. doi:10.1038/sj.onc.1208597

Zhao S, Shetty J, Hou L, Delcher A, Zhu B, Osoegawa K, de Jong P, Nierman WC, Strausberg RL, Fraser CM. 2004. Human, mouse, and rat genome large-scale rearrangements: stability versus speciation. *Genome Res* **14:** 1851–1860. doi:10.1101/gr.2663304

Cite this article as *Cold Spring Harb Perspect Med* doi: 10.1101/cshperspect.a041328

The Power and Promise of Patient-Derived Xenografts of Human Breast Cancer

Michael T. Lewis[1] and Carlos Caldas[2]

[1]Baylor College of Medicine, The Lester and Sue Smith Breast Center, Departments of Molecular and Cellular Biology and Radiology, Baylor College of Medicine, Houston, Texas 77030, USA

[2]Cancer Research UK Cambridge Institute, University of Cambridge, Li Ka Shing Centre, Cambridge CB2 0RE, United Kingdom

Correspondence: mtlewis@bcm.edu; carlos.caldas@cruk.cam.ac.uk

In 2016, a group of researchers engaged in the development of patient-derived xenografts (PDXs) of human breast cancer provided a comprehensive review of the state of the field. In that review, they summarized the clinical problem that PDXs might address, the technical approaches to their generation (including a discussion of host animals and transplant conditions tested), and presented transplantation success (take) rates across groups and across transplantation conditions. At the time, there were just over 500 unique PDX models created by these investigators representing all three clinically defined subtypes (ER[+], HER2[+], and TNBC). Today, many of these PDX resources have at least doubled in size, and several more PDX development groups now exist, such that there may be well upward of 1000 PDX models of human breast cancer in existence worldwide. They also presented a series of open questions for the field. Many of these questions have been addressed. However, several remain open, or only partially addressed. Herein, we revisit these questions, and recount the progress that has been made in a number of areas with respect to generation, characterization, and use of PDXs in translational research, and re-present questions that remain open. These open questions, and others, are now being addressed not only by individual investigators, but also large, well-funded consortia including the PDXNet program of the National Cancer Institute in the United States, and the EuroPDX Consortium, an organization of PDX developers across Europe. Finally, we discuss the new opportunities in PDX-based research.

There is great and growing interest in using patient-derived xenograft (PDX) models of human breast cancer as both an experimental platform for basic science, and as a drug development and testing platform for translational science. PDX models have proven to be remarkably stable with respect to histological phenotype, growth characteristics, biomarker expression, genomics, transcriptomics, and proteomics over time (passage in host mice). They have also proven themselves to recapitulate, to a large extent, the biology of the tumors-of-origin at all of these levels, as well as with respect to their responses to treatment, to the extent

---

Cite this article as *Cold Spring Harb Perspect Med* doi: 10.1101/cshperspect.a041329

that this question has been investigated thus far.

These biologically and translationally relevant features have led to rapid adoption in research laboratories worldwide, typically at the expense of the use of long-established breast cancer cell lines in two-dimensional and three-dimensional growth conditions. These cell lines show significant limitations, in many cases, even when grown in four dimensions as cell line xenografts, that lessen their relevance for certain applications such as drug evaluation. In fact, variability from laboratory to laboratory and isolate to isolate (e.g., Nugoli et al. 2003; Prinz et al. 2011; Ben-David et al. 2019) impacts drug responses. Perhaps it should not be surprising that as much as 95% of all cell-line-based results are not translatable clinically (Johnson et al. 2001; Hait 2010). That said, their experimental flexibility and ease of use has allowed for their continued employment in many molecular and genetic studies, with valuable insights into mechanism(s) of signaling and gene regulation.

Despite their attractiveness as an experimental platform, PDXs are not without their own limitations. These limitations can be significant, and indeed hamper adoption by many groups who may lack the time, space, personnel, training, or funding to conduct PDX-based research on a meaningful scale. Fortunately, most PDX development groups have proven more than willing to collaborate with such groups to mutual benefit. Indeed, the combined expertise of groups with domain knowledge in the biological question asked, and those with domain knowledge in the biology, behavior, and use of PDX models at scale, has allowed the field to gain insights into questions that might not have been gained otherwise if the two types of groups had not worked together collaboratively.

USE OF PDXs BY PHARMACEUTICAL COMPANIES AND THE RISE OF THE CONTRACT RESEARCH ORGANIZATION (CRO)

Several pharmaceutical companies have embraced PDXs, and models derived from them (e.g., PDX-derived organoids [PDXOs] and PDX-derived tumor cells [PDTCs] grown in either 2D or 3D on scaffolds or in gel matrices) as experimental platforms for screening drugs and evaluating their efficacy in vivo (e.g., Gao et al. 2015; Gu et al. 2015; Bruna et al. 2016 and subsequent publications). While some pharmaceutical companies have opted to conduct their PDX-based research in house, others have chosen to conduct their research with the assistance of a growing number of CROs. Unfortunately for the field, data generated either by the companies themselves, or by the CROs they engage, are rarely made public. The reasons for this can be trivial or compelling. For a trivial example, if an agent in question failed to have the desired efficacy in the PDX models chosen, it would be understandable that a company may wish to keep this result to themselves. Conversely, it may be critical that data be held in house. For example, if the agent in question had outstanding efficacy, a company may wish to exploit these data to justify clinical trials and drug approval efforts without the possibility of being compromised by exploitation of those data by the research community at large.

Regardless of the reasons for reluctance to release results, data such as these do have a shelf life with respect to their usefulness in drug development efforts. Drug candidates that fail to meet their clinical end points in trials (again for good, or trivial reasons) are supplanted with the next generation of the compound, or are abandoned for other reasons unrelated to their efficacy (e.g., side-effect profile, cost of production, order of market entry, etc.). New drugs from competitors may replace a given drug in the market. Yet still, such data have value to the research community. For example, if a particular agent was effective in a selected set of PDXs, such PDXs could be chosen for subsequent evaluation of agents with a similar mechanism of activity, or because a novel agent had the same molecular target. Such PDXs could also be chosen for use in experiments involving tool compounds (compounds that have biological effects and known targets, but that are not being developed as drugs) for investigation of molecular-genetic mechanisms, particularly those of resistance.

Pharmaceutical companies are currently taking part in centralized drug distribution programs such as the Cancer Therapy Evaluation Program (CTEP), in which drugs are contributed to the program by companies, and then distributed to academic investigators for use in their research, typically with the hope of developing a clinical trial. A useful companion program would be a centralized repository of unpublished PDX-based drug treatment results conducted, or commissioned, by participating companies. Such data could be curated and displayed in a dedicated web portal for use by the research community as a whole. In this context, even negative data have value in that such data might forewarn researchers and perhaps prevent the waste of time and money by grant-funded investigators who might repeat the experiment with the same negative outcome.

OPEN QUESTIONS (ca. 2016), AND SOME ANSWERS

Open questions discussed in 2016 had to do mainly with the uncertainty as to whether PDXs were truly superior to long-established cell lines for translational biology, the degree to which PDXs accurately reflected the biology of the tumors-of-origin, and the uncertainty of exactly how best to generate, characterize, and "credential" PDX models such that they remain useful over time.

While full or partial answers to several of these questions were presented in the 2016 review itself, a few remain unanswered to varying degrees, and new questions have arisen as the community gains experience. In the intervening 6 years since publication, new data related to these questions has been published that increase confidence in the utility and power of PDXs for translational research. For those questions addressed adequately in the review, we will not discuss these in detail here. Rather, we will discuss any new data relevant to the questions posed originally, and focus our attention on lingering open questions, as well as new questions that have arisen.

Which immunocompromised mouse model should be used for most efficient PDX generation while still retaining the highest possible fidelity with the human disease? To an extent, the jury is still out on this question. While data were summarized previously to demonstrate that, under otherwise comparable conditions and patient tumor population, NOD/SCID, SCID/Bg, NSG, and NRG mice performed equivalently with respect to the overall take rate (Dobrolecki et al. 2016). That said, no head-to-head comparison using identical source material for transplantation has been published to demonstrate whether one host is superior to the other, or indeed to any other host strain used by any other groups. As such, it remains possible that one immunocompromised host may allow growth of one class of tumors, while the other performs better for a different class of tumors, although overall their proficiency for supporting PDX growth is equivalent.

Potentially arguing against this idea are recent clustering results of RNA-seq and proteomic data from PDXs developed and maintained in SCID/Bg (Dobrolecki et al. 2016), versus PDXs developed in NSG (Guillen et al. 2022) but maintained in SCID/Bg (all NSG-grown PDXs grew in SCID/Bg) that show considerable intermingling of tumor gene expression patterns, suggesting that there is no overt selection of one set of tumors over another conferred by the nature of these two host strains (Petrosyan et al. 2023).

What is the best transplantation site? Like the question above related to host superiority, there has been remarkably little systematic attention paid to this important question in the literature, and what is in the literature is by no means conclusive. Ultimately, the answer seems to be "…well, that depends."

With respect to available transplantation sites for PDX development and use, tissue may be transplanted orthotopically within the mammary fat pad (either the intact [IFP], or epithelium-free "cleared" fat pad [CFP]) (Deome et al. 1959), injected into the mammary ducts themselves (Behbod et al. 2009; Richard et al. 2016; Sflomos et al. 2021), or heterotopically at other sites (aka "ectopic"), including subcutaneously (flank or subscapular fat), under the renal capsule, or the anterior chamber of the eye (neither the renal capsule nor the eye have been used for establishment thus far) (Dobrolecki et al. 2016).

There are several studies comparing orthotopic and heterotopic growth of mouse, dog, and human cell line–derived xenografts (CDXs). However, there are remarkably few studies with PDXs. Some cell line studies show differences for a given measurement/observation in one direction, while other studies show the opposite. For example, among other studies using mouse 4T1 cells (Balb/c), orthotopic tumors grew faster with more uniformity, had higher blood vessel densities, and showed increased metastatic behavior compared to cells grown subcutaneously (Zhang et al. 2018). Similar results were obtained using MDA-MB-231 cells (Ho et al. 2012), where blood vessel permeability was also increased in orthotopic tumors. For 4T1, the mammary fat pad supported higher T-cell-dependent induction of mediators of inflammation to enhance metastasis (Takahashi et al. 2015). Using E0771 cells (C57BL/6 mice), enhanced growth in orthotopic tumors was dependent on a metabolic interaction with adipose tissue in the mammary fat pad (Micallef et al. 2021). In orthotopic versus subcutaneous transplantation using four breast cancer cell lines (MCF7, ZR75-1, MDA-MB-231, and MCF10Ca1h), these investigators observed differences in final tumor size (orthotopic larger) and in promoter methylation for several genes. Finally, there were differences in vascularization, with the orthotopic site more highly vascularized than the subcutaneous location (Fleming et al. 2010).

On the flip side of this discussion using CDXs, with two inflammatory breast cancer cell lines IPC-366 (canine) and SUM149 (human), Caceres et al. showed that there were significant differences in growth rate (heterotopic faster) and metastasis rate (heterotopic higher), but only using the IPC-366 cell line, with no significant differences in histology steroid hormone secretion between the two sites for either (Caceres et al. 2021). There was a similar observation with human MCF7 cells with orthotopic tumors having a lower proportion of functional blood vessels than subcutaneous (density was not different) (Fung et al. 2015). These tumors were treated subsequently with common chemotherapeutics (paclitaxel, 5FU, doxorubicin), which resulted in an increase in functional blood vessels at both the orthotopic and heterotopic sites. Thus, the differences between the two sites may be model- and manipulation-dependent.

An important caveat to most established cell line experiments is that the results cannot be compared with what existed in the tumor-of-origin. In other words, perhaps a cell line model grows faster, looks mesenchymal, and metastasizes more from the orthotopic site, but the tumor-of-origin is actually slow growing, looks more epithelial, and the patient showed little evidence of metastasis at the time the cell line was made. Such a distinction might disqualify the cell line as an accurate reflection of the tumor-of-origin. Of course, most available cell lines were produced using cells either from pleural effusions or ascites, so these patients were definitively metastatic.

Many investigators believe the orthotopic site in the inguinal (#4) mouse mammary gland is better than any other site in the mouse with respect to recapitulation of the biology of the tumor-of-origin. Indeed, over a decade ago, the #4 gland was shown to allow better cell line and PDX growth than the thoracic glands (#2 and #3) (Fleming et al. 2010), and conferred a different response to etoposide pretreatment with tumor volume decreased in the thoracic glands, whereas the inguinal glands showed no effect of the treatment.

Subsequently, with respect to establishment of PDXs, several groups have now shown similar rates of engraftment ("take rate"). In addition, mRNA and protein expression in medium-to-large PDX collections established and maintained either orthotopically or subcutaneously, correlate very closely with those of the tumor-of-origin (DeRose et al. 2011; Eirew et al. 2015; Bruna et al. 2016; Dobrolecki et al. 2016, and references therein). Further, the histology and clonal heterogeneity of the parental tumor are generally retained in both sites (Zhang et al. 2013; Bruna et al. 2016). These results would suggest that the two sites are roughly equivalent to one another as sites for engraftment.

With respect to recapitulation of tumor-of-origin response to therapeutics, or the ability of PDX-based results to inform/validate patient stratification in clinical trials, there are less data. From these limited data, is does appear that orthotopic transplants show excellent recapitulation of treatment responses to taxanes and perhaps other agents (DeRose et al. 2011; Zhang et al. 2013; Guillen et al. 2022; Petrosyan et al. 2023). It also appears that both subcutaneous and/or orthotopic PDX models can be used to validate clinically derived biomarkers (Pellegrino et al. 2022), to inform patient treatment (Guillen et al. 2022), and perhaps ultimately inform patient stratification for inclusion into prospective clinical trials.

That said, there are some data to suggest that, as for CDXs, the two sites may not, in fact, be fully equivalent for PDX growth. In a recent small-scale, head-to-head study, tissue from 11 donors were implanted into NSG mice both subcutaneously and orthotopically (Okano et al. 2020). Of the 11 donor tumors, six were ER$^+$ and five were TNBC. As could be expected, only one of the six ER$^+$ yielded a PDX, while four of the five TNBC samples yielded a PDX (presumably at both sites, although not stated explicitly). Those grown at the orthotopic site were larger than those grown subcutaneously. These tumors were then retransplanted to passage 3 (P3) to both sites to evaluate re-engraftment rates. Those grown orthotopically showed higher rates of re-engraftment, and faster growth rates corroborated by Ki67 staining, although no obvious histological differences were shown. Unfortunately, the tumors-of-origin were not evaluated for comparison.

As of this writing, one author favors the orthotopic site, and one author favors the subcutaneous site, with no empirical evidence to sway either author one way or the other. That said, the long and the short of it is, if transplantation site is a potential concern, the relevant experiments should be done to determine the importance of transplantation site for behavior of a given model and for a given purpose. Further, results should be compared with observations in the

tumor-of-origin from the patient whenever feasible. Once satisfied, one can decide firmly which transplantation site to use.

If using the mammary fat pad, is it necessary to eliminate endogenous mouse epithelium from the mammary fat pad (i.e., "clear" the fat pad) for efficient orthotopic transplantation? In short, no. In fact, the presence of the endogenous mouse mammary epithelium appears to enhance the take rate overall. In other words, it cannot hurt, and it might help (Dobrolecki et al. 2016). Not clearing also saves time and money, both of which are required for PDX work given the time required to conduct a PDX-based experiment, and the overall higher expenses associated with PDX work.

Is estradiol supplementation necessary? If so, what is the most effective way to deliver it? The answer to this question appears to be "that depends." There are convincing data from multiple groups that E2 supplementation is essential for the growth of some ER$^+$ PDXs, and that supplementation may actually be required for the growth of TNBC when transplanted into the mammary fat pad (Iyer et al. 2012; Zhang et al. 2013; Dobrolecki et al. 2016, and references therein). However, using ductal injections of tumor cells, this requirement for supplementation may not be necessary (Fiche et al. 2019). If generalizable, intraductal injection represents a significant technical breakthrough for the generation and growth of ER$^+$ PDXs.

How can lower grade tumors, ductal carcinoma in situ (DCIS), and normal tissue be grown with higher efficiency? While little to no progress has been made with respect to growing normal human mammary tissue, excellent strides have been made in efforts to grow lower grade tumors. An interesting development that requires careful evaluation is, again, the use of intraductal injection for growth of ductal carcinoma in situ (DCIS) (Hong et al. 2022), ER$^+$ breast cancers (Richard et al. 2016; Fiche et al. 2019), lobular carcinomas (Sflomos et al. 2021), and apocrine tumors (Richard et al. 2016). DCIS essentially does not grow using either subcutaneous or orthotopic (fat pad) transplantation. However, intraductal injection does allow growth of at least some DCIS. ER$^+$ breast cancers and less studied lobular breast cancers are notoriously difficult to engraft, but recent success using intraductal injection appears to be changing that dynamic for the better (Behbod et al. 2009; Fiche et al. 2019; Sflomos et al. 2021; Hong et al. 2022). It remains possible that improvements in humanization techniques may also improve take rates, as these methods are only now coming on line (see below).

It is important to note that, in general, tumors grown using the intraductal injection approach tend to remain localized within the duct (similar to ductal carcinoma in situ) and thus do not lend themselves to drug treatment studies as conducted typically on 3D tumors grown in the mammary fat pad, which are easily measured.

Can "humanization" of immunocompromised mice enhance take rate and improve translational relevance? A known limitation of PDX models is the lack of a fully functional immune system in the host mouse. As a general premise, the mouse mammary stroma (or subcutaneous space) is understandably different from human. Hence, there is the effort to "humanize" immunocompromised hosts. That said, the microenvironment surrounding, intermingling, and interacting with the PDX epithelial cells recapitulates much of the biology of the stroma in human tumors.

As discussed previously (Dobrolecki et al. 2016), there are multiple levels of "humanization" of immunocompromised host mice. There are essentially three main types: (1) transgenic or knockin expression of human genes, particularly those encoding ligands for which the mouse counterpart fails to engage the human receptor, or does not exist in the mouse genome, (2) introduction of human cells of selected types, including immortalized fibroblasts, T cells, B cells, etc., and (3) immune system reconstitution using blood- or bone marrow–derived hematopoietic stem cells. These humanization methods were well reviewed recently in Jin et al. (2021).

With respect to introduction of human genes, this method can reportedly yield enhanced results. For example, knockin of the human prolactin gene (the mouse protein does not activate the human receptor [Utama et al. 2006]

reportedly increases the take rate for ER^+ breast cancer from a paltry 10% to a dramatically improved 43% [Sun et al. 2021a,b]). However, this method may act as a double-edged sword with respect to $HER2^+$ PDX where it may prove to inhibit their growth (Hachim et al. 2019). Thus, this method of humanization may need to be finely tailored for each clinical subtype of breast cancer, and indeed each molecular subgroup within each subtype.

Other genes-of-interest have not been evaluated fully. For example, IL-6 (the mouse counterpart of which does not stimulate the human IL-6 receptor) and IL-8 (a gene for which there is no true counterpart in mice, although there are genes that show homology). Thus, there is ample room for study in this area. Both of these genes are well known to affect epithelial cell function.

With respect to the introduction of human cells for either PDX establishment or therapeutics, despite their attractiveness in theory, implementation into common practice has met with some technical and biological challenges. However, there has been some recent successes (e.g., Byrd et al. 2018; Nalawade et al. 2021; Scherer et al. 2021; for review, see Xie et al. 2020), particularly with respect to the investigation of immuno-oncology agents, and CAR T-cell therapies, as well as adoptive T- and NK-cell therapies (for review, see Dees et al. 2020; Jin et al. 2021; Qiu et al. 2022). In addition, there is evidence that modulation of the immune system using autologous dendritic cell transplantation together with certain chemotherapies such as anthracycline-cyclophosphamide treatment can restore T-cell function and promote immunogenic cell death (Bernal-Estévez et al. 2021), thereby potentially enhancing chemotherapy response. This will likely be one of the most active areas of breast cancer research in the coming years.

As new host strains and as new and refined methods for humanization are developed, PDXs will become an even more powerful platform for understanding the role(s) of the human immune system in regulation of tumor growth and behavior, as well as for the development of novel agents that engage, or alter, the immune system to enhance antitumor activities.

To what degree do PDX models truly recapitulate the biology of the tumor-of-origin in the patient, particularly with respect to treatment response? This has been a very active area of research in recent years, for if PDXs do not recapitulate a vast majority of patient biology, their utility as experimental models would be quite limited—and hardly worth the effort and expense to use them.

Thankfully, PDXs have thus far shown comparable patterns of intratumoral cellular heterogeneity (Georgopoulou et al. 2021), as well as conservation of genomic copy-number alterations and mutations consistent with the tumor-of-origin. They also show remarkable genomic stability over multiple passages (at least 20) in mice (Kerbel et al. 2013; Zhang et al. 2013, 2014; Sun et al. 2021a,b; Woo et al. 2021; Petrosyan et al. 2023). Further, their transcriptomic and proteomic patterns of gene expression not only show a high degree of concordance with the tumor-of-origin directly from the patient, but PDX gene expression at both the transcriptomic and proteomic levels is directly comparable to those contained in the TCGA and CPTAC repositories (Kerbel et al. 2013; Petrosyan et al. 2023) where they represent all of the clinical subtypes, and many of the molecular subtypes, particularly in the basal subtype (mainly TNBC).

Perhaps most importantly, several studies in breast and other cancers have recently shown that responses of PDXs to a select number of therapeutic agents are qualitatively similar to those of the tumor-of-origin in the patient when challenged with the same or similar agent (e.g., Zhang et al. 2013; Izumchenko et al. 2017; Lazzari et al. 2019; Ice et al. 2020; Kang et al. 2020; Petrosyan et al. 2023, and references therein). This quality has even allowed translation of a discovery in PDX-derived organoids to be validated in the corresponding PDXs, and translated to the corresponding patient in "real time" (Guillen et al. 2022).

Are PDXs really any better than xenografts made using long-established cell lines? While it has been shown that, with the exception of the luminal A molecular subtype, breast cancer cell lines do represent the spectrum of human tu-

mors to a degree (Prat et al. 2013), it is well-established that cell lines grown in 2D versus 3D conditions have dramatically different gene expression patterns (Kenny et al. 2007) as well as dramatically different responses to various therapeutic agents (Imamura et al. 2015), with 3D methods showing a closer resemblance to in vivo grown cells (Birgersdotter et al. 2005) as well as better recapitulation of patient biology than 2D with respect to gene expression (Guillen et al. 2022; Ozkan et al. 2022).

What is less clear is whether traditional cell lines grown in 4D (mice) represent the spectrum of human disease, and their respective tumors-of-origin, as well as do PDXs (Sun et al. 2021a,b; Woo et al. 2021). While the former can be addressed with time and money, and has been addressed to an extent (Riaz et al. 2014), the latter cannot, in most cases, as the source material and clinical information required to do so no longer exist, and in vivo growth conditions for many traditional cell lines have not been found.

Unlike traditional cell lines that have drifted genetically over time, and were altered differently from laboratory to laboratory, PDXs have thus far shown remarkable genomic stability over multiple passages (at least 20) in mice (Zhang et al. 2013; Woo et al. 2021). Despite early indications that PDXs may show consistent mouse-specific genomic alterations, they do not appear to do so (Ben-David et al. 2017; Woo et al. 2021). Thus, because PDXs retain biology and behavior of their tumors-of-origin, and traditional cell lines cannot be demonstrated to do so in most cases, PDXs can be considered superior. Recent success in generating PDX-derived organoid lines and cell lines will likely result in the ultimate abandonment of long-established cell lines in favor of these PDX-derived cell lines, which in many cases can be evaluated directly in comparison with their tumors-of-origin.

NEW QUESTIONS AND LIMITATIONS OF BREAST CANCER PDXs

So why are some tumor types more difficult to grow as PDXs than others? Is this due to intrinsic features of the tumor epithelium? Are we doing something wrong technically? Or are we just missing something important in the mouse that is present in the human. More than likely, the reason will be a combination of these three possibilities, and others not mentioned. For example, certain tumors may be inherently unable to adapt to the mouse microenvironment and will either linger, but not grow, as has been observed (Zhang et al. 2013), or they may simply die. Others may possess features that induce what is left of the mouse immune system (e.g., macrophages) to kill it. Although unlikely, a new surgical and/or implantation technique may need to be developed that imposes less stress on the tumor between excision and transplantation. Almost certainly we are missing something, or many things, that are important. As mentioned above, there are mouse hormones (PRL), and cytokines (IL-6) that are incapable of activating the human receptors for them. Further, there are genes that exist in humans that do not exist in mice (IL-8) that are known to contribute to tumor biology. We are almost certainly missing other things that we know about, such as the tumor-promoting actions of the immune system in some cases, and other thing about which we know nothing.

To what degree do PDX collections reflect the racial/ethnic diversity of patients? In short, poorly. While there are a few PDX collections that do have some racial/ethnic diversity (Dobrolecki et al. 2016), as a community, we are falling well short of the mark. Sorely lacking are, for example, Asians (although there may be collections in Asia of which the authors are unaware, but would like to be), Pacific Islanders, continental Africans, as well as Native Americans and other indigenous populations around the world.

To correct this situation will take a concerted effort by PDX groups worldwide to identify these populations, reach out to them, and, if possible, establish the infrastructure and logistics to recruit patients, treat them, and get their tissue into the hands of a PDX developer. While difficult, such a workflow was implemented by Dr. Max Wicha to establish a number of PDX models from women in Ghana (Jiagge et al. 2016, 2021). The resulting PDXs showed ethnicity-associated gene-expression patterns distinct for women with African ancestry, and allowed identification of an ethnicity-specific immune cell signature in TNBC that may be targetable.

What are the strengths and limitations of PDXs for molecular pharmacology and drug evaluation? The strengths of PDX models as a platform for drug candidate evaluation are now very well documented (Collins and Lang 2018), as evidenced by the growing number of publications in which PDX models are used to perform these evaluations, and the growing number of clinical trials initiated that are supported by PDX-based results. This strength is also evidenced by the growing requirement for use of PDX-based data in publications, grant applications, clinical trial designs, and regulatory agency approvals of drugs for use against breast cancer, for which data using established cell lines would have been acceptable previously.

This said, there is evidence that the range of PDX models that can be established is related to the immunological state (i.e., "hot" vs. "cold" of the parental tumor in the patient such that there is a distinct bias toward engraftment of immunologically "cold" tumors as PDXs) (Fig. 1; Petrosyan et al. 2022).

It is less clear whether PDXs can be used effectively for drug target identification, but they are showing some promise (Huang et al. 2017; Sun et al. 2021). What is even less clear is whether cohorts of PDXs can be used as surrogates for clinical cohorts for the development of response and resistance signatures, although again PDXs are showing some promise here too (Matossian et al. 2019; Guo et al. 2020; Petrosyan et al. 2023).

Where PDXs fall well short of the mark is with the development and testing of immuno-oncology agents, although humanization efforts may address this issue, in part, for certain targets.

SOME NEW OPPORTUNITIES

Metastasis

A subset of PDXs has been shown not only to metastasize to organ sites commonly affected in

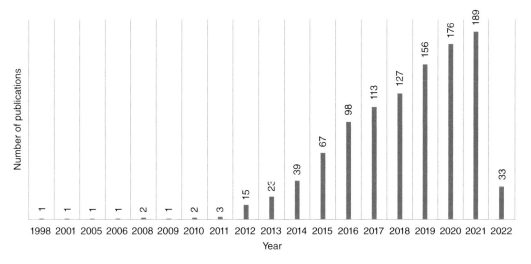

Figure 1. Publications with breast patient–derived xenograft in their title or abstract by year.

patients (Zhang et al. 2013; Giuliano et al. 2015), but also to recapitulate the patterns of metastasis observed in the patient-of-origin (DeRose et al. 2011; for review, see Roarty and Echeverria 2021). It is not yet known which mouse model should be used to have the best representation of PDX-matched patient biology, but NSG appears to be a good place to begin. With the availability of proteogenomic data on a large number of PDXs, there should be a wealth of information to emerge from studies of metastasis (e.g., Sprouffske et al. 2020).

Treatment Resistance, Dormancy, and Recurrence

Emerging data in multiple PDXs using various agents shows that resistance is common, although some PDXs do respond (Powell et al. 2020; Savage et al. 2020). Why some respond, and others do not, can perhaps be addressed in cohorts of PDXs, which may allow accurate prediction of response, not only in PDXs, but patients as well (Petrosyan et al. 2023). Even in those that apparently respond completely, some residual tumor cells remain and lay dormant for weeks to months before they regrow, and regrowth appears to be associated with a metabolic shift that might be exploited therapeutically (Echeverria et al. 2019).

Related to the question of metastasis, the reproducibility of recurrence in PDX models after different chemotherapy regimens and/or targeted agents, and the ability to conduct multiple experiments in the same PDX, offers a remarkable opportunity to understand why breast cancers recur when they do, and the underlying processes by which recurrence happens.

Clonal Heterogeneity and Tumor Evolution

Related to the question of treatment resistance is the presence of cell heterogeneity within individual tumors that influences apparent response, as well as the evolution of tumors over time, and in response to treatment (Nguyen et al. 2014; Eirew et al. 2015; Kim et al. 2018; Menghi et al. 2022). The advent of single cell sequencing and barcoding techniques should allow breakthrough understanding of these various cell types, and how best to target them for elimination of tumors. Indeed, recent barcoding experiments suggest that most disseminated tumor cells do not have the capacity to regenerate metastatic tumors (Merino et al. 2019). Further, upon treatment with cisplatin, about 50% of such disseminated clones could be eliminated at distal sites, yet the clonal diversity in the locally recurrent tumor remained unchanged (Merino et al. 2019).

CONCLUDING REMARKS

Despite their potential shortcomings with respect to some experimental approaches, PDXs are beginning to prove themselves as some of the most translationally relevant models of breast and other cancers in existence. The next 5 to 10 years will determine exactly how relevant these models are as more larger-scale studies are published, especially in the context of coclinical trials in which a cohort of PDX-bearing mice are treated, to the greatest extent possible, just as a corresponding patient cohort are being treated in a clinical trial. Ideally, the PDXs and patient cohorts would show a high degree of consistency and comparability between them to bring the "bench-to-bedside (and back again)" goal for translational research full circle. Thus far, all indications are that PDXs will be quite relevant for the foreseeable future. Powerful and promising indeed.

REFERENCES

Behbod F, Kittrell FS, LaMarca H, Edwards D, Kerbawy S, Heestand JC, Young E, Mukhopadhyay P, Yeh HW, Allred DC, et al. 2009. An intraductal human-in-mouse transplantation model mimics the subtypes of ductal carcinoma in situ. *Breast Cancer Res* **11**: R66. doi:10.1186/bcr2358

Ben-David U, Ha G, Tseng YY, Greenwald NF, Oh C, Shih J, McFarland JM, Wong B, Boehm JS, Beroukhim R, et al. 2017. Patient-derived xenografts undergo mouse-specific tumor evolution. *Nat Genet* **49**: 1567–1575. doi:10.1038/ng.3967

Ben-David U, Beroukhim R, Golub TR. 2019. Genomic evolution of cancer models: perils and opportunities. *Nat Rev Cancer* **19**: 97–109. doi:10.1038/s41568-018-0095-3

Bernal-Estévez DA, Ortíz Barbosa MA, Ortíz-Montero P, Cifuentes C, Sánchez R, Parra-López CA. 2021. Autologous dendritic cells in combination with chemotherapy restore responsiveness of T cells in breast cancer patients: a single-arm phase I/II trial. *Front Immunol* **12**: 669965. doi:10.3389/fimmu.2021.669965

Birgersdotter A, Sandberg R, Ernberg I. 2005. Gene expression perturbation in vitro—a growing case for three-dimensional (3D) culture systems. *Semin Cancer Biol* **15**: 405–412. doi:10.1016/j.semcancer.2005.06.009

Bruna A, Rueda OM, Greenwood W, Batra AS, Callari M, Batra RN, Pogrebniak K, Sandoval J, Cassidy JW, Tufegdzic-Vidakovic A, et al. 2016. A biobank of breast cancer explants with preserved intra-tumor heterogeneity to screen anticancer compounds. *Cell* **167**: 260–274.e22. doi:10.1016/j.cell.2016.08.041

Byrd TT, Fousek K, Pignata A, Szot C, Samaha H, Seaman S, Dobrolecki L, Salsman VS, Oo HZ, Bielamowicz K, et al. 2018. TEM8/ANTXR1-specific CAR T cells as a targeted therapy for triple-negative breast cancer. *Cancer Res* **78**: 489–500. doi:10.1158/00085472.CAN-16-1911

Caceres S, Alonso-Diez A, Crespo B, Peña L, Illera MJ, Silvan G, de Andres PJ, Illera JC. 2021. Tumor growth progression in ectopic and orthotopic xenografts from inflammatory breast cancer cell lines. *Vet Sci* **8**: 194. doi:10.3390/vetsci8090194

Collins AT, Lang SH. 2018. A systematic review of the validity of patient derived xenograft (PDX) models: the implications for translational research and personalised medicine. *PeerJ* **6**: e5981. doi:10.7717/peerj.5981

Dees S, Ganesan R, Singh S, Grewal IS. 2020. Emerging CAR-T cell therapy for the treatment of triple-negative breast cancer. *Mol Cancer Ther* **19**: 2409–2421. doi:10.1158/15357163.MCT-20-0385

Deome KB, Faulkin LJ Jr, Bern HA, Blair PB. 1959. Development of mammary tumors from hyperplastic alveolar nodules transplanted into gland-free mammary fat pads of female C3H mice. *Cancer Res* **19**: 515–520.

DeRose YS, Wang G, Lin YC, Bernard PS, Buys SS, Ebbert MT, Factor R, Matsen C, Milash BA, Nelson E, et al. 2011. Tumor grafts derived from women with breast cancer authentically reflect tumor pathology, growth, metastasis and disease outcomes. *Nat Med* **17**: 1514–1520. doi:10.1038/nm.2454

Dobrolecki LE, Airhart SD, Alferez DG, Aparicio S, Behbod F, Bentires-Alj M, Brisken C, Bult CJ, Cai S, Clarke RB, et al. 2016. Patient-derived xenograft (PDX) models in basic and translational breast cancer research. *Cancer Metastasis Rev* **35**: 547–573. doi:10.1007/s10555-016-9653-x

Echeverria GV, Ge Z, Seth S, Zhang X, Jeter-Jones S, Zhou X, Cai S, Tu Y, McCoy A, Peoples M, et al. 2019. Resistance to neoadjuvant chemotherapy in triple-negative breast cancer mediated by a reversible drug-tolerant state. *Sci Transl Med* **11**: eaav0936. doi:10.1126/scitranslmed.aav0936

Eirew P, Steif A, Khattra J, Ha G, Yap D, Farahani H, Gelmon K, Chia S, Mar C, Wan A, et al. 2015. Dynamics of genomic clones in breast cancer patient xenografts at single-cell resolution. *Nature* **518**: 422–426. doi:10.1038/nature13952

Fiche M, Scabia V, Aouad P, Battista L, Treboux A, Stravodimou A, Zaman K, Dormoy V, Ayyanan A, et al. 2019. Intraductal patient-derived xenografts of estrogen receptor α-positive breast cancer recapitulate the histopathological spectrum and metastatic potential of human lesions. *J Pathol* **247**: 287–292. doi:10.1002/path.5200

Fleming JM, Miller TC, Meyer MJ, Ginsburg E, Vonderhaar BK. 2010. Local regulation of human breast xenograft models. *J Cell Physiol* **224**: 795–806. doi:10.1002/jcp.22190

Fung AS, Lee C, Yu M, Tannock IF. 2015. The effect of chemotherapeutic agents on tumor vasculature in subcutaneous and orthotopic human tumor xenografts. *BMC Cancer* **15**: 112. doi:10.1186/s12885-015-1091-6

Gao H, Korn JM, Ferretti S, Monahan JE, Wang Y, Singh M, Zhang C, Schnell C, Yang G, Zhang Y, et al. 2015. High-throughput screening using patient-derived tumor xenografts to predict clinical trial drug response. *Nat Med* **21**: 1318–1325. doi:10.1038/nm.3954

Georgopoulou D, Callari M, Rueda OM, Shea A, Martin A, Giovannetti A, Qosaj F, Dariush A, Chin SF, Carnevalli LS, et al. 2021. Landscapes of cellular phenotypic diversity in breast cancer xenografts and their impact on drug response. *Nat Commun* **12**: 1998. doi:10.1038/s41467-021-22303-z

Giuliano M, Herrera S, Christiny P, Shaw C, Creighton CJ, Mitchell T, Bhat R, Zhang X, Mao S, Dobrolecki LE, et al. 2015. Circulating and disseminated tumor cells from breast cancer patient-derived xenograft-bearing mice as a novel model to study metastasis. *Breast Cancer Res* **17**: 3. doi:10.1186/s13058-014-0508-5

Gu Q, Zhang B, Sun H, Xu Q, Tan Y, Wang G, Luo Q, Xu W, Yang S, Li J, et al. 2015. Genomic characterization of a large panel of patient-derived hepatocellular carcinoma xenograft tumor models for preclinical development. *Oncotarget* **6**: 20160–20176. doi:10.18632/oncotarget.3969

Guillen KP, Fujita M, Butterfield AJ, Scherer SD, Bailey MH, Chu Z, DeRose YS, Zhao L, Cortes-Sanchez E, Yang CH, et al. 2022. A human breast cancer-derived xenograft and organoid platform for drug discovery and precision oncology. *Nat Cancer* **3**: 232–250. doi:10.1038/s43018-022-00337-6

Guo Z, Primeau T, Luo J, Zhang C, Sun H, Hoog J, Gao F, Huang S, Edwards DP, Davies SR, et al. 2020. Proteomic resistance biomarkers for PI3K inhibitor in triple negative breast cancer patient-derived xenograft models. *Cancers (Basel)* **12**: 3857. doi:10.3390/cancers12123857

Hachim IY, López-Ozuna VM, Hachim MY, Lebrun JJ, Ali S. 2019. Prolactin hormone exerts anti-tumorigenic effects in HER-2 overexpressing breast cancer cells through regulation of stemness. *Stem Cell Res* **40**: 101538. doi:10.1016/j.scr.2019.101538

Hait WN. 2010. Anticancer drug development: the grand challenges. *Nat Rev Drug Discov* **9**: 253–254. doi:10.1038/nrd3144

Ho KS, Poon PC, Owen SC, Shoichet MS. 2012. Blood vessel hyperpermeability and pathophysiology in human tumour xenograft models of breast cancer: a comparison of ectopic and orthotopic tumours. *BMC Cancer* **12**: 579. doi:10.1186/1471-2407-12-579

Hong Y, Limback D, Elsarraj HS, Harper H, Haines H, Hansford H, Ricci M, Kaufman C, Wedlock E, Xu M, et al. 2022. Mouse-INtraDuctal (MIND): an in vivo model for studying the underlying mechanisms of DCIS malignancy. *J Pathol* **256**: 186–201. doi:10.1002/path.5820

Huang KL, Li S, Mertins P, Cao S, Gunawardena HP, Ruggles KV, Mani DR, Clauser KR, Tanioka M, Usary J, et al. 2017. Proteogenomic integration reveals therapeutic targets in breast cancer xenografts. *Nat Commun* **8**: 14864. doi:10.1038/ncomms14864

Ice RJ, Chen M, Sidorov M, Le Ho T, Woo RWL, Rodriguez-Brotons A, Luu T, Jian D, Kim KB, Leong SP, et al. 2020. Drug responses are conserved across patient-derived xenograft models of melanoma leading to identification of novel drug combination therapies. *Br J Cancer* **122**: 648–657. doi:10.1038/s41416-019-0696-y

Imamura Y, Mukohara T, Shimono Y, Funakoshi Y, Chayahara N, Toyoda M, Kiyota N, Takao S, Kono S, Nakatsura T, et al. 2015. Comparison of 2D- and 3D-culture models as drug-testing platforms in breast cancer. *Oncol Rep* **33**: 1837–1843. doi:10.3892/or.2015.3767

Iyer V, Klebba I, McCready J, Arendt LM, Betancur-Boissel M, Wu MF, Zhang X, Lewis MT, Kuperwasser C. 2012. Estrogen promotes ER-negative tumor growth and angiogenesis through mobilization of bone marrow-derived monocytes. *Cancer Res* **72**: 2705–2713. doi:10.1158/0008-5472.CAN-11-3287

Izumchenko E, Paz K, Ciznadija D, Sloma I, Katz A, Vasquez-Dunddel D, Ben-Zvi I, Stebbing J, McGuire W, Harris W, et al. 2017. Patient-derived xenografts effectively capture responses to oncology therapy in a heterogeneous cohort of patients with solid tumors. *Ann Oncol* **28**: 2595–2605. doi:10.1093/annonc/mdx416

Jiagge E, Oppong JK, Bensenhaver J, Aitpillah F, Gyan K, Kyei I, Osei-Bonsu E, Adjei E, Ohene-Yeboah M, Toy K, et al. 2016. Breast cancer and African ancestry: lessons learned at the 10-year anniversary of the Ghana-Michigan Research Partnership and International Breast Registry. *J Glob Oncol* **2**: 302–310. doi:10.1200/JGO.2015.002881

Jiagge EM, Ulintz PJ, Wong S, McDermott SP, Fossi SI, Suhan TK, Hoenerhoff MJ, Bensenhaver JM, Salem B, Dziubinski M, et al. 2021. Multiethnic PDX models predict a possible immune signature associated with TNBC of African ancestry. *Breast Cancer Res Treat* **186**: 391–401. doi:10.1007/s10549-021-06097-8

Jin KT, Du WL, Lan HR, Liu YY, Mao CS, Du JL, Mou XZ. 2021. Development of humanized mouse with patient-derived xenografts for cancer immunotherapy studies: a comprehensive review. *Cancer Sci* **112**: 2592–2606. doi:10.1111/cas.14934

Johnson JI, Decker S, Zaharevitz D, Rubinstein LV, Venditti JM, Schepartz S, Kalyandrug S, Christian M, Arbuck S, Hollingshead M, et al. 2001. Relationships between drug activity in NCI preclinical in vitro and in vivo models and early clinical trials. *Br J Cancer* **84**: 1424–1431. doi:10.1054/bjoc.2001.1796

Kang HN, Kim JH, Park AY, Choi JW, Lim SM, Kim J, Shin EJ, Hong MH, Pyo KH, Yun MR, et al. 2020. Establishment and characterization of patient-derived xenografts as paraclinical models for head and neck cancer. *BMC Cancer* **20**: 316. doi:10.1186/s12885-020-06786-5

Kenny PA, Lee GY, Myers CA, Neve RM, Semeiks JR, Spellman PT, Lorenz K, Lee EH, Barcellos-Hoff MH, Petersen OW, et al. 2007. The morphologies of breast cancer cell lines in three-dimensional assays correlate with their profiles of gene expression. *Mol Oncol* **1**: 84–96. doi:10.1016/j.molonc.2007.02.004

Kerbel RS, Guerin E, Francia G, Xu P, Lee CR, Ebos JM, Man S. 2013. Preclinical recapitulation of antiangiogenic drug clinical efficacies using models of early or late stage breast cancer metastasis. *Breast* **22** (Suppl 2): S57–S65. doi:10.1016/j.breast.2013.07.011

Kim H, Kumar P, Menghi F, Noorbakhsh J, Cerveira E, Ryan M, Zhu Q, Ananda G, George J, Chen HC, et al. 2018. High-resolution deconstruction of evolution induced by chemotherapy treatments in breast cancer xenografts. *Sci Rep* **8**: 17937. doi:10.1038/s41598-018-36184-8

Lazzari L, Corti G, Picco G, Isella C, Montone M, Arcella P, Durinikova E, Zanella ER, Novara L, Barbosa F, et al. 2019. Patient-derived xenografts and matched cell lines identify pharmacogenomic vulnerabilities in colorectal cancer. *Clin Cancer Res* **25**: 6243–6259. doi:10.1158/1078-0432.CCR-18-3440

Matossian MD, Burks HE, Elliott S, Hoang VT, Bowles AC, Sabol RA, Wahba B, Anbalagan M, Rowan B, Abazeed ME, et al. 2019. Drug resistance profiling of a new triple negative breast cancer patient-derived xenograft model. *BMC Cancer* **19:** 205. doi:10.1186/s12885-019-5401-2

Menghi F, Banda K, Kumar P, Straub R, Dobrolecki L, Rodriguez IV, Yost SE, Chandok H, Radke MR, Somlo G, et al. 2022. Genomic and epigenomic *BRCA* alterations predict adaptive resistance and response to platinum-based therapy in patients with triple-negative breast and ovarian carcinomas. *Sci Transl Med* **14:** eabn1926. doi:10.1126/scitranslmed.abn1926

Merino D, Weber TS, Serrano A, Vaillant F, Liu K, Pal B, Di Stefano L, Schreuder J, Lin D, Chen Y, et al. 2019. Barcoding reveals complex clonal behavior in patient-derived xenografts of metastatic triple negative breast cancer. *Nat Commun* **10:** 766. doi:10.1038/s41467-019-08595-2

Micallef P, Wu Y, Bauzá-Thorbrügge M, Chanclón B, Vujičić M, Peris E, Ek CJ, Wernstedt Asterholm I. 2021. Adipose tissue–breast cancer crosstalk leads to increased tumor lipogenesis associated with enhanced tumor growth. *Int J Mol Sci* **22:** 11881. doi:10.3390/ijms222111881

Nalawade SA, Shafer P, Bajgain P, McKenna MK, Ali A, Kelly L, Joubert J, Gottschalk S, Watanabe N, Leen A, et al. 2021. Selectively targeting myeloid-derived suppressor cells through TRAIL receptor 2 to enhance the efficacy of CAR T cell therapy for treatment of breast cancer. *J Immunother Cancer* **9:** e003237. doi:10.1136/jitc-2021-003237

Nguyen LV, Cox CL, Eirew P, Knapp DJ, Pellacani D, Kannan N, Carles A, Moksa M, Balani S, Shah S, et al. 2014. DNA barcoding reveals diverse growth kinetics of human breast tumour subclones in serially passaged xenografts. *Nat Commun* **5:** 5871. doi:10.1038/ncomms6871

Nugoli M, Chuchana P, Vendrell J, Orsetti B, Ursule L, Nguyen C, Birnbaum D, Douzery EJ, Cohen P, Theillet C. 2003. Genetic variability in MCF-7 sublines: evidence of rapid genomic and RNA expression profile modifications. *BMC Cancer* **3:** 13. doi:10.1186/1471-2407-3-13

Okano M, Oshi M, Butash A, Okano I, Saito K, Kawaguchi T, Nagahashi M, Kono K, Ohtake T, Takabe K. 2020. Orthotopic implantation achieves better engraftment and faster growth than subcutaneous implantation in breast cancer patient-derived xenografts. *J Mammary Gland Biol Neoplasia* **25:** 27–36. doi:10.1007/s10911-020-09442-7

Özkan H, Öztürk DG, Korkmaz G. 2022. Transcriptional factor repertoire of breast cancer in 3D cell culture models. *Cancers (Basel)* **14:** 1023. doi:10.3390/cancers14041023

Pellegrino B, Herencia-Ropero A, Llop-Guevara A, Pedretti F, Moles-Fernández A, Viaplana C, Villacampa G, Guzmán M, Rodríguez O, Grueso J, et al. 2022. Preclinical in vivo validation of the RAD51 test for identification of homologous recombination-deficient tumors and patient stratification. *Cancer Res* **82:** 1646–1657. doi:10.1158/0008-5472.CAN-21-2409

Petrosyan V, Dobrolecki LE, LaPlante EL, Srinivasan RR, Bailey MH, Welm AL, Welm BE, Lewis MT, Milosavljevic A. 2022. Immunologically "cold" triple negative breast cancers engraft at a higher rate in patient derived xeno-

grafts. *NPJ Breast Cancer* **8:** 104. doi:10.1038/s41523-022-00476-0

Petrosyan V, Dobrolecki LE, Thistlethwaite L, Lewis AN, Sallas C, Srinivasan RR, Lei JT, Kovacevic V, Obradovic P, Ellis MJ, et al. 2023. Identifying biomarkers of differential chemotherapy response in TNBC patient-derived xenografts with a CTD/WGCNA approach. *iScience* **26:** 105799. doi:10.1016/j.isci.2022.105799

Powell RT, Redwood A, Liu X, Guo L, Cai S, Zhou X, Tu Y, Zhang X, Qi Y, Jiang Y, et al. 2020. Pharmacologic profiling of patient-derived xenograft models of primary treatment-naive triple-negative breast cancer. *Sci Rep* **10:** 17899. doi:10.1038/s41598-020-74882-4

Prat A, Karginova O, Parker JS, Fan C, He X, Bixby L, Harrell JC, Roman E, Adamo B, Troester M, et al. 2013. Characterization of cell lines derived from breast cancers and normal mammary tissues for the study of the intrinsic molecular subtypes. *Breast Cancer Res Treat* **142:** 237–255. doi:10.1007/s10549-013-2743-3

Prinz F, Schlange T, Asadullah K. 2011. Believe it or not: how much can we rely on published data on potential drug targets? *Nat Rev Drug Discov* **10:** 712. doi:10.1038/nrd3439-c1

Qiu D, Zhang G, Yan X, Xiao X, Ma X, Lin S, Wu J, Li X, Wang W, Liu J, et al. 2022. Prospects of immunotherapy for triple-negative breast cancer. *Front Oncol* **11:** 797092. doi:10.3389/fonc.2021.797092

Riaz M, Setyono-Han B, Timmermans MA, Trapman AM, Bolt-de Vries J, Hollestelle A, Janssens RC, Look MP, Schutte M, Foekens JA, et al. 2014. Growth and metastatic behavior of molecularly well-characterized human breast cancer cell lines in mice. *Breast Cancer Res Treat* **148:** 19–31. doi:10.1007/s10549-014-3142-0

Richard E, Grellety T, Velasco V, MacGrogan G, Bonnefoi H, Iggo R. 2016. The mammary ducts create a favourable microenvironment for xenografting of luminal and molecular apocrine breast tumours. *J Pathol* **240:** 256–261. doi:10.1002/path.4772

Roarty K, Echeverria GV. 2021. Laboratory models for investigating breast cancer therapy resistance and metastasis. *Front Oncol* **11:** 645698. doi:10.3389/fonc.2021.645698

Savage P, Pacis A, Kuasne H, Liu L, Lai D, Wan A, Dankner M, Martinez C, Muñoz-Ramos V, Pilon V, et al. 2020. Chemogenomic profiling of breast cancer patient-derived xenografts reveals targetable vulnerabilities for difficult-to-treat tumors. *Commun Biol* **3:** 310. doi:10.1038/s42003-020-1042-x

Scherer SD, Riggio AI, Haroun F, DeRose YS, Ekiz HA, Fujita M, Toner J, Zhao L, Li Z, Oesterreich S, et al. 2021. An immune-humanized patient-derived xenograft model of estrogen-independent, hormone receptor positive metastatic breast cancer. *Breast Cancer Res* **23:** 100. doi:10.1186/s13058-021-01476-x

Sflomos G, Battista L, Aouad P, De Martino F, Scabia V, Stravodimou A, Ayyanan A, Ifticene Treboux A, Bucher P, Fiche M, et al. 2021. Intraductal xenografts show lobular carcinoma cells rely on their own extracellular matrix and LOXL1. *EMBO Mol Med* **13:** e13180. doi:10.15252/emmm.202013180

Sprouffske K, Kerr G, Li C, Prahallad A, Rebmann R, Waehle V, Naumann U, Bitter H, Jensen MR, Hofmann F, et al.

2020. Genetic heterogeneity and clonal evolution during metastasis in breast cancer patient-derived tumor xenograft models. *Comput Struct Biotechnol J* **18:** 323–331. doi:10.1016/j.csbj.2020.01.008

Sun H, Cao S, Mashl RJ, Mo CK, Zaccaria S, Wendl MC, Davies SR, Bailey MH, Primeau TM, Hoog J, et al. 2021a. Comprehensive characterization of 536 patient-derived xenograft models prioritizes candidates for targeted treatment. *Nat Commun* **12:** 5086. doi:10.1038/s41467-021-25177-3

Sun Y, Yang N, Utama FE, Udhane SS, Zhang J, Peck AR, Yanac A, Duffey K, Langenheim JF, Udhane V, et al. 2021b. NSG-Pro mouse model for uncovering resistance mechanisms and unique vulnerabilities in human luminal breast cancers. *Sci Adv* **7:** eabc8145. doi:10.1126/sciadv.abc8145

Takahashi K, Nagai N, Ogura K, Tsuneyama K, Saiki I, Irimura T, Hayakawa Y. 2015. Mammary tissue microenvironment determines T cell-dependent breast cancer-associated inflammation. *Cancer Sci* **106:** 867–874. doi:10.1111/cas.12685

Utama FE, LeBaron MJ, Neilson LM, Sultan AS, Parlow AF, Wagner KU, Rui H. 2006. Human prolactin receptors are insensitive to mouse prolactin: implications for xenotransplant modeling of human breast cancer in mice. *J Endocrinol* **188:** 589–601. doi:10.1677/joe.1.06560

Woo XY, Giordano J, Srivastava A, Zhao ZM, Lloyd MW, de Bruijn R, Suh YS, Patidar R, Chen L, Scherer S, et al. 2021. Conservation of copy number profiles during engraftment and passaging of patient-derived cancer xenografts. *Nat Genet* **53:** 86–99. doi:10.1038/s41588-020-00750-6

Xie Y, Hu Y, Zhou N, Yao C, Wu L, Liu L, Chen F. 2020. CAR T-cell therapy for triple-negative breast cancer: where we are. *Cancer Lett* **491:** 121–131. doi:10.1016/j.canlet.2020.07.044

Zhang X, Claerhout S, Prat A, Dobrolecki LE, Petrovic I, Lai Q, Landis MD, Wiechmann L, Schiff R, Giuliano M, et al. 2013. A renewable tissue resource of phenotypically stable, biologically and ethnically diverse, patient-derived human breast cancer xenograft models. *Cancer Res* **73:** 4885–4897. doi:10.1158/0008-5472.CAN-12-4081

Zhang H, Cohen AL, Krishnakumar S, Wapnir IL, SVeeriah S, Deng G, Coram MA, Piskun CM, Longacre TA, Herrler M, et al. 2014. Patient-derived xenografts of triple-negative breast cancer reproduce molecular features of patient tumors and respond to mTOR inhibition. *Breast Cancer Res* **16:** R36. doi:10.1186/bcr3640

Zhang Y, Zhang GL, Sun X, Cao KX, Ma C, Nan N, Yang GW, Yu MW, Wang XM. 2018. Establishment of a murine breast tumor model by subcutaneous or orthotopic implantation. *Oncol Lett* **15:** 6233–6240. doi:10.3892/ol.2018.8113

Multidimensional Imaging of Breast Cancer

Anne C. Rios,[1,2] Jacco van Rheenen,[2,3] and Colinda L.G.J. Scheele[4,5]

[1]Princess Máxima Center for Pediatric Oncology, 3584 CS Utrecht, The Netherlands

[2]Oncode Institute, 3521 AL Utrecht, The Netherlands

[3]Department of Molecular Pathology, The Netherlands Cancer Institute, 1066 CX Amsterdam, The Netherlands

[4]Laboratory for Intravital Imaging and Dynamics of Tumor Progression, VIB Center for Cancer Biology, KU Leuven, 3000 Leuven, Belgium

[5]Department of Oncology, KU Leuven, 3000 Leuven, Belgium

Correspondence: Colinda.Scheele@kuleuven.be

Breast cancer is a pathological condition characterized by high morphological and molecular heterogeneity. Not only the breast cancer cells, but also their tumor micro-environment consists of a multitude of cell types and states, which continuously evolve throughout progression of the disease. To understand breast cancer evolution within this complex environment, in situ analysis of breast cancer and their co-evolving cells and structures in space and time are essential. In this review, recent technical advances in three-dimensional (3D) and intravital imaging of breast cancer are discussed. Moreover, we highlight the resulting new knowledge on breast cancer biology obtained through these innovative imaging technologies. Finally, we discuss how multidimensional imaging technologies can be integrated with molecular profiling to understand the full complexity of breast cancer and the tumor micro-environment during tumor progression and treatment response.

Recent exciting developments in multidimensional imaging enabled the large-scale analysis of the heterogeneity of breast cancer, including the spatial distribution of the multitude of cell types and states present in tumors. These analyses identified the presence of many populations of intertwined cancer and microenvironmental cells with various cellular shapes that are intricately organized in 3D. Importantly, this organization is not static and can evolve over time with cells actively moving around. Spatial and dynamic analyses are, therefore, crucial to not only profile the cellular, but also this structural composition to fully appreciate the significance of cell localization, interconnection, morphology, and behavior for cancer progression and treatment design. Advances in tissue clearing (Richardson and Lichtman 2015; Almagro et al. 2021) coupled with light sheet (Reynaud et al. 2015), confocal (Jonkman et al. 2020), and multiphoton (Andresen et al. 2009) microscopy technologies allow for the observation of complex structures within large intact tissue specimens and led to discoveries relevant to breast development (Dawson et al. 2021), including mammary stem cell biology (Rios et al. 2014;

Davis et al. 2016; Scheele et al. 2017) and interplay with the microenvironment (Dawson et al. 2020; Lloyd-Lewis 2020). By providing the 3D context of tissue architecture and cellular composition, and even subcellular organization of healthy and diseased tissue, these techniques can also uniquely unveil spatiodynamic traits of cancer (Fig. 1A; van Ineveld et al. 2021). Moreover, an increased number of sample preparation methods designed for preserving endogenous fluorescence (Li et al. 2019; Rios et al. 2019; Messal et al. 2021a), performing large sample immunolabeling (Renier et al. 2014; Ku et al. 2020) and moving toward multiplex imaging (Murray et al. 2015; Goltsev et al. 2018; Stoltzfus et al. 2020; van Ineveld et al. 2021; Seo et al. 2022), allowed adaptation of 3D imaging from application in fluorescently engineered animal models to studying biological processes in human samples. Here, we review recent advances in sample preparation and multidimensional imaging technologies and the resulting new knowledge on breast tumor biology. Furthermore, we discuss expected future contributions and novel technological developments that will transcend the application of multidimensional imaging in the field of breast cancer research.

CLARIFYING BREAST CANCER TISSUE

Most tissues and tumors contain a complex mixture of water, proteins, and lipids, each of which have different refractive indices (RIs), which makes them inherently opaque. The RI mismatch between the different constituents induces light scattering and, thereby, introduces loss of fluorescent signal when performing imaging in intact tissues. The breast is especially difficult to image because of its lipid-rich structure with epithelial mammary ducts buried deep within adipose tissue that strongly reflects light (Dawson and Visvader 2021). Furthermore, breast tumors change in density, as compared to surrounding healthy tissue, and are often characterized by a dense fibrotic tumor stroma, due to accumulation of extracellular matrix (ECM) proteins, such as collagen and fibronectin (Fabiano et al. 2022). To achieve complex 3D images of intact breast tissue and tumors, dedi-

cated protocols are required to render breast tumor tissue transparent, while keeping the original tissue structure and composition intact. In this section, we summarize advances in tissue clearing and highlight the most suitable methods for breast cancer applications.

Over the last decade, various chemical strategies have been developed to render organs translucent to reduce light scattering and increase imaging depth. Pioneering studies have primarily focused on deep brain imaging with the development of clearing agents that preserve brain architecture and cellular composition for analysis of intact neural networks (Chung et al. 2013). Since then, over 60 different clearing methods have been described and applied to diverse organs and biomedical research fields, including developmental and stem cell biology, immunology, and oncology (Almagro et al. 2021). They depend on two main categories of clearing agents: solvent-based solutions that remove water and lipids, while homogenizing the tissue RI, and aqueous solutions that simply match the RI of the tissue, for some methods including additional lipid removal (Richardson and Lichtman 2015). Solvent-based clearing agents, such as the DISCO family (Ertürk et al. 2012; Renier et al. 2014; Pan et al. 2016; Chi et al. 2018; Perin et al. 2019; Qi et al. 2019), are most effective in achieving transparency and have, therefore, been of particular interest for macroscale 3D imaging. However, they can cause tissue shrinkage, structural deformation, and tend to quench native fluorescence (Lloyd-Lewis et al. 2016; Pan et al. 2016). Aqueous-based agents, in particular FUnGI (Rios et al. 2019), Ce3D (Li et al. 2017), FLASH (Messal et al. 2021a), CUBIC, and SeeDB (Davis et al. 2016; Lloyd-Lewis et al. 2016), are well-suited for high-resolution single-cell 3D imaging of the mammary gland, as well as breast tumors (Rios et al. 2019; Messal et al. 2021a). The sample volume and desired resolution (e.g., macroscale, [sub]cellular) will determine the clearing method of choice and subsequent imaging technique best suited for visualization of the cleared specimen (for further reading on tissue clearing and method selection, please refer to Almagro et al. 2021; van Ineveld et al. 2022).

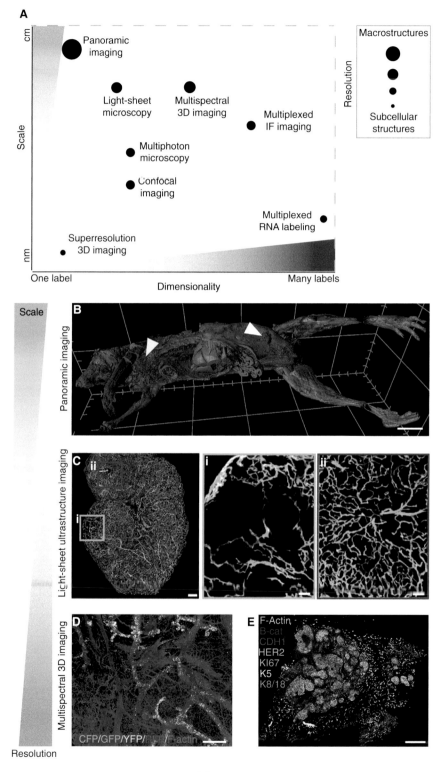

Figure 1. (*See following page for legend.*)

THE MACROSCOPIC LANDSCAPE OF BREAST CANCER

Light-sheet microscopy presents an excellent system for imaging large samples, such as entire mice and recently even human organs (Zhao et al. 2020). Through selective plane illumination, fluorescent molecules are excited together in an entire sheet and detected simultaneously, allowing increased scan speed acquisition compared to classical, laser point scanning confocal microscopy. Although recent advances have achieved subcellular resolution imaging (Chakraborty et al. 2019), common light-sheet microscopy uses relatively low numerical aperture objectives in detriment of some resolution to achieve a large field of view. As a result, it is especially suited for studying macroscopic structures, such as blood and lymphatic vasculature to fully appreciate the complex organization of these networks in 3D (Fig. 1A–C). It has, thereby, provided key information on how these structures within the tumor microenvironment (TME) can be actively remodeled by the tumor cells (Liu et al. 2013; Lin et al. 2016; Chen et al. 2019). For example, it has uniquely revealed the chaotic and immature tumor angioarchitecture, identifying tumor areas with poorly perfused microvasculature providing a cause for frequently observed intratumoral hypoxia, a feature that often facilitates cancer cell evasion (Fig. 1C; Dobosz et al. 2014). Applied in animal models, volumetric imaging of blood vasculature has also helped to assess the efficacy of therapy delivery within mammary carcinoma (Lee et al. 2019a). 3D visualization of systemic antibody drug delivery, including the well-known immune checkpoint inhibitor anti-PD-L1 (programmed death-ligand 1), identified increased permeability of capillaries when compared to blood vessels that were supported by vascular fibroblasts and smooth muscle cells (Lee et al. 2019a). In addition, targeted anti-HER2 antibody was found to be sequestered into periphery nests within HER2$^+$ tumors, potentially by binding to target antigens, as control nontargeted antibodies easily extravagated into the tumor parenchyma (Lee et al. 2019a). These findings might explain fluctuating responses to systemic treatment that have been linked to perfusion restraints (Dobosz et al. 2014). Light-sheet 3D imaging also uniquely identified a novel crossover pathway between lymph and blood vasculature, with breast metastasizing cells exiting the lymph node by invading local lymph node blood vessels, rather than using efferent lymphatic vessels (Brown et al. 2018).

Figure 1. 3D imaging technologies to uncover the spatial traits of breast cancer progression. (*A*) 3D diagram demonstrating the features of diverse commonly used imaging modalities to study intact breast cancer samples. The horizontal axis represents the image dimensionality (number of labels typically used to visualize different structures, cells, or molecules), the vertical axis represents the imaging scale ranging from nanometers (nm) to centimeters (cm), and the size of the dots represents image resolution ranging from subcellular (small dots) to macrostructural (large dots). Note that none of the currently available imaging technologies reach a combination of large-scale high dimensionality and high resolution. (IF) Immunofluorescence. (*B*) An example of panoramic imaging of the full intact mouse body using whole-body clearing and light-sheet imaging, which can be used to detect small cell numbers, such as micrometastases, in large volumes, such as the entire mouse. Scale bar, 2 cm. (Photo in *B* is reprinted from Pan et al. 2019 with permission from Elsevier © 2019.) (*C*) 3D light-sheet imaging of macrostructural components in primary breast tumors highlighting high spatial heterogeneity in vascular architecture including poorly perfused regions (region i) and well-vascularized regions (region ii) within the same tumor sample. Scale bars, 250 µm (overview panel) and 100 µm (zoom regions). (Photo *C* is reprinted from Dobosz et al. 2014 under the terms of a Creative Commons CC-BY-NC-ND license.) (*D*) Whole-mount 3D confocal imaging of an *Elf5-rtTA/TetO-cre/Pten^fl/fl/p53^fl/fl/R26R-Confetti* mammary gland with F-actin (blue) in a precancerous stage. Mutant cells are labeled with CFP, GFP, YFP, or RFP. Scale bar, 500 µm. (Photo in *D* is reprinted from Rios et al. 2019 with permission from Elsevier © 2019.) (*E*) Representative 3D image of breast tumor tissue derived from xenotransplanted human breast tumor organoids, immunolabeled with seven different markers and imaged using multispectral confocal imaging. Scale bar, 300 µm. (Photo *E* is reprinted from van Ineveld et al. 2021 with permission from the authors © 2021, who hold full copyright under exclusive license to Springer Nature America.)

This key observation shed new light on the debate about whether lymph nodes could be an active dissemination route and which preferential route tumor cells use to reach blood circulation from lymph nodes. It might have implications for treatment decisions, as tumor cells located adjacent to, or within blood vessels, are expected to be more predictive of poor prognosis, compared to potentially larger lymph node tumor cell deposits distant from blood vessels (Dart 2018; Tjan-Heijnen and Viale 2018). With breast cancer being a key example of tumors that constantly fail to respond to angiogenesis-targeted therapy for improving survival outcomes (Ayoub et al. 2022), macroscale 3D imaging technology might be of great interest for testing novel angiogenic therapies to normalize tumor vasculature and study their impact on drug delivery as well as metastatic dissemination.

Next to the above-described evaluation of specific metastatic routes and local drug delivery, recent technology advances in panoptic 3D imaging now provide an unprecedented readout to assess metastatic tumor burden alongside therapy distribution, not only at the tumor site, but also at the whole-body level. The potent clearing methodologies, CUBIC (Susaki et al. 2014) and next generation of 3DISCO (such as uDISCO [Pan et al. 2016] and subsequently vDISCO [Pan et al. 2019]), together with further-advanced light-sheet microscopes and immunolabeling protocols, have enabled detection of cells in small quantities, while scanning massive volumes to image entire mice (Kubota et al. 2017; Cai et al. 2019). Using 13 mouse models and nine cancer cell lines, including the human metastatic breast cancer line, MDA-MB-231, Kubota et al. (2017) used CUBIC to quantify metastasis at the whole-mouse body level and assess their response to chemotherapy. The unprecedented scale of whole-body imaging poses significant challenges for handling and analysis of large imaging data sets, necessitating the development of deep learning–based quantification platforms (Fig. 1B). The DeepMACT analysis pipeline (Pan et al. 2019) allows mapping of individual metastasis in whole-animal breast cancer models. This uniquely revealed a previously unrecognized micrometastasis pattern at numerous sites

across the body, but also only a partial degree of overlap with the distribution of a monoclonal antibody therapy, suggesting incomplete targeting (Pan et al. 2019). Building onto this seminal work, whole-body distribution mapping of novel targeted or cellular therapies is a promising avenue that could uniquely be addressed by these 3D panoramic imaging technologies. Further automation of these imaging pipelines could deliver an invaluable tool for translational studies to identify the most promising therapies prior to clinical testing (for an extensive overview of 3D image analyses pipelines, we refer to van Ineveld et al. 2022). Together, this body of work demonstrates the value of 3D imaging to unravel the macroscopic landscape of breast cancer across the full trajectory from tumor progression to end-stage metastasis, as well as during treatment response.

MAPPING BREAST CANCER HETEROGENEITY

Confocal or two-photon microscopes offer mainstream systems for large-scale 3D imaging with higher resolution, as compared to light-sheet technology, yet at the expense of some imaging depth and speed. Therefore, recent efforts have been directed at obtaining easy-to-use protocols for routine 3D imaging of intact tumors with clearing methodologies that preserve epitope integrity, limit distortion of tissue, and allow for rapid sample preparation (Li et al. 2017, 2019; Rios et al. 2019; Messal et al. 2021a). These protocols are compatible with visualization of native fluorescence and, thereby, enable visualization of breast cancer progression in fluorescently engineered animal or xenograft models. Multicolored fluorescent lineage-tracing approaches, such as confetti, enable tracking of individual cell fates throughout the progression of cancer (Kretzschmar and Watt 2012) and have allowed prediction of clonal relationships between tumor cells in vivo and, thereby, the modeling of tumor-wide clonal contributions (Fig. 1D; Lamprecht et al. 2017; Yanai et al. 2017; Rios et al. 2019; Tang et al. 2019; Tiede et al. 2021; Yum et al. 2021). For instance, LSR-3D (large-scale single-cell resolution-3D) imaging of confetti lineage-tracing breast cancer models revealed profound clonal restriction during breast cancer progression (Rios et al. 2019), a phenomenon also observed in other cancers (Vermeulen et al. 2013; Snippert et al. 2014; Brown et al. 2017; Ying et al. 2018; Bruens et al. 2020). Integration with clonal RNA sequencing and immunolabeling revealed the molecular heterogeneity of these predominant stem- or progenitor cell–derived clones, with cellular phenotypes varying along a spectrum of reversible states in between epithelial and mesenchymal fate (Fig. 1D; Rios et al. 2019). Intratumoral heterogeneity also manifests as differences in cell proliferation within invasive carcinomas derived from different subtypes of breast cancer, including luminal A-, luminal B-, and HER2-enriched human breast cancer. Interestingly, in vivo clonal tracing demonstrates that cancer cell populations efficiently targeted by chemotherapy are mainly slow-proliferative clones, whereas chemotherapy fails to repress fast-proliferative clones, suggesting that intratumoral heterogeneity of tumor cell proliferation may enhance the tumor's resistance to chemotherapy (Tiede et al. 2021). Altogether, this indicates inherent tumor plasticity and shows that many clonal evolution routes can lead to effective cancer progression and treatment resistance (Rios et al. 2019; Tiede et al. 2021). Multidimensional imaging is also relevant for gaining knowledge in metastasis. Deep multiphoton microscopy combined with an integrative analysis pipeline, SMART 3D, revealed the spatial landscape of experimentally induced brain metastasis derived from breast cancer cells and again identified proliferative differences in the metastases, as well as specific metastasis-associated astrogliosis (Guldner et al. 2016).

Altogether, these findings show the power of confocal/multiphoton 3D imaging to visualize and quantify global spatial and clonal changes occurring during cancer progression and reveal key cellular processes, such as proliferation and epithelial-to-mesenchymal transition (EMT), that drive tumor heterogeneity. In other types of cancers, these 3D imaging techniques have, furthermore, identified physical constraints, such as epithelial curvature, to not only dictate morphology, but also predict aggressive tumor behavior, such as invasion (Messal et al. 2019; Fiore et al. 2020). This type of analysis might be of great interest for early-stage ductal carcinoma in situ (DCIS), accounting for 25% of all diagnosed breast cancers that can be considered noninvasive or pre-invasive breast cancer. They often associate with a high survival rate, but due to their heterogeneous pathological traits are clinically challenging in terms of the therapeutic decision whether to offer less aggressive treatment (Farante et al. 2022). As there is currently no method to predict which of the DCIS will become invasive, multidimensional analysis of the biophysical properties of DCIS could potentially help to predict invasiveness and, thereby, guide treatment management in the near future.

RESOLVING TME INTERPLAY IN BREAST CANCER

Multidimensional imaging contributed significantly to the identification of tumor heterogeneity including the complex TME. This is of particular importance for the immune compartment that has been increasingly recognized for its role in tumor progression and is, thereby, an important focus of ongoing therapy development (Hegde and Chen 2020). Indeed, 3D imaging has been applied to spatially map diverse immune cell subsets in breast tumors (Lee et al. 2019b), as well as the expression of immune regulatory molecules (Lee et al. 2019a). Multispectral imaging that increases the number of markers that can be visualized by spectrally resolving numerous fluorophores offers a methodology of choice for simultaneously analyzing diverse immune and cancer cell types and states. The Opal method, although performed in 2D, allows recording of up to eight markers in a single acquisition. This methodology was used to spatially decode the TME immune composition of breast tumors, and helped identify differences in immune infiltration and immunomodulatory molecules PDL-1 and CTL4 in BRCA1-deficient triple-negative breast cancer (TNBC) tumors and provided strong rationale for clinical studies evaluating the efficacy of combined immune checkpoint inhibitors in BRCA-associated TNBC (Nolan et al. 2017). This clearly shows the clinical relevance of spatially resolving the complexity of tumors and their TME by visualizing multiple molecular markers at once. In recent years, several protocols have been developed for multispectral 3D imaging with a high number of markers (Valm et al. 2017; Coutu et al. 2018; Li et al. 2019; van Ineveld et al. 2021). mLSR-3D, for example, implements "on the fly" linear unmixing for single-scan acquisition of eight spectrally resolved fluorophores and enables fast and light-efficient high-dimensional imaging (van Ineveld et al. 2021). Combined with a deep learning–based segmentation pipeline, STAPL-3D, hundreds of molecular and spatial features can be extracted from millions of cells in 3D imaging data sets, allowing omics-like analysis of imaging data (van Ineveld et al. 2021). The power of this 3D imaging and computational advance was demonstrated by the identification of novel and rare tumor cell subsets in pediatric kidney Wilms tumor, but also characterization of intratumoral heterogeneity that can be modeled in breast cancer organoids and derived xenograft models (Fig. 1E; van Ineveld et al. 2021). This showcases the already highly relevant resolving capacity of multispectral 3D microscopy to capture the complex cellular heterogeneity of tumors by overcoming the traditional limit of four to five markers imaged at the time. This can be further expanded with ongoing advancements that are expected to continue to sharply increase the number of markers that can be imaged (Seo et al. 2022).

Another important imaging trend for increasing molecular resolution is represented by multiplex fluorescence imaging, a methodology that implements sequential rounds of protein or RNA labeling and microscopy acquisition (Junker et al. 2014; Chen et al. 2017; Karaiskos et al. 2017). RNA labeling currently enables the highest in situ molecular resolution (Murray et al. 2015; Wang et al. 2018b; Alon et al. 2021). Applied to a core biopsy from breast cancer metastasis, Exseq (Expansion sequencing), for instance, elucidated the relationship between tumor cells and fibroblasts in the hypoxic tumor environment (Alon et al. 2021). By providing spatial profiling of 296 genes within 2395 cells, this technology showed that tumor cells overexpress HIF1A, a molecule involved in tumor maintenance, when in close proximity to HSPG2 positive fibroblasts (Alon et al. 2021). Furthermore, *S100A8*, a regulator of inflammatory immune responses and known biomarker for potential relapse or late progression of breast cancer (Wang et al. 2018a; Zhong et al. 2018), was found to be overexpressed fourfold in a subset of B cells, when they are located near EGFR-positive tumor cells. This shows the power of enhanced spatial molecular mapping for identification of subtle changes in cellular states, for example, in immune cells, according to their distance to other cell types such as tumor cells (Alon et al. 2021). However, for larger specimens, the multiplexing experimental procedure can become incredibly time consuming and complex, as sample coregistration is required to integrate data from each sequential imaging round in 3D. STARmap, for example, reports visualization of 1020 genes in 5 days on thin sections, but this number drastically reduces to 28-in-thick samples, while almost doubling experimental duration (Wang et al. 2018b).

Overall, with both multiplexed and multispectral imaging still rapidly evolving, we see a clear application for high-dimensional 3D imaging in probing the heterogeneity of both the tumor itself, as well as its cellular environment and, importantly, the interaction between those two. Its power for drawing connections between spatial cellular organizations within tumors and how those relate to different cell types and states has already been demonstrated and might eventually be linked to clinical outcomes. To fully explore this clinical predictive power and start using it as a diagnostic tool, 3D imaging of archival samples stored in patient biobanks will be the next step forward, as discussed in the following section.

3D VISUALIZATION OF BREAST CANCER ARCHIVAL TISSUE

Pathological formalin-fixed paraffin-embedded (FFPE) samples are current practice for diagnosis and, therefore, stored alongside relevant clinical data. This represents a widely available resource that is for now not fully exploited for its clinical and research potential by routine 2D sectional imaging (Haddad et al. 2021). Indeed, standard histological practices only allow for the observation of a few slices per sample, representing a fraction of the entire sample, thereby providing limited information. As a result, the full spatial heterogeneity in complex tissues, such as breast cancer samples, cannot be studied. This deficiency is underscored by current 2D methods for histological grading of primary tumors leading to inconsistency and misdiagnosis (Kruskal et al. 1997; Brown et al. 2010; Catalona et al. 2017; Epstein 2018; Ahdoot et al. 2020). For breast cancer, it has been shown that chemotherapy-induced histological changes associate with false-negative sentinel lymph nodes and, therefore, might be misleading for monitoring treatment effects on late metastatic progression (Brown et al. 2010). Many laboratories have begun to explore 3D histopathology of FFPE primary patient material for tumor grading (van Royen et al. 2016) or to assign novel treatment options, such as immunotherapy (Si et al. 2019).

3D imaging of full FFPE tissue blocks can be achieved by 3D reconstruction of serial acquired 2D sections. The advantages of this approach include full compatibility with current tissue-staining protocols and a vast number of antibodies validated for immunohistochemistry. 3D histological reconstructions of DCIS have identified structural dissimilarities between benign lobules and DCIS, offering an improved structural analysis to potentially discriminate noncancerous-resembling lobular structures from early-stage breast cancer manifestations (Booth et al. 2015). However, 3D rendering from serial sectioning is tremendously laborious and time consuming, as for large volumes thousands of slides must be processed and imaged. Automated staining machines and slide scanners can potentially relieve some of this burden, but still artefacts can arise, due to tissue deformations associated with sectioning. Protocols for 3D imaging of intact FFPE specimens can prevent these drawbacks and are currently emerging based on similar principles of optical clearing, as described for freshly fixed tissue. Furthermore, these protocols are similarly compatible with immunolabeling and confocal or light-sheet microscopy (Rios et al. 2016; Nojima et al. 2017; Chen et al. 2019; Yoshizawa et al. 2020; van Ineveld et al., in press). 3D imaging of breast cancer FFPE tissue, revealed spatial variation in the cell-cycling profile of tumor cells throughout the tumor, which was not detectable in FFPE 2D sections (Tanaka et al. 2018). Further clinical application of 3D archival tissue imaging was demonstrated by DIPCO, an iDISCO-based protocol for FFPE tissue, shown to have better accuracy compared to 2D histological analysis for diagnosis and stratification of patient prognosis (Tanaka et al. 2017, 2018). In addition, an updated protocol including in situ hybridization for RNA detection, further improving the molecular phenotyping capabilities of the technique (Tanaka et al. 2020). This revealed in TNBC, a disease subtype strongly associated with poor prognosis, a higher density of spatial niches enriched with cancer stem-like cells, a specific cancer cell population known to fuel breast cancer progression and, so far, difficult to therapeutically target (Tanaka et al. 2020). Further implementation of such advancements for breast cancer might be of great interest for treatment guidance and monitoring outcomes, as discussed previously for DCIS and sentinel lymph nodes.

The wealth of biologically and clinically relevant data obtained from FFPE tissues can be further expanded by superresolution imaging. This allows differentiation malignant macrostructures beyond the diffraction limit (Weiss 2000) and can achieve below 1 nm or 20 nm resolution, depending on the specific method (Rust et al. 2006; Jones 2012). Applied to clinical FFPE sections of cancerous tissue, it revealed 3D macrostructures resolvable to a depth of 800 nm (Creech et al. 2017). HER2-receptor expression, an important biomarker for breast cancer, could be accurately visualized on membrane protrusions in FFPE breast cancer biopsies. These were previously shown to be implicated in persistent localization and signaling of HER2 in breast cancer cells in vitro (Jeong et al. 2016), which demonstrates that similar HER2-enriched protrusions appear to exist in vivo and may play a functional role in breast tumor biology (Creech et al. 2017). Thus, although for now limited to very thin sections, superresolution

Cite this article as *Cold Spring Harb Perspect Med* doi: 10.1101/cshperspect.a041330

imaging may offer important functional insights into disease mechanisms and diagnosis, by precisely mapping structural molecular arrangements within cancer cells. Together with 3D imaging technology advances for archival samples, this can create a very complete picture of breast tumor characteristics associated with poor prognosis and treatment response.

This body of literature already demonstrates enhanced discriminative and, thereby, clinical potential of 3D archival tissue imaging. Its potential will further expand with the number of molecules and, thus, cell types and macrostructures, that can be visualized. Advances in label-free imaging approaches, including THG (Third Harmonic Generation), SHG (Second Harmonic Generation), and FTIR (Fourier-transform infrared spectroscopy) (Ren et al. 2017; Rivenson et al. 2019; Gavgiotaki et al. 2020), can further help to achieve this goal. They have already been applied to 2D tissue sections to discriminate morphology differences between malignant and healthy breast tissue, as well different grades of the disease (Gavgiotaki et al. 2020). These approaches might offer easy-to-implement alternatives to antibody labeling of thick FFPE tissues. Another exciting avenue is highly multiplexed molecular and cellular mapping using multidimensional imaging approaches, including 3D imaging mass cytometry (di Palma and Bodenmiller 2015; Kuett et al. 2021). It has the important potential of adding high-dimensional molecular classification for cancer and microenvironmental cells to histological H&E scoring of FFPE samples, thereby identifying novel biomarkers to refine diagnosis or guide treatment decision, especially for immunotherapy, the largest current body of development in cancer therapy.

INTRAVITAL MICROSCOPY TO STUDY BREAST TUMOR DYNAMICS

3D imaging of fixed breast tissues provides a snapshot of the architecture and the different cellular and noncellular components, but does not provide information on the dynamic interplay between the various components and their subsequent fate. To uncover cellular and architectural dynamics, intravital microscopy (IVM) approaches have been developed. To obtain high-resolution images deep inside tissues (up to 1 mm) while minimizing photo damage, multiphoton imaging technologies have been the preferred choice for IVM of breast tissues (Bakker et al. 2022; Choe et al. 2022). To get optical access beyond the imaging depth of the multiphoton microscope, the mammary gland can be surgically exposed (often referred to as the skin-flap approach) to allow for longitudinal imaging over several hours (Ewald et al. 2011; Harney et al. 2016; Dawson et al. 2021; Messal et al. 2021b). To get visual access to tissues over multiple weeks, permanent imaging windows were developed, which are surgically implanted on top of the anatomical location of interest, which can be the primary mammary tumor site (mammary imaging window) (Kedrin et al. 2008; Jacquemin et al. 2021; Messal et al. 2021b; Maiorino et al. 2022; Mourao et al. 2022) or the metastatic site (bone marrow, liver, lung, brain) (Fig. 2; Ritsma et al. 2013; Alieva et al. 2014; Entenberg et al. 2018).

To obtain contrast between different components in breast tissues (e.g., epithelial cells, stromal cells, immune cells, ECM), it is crucial to fluorescently label components of interest. For most IVM studies, this is done either by inducible reporter constructs in genetically engineered mouse models, transplantation of fluorescently labeled cells, intravenous or local injection of fluorescently labeled antibodies raised against cell- or matrix-specific epitopes, or injection of dyes, such as dextran or lectin, to label the vasculature. Of note, a recent report underscored that expression of fluorescent proteins or any other xenobiotic marker genes may elicit immunogenicity leading to attenuation of tumor growth and progression (Grzelak et al. 2022), underscoring the importance of evaluating the effects of experimental design in imaging studies (Day et al. 2022). When using multiphoton imaging, several label-free imaging approaches can be used to visualize structural components of the tissue environment, such as collagen I, fat cells, muscles, and nerve bundles (Friedl et al. 2007; Weigelin et al. 2016; You et al. 2018). For each

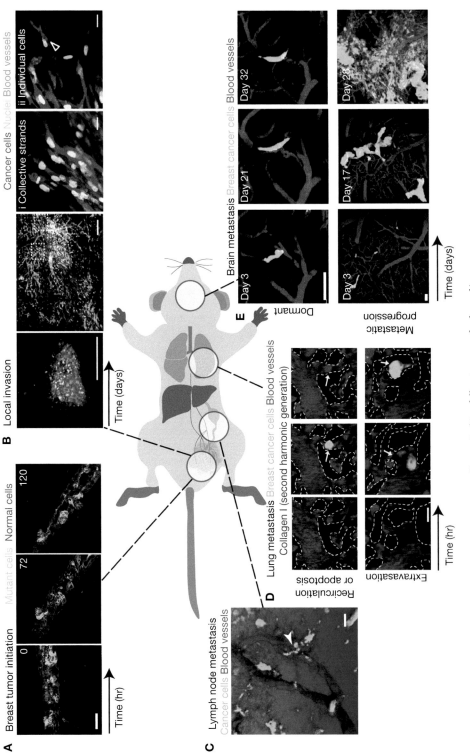

Figure 2. (*See following page for legend.*)

process that will be imaged in vivo, the experimental setup needs to be optimized. For instance, fast events such as cancer cell migration, cancer cell–immune cell interactions, or intravasation will require IVM time-lapse imaging with a short time interval of seconds or minutes, whereas the study of cancer cell growth dynamics will require multiday IVM through a chronic imaging window with a time interval of days or even weeks. In this part of the review, we discuss how IVM (3D imaging over time) has contributed to our understanding of the formation and progression of breast cancer.

IMAGING BREAST TUMOR INITIATION AND PROGRESSION

IVM has been used to study the cellular dynamics of all stages of breast tumor formation. For example, using an inducible model of β-catenin activation combined with lineage tracing and real-time IVM, it was elegantly shown that luminal and basal epithelial cells rearrange upon sustained Wnt signaling (Fig. 2A; Lloyd-Lewis et al. 2022). These altered dynamics eventually led to the formation of hyperplastic lesions with squamous features, irrespective of the cell lineage of origin (Lloyd-Lewis et al. 2022). Using a similar longitudinal IVM approach combined with multicolor confetti lineage tracing at a later state of mammary tumor development, it was shown

that primary mammary tumors display a large degree of plasticity (Zomer et al. 2013). Following the same cells and their progeny over multiple weeks indicated that the vast majority of the mammary tumor cells have a short lifetime (<wk) and do not contribute to tumor growth. Only a small subset of labeled cancer cells remained and proliferated leading to outgrowth of large clones as also observed by 3D microscopy of fixed tumor tissues (Rios et al. 2019). This indicates that these cancer cells represent the cancer stem cell (CSC) population and fuel tumor progression. Interestingly, IVM over multiple weeks showed that cells can gain or lose the CSC clonogenic properties, indicating widespread phenotypic plasticity in primary mammary tumors (Zomer et al. 2013). Together, these studies on primary breast lesion dynamics uncover a role for cell-state plasticity already starting from the first stages of tumorigenesis.

IMAGING BREAST CANCER CELL ESCAPE

Over the past years, IVM revealed several mechanisms on how cells escape from primary mammary tumors. First, cells need to detach from neighboring cells and migrate toward vessels. IVM has revealed several modes of in vivo migration including collective strand invasion (Fig. 2B; Ilina et al. 2018), cellular streaming (Patsialou et al. 2013; Leung et al. 2017), and individual

Figure 2. Dynamic intravital microscopy to uncover breast cancer dynamics at a cellular level. (*A*) Intravital imaging of mutant mammary cells revealed the first steps of tumorigenesis in the mammary gland where β-catenin activation drives the formation of hyperplastic regions with squamous differentiation within a few days. Scale bar, 50 μm. (Photos in *A* are reprinted from Lloyd-Lewis et al. 2022 under the terms of a Creative Commons Attribution 4.0 International license (CC BY 4.0).) (*B*) In vivo imaging of mammary tumor progression revealed several modes of invasion, including collective strand invasion (panel i) and single-cell invasion (panel ii). Scale bars, 200 μm (overviews) and 25 μm (panels i and ii). (Photos in *B* are reprinted from Ilina et al. 2018 under the terms of a Creative Commons CC-BY license.) (*C*) In vivo detection of metastatic breast cancer cells in the vasculature of the lymph node elucidating a role for the lymph node as a cancer cell hub prior to further distant metastatic spreading. Scale bar, 25 μm. (Photo in *C* is reprinted from Pereira et al. 2018 with permission from American Association for the Advancement of Science © 2018.) (*D*) Time-lapse intravital imaging of the lung vasculature allows behavioral tracking of disseminated tumors cells upon arrival in the lungs, including recirculation or apoptosis (*top* panels) or extravasation (*bottom* panels). Scale bar, 15 μm. (Photos in *D* are reprinted from Borriello et al. 2022 under the terms of a Creative Commons Attribution 4.0 International License (CC BY 4.0).) (*E*) Multiday intravital microscopy of metastatic breast cancer cells in the brain revealed that the rate-limiting step for metastatic progression is to overcome the dormancy-inducing microenvironment. Scale bars, 40 μm. (Photos in *E* are reprinted from Dai et al. 2022 with permission from the authors who hold the exclusive copyright license for this work published by Springer Nature.)

cell movement (Zomer et al. 2015; Beerling et al. 2016). Cell motility can be driven by diverse intracellular mechanisms or by cues in the TME. IVM of mammary tumors combined with in silico modeling recently showed that the mode of migration may be determined by a combination of ECM confinement and cell–cell junction stability, where highly confined tumor neighborhoods drive collective migration irrespective of cell–cell junction stability and single-cell escape occurs in areas with free space (Ilina et al. 2020). A systematic analysis of migratory cells in mammary carcinoma revealed that motile mammary tumor cells show differences in their migration speed. Slow migratory mammary tumor cells expressed CSC markers and had the capacity to degrade the ECM and disseminate (Sharma et al. 2021), whereas the faster migratory mammary tumor cells lacked these capabilities (Gligorijevic et al. 2014). The important role of matrix composition and density was underscored by dynamic IVM of the collagen network (using second harmonics generation) upon down-regulation of SERPIN E2, an extracellular protease inhibitor known to be involved in promoting breast cancer metastasis. SERPIN E2 knockdown or inhibition led to a cascade of changes in the TME, eventually leading to the deposition of a dense collagen network around the primary mammary tumor (Smirnova et al. 2016). Interestingly, this excessive matrix deposition showed a clear inhibitory effect both on local invasion as well as on distant metastasis formation (Smirnova et al. 2016).

Both during intravasation (entry into the bloodstream) at the primary tumor site and extravasation (exit from the bloodstream into a secondary organ), cancer cells need to cross the endothelial cell barrier. IVM has elucidated several mechanisms by which cancer cells can enter or exit the bloodstream. IVM showed that intravasation of disseminating primary breast cancer cells into the bloodstream can be facilitated by perivascular macrophages. The close interaction between the macrophage and endothelial cells leads to transient and local disconnection of the endothelial cell–cell junctions, which in turn facilitates cancer cell intravasation and dissemination (Harney et al. 2015; Kara-

giannis et al. 2017). Moreover, it was shown that the cancer cells interacting with these macrophages expressed CSC markers (Sharma et al. 2021), indicating that CSCs have the capacity to intravasate. IVM time-lapse imaging of intravasation revealed that invading breast cancer cells form invadopodia, which are dynamic and actin-rich protrusions. Invadopodia are assembled and disassembled in response to diverse external stimuli, such as integrin signaling (Paz et al. 2014) or chemotactic factors (Williams et al. 2019), and inhibition of invadopodia formation blocks extravasation and metastatic outgrowth (Leong et al. 2014). Upon intravasation into the bloodstream, IVM revealed the mechanisms by which circulating tumor cell (CTC) clusters may arise. Using time-lapse imaging of diverse patient-derived xenograft and spontaneous mouse models, it was shown that migrating tumor cells either cluster together near the vasculature or intravasate as individual cells and subsequently aggregate in the bloodstream. Tumor cell clustering was shown to depend on CD44 expression, which mediates intercellular interactions within the tumor cell aggregates with increased capacity to seed lung metastases (Liu et al. 2019).

THE ROLE OF CANCER CELL PLASTICITY DURING TUMOR PROGRESSION

Primary mammary tumor cells display a high degree of plasticity between epithelial and mesenchymal states to respond to changing environments. Using IVM of primary mammary tumors with endogenously labeled E-cadherin, it was shown that mammary tumor cells dynamically lose or gain E-cadherin expression during the metastatic cascade (Beerling et al. 2016). The majority of the migratory cancer cells within the primary tumor transiently lost their membranous E-cadherin expression, while this was always retained in nonmigrating E-cadherin-positive cancer cells. Moreover, the motile cancer cells adopted a mesenchymal expression profile, suggesting that an EMT preceded local invasion and intravasation (Beerling et al. 2016). This EMT phenotype was shown to be specifically induced in close proximity to blood vessels

(Zhao et al. 2016). Importantly, a recent study using novel reporters for EMT lineage tracing showed that mammary tumor cells mostly transition between an epithelial and partial EMT state (Lüönd et al. 2021). Moreover, it was shown that mammary tumor cells with partial EMT show increased collective migration, and the ability to metastasize and grow out in the lungs, whereas mammary tumor cells undergoing full EMT were static, lost their plasticity, and failed to establish metastases (Bornes et al. 2019; Lüönd et al. 2021). Indeed, upon intravasation and arrival in the lungs, it was shown that metastatic tumor cells quickly reverted to an epithelial state (Beerling et al. 2016). Together, these studies underscore the importance of EMT plasticity during the metastatic cascade in breast cancer.

IMAGING BREAST TUMOR METASTASES

IVM of spontaneous metastases is challenging as it is impossible to predict where and when primary mammary tumor cells will arrive in the secondary organ and grow out into metastatic lesions. Therefore, experimental metastasis models are often used to study the dynamics of cancer cell extravasation and metastatic outgrowth, which are obtained by direct injection of the primary breast cancer cells into the bloodstream, such as by tail vein injection to obtain lung metastases, mesenteric vein injection to obtain liver metastases, or by intracardiac injection to obtain brain metastases. Recently, a new permanent lung imaging window was designed to allow for longitudinal visualization of the lung (Entenberg et al. 2018), thereby enabling the assessment of spontaneous lung metastasis formation using diverse cell lines of breast cancer directly derived from the primary tumor. Direct comparison of the in vivo dynamics of experimental and spontaneous metastases in the lung revealed significant behavioral differences, where spontaneous disseminating cancer cells showed increased levels of retention in the lung vasculature, faster extravasation from the lung vessels, and increased survival after extravasation when compared to experimental metastatic cells (Fig. 2D; Borriello et al. 2022). Mechanisti-

cally, it was shown that spontaneously disseminating cells represent a subset of cancer cells that concomitantly expressed increased levels of stem cell markers, such as SOX9, and markers for cellular dormancy, such as NRF2F1 (Borriello et al. 2022). Using an innovative hypoxia-inducing nano-intravital device (Williams et al. 2016), it was shown that a hypoxic environment in primary mammary tumors can induce local dormancy (Fluegen et al. 2017). Although the ability to intravasate was not changed in these hypoxia-induced dormant tumor cells, their dormant state, but not their hypoxic state, was retained upon seeding in the lungs. This indicates that a hypoxic environment at the primary tumor site may prime cells for into a dormant state at the distant site for the long term.

Tumor cell dormancy was also shown to play an important role during breast cancer brain metastases initiation and outgrowth. Using multiday IVM through a cranial imaging window in an experimental model of TNBC brain metastasis, it was demonstrated that the rate-limiting step to form brain metastasis is not the capacity for extravasation, but rather the capacity to overcome a dormancy-inducing microenvironment (Fig. 2E; Dai et al. 2022). Proliferative and expanding brain metastasis were situated in a vascular niche devoid of astrocytes, whereas dormant TNBC cells were associated with vessels that directly associated with astrocytes. Further analyses revealed that astrocytes prevent TNBC proliferation by deposition of laminin-211, a component of the parenchymal basement membrane. Laminin-211 is sensed by the TNBC through dystroglycan, which in turn negatively regulates transcriptional activator Yes associated protein (YAP), thereby inhibiting proliferation in these metastatic TNBC cells (Dai et al. 2022). A similar dormancy-sustaining mechanism by the ECM was described for disseminated tumor cells in the lung, where deposition of collagen type III was shown to be required to retain a dormant state (di Martino et al. 2022). Reactivation of dormant cells could be achieved both by disruption of the collagen III architecture and by changes in the collagen III abundance (di Martino et al. 2022). In addition to the local tissue architecture, other extrinsic factors were identified as stimuli-reactivating dormant metastatic breast cancer cells. In models of experimental and spontaneous breast cancer lung metastases, sustained inflammation, for example, induced by exposure to tobacco smoke, was identified as one of these triggers (Park et al. 2016; Albrengues et al. 2018). IVM revealed that inflammation resulted in high infiltration of neutrophils in the lungs, which in turn resulted in dormant cancer cells that reentered the cell cycle. Interestingly, the effect of the neutrophils was not exerted by the immune cells directly, but in an indirect manner through the deposition of neutrophil extracellular traps (NETs). NETs are DNA scaffolds in the extracellular space associated with proteases, which in turn were shown to facilitate proteolytic ECM remodeling and cleavage of laminin-111 to trigger cancer cell proliferation (Albrengues et al. 2018). Together, these experiments reveal a local and temporal control mechanism of tumor cell dormancy by structural and cellular components of the host tissues. Sustaining these natural barriers could potentially prevent outgrowth of disseminated cancer cells into overt metastases.

In addition to blood-borne metastases, two hallmark papers recently provided evidence for a hybrid route in which lymph node metastases serve as a hub prior to further dissemination to distant organs using either spontaneous metastases models or experimental lymph node metastases of breast cancer by intralymphatic infusion (Fig. 2C; Brown et al. 2018; Pereira et al. 2018). To study the dynamics of spontaneous lymph node metastases, diverse cancer models (including mammary tumor cells) expressing a photoconvertible protein were orthotopically transplanted (Pereira et al. 2018). Cancer cells that spontaneously metastasized to the lymph node were photoconverted (from green to red) to determine the origin of distant metastases arising in the lungs (i.e., nonconverted [green] metastatic lesions would indicate direct transit from the primary tumor, whereas converted [red] metastatic lesions would indicate a hybrid route via the lymph node). Interestingly, both red and green lung metastasis were detected indicating that both routes are used by cancer cells to reach the distant organs. Next, using time-lapse IVM, it was shown that cancer cells disseminating through the lymphatics enter the lymph nodes via the subcapsular sinus and migrate toward the cortex of the lymph nodes, where they associated with the endothelial cells and the resident dendritic cells (Pereira et al. 2018). Similar spatial dynamics were observed in static time-point analyses after intralymphatic infusion of breast cancer cells (Brown et al. 2018), where the high endothelial venules, the main point of entry for incoming lymphocytes, were identified as the main point of exit for the cancer cells to seed to distant organs (Brown et al. 2018). Intravasation of breast cancer cells into the lymphatics was shown to correlate with Ezrin expression, and IVM of breast cancer cell behavior after systemic Ezrin inhibition revealed impaired cell migration and a reduction in metastatic burden in the lymph nodes and lungs (Ghaffari et al. 2019). Recently, the role of lymph node colonization prior to distant metastasis formation was shown to extend beyond a mere transit route. Instead, it was shown that cancer cells colonizing the lymph node are able to evade the immune system, activate regulatory T cells, and actively induce systemic tumor-specific immune tolerance to facilitate further metastatic spread (Reticker-Flynn et al. 2022). Together, these studies identify the lymph nodes as an initial hub prior to systemic spreading using a powerful combination of dynamic IVM and static ex vivo imaging.

IMAGING OF DRUG RESPONSE DYNAMICS IN BREAST CANCER MODELS

To successfully treat breast cancer, it is important to know the in vivo biodistribution, kinetics, and mode of action of a potential drug. Both drug delivery and dynamics are hard to predict based on ex vivo analysis, and, specifically in the case of a heterogeneous response, a cellular resolution is required to understand where and when a potential anticancer drug exerts its function. IVM is perfectly suited to visualize drug distribution and dynamics, and several IVM studies have helped to elucidate drug dynamics and mode of action. For instance, when using IVM, it was shown that Parp inhibitors reach their cellular targets within seconds to minutes in the in vivo setting at a sufficiently high concentration, suggesting that resistance is not caused by an inefficient bioavailability of the drug (Thurber et al. 2013). Moreover, the TME is an important player in drug responses that can only be assessed in the unperturbed in vivo setting (Miller and Weissleder 2017). For example, in the case of monoclonal antibodies targeting HER2-positive breast cancers (trastuzumab), IVM revealed that, at the first 24 hours after injection, the HER2-overexpressing breast cancer cells were reached. However, soon after, this distribution shifted toward predominant accumulation of trastuzumab in the tumor-associated macrophages (Li et al. 2020). A similar drug-sequestering effect of the tumor-associated macrophages was observed in tumors treated with anti-PD1 antibodies used to target and activate intratumoral CD8[+] T cells (Arlauckas et al. 2017). Although an immediate binding of anti-PD1 to the T cells within the tumor was observed, this interaction was transient as tumor-associated macrophages captured these antibodies from the T cells within a few minutes (Arlauckas et al. 2017). Anti-PD1 engagement could be prolonged using a blockade of the Fcγ receptors, indicating that a combination therapy of anti-PD1 with Fcγ-receptor blockade could be a way to improve immunotherapy (Arlauckas et al. 2017). Also, in doxorubicin-treated mammary tumors, IVM demonstrated that the TME plays an important role, where differential regulation of vascular permeability and myeloid cell infiltration were shown to be key determinants of effective drug response (Nakasone et al. 2012).

Many breast cancer patients eventually develop resistance to therapy. To better understand the dynamics of drug-resistance, IVM has been used as a tool. For instance, IVM combined with fluorescent analogs of eribulin, a microtubule-targeting agent, identified that resistance was spatially determined by distance of the cancer cells to the 3D tumor vasculature and the expression of drug efflux proteins (Laughney et al. 2014). A promising class of anticancer drugs are therapeutic nanoparticles, which not only lead to reduced toxicity, but also show improved bioavailability. However, efficient delivery to the tumor site to maximize their therapeutic effect is still a hurdle in clinical practice. Using a combination of IVM and modeling, it was shown that a single, low dose of local radiation can be used to boost nanoparticle influx into the breast tumor (Miller et al. 2017). Local radiation leads to a transient increase in vascular permeability, which in turn drives extravasation of nanoparticles and tumor-associated macrophages, which took up the therapeutic nanoparticles, into the tumor. Several other approaches have indeed used this tendency of tumor-associated immune cells to take up the nanoparticles. Both tumor-associated macrophages (Rodell et al. 2018), neutrophils (Chu et al. 2017; Naumenko et al. 2020), or myeloid cells were used as "Trojan horses" to improve the delivery of therapeutic nanoparticles specifically toward the tumor site (Lin et al. 2020).

To study the molecular dynamics together with the spatiotemporal dynamics of drug target vulnerabilities, several biosensor mouse models were developed (Erami et al. 2016; Nobis et al. 2017, 2018). Recently, a Rac1-FRET biosensor mouse was generated to study the potential of Rac1-inhibition in breast cancer. Rac1, a small GTPase and regulator of the actin cytoskeleton, may play a key role in the major steps of metastasis, such as cell motility, invasion, and EMT–MET (mesenchymal-to-epithelial transition) plasticity during extravasation and distant colonization (Floerchinger et al. 2021). IVM of MMTV-PyMT and MMTV-Her2 mouse models expressing the Rac1-FRET biosensor in primary breast tumors revealed that Rac1 activity is specifically up regulated in proximity to tumor vasculature. Pharmacodynamic monitoring of Rac1 inhibition through a mammary imaging window showed a temporal decrease in Rac activity and a reduction in intratumoral migration in the short term, as well as increased survival and decreased lung metastasis on the long term (Floerchinger et al. 2021). Taken together, IVM has proven to be a valuable tool not only to visualize biodistribution of diverse anticancer drugs, but also to elucidate the spatial and temporal heterogeneity of drug efficacy in the intact in vivo environment.

CONCLUDING REMARKS AND FUTURE DIRECTIONS

Due to advances in sample preparation protocols and microscope technology, we can now achieve incredible cellular detail in large tissue volumes, demonstrating an invaluable asset for 3D and intravital imaging of breast cancer. Novel reporter model systems combined with innovative imaging technologies now allow for visualization of diverse aspects of the cancer cells and their TME beyond cell migration or proliferation, such as the molecular signaling dynamics or tumor metabolism, in a multiplexed way (Zhu et al. 2017; Madonna et al. 2021). With accompanying advances in artificial intelligence and large data handling, a multitude of parameters can be extracted from those imaging data sets. A recent study exemplified the power of combining imaging with large-scale analyses to extract the behavioral identity and states of immune cells in their native environments (Crainiciuc et al. 2022). The combination of dynamic imaging and large-scale or automated analyses offers a unique opportunity to correlate tumoral heterogeneity in space and time with clinical outcomes such as cancer progression or treatment response. Finally, recent developments allow for noninvasive monitoring of tumor growth dynamics in preclinical models of breast cancer at a cellular resolution using a noninvasive and vacuum stabilized imaging window, allowing for dynamic repositioning of the imaging field-of-view (Ozturk et al. 2021). Using a similar noninvasive approach with a miniaturized microscope, MediSCAPE, has demonstrated live imaging of oral mucosa in healthy volunteers, thereby paving the way for live imaging in humans (Patel et al. 2022). Such advancements hold promise that, in the coming years, live imaging could be further developed for real-time diagnosis and even incorporate dynamic information that might lead to new biomarkers of cancer aggressiveness.

Another way for deep exploration of breast cancer biology at the patient population scale is provided by recent advancements in human breast tissue in vitro modeling and development of patient-derived breast cancer organoid biobanks (Sachs et al. 2018; Rosenbluth et al. 2020; Dekkers et al. 2021) combined with 3D imaging. For instance, a combination of organoid culture and imaging can provide a powerful readout for testing the efficacy of immunotherapies, such as engineered T cells in coculture assays with cancer organoids (for a comprehensive review on organoid imaging, refer to Rios and Clevers 2018; Lukonin et al. 2021). This potential was demonstrated by applying patient-derived cancer organoids and 3D live imaging to reveal a cancer-metabolome-sensing T cell immunotherapy with broad targeting efficacy across several subtypes of breast cancer (Dekkers et al. 2022). With the deployment of BEHAV3D, a tailored 3D imaging-based framework for organoid technology, the study further revealed mode-of-action of diverse cellular immunotherapies and the influence of patient-tumor variability on their functioning. Moreover, efforts dedicated to integrating imaging with other omics data sets (RNAseq, DNAseq, ATACseq, etc.), will further enhance the dimensionality retrieved from a single biological specimen (Rodriques et al. 2019; Cang and Nie 2020; Liu et al. 2020; Mantri et al. 2021; Payne et al. 2021; Zanfardino et al. 2021). Thus, we can expect that combining advancements in human model engineering with tailored imaging modalities, integrated with behavioral phenotyping and cell-type identification at the single-cell level, will move the field of breast cancer toward precision medicine and accelerated drug development in the coming years (Rios and Clevers 2018; Dekkers et al. 2022). These future directions will greatly aid in understanding the underlying molecular signatures responsible for complex cellular heterogeneity of breast tumors and their microenvironment identified by imaging as well as the heterogenous composition and mode of action of cellular cancer therapies.

ACKNOWLEDGMENTS

Due to the limited number of papers that we could discuss and cite in this review, we focused on the latest literature for multidimensional imaging in breast cancer. We apologize to all authors of key papers in the field that were not recent enough to include in this review. J.v.R. was supported by the Netherlands Organization of Scientific Research NWO (VICI 09150182110004), CancerGenomics.nl (Netherlands Organisation for Scientific Research) program, and the Doctor Josef Steiner Foundation (to J.v.R). CLGJS was supported by an EMBO postdoctoral fellowship (grant ALTF-1035-2020), the FEBS excellence award, and an Excellence of Science (EOS) grant (project ID: 40007532) of Fonds Wetenschappelijk Onderzoek-Le Fonds de la Recherche Scientifique (FWO-FNRS). A.C.R. received funding from the European Research Council (ERC) under the European Union's Horizon 2020 research and innovation programme (grant agreement No. 804412). This work was financially supported by the Princess Máxima Center for Pediatric Oncology and St. Baldrick's Robert J. Arceci International Innovation award.

REFERENCES

Ahdoot M, Wilbur AR, Reese SE, Lebastchi AH, Mehralivand S, Gomella PT, Bloom J, Gurram S, Siddiqui M, Pinsky P, et al. 2020. MRI-targeted, systematic, and combined biopsy for prostate cancer diagnosis. *N Engl J Med* **382:** 917–928. doi:10.1056/NEJMoa1910038

Albrengues J, Shields MA, Ng D, Park CG, Ambrico A, Poindexter ME, Upadhyay P, Uyeminami DL, Pommier A, Küttner V, et al. 2018. Neutrophil extracellular traps produced during inflammation awaken dormant cancer cells in mice. *Science* **361:** eaao4227. doi:10.1126/science.aao4227

Alieva M, Ritsma L, Giedt RJ, Weissleder R, van Rheenen J. 2014. Imaging windows for long-term intravital imaging: general overview and technical insights. *Intravital* **3:** e29917. doi:10.4161/intv.29917

Almagro J, Messal HA, Zaw Thin M, van Rheenen J, Behrens A. 2021. Tissue clearing to examine tumour complexity in three dimensions. *Nat Rev Cancer* **21:** 718–730. doi:10.1038/s41568-021-00382-w

Alon S, Goodwin DR, Sinha A, Wassie AT, Chen F, Daugharthy ER, Bando Y, Kajita A, Xue AG, Marrett K, et al. 2021. Expansion sequencing: spatially precise in situ transcriptomics in intact biological systems. *Science* **371:** eaax2656. doi:10.1126/science.aax2656

Andresen V, Alexander S, Heupel WM, Hirschberg M, Hoffman RM, Friedl P. 2009. Infrared multiphoton microscopy: subcellular-resolved deep tissue imaging. *Curr Opin Biotechnol* **20:** 54–62. doi:10.1016/j.copbio.2009.02.008

Arlauckas SP, Garris CS, Kohler RH, Kitaoka M, Cuccarese MF, Yang KS, Miller MA, Carlson JC, Freeman GJ, Anthony RM, et al. 2017. In vivo imaging reveals a tumor-associated macrophage-mediated resistance pathway in anti-PD-1 therapy. *Sci Transl Med* **9:** eaal3604. doi:10.1126/scitranslmed.aal3604

Ayoub NM, Jaradat SK, Al-Shami KM, Alkhalifa AE. 2022. Targeting angiogenesis in breast cancer: current evidence and future perspectives of novel anti-angiogenic approaches. *Front Pharmacol* **13:** 485. doi:10.3389/fphar.2022.838133

Bakker GJ, Weischer S, Ferrer Ortas J, Heidelin J, Andresen V, Beutler M, Beaurepaire E, Friedl P. 2022. Intravital deep-tumor single-beam 3-photon, 4-photon, and harmonic microscopy. *eLife* **11:** e63776. doi:10.7554/eLife.63776

Beerling E, Seinstra D, de Wit E, Kester L, van der Velden D, Maynard C, Schäfer R, van Diest P, Voest E, van Oudenaarden A, et al. 2016. Plasticity between epithelial and mesenchymal states unlinks EMT from metastasis-enhancing stem cell capacity. *Cell Rep* **14:** 2281–2288. doi:10.1016/j.celrep.2016.02.034

Booth ME, Treanor D, Roberts N, Magee DR, Speirs V, Hanby AM. 2015. Three-dimensional reconstruction of ductal carcinoma in situ with virtual slides. *Histopathology* **66:** 966–973. doi:10.1111/his.12561

Bornes L, van Scheppingen RH, Beerling E, Schelfhorst T, Ellenbroek SIJ, Seinstra D, van Rheenen J. 2019. Fsp1-mediated lineage tracing fails to detect the majority of disseminating cells undergoing EMT. *Cell Rep* **29:** 2565–2569.e3. doi:10.1016/j.celrep.2019.10.107

Borriello L, Coste A, Traub B, Sharma VP, Karagiannis GS, Lin Y, Wang Y, Ye X, Duran CL, Chen X, et al. 2022. Primary tumor associated macrophages activate programs of invasion and dormancy in disseminating tumor cells. *Nat Commun* **13:** 626. doi:10.1038/s41467-022-28076-3

Brown AS, Hunt KK, Shen J, Huo L, Babiera GV, Ross MI, Meric-Bernstam F, Feig BW, Kuerer HM, Boughey JC, et al. 2010. Histologic changes associated with false-negative sentinel lymph nodes after preoperative chemotherapy in patients with confirmed lymph node-positive breast cancer before treatment. *Cancer* **116:** 2878–2883. doi:10.1002/cncr.25066

Brown S, Pineda CM, Xin T, Boucher J, Suozzi KC, Park S, Matte-Martone C, Gonzalez DG, Rytlewski J, Beronja S, et al. 2017. Correction of aberrant growth preserves tissue homeostasis. *Nature* **548:** 334–337. doi:10.1038/nature23304

Brown M, Assen FP, Leithner A, Abe J, Schachner H, Asfour G, Bago-Horvath Z, Stein J, Uhrin P, Sixt M, et al. 2018. Lymph node blood vessels provide exit routes for metastatic tumor cell dissemination in mice. *Science* **359:** 1408–1411. doi:10.1126/science.aal3662

Bruens L, Ellenbroek SIJ, Suijkerbuijk SJE, Azkanaz M, Hale AJ, Toonen P, Flanagan DJ, Sansom OJ, Snippert HJ, van Rheenen J. 2020. Calorie restriction increases the number of competing stem cells and decreases mutation retention in the intestine. *Cell Rep* **32:** 107937. doi:10.1016/j.celrep.2020.107937

Cai R, Pan C, Ghasemigharagoz A, Todorov MI, Förstera B, Zhao S, Bhatia HS, Parra-Damas A, Mrowka L, Theodorou D, et al. 2019. Panoptic imaging of transparent mice reveals whole-body neuronal projections and skull-meninges connections. *Nat Neurosci* **22:** 317–327. doi:10.1038/s41593-018-0301-3

Cang Z, Nie Q. 2020. Inferring spatial and signaling relationships between cells from single cell transcriptomic data. *Nat Commun* **11**: 2084. doi:10.1038/s41467-020-15968-5

Catalona WJ, Richie JP, Ahmann FR, Hudson MA, Scardino PT, Flanigan RC, DeKernion JB, Ratliff TL, Kavoussi LR, Dalkin BL, et al. 2017. Comparison of digital rectal examination and serum prostate specific antigen in the early detection of prostate cancer: results of a multicenter clinical trial of 6,630 men. *J Urol* **197**: S200–S207. doi:10.1016/j.juro.2016.10.073

Chakraborty T, Driscoll MK, Jeffery E, Murphy MM, Roudot P, Chang BJ, Vora S, Wong WM, Nielson CD, Zhang H, et al. 2019. Light-sheet microscopy of cleared tissues with isotropic, subcellular resolution. *Nature Methods* **16**: 1109–1113. doi:10.1038/s41592-019-0615-4

Chen J, Suo S, Tam PP, Han JDJ, Peng G, Jing N. 2017. Spatial transcriptomic analysis of cryosectioned tissue samples with Geo-seq. *Nat Protoc* **12**: 566–580. doi:10.1038/nprot.2017.003

Chen Y, Shen Q, White SL, Gokmen-Polar Y, Badve S, Goodman LJ. 2019. Three-dimensional imaging and quantitative analysis in CLARITY processed breast cancer tissues. *Sci Rep* **9**: 5624. doi:10.1038/s41598-019-41957-w

Chi J, Crane A, Wu Z, Cohen P. 2018. Adipo-Clear: a tissue clearing method for three-dimensional imaging of adipose tissue. *J Vis Exp* **2018**: 58271.

Choe K, Hontani Y, Wang T, Hebert E, Ouzounov DG, Lai K, Singh A, Béguelin W, Melnick AM, Xu C. 2022. Intravital three-photon microscopy allows visualization over the entire depth of mouse lymph nodes. *Nat Immunol* **23**: 330–340. doi:10.1038/s41590-021-01101-1

Chu D, Dong X, Zhao Q, Gu J, Wang Z. 2017. Photosensitization priming of tumor microenvironments improves delivery of nanotherapeutics via neutrophil infiltration. *Adv Mater* **29**: 1701021. doi:10.1002/adma.201701021

Chung K, Wallace J, Kim SY, Kalyanasundaram S, Andalman AS, Davidson TJ, Mirzabekov JJ, Zalocusky KA, Mattis J, Denisin AK, et al. 2013. Structural and molecular interrogation of intact biological systems. *Nature* **497**: 332–337. doi:10.1038/nature12107

Coutu DL, Kokkaliaris KD, Kunz L, Schroeder T. 2018. Multicolor quantitative confocal imaging cytometry. *Nat Methods* **15**: 39–46. doi:10.1038/nmeth.4503

Crainiciuc G, Palomino-Segura M, Molina-Moreno M, Sicilia J, Aragones DG, Li JLY, Madurga R, Adrover JM, Aroca-Crevillén A, Martin-Salamanca S, et al. 2022. Behavioural immune landscapes of inflammation. *Nature* **601**: 415–421. doi:10.1038/s41586-021-04263-y

Creech MK, Wang J, Nan X, Gibbs SL. 2017. Superresolution imaging of clinical formalin fixed paraffin embedded breast cancer with single molecule localization microscopy. *Sci Rep* **7**: 40766. doi:10.1038/srep40766

Dai J, Cimino PJ, Gouin KH, Grzelak CA, Barrett A, Lim AR, Long A, Weaver S, Saldin LT, Uzamere A, et al. 2022. Astrocytic laminin-211 drives disseminated breast tumor cell dormancy in brain. *Nat Cancer* **3**: 25–42.

Dart A. 2018. Take a left here. *Nat Rev Cancer* **18**: 337. doi:10.1038/s41568-018-0011-x

Davis FM, Lloyd-Lewis B, Harris OB, Kozar S, Winton DJ, Muresan L, Watson CJ. 2016. Single-cell lineage tracing in the mammary gland reveals stochastic clonal dispersion of stem/progenitor cell progeny. *Nat Commun* **7**: 13053. doi:10.1038/ncomms13053

Dawson CA, Visvader JE. 2021. The cellular organization of the mammary gland: insights from microscopy. *J Mammary Gland Biol Neoplasia* **26**: 71–85. doi:10.1007/s10911-021-09483-6

Dawson CA, Pal B, Vaillant F, Gandolfo LC, Liu Z, Bleriot C, Ginhoux F, Smyth GK, Lindeman GJ, Mueller SN, et al. 2020. Tissue-resident ductal macrophages survey the mammary epithelium and facilitate tissue remodelling. *Nat Cell Biol* **22**: 546–558. doi:10.1038/s41556-020-0505-0

Dawson CA, Mueller SN, Lindeman GJ, Rios AC, Visvader JE. 2021. Intravital microscopy of dynamic single-cell behavior in mouse mammary tissue. *Nat Protoc* **16**: 1907–1935. doi:10.1038/s41596-020-00473-2

Day CP, Pérez-Guijarro E, Lopès A, Goldszmid RS, Murgai M, Wakefield L, Merlino G. 2022. Recognition of observer effect is required for rigor and reproducibility of preclinical animal studies. *Cancer Cell* **40**: 231–232. doi:10.1016/j.ccell.2022.01.015

Dekkers JF, van Vliet EJ, Sachs N, Rosenbluth JM, Kopper O, Rebel HG, Wehrens EJ, Piani C, Visvader JE, Verissimo CS, et al. 2021. Long-term culture, genetic manipulation and xenotransplantation of human normal and breast cancer organoids. *Nat Protoc* **16**: 1936–1965. doi:10.1038/s41596-020-00474-1

Dekkers JF, Alieva M, Cleven A, Keramati F, Wezenaar AKL, van Vliet EJ, Puschhof J, Brazda P, Johanna I, Meringa AD, et al. 2022. Uncovering the mode of action of engineered T cells in patient cancer organoids. *Nat Biotechnol* doi:10.1038/s41587-022-01397-w

di Martino JS, Nobre AR, Mondal C, Taha I, Farias EF, Fertig EJ, Naba A, Aguirre-Ghiso JA, Bravo-Cordero JJ. 2022. A tumor-derived type III collagen-rich ECM niche regulates tumor cell dormancy. *Nat Cancer* **3**: 90–107. doi:10.1038/s43018-021-00291-9

di Palma S, Bodenmiller B. 2015. Unraveling cell populations in tumors by single-cell mass cytometry. *Curr Opin Biotechnol* **31**: 122–129. doi:10.1016/j.copbio.2014.07.004

Dobosz M, Ntziachristos V, Scheuer W, Strobel S. 2014. Multispectral fluorescence ultramicroscopy: three-dimensional visualization and automatic quantification of tumor morphology, drug penetration, and antiangiogenic treatment response. *Neoplasia* **16**: 1–13. doi:10.1593/neo.131848

Entenberg D, Voiculescu S, Guo P, Borriello L, Wang Y, Karagiannis GS, Jones J, Baccay F, Oktay M, Condeelis J. 2018. A permanent window for the murine lung enables high-resolution imaging of cancer metastasis. *Nat Methods* **15**: 73–80. doi:10.1038/nmeth.4511

Epstein JI. 2018. Prostate cancer grading: a decade after the 2005 modified system. *Mod Pathol* **31**: S47–S63. doi:10.1038/modpathol.2017.133

Erami Z, Herrmann D, Warren SC, Nobis M, McGhee EJ, Lucas MC, Leung W, Reischmann N, Mrowinska A, Schwarz JP, et al. 2016. Intravital FRAP imaging using an E-cadherin-GFP mouse reveals disease- and drug-dependent dynamic regulation of cell-cell junctions in live tissue. *Cell Rep* **14:** 152–167. doi:10.1016/j.celrep.2015.12.020

Ertürk A, Becker K, Jährling N, Mauch CP, Hojer CD, Egen JG, Hellal F, Bradke F, Sheng M, Dodt HU. 2012. Three-dimensional imaging of solvent-cleared organs using 3DISCO. *Nat Protoc* **7:** 1983–1995. doi:10.1038/nprot.2012.119

Ewald AJ, Werb Z, Egeblad M. 2011. Preparation of mice for long-term intravital imaging of the mammary gland. *Cold Spring Harb Protocols* **2011:** pdb.prot5562.

Fabiano E, Zhang J, Reinhart-King CA. 2022. Tissue density in the progression of breast cancer: bedside to bench and back again. *Curr Opin Biomed Eng* **22:** 100383. doi:10.1016/j.cobme.2022.100383

Farante G, Toesca A, Magnoni F, Lissidini G, Vila J, Mastropasqua M, Viale G, Penco S, Cassano E, Lazzeroni M, et al. 2022. Advances and controversies in management of breast ductal carcinoma in situ (DCIS). *Eur J Surg Oncol* **48:** 736–741. doi:10.1016/j.ejso.2021.10.030

Fiore VF, Krajnc M, Quiroz FG, Levorse J, Pasolli HA, Shvartsman SY, Fuchs E. 2020. Mechanics of a multilayer epithelium instruct tumour architecture and function. *Nature* **585:** 433–439. doi:10.1038/s41586-020-2695-9

Floerchinger A, Murphy KJ, Latham SL, Warren SC, McCulloch AT, Lee YK, Stoehr J, Mélénec P, Guaman CS, Metcalf XL, et al. 2021. Optimizing metastatic-cascade-dependent Rac1 targeting in breast cancer: guidance using optical window intravital FRET imaging. *Cell Rep* **36:** 109689. doi:10.1016/j.celrep.2021.109689

Fluegen G, Avivar-Valderas A, Wang Y, Padgen MR, Williams JK, Nobre AR, Calvo V, Cheung JF, Bravo-Cordero JJ, Entenberg D, et al. 2017. Phenotypic heterogeneity of disseminated tumour cells is preset by primary tumour hypoxic microenvironments. *Nat Cell Biol* **19:** 120–132. doi:10.1038/ncb3465

Friedl P, Wolf K, Harms G, von Andrian UH. 2007. Biological second and third harmonic generation microscopy. *Curr Protoc Cell Biol* **4:** 4.15.

Gavgiotaki E, Filippidis G, Tsafas V, Bovasianos S, Kenanakis G, Georgoulias V, Tzardi M, Agelaki S, Athanassakis I. 2020. Third harmonic generation microscopy distinguishes malignant cell grade in human breast tissue biopsies. *Sci Rep* **10:** 11055. doi:10.1038/s41598-020-67857-y

Ghaffari A, Hoskin V, Turashvili G, Varma S, Mewburn J, Mullins G, Greer PA, Kiefer F, Day AG, Madarnas Y, et al. 2019. Intravital imaging reveals systemic ezrin inhibition impedes cancer cell migration and lymph node metastasis in breast cancer. *Breast Cancer Res* **21:** 12. doi:10.1186/s13058-018-1079-7

Gligorijevic B, Bergman A, Condeelis J. 2014. Multiparametric classification links tumor microenvironments with tumor cell phenotype. *PLoS Biol* **12:** e1001995. doi:10.1371/journal.pbio.1001995

Goltsev Y, Samusik N, Kennedy-Darling J, Bhate S, Hale M, Vazquez G, Black S, Nolan GP. 2018. Deep profiling of mouse splenic architecture with CODEX multiplexed imaging. *Cell* **174:** 968–981.e15. doi:10.1016/j.cell.2018.07.010

Grzelak CA, Goddard ET, Lederer EE, Rajaram K, Dai J, Shor RE, Lim AR, Kim J, Beronja S, Funnell APW, et al. 2022. Elimination of fluorescent protein immunogenicity permits modeling of metastasis in immune-competent

settings. *Cancer Cell* **40:** 1–2. doi:10.1016/j.ccell.2021.11.004

Guldner IH, Yang L, Cowdrick KR, Wang Q, Alvarez Barrios WV, Zellmer VR, Zhang Y, Host M, Liu F, Chen DZ, et al. 2016. An integrative platform for three-dimensional quantitative analysis of spatially heterogeneous metastasis landscapes. *Sci Rep* **6:** 24201. doi:10.1038/srep24201

Haddad TS, Friedl P, Farahani N, Treanor D, Zlobec I, Nagtegaal I. 2021. Tutorial: methods for three-dimensional visualization of archival tissue material. *Nat Protoc* **16:** 4945–4962. doi:10.1038/s41596-021-00611-4

Harney AS, Arwert EN, Entenberg D, Wang Y, Guo P, Qian BZZ, Oktay MH, Pollard JW, Jones JG, Condeelis JS. 2015. Real-time imaging reveals local, transient vascular permeability, and tumor cell intravasation stimulated by TIE2hi macrophage–derived VEGFA. *Cancer Discov* **5:** 932–943. doi:10.1158/2159-8290.CD-15-0012

Harney AS, Wang Y, Condeelis JS, Entenberg D. 2016. Extended time-lapse intravital imaging of real-time multicellular dynamics in the tumor microenvironment. *J Vis Exp* **112:** 54042.

Hegde PS, Chen DS. 2020. Top 10 challenges in cancer immunotherapy. *Immunity* **52:** 17–35. doi:10.1016/j.immuni.2019.12.011

Ilina O, Campanello L, Gritsenko PG, Vullings M, Wang C, Bult P, Losert W, Friedl P. 2018. Intravital microscopy of collective invasion plasticity in breast cancer. *Dis Model Mech* **11:** dmm034330. doi:10.1242/dmm.034330

Ilina O, Gritsenko PG, Syga S, Lippoldt J, la Porta CAM, Chepizhko O, Grosser S, Vullings M, Bakker GJ, Starruß J, et al. 2020. Cell–cell adhesion and 3D matrix confinement determine jamming transitions in breast cancer invasion. *Nat Cell Biol* **22:** 1103–1115. doi:10.1038/s41556-020-0552-6

Jacquemin G, Benavente-Diaz M, Djaber S, Bore A, Dangles-Marie V, Surdez D, Tajbakhsh S, Fre S, Lloyd-Lewis B. 2021. Longitudinal high-resolution imaging through a flexible intravital imaging window. *Sci Adv* **7:** eabg7663. doi:10.1126/sciadv.abg7663

Jeong J, VanHouten JN, Dann P, Kim W, Sullivan C, Yu H, Liotta L, Espina V, Stern DF, Friedman PA, et al. 2016. PMCA2 regulates HER2 protein kinase localization and signaling and promotes HER2-mediated breast cancer. *Proc Natl Acad Sci* **113:** E282–E290. doi:10.1073/pnas.1516138113

Jones CG. 2012. Scanning electron microscopy: preparation and imaging for SEM. *Methods Mol Biol* **915:** 1–20. doi:10.1007/978-1-61779-977-8_1

Jonkman J, Brown CM, Wright GD, Anderson KI, North AJ. 2020. Tutorial: guidance for quantitative confocal microscopy. *Nat Protoc* **15:** 1585–1611. doi:10.1038/s41596-020-0313-9

Junker JP, Noël ES, Guryev V, Peterson KA, Shah G, Huisken J, McMahon AP, Berezikov E, Bakkers J, van Oudenaarden A. 2014. Genome-wide RNA tomography in the zebrafish embryo. *Cell* **159:** 662–675. doi:10.1016/j.cell.2014.09.038

Karagiannis GS, Pastoriza JM, Wang Y, Harney AS, Entenberg D, Pignatelli J, Sharma VP, Xue EA, Cheng E, D'Alfonso TM, et al. 2017. Neoadjuvant chemotherapy induces breast cancer metastasis through a TMEM-mediated mechanism. *Sci Transl Med* **9:** eaan0026. doi:10 .1126/scitranslmed.aan0026

Karaiskos N, Wahle P, Alles J, Boltengagen A, Ayoub S, Kipar C, Kocks C, Rajewsky N, Zinzen RP. 2017. The *Drosophila* embryo at single-cell transcriptome resolution. *Science* **358:** 194–199. doi:10.1126/science.aan 3235

Kedrin D, Gligorijevic B, Wyckoff J, Verkhusha V, Condeelis J, Segall JE, van Rheenen J. 2008. Intravital imaging of metastatic behavior through a mammary imaging window. *Nat Methods* **5:** 1019–1021. doi:10.1038/nmeth .1269

Kretzschmar K, Watt FM. 2012. Lineage tracing. *Cell* **148:** 33–45. doi:10.1016/j.cell.2012.01.002

Kruskal JB, Kane RA, Sentovich SM, Longmaid HE. 1997. Pitfalls and sources of error in staging rectal cancer with endorectal US. *Radiographics* **17:** 609–626. doi:10.1148/ radiographics.17.3.9153700

Ku T, Guan W, Evans NB, Sohn CH, Albanese A, Kim JG, Frosch MP, Chung K. 2020. Elasticizing tissues for reversible shape transformation and accelerated molecular labeling. *Nat Methods* **17:** 609–613. doi:10.1038/s41592-020-0823-y

Kubota SI, Takahashi K, Nishida J, Morishita Y, Ehata S, Tainaka K, Miyazono K, Ueda HR. 2017. Whole-body profiling of cancer metastasis with single-cell resolution. *Cell Rep* **20:** 236–250. doi:10.1016/j.celrep.2017.06.010

Kuett L, Catena R, Özcan A, Plüss A, Ali HR, Al Sa'd M, Alon S, Aparicio S, Battistoni G, Balasubramanian S, et al. 2021. Three-dimensional imaging mass cytometry for highly multiplexed molecular and cellular mapping of tissues and the tumor microenvironment. *Nat Cancer* **3:** 122–133.

Lamprecht S, Schmidt EM, Blaj C, Hermeking H, Jung A, Kirchner T, Horst D. 2017. Multicolor lineage tracing reveals clonal architecture and dynamics in colon cancer. *Nat Commun* **8:** 1406. doi:10.1038/s41467-017-00976-9

Laughney AM, Kim E, Sprachman MM, Miller MA, Kohler RH, Yang KS, Orth JD, Mitchison TJ, Weissleder R. 2014. Single-cell pharmacokinetic imaging reveals a therapeutic strategy to overcome drug resistance to the microtubule inhibitor eribulin. *Sci Transl Med* **6:** 261ra152. doi:10 .1126/scitranslmed.3009318

Lee SS, Bindokas VP, Kron SJ. 2019a. Multiplex three-dimensional mapping of macromolecular drug distribution in the tumor microenvironment. *Mol Cancer Ther* **18:** 213–226. doi:10.1158/1535-7163.MCT-18-0554

Lee SSY, Bindokas VP, Lingen MW, Kron SJ. 2019b. Nondestructive, multiplex three-dimensional mapping of immune infiltrates in core needle biopsy. *Lab Invest* **99:** 1400–1413. doi:10.1038/s41374-018-0156-y

Leong HS, Robertson AE, Stoletov K, Leith SJ, Chin CA, Chien AE, Hague MN, Ablack A, Carmine-Simmen K, McPherson VA, et al. 2014. Invadopodia are required for cancer cell extravasation and are a therapeutic target for metastasis. *Cell Rep* **8:** 1558–1570. doi:10.1016/j.celrep .2014.07.050

Leung E, Xue A, Wang Y, Rougerie P, Sharma VP, Eddy R, Cox JJ, Condeelis J. 2017. Blood vessel endothelium-directed tumor cell streaming in breast tumors requires the HGF/C-Met signaling pathway. *Oncogene* **36:** 2680–2692. doi:10.1038/onc.2016.421

Li W, Germain RN, Gerner MY. 2017. Multiplex, quantitative cellular analysis in large tissue volumes with clearing-enhanced 3D microscopy (Ce3D). *Proc Natl Acad Sci* **114:** E7321–E7330.

Li W, Germain RN, Gerner MY. 2019. High-dimensional cell-level analysis of tissues with Ce3D multiplex volume imaging. *Nat Protoc* **14:** 1708–1733. doi:10.1038/s41596-019-0156-4

Li R, Attari A, Prytyskach M, Garlin MA, Weissleder R, Miller MA. 2020. Single-cell intravital microscopy of trastuzumab quantifies heterogeneous in vivo kinetics. *Cytometry A* **97:** 528–539. doi:10.1002/cyto.a.23872

Lin PY, Peng SJ, Shen CN, Pasricha PJ, Tang SC. 2016. PanIN-associated pericyte, glial, and islet remodeling in mice revealed by 3D pancreatic duct lesion histology. *Am J Physiol Gastrointest Liver Physiol* **311:** G412–G422. doi:10.1152/ajpgi.00071.2016

Lin Q, Fathi P, Chen X. 2020. Nanoparticle delivery in vivo: a fresh look from intravital imaging. *eBioMedicine* **59:** 102958.

Liu YA, Pan ST, Hou YC, Shen MY, Peng SJ, Tang SC, Chung YC. 2013. 3-D visualization and quantitation of microvessels in transparent human colorectal carcinoma. *PLoS ONE* **8:** e81857. doi:10.1371/journal.pone .0081857

Liu X, Taftaf R, Kawaguchi M, Chang YF, Chen W, Entenberg D, Zhang Y, Gerratana L, Huang S, Patel DB, et al. 2019. Homophilic CD44 interactions mediate tumor cell aggregation and polyclonal metastasis in patient-derived breast cancer models. *Cancer Discov* **9:** 96–113. doi:10 .1158/2159-8290.CD-18-0065

Liu Y, Yang M, Deng Y, Su G, Enninful A, Guo CC, Tebaldi T, Zhang D, Kim D, Bai Z, et al. 2020. High-spatial-resolution multi-omics sequencing via deterministic barcoding in tissue. *Cell* **183:** 1665–1681.e18. doi:10.1016/j.cell .2020.10.026

Lloyd-Lewis B. 2020. Multidimensional imaging of mammary gland development: a window into breast form and function. *Front Cell Dev Biol* **8:** 203. doi:10.3389/fcell .2020.00203

Lloyd-Lewis B, Davis FM, Harris OB, Hitchcock JR, Lourenco FC, Pasche M, Watson CJ. 2016. Imaging the mammary gland and mammary tumours in 3D: optical tissue clearing and immunofluorescence methods. *Breast Cancer Res* **18:** 1–17.

Lloyd-Lewis B, Gobbo F, Perkins M, Jacquemin G, Huyghe M, Faraldo MM, Fre S. 2022. In vivo imaging of mammary epithelial cell dynamics in response to lineage-biased Wnt/β-catenin activation. *Cell Rep* **38:** 110461. doi:10 .1016/j.celrep.2022.110461

Lukonin I, Zinner M, Liberali P. 2021. Organoids in image-based phenotypic chemical screens. *Exp Mol Med* **53:** 1495–1502. doi:10.1038/s12276-021-00641-8

Lüönd F, Sugiyama N, Bill R, Bornes L, Hager C, Tang F, Santacroce N, Beisel C, Ivanek R, Bürglin T, et al. 2021. Distinct contributions of partial and full EMT to breast cancer malignancy. *Dev Cell* **56:** 3203–3221.e11. doi:10 .1016/j.devcel.2021.11.006

Madonna MC, Duer JE, Lee J, Williams J, Avsaroglu B, Zhu C, Deutsch R, Wang R, Crouch BT, Hirschey MD, et al. 2021. In vivo optical metabolic imaging of long-chain fatty acid uptake in orthotopic models of triple-negative breast cancer. *Cancers (Basel)* **13**: 1–20.

Maiorino L, Shevik M, Adrover JM, Han X, Georgas E, Wilkinson JE, Seidner H, Foerschner L, Tuveson DA, Qin YX, et al. 2022. Longitudinal intravital imaging through clear silicone windows. *J Vis Exp* doi:10.3791/62757

Mantri M, Scuderi GJ, Abedini-Nassab R, Wang MFZ, McKellar D, Shi H, Grodner B, Butcher JT, de Vlaminck I. 2021. Spatiotemporal single-cell RNA sequencing of developing chicken hearts identifies interplay between cellular differentiation and morphogenesis. *Nat Commun* **12**: 1771. doi:10.1038/s41467-021-21892-z

Messal HA, Alt S, Ferreira RMM, Gribben C, Wang VMY, Cotoi CG, Salbreux G, Behrens A. 2019. Tissue curvature and apicobasal mechanical tension imbalance instruct cancer morphogenesis. *Nature* **566**: 126–130. doi:10.1038/s41586-019-0891-2

Messal HA, Almagro J, Zaw Thin M, Tedeschi A, Ciccarelli A, Blackie L, Anderson KI, Miguel-Aliaga I, van Rheenen J, Behrens A. 2021a. Antigen retrieval and clearing for whole-organ immunofluorescence by FLASH. *Nat Protoc* **16**: 239–262. doi:10.1038/s41596-020-00414-z

Messal HA, van Rheenen J, Scheele CLGJ. 2021b. An intravital microscopy toolbox to study mammary gland dynamics from cellular level to organ scale. *J Mammary Gland Biol and Neoplasia* **26**: 9–27. doi:10.1007/s10911-021-09487-2

Miller MA, Weissleder R. 2017. Imaging of anticancer drug action in single cells. *Nat Rev Cancer* **17**: 399–414. doi:10.1038/nrc.2017.41

Miller MA, Chandra R, Cuccarese MF, Pfirschke C, Engblom C, Stapleton S, Adhikary U, Kohler RH, Mohan JF, Pittet MJ, et al. 2017. Radiation therapy primes tumors for nanotherapeutic delivery via macrophage-mediated vascular bursts. *Sci Transl Med* **9**: eaal0225.

Mourao L, Ciwinska M, van Rheenen J, Scheele CLGJ. 2022. Longitudinal intravital microscopy using a mammary imaging window with replaceable lid. *J Vis Exp* doi:10.3791/63326

Murray E, Cho JH, Goodwin D, Ku T, Swaney J, Kim SY, Choi H, Park YG, Park JY, Hubbert A, et al. 2015. Simple, scalable proteomic imaging for high-dimensional profiling of intact systems. *Cell* **163**: 1500–1514. doi:10.1016/j.cell.2015.11.025

Nakasone ES, Askautrud HA, Kees T, Park JHJH, Plaks V, Ewald AJ, Fein M, Rasch MG, Tan YX, Qiu J, et al. 2012. Imaging tumor–stroma interactions during chemotherapy reveals contributions of the microenvironment to resistance. *Cancer Cell* **21**: 488–503. doi:10.1016/j.ccr.2012.02.017

Naumenko V, Nikitin A, Garanina A, Melnikov P, Vodopyanov S, Kapitanova K, Potashnikova D, Vishnevskiy D, Alieva I, Ilyasov A, et al. 2020. Neutrophil-mediated transport is crucial for delivery of short circulating magnetic nanoparticles to tumors. *Acta Biomater* **104**: 176–187. doi:10.1016/j.actbio.2020.01.011

Nobis M, Herrmann D, Warren SC, Kadir S, Leung W, Killen M, Magenau A, Stevenson D, Lucas MC, Reischmann N, et al. 2017. A RhoA-FRET biosensor mouse for intravital imaging in normal tissue homeostasis and disease contexts. *Cell Rep* **21**: 274–288. doi:10.1016/j.celrep.2017.09.022

Nobis M, Warren SC, Lucas MC, Murphy KJ, Herrmann D, Timpson P. 2018. Molecular mobility and activity in an intravital imaging setting—implications for cancer progression and targeting. *J Cell Sci* **131**: jcs206995. doi:10.1242/jcs.206995

Nojima S, Susaki EA, Yoshida K, Takemoto H, Tsujimura N, Iijima S, Takachi K, Nakahara Y, Tahara S, Ohshima K, et al. 2017. CUBIC pathology: three-dimensional imaging for pathological diagnosis. *Sci Rep* **7**: 9269. doi:10.1038/s41598-017-09117-0

Nolan E, Savas P, Policheni AN, Darcy PK, Vaillant F, Mintoff CP, Dushyanthen S, Mansour M, Pang JB, Fox SB, et al. 2017. Combined immune checkpoint blockade as a therapeutic strategy for *BRCA1*-mutated breast cancer. *Sci Transl Med* **9**: eaal492. doi:10.1126/scitranslmed.aal4922

Ozturk MS, Montero MG, Wang L, Chaible LM, Jechlinger M, Prevedel R. 2021. Intravital mesoscopic fluorescence molecular tomography allows non-invasive in vivo monitoring and quantification of breast cancer growth dynamics. *Commun Biol* **4**: 556. doi:10.1038/s42003-021-02063-8

Pan C, Cai R, Quacquarelli FP, Ghasemigharagoz A, Lourbopoulos A, Matryba P, Plesnila N, Dichgans M, Hellal F, Ertürk A, et al. 2016. Shrinkage-mediated imaging of entire organs and organisms using uDISCO. *Nat Methods* **13**: 859–867. doi:10.1038/nmeth.3964

Pan C, Schoppe O, Parra-Damas A, Cai R, Todorov MI, Gondi G, von Neubeck B, Böğürcü-Seidel N, Seidel S, Sleiman K, et al. 2019. Deep learning reveals cancer metastasis and therapeutic antibody targeting in the entire body. *Cell* **179**: 1661–1676.e19. doi:10.1016/j.cell.2019.11.013

Park J, Wysocki RW, Amoozgar Z, Maiorino L, Fein MR, Jorns J, Schott AF, Kinugasa-Katayama Y, Lee Y, Won NH, et al. 2016. Cancer cells induce metastasis-supporting neutrophil extracellular DNA traps. *Sci Transl Med* **8**: 361ra138.

Patel KB, Liang W, Casper MJ, Voleti V, Li W, Yagielski AJ, Zhao HT, Perez Campos C, Lee GS, Liu JM, et al. 2022. High-speed light-sheet microscopy for the in-situ acquisition of volumetric histological images of living tissue. *Nat Biomed Eng* **6**: 569–583. doi:10.1038/s41551-022-00849-7

Patsialou A, Bravo-Cordero JJ, Wang Y, Entenberg D, Liu H, Clarke M, Condeelis JS. 2013. Intravital multiphoton imaging reveals multicellular streaming as a crucial component of in vivo cell migration in human breast tumors. *Intravital* **2**: e25294. doi:10.4161/intv.25294

Payne AC, Chiang ZD, Reginato PL, Mangiameli SM, Murray EM, Yao CC, Markoulaki S, Earl AS, Labade AS, Jaenisch R, et al. 2021. In situ genome sequencing resolves DNA sequence and structure in intact biological samples. *Science* **371**: eaay3446. doi:10.1126/science.aay3446

Paz H, Pathak N, Yang J. 2014. Invading one step at a time: the role of invadopodia in tumor metastasis. *Oncogene* **33**: 4193–4202. doi:10.1038/onc.2013.393

Pereira ER, Kedrin D, Seano G, Gautier O, Meijer EFJ, Jones D, Chin SM, Kitahara S, Bouta EM, Chang J, et al. 2018. Lymph node metastases can invade local blood vessels, exit the node, and colonize distant organs in mice. *Science* **359**: 1403–1407. doi:10.1126/science.aal3622

Perin P, Voigt FF, Bethge P, Helmchen F, Pizzala R. 2019. iDISCO⁺ for the study of neuroimmune architecture of the rat auditory brainstem. *Front Neuroanat* **13**: 15. doi:10.3389/fnana.2019.00015

Qi Y, Yu T, Xu J, Wan P, Ma Y, Zhu J, Li Y, Gong H, Luo Q, Zhu D. 2019. FDISCO: advanced solvent-based clearing method for imaging whole organs. *Archivio di Studi Urbani e Regionali* **48**: 1–14. doi:10.1126/sciadv.aau8355

Ren J, Choi H, Chung K, Bouma BE. 2017. Label-free volumetric optical imaging of intact murine brains. *Sci Rep* **7**: 46306. doi:10.1038/srep46306

Renier N, Wu Z, Simon DJ, Yang J, Ariel P, Tessier-Lavigne M. 2014. iDISCO: a simple, rapid method to immunolabel large tissue samples for volume imaging. *Cell* **159**: 896–910. doi:10.1016/j.cell.2014.10.010

Reticker-Flynn NE, Zhang W, Belk JA, Basto PA, Escalante NK, Pilarowski GOW, Bejnood A, Martins MM, Kenkel JA, Linde IL, et al. 2022. Lymph node colonization induces tumor-immune tolerance to promote distant metastasis. *Cell* **185**: 1924–1942.e23. doi:10.1016/j.cell.2022.04.019

Reynaud EG, Peychl J, Huisken J, Tomancak P. 2015. Guide to light-sheet microscopy for adventurous biologists. *Nat Methods* **12**: 30–34. doi:10.1038/nmeth.3222

Richardson DS, Lichtman JW. 2015. Clarifying tissue clearing. *Cell* **162**: 246–257. doi:10.1016/j.cell.2015.06.067

Rios AC, Clevers H. 2018. Imaging organoids: a bright future ahead. *Nat Methods* **15**: 24–26. doi:10.1038/nmeth.4537

Rios AC, Fu NY, Lindeman GJ, Visvader JE. 2014. In situ identification of bipotent stem cells in the mammary gland. *Nature* **506**: 322–327. doi:10.1038/nature12948

Rios AC, Fu NY, Jamieson PR, Pal B, Whitehead L, Nicholas KR, Lindeman GJ, Visvader JE. 2016. Essential role for a novel population of binucleated mammary epithelial cells in lactation. *Nat Commun* **7**: 11400. doi:10.1038/ncomms11400

Rios AC, Capaldo BD, Vaillant F, Pal B, van Ineveld R, Dawson CA, Chen Y, Nolan E, Fu NY, Group D, et al. 2019. Intraclonal plasticity in mammary tumors revealed through large-scale single-cell resolution 3D imaging. *Cancer Cell* **35**: 618–632.e6. doi:10.1016/j.ccell.2019.02.010

Ritsma L, Steller EJA, Ellenbroek SIJ, Kranenburg O, Borel Rinkes IHM, van Rheenen J. 2013. Surgical implantation of an abdominal imaging window for intravital microscopy. *Nat Protoc* **8**: 583–594. doi:10.1038/nprot.2013.026

Rivenson Y, Wang H, Wei Z, de Haan K, Zhang Y, Wu Y, Günaydin H, Zuckerman JE, Chong T, Sisk AE, et al. 2019. Virtual histological staining of unlabelled tissue-autofluorescence images via deep learning. *Nat Biomed Eng* **3**: 466–477. doi:10.1038/s41551-019-0362-y

Rodell CB, Arlauckas SP, Cuccarese MF, Garris CS, Li R, Ahmed MS, Kohler RH, Pittet MJ, Weissleder R. 2018. TLR7/8-agonist-loaded nanoparticles promote the polarization of tumour-associated macrophages to enhance cancer immunotherapy. *Nat Biomed Eng* **2**: 578–588. doi:10.1038/s41551-018-0236-8

Rodriques SG, Stickels RR, Goeva A, Martin CA, Murray E, Vanderburg CR, Welch J, Chen LM, Chen F, Macosko EZ. 2019. Slide-seq: a scalable technology for measuring genome-wide expression at high spatial resolution. *Science* **363**: 1463–1467. doi:10.1126/science.aaw1219

Rosenbluth JM, Schackmann RCJ, Gray GK, Selfors LM, Li CM-C, Boedicker M, Kuiken HJ, Richardson A, Brock J, Garber J, et al. 2020. Organoid cultures from normal and cancer prone human breast tissues preserve complex epithelial lineages. *Nat Commun* **11**: 1711. doi:10.1038/s41467-020-15548-7

Rust MJ, Bates M, Zhuang X. 2006. Sub-diffraction-limit imaging by stochastic optical reconstruction microscopy (STORM). *Nat Methods* **3**: 793–795. doi:10.1038/nmeth929

Sachs N, de Ligt J, Kopper O, Gogola E, Bounova G, Weeber F, Balgobind AV, Wind K, Gracanin A, Begthel H, et al. 2018. A living biobank of breast cancer organoids captures disease heterogeneity. *Cell* **172**: 373–386.e10. doi:10.1016/j.cell.2017.11.010

Scheele CLGJ, Hannezo E, Muraro MJ, Zomer A, Langedijk NSM, van Oudenaarden A, Simons BD, van Rheenen J. 2017. Identity and dynamics of mammary stem cells during branching morphogenesis. *Nature* **542**: 313–317. doi:10.1038/nature21046

Seo J, Sim Y, Kim J, Kim H, Cho I, Nam H, Yoon YG, Chang JB. 2022. PICASSO allows ultra-multiplexed fluorescence imaging of spatially overlapping proteins without reference spectra measurements. *Nat Commun* **13**: 2475. doi:10.1038/s41467-022-30168-z

Sharma VP, Tang B, Wang Y, Duran CL, Karagiannis GS, Xue EA, Entenberg D, Borriello L, Coste A, Eddy RJ, et al. 2021. Live tumor imaging shows macrophage induction and TMEM-mediated enrichment of cancer stem cells during metastatic dissemination. *Nat Commun* **12**: 7300. doi:10.1038/s41467-021-27308-2

Si Y, Merz SF, Jansen P, Wang B, Bruderek K, Altenhoff P, Mattheis S, Lang S, Gunzer M, Klode J, et al. 2019. Multidimensional imaging provides evidence for down-regulation of T cell effector function by MDSC in human cancer tissue. *Sci Immunol* **4**: eaaw9159.

Smirnova T, Bonapace L, MacDonald G, Kondo S, Wyckoff J, Ebersbach H, Fayard B, Doelemeyer A, Coissieux M-M, Heideman MR, et al. 2016. Serpin E2 promotes breast cancer metastasis by remodeling the tumor matrix and polarizing tumor associated macrophages. *Oncotarget* **7**: 82289–82304. doi:10.18632/oncotarget.12927

Snippert HJ, Schepers AG, Es JH, Simons BD, Clevers H. 2014. Biased competition between Lgr5 intestinal stem cells driven by oncogenic mutation induces clonal expansion. *EMBO Rep* **15**: 62–69. doi:10.1002/embr.201337799

Stoltzfus CR, Filipek J, Gern BH, Olin BE, Leal JM, Wu Y, Lyons-Cohen MR, Huang JY, Paz-Stoltzfus CL, Plumlee CR, et al. 2020. CytoMAP: a spatial analysis toolbox reveals features of myeloid cell organization in lymphoid tissues. *Cell Rep* **31**: 107523. doi:10.1016/j.celrep.2020.107523

Susaki EA, Tainaka K, Perrin D, Kishino F, Tawara T, Watanabe TM, Yokoyama C, Onoe H, Eguchi M, Yamaguchi S, et al. 2014. Whole-brain imaging with single-cell resolution using chemical cocktails and computational analysis. *Cell* **157**: 726–739. doi:10.1016/j.cell.2014.03.042

Tanaka N, Kanatani S, Tomer R, Sahlgren C, Kronqvist P, Kaczynska D, Louhivuori L, Kis L, Lindh C, Mitura P, et al. 2017. Whole-tissue biopsy phenotyping of three-dimensional tumours reveals patterns of cancer heterogeneity. *Nat Biomed Eng* **1:** 796–806. doi:10.1038/s41551-017-0139-0

Tanaka N, Kaczynska D, Kanatani S, Sahlgren C, Mitura P, Stepulak A, Miyakawa A, Wiklund P, Uhlén P. 2018. Mapping of the three-dimensional lymphatic microvasculature in bladder tumours using light-sheet microscopy. *Br J Cancer* **118:** 995–999. doi:10.1038/s41416-018-0016-y

Tanaka N, Kanatani S, Kaczynska D, Fukumoto K, Louhivuori L, Mizutani T, Kopper O, Kronqvist P, Robertson S, Lindh C, et al. 2020. Three-dimensional single-cell imaging for the analysis of RNA and protein expression in intact tumour biopsies. *Nat Biomed Eng* **4:** 875–888. doi:10.1038/s41551-020-0576-z

Tang YJ, Huang J, Tsushima H, Ban GI, Zhang H, Oristian KM, Puviindran V, Williams N, Ding X, Ou J, et al. 2019. Tracing tumor evolution in sarcoma reveals clonal origin of advanced metastasis. *Cell Rep* **28:** 2837–2850.e5. doi:10.1016/j.celrep.2019.08.029

Thurber GM, Yang KS, Reiner T, Kohler RH, Sorger P, Mitchison T, Weissleder R. 2013. Single-cell and subcellular pharmacokinetic imaging allows insight into drug action in vivo. *Nat Commun* **4:** 1–10. doi:10.1038/ncomms2506

Tiede S, Kalathur RKR, Lüönd F, von Allmen L, Szczerba BM, Hess M, Vlajnic T, Müller B, Canales Murillo J, Aceto N, et al. 2021. Multi-color clonal tracking reveals intrastage proliferative heterogeneity during mammary tumor progression. *Oncogene* **40:** 12–27. doi:10.1038/s41388-020-01508-4

Tjan-Heijnen V, Viale G. 2018. The lymph node and the metastasis. *N Engl J Med* **378:** 2045–2046. doi:10.1056/NEJMcibr1803854

Valm AM, Cohen S, Legant WR, Melunis J, Hershberg U, Wait E, Cohen AR, Davidson MW, Betzig E, Lippincott-Schwartz J. 2017. Applying systems-level spectral imaging and analysis to reveal the organelle interactome. *Nature* **546:** 162–167. doi:10.1038/nature22369

van Ineveld RL, Kleinnijenhuis M, Alieva M, de Blank S, Barrera Roman M, van Vliet EJ, Martínez Mir C, Johnson HR, Bos FL, Heukers R, et al. 2021. Revealing the spatiophenotypic patterning of cells in healthy and tumor tissues with mLSR-3D and STAPL-3D. *Nat Biotechnol* **39:** 1239–1245. doi:10.1038/s41587-021-00926-3

van Ineveld RL, Ee E, van Vliet J, Wehrens EJ, Alieva M, Rios AC. 2022. 3D imaging for driving cancer discovery. *EMBO J* **41:** e109675.

van Ineveld RL, Collot R, Barrera Roman M, et al. (in press).

van Royen ME, Verhoef EI, Kweldam CF, van Cappellen WA, Kremers GJ, Houtsmuller AB, van Leenders GJ. 2016. Three-dimensional microscopic analysis of clinical prostate specimens. *Histopathology* **69:** 985–992. doi:10.1111/his.13022

Vermeulen L, Morrissey E, van der Heijden M, Nicholson AM, Sottoriva A, Buczacki S, Kemp R, Tavaré S, Winton DJ. 2013. Defining stem cell dynamics in models of intestinal tumor initiation. *Science* **342:** 995–998. doi:10.1126/science.1243148

Wang D, Liu G, Wu B, Chen L, Zeng L, Pan Y. 2018a. Clinical significance of elevated S100A8 expression in breast can-

cer patients. *Front Oncol* **8:** 496. doi:10.3389/fonc.2018.00496

Wang X, Allen WE, Wright MA, Sylwestrak EL, Samusik N, Vesuna S, Evans K, Liu C, Ramakrishnan C, Liu J, et al. 2018b. Three-dimensional intact-tissue sequencing of single-cell transcriptional states. *Science* **361:** eaat5691. doi:10.1126/science.aat5691

Weigelin B, Bakker GJ, Friedl P. 2016. Third harmonic generation microscopy of cells and tissue organization. *J Cell Sci* **129:** 245–255.

Weiss S. 2000. Shattering the diffraction limit of light: a revolution in fluorescence microscopy? *Proc Natl Acad Sci* **97:** 8747–8749. doi:10.1073/pnas.97.16.8747

Williams JK, Entenberg D, Wang Y, Avivar-Valderas A, Padgen M, Clark A, Aguirre-Ghiso JA, Castracane J, Condeelis JS. 2016. Validation of a device for the active manipulation of the tumor microenvironment during intravital imaging. *Intravital* **5:** e1182271. doi:10.1080/21659087.2016.1182271

Williams KC, Cepeda MA, Javed S, Searle K, Parkins KM, Makela A, Hamilton AM, Soukhtehzari S, Kim Y, Tuck AB, et al. 2019. Invadopodia are chemosensing protrusions that guide cancer cell extravasation to promote brain tropism in metastasis. *Oncogene* **38:** 3598–3615. doi:10.1038/s41388-018-0667-4

Yanai H, Atsumi N, Tanaka T, Nakamura N, Komai Y, Omachi T, Tanaka K, Ishigaki K, Saiga K, Ohsugi H, et al. 2017. Intestinal cancer stem cells marked by Bmi1 or Lgr5 expression contribute to tumor propagation via clonal expansion. *Sci Rep* **7:** 41838. doi:10.1038/srep41838

Ying Z, Sandoval M, Beronja S. 2018. Oncogenic activation of PI3K induces progenitor cell differentiation to suppress epidermal growth. *Nat Cell Biol* **20:** 1256–1266. doi:10.1038/s41556-018-0218-9

Yoshizawa T, Hong SM, Jung D, Noë M, Kiemen A, Wu PH, Wirtz D, Hruban RH, Wood LD, Oshima K. 2020. Three-dimensional analysis of extrahepatic cholangiocarcinoma and tumor budding. *J Pathol* **251:** 400–410. doi:10.1002/path.5474

You S, Tu H, Chaney EJ, Sun Y, Zhao Y, Bower AJ, Liu YZ, Marjanovic M, Sinha S, Pu Y, et al. 2018. Intravital imaging by simultaneous label-free autofluorescence-multiharmonic microscopy. *Nat Commun* **9:** 2125. doi:10.1038/s41467-018-04470-8

Yum MK, Han S, Fink J, Wu SHS, Dabrowska C, Trendafilova T, Mustata R, Chatzeli L, Azzarelli R, Pshenichnaya I, et al. 2021. Tracing oncogene-driven remodeling of the intestinal stem cell niche. *Nature* **594:** 442–447. doi:10.1038/s41586-021-03605-0

Zanfardino M, Castaldo R, Pane K, Affinito O, Aiello M, Salvatore M, Franzese M. 2021. MuSA: a graphical user interface for multi-OMICs data integration in radiogenomic studies. *Sci Rep* **11:** 1550. doi:10.1038/s41598-021-81200-z

Zhao Z, Zhu X, Cui K, Mancuso J, Federley R, Fischer K, Teng G, Mittal V, Gao D, Zhao H, et al. 2016. In vivo visualization and characterization of epithelial–mesenchymal transition in breast tumors. *Cancer Res* **76:** 2094–2104. doi:10.1158/0008-5472.CAN-15-2662

Zhao S, Todorov MI, Cai R, Maskari RA, Steinke H, Kemter E, Mai H, Rong Z, Warmer M, Stanic K, et al. 2020.

Cellular and molecular probing of intact human organs. *Cell* **180:** 796–812.e19. doi:10.1016/j.cell.2020.01 .030

Zhong JM, Li J, Kang AD, Huang SQ, Liu WB, Zhang Y, Liu ZH, Zeng L. 2018. Protein S100-A8: a potential metastasis-associated protein for breast cancer determined via iTRAQ quantitative proteomic and clinicopathological analysis. *Oncol Lett* **15:** 5285–5293.

Zhu C, Martinez AF, Martin HL, Li M, Crouch BT, Carlson DA, Haystead TAJ, Ramanujam N. 2017. Near-simultaneous intravital microscopy of glucose uptake and mito-chondrial membrane potential, key endpoints that reflect major metabolic axes in cancer. *Sci Rep* **7:** 13772.

Zomer A, Ellenbroek SIJ, Ritsma L, Beerling E, Vrisekoop N, van Rheenen J. 2013. Brief report: intravital imaging of cancer stem cell plasticity in mammary tumors. *Stem Cells* **31:** 602–606. doi:10.1002/stem.1296

Zomer A, Maynard C, Verweij FJ, Kamermans A, Schäfer R, Beerling E, Schiffelers RM, de Wit E, Berenguer J, Ellenbroek SIJ, et al. 2015. In vivo imaging reveals extracellular vesicle-mediated phenocopying of metastatic behavior. *Cell* **161:** 1046–1057. doi:10.1016/j.cell.2015.04.042

Cite this article as *Cold Spring Harb Perspect Med* doi: 10.1101/cshperspect.a041330

Spatial Biology of Breast Cancer

H. Raza Ali[1] and Robert B. West[2]

[1]Cancer Research UK Cambridge Institute, University of Cambridge, Li Ka Shing Centre, Cambridge CB2 0RE, United Kingdom

[2]Department of Pathology, Stanford University Medical Center, Stanford, California 94305, USA

Correspondence: raza.ali@cruk.cam.ac.uk; rbwest@stanford.edu

Spatial findings have shaped on our understanding of breast cancer. In this review, we discuss how spatial methods, including spatial transcriptomics and proteomics and the resultant understanding of spatial relationships, have contributed to concepts regarding cancer progression and treatment. In addition to discussing traditional approaches, we examine how emerging multiplex imaging technologies have contributed to the field and how they might influence future research.

Spatial concepts have helped define our understanding of breast cancer. Carcinoma in situ, invasion, pagetoid spread, and metastasis, for example, describe critical stages in breast disease that are defined by spatial context. These spatial concepts are born from the methods we have developed to study breast cancer. As researchers, science and medicine have developed more sophisticated methods to observe disease, our insight into the processes have increased coordinately. We are currently in a golden age of spatial methods. These emerging approaches give both new resolution and added dimension to our understanding of cancer.

Historically, breast cancer research has focused on the neoplastic cells themselves. This aligns with the concept that cancer is a genetic disease (Nowell 1976). However, it may be more accurate to say that neoplasia is a genetic disease, whereas cancer is a combination of genetic and spatial events, which are influenced by a host of cells and factors in the tumor microenvironment (TME). It is understood that the microenvironment contains stromal cells, immune cells, and others (Azizi et al. 2018; Wagner et al. 2019), but it is unclear what biology the spatial relationships confer and how they influence disease. Moreover, in the breast (and in other organ systems) there are nonneoplastic epithelial cells that intimately interact with the neoplastic process (e.g., myoepithelial cells and neuroendocrine cells) that likely play a critical role in neoplasia progression but are understudied. New technologies to understand spatial relationships are at our disposal and a raft of clinically significant spatial questions await to be answered.

Here, we discuss how spatial relationships underpin breast cancer progression, treatment, and relapse. We will describe how histopathology has defined our understanding of discrete stages in disease progression based largely on spatial landmarks, explain how recent work shows that cancer genomes are sculpted by spatial context (Casasent et al. 2018; Danenberg et al. 2022), and discuss how both genetic and nongenetic spatial heterogeneity in established tumors may underpin treatment resistance and eventual relapse (Hwang et al. 2022). We describe how emerging technologies for highly multiplexed tissue imaging are shaping novel concepts such as multicellular functional units in situ, and the potential for these approaches to achieve personalized medicine for novel treatments such as immunotherapy (Lewis et al. 2021). Throughout, we will anchor our discussion in established breast pathology and conclude by speculating on how novel, complex technologies could be reconciled with traditional approaches.

SPATIAL IMPLICATIONS FOR CLINICAL DISEASE PROGRESSION

Spatial findings have a large influence on our understanding of both clinical and biological significance in breast cancer. Light microscopy with a hematoxylin and eosin (H&E) stain has been the main driver of spatial observations for several decades. Fundamental questions of cancer spread and growth were unclear before microscopy (Grange et al. 2002). Spatial concepts such as carcinoma in situ and Paget's disease were characterized microscopically (Broders 1932) and helped generate an understanding of cancer development. One of the most important clinical observations is the progression to invasion, as defined by the presence of neoplastic cells outside of the epithelial compartment. It is a critical clinical step that correlates with a significantly higher risk of morbidity and mortality (Pinder and Ellis 2003). The 5-year survival of women with invasive breast carcinoma (IBC) versus ductal carcinoma in situ (DCIS) is significantly different, as is the likelihood of metastatic disease (Narod et al. 2015). Treatment options for IBC often include systemic therapy that carries its own morbidity. In contrast, DCIS treatment is largely local and the risk for death by disease is exceptionally low without progression to IBC. Determining whether neoplastic cells are invasive or not by H&E light microscopy is one of the most important and effective clinical tests in the treatment of breast cancer. Remarkably, molecular attempts to identify genomics features that distinguish DCIS and IBC have not been found. Basic proteomic testing, with bright field immunohistochemistry (IHC), does contribute through the identification of mature myoepithelial cells in preinvasive processes (Pinder and Ellis 2003). Nevertheless, these clinical differences exist despite there being very little difference in the genomic and proteomic composition of DCIS and IBC cells (Newburger et al. 2013; Kim et al. 2015; Krøigård et al. 2015). The two entities principally differ in their spatial context.

SPATIAL HETEROGENEITY OF CLINICAL MOLECULAR BIOMARKERS

Clinical molecular biomarkers are best evaluated in spatial context. Estrogen receptor (ER) expression, for example, was shown in large meta-analyses to predict the efficacy of tamoxifen. But at that time, its expression was measured using lysed tissue by a radiolabeled ligand-binding assay (LBA) (Early Breast Cancer Trialists' Collaborative Group 1998). Although this approach was shown to be predictive of response to treatment, it was far from ideal because of the limitations of an assay for which tissue must be homogenized: it did not reveal which cells expressed ER and had limited sensitivity that depended on overall tissue composition. Normal breast epithelial cells, those in benign and preinvasive lesions as well as some stromal cells, can all express ER at high levels for instance and they, not invasive tumor cells, could be the source of ER in some cases deemed positive by LBA. Conversely, a tumor characterized by sparse ER-positive tumor cells located in a cellular stroma with extensive lymphocytic infiltration may not, in aggregate, contain enough ER to be identified as positive by an assay based on bulk tissue measurements such as LBA. The coming of IHC heralded a revolution in spatial biomarker analysis, and quickly replaced LBA as the assay of choice for ER status (Harvey et al. 1999). By integrating a molecular readout with spatial location, pathologists could determine whether a tumor expressed ER with unprecedented (single-cell) sensitivity and specificity. When therapy targeting human epidermal growth factor receptor 2 (HER2) was developed, IHC was a natural choice for identifying which patients were likely to respond (Slamon et al. 1987, 1989). Presently, routine analysis of HER2 protein expression and *ERBB2* copy number both rely on approaches that account for spatial context (IHC and fluorescence in situ hybridization [FISH], respectively) (Wolff et al. 2013). Another advantage of spatial analysis is that heterogeneous expression patterns can be readily identified. Among HER2-positive tumors, for example, it is known that a rare subset shows subclonal HER2 overexpression that may have implications for treatment response and risk of relapse; these cases are easily flagged by HER2 IHC (Ng et al. 2015). Cell type and abundance are therefore both important for accurate patient stratification by molecular biomarkers, and pathologists can reliably ascertain these features using high-resolution spatial assays that preserve tissue morphology.

The characteristic heterogeneity of breast cancer remains a challenge in both the laboratory and the clinic. Most past efforts have focused on resolving intertumor heterogeneity but, by overlooking intratumor heterogeneity, these efforts forgo this important source of phenotypic variation. Past approaches aimed at tackling intertumor heterogeneity relied on bulk omics methodology (Kristensen et al. 2014). The intrinsic subtypes of breast cancer, for example, were discovered using bulk tumor transcriptomes (Perou et al. 2000). Similarly, integrating bulk genomic and transcriptomic profiles yielded the 10 integrative clusters of breast cancer based on patterns of driver copy-number aberrations (Curtis et al. 2012). Clinical subtypes of breast cancer, particularly triple-negative tumors, have been further subdivided based on transcriptomics (Lehmann et al. 2011). Among the motivations for discovering tumor subtypes is the idea that new entities will prove more biologically homogeneous and will therefore be characterized by distinct clinical outcomes. A complementary approach based on similar technology has been to identify prognostic gene signatures composed of a collection of genes that, together, predict outcome among a subgroup of breast cancer patients (Paik et al. 2004). Several signatures have been described

that address the common clinical problem of determining the probable benefit a given patient will derive from systemic therapy. Some, such as OncotypeDX (Paik et al. 2004), are widely recommended in treatment guidelines to aid such decisions, and large randomized trials have been conducted to test their clinical utility. Both tumor subtyping efforts and prognostic gene signatures have shed light on breast cancer heterogeneity and aid clinical decision making. Being based on bulk tissue assays, however, they are subject to limitations because spatial context is not considered. Tissue composition could, for example, be a source of false positives or negatives. In some tumors, it may be that extensive in situ disease overwhelms signal from the minority invasive component, or that an indolent low-grade tumor associated with rare foci of high grade in situ disease is assigned a higher risk score than would otherwise be the case. The source of RNA for some genes can also vary wildly and can also lead to misclassification. For example, *BCL2* expression by tumor cells known to be highly prognostic in breast cancer (Dawson et al. 2010; Ali et al. 2012) is among the genes that comprise the OncotypeDX signature (Paik et al. 2004), but is also highly expressed by B cells. Tumors with extensive lymphocytic infiltration may express *BCL2* at comparable or higher levels as those that contain *BCL2*-expressing cancer cells. But in the case of OncotypeDX, it is not clear which case is dominant or which qualified *BCL2* for inclusion in the signature. These examples show bulk omics approaches neglect spatial context, but that it is critical for accurate biomarker analyses.

SPATIAL HETEROGENEITY OF GENOMIC FEATURES

Spatial approaches also offer the possibility of better understanding tumor genomic evolution. Deep sequencing has revealed that breast tumors comprise clonal populations with variable phylogenetic structure (Shah et al. 2012; Yates et al. 2015). Multiregion sequencing, by sampling tissue regions separated in space, has better characterized the clonal structure of renal and lung cancers (Gerlinger et al. 2012, 2014; Jamal-Hanjani et al. 2017) and breast cancer (Yates et al. 2015).

In breast cancer, spatial analyses to characterize tumor evolution have focused on two scenarios: transition from in situ to invasive disease, and the selective effect of neoadjuvant treatment. FISH, in conjunction with immunofluorescence (IF), has been used to profile paired in situ and invasive carcinoma (Park et al. 2010). Diversity statistics based on FISH showed cancer clonal populations differed between tumor subtypes, and, in some cases, between in situ and adjacent invasive carcinoma (Park et al. 2010). Coupling laser capture microdissection with a single-nucleus genome sequencing protocol as topographic single-cell sequencing (TSCS), again in paired in situ and invasive carcinoma, has shed light on whether the transition from an intraductal environment into breast stroma is an evolutionary bottleneck (Casasent et al. 2018). Using TSCS to precisely characterize copy-number aberrations genome-wide in single cells from synchronous in situ and invasive carcinomas showed that the phylogenetic structure of preinvasive cells and invasive cancer cells was best explained by a multiclonal model of invasion, arguing against the idea of a genetic evolutionary bottleneck (Casasent et al. 2018). It is, however, known that in situ carcinoma cells can exhibit distinct phenotypic profiles in comparison to their invasive counterparts, such as more frequent HER2 overexpression, suggesting that adaptation to the breast stroma may be nongenetic (Bergholtz et al. 2020; Strand et al. 2022). Another paired analysis of in situ and invasive carcinoma, spanning spatial and bulk genomic data, showed that immune composition significantly differed, however (Gil Del Alcazar et al. 2017). Invasive carcinoma, for instance, contained fewer activated cytotoxic T cells and lower T-cell receptor clonotype diversity (Gil Del Alcazar et al. 2017). These findings raise the possibility that immune escape is required for breast cancer progression but that the dominant mechanisms may not be mediated by copy-number aberrations.

Recent work coupling high-depth genome sequencing and in situ sequencing has resolved tumor subclones in situ at high resolution (Lomakin et al. 2022). By first building a clonal phylogeny using bulk sequencing data, it was possible to select probes for in situ RNA profiling that

defined branch points in the phylogenetic tree and were therefore highly discriminatory of tumor subclones. Applying this approach to the question of DCIS to invasive carcinoma progression revealed the unexpected finding that clonal separation in space is often constrained by microanatomy with, for example, contiguous clonal expansion within ducts (Lomakin et al. 2022).

Although the drivers remain obscure, these studies show that breast tumor evolution is sculpted by spatial context.

SPATIAL IMPLICATIONS FOR TREATMENT RESPONSE

Treatment is a major selection pressure on cancer cells (Nowell 1976; Aparicio and Caldas 2013). Those that endure treatment, and ultimately seed sites of relapse, employ both genetic and nongenetic mechanisms. It is becoming clear that, in addition to cell-intrinsic resistance mechanisms, cancer cells can resist treatment by extrinsic mechanisms contingent on their spatial context. Diverse TMEs can alter proximity to vasculature, relationships to the extracellular matrix (ECM), and immune predation, inevitably changing the landscape of sensitivity to therapy.

Breast cancer is an opportune setting in which to study the basis of treatment resistance because neoadjuvant therapy is common and provides snapshots of a tumor before and after treatment. To investigate the influence of cellular spatial context on cancer cell persistence following neoadjuvant treatment, the Polyak group have used a combination of IF and in situ hybridization (for key copy-number alterations and one hotspot mutation) in paired pre- and posttreatment samples (Janiszewska et al. 2015). Coupling these approaches to quantitative image analysis allowed for generation of "topology maps" to investigate the relationships between cellular phenotype, genotype, and spatial context (Janiszewska et al. 2015). They found that tumor cells tend to cluster according to their phenotype rather than their genotype. By comparing the spatial dispersion of *PIK3CA*-mutant cells before and after treatment, the group also speculate that *PIK3CA* confers greater migratory capacity (Janiszewska et al. 2015). These studies show that chemotherapy impacts both the overall cancer cell population but also their distribution in space uncovering spatial context as a determinant of response. More recent work based on multiplexed IF corroborates the idea that cellular spatial context can impact the ability of cancer cells to withstand treatment. In the setting of triple-negative breast cancer (TNBC), using both animal models and primary human tissues, clusters of quiescent cancer cells are found to resist the effect of immunotherapy (Baldominos et al. 2022). These cell clusters are characterized by hypoxia, glycolysis, and a distinctive TME comprising exhausted T cells and immunosuppressive fibroblasts. This work shows that intrinsic and extrinsic mechanisms of treatment resistance can be correlated and may by inextricably linked (Baldominos et al. 2022). Multicellular spatial context is increasingly recognized as a determinant of treatment response and resistance in breast cancer. Additional work using higher resolution analyses in well-designed clinical studies is likely to reveal the extent of its influence and suggest means of disrupting TMEs that shield cells from the toxic effects of treatment.

MULTIPLEXED IMAGING APPROACHES

Spatial technologies profile either proteins or transcripts in situ and are both complementary and competing (Lewis et al. 2021). Underpinning these technologies is a rapidly developing field of image analysis. The methodologic details have been thoroughly described elsewhere (Lewis et al. 2021). In their current state, key differences in the methods are cellular resolution, multiplexing capacity, sensitivity, throughput, and ease of use. Proteomic approaches can achieve subcellular resolution and enable the classification of all cells and quantitative measurements of physical relationships between cells (Danenberg et al. 2022; Risom et al. 2022). The study by Risom et al. (2022) was able to define subductal compartment and distinguish between tightly associated cell types, such as myoepithelial cells and luminal cells. In contrast, spatial transcriptomics generates data on regions of interest, which may include several cells (Wu et al. 2021). Cell-type composition must be estimated, or deconvo-

luted, as the sequencing-based spatial transcriptomics does not have single-cell resolution. This necessarily reduces the types of observations that can be made regarding cell-to-cell interactions.

Transcriptomic methods can measure between 100 and 10,000+ unique transcripts dependent on the method (RNA extraction and sequencing vs. in situ hybridization vs. in situ sequencing). The sensitivity of these approaches can be quite good but diminishes with the quality of the tissue. Unlike RNA-based methods, proteomic assays rely on antibodies, which are finite and complex in their biology. Antibodies may have high binding affinity that may be altered by the necessary direct conjugation step. As such, antibody selection and planning of the panel is a unique and critical step in this analysis. At present, proteomic methods measure $\sim$10–100 features, in part dependent on the antibody panel selected (Hickey et al. 2022). New multiplexed antibody-based methods can be organized based on the signaling chemistry labeling the antibody: using traditional variations in fluorescence, oligonucleotide-barcoding, or mass spectrometry. The latter employs metal tags, liberated by either ion beam or laser (Angelo et al. 2014; Giesen et al. 2014). The diversity mainly relates to the process of labeling the primary antibody and this results in choices in sensitivity, throughput, and ease of use. Throughput and general ease of use (complexity of process and cost) varies widely and can greatly influence the type of experiments that can be performed.

STUDY DESIGN AND ANALYSIS CHALLENGES

Study design and analysis can be partitioned into three phases: preanalytic, analytic, and postanalytic concepts.

A common approach to studying human samples is to correlate results with longitudinal observations. Typically, this requires the study of tissues that have been archived for an extended period. A technical issue associated with this approach is the inherent tissue quality and the preservation of macromolecules. Factors that influence this include the initial tissue fixation process, the storage conditions, and the length of time between tissue acquisition and data generation (DiVito et al. 2004; Camp et al. 2008). For example, cut sections often have significant degradation of RNA and protein over several weeks or months, whereas tissue in blocks may have reasonably well-preserved protein and RNA over years. RNA quality is typically more sensitive to these factors than protein quality (DiVito et al. 2004). However, generally small fragments of RNA, ~100–200 nucleotides in length, can be measured when longer fragments have been degraded (Beck et al. 2010; Sweeney et al. 2013; Foley et al. 2019).

Perhaps the least discussed aspect of spatial technologies is tissue annotation. Selecting a region of interest is a challenging problem in many situations. Oftentimes the experimental study requires that studies be done on tissue microarrays (Kononen et al. 1998). The limited extent of a sample represented by a core on a tissue microarray can make ascertainment of the correct region of interest difficult.

In contrast to RNA studies in which the probe or sequencing approach is very reproducible, epitope-based approaches are challenged with antibodies in which sensitivity and specificity can be poorly defined or may change from batch to batch (Brennan et al. 2010; Bodenmiller 2016). There has been extensive work to adequately control antibody binding, but for antibodies that are not commonly used this is not always the case (Bordeaux et al. 2010). A second consideration in multiplex staining is the performance of a variety of antibodies using a single set of conditions. This includes antibody retrieval, incubation buffer, and blocking reagents (Bordeaux et al. 2010). All these features may change the performance of the antibody rendering prior workup of antibody sensitivity and specificity obsolete.

There are guidelines for best practices for data analysis of highly multiplexed tissue images (Schapiro et al. 2022). Processing of highly multiplexed images is a critical postanalytic step that can profoundly influence data interpretation. The questions an investigator is seeking to answer should drive the exact approach taken. Early image-processing steps include reducing technical noise (Lu et al. 2023) and aggregation of pixel-level data at different resolutions (Keren et al. 2018). Each step can be achieved by different methods and each method tuned by different parameters, each impacting the final readout.

In many of these approaches, cell segmentation is a critical aspect of assigning biology to individual cells. Although the nucleus can be used as a seed for the segmentation process, the variety of cell shapes and sizes can pose challenges for accurate segmentation and can possibly lead to underappreciation of cell-to-cell interactions. On the other hand, cells that are closely approximated, such as myoepithelial cells and luminal cells of the breast, can be difficult to distinguish. Creative approaches in cell segmentation are required to accurately localize signals to the correct cell. Although this problem would be well recognized within our current understanding of epithelial phenotypic states, it is a significant problem in our efforts to understand new epithelial states. Solutions will likely require coordination of nonspatial techniques such as single-cell RNA (scRNA) sequencing onto spatial methods. Cell-segmentation fidelity has been improved by the development of deep-learning models designed for tissue sections (Sirinukunwattana et al. 2016; Hollandi et al. 2020; Greenwald et al. 2022), including multiplexed image data, and is likely to improve further as more diverse training data is accumulated.

Similarly, different approaches can be adopted to phenotyping single segmented cells from traditional gating to unsupervised data-driven clustering (Levine et al. 2015; Van Gassen et al. 2015). Clustering algorithms can fall prey to errors dependent on technical noise. This noise may be represented by poor signal quality or the confusion between two closely approximated cells. This is like the problems encountered in scRNA-seq in which cells that are dying demonstrate low read quality and technical artifacts including cell doublets that can lead to confusion of cell type, or even the impression of gain of a new phenotype. Interestingly, some groups have taken advantage of incomplete single-cell dissociation (doublets or more) to systematically characterize cell–cell interactions (Boisset et al. 2018; Andrews et al. 2021). An advantage of tissue-based data as compared to single-cell suspensions, however, is that the validity of derived phenotypes can be compared against known cell morphology and spatial context (Danenberg et al. 2022).

EMERGING PRINCIPLES OF SPATIAL BIOLOGY IN BREAST CANCER

The histopathology of breast tumors has long revealed their heterogeneous cellular composition. This diversity, shared by both cancer cells and those of the TME, characterizes the heterocellular ecosystem of interacting cells that comprise tumors (Azizi et al. 2018; Wagner et al. 2019). Our understanding of tumors as dynamic ecosystems has been advanced by several orthogonal methods, all of which aim to enumerate intratumoral cellular populations, but which differ in their power to resolve different cell types. Computational pathology, for example, has been used to quantify tumor composition according to coarse cell categories (epithelial, lymphocyte, and stromal) revealing associations with treatment response (Ali et al. 2016a, 2017) and evolution (AbdulJabbar et al. 2020). To estimate cellular composition in greater detail across large sample collections, methods for deconvolving bulk gene expression data have been developed (Newman et al. 2015, 2019). Use of these methods has revealed key immune cell types associated with outcome across tumors including breast cancer (Ali et al. 2016b) and, more recently, has revealed multicellular TME communities in cancer and their clinical associations (Luca et al. 2021). Employing more sophisticated machine learning, tools such as EcoTyper now enable analysis of a myriad of TME states from bulk transcriptomic data (Luca et al. 2021). These results provide insight into TME biology and can be prognostic for tumor behavior (Gentles et al. 2015). It also enables a more nuanced understanding of how the neoplastic cells interact or evade the host response to growth and invasion.

But deconvolving mixed signals is challenging, contingent on prior knowledge of cell transcriptional states and may be limited in its detection of rare cell subpopulations that contribute little to an average signal. These limitations can be addressed by dissociating tumors into single-cell suspensions and subjecting cells to high-dimensional analysis (Azizi et al. 2018; Wagner et al. 2019). A series of studies have taken this approach to precisely map cell types in breast cancer. The complexity of immune cell states was revealed by scRNA-seq that showed both T cells and macrophages are characterized by continuums of activation rather than discrete states (Azizi et al. 2018). A large analysis using suspension mass cytometry investigated the relationship between tumor cell phenotypes and immune cell composition. Cancer cell phenotypes that were dominant and particular to tumors were uncovered by this approach, showing the remarkable plasticity of which they are capable (Wagner et al. 2019). In addition, high-grade tumors were characterized by PD-L1$^+$ macrophages and exhausted T cells (Wagner et al. 2019). Integration of scRNA-seq and epitope-based analysis by CITE-seq, later corroborated the significance of PD-L1$^+$ macrophage populations, finding that their abundance is linked to clinical outcome (Wu et al. 2021). In all, these studies demonstrate that breast tumors contain diverse interacting cells that are interdependent and likely drive disease progression. A major limitation of these approaches, however, is that they do not account for the multicellular spatial organization of tumors.

MULTICELLUAR SPATIAL ORGANIZATION BY MULTIPLEXED IMAGING

Single-cell atlases of breast cancer have provided a census of the cellular phenotypes that populate tumors, but these approaches do not reveal the landscape in which cells are resident. Several studies of solid tumors including breast cancer have sought to place cell phenotypes in their native spatial context using multiplexed imaging of tumor tissues (Keren et al. 2018; Ali et al. 2020; Jackson et al. 2020; Danenberg et al. 2022). These studies have shown that tumors are highly structured ecosystems characterized by phenotypic transitions at tissue planes and organized multicellular structures. For example, multiplexed imaging revealed that the tumor–stroma interface is characterized by complex multicellular structures and that the location of infiltrating leukocytes distinguishes triple-negative tumors as being compartmentalized (separated) or mixed (infiltrated) (Keren et al. 2018). Further spatial organization of the breast TME was revealed by a data-driven approach where cells clustered in space were grouped using social network analysis and classified based on the connectivity of their constituent cell phenotypes (Danenberg et al. 2022). The 10 recurrent multicellular structures discovered by the approach were associated with distinct genomic profiles and clinical outcome (Danenberg et al. 2022). Exhausted and regulatory T cells, for example, co-occurred in "suppressed expansion" structures that were the largest and most heterogeneous of all TME structures discovered. They also contained most proliferating T cells suggesting they acted as a functional niche enabling differential cell activation (Danenberg et al. 2022).

Multiplexed imaging has also been used to classify breast tumors into "single-cell pathology" subtypes that account for both the tumor and TME (Jackson et al. 2020). A key aim of multiplexed imaging of cancer pathology has been to identify multicellular structures that correspond to known tissue organization and align with underlying function. How best to account for spatial cell context to address this aim is unclear but methods include use of a sliding window of fixed size (Schürch et al. 2020) and representation of cell–cell interactions as a spatial network (Danenberg et al. 2022). The purpose of these methods is to integrate phenotypic and contextual features to classify cells. Network analysis has been extended by use of "message passing" to account for the phenotypic features of neighboring cells (Kim et al. 2022). These approaches reveal higher-order tissue structure and cellular neighborhoods that could represent discrete functional units. Placing cell phenotypes in spatial context, while accounting for their activation state, has linked structure and function with unprecedented resolution. These approaches have the potential to improve patient stratification and by imaging tissues from window-of-opportunity trials (Marron et al. 2022), for example, to elucidate the cellular basis of response and resistance. These studies have revealed that the TME in breast cancer is not uniform but is a mosaic of diverse structures varying in cellular composition and spatial context.

An example of the complexity of the TME and the spatial methods that can uncover it is a study by Risom et al. (2022) that examines DCIS by multi-ion beam imaging (MIBI). A combination of epithelial, stromal, and immune protein markers was used to assess the TME in a case series with mature follow-up. Using an unsupervised approach to identify new cell subtypes and states, the authors identified a myoepithelial population that expresses E-cadherin, which is associated with a higher risk for ipsilateral IBC recurrence. Previously, myoepithelial cells have been difficult to study because of their proximity to luminal cells. The highly multiplexed approach to protein expression in this study allowed for the discovery of new cell states while also having the throughput to make clinical observations with large numbers of samples. Moreover, this study demonstrates that technologies such as MIBI can reduce the TME to cell types and states identified within functional compartments and cellular relationships and neighborhoods quantitatively defined (Risom et al. 2022).

FUTURE/OUTLOOK

Astronomers examine distant bodies to gain insight into historical events by building stronger telescopes that can image features farther into space. With these approaches, scientists can make observations and conclusions on how the universe has evolved rather than simply observing the somewhat static events of the present. We can undertake a similar operation in our studies of human tissue. An archival sample represents the pathology of disease at the fixed time of resection. Although the invasive carcinoma is the clear entity of clinical significance in patient care, clonally related precursor lesions, such as DCIS and benign breast disease (BBD), can give us insight into the events leading to the invasive carcinoma development. Often, these events or residua of these events are evident adjacent to the invasive carcinoma in the same clinical specimen. A significant challenge in studying these events is their microscopic size. Here, we can appreciate the power of specialized measurements, both transcriptomic and proteomic, in their ability to make observations on cellular composition and expression returning spatial data and single-cell data. An extraordinary range of spatial technologies, of increasing parameters and resolution, are becoming widely available. Integrating these data will present a major challenge in future studies. Adapting mosaic or dictionary integration methods developed for single-cell data to spatial data, accounting for the added dimension of space, may prove a promising strategy (Ghazanfar et al. 2023; Hao et al. 2023).

One important question is what changes occur at the early stages of neoplasia (Rozenblatt-Rosen et al. 2020)? Epithelial heterogeneity, interactions with myoepithelial cells and ECM, and immune surveillance are likely to be important in these formative stages (Risom et al. 2022). It is tempting to assume that the changes at the DCIS stage are simply less well defined at earlier stages but essentially the same. This is possible, but it ignores that DCIS and BBD may have different growth challenges. Like DCIS, some BBD grow into the ductal lumen and may have proapoptotic stress associated with loss of contact with the ECM (anoikis). Some DCIS lesions are clearly under additional stresses as shown by frequent central necrosis (van de Vijver et al. 1988). These may include hypoxia and immune predation. In addition, the genomic conditions are quite different. DCIS frequently bears mutations that destabilize the genome and has genomic complexity comparable to invasive cancer. BBD, in contrast, has fewer genomic events and is enriched for mutations associated with survival (e.g., *PIK3CA* mutations). Spatial studies will help us understand changing cellular interactions that contribute to this transition.

Another critical aspect of studying breast precancers is the clonal relationships between entities at different stages of progression. In many instances, it is not readily apparent whether precancers are related to one another and to clinically significant invasive lesions. Methods for studying expression are rapidly emerging and will be quite robust in the study of progression. However, similar methods for evaluating genomics relationships are at an early stage and lack comparable resolution. Sequencing in conjunction with microdissection begins an important advance. However, techniques that are more flexible, robust, and with higher throughput are needed. The combination of multiplex proteomic and transcriptomic imaging provides a tremendous amount of data on gene expression and cell composition. Spatial assessment of the genomics changes in breast neoplasia is, however, missing. Genomic heterogeneity has been long recognized as an important aspect in breast neoplasia development. Studies have examined genomic heterogeneity in invasive, metastatic, and precancerous neoplasia. However, the identification of morphologic and microstage changes in evolution have been understudied. Assays that can combine the proteomic and transcriptomic observations with spatial resolution will reveal insights into neoplasia evolution and add a new dimension to our understanding of breast neoplasia development.

CONCLUSIONS

When and how spatial observations will make their way into clinics is unclear. Spatial methods are discovering many new subtypes of cells and tissue features. It remains to be seen whether the clinical significance of observations made with highly multiplex spatial analyses will require such a degree of multiplexed observations or whether a less complicated derivative clinical test could be used instead. Simpler approaches exist, if not currently used routinely in the clinic, such as small multiplex IF panels and H&E image analysis. H&E image analysis is promising, not least because H&E remains the standard for clinical care and is widely available. Assessing whether multiplex observations can be reduced to a more pragmatic representation will be an important topic of research in the coming years.

With spatial observations comes an appreciation of TME biology. It is possible that the emerging ecology-aware concepts might impact clinical care by identifying features that improve the predictive value of existing therapeutic companion tests. Receptor–ligand and other cell-to-cell communication events are, for example, likely to become more important as immunotherapies become more widespread. Understanding the evolution and ecology of cancer cell population dynamics, not simply within the contemporary clinical specimen, but with longitudinal sampling, may provide critical insights into clinical management.

Highly multiplexed imaging technologies have brought breast histopathology to the fore and underscore the importance of spatial context as a determinant of tumor biology.

REFERENCES

AbdulJabbar K, Raza SEA, Rosenthal R, Jamal-Hanjani M, Veeriah S, Akarca A, Lund T, Moore DA, Salgado R, Al Bakir M, et al. 2020. Geospatial immune variability illuminates differential evolution of lung adenocarcinoma. *Nat Med* **26**: 1054–1062. doi:10.1038/s41591-020-0900-x

Ali HR, Dawson SJ, Blows FM, Provenzano E, Leung S, Nielsen T, Pharoah PD, Caldas C. 2012. A Ki67/BCL2 index based on immunohistochemistry is highly prognostic in ER-positive breast cancer. *J Pathol* **226**: 97–107. doi:10.1002/path.2976

Ali HR, Dariush A, Provenzano E, Bardwell H, Abraham JE, Iddawela M, Vallier AL, Hiller L, Dunn JA, Bowden SJ, et al. 2016a. Computational pathology of pre-treatment biopsies identifies lymphocyte density as a predictor of response to neoadjuvant chemotherapy in breast cancer. *Breast Cancer Res* **18**: 21. doi:10.1186/s13058-016-0682-8

Ali HR, Chlon L, Pharoah PD, Markowetz F, Caldas C. 2016b. Patterns of immune infiltration in breast cancer and their clinical implications: a gene-expression-based retrospective study. *PLoS Med* **13**: e1002194. doi:10.1371/journal.pmed.1002194

Ali HR, Dariush A, Thomas J, Provenzano E, Dunn J, Hiller L, Vallier AL, Abraham J, Piper T, Bartlett JMS, et al. 2017. Lymphocyte density determined by computational pathology validated as a predictor of response to neoadjuvant chemotherapy in breast cancer: secondary analysis of the ARTemis trial. *Ann Oncol* **28**: 1832–1835. doi:10.1093/annonc/mdx266

Ali HR, Jackson HW, Zanotelli VRT, Danenberg E, Fischer JR, Bardwell H, Provenzano E; CRUK IMAXT Grand Challenge Team; Rueda OM, Chin SF, et al. 2020. Imaging mass cytometry and multiplatform genomics define the phenogenomic landscape of breast cancer. *Nat Cancer* **1**: 163–175. doi:10.1038/s43018-020-0026-6

Andrews N, Serviss JT, Geyer N, Andersson AB, Dzwonkowska E, Šutevski I, Heijboer R, Baryawno N, Gerling M, Enge M. 2021. An unsupervised method for physical cell interaction profiling of complex tissues. *Nat Methods* **18**: 912–920. doi:10.1038/s41592-021-01196-2

Angelo M, Bendall SC, Finck R, Hale MB, Hitzman C, Borowsky AD, Levenson RM, Lowe JB, Liu SD, Zhao S, et al. 2014. Multiplexed ion beam imaging of human breast tumors. *Nat Med* **20**: 436–442. doi:10.1038/nm.3488

Aparicio S, Caldas C. 2013. The implications of clonal genome evolution for cancer medicine. *N Engl J Med* **368**: 842–851. doi:10.1056/NEJMra1204892

Azizi E, Carr AJ, Plitas G, Cornish AE, Konopacki C, Prabhakaran S, Nainys J, Wu K, Kiseliovas V, Setty M, et al. 2018. Single-cell map of diverse immune phenotypes in the breast tumor microenvironment. *Cell* **174**: 1293–1308.e36. doi:10.1016/j.cell.2018.05.060

Baldominos P, Barbera-Mourelle A, Barreiro O, Huang Y, Wight A, Cho JW, Zhao X, Estivill G, Adam I, Sanchez X, et al. 2022. Quiescent cancer cells resist T cell attack by forming an immunosuppressive niche. *Cell* **185**: 1694–1708.e19. doi:10.1016/j.cell.2022.03.033

Beck AH, Weng Z, Witten DM, Zhu S, Foley JW, Lacroute P, Smith CL, Tibshirani R, van de Rijn M, Sidow A, et al. 2010. 3′-end sequencing for expression quantification (3SEQ) from archival tumor samples. *PLoS ONE* **5**: e8768. doi:10.1371/journal.pone.0008768

Bergholtz H, Lien TG, Swanson DM, Frigessi A, Oslo Breast Cancer Research Consortium (OSBREAC), Daidone MG, Tost J, Wärnberg F, Sørlie T. 2020. Contrasting DCIS and invasive breast cancer by subtype suggests basal-like DCIS as distinct lesions. *NPJ Breast Cancer* **6**: 26. doi:10.1038/s41523-020-0167-x

Bodenmiller B. 2016. Multiplexed epitope-based tissue imaging for discovery and healthcare applications. *Cell Syst* **2**: 225–238. doi:10.1016/j.cels.2016.03.008

Boisset JC, Vivié J, Grün D, Muraro MJ, Lyubimova A, van Oudenaarden A. 2018. Mapping the physical network of cellular interactions. *Nat Methods* **15**: 547–553. doi:10.1038/s41592-018-0009-z

Bordeaux J, Welsh A, Agarwal S, Killiam E, Baquero M, Hanna J, Anagnostou V, Rimm D. 2010. Antibody validation. *BioTechniques* **48**: 197–209. doi:10.2144/000113382

Brennan D, O'Connor D, Rexhepaj E, Ponten F, Gallagher W. 2010. Antibody-based proteomics: fast-tracking molecular diagnostics in oncology. *Nat Rev Cancer* **10**: 605–617. doi:10.1038/nrc2902

Broders AC. 1932. Carcinoma in situ contrasted with benign penetrating epithelium. *J Am Med Assoc* **99**: 1670–1674. doi:10.1001/jama.1932.02740720024007

Camp RL, Neumeister V, Rimm DL. 2008. A decade of tissue microarrays: progress in the discovery and validation of cancer biomarkers. *J Clin Oncol* **26**: 5630–5637. doi:10.1200/JCO.2008.17.3567

Casasent AK, Schalck A, Gao R, Sei E, Long A, Pangburn W, Casasent T, Meric-Bernstam F, Edgerton ME, Navin NE. 2018. Multiclonal invasion in breast tumors identified by topographic single cell sequencing. *Cell* **172**: 205–217.e12. doi:10.1016/j.cell.2017.12.007

Curtis C, Shah SP, Chin SF, Turashvili G, Rueda OM, Dunning MJ, Speed D, Lynch AG, Samarajiwa S, Yuan Y, et al. 2012. The genomic and transcriptomic architecture of 2,000 breast tumours reveals novel subgroups. *Nature* **486**: 346–352. doi:10.1038/nature10983

Danenberg E, Bardwell H, Zanotelli VRT, Provenzano E, Chin SF, Rueda OM, Green A, Rakha E, Aparicio S, Ellis IO, et al. 2022. Breast tumor microenvironment structures are associated with genomic features and clinical outcome. *Nat Genet* **54**: 660–669. doi:10.1038/s41588-022-01041-y

Dawson SJ, Makretsov N, Blows FM, Driver KE, Provenzano E, Le Quesne J, Baglietto L, Severi G, Giles GG, McLean CA, et al. 2010. BCL2 in breast cancer: a favourable prognostic marker across molecular subtypes and independent of adjuvant therapy received. *Br J Cancer* **103**: 668–675. doi:10.1038/sj.bjc.6605736

DiVito KA, Charette LA, Rimm DL, Camp RL. 2004. Long-term preservation of antigenicity on tissue microarrays. *Lab Invest* **84**: 1071–1078. doi:10.1038/labinvest.3700131

Early Breast Cancer Trialists' Collaborative Group. 1998. Tamoxifen for early breast cancer: an overview of the randomised trials. *Lancet* **351**: 1451–1467. doi:10.1016/S0140-6736(97)11423-4

Foley JW, Zhu C, Jolivet P, Zhu SX, Lu P, Meaney MJ, West RB. 2019. Gene expression profiling of single cells from archival tissue with laser-capture microdissection and Smart-3SEQ. *Genome Res* **29**: 1816–1825. doi:10.1101/gr.234807.118

Gentles AJ, Newman AM, Liu CL, Bratman SV, Feng W, Kim D, Nair VS, Xu Y, Khuong A, Hoang CD, et al. 2015. The prognostic landscape of genes and infiltrating immune cells across human cancers. *Nat Med* **21**: 938–945. doi:10.1038/nm.3909

Gerlinger M, Rowan AJ, Horswell S, Math M, Larkin J, Endesfelder D, Gronroos E, Martinez P, Matthews N, Stewart A, et al. 2012. Intratumor heterogeneity and branched evolution revealed by multiregion sequencing. *N Engl J Med* **366**: 883–892. doi:10.1056/NEJMoa1113205

Gerlinger M, Horswell S, Larkin J, Rowan AJ, Salm MP, Varela I, Fisher R, McGranahan N, Matthews N, Santos CR, et al. 2014. Genomic architecture and evolution of clear cell renal cell carcinomas defined by multiregion sequencing. *Nat Genet* **46**: 225–233. doi:10.1038/ng.2891

Ghazanfar S, Guibentif C, Marioni JC. 2023. Stabilized mosaic single-cell data integration using unshared features. *Nat Biotechnol* doi:10.1038/s41587-023-01766-z

Giesen C, Wang HA, Schapiro D, Zivanovic N, Jacobs A, Hattendorf B, Schüffler PJ, Grolimund D, Buhmann JM, Brandt S, et al. 2014. Highly multiplexed imaging of tumor tissues with subcellular resolution by mass cytometry. *Nat Methods* **11**: 417–422. doi:10.1038/nmeth.2869

Gil Del Alcazar CR, Huh SJ, Ekram MB, Trinh A, Liu LL, Beca F, Zi X, Kwak M, Bergholtz H, Su Y, et al. 2017. Immune escape in breast cancer during in situ to invasive carcinoma transition. *Cancer Discov* **7**: 1098–1115. doi:10.1158/2159-8290.cd-17-0222

Grange JM, Stanford JL, Stanford CA. 2002. Campbell De Morgan's "Observations on Cancer," and their relevance today. *J R Soc Med* **95**: 296–299. doi:10.1177/014107680209500609

Greenwald NF, Miller G, Moen E, Kong A, Kagel A, Dougherty T, Fullaway CC, McIntosh BJ, Leow KX, Schwartz MS, et al. 2022. Whole-cell segmentation of tissue images with human-level performance using large-scale data annotation and deep learning. *Nat Biotechnol* **40**: 555–565. doi:10.1038/s41587-021-01094-0

Hao Y, Stuart T, Kowalski MH, Choudhary S, Hoffman P, Hartman A, Srivastava A, Molla G, Madad S, Fernandez-Granda C, et al. 2023. Dictionary learning for integrative, multimodal and scalable single-cell analysis. *Nat Biotechnol* doi:10.1038/s41587-023-01767-y

Harvey JM, Clark GM, Osborne CK, Allred DC. 1999. Estrogen receptor status by immunohistochemistry is superior to the ligand-binding assay for predicting response to adjuvant endocrine therapy in breast cancer. *J Clin Oncol* **17**: 1474–1481. doi:10.1200/JCO.1999.17.5.1474

Hickey JW, Neumann EK, Radtke AJ, Camarillo JM, Beuschel RT, Albanese A, McDonough E, Hatler J, Wiblin AE, Fisher J, et al. 2022. Spatial mapping of protein composition and tissue organization: a primer for multiplexed antibody-based imaging. *Nat Methods* **19**: 284–295. doi:10.1038/s41592-021-01316-y

Hollandi R, Szkalisity A, Toth T, Tasnadi E, Molnar C, Mathe B, Grexa I, Molnar J, Balind A, Gorbe M, et al. 2020. NucleAIzer: a parameter-free deep learning framework for nucleus segmentation using image style transfer. *Cell Syst* **10**: 453–458.e6. doi:10.1016/j.cels.2020.04.003

Hwang WL, Jagadeesh KA, Guo JA, Hoffman HI, Yadollahpour P, Reeves JW, Mohan R, Drokhlyansky E, Van Wittenberghe N, Ashenberg O, et al. 2022. Single-nucleus and spatial transcriptome profiling of pancreatic cancer identifies multicellular dynamics associated with neoadjuvant treatment. *Nat Genet* **54**: 1178–1191. doi:10.1038/s41588-022-01134-8

Jackson HW, Fischer JR, Zanotelli VRT, Ali HR, Mechera R, Soysal SD, Moch H, Muenst S, Varga Z, Weber WP, et al. 2020. The single-cell pathology landscape of breast cancer. *Nature* **578**: 615–620. doi:10.1038/s41586-019-1876-x

Jamal-Hanjani M, Wilson GA, McGranahan N, Birkbak NJ, Watkins TBK, Veeriah S, Shafi S, Johnson DH, Mitter R, Rosenthal R, et al. 2017. Tracking the evolution of non-small-cell lung cancer. *N Engl J Med* **376**: 2109–2121. doi:10.1056/NEJMoa1616288

Janiszewska M, Liu L, Almendro V, Kuang Y, Paweletz C, Sakr RA, Weigelt B, Hanker AB, Chandarlapaty S, King TA, et al. 2015. In situ single-cell analysis identifies heterogeneity for PIK3CA mutation and HER2 amplification in HER2-positive breast cancer. *Nat Genet* **47**: 1212–1219. doi:10.1038/ng.3391

Keren L, Bosse M, Marquez D, Angoshtari R, Jain S, Varma S, Yang SR, Kurian A, Van Valen D, West R, et al. 2018. A structured tumor-immune microenvironment in triple negative breast cancer revealed by multiplexed ion beam imaging. *Cell* **174**: 1373–1387.e19. doi:10.1016/j.cell.2018.08.039

Kim SY, Jung SH, Kim MS, Baek IP, Lee SH, Kim TM, Chung YJ, Lee SH. 2015. Genomic differences between pure ductal carcinoma in situ and synchronous ductal carcinoma in situ with invasive breast cancer. *Oncotarget* **6**: 7597–7607. doi:10.18632/oncotarget.3162

Kim J, Rustam S, Mosquera JM, Randell SH, Shaykhiev R, Rendeiro AF, Elemento O. 2022. Unsupervised discovery of tissue architecture in multiplexed imaging. *Nat Methods* **19**: 1653–1661. doi:10.1038/s41592-022-01657-2

Kononen J, Bubendorf L, Kallioniemi A, Bärlund M, Schraml P, Leighton S, Torhorst J, Mihatsch MJ, Sauter G, Kallioniemi OP. 1998. Tissue microarrays for high throughput molecular profiling of tumor specimens. *Nat Med* **4**: 844–847. doi:10.1038/nm0798-844

Kristensen VN, Lingjærde OC, Russnes HG, Vollan HK, Frigessi A, Børresen-Dale AL. 2014. Principles and methods of integrative genomic analyses in cancer. *Nat Rev Cancer* **14**: 299–313. doi:10.1038/nrc3721

Krøigård AB, Larsen MJ, Lænkholm AV, Knoop AS, Jensen JD, Bak M, Mollenhauer J, Kruse TA, Thomassen M. 2015. Clonal expansion and linear genome evolution through breast cancer progression from pre-invasive stages to asynchronous metastasis. *Oncotarget* **6:** 5634–5649. doi:10.18632/oncotarget.3111

Lehmann BD, Bauer JA, Chen X, Sanders ME, Chakravarthy AB, Shyr Y, Pietenpol JA. 2011. Identification of human triple-negative breast cancer subtypes and preclinical models for selection of targeted therapies. *J Clin Invest* **121:** 2750–2767. doi:10.1172/jci45014

Levine JH, Simonds EF, Bendall SC, Davis KL, Amir e-A, Tadmor MD, Litvin O, Fienberg HG, Jager A, Zunder ER, et al. 2015. Data-driven phenotypic dissection of AML reveals progenitor-like cells that correlate with prognosis. *Cell* **162:** 184–197. doi:10.1016/j.cell.2015.05.047

Lewis SM, Asselin-Labat ML, Nguyen Q, Berthelet J, Tan X, Wimmer VC, Merino D, Rogers KL, Naik SH. 2021. Spatial omics and multiplexed imaging to explore cancer biology. *Nat Methods* **18:** 997–1012. doi:10.1038/s41592-021-01203-6

Lomakin A, Svedlund J, Strell C, Gataric M, Shmatko A, Rukhovich G, Park JS, Ju YS, Dentro S, Kleshchevnikov V, et al. 2022. Spatial genomics maps the structure, nature and evolution of cancer clones. *Nature* **611:** 594–602. doi:10.1038/s41586-022-05425-2

Lu P, Oetjen KA, Bender DE, Ruzinova MB, Fisher DAC, Shim KG, Pachynski RK, Brennen WN, Oh ST, Link DC, et al. 2023. IMC-Denoise: a content aware denoising pipeline to enhance imaging mass cytometry. *Nat Commun* **14:** 1601. doi:10.1038/s41467-023-37123-6

Luca BA, Steen CB, Matusiak M, Azizi A, Varma S, Zhu C, Przybyl J, Espín-Pérez A, Diehn M, Alizadeh AA, et al. 2021. Atlas of clinically distinct cell states and ecosystems across human solid tumors. *Cell* **184:** 5482–5496.e28. doi:10.1016/j.cell.2021.09.014

Marron TU, Galsky MD, Taouli B, Fiel MI, Ward S, Kim E, Yankelevitz D, Doroshow D, Guttman-Yassky E, Ungar B, et al. 2022. Neoadjuvant clinical trials provide a window of opportunity for cancer drug discovery. *Nat Med* **28:** 626–629. doi:10.1038/s41591-022-01681-x

Narod SA, Iqbal J, Giannakeas V, Sopik V, Sun P. 2015. Breast cancer mortality after a diagnosis of ductal carcinoma in situ. *JAMA Oncol* **1:** 888–896. doi:10.1001/jamaoncol.2015.2510

Newburger DE, Kashef-Haghighi D, Weng Z, Salari R, Sweeney RT, Brunner AL, Zhu SX, Guo X, Varma S, Troxell ML, et al. 2013. Genome evolution during progression to breast cancer. *Genome Res* **23:** 1097–1108. doi:10.1101/gr.151670.112

Newman AM, Liu CL, Green MR, Gentles AJ, Feng W, Xu Y, Hoang CD, Diehn M, Alizadeh AA. 2015. Robust enumeration of cell subsets from tissue expression profiles. *Nat Methods* **12:** 453–457. doi:10.1038/nmeth.3337

Newman AM, Steen CB, Liu CL, Gentles AJ, Chaudhuri AA, Scherer F, Khodadoust MS, Esfahani MS, Luca BA, Steiner D, et al. 2019. Determining cell type abundance and expression from bulk tissues with digital cytometry. *Nat Biotechnol* **37:** 773–782. doi:10.1038/s41587-019-0114-2

Ng CK, Martelotto LG, Gauthier A, Wen HC, Piscuoglio S, Lim RS, Cowell CF, Wilkerson PM, Wai P, Rodrigues DN, et al. 2015. Intra-tumor genetic heterogeneity and alternative driver genetic alterations in breast cancers with heterogeneous HER2 gene amplification. *Genome Biol* **16:** 107. doi:10.1186/s13059-015-0657-6

Nowell PC. 1976. The clonal evolution of tumor cell populations. *Science* **194:** 23–28. doi:10.1126/science.959840

Paik S, Shak S, Tang G, Kim C, Baker J, Cronin M, Baehner FL, Walker MG, Watson D, Park T, et al. 2004. A multigene assay to predict recurrence of tamoxifen-treated, node-negative breast cancer. *N Engl J Med* **351:** 2817–2826. doi:10.1056/NEJMoa041588

Park SY, Gönen M, Kim HJ, Michor F, Polyak K. 2010. Cellular and genetic diversity in the progression of in situ human breast carcinomas to an invasive phenotype. *J Clin Invest* **120:** 636–644. doi:10.1172/jci40724

Perou CM, Sørlie T, Eisen MB, van de Rijn M, Jeffrey SS, Rees CA, Pollack JR, Ross DT, Johnsen H, Akslen LA, et al. 2000. Molecular portraits of human breast tumours. *Nature* **406:** 747–752. doi:10.1038/35021093

Pinder SE, Ellis IO. 2003. The diagnosis and management of pre-invasive breast disease: ductal carcinoma in situ (DCIS) and atypical ductal hyperplasia (ADH)—current definitions and classification. *Breast Cancer Res* **5:** 254–257. doi:10.1186/bcr623

Risom T, Glass DR, Averbukh I, Liu CC, Baranski A, Kagel A, McCaffrey EF, Greenwald NF, Rivero-Gutiérrez B, Strand SH, et al. 2022. Transition to invasive breast cancer is associated with progressive changes in the structure and composition of tumor stroma. *Cell* **185:** 299–310.e18. doi:10.1016/j.cell.2021.12.023

Rozenblatt-Rosen O, Regev A, Oberdoerffer P, Nawy T, Hupalowska A, Rood JE, Ashenberg O, Cerami E, Coffey RJ, Demir E, et al. 2020. The human tumor atlas network: charting tumor transitions across space and time at single-cell resolution. *Cell* **181:** 236–249. doi:10.1016/j.cell.2020.03.053

Schapiro D, Yapp C, Sokolov A, Reynolds SM, Chen YA, Sudar D, Xie Y, Muhlich J, Arias Camison R, Arena S, et al. 2022. MITI minimum information guidelines for highly multiplexed tissue images. *Nat Methods* **19:** 262–267. doi:10.1038/s41592-022-01415-4

Schürch CM, Bhate SS, Barlow GL, Phillips DJ, Noti L, Zlobec I, Chu P, Black S, Demeter J, McIlwain DR, et al. 2020. Coordinated cellular neighborhoods orchestrate antitumoral immunity at the colorectal cancer invasive front. *Cell* **182:** 1341–1359.e19. doi:10.1016/j.cell.2020.07.005

Shah SP, Roth A, Goya R, Oloumi A, Ha G, Zhao Y, Turashvili G, Ding J, Tse K, Haffari G, et al. 2012. The clonal and mutational evolution spectrum of primary triple-negative breast cancers. *Nature* **486:** 395–399. doi:10.1038/nature10933

Sirinukunwattana K, Ahmed Raza SE, Tsang Y-W, Snead DR, Cree IA, Rajpoot NM. 2016. Locality sensitive deep learning for detection and classification of nuclei in routine colon cancer histology images. *IEEE Trans Med Imaging* **35:** 1196–1206. doi:10.1109/tmi.2016.2525803

Slamon DJ, Clark GM, Wong SG, Levin WJ, Ullrich A, McGuire WL, et al. 1987. Human breast cancer: correlation of relapse and survival with amplification of the HER-2/*neu* oncogene. *Science* **235:** 177–182. doi:10.1126/science.3798106

Slamon DJ, Godolphin W, Jones LA, Holt JA, Wong SG, Keith DE, Levin WJ, Stuart SG, Udove J, Ullrich A, et al. 1989. Studies of the HER-2/*neu* proto-oncogene in human breast and ovarian cancer. *Science* **244:** 707–712. doi:10.1126/science.2470152

Strand SH, Rivero-Gutiérrez B, Houlahan KE, Seoane JA, King LM, Risom T, Simpson LA, Vennam S, Khan A, Cisneros L, et al. 2022. Molecular classification and biomarkers of clinical outcome in breast ductal carcinoma in situ: analysis of TBCRC 038 and RAHBT cohorts. *Cancer Cell* **40:** 1521–1536.e7. doi:10.1016/j.ccell.2022.10.021

Sweeney RT, Zhang B, Zhu SX, Varma S, Smith KS, Montgomery SB, van de Rijn M, Zehnder J, West RB. 2013. Desktop transcriptome sequencing from archival tissue to identify clinically relevant translocations. *Am J Surg Pathol* **37:** 796–803. doi:10.1097/PAS.0b013e31827ad9b2

van de Vijver MJ, Peterse JL, Mooi WJ, Wisman P, Lomans J, Dalesio O, Nusse R. 1988. Neu-protein overexpression in breast cancer. Association with comedo-type ductal carcinoma in situ and limited prognostic value in stage II breast cancer. *N Engl J Med* **319:** 1239–1245. doi:10.1056/nejm198811103191902

Van Gassen S, Callebaut B, Van Helden MJ, Lambrecht BN, Demeester P, Dhaene T, Saeys Y. 2015. FlowSOM: using self-organizing maps for visualization and interpretation of cytometry data. *Cytometry A* **87:** 636–645. doi:10.1002/cyto.a.22625

Wagner J, Rapsomaniki MA, Chevrier S, Anzeneder T, Langwieder C, Dykgers A, Rees M, Ramaswamy A, Muenst S, Soysal SD, et al. 2019. A single-cell atlas of the tumor and immune ecosystem of human breast cancer. *Cell* **177:** 1330–1345.e18. doi:10.1016/j.cell.2019.03.005

Wolff AC, Hammond ME, Hicks DG, Dowsett M, McShane LM, Allison KH, Allred DC, Bartlett JM, Bilous M, Fitzgibbons P, et al. 2013. Recommendations for human epidermal growth factor receptor 2 testing in breast cancer: American society of clinical oncology/college of American pathologists clinical practice guideline update. *J Clin Oncol* **31:** 3997–4013. doi:10.1200/JCO.2013.50.9984

Wu SZ, Al-Eryani G, Roden DL, Junankar S, Harvey K, Andersson A, Thennavan A, Wang C, Torpy JR, Bartonicek N, et al. 2021. A single-cell and spatially resolved atlas of human breast cancers. *Nat Genet* **53:** 1334–1347. doi:10.1038/s41588-021-00911-1

Yates LR, Gerstung M, Knappskog S, Desmedt C, Gundem G, Van Loo P, Aas T, Alexandrov LB, Larsimont D, Davies H, et al. 2015. Subclonal diversification of primary breast cancer revealed by multiregion sequencing. *Nat Med* **21:** 751–759. doi:10.1038/nm.3886

breast cancer subtypes; and identify emerging opportunities for modulating the immune microenvironment with the goal of eliminating tumors.

IMMUNE CELLS AND MECHANISMS IN BREAST CANCER

The breast cancer TIME is highly heterogeneous, varies between and within clinical subtypes, and contains immune cells from both lymphoid and myeloid origins (Yuan et al. 2021). T cells, B cells, and natural killer (NK) cells constitute major lymphoid lineages, while macrophages, monocytes, dendritic cells (DCs), and neutrophils comprise the predominant myeloid populations found in the breast cancer TIME. These cell types each exhibit heterogeneous functional states that can directly eliminate tumors or actively modulate the immune response to establish a protumorigenic or an antitumorigenic niche. The spatial composition and complexity of the TIME is increasingly recognized to distinguish breast cancers and their outcomes. The complexity of the TIME in breast cancer raises both new opportunities in disease management and challenges in developing appropriate research models. Here, we will summarize key findings from clinical and preclinical studies on the involvement of various immune cell types in regulating the outcome in breast cancer.

Cytotoxic and Helper T Cells

The role of T lymphocytes has been a predominant focus in breast cancer research. T cells are key effectors of adaptive immunity. CD8$^+$ cytotoxic T lymphocytes (CTLs) directly induce tumor cell death, while CD4$^+$ helper T cells orchestrate the functions of other immune cells within the TIME. CTL-mediated killing of a cancer cell involves antigen-specific recognition followed by the secretion of cytotoxic molecules including granzymes and perforin, as well as release of proinflammatory effector cytokines such as interferon γ (IFN-γ) (Barry and Bleackley 2002). Although T cells are the predominant tumor-infiltrating lymphocytes (TILs) in breast cancer (Ruffell et al. 2012), both numbers and composition of this infiltrate vary within and between tumors. Studies of large retrospective cohorts of breast cancer patients have shown that increased CD8$^+$ T cells are generally associated with good outcome in breast cancer patients (Ali et al. 2014a). However, studies have shown that the infiltration of CD8$^+$ T cells throughout the tumor bed or their retention in the stroma might differentially impact cancer outcomes, highlighting the importance of understanding the breast cancer immune contexture (Ali et al. 2014a; Khoury et al. 2018). Analyses of patient samples at the single-cell level, albeit with smaller cohorts, have started to demonstrate this immune heterogeneity in breast cancer. For example, examination of eight breast tumors suggested a higher heterogeneity in tumor-infiltrating immune cells from lymphoid and myeloid lineages when compared to matched normal breast tissue (Azizi et al. 2017). Furthermore, this work implies that T-cell clonality and phenotype are uniquely shaped by the tissue context, and that immune biomarkers in blood may not represent an accurate snapshot of the TIME. Notably, in breast tumors with a high-TIL content, a subset of CD8$^+$ T cells exhibited a tissue-resident memory (TRM) phenotype characterized by CD103 positivity in addition to high expression of immune checkpoint and effector molecules (Savas et al. 2018). Further, the TRM gene-expression signature derived from CD8/CD103 dual positive cells more strongly predicted survival for breast cancer patients than did CD8 expression alone, suggesting that TRM cells may play a key role in cancer immunosurveillance and control. Interestingly, the development of the TRM phenotype in this cohort did not rely on pretreatment with immune checkpoint blockade (ICB) or neo-adjuvant chemotherapy. While the TRM phenotype was not exclusive to patients experiencing long-term survival in breast cancer, analysis of data from melanoma patients treated with ICB indicated that the TRM signature is enriched at baseline in responders. Furthermore, TRM-specific genes were significantly up-regulated during treatment with anti-PD1 ICB, suggesting that TRM may be key effectors of ICB response (Savas et al. 2018).

Cite this article as *Cold Spring Harb Perspect Med* doi: 10.1101/cshperspect.a041324

The importance of the composition of CD8[+] T-cell infiltrate in combating tumor progression was also shown in a study examining the dynamics of the immune landscape during progression from precancerous breast lesions, ductal carcinoma in situ (DCIS), to invasive ductal carcinoma (IDC) (Gil Del Alcazar et al. 2017). This work revealed enhanced diversity and activation of CD8[+] CTLs at microinvasive DCIS transition stages, whereas, in more advanced IDCs from the same patients, these cells exhibited exhaustion with a lower activation state and a less diverse T-cell receptor (TCR) repertoire. Thus, both the quantity and quality of tumor-infiltrating CTLs play an important role in controlling tumor progression. Critical roles of CD8[+] CTLs in antitumor immunity have also been functionally confirmed in preclinical breast cancer models, where depletion of these cells leads to tumor progression (Gómez-Aleza et al. 2020; Lai et al. 2021; Li et al. 2021). In parallel, CD8[+] CTLs have been shown to play important roles at distant metastatic sites. This was initially suggested by studies showing that breast cancer cells disseminate early during tumor development, but immunosurveillance at distant metastatic loci mediated mainly by CD8[+] T cells prevents metastatic outgrowth (Eyles et al. 2010). A recent study identified CD39[+]PD1[+]CD8[+] T cells in primary tumors as well as dormant metastatic nodules in mice (Tallón de Lara et al. 2021). These cells kept tumors dormant through secretion of TNF-α and IFN-γ and blocked metastatic outgrowth when adoptively transferred prior to intravenous tumor cell injection in an experimental metastasis model. These studies suggest that CD8[+] T cells are key players in the immune system for controlling tumor progression and metastasis.

Compared to CTLs, the involvement of CD4[+] helper T cells in breast cancer immunosurveillance appears to be more complex. A study modeling breast cancer using two distinct mouse strains (BALB/c and C57BL/6) revealed that CD4[+] T cells have tumor-promoting effects and have negative prognostic associations with disease outcome in patients (Huang et al. 2015). Supporting this view, in an "immune-humanized" mouse model of triple-negative breast cancer (TNBC), CD4[+] T cells were shown to infiltrate tumors and differentiate into immunosuppressive regulatory T cells (Tregs), which promote tumor growth through their immunesuppressive function (Su et al. 2017). Here, blocking the recruitment of naive CD4[+] T cells into tumors reduced the ability of Tregs to accumulate and suppressed tumor growth. Increased Treg infiltration was also shown during progression from DCIS to IDC concomitant with CTL exhaustion (Gil Del Alcazar et al. 2017). On the other hand, in an HER2[+] mouse model of breast cancer, treatment with an antibody-drug conjugate (ADC) targeting HER2 (trastuzumab-emtansine) elicited a strong antitumor immune response and led to an increase in Tregs in animals responding to treatment (Müller et al. 2015). However, in this case, the increase in Tregs was concomitant with a higher level of TILs with an antitumor T helper 1 (Th1) phenotype, which uniquely rendered tumors susceptible to anti-CTLA-4/PD-1 ICB immunotherapy. ICB therapy takes advantage of blocking inhibitory signals for T cells, such as those conveyed through the CTLA-4 or PD-1 receptors (Buchbinder and Desai 2016).

The Th1/Th2 paradigm for helper T-cell activation was defined as early as 1986 based on the cellular secretory phenotype (Mosmann et al. 1986), and it has been used extensively for studying the immune response, although some T-cell phenotypes may not exactly fit into these polarized classes (Muraille and Leo 1998). Th1 cells primarily secrete proinflammatory cytokines IL-12, TNF, and IFN-γ, which are important for antitumor immunity, while Th2 cells are characterized by the production of IL-4, IL-5, and IL-10, which are typically anti-inflammatory and counterproductive for antitumor immunity (Kim and Cantor 2014). Tregs can produce some of the same anti-inflammatory cytokines to modulate the immune response. Interestingly, Tregs were shown to be important for achieving a therapeutic window in which antitumor immunity can occur in the absence of autoimmunity, in the context of ICB (Müller et al. 2015). The importance of Th1 CD4[+] T cells in breast cancer was further supported by identification of a stem-like, Th1-skewed CD4[+] T-cell population that was essential for eliminating tumors that had metastasized to lungs in the PyMT mouse model of breast cancer (Fig. 1; Lai et al. 2021). On the other hand, another study found no significant association between metastasis-free survival and bulk *CD4* or *Foxp3* (key transcription factor for Tregs) expression within human TNBC (Schmidt et al. 2018), suggesting that bulk tumor mRNA analysis approaches may not be optimal for untangling complex relationships within the TIME.

NK Cells

NK cells are innate lymphoid cells whose responses are controlled by a set of activating and inhibitory immune receptors. NK cells can be activated through a "missing-self" mechanism where the lack of MHC-I on the target cell surface results in failure to engage inhibitory NK receptors, leading to release of cytotoxic mole-

cules. Thus, NK cells are critical for elimination of cancer cells that evade CTL recognition by down-regulating MHC-I expression (Guillerey et al. 2016). Multiple studies have shown that NK cells play an important antitumorigenic role in breast cancer, and NK cell dysfunction is associated with advanced disease. For instance, locally advanced breast cancer samples contain fewer numbers of NK cells compared to healthy breast tissue, and neoadjuvant chemotherapy was shown to increase NK cell numbers particularly in patients with pathological responses to treatment (Verma et al. 2015). Similarly, enhanced peripheral NK cell activity correlates with the disappearance of axillary node metastases in breast cancer after neoadjuvant chemotherapy (Kim et al. 2019). Furthermore, a small clinical trial involving high dose chemotherapy with subsequent autologous hematopoietic stem cell transplantation indicated that residual disease can be eliminated upon adoptive immuno-

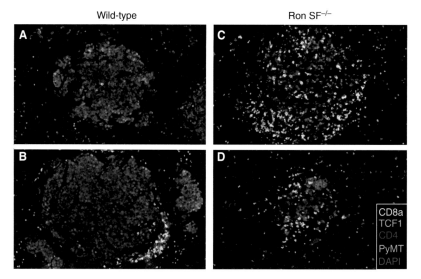

Figure 1. Mammary-selective overexpression of polyomavirus middle-T antigen (MMTV-PyMT) is a commonly used preclinical research model for breast cancer. These multicolor immunofluorescence images show metastatic lesions in mouse lungs 2 weeks after injecting MMTV-PyMT cells intravenously as part of an experimental metastasis assay. (*A,B*) Metastatic tumors in wild-type (FVB) mouse hosts are poorly infiltrated by immune cells (*A*) or exhibit an immune-excluded phenotype (*B*), similar to what is often observed in human metastatic breast cancer. (*C,D*) Genetic deletion of a negative immune regulator called "short-form Ron kinase" (Ron SF$^{-/-}$) results in swarming of metastatic tumors in the lungs by CD4$^+$ and CD8$^+$ T cells. Notably, these cells expressed TCF1, a transcription factor associated with increased effector and stem-like features (Lai et al. 2021). The robust infiltration of tumors with TCF1-expressing CD4$^+$ and CD8$^+$ T cells led to a lower tumor burden and long-term survival in mice. These findings suggest that preclinical research models are invaluable for identifying and validating new immunotherapy targets in breast cancer.

therapy with IL-2 and NK cells in metastatic breast cancer (deMagalhaes-Silverman et al. 2000). Unsurprisingly, tumors can develop various mechanisms to evade NK-mediated cytotoxicity, including altered expression of NK cell receptors and ligands and secretion of immunosuppressive cytokines such as TGF-β (Mamessier et al. 2011; Slattery et al. 2021). Interestingly, NK cells in the peripheral blood of advanced breast cancer patients were shown to be enriched in immature and noncytotoxic subsets (Mamessier et al. 2013), suggesting that the tumor-derived factors can shape NK phenotypes. A recent study indicated that such factors not only can inhibit NK cell cytotoxicity, but they can also educate NK cells to a prometastatic state (Chan et al. 2020). Interestingly, protumorigenic effects of tumor-educated NK cells were blocked by antibodies specific to TIGIT, a negative immune checkpoint receptor that is currently being targeted in clinical trials (Chan et al. 2020). In addition to TIGIT, PD-1 can also be expressed on NK cells in tumor-draining lymph nodes (Frazao et al. 2019), opening interesting avenues for targeting the NK cell axis using ICB therapy in cancer.

Macrophages/Monocytes

Macrophages are highly diverse innate immune cells of the mononuclear phagocyte system and play critical roles in mediating the balance between cancer progression and antitumor immunity. Macrophage accumulation in breast tumors is often associated with poor survival in patients; however, these analyses rely heavily on general markers of bulk macrophages such as CD68. In reality, macrophages are much more diverse and can adopt both pro- and antitumorigenic identities. Historically, macrophages have been described to exist along a phenotypic continuum that ascribes a polarized activation status as either "M1" or "M2" (Mills et al. 2000). M1 macrophages are classically activated, often expressing proinflammatory markers including inducible nitric oxide synthase (iNOS) or CD80/CD86, and in cancer these are considered to be antitumorigenic. In contrast, M2-macrophages are alternatively activated, and express anti-

inflammatory markers, CD163, Arginase-1 (Arg1) or CD206, and exhibit protumorigenic behavior. Many studies that describe tumor-associated macrophages (TAMs) assume an M2-like activation status; however, it is now appreciated that macrophages can adopt features of both M1 and M2 cells simultaneously, particularly when making comparisons between in vitro and in vivo models (Orecchioni et al. 2019). Therefore, this polarization paradigm has been dropped in favor of a broader appreciation for macrophage diversity and heterogeneity within tumors (Murray et al. 2014; Ginhoux et al. 2016).

TAMs are known to be phenotypically and ontogenically diverse (Lavin et al. 2015). Tissue-resident macrophages arise from the embryonic yolk sac during development and their replenishing ability is tissue-dependent (Ginhoux and Guilliams 2016). On the other hand, monocyte-derived macrophages are recruited from the bone marrow in response to inflammatory cues and can evolve from two developmentally distinct precursors (granulocyte-monocyte progenitors [GMPs] or monocyte-DC progenitors [MDPs]) (Yáñez et al. 2015, 2017). Within the fatty breast tissue, macrophages also display spatial and functional diversity in their activation status in accordance with the metabolic status of adipose tissue. Most macrophages within the breast adipose tissue, particularly in women with a high body mass index (BMI), are associated with adipocyte hypertrophy, where they form inflammatory crown-like structures (CLSs) that are associated with elevated breast cancer risk, distant recurrence, and cancer mortality (Iyengar et al. 2016; Koru Sengul et al. 2016; Carter et al. 2018). Adding further complexity to macrophage diversity, CLS macrophages are lipid-laden and exhibit unique features, including high proliferation and expression of CD11c, CD9, or TREM2 (Patsouris et al. 2008; Amano et al. 2014; Hill et al. 2018; Jaitin et al. 2019). Moreover, the mammary gland undergoes various remodeling events during adult life, for example during lactation, which relies heavily on macrophage phagocytosis during postlactational involution (O'Brien et al. 2012). Therefore, macrophages within breast tumors are highly diverse despite many studies using broadly defining markers to study their functional contribution to breast cancer.

In preclinical models of breast cancer, macrophages represent the predominant myeloid component of the TIME, and macrophage depletion is sufficient to blunt tumor progression and reinvigorate antitumor immunity by the adaptive immune system (DeNardo et al. 2011; Strachan et al. 2013; Franklin et al. 2014; Ruffell et al. 2014). In the widely studied MMTV-PyMT model, at least two distinct macrophage populations have been described, including CD11b^{hi} MHC-II^{hi} tissue-resident macrophages with low proliferative capacity, and CD11b^{lo} MHC-II^{hi} macrophages that are more proliferative and are directly associated with the tumor (Dawson et al. 2020). Both macrophage subsets appear to be peripherally derived and replenished by inflammatory Ly6C^{hi} monocytes, which express CCR2 and CX3CR1 (Geissmann et al. 2003; Franklin et al. 2014; Dawson et al. 2020); however, it is unclear whether Ly6C^{hi} GMP- versus MDP-derived monocytes differentially contribute to these macrophage pools (although their high expression of MHC-II might suggest so) (Yáñez et al. 2017). Macrophage recruitment to breast tumors appears to be in part dictated by spatial features of the tumor microenvironment, such as hypoxia (Movahedi et al. 2010). Indeed, mechanistic studies have explored the profound effect of nutrient gradients on macrophage localization and activation states (Carmona-Fontaine et al. 2017; Gilmore et al. 2021), as well as the consequences of macrophage metabolism on nutrient partitioning between other cells within the tumor microenvironment, such as activated T cells (Vitale et al. 2019; Reinfeld et al. 2021). The robust heterogeneity of macrophages in cancer suggests there may be untapped therapeutic potential in reprogramming or targeting specific subsets of these cells in cancer.

Neutrophils

Similar to macrophages, there is a growing appreciation for the heterogeneity of neutrophils both in steady-state and inflammatory contexts

(Ng et al. 2019; Hedrick and Malanchi 2022). Neutrophils are highly cytotoxic polymorphonuclear granulocytes that act as first responders during inflammation and comprise up to 70% of all peripheral leukocytes in humans. Although studies have historically viewed neutrophils as short-lived phenotypically homogenous cells, emerging research is demonstrating significant functional diversity based on physiologic context (e.g., tissue type [Casanova-Acebes et al. 2018; Ballesteros et al. 2020; Xie et al. 2020], time of day [Casanova-Acebes et al. 2013, 2018; Adrover et al. 2019, 2020], sex [Blazkova et al. 2017; Gupta et al. 2020; Lu et al. 2021], age [Wenisch et al. 2000; Hazeldine et al. 2014; Tomay et al. 2018; Lu et al. 2021], and microbiome [Clarke et al. 2010; Balmer et al. 2014; Deshmukh et al. 2014; Khosravi et al. 2014; Zhang et al. 2015]) as well as pathologic context (e.g., cancer-related surgical stress or infection [Cools-Lartigue et al. 2013; Tohme et al. 2016; Najmeh et al. 2017], smoking [Hosseinzadeh et al. 2016; Albrengues et al. 2018; Wculek et al. 2020b; Tyagi et al. 2021], and obesity [Xia et al. 2011; Nagareddy et al. 2014; McDowell et al. 2021]). Neutrophils develop in the bone marrow from GMPs and are replenished diurnally through granulopoiesis (Evrard et al. 2018). It is estimated that humans produce up to 1 billion new neutrophils per kg per day at steady state and up to 10 billion during inflammatory settings (Ley et al. 2018). Normally, neutrophils are released from the bone marrow as fully mature effector cells in response to granulocyte macrophage (GM)/granulocyte-colony stimulating factor (G-CSF) and CXCR2 chemokines such as IL-8; however, tumors chronically produce high levels of these factors (Kowanetz et al. 2010; Casbon et al. 2015; Coffelt et al. 2015; Wculek and Malanchi 2015; Hsu et al. 2019), causing neutrophils to be prematurely released as committed progenitors (Evrard et al. 2018; Zhu et al. 2018). Within tumors, multiple neutrophil identities have been described (Zilionis et al. 2019), although in breast cancer, similar in-depth characterization of neutrophil heterogeneity has yet to be fully defined.

Many studies on neutrophils and breast cancer have focused on their contribution to metastatic disease. In the MMTV-PyMT and KEP (K14cre;Cdh1$^{F/F}$;Trp53$^{F/F}$) preclinical models, neutrophil accumulation has been shown to precede the arrival of tumor cells within the premetastatic niche, where they promote colonization and outgrowth, such that neutrophil depletion with anti-Ly6G antibodies is sufficient to blunt metastatic progression (Coffelt et al. 2015; Wculek and Malanchi 2015). In the D2.0R breast tumor dormancy model, neutrophils contribute to awakening of dormant metastases within the lung by releasing neutrophil extracellular DNA traps (NETs) in response to inflammatory triggers within the host (e.g., tobacco, acute infection) (Park et al. 2016; Albrengues et al. 2018). Similarly, in the 4T1 model, it has been shown that neutrophils can cluster with circulating tumor cells within the periphery to promote cell-cycle progression during metastasis (Szczerba et al. 2019), while simultaneously blocking NK-mediated antitumor immune surveillance and facilitating extravasation (Spiegel et al. 2016). Emerging work now suggests that distinct genetic drivers of breast cancer may predispose to neutrophilic inflammation and facilitate metastasis. Notably, a comparison of 16 genetic mouse models of breast cancer demonstrated that loss of p53 within cancer cells and subsequent secretion of WNT ligands is a trigger for macrophage-dependent neutrophil infiltration and prometastatic functions (Wellenstein et al. 2019). Similar to macrophages, neutrophil metabolism has also been shown to play an important role in influencing antitumor immunity; although neutrophils are largely glycolytic, they adapt to the glucose-restricted microenvironment by engaging alternative metabolic pathways to sustain a protumoral phenotype in 4T1 breast cancer models (Rice et al. 2018; Hsu et al. 2019; Ombrato et al. 2019; Li et al. 2020). These studies shed light on the multifaceted effects of neutrophils at various stages of the metastatic cascade, and functional interdependency of cells within the TIME. However, because of their strong circadian rhythmicity, it is likely that developing neutrophil-targeted immunotherapies will require chronopharmacological considerations, thus making these cells challenging to target clinically.

Dendritic Cells

Derived from lymphoid and myeloid origins, DCs are key players responsible for bridging the innate and adaptive immunity. DCs are present in peripheral tissues and lymphoid organs, and they specialize in presenting extracellular and intracellular antigens to prime T-cell responses (Collin et al. 2013; Lee et al. 2017). Three major DC subsets were defined in humans based on distinct expression patterns of transcription factors IFN regulatory factor 4 (IRF4) and IRF8: plasmacytoid DCs (pDCs, IRF4$^+$IRF8$^+$), conventional DC1 (cDC1, IRF4$^-$IRF8$^+$), and conventional DC2 (cDC2, IRF4$^+$IRF8$^-$) (Guilliams et al. 2016; Collin and Bigley 2018). In addition to these, specialized DCs can participate in the immune response including Langerhans cells in squamous epithelial tissues and monocyte-derived inflammatory DCs (Mo-DCs) (Collin and Bigley 2018). As professional antigen-presenting cells, DCs can phagocytose extracellular antigens and present them on MHC-II molecules for priming CD4$^+$ helper T-cell responses (Cabeza-Cabrerizo et al. 2021). However, one of the key features of DCs in the context of tumor immunity is antigen cross-presentation in which antigens from dying tumor cells can be internalized and loaded onto MHC-I molecules to induce CD8$^+$ killer T-cell responses (Albert et al. 1998; Joffre et al. 2012). Antigen presentation capabilities of DCs are regulated by costimulatory ligands expressed on their surfaces, and a recent study in melanoma showed that antitumor functions of DCs may be circadian in nature through the regulation of CD80 expression (Wang et al. 2023). While most DC subsets are evolutionarily conserved across multiple mammals, the functional capacities of these cells and their surface markers may differ between species (Vu Manh et al. 2015; Guilliams et al. 2016). Studies in mice and humans have shown that DCs can exhibit a range of activation phenotypes and regulate immunological tolerance as well as inflammatory responses (Mellman and Steinman 2001; Anderson et al. 2017). Particularly, immature DCs presenting self-antigens can induce immunologic tolerance, whereas non-self-antigen-loaded mature DCs can assume a migratory phenotype to initiate a cascade of immuno-stimulatory reactions in lymphoid organs. The critical roles of DCs in regulating immunity have generated interest in modulating their functions for an effective antitumor immune response (Wculek et al. 2020a). Various thera-peutics targeting DCs and/or their activation phenotypes have been clinically approved in-cluding tumor antigen-pulsed DC vaccine in prostate cancer (Kantoff et al. 2010), GM-CSF-expressing oncolytic virus in melanoma (Bom-mareddy et al. 2017), and TLR7/8 ligands in nonmelanoma skin cancers (Drobits et al. 2012). There are no DC-targeting therapies ap-proved for breast cancer yet, but studies indicate that DCs can be dysregulated in this disease, suggesting possible intervention points.

DCs isolated from the peripheral blood of lymph nodes of breast cancer patients were shown to be dysfunctional and unable to induce T-cell activation in mixed lymphocyte reaction experiments (Satthaporn et al. 2004; Gervais et al. 2005). In this study, DCs from a wide range of breast cancer subtypes expressed lower levels of MHC-II and costimulatory molecules sug-gesting that they exist in an immature state. Oth-er groups examined DCs within the TIME and revealed that DCs within the tumor bed have an immature phenotype, whereas mature DCs are located at the peritumoral sites (Bell et al. 1999; Coventry et al. 2002). A more recent study focus-ing on intratumoral DC subsets in luminal breast cancer and TNBC showed subset-specific gene-expression patterns when compared to DCs from the uninvolved tissues (Michea et al. 2018). Interestingly, the vast majority of differ-entially up-regulated genes in cancer-infiltrating DCs were found to be subset-specific, suggesting that there is not a strong tissue-specific gene-expression pattern shared among DC popula-tions within breast cancer. Importantly, DCs in TNBC were enriched for immunological signa-tures including the IFN pathway, whereas DCs in luminal cancer were enriched for wound healing and extracellular matrix reorganization path-ways (Michea et al. 2018). In addition to tran-scriptomic signatures, the composition of DCs within the breast TIME can have differential im-pact on clinical outcomes. For instance, DCs ex-pressing a CD1a invariant MHC-I molecule were shown to be associated with reduced metastasis and longer survival (Coventry and Morton 2003; Giorello et al. 2021; Szpor et al. 2021). On the other hand, infiltration of breast tumors with CD123+ pDCs correlated with poor relapse-free and overall survival (Treilleux et al. 2004). In a later study, the negative effects of tumor-associated pDCs were attributed to Treg activa-tion, and intratumoral administration of TLR7 ligand was shown to reverse pDC-mediated tu-mor promotion (le Mercier et al. 2013). These findings suggest that cancer subtype, spatial fea-tures within the tumor, and the context of the TIME can have a significant impact on the DC phenotypes in breast cancer. Indeed, the precon-ditioning of the DCs and/or the tumor microen-vironment can be an effective approach to strengthen antitumor immunity. For instance, tumor-antigen-pulsed cDC1-type polarized DCs were shown to secrete high levels of IL-12 and expand tumor-specific CD4+ and CD8+ T cells in HER2+ breast cancer in clinical trials (Czerniecki et al. 2007). Similarly, coadministra-tion of antigen-pulsed DCs and cytokine-in-duced killer (CIK) cells (CD3+CD56+ T cells generated in vitro from peripheral blood lym-phocytes using anti-CD3 monoclonal antibody, IL-2, IL-1α, and IFN-γ) enhanced antitumor im-mune responses and prolonged survival in mul-tiple studies (Wang et al. 2014; Shevchenko et al. 2020). Taken together, although the mecha-nisms by which DCs participate in the regulation of antitumor immunity remain to be fully eluci-dated, there is an active interest for modulating DCs as a cancer immunotherapy.

HIGH-THROUGHPUT APPROACHES FOR CHARACTERIZING THE BREAST CANCER TIME

The last few decades have witnessed an explosive growth in the amount of cancer genomic data. Historically, bulk tissue gene-expression micro-arrays, transcriptome analyses via bulk RNA se-quencing, and mutation profiling through whole-exome sequencing have uncovered sever-al complexities in breast cancer. The META-BRIC microarray study, involving nearly 2000 primary tumor samples, was among the first comprehensive characterization efforts in breast cancer (Curtis et al. 2012). This work integrated copy number analysis with gene-expression profiling and identified *cis*- and *trans*-acting genomic aberrations. This high dimensional data defined novel patient subgroups with varying prognoses and intratumoral immune involvement. Shortly after the landmark METABRIC study, a comprehensive analysis of 825 breast cancer samples was released by the TCGA network, which involved DNA copy number arrays, reverse phase protein arrays, DNA methylation and exome sequencing, and mRNA and miRNA-expression profiling (Koboldt et al. 2012). Both studies reported overlapping and complementary findings on two independent data sets and defined the molecular characteristics of human breast cancer. Many subsequent studies have used these data sets and found tumor-specific immune signatures for predicting survival and immunotherapy response in cancer (Charoentong et al. 2017; Ock et al. 2017; Auslander et al. 2018; Cristescu et al. 2018).

Although the data gathered from these bulk tissue analyses have been indispensable for new discoveries and are still heavily used, findings can be difficult to attribute to particular cell types or individual phenotypic states. With the advent of single-cell analysis technologies, it is now clear that the TIME contains highly diverse immune lineages with considerable functional heterogeneity. For example, as discussed above, several landmark studies using single-cell transcriptomics in breast cancer have led to new discoveries that have deepened our understanding of TAM and T-cell biology, as well as the heterogeneity of malignant cells (Azizi et al. 2017; Lai et al. 2021; Pal et al. 2021). Others have further expanded our knowledge about the breast cancer TIME by interrogating the single-cell spatial proteome through high-dimensional mass cytometry approaches (Schulz et al. 2018; Ali et al. 2020; Jackson et al. 2020). These studies have demonstrated that cellular neighborhoods in breast cancer are predictive of patient outcomes, and that spatial examination of the TIME can better segregate patient subsets compared to current clinical classifications (Ali et al. 2020; Jackson et al. 2020).

Taken together, integration of multiple technologies has enabled an unprecedented understanding of the breast cancer TIME. The increasing availability of these high-dimensional data sets will also continue to fuel the development of visual analysis interfaces and dashboards to further facilitate data sharing and dissemination (Fig. 2). Importantly, iterative analyses of large amounts of data from these studies using sophisticated computational methods, such as machine learning and deep learning, will surely open new avenues for research in the years ahead.

IMMUNE LANDSCAPES OF BREAST CANCER SUBTYPES

Breast cancer has been clinically characterized as estrogen and progesterone receptor-positive (ER^+, PR^+, or hormone receptor-positive, HR^+), human epidermal growth factor receptor 2-positive ($HER2^+$), or the TNBC subset, which lacks expression of ER, PR, and HER2. Examination of gene-expression patterns in resected tumors demonstrated the heterogeneity beyond these classifications and indicated that breast cancer can be divided into at least five prognostically variant subtypes, including luminal-A, luminal-B, basal, $HER2^+$, and normal breast-like (Perou et al. 2000; Sørlie et al. 2003). ER^+ tumors are characterized by a relatively high expression of genes associated with the luminal subtype, while TNBC tumors are mostly basal in nature (Perou et al. 2000). Analyses of breast tumors from patients and preclinical research models have revealed that the tumor subtype affects the composition of immune infiltration. Although the prevalence of TILs is higher in $HER2^+$ and TNBC subtypes (Ali et al. 2014a), TIL infiltration and abundance is a prognostic biomarker for all breast cancer subtypes and an independent predictor of response to neoadjuvant chemotherapy (Loi et al. 2013). Guidelines are now established by the international TILs working group on scoring TILs in breast cancer (El Bairi et al. 2021), and in 2019 the WHO officially listed TILs as one of the biomarkers for the clinicopathological analysis of breast cancer with efforts to include TIL scores into practice to further define disease subsets.

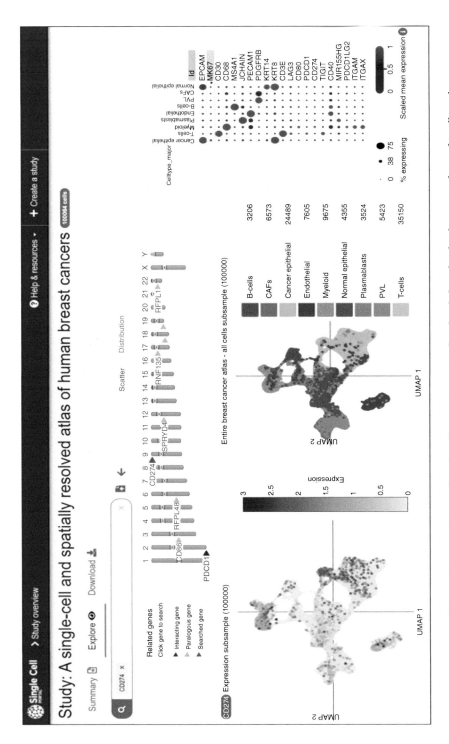

Figure 2. The rapid accumulation of single-cell transcriptomics data has led to the development of user-friendly web interfaces to facilitate data exploration and dissemination. This screenshot was obtained from the Broad Institute's Single Cell Portal (singlecell.broadinstitute.org), which hosts 436 single-cell RNA sequencing data sets at the time of this writing, and at least 10 of these data sets were derived from breast cancer studies. The Single Cell Portal enables users to interact with complex multidimensional data in a point-and-click manner to visualize single-cell clusters and expression of genes of interest. In this example, more than 100,000 cells from 26 different breast tumor samples are shown based on Wu et al. (2021). Such visual analysis solutions are key for bridging the gap between disciplines and increasing data usability.

HR⁺ Breast Cancer

Historically, HR$^+$ breast tumors have been considered immunologically cold as there are few T cells associated with these tumors (Loi et al. 2014). However, other immune cells can be associated with the TIME in HR$^+$ breast cancer. In an analysis of over 11,000 HR$^+$ breast tumors, the immune parameter that correlated most significantly with poor clinical outcome was the presence of TAMs (Ali et al. 2016; Bense et al. 2016). Protumorigenic M2-like TAMs have been shown to induce endocrine therapy resistance in HR$^+$ breast cancer cells in vitro and in vivo through NF-κB- and IL-6-dependent signaling pathways (Castellaro et al. 2019). In contrast, a higher fraction of M1-like TAMs in HR$^+$ breast cancer (which can mediate phagocytosis and cross-presentation of antigen to T cells) correlated with a higher pathologic complete response (pCR) rate as well as prolonged disease-free survival and overall survival (Bense et al. 2016). Macrophage depletion has shown promise in solid tumors including breast cancer (Wesolowski et al. 2019), and identifying strategies to harness the antitumor potential of macrophages may offer new potential opportunities for the treatment of HR$^+$ breast cancer. Due to the low abundance of CD8$^+$ T cells in HR$^+$ tumors, this subtype is unlikely to benefit from current ICB therapies that aim to reinvigorate a previously existing but ineffective T-cell response. However, since dying cells can elicit an enhanced immune response through release of tumor antigens, coupling therapies for HR$^+$ tumors, such as CDK4/6 inhibitors, together with ICB agents may be a useful therapeutic approach (Zhang et al. 2019).

HER2⁺ Breast Cancer

HER2$^+$ breast cancer is a heterogeneous disease with HR$^+$/HER2$^+$ and HR$^-$/HER2$^+$ subsets. Among different therapeutic approaches, immunotherapies represent a relevant option for HER2$^+$ breast cancer patients, both in the adjuvant and metastatic setting. The concept that TILs are a good prognostic indicator for HER2$^+$ breast cancer is generally accepted (Den-

kert et al. 2018), although the median TIL levels are lower in HR$^+$/HER2$^+$ tumors than in HR$^-$/HER2$^+$ tumors (Solinas et al. 2017). Conversely, elevated immunoregulatory cells, such as Tregs and M2-like macrophages, have been associated with poor prognosis (Bates et al. 2006; Mahmoud et al. 2012). HER2-targeting monoclonal antibodies such as trastuzumab and pertuzumab have dramatically improved the clinical outcome in HER2$^+$ breast cancer by, in part, activating the immune response through antibody-dependent cellular cytotoxicity (Demonty et al. 2007). This implies that HER2$^+$ tumors that progress on these now standard-of-care therapies, may already have protumorigenic immune microenvironments. Murine models of HER2$^+$ breast cancer have been useful for understanding the mechanisms of trastuzumab resistance and characterizing the immune cell dynamics during ineffective antitumor immunity. For example, recent work involving mammary-specific expression of Neu, the rat homolog of HER2 (MMTV-Neu model), tested the efficacy of combining trastuzumab and CDK4/6 inhibitors and showed an enrichment of an immunosuppressive myeloid cell population in tumors that are resistant to the therapy (Wang et al. 2019a). In a search to find efficacious drug combinations, another study involving a similar model reported that PD1-blocking antibodies augment anti-HER2 therapy responses (Stagg et al. 2011). Many clinical trials are now underway to test immune therapies in HER2-positive tumors and are covered in companion literature (see Emens and Loi 2023).

TNBC Subtype

TNBC tumors are associated with the highest TILs of all breast cancer subtypes, likely reflecting the higher mutational complexity of TNBC (Karaayvaz et al. 2018; Bao et al. 2021). Although great patient-to-patient heterogeneity exists within TNBC, leukocytes can comprise more than 60% of all cells, and 40%–60% of the tumor immune infiltrate is composed of T cells (Gil Del Alcazar et al. 2017). To define their immunological state, solid tumors have been categorized into three groups: "immune-desert" (cold) tumors largely devoid of lymphocytes, "immune-excluded" where lymphocytes are present in the peritumoral stroma only, and "immune-infil-trated/inflamed" (hot) tumors (Chen and Mellman 2017). Among these, immune-cold breast cancers are characterized by the worst surviv-ability, immune-excluded tumors exhibit somewhat poor prognosis, and the immune-hot tumors are associated with better outcomes (Ali et al. 2014a,b). These observations suggest that where the immune cells are found in the tumor matters, and they provide an important insight as to why immune signatures alone from bulk gene-expression data from tumor biopsies may be confounding (Tofigh et al. 2014; Gruosso et al. 2019). Nevertheless, bulk gene-expression data are more readily available in the literature at this time and efforts to decipher the complexity of TIME in these data sets continue. For instance, a landmark study investigating bulk gene-expression data from The Cancer Genome Atlas (TCGA) across different tumor types defined six unique immune signatures (C1–6) within human cancers (Thorsson et al. 2018). Among these, the basal molecular subtype of breast cancer, which corresponds to the majority of TNBC, was shown to be enriched in C1 (wound healing, 60.5%) and C2 (IFN-γ-dominant, 34%) states (Thorsson et al. 2018), indicating the heterogeneity of the immune infiltration in TNBC.

A spatial examination of the TIME has provided a further understanding of immunological heterogeneity within the tumor architecture of TNBC (Fig. 3), which can have a significant impact on outcomes. Spatial enrichment analyses show "immune-infiltrated/inflamed" TNBC tumors in which $Gzmb^+CD8^+$ cytotoxic T cells are distributed throughout the tumor bed and exhibit a type 1 IFN signature, M1-like macrophages, and elevated expression of immune inhibitory molecules PD-L1 and indoleamine 2,3-dioxygenase (IDO) have a better prognosis than TNBC tumors in which immune cells are separated from the cancer cells (Gruosso et al. 2019). These "immune-excluded" tumors displayed hallmarks of exhausted $CD8^+$ T cells, elevated levels of protumorigenic M2-like macrophages, as well as neutrophils and $IL-17^+$ T cells, which is consistent with the promotion of metastatic spread in the PyMT murine model

of breast cancer by IL-17-expressing γδT cells (Coffelt et al. 2015). However, using imaging mass cytometry and machine learning, another study categorized TNBC tumors into "mixed" and "compartmentalized" subsets in which immune cells are found within the tumor or spatially separated, respectively, and reported that a subset of "compartmentalized" tumors has improved survival (Keren et al. 2018). Notably, the expression patterns of immune checkpoint molecules differed between these spatial configurations. In particular, PD-1/PD-L1 interactions occurred between $CD4^+$ T cells and macrophages in compartmentalized tumors, while they were more frequent between $CD8^+$ CTLs and tumor cells in mixed tumors (Keren et al. 2018). The heterogeneous spatial composition of the TIME and its correlation with differential outcomes were recently shown by the ARTEMIS clinical trial (NCT02276443) designed to assess response to initial neoadjuvant therapy using diagnostic imaging followed by personalized targeted therapy in TNBC. The authors reported a shorter physical distance between malignant cells and $CD3^+$ T cells in patients experiencing pCR compared to patients with residual cancer burden (Yam et al. 2021). Interestingly, patients responding to neoadjuvant chemotherapy tended to have more clonal intratumoral T-cell populations and higher PD-L1 positivity prior to treatment in this study. These data suggest that the cellular and spatial context is an important determinant within the TIME. It is important to note that the choice of the analytical method can have a strong impact on the predictive molecular signatures. At the other end of the spectrum of intratumoral inflammation, "immune-cold" TNBC is characterized by the absence of $CD8^+$ T cells, increased fibrosis, TGF-β-regulated gene-expression signature, M2-polarized macrophages, as well as elevated expression of the immunosuppressive PD-L1 family member, B7-H4, but not PD-L1 (Altan et al. 2018; Gruosso et al. 2019). In contrast, B7-H4 was absent in inflamed PD-L1-positive tumors, suggesting this could be a novel immunotherapy target in TNBC (Gruosso et al. 2019; Li et al. 2019b).

Large-scale immune classification from transcriptomic data from a cohort of 299 patients (METABRIC study) supported the notion that

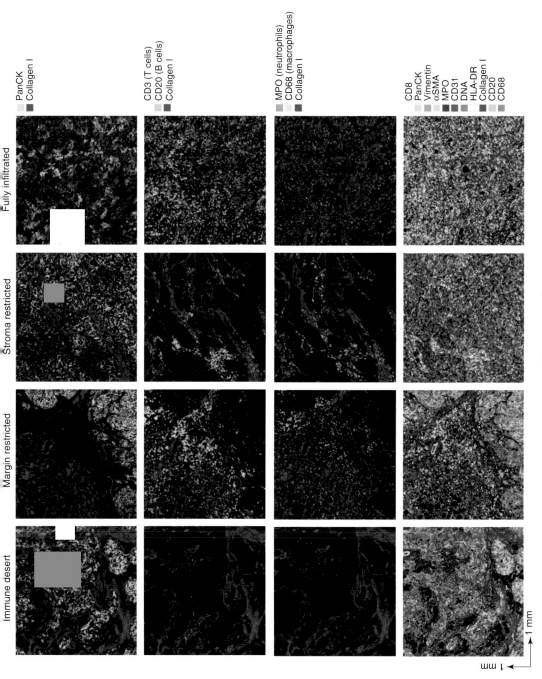

Figure 3. (*See following page for legend.*)

TNBC can be subdivided into these three TIME categories based on immune-specific gene-expression profiles to predict survival and sensitivity to PD-1 blockade (Bareche et al. 2020; Hu et al. 2021). Interestingly, multiple studies reported the presence of tertiary lymphoid structures, organized lymph node–like structures at the peritumoral regions, and their positive association with survival in TNBC (Martinet et al. 2011; Gu-Trantien et al. 2013). Recent studies used imaging mass cytometry to further dissect the features of intratumoral lymphoid and vascular structures and contributed to the complex picture of the TIME in breast cancer (Danenberg et al. 2022).

Despite success in other cancer types, ICB is not yet optimally effective in TNBC, and various strategies have been proposed to boost antitumor immune mechanisms (Vonderheide et al. 2017). Additionally, efforts are underway to more accurately stratify patients with inflamed tumors to obtain better immunotherapy outcomes. Importantly, a recent clinical trial indicated that preconditioning with cisplatin or doxorubicin can lead to up-regulation of inflammatory genes and improve anti-PD-1 responses in TNBC, suggesting that the TIME can be rewired to elicit an effective antitumor state (Voorwerk et al. 2019). Furthermore, several ADCs are currently approved or being evaluated in the clinic for the treatment of breast cancer (Nagayama et al. 2020). In these studies, different types of cytotoxic payloads are conjugated to antibodies against molecular targets, including HER2, TROP-2, LIV-1, and EGFR, that are variably expressed in TNBC. It is possible that increasing tumor cell death using ADCs may modulate the TIME in a way that makes immunotherapy more effective. Taken together, it is becoming increasingly clear that the highly heterogeneous TIME in TNBC provides both opportunities and challenges for reaching the full potential of immunotherapy.

EMERGING OPPORTUNITIES TO MODULATE THE TIME IN BREAST CANCER

Immune Checkpoint Inhibitors

As we better understand the immunoregulatory mechanisms in breast cancer, we are becoming able to devise new therapeutic approaches to improve patient outcomes. The clinical success of ICB therapies such as anti-PD-1 and anti-CTLA-4 in a few cancer types has placed immunotherapy in the center of attention for cancer therapy in general (Callahan et al. 2016). PD-1 and CTLA-4 immune checkpoint receptors are up-regulated in activated T cells and their engagement with the ligands PD-L1/PD-L2 and CD80/CD86, respectively, results in the inhibition of T-cell functions (Wei et al. 2018). Antibody-mediated blockade of these receptors increases T-cell infiltration into tumors and the production of immune effector molecules (Curran et al. 2010). Although ICB can lead to complete tumor clearance in some cases, the majority of patients do not benefit from therapy (Wolchok et al. 2017). In the context of breast cancer, ICB is approved only for TNBC but still has minimal efficacy in that subtype (Villacampa et al. 2022). A phase 1b clinical trial testing the PD-1 inhibitor pembrolizumab in TNBC reported an 18.5% overall response rate, even though patients were enrolled based on PD-L1 expression in the stroma or tumor (Nanda et al. 2016). Among the key challenges of ICB therapy in breast cancer are nonexistent or modest TIL levels in the "immune-cold" tumors, low tumor antigenicity, and the presence of compensatory immunoregulatory mechanisms in the TIME (Vonderheide et al. 2017). There are many current studies now investigating the effects of combining ICB with other anticancer agents and targeting other immune-modulatory processes in breast cancer, some of which are discussed below. For further information on

Figure 3. Highly multiplexed imaging of the breast tumor microenvironment reveals the complexity of the immune infiltration. Representative images from imaging mass cytometry (IMC) on human breast tumor tissues from four patients (one patient per column). Each column depicts a patient tumor (1 mm^2 area) from the indicated histological immune subtype (as defined in Gruosso et al. 2019). Each row represents different combinations of markers within each image (*right* color legends).

clinical breast cancer immunotherapy, we refer the reader to Emens and Loi (2023).

Myeloid-Targeted Therapies

In light of the profound contributions of myeloid cells to the progression and metastasis of solid tumors, there is an emerging interest in developing immunotherapies that target specific subsets of these cells. Notably, CSF-1R inhibitors have been used extensively both in preclinical models and in patients to deplete or reprogram macrophages in cancer. Blockade of CSF-1R signaling has been proven to have multiple benefits, including blunting tumor growth and metastasis (Strachan et al. 2013; Klemm et al. 2021) and enhancing chemotherapeutic response (DeNardo et al. 2011; Ruffell et al. 2014; Olson et al. 2017). Building on these findings, studies have now begun to explore the therapeutic potential of targeting specific macrophage ontogenies in cancer. For example, targeting TAMs in breast cancer models is effective to restore antitumor adaptive immune responses, whereas targeting resident mammary macrophages is not (Franklin et al. 2014). Similarly, anti-CD49d has been identified as a putative approach to target monocyte-derived macrophages in mouse models of breast cancer metastasis to brain, while leaving the nonreplenishing tissue-resident microglia population intact (Bowman et al. 2016). As an alternative approach to depleting specific macrophage populations, studies have also developed methods to reprogram macrophages to adopt an antitumorigenic identity. For example, class IIa HDAC inhibitors can reprogram monocytes and macrophages to limit tumor progression in preclinical breast cancer models, while leaving lymphocyte functions intact (Lobera et al. 2013; Guerriero et al. 2017). Class IIa HDAC inhibition elicits a robust antitumor effect by enhancing macrophage phagocytosis and synergizes with chemotherapy and ICB. These discoveries and others highlight the potential value in reprogramming the immunosuppressive myeloid niche to leverage existing standard-of-care treatments or immunotherapies.

Similarly, there is growing interest in targeting neutrophils in the context of cancer, although there are complex considerations. First, neutropenia is often a consequence of systemic chemotherapy, and it is treated clinically with recombinant G-CSF to boost white blood cell counts. Although necessary to avoid serious risks from infection, it is unclear whether G-CSF supplementation has counterproductive effects in cancer patients by stimulating neutrophil production and activation. Fortunately, retrospective analysis of breast cancer patients with brain metastasis has shown that G-CSF treatment does not exacerbate metastatic disease (Fujii et al. 2021); however, this has yet to be expanded to additional metastatic contexts, particularly lung or liver where the majority of preclinical work has been done. Second, although preclinical models have focused on neutralizing antibodies against the canonical neutrophil marker Ly6G, this causes rapid and robust neutropenia, which is not translatable to patients. Therefore, therapies that target protumorigenic neutrophil functions are receiving more attention, such as NET inhibitors like recombinant DNase1. Although these have not yet been evaluated in cancer patients, emerging clinical trials are now using NET inhibitors for the first time in patients with severe COVID-19 (NCT04409925, NCT04359654, NCT04445285, NCT04541979, NCT04432987, NCT04402944). Finally, given the robust fluctuations in neutrophil frequency and effector function in response to the circadian clock (Casanova-Acebes et al. 2018; Adrover et al. 2019), it is likely that chronopharmacology-based approaches will need to be considered for the development of neutrophil-targeted therapies in the clinical setting.

Emerging Targets and Strategies to Reinvigorate Antitumor Immunity

Resistance to chemotherapy and metastasis are recognized as main factors of mortality in breast cancer. Studies have reported that up to 20% of treatment-naive patients show response to single-agent anti PD-L1 therapy (Adams et al. 2019), which is similar to first-line chemotherapy response rates in TNBC (Li et al. 2019a). This raises an interesting point about potential immune-mediated mechanisms of tumor clearance in chemotherapy. TNBC patients that respond to PD-L1 ICB are predicted to be those that display an inflamed TIME, which reflects 30%–40% of TNBC patients (Bareche et al. 2020; Hu et al. 2021), and it remains to be determined whether specific chemotherapeutic agents can potentiate immunotherapy.

The TIME is a complex network of internal and external components that are sculpted during tumor immune evolution (Finak et al. 2008). Recent multiplex and single-cell analyses of immune microenvironments demonstrate that T cells within the TIME can be found in transitionary activation states as well as terminal differentiation and exhaustion associated with signatures of hypoxia (Azizi et al. 2017). Moreover, many immune cells display covariant expression patterns for the costimulatory and coinhibitory molecules, which suggests targeting one alone may not be a sufficient immunotherapy approach. For example, Foxp3$^+$ Tregs that express coinhibitory CTLA-4 along with other coinhibitory molecules, TIGIT, and the costimulatory receptor GITR have been demonstrated to selectively inhibit proinflammatory Th1 and Th17 but not Th2 responses (Joller et al. 2014). Similarly, macrophage populations are demonstrated along a continuum, with enrichment in M2-like macrophages for the scavenger receptor MARCO and coinhibitory receptor B7-H3 in humans and mouse models (Kos et al. 2022). The B7 family of immune checkpoint molecules, of which PD-L1 is a member (B7-H1), has been recognized as crucial modulators of adaptive immunity. However, members of this family, such as the coinhibitory molecules B7-H3 and B7-H4, can be expressed both in immune and tumor cells and often correlate with immune-cold TNBC subtypes (Altan et al. 2018; Gruosso et al. 2019; Bareche et al. 2020). Thus, these molecules may both promote tumor-intrinsic mechanisms of progression in addition to stimulating an immunosuppressive tumor microenvironment (Podojil and Miller 2017).

In addition to complexity within the immune cell subsets, studies have shown that human breast cancer cells can aberrantly express genes classically thought to be immune cell–specific, such as components of class-II antigen presentation machinery (MHC-II), and this can correlate with favorable prognosis and long-term disease-free survival (Andres et al. 2016). Mechanistic investigations in mice showed that MHC-II expression in breast cancer cells promote antigen presentation to CD4$^+$ T cells and elicitation of Th1 response leading to tumor rejection (Mortara et al. 2006). Hence, accurately stratifying patients for ICB and improving the therapeutic outcomes require a fuller understanding of the complex TIME in breast cancer patients as well as tumor cell capacity to display immunogenic antigens.

It is increasingly appreciated that targeted therapies can also modulate the TIME and thus can be considered as unorthodox immunotherapeutics (Petroni et al. 2020). For instance, CDK4/6 inhibitors intended for cell-cycle arrest can lead to immunostimulation through multiple mechanisms. One of these mechanisms may involve induction of cancer cell senescence, which is characterized by secretion of chemokines and expression of ligands responsible for activating NK and T cells (Textor et al. 2011; Iannello et al. 2013). CDK4/6 inhibitors can also lead to up-regulation of antigen presentation machinery and improve ICB responses in a senescence-independent manner (Goel et al. 2017; Schaer et al. 2018). In addition to effects on tumor cells, cell-cycle inhibitors can directly regulate immune cells within the TIME. Studies in murine breast and lung cancer models have shown that CDK4/6 inhibition can block Treg functions and enhance antitumor T-cell responses (Deng et al. 2018; Schaer et al. 2018). These findings indicate that targeted therapeutics have multiple mechanisms of action and can potentially be combined with other immunotherapeutics for improved outcomes in cancer.

Considerations for Clinical Translation of Emerging TIME Therapies

As emerging immunotherapies continue their development in the preclinical setting, an important hurdle that we will face is the practicality of translating treatments to the clinic where patients are immunologically diverse. This is akin to comparative studies conducted between wild mice and laboratory mice, which demonstrated a striking enhancement in immune activity in wild mice driven by a high degree of environmental pathogen exposure (Abolins et al. 2017). These seminal discoveries called into question the relevance of laboratory mice to study immunological diseases, where experimental conditions are more sterile and all physiologic variation is tightly controlled (e.g., diet, exercise, temperature, genetics, etc). Indeed, recent clinical discoveries have shed light on how immunological diversity among humans impacts cancer immunotherapy. For example, the gut microbiome, which is heavily influenced by diet and other lifestyle factors, is emerging as a major regulator of ICB efficacy both in mouse (Gopalakrishnan et al. 2018; Matson et al. 2018; Routy et al. 2018) and human (McQuade et al. 2018; Baruch et al. 2021) disease settings. Similarly, preclinical and epidemiological studies have shown that obesity is associated with an unexpected improvement in response to ICB (McQuade et al. 2018; Wang et al. 2019b; Kichenadasse et al. 2020), owing to its chronic inflammatory effects that are otherwise usually harmful. Similar observations have been made in aging studies, where it has been discovered that older melanoma patients respond to ICB better than younger patients, which may be due to a higher CD8$^+$/Treg cell ratio (Kugel et al. 2018; Fane and Weeraratna 2020). Additionally, various aspects of lifestyle, such as stress or circadian disturbances, also affect immunological responses and cancer outcomes. For example, stress hormones can lead to activation of neutrophils, which can awaken dormant tumor cells and potentially act as a trigger for recurrence (Perego et al. 2020). Taken together, these findings raise the question of whether epigenetic reprogramming of immune cells via trained immunity—uniquely shaped by a lifetime of pathogen exposures, vaccinations, chronic diseases, diet, and other lifestyle factors within each person—explains why differential responses to cancer immunotherapies have been so challenging to predict at the individual level (Christ et al. 2018; Kaufmann et al. 2018; Mitroulis et al. 2018; Chavakis et al. 2019; Kalafati et al. 2020; Netea et al. 2020). They also highlight the importance of understanding disparities in cancer outcomes through an intersectional lens, by taking into consideration age, gender identity, racialization, weight, socioeconomic status, and other factors that may influence treatment access and efficacy (Murthy et al. 2004; Brady and Weeraratna 2020; Carpten et al. 2021).

FUTURE DIRECTIONS

Developing technologies have provided us with an unprecedented view of the breast cancer TIME. Not surprisingly, as we examine more patient samples at an ever-increasing resolution, we are discovering previously unappreciated complexities of breast cancer and how the immune system participates in disease control and progression. Although bulk transcriptomic and genomic analyses have equipped us with essential insights about disease mechanisms, the tumor immunology arena is undoubtedly shifting toward interrogating the TIME at single-cell resolution in situ (Giesen et al. 2014; Ali et al. 2020; Jackson et al. 2020; Andersson et al. 2021; Wu et al. 2021; Danenberg et al. 2022), and in three dimensions (Kuett et al. 2022). Technology advancements are now building toward four dimensions (Crainiciuc et al. 2022). Such approaches will reveal critical interactions between tumor and immune cells, which confer differential disease outcomes. Notably, the analysis of high-dimensional data obtained in these experiments requires the development of sophisticated algorithms that necessitate the integration of artificial intelligence and machine learning into biomedical research (Butler et al. 2018; Chen et al. 2018; Bao et al. 2021). On this front, the availability of a wide collection of large data sets of patient samples with clinical outcome information will be important for revealing the strongest predictors as well as validating molecular targets to effectively combat breast cancer. In addition to examining the TIME, more sophisticated assessment of corresponding systemic factors such as serum biomarkers can further help distinguish patient subsets with different risk profiles. Defining the key features of the local and systemic immune dysregulation in cancer can also help us classify the disease into differentially actionable groups, prioritize treatment, and aid clinical trial design.

Complementing clinical research initiatives, it remains important to study breast cancer in preclinical animal models where mechanistic experimentation is possible. Many of the key immune cells and mechanisms are conserved between mouse and human, as suggested by comparable antitumor immune responses during CTLA-4 blockade between the two species (Leach et al. 1996; Wolchok et al. 2017). This mechanistic conservation was further supported by a study comparing human and mouse immune responses during sepsis, although this study also identified some species-specific effects in inflammation (Godec et al. 2016). Thus, differences between mouse and human immune systems must be considered when interpreting findings from preclinical studies. For example, there may be value in incorporating more immunological diverse preclinical models in cancer research, such as outbred or wild mice, rather than relying on inbred pathogen-free models (Abolins et al. 2017). A study involving outbred mice that express human HER2 recently revealed that MHC-IB-dependent NK cell responses can counteract Treg functions (Wei et al. 2020). Alternatively, to develop clinically relevant research models, other groups have created immune-humanized mice in which tumors are established simultaneously with the human immune system in immunocompromised mice. Humanized mice can be created through multiple methods including injecting mature immune cells, or transferring hematopoietic stem cells alone or in combination with supportive cytokines and lymphoid organ fragments (Covassin et al. 2013; Katano et al. 2015; Jespersen et al. 2017; Scherer et al. 2021). While they can be overly reductionist, immune-humanized mouse models are valuable for studying specific cellular interactions and answering precisely defined research questions. Since immune-humanized mouse models offer a relatively tractable system for experimental manipulations, they can be useful for testing new therapies. Taken together, validating the findings from preclinical models will continue to be important for the next generation of discoveries in breast cancer immunology.

Finally, we would like to provide a few humble opinions on how to further increase the im-

pact of breast cancer research. First, as almost all approved therapies have been developed as a result of investigating the basic mechanisms that underpin biology, we are confident that our ever-increasing collective knowledge will continue to be a driver of new advancements in the clinic. Notably, the most effective research initiatives are the ones combining sophisticated preclinical models with clinical research infrastructures to establish a translational research ecosystem. Thus, it is essential to foster connections between the bench and the bedside by establishing environments conducive to collaborations. Second, expanding biobanks with patient samples and data collected throughout the spectrum of clinical care will be essential for uncovering yet unknown mechanisms of immune dysfunction in cancer. However, as new approaches and technologies emerge, we may have to rethink how the patient tissues are acquired and stored for future analyses. Traditionally, pathological assessments and bulk transcriptomic/genomic analyses were performed on formalin-fixed, paraffin-embedded (FFPE) tumor samples. While paraffin embedding of fixed tissues can have advantages for histology and long-term storage, it may not allow interrogating the TIME in its native spatial configuration or at single-cell resolution. It is becoming increasingly clear that both the quality and quantity of the immune infiltrate can have a significant impact on cancer outcomes, and, thus, preserving the features of collected tissues will be important for future investigations. Third, a multiparameter fingerprinting of the host milieu may inform risk assessment and help devise personalized clinical trials. Often, patients are enrolled in clinical trials based on the expression of a specific immunotherapy target in the stroma or the tumor bed. While it is sensible to test an immunotherapy agent on patients with a clear expression of the specific molecular target such as PD-L1, many of these "good candidates" still fail to respond to the therapy. With future research, we may be able to shift from this suboptimal categorization approach to qualifying patients for treatment based on specific molecular and cellular fingerprints to achieve the best therapeutic outcomes. A remaining challenge in this context is the heterogeneity of cancer cells and the infiltrating immune cells, which makes it difficult to find parallels even among a large number of patients. Eventually, this can pose an obstacle for conducting clinical trials at high statistical power, perhaps suggesting new statistical methods might also be needed beyond what has been traditionally used.

Last but not least, clinical research would not be possible without the consent and participation of patients and their families. In this context, a critical responsibility befalls both basic scientists and physicians to educate the public about the importance of basic and translational research and clinical trials. To improve patient participation and compliance, in addition to increasing scientific literacy across all walks of life, we must also recognize preexisting biases and disparities in healthcare access. This means looking at patient populations from a holistic perspective and considering other social determinants of health, including socioeconomic status, racial background, social networks, and environmental factors to improve outcomes. Furthermore, to make the new findings applicable to a wide range of patients, research designs must be diversified to account for the effects of age, race, and gender. Analyzing samples from a diverse group of patients can help us develop more relevant biomarkers and discover new therapeutic targets, which would, in turn, improve overall outcomes by creating positive public feedback for future research initiatives. Taken together, we believe that advanced technologies in basic science and robust clinical research infrastructures will continue to reveal new facets in breast cancer immunology and lead to the development of next-generation therapeutics to improve outcomes in breast cancer.

ACKNOWLEDGMENTS

The authors are grateful for technical support from core facilities from the Rosalind and Morris Goodman Cancer Institute (GCI) and Life Sciences Complex at McGill University, including the Single Cell and Imaging Mass Cytometry Platform (SCIMAP) and Histology core facility. The authors acknowledge financial support from the McGill Interdisciplinary Initiative in Infection and Immunity, the Terry Fox Research Institute (D.F.Q.); the Quebec Cancer Consortium, the Ministère de l'Économie l'Innovation et de l'Énergie du Québec through the Fonds d'accélération des collaborations en santé (M.P., D.F.Q.); Susan G. Komen and Breast Cancer Research Foundations (A.L.W.); and the Scientific and Technological Research Council of Turkey (H.A.E.). We also thank Dr. S.-C. Alicia Lai for the images in Figure 1.

REFERENCES

*Reference is also in this subject collection.

Abolins S, King EC, Lazarou L, Weldon L, Hughes L, Drescher P, Raynes JG, Hafalla JCR, Viney ME, Rile EM. 2017. The comparative immunology of wild and laboratory mice, *Mus musculus domesticus*. *Nat Commun* 8: 1–13. doi:10.1038/ncomms14811

Adams S, Schmid P, Rugo HS, Winer EP, Loirat D, Awada A, Cescon DW, Iwata H, Campone M, Nanda R, et al. 2019. Pembrolizumab monotherapy for previously treated metastatic triple-negative breast cancer: cohort A of the phase II KEYNOTE-086 study. *Ann Oncol* 30: 397–404. doi:10.1093/annonc/mdy517

Adrover JM, Del Fresno C, Crainiciuc G, Cuartero MI, Casanova-Acebes M, Weiss LA, Huerga-Encabo H, Silvestre-Roig C, Rossaint J, Cossío I, et al. 2019. A neutrophil timer coordinates immune defense and vascular protection. *Immunity* 50: 390–402.e10. doi:10.1016/j.immuni.2019.01.002

Adrover JM, Aroca-Crevillén A, Crainiciuc G, Ostos F, Rojas-Vega Y, Rubio-Ponce A, Cilloniz C, Bónzon-Kulichenko E, Calvo E, Rico D, et al. 2020. Programmed "disarming" of the neutrophil proteome reduces the magnitude of inflammation. *Nat Immunol* 21: 135–144. doi:10.1038/s41590-019-0571-2

Albert ML, Sauter B, Bhardwaj N. 1998. Dendritic cells acquire antigen from apoptotic cells and induce class I-restricted CTLs. *Nature* 392: 86–89. doi:10.1038/32183

Albrengues J, Shields MA, Ng D, Park CG, Ambrico A, Poindexter ME, Upadhyay P, Uyemami DL, Pommier A, Küttner V, et al. 2018. Neutrophil extracellular traps produced during inflammation awaken dormant cancer cells in mice. *Science* 361: eaao4227. doi:10.1126/science.aao4227

Ali HR, Provenzano E, Dawson SJ, Blows FM, Liu B, Shah M, Earl HM, Poole CJ, Hiller L, Dunn JA, et al. 2014a. Association between CD8+ T-cell infiltration and breast cancer survival in 12,439 patients. *Ann Oncol* 25: 1536–1543. doi:10.1093/annonc/mdu191

Ali HR, Rueda OM, Chin SF, Curtis C, Dunning MJ, Aparicio SAJR, Caldas C. 2014b. Genome-driven integrated classification of breast cancer validated in over 7,500 samples. *Genome Biol* 15: 431. doi:10.1186/s13059-014-0431-1

Ali HR, Chlon L, Pharoah PDP, Markowetz F, Caldas C. 2016. Patterns of immune infiltration in breast cancer and their clinical implications: a gene-expression-based retrospective study. *PLoS Med* **13**: e1002194. doi:10.1371/journal.pmed.1002194

Ali HR, Jackson HW, Zanotelli VRT, Danenberg E, Fischer JR, Bardwell H, Provenzano E, Rueda OM, Chin S-F, Aparicio S, et al. 2020. Imaging mass cytometry and multiplatform genomics define the phenogenomic landscape of breast cancer. *Nat Cancer* **1**: 163–175. doi:10.1038/s43018-020-0026-6

Altan M, Kidwell KM, Pelekanou V, Carvajal-Hausdorf DE, Schalper KA, Toki MI, Thomas DG, Sabel MS, Hayes DF, Rimm DL. 2018. Association of B7-H4, PD-L1, and tumor infiltrating lymphocytes with outcomes in breast cancer. *NPJ Breast Cancer* **4**: 40. doi:10.1038/s41523-018-0095-1

Amano SU, Cohen JL, Vangala P, Tencerova M, Nicoloro SM, Yawe JC, Shen Y, Czech MP, Aouadi M. 2014. Local proliferation of macrophages contributes to obesity-associated adipose tissue inflammation. *Cell Metab* **19**: 162–171. doi:10.1016/j.cmet.2013.11.017

Anderson DA III, Murphy KM, Briseño CG, Leonard WJ, Schreiber RD. 2017. Development, diversity, and function of dendritic cells in mouse and human. *Cold Spring Harb Perspect Biol* **10**: a028613.

Andersson A, Larsson L, Stenbeck L, Salmén F, Ehinger A, Wu SZ, Al-Eryani G, Roden D, Swarbrick A, Borg Å, et al. 2021. Spatial deconvolution of HER2-positive breast cancer delineates tumor-associated cell type interactions. *Nat Commun* **12**: 6012. doi:10.1038/s41467-021-26271-2

Andres F, Yufeng L, Dongquan C, William EG, Katherine LU, Natalie DM, Erinn DK, Todd CB, Christos V, Donald JB, et al. 2016. Expression of the MHC class II pathway in triple-negative breast cancer tumor cells is associated with a good prognosis and infiltrating lymphocytes. *Cancer Immunol Res* **4**: 390–399. doi:10.1158/2326-6066.CIR-15-0243

Auslander N, Zhang G, Lee JS, Frederick DT, Miao B, Moll T, Tian T, Wei Z, Madan S, Sullivan RJ, et al. 2018. Robust prediction of response to immune checkpoint blockade therapy in metastatic melanoma. *Nat Med* **24**: 1545–1549. doi:10.1038/s41591-018-0157-9

Azizi E, Carr AJ, Plitas G, Cornish AE, Konopacki C, Prabhakaran S, Nainys J, Wu K, Kiseliovas V, Setty M, et al. 2017. Single-cell map of diverse immune phenotypes driven by the tumor microenvironment. *Cell* **174**: 221994.

Ballesteros I, Rubio-Ponce A, Genua M, Lusito E, Kwok I, Fernández-Calvo G, Khoyratty TE, van Grinsven E, González-Hernández S, Nicolás-Ávila JA, et al. 2020. Co-option of neutrophil fates by tissue environments. *Cell* **183**: 1282–1297.e18. doi:10.1016/j.cell.2020.10.003

Balmer ML, Schürch CM, Saito Y, Geuking MB, Li H, Cuenca M, Kovtonyuk LV, McCoy KD, Hapfelmeier S, Ochsenbein AF, et al. 2014. Microbiota-derived compounds drive steady-state granulopoiesis via MyD88/TICAM signaling. *J Immunol* **193**: 5273–5283. doi:10.4049/jimmunol.1400762

Bao X, Shi R, Zhao T, Wang Y, Anastasov N, Rosemann M, Fang W. 2021. Integrated analysis of single-cell RNA-seq and bulk RNA-seq unravels tumour heterogeneity plus M2-like tumour-associated macrophage infiltration and aggressiveness in TNBC. *Cancer Immunol Immunother* **70**: 189–202. doi:10.1007/s00262-020-02669-7

Bareche Y, Buisseret L, Gruosso T, Girard E, Venet D, Dupont F, Desmedt C, Larsimont D, Park M, Rothé F, et al. 2020. Unraveling triple-negative breast cancer tumor microenvironment heterogeneity: towards an optimized treatment approach. *J Natl Cancer Inst* **112**: 708–719. doi:10.1093/jnci/djz208

Barry M, Bleackley RC. 2002. Cytotoxic T lymphocytes: all roads lead to death. *Nat Rev Immunol* **2**: 401–409. doi:10.1038/nri819

Baruch EN, Youngster I, Ben-Betzalel G, Ortenberg R, Lahat A, Katz L, Adler K, Dick-Necula D, Raskin S, Bloch N, et al. 2021. Fecal microbiota transplant promotes response in immunotherapy-refractory melanoma patients. *Science* **371**: 602–609. doi:10.1126/science.abb5920

Bates GJ, Fox SB, Han C, Leek RD, Garcia JF, Harris AL, Banham AH. 2006. Quantification of regulatory T cells enables the identification of high-risk breast cancer patients and those at risk of late relapse. *J Clin Oncol* **24**: 5373–5380. doi:10.1200/JCO.2006.05.9584

Bell D, Chomarat P, Broyles D, Netto G, Harb GM, Lebecque S, Valladeau J, Davoust J, Palucka KA, Banchereau J. 1999. In breast carcinoma tissue, immature dendritic cells reside within the tumor, whereas mature dendritic cells are located in peritumoral areas. *J Exp Med* **190**: 1417–1426. doi:10.1084/jem.190.10.1417

Bense RD, Sotiriou C, Piccart-Gebhart MJ, Haanen JBAG, van Vugt MATM, de Vries EGE, Schroder CP, Fehrmann RSN. 2016. Relevance of tumor-infiltrating immune cell composition and functionality for disease outcome in breast cancer. *J Natl Cancer Inst* **109**: djw192. doi:10.1093/jnci/djw192

Blazkova J, Gupta S, Liu Y, Gaudilliere B, Ganio EA, Bolen CR, Saar-Dover R, Fragiadakis GK, Angst MS, Hasni S, et al. 2017. Multicenter systems analysis of human blood reveals immature neutrophils in males and during pregnancy. *J Immunol* **198**: 2479–2488. doi:10.4049/jimmunol.1601855

Bommareddy PK, Patel A, Hossain S, Kaufman HL. 2017. Talimogene laherparepvec (T-VEC) and other oncolytic viruses for the treatment of melanoma. *Am J Clin Dermatol* **18**: 1–15. doi:10.1007/s40257-016-0238-9

Bowman RL, Klemm F, Akkari L, Pyonteck SM, Sevenich L, Quail DF, Dhara S, Simpson K, Gardner EE, Iacobuzio-Donahue CA, et al. 2016. Macrophage ontogeny underlies differences in tumor-specific education in brain malignancies. *Cell Rep* **17**: 2445–2459. doi:10.1016/j.celrep.2016.10.052

Brady DC, Weeraratna AT. 2020. The race toward equity: increasing racial diversity in cancer research and cancer care. *Cancer Discov* **10**: 1451–1454. doi:10.1158/2159-8290.CD-20-1193

Buchbinder EI, Desai A. 2016. CTLA-4 and PD-1 pathways: similarities, differences, and implications of their inhibition. *Am J Clin Oncol* **39**: 98–106. doi:10.1097/COC.0000000000000239

Butler A, Hoffman P, Smibert P, Papalexi E, Satija R. 2018. Integrating single-cell transcriptomic data across different conditions, technologies, and species. *Nat Biotechnol* **36**: 411–420. doi:10.1038/nbt.4096

Cabeza-Cabrerizo M, Cardoso A, Minutti CM, Pereira da Costa M, Reis e Sousa C. 2021. Dendritic cells revisited. *Annu Rev Immunol* **39:** 131–166.

Callahan MK, Postow MA, Wolchok JD. 2016. Targeting T cell co-receptors for cancer therapy. *Immunity* **44:** 1069–1078. doi:10.1016/j.immuni.2016.04.023

Carmona-Fontaine C, Deforet M, Akkari L, Thompson CB, Joyce JA, Xavier JB. 2017. Metabolic origins of spatial organization in the tumor microenvironment. *Proc Natl Acad Sci* **114:** 2934–2939. doi:10.1073/pnas.1700600114

Carpten JD, Fashoyin-Aje L, Garraway LA, Winn R. 2021. Making cancer research more inclusive. *Nat Rev Cancer* **21:** 613–618. doi:10.1038/s41568-021-00369-7

Carter JM, Hoskin TL, Pena MA, Brahmbhatt R, Winham SJ, Frost MH, Stallings-Mann M, Radisky DC, Knutson KL, Visscher DW, et al. 2018. Macrophagic "crown-like structures" are associated with an increased risk of breast cancer in benign breast disease. *Cancer Prev Res (Phila)* **11:** 113–119. doi:10.1158/1940-6207.CAPR-17-0245

Casanova-Acebes M, Pitaval C, Weiss LA, Nombela-Arrieta C, Chèvre R, A-González N, Kunisaki Y, Zhang D, van Rooijen N, Silberstein LE, et al. 2013. Rhythmic modulation of the hematopoietic niche through neutrophil clearance. *Cell* **153:** 1025–1035. doi:10.1016/j.cell.2013.04.040

Casanova-Acebes M, Nicolás-Ávila JA, Li JL, García-Silva S, Balachander A, Rubio-Ponce A, Weiss LA, Adrover JM, Burrows K, A-González N, et al. 2018. Neutrophils instruct homeostatic and pathological states in naive tissues. *J Exp Med* **215:** 2778–2795. doi:10.1084/jem.20181468

Casbon AJ, Reynaud D, Park C, Khuc E, Gan DD, Schepers K, Passegué E, Werb Z. 2015. Invasive breast cancer reprograms early myeloid differentiation in the bone marrow to generate immunosuppressive neutrophils. *Proc Natl Acad Sci* **112:** E566–E575. doi:10.1073/pnas.1424927112

Castellaro AM, Rodriguez-Baili MC, di Tada CE, Gil GA. 2019. Tumor-associated macrophages induce endocrine therapy resistance in ER⁺ breast cancer cells. *Cancers (Basel)* **11:** 189. doi:10.3390/cancers11020189

Chan IS, Knútsdóttir H, Ramakrishnan G, Padmanaban V, Warrier M, Ramirez JC, Dunworth M, Zhang H, Jaffee EM, Bader JS, et al. 2020. Cancer cells educate natural killer cells to a metastasis-promoting cell state. *J Cell Biol* **219:** e202001134. doi:10.1083/jcb.202001134

Charoentong P, Finotello F, Angelova M, Mayer C, Efremova M, Rieder D, Hackl H, Trajanoski Z. 2017. Pan-cancer immunogenomic analyses reveal genotype-immunophenotype relationships and predictors of response to checkpoint blockade. *Cell Rep* **18:** 248–262. doi:10.1016/j.celrep.2016.12.019

Chavakis T, Mitroulis I, Hajishengallis G. 2019. Hematopoietic progenitor cells as integrative hubs for adaptation to and fine-tuning of inflammation. *Nat Immunol* **20:** 802–811. doi:10.1038/s41590-019-0402-5

Chen DS, Mellman I. 2017. Elements of cancer immunity and the cancer-immune set point. *Nature* **541:** 321–330. doi:10.1038/nature21349

Chen Z, Quan L, Huang A, Zhao Q, Yuan Y, Yuan X, Shen Q, Shang J, Ben Y, Qin FXF, et al. 2018. seq-ImmuCC: cell-centric view of tissue transcriptome measuring cellular compositions of immune microenvironment from mouse RNA-seq data. *Front Immunol* **9:** 1–11. doi:10.3389/fimmu.2018.01286

Christ A, Günther P, Lauterbach MAR, Duewell P, Biswas D, Pelka K, Scholz CJ, Oosting M, Haendler K, Baßler K, et al. 2018. Western diet triggers NLRP3-dependent innate immune reprogramming. *Cell* **172:** 162–175.e14. doi:10.1016/j.cell.2017.12.013

Clarke TB, Davis KM, Lysenko ES, Zhou AY, Yu Y, Weiser JN. 2010. Recognition of peptidoglycan from the microbiota by Nod1 enhances systemic innate immunity. *Nat Med* **16:** 228–231. doi:10.1038/nm.2087

Coffelt SB, Kersten K, Doornebal CW, Weiden J, Vrijland K, Hau CS, Verstegen NJM, Ciampricotti M, Hawinkels L, Jonkers J, et al. 2015. IL-17-producing γδ T cells and neutrophils conspire to promote breast cancer metastasis. *Nature* **522:** 345–348. doi:10.1038/nature14282

Collin M, Bigley V. 2018. Human dendritic cell subsets: an update. *Immunology* **154:** 3–20. doi:10.1111/imm.12888

Collin M, Mcgovern N, Haniffa M. 2013. Human dendritic cell subsets. *Immunology* **140:** 22–30. doi:10.1111/imm.12117

Cools-Lartigue J, Spicer J, McDonald B, Gowing S, Chow S, Giannias B, Bourdeau F, Kubes P, Ferri L. 2013. Neutrophil extracellular traps sequester circulating tumor cells and promote metastasis. *J Clin Invest* **123:** 3446–3458. doi:10.1172/JCI67484

Covassin L, Jangalwe S, Jouvet N, Laning J, Burzenski L, Shultz LD, Brehm MA. 2013. Human immune system development and survival of non-obese diabetic (NOD)-scid IL2rγ^null (NSG) mice engrafted with human thymus and autologous haematopoietic stem cells. *Clin Exp Immunol* **174:** 372–388. doi:10.1111/cei.12180

Coventry BJ, Morton J. 2003. CD1a-positive infiltrating-dendritic cell density and 5-year survival from human breast cancer. *Br J Cancer* **89:** 533–538. doi:10.1038/sj.bjc.6601114

Coventry BJ, Lee PL, Gibbs D, Hart DNJ. 2002. Dendritic cell density and activation status in human breast cancer—CD1a, CMRF-44, CMRF-56 and CD-83 expression. *Br J Cancer* **86:** 546–551. doi:10.1038/sj.bjc.6600132

Crainiciuc G, Palomino-Segura M, Molina-Moreno M, Sicilia J, Aragones DG, Li JLY, Madurga R, Adrover JM, Aroca-Crevillén A, Martin-Salamanca S, et al. 2022. Behavioural immune landscapes of inflammation. *Nature* **601:** 415–421. doi:10.1038/s41586-021-04263-y

Cristescu R, Mogg R, Ayers M, Albright A, Murphy E, Yearley J, Sher X, Liu XQ, Lu H, Nebozhyn M, et al. 2018. Pan-tumor genomic biomarkers for PD-1 checkpoint blockade-based immunotherapy. *Science* **362:** eaar3593. doi:10.1126/science.aar3593

Curran MA, Montalvo W, Yagita H, Allison JP. 2010. PD-1 and CTLA-4 combination blockade expands infiltrating T cells and reduces regulatory T and myeloid cells within B16 melanoma tumors. *Proc Natl Acad Sci* **107:** 4275–4280. doi:10.1073/pnas.0915174107

Curtis C, Shah SP, Chin SF, Turashvili G, Rueda OM, Dunning MJ, Speed D, Lynch AG, Samarajiwa S, Yuan Y, et al. 2012. The genomic and transcriptomic architecture of 2,000 breast tumours reveals novel subgroups. *Nature* **486:** 346–352. doi:10.1038/nature10983

Czerniecki BJ, Koski GK, Koldovsky U, Xu S, Cohen PA, Mick R, Nisenbaum H, Pasha T, Xu M, Fox KR, et al.

2007. Targeting HER-2/*neu* in early breast cancer development using dendritic cells with staged interleukin-12 burst secretion. *Cancer Res* **67:** 1842–1852. doi:10.1158/0008-5472.CAN-06-4038

Danenberg E, Bardwell H, Zanotelli VRT, Provenzano E, Chin SF, Rueda OM, Green A, Rakha E, Aparicio S, Ellis IO, et al. 2022. Breast tumor microenvironment structures are associated with genomic features and clinical outcome. *Nat Genet* **54:** 660–669. doi:10.1038/s41588-022-01041-y

Dawson CA, Pal B, Vaillant F, Gandolfo LC, Liu Z, Bleriot C, Ginhoux F, Smyth GK, Lindeman GJ, Mueller SN, et al. 2020. Tissue-resident ductal macrophages survey the mammary epithelium and facilitate tissue remodelling. *Nat Cell Biol* **22:** 546–558. doi:10.1038/s41556-020-0505-0

deMagalhaes-Silverman M, Donnenberg A, Lembersky B, Elder E, Lister J, Rybka W, Whiteside T, Ball E. 2000. Posttransplant adoptive immunotherapy with activated natural killer cells in patients with metastatic breast cancer. *J Immunother* **23:** 154–160. doi:10.1097/00002371-200001000-00018

Demonty G, Bernard-Marty C, Puglisi F, Mancini I, Piccart M. 2007. Progress and new standards of care in the management of HER-2 positive breast cancer. *Eur J Cancer* **43:** 497–509. doi:10.1016/j.ejca.2006.10.020

DeNardo DG, Brennan DJ, Rexhepaj E, Ruffell B, Shiao SL, Madden SF, Gallagher WM, Wadhwani N, Keil SD, Junaid SA, et al. 2011. Leukocyte complexity predicts breast cancer survival and functionally regulates response to chemotherapy. *Cancer Discov* **1:** 54–67. doi:10.1158/2159-8274.CD-10-0028

Deng J, Wang ES, Jenkins RW, Li S, Dries R, Yates K, Chhabra S, Huang W, Liu H, Aref AR, et al. 2018. CDK4/6 inhibition augments antitumor immunity by enhancing T-cell activation. *Cancer Discov* **8:** 216–233. doi:10.1158/2159-8290.CD-17-0915

Denkert C, von Minckwitz G, Darb-Esfahani S, Lederer B, Heppner BI, Weber KE, Budczies J, Huober J, Klauschen F, Furlanetto J, et al. 2018. Tumour-infiltrating lymphocytes and prognosis in different subtypes of breast cancer: a pooled analysis of 3771 patients treated with neoadjuvant therapy. *Lancet Oncol* **19:** 40–50. doi:10.1016/S1470-2045(17)30904-X

Deshmukh HS, Liu Y, Menkiti OR, Mei J, Dai N, O'Leary CE, Oliver PM, Kolls JK, Weiser JN, Worthen GS. 2014. The microbiota regulates neutrophil homeostasis and host resistance to *Escherichia coli* K1 sepsis in neonatal mice. *Nat Med* **20:** 524–530. doi:10.1038/nm.3542

Drobits B, Holcmann M, Amberg N, Swiecki M, Grundtner R, Hammer M, Colonna M, Sibilia M. 2012. Imiquimod clears tumors in mice independent of adaptive immunity by converting pDCs into tumor-killing effector cells. *J Clin Invest* **122:** 575–585. doi:10.1172/JCI61034

El Bairi K, Haynes HR, Blackley E, Fineberg S, Shear J, Turner S, de Freitas JR, Sur D, Amendola LC, Gharib M, et al. 2021. The tale of TILs in breast cancer: a report from the international immuno-oncology biomarker working group. *NPJ Breast Cancer* **7:** 150. doi:10.1038/s41523-021-00346-1

* Emens LA, Loi S. 2023. Immunotherapy approaches for breast cancer patients in 2023. *Cold Spring Harb Perspect Med* doi:10.1101/cshperspect.a041332

Evrard M, Kwok IWH, Chong SZ, Teng KWW, Becht E, Chen J, Sieow JL, Penny HL, Ching GC, Devi S, et al. 2018. Developmental analysis of bone marrow neutrophils reveals populations specialized in expansion, trafficking, and effector functions. *Immunity* **48:** 364–379.e8. doi:10.1016/j.immuni.2018.02.002

Eyles J, Puaux AL, Wang X, Toh B, Prakash C, Hong M, Tan TG, Zheng L, Ong LC, Jin Y, et al. 2010. Tumor cells disseminate early, but immunosurveillance limits metastatic outgrowth, in a mouse model of melanoma. *J Clin Invest* **120:** 2030–2039. doi:10.1172/JCI42002

Fane M, Weeraratna AT. 2020. How the ageing microenvironment influences tumour progression. *Nat Rev Cancer* **20:** 89–106. doi:10.1038/s41568-019-0222-9

Finak G, Bertos N, Pepin F, Sadekova S, Souleimanova M, Zhao H, Chen H, Omeroglu G, Meterissian S, Omeroglu A, et al. 2008. Stromal gene expression predicts clinical outcome in breast cancer. *Nat Med* **14:** 518–527. doi:10.1038/nm1764

Franklin RA, Liao W, Sarkar A, Kim MV, Bivona MR, Liu K, Pamer EG, Li MO. 2014. The cellular and molecular origin of tumor-associated macrophages. *Science* **344:** 921–925. doi:10.1126/science.1252510

Frazao A, Messaoudene M, Nunez N, Dulphy N, Roussin F, Sedlik C, Zitvogel L, Piaggio E, Toubert A, Caignard A. 2019. CD16$^+$NKG2A^{high} natural killer cells infiltrate breast cancer–draining lymph nodes. *Cancer Immunol Res* **7:** 208–218. doi:10.1158/2326-6066.CIR-18-0085

Fujii T, Rehman H, Chung SY, Shen J, Newman J, Wu V, Hines A, Azimi-Nekoo E, Fayyaz F, Lee M, et al. 2021. Treatment with granulocyte-colony stimulating factor (G-CSF) is not associated with increased risk of brain metastasis in patients with de novo stage IV breast cancer. *J Cancer* **12:** 5687–5692. doi:10.7150/jca.63159

Geissmann F, Jung S, Littman DR. 2003. Blood monocytes consist of two principal subsets with distinct migratory properties. *Immunity* **19:** 71–82. doi:10.1016/S1074-7613(03)00174-2

Gervais A, Levêque J, Bouet-Toussaint F, Burtin F, Lesimple T, Sulpice L, Patard JJ, Genetet N, Catros-Quemener V. 2005. Dendritic cells are defective in breast cancer patients: a potential role for polyamine in this immunodeficiency. *Breast Cancer Res* **7:** 1–10. doi:10.1186/bcr1001

Giesen C, Wang HAO, Schapiro D, Zivanovic N, Jacobs A, Hattendorf B, Schüffler PJ, Grolimund D, Buhmann JM, Brandt S, et al. 2014. Highly multiplexed imaging of tumor tissues with subcellular resolution by mass cytometry. *Nat Methods* **11:** 417–422. doi:10.1038/nmeth.2869

Gil Del Alcazar CR, Huh SJ, Ekram MB, Trinh A, Liu LL, Beca F, Zi X, Kwak M, Bergholtz H, Su Y, et al. 2017. Immune escape in breast cancer during in situ to invasive carcinoma transition. *Cancer Discov* **7:** 1098–1115. doi:10.1158/2159-8290.CD-17-0222

Gilmore AC, Flaherty SJ, Somasundaram V, Scheiblin DA, Lockett SJ, Wink DA, Heinz WF. 2021. An in vitro tumorigenesis model based on live-cell-generated oxygen and nutrient gradients. *Commun Biol* **4:** 477. doi:10.1038/s42003-021-01954-0

Ginhoux F, Guilliams M. 2016. Tissue-resident macrophage ontogeny and homeostasis. *Immunity* **44:** 439–449. doi:10.1016/j.immuni.2016.02.024

Ginhoux F, Schultze JL, Murray PJ, Ochando J, Biswas SK. 2016. New insights into the multidimensional concept of macrophage ontogeny, activation and function. *Nat Immunol* **17:** 34–40. doi:10.1038/ni.3324

Giorello MB, Matas A, Marenco P, Davies KM, Borzone FR, de Calcagno ML, García-Rivello H, Wernicke A, Martinez LM, Labovsky V, et al. 2021. CD1a- and CD83-positive dendritic cells as prognostic markers of metastasis development in early breast cancer patients. *Breast Cancer (Auckl)* **28:** 1328–1339. doi:10.1007/s12282-021-01270-9

Godec J, Tan Y, Liberzon A, Tamayo P, Bhattacharya S, Butte AJ, Mesirov JP, Haining WN. 2016. Compendium of immune signatures identifies conserved and species-specific biology in response to inflammation. *Immunity* **44:** 194–206. doi:10.1016/j.immuni.2015.12.006

Goel S, Decristo MJ, Watt AC, Brinjones H, Sceneay J, Li BB, Khan N, Ubellacker JM, Xie S, Metzger-Filho O, et al. 2017. CDK4/6 inhibition triggers anti-tumour immunity. *Nature* **548:** 471–475. doi:10.1038/nature23465

Gómez-Aleza C, Nguyen B, Yoldi G, Ciscar M, Barranco A, Hernández-Jiménez E, Maetens M, Salgado R, Zafeiroglou M, Pellegrini P, et al. 2020. Inhibition of RANK signaling in breast cancer induces an anti-tumor immune response orchestrated by CD8+ T cells. *Nat Commun* **11:** 1–18. doi:10.1038/s41467-020-20138-8

Gopalakrishnan V, Spencer CN, Nezi L, Reuben A, Andrews MC, Karpinets TV, Prieto PA, Vicente D, Hoffman K, Wei SC, et al. 2018. Gut microbiome modulates response to anti-PD-1 immunotherapy in melanoma patients. *Science* **359:** 97–103. doi:10.1126/science.aan4236

Gruosso T, Gigoux M, Manem VSK, Bertos N, Zuo D, Perlitch I, Saleh SMI, Zhao H, Souleimanova M, Johnson RM, et al. 2019. Spatially distinct tumor immune microenvironments stratify triple-negative breast cancers. *J Clin Invest* **129:** 1785–1800. doi:10.1172/JCI96313

Guerriero JL, Sotayo A, Ponichtera HE, Castrillon JA, Pourzia AL, Schad S, Johnson SF, Carrasco RD, Lazo S, Bronson RT, et al. 2017. Class IIa HDAC inhibition reduces breast tumours and metastases through anti-tumour macrophages. *Nature* **543:** 428–432. doi:10.1038/nature21409

Guillerey C, Huntington ND, Smyth MJ. 2016. Targeting natural killer cells in cancer immunotherapy. *Nat Immunol* **17:** 1025–36. doi:10.1038/ni.3518

Guilliams M, Dutertre CA, Scott CL, McGovern N, Sichien D, Chakarov S, van Gassen S, Chen J, Poidinger M, de Prijck S, et al. 2016. Unsupervised high-dimensional analysis aligns dendritic cells across tissues and species. *Immunity* **45:** 669–684. doi:10.1016/j.immuni.2016.08.015

Gupta S, Nakabo S, Blanco LP, O'Neil LJ, Wigerblad G, Goel RR, Mistry P, Jiang K, Carmona-Rivera C, Chan DW, et al. 2020. Sex differences in neutrophil biology modulate response to type I interferons and immunometabolism. *Proc Natl Acad Sci* **117:** 16481–16491. doi:10.1073/pnas.2003603117

Gu-Trantien C, Loi S, Garaud S, Equeter C, Libin M, De Wind A, Ravoet M, Le Buanec H, Sibille C, Manfouo-Foutsop G, et al. 2013. CD4+ follicular helper T cell infil-tration predicts breast cancer survival. *J Clin Invest* **123:** 2873–2892. doi:10.1172/JCI67428

Hazeldine J, Harris P, Chapple IL, Grant M, Greenwood H, Livesey A, Sapey E, Lord JM. 2014. Impaired neutrophil extracellular trap formation: a novel defect in the innate immune system of aged individuals. *Aging Cell* **13:** 690–698. doi:10.1111/acel.12222

Hedrick CC, Malanchi I. 2022. Neutrophils in cancer: heterogeneous and multifaceted. *Nat Rev Immunol* **22:** 173–187. doi:10.1038/s41577-021-00571-6

Hill DA, Lim HW, Kim YH, Ho WY, Foong YH, Nelson VL, Nguyen HCB, Chegireddy K, Kim J, Habertheuer A, et al. 2018. Distinct macrophage populations direct inflammatory versus physiological changes in adipose tissue. *Proc Natl Acad Sci* **115:** E5096–E5105.

Hosseinzadeh A, Thompson PR, Segal BH, Urban CF. 2016. Nicotine induces neutrophil extracellular traps. *J Leukoc Biol* **100:** 1105–1112. doi:10.1189/jlb.3AB0815-379RR

Hsu BE, Tabariès S, Johnson RM, Andrzejewski S, Senecal J, Lehuédè C, Annis MG, Ma EH, Völs S, Ramsay L, et al. 2019. Immature low-density neutrophils exhibit metabolic flexibility that facilitates breast cancer liver metastasis. *Cell Rep* **27:** 3902–3915.e6. doi:10.1016/j.celrep.2019.05.091

Hu S, Qu X, Jiao Y, Hu J, Wang B. 2021. Immune classification and immune landscape analysis of triple-negative breast cancer. *Front Genet* **12:** 2154.

Huang Y, Ma C, Zhang Q, Ye J, Wang F, Zhang Y, Hunborg P, Varvares MA, Hoft DF, Hsueh EC, et al. 2015. CD4+ and CD8+ T cells have opposing roles in breast cancer progression and outcome. *Oncotarget* **6:** 17462–17478. doi:10.18632/oncotarget.3958

Iannello A, Thompson TW, Ardolino M, Lowe SW, Raulet DH. 2013. p53-dependent chemokine production by senescent tumor cells supports NKG2D-dependent tumor elimination by natural killer cells. *J Exp Med* **210:** 2057–2069. doi:10.1084/jem.20130783

Iyengar NM, Zhou XK, Gucalp A, Morris PG, Howe LR, Giri DD, Morrow M, Wang H, Pollak M, Jones LW, et al. 2016. Systemic correlates of white adipose tissue inflammation in early-stage breast cancer. *Clin Cancer Res* **22:** 2283–2289. doi:10.1158/1078-0432.CCR-15-2239

Jackson HW, Fischer JR, Zanotelli VRT, Ali HR, Mechera R, Soysal SD, Moch H, Muenst S, Varga Z, Weber WP, et al. 2020. The single-cell pathology landscape of breast cancer. *Nature* **578:** 615–620. doi:10.1038/s41586-019-1876-x

Jaitin DA, Adlung L, Thaiss CA, Weiner A, Li B, Descamps H, Lundgren P, Bleriot C, Liu Z, Deczkowska A, et al. 2019. Lipid-associated macrophages control metabolic homeostasis in a Trem2-dependent manner. *Cell* **178:** 686–698.e14. doi:10.1016/j.cell.2019.05.054

Jespersen H, Lindberg MF, Donia M, Söderberg EMV, Andersen R, Keller U, Ny L, Svane IM, Nilsson LM, Nilsson JA. 2017. Clinical responses to adoptive T-cell transfer can be modeled in an autologous immune-humanized mouse model. *Nat Commun* **8:** 1–10. doi:10.1038/s41467-017-00786-z

Joffre OP, Segura E, Savina A, Amigorena S. 2012. Cross-presentation by dendritic cells. *Nat Rev Immunol* **12:** 557–569. doi:10.1038/nri3254

Joller N, Lozano E, Burkett PR, Patel B, Xiao S, Zhu C, Xia J, Tan TG, Sefik E, Yajnik V, et al. 2014. Treg cells expressing the coinhibitory molecule TIGIT selectively inhibit proinflammatory Th1 and Th17 cell responses. *Immunity* **40:** 569–581. doi:10.1016/j.immuni.2014.02.012

Kalafati L, Kourtzelis I, Schulte-Schrepping J, Li X, Hatzioannou A, Grinenko T, Hagag E, Sinha A, Has C, Dietz S, et al. 2020. Innate immune training of granulopoiesis promotes anti-tumor activity. *Cell* **183:** 771–785.e12. doi:10.1016/j.cell.2020.09.058

Kantoff PW, Higano CS, Shore ND, Berger ER, Small EJ, Penson DF, Redfern CH, Ferrari AC, Dreicer R, Sims RB, et al. 2010. Sipuleucel-T immunotherapy for castration-resistant prostate cancer. *N Engl J Med* **363:** 411–422. doi:10.1056/NEJMoa1001294

Karaayvaz M, Cristea S, Gillespie SM, Patel AP, Mylvaganam R, Luo CC, Specht MC, Bernstein BE, Michor F, Ellisen LW. 2018. Unravelling subclonal heterogeneity and aggressive disease states in TNBC through single-cell RNA-seq. *Nat Commun* **9:** 3588. doi:10.1038/s41467-018-06052-0

Katano I, Takahashi T, Ito R, Kamisako T, Mizusawa T, Ka Y, Ogura T, Suemizu H, Kawakami Y, Ito M. 2015. Predominant development of mature and functional human NK cells in a novel human IL-2-producing transgenic NOG mouse. *J Immunol* **194:** 3513–3525. doi:10.4049/jimmunol.1401323

Kaufmann E, Sanz J, Dunn JL, Khan N, Mendonça LE, Pacis A, Tzelepis F, Pernet E, Dumaine A, Grenier JC, et al. 2018. BCG educates hematopoietic stem cells to generate protective innate immunity against tuberculosis. *Cell* **172:** 176–190.e19. doi:10.1016/j.cell.2017.12.031

Keren L, Bosse M, Marquez D, Angoshtari R, Jain S, Varma S, Yang SR, Kurian A, van Valen D, West R, et al. 2018. A structured tumor-immune microenvironment in triple negative breast cancer revealed by multiplexed ion beam imaging. *Cell* **174:** 1373–1387.e19. doi:10.1016/j.cell.2018.08.039

Khosravi A, Yáñez A, Price JG, Chow A, Merad M, Goodridge HS, Mazmanian SK. 2014. Gut microbiota promote hematopoiesis to control bacterial infection. *Cell Host Microbe* **15:** 374–381. doi:10.1016/j.chom.2014.02.006

Khoury T, Nagrale V, Opyrchal M, Peng X, Wang D, Yao S. 2018. Prognostic significance of stromal versus intratumoral infiltrating lymphocytes in different subtypes of breast cancer treated with cytotoxic neoadjuvant chemotherapy. *Appl Immunohistochem Mol Morphol* **26:** 523–532. doi:10.1097/PAI.0000000000000466

Kichenadasse G, Miners JO, Mangoni AA, Rowland A, Hopkins AM, Sorich MJ. 2020. Association between body mass index and overall survival with immune checkpoint inhibitor therapy for advanced non–small cell lung cancer. *JAMA Oncol* **6:** 512–518. doi:10.1001/jamaoncol.2019.5241

Kim HJ, Cantor H. 2014. CD4 T-cell subsets and tumor immunity: the helpful and the not-so-helpful. *Cancer Immunol Res* **2:** 91–98. doi:10.1158/2326-6066.CIR-13-0216

Kim R, Kawai A, Wakisaka M, Funaoka Y, Yasuda N, Hidaka M, Morita Y, Ohtani S, Ito M, Arihiro K. 2019. A potential role for peripheral natural killer cell activity induced by preoperative chemotherapy in breast cancer patients.

Cancer Immunol Immunother **68:** 577–585. doi:10.1007/s00262-019-02305-z

Klemm F, Möckl A, Salamero-Boix A, Alekseeva T, Schäffer A, Schulz M, Niesel K, Maas RR, Groth M, Elie BT, et al. 2021. Compensatory CSF2-driven macrophage activation promotes adaptive resistance to CSF1R inhibition in breast-to-brain metastasis. *Nat Cancer* **2:** 1086–1101. doi:10.1038/s43018-021-00254-0

Koboldt DC, Fulton RS, McLellan MD, Schmidt H, Kalicki-Veizer J, McMichael JF, Fulton LL, Dooling DJ, Ding L, Mardis ER, et al. 2012. Comprehensive molecular portraits of human breast tumours. *Nature* **490:** 61–70. doi:10.1038/nature11412

Koru-Sengul T, Santander AM, Miao F, Sanchez LG, Jorda M, Glück S, Ince TA, Nadji M, Chen Z, Penichet ML, et al. 2016. Breast cancers from black women exhibit higher numbers of immunosuppressive macrophages with proliferative activity and of crown-like structures associated with lower survival compared to non-black Latinas and Caucasians. *Breast Cancer Res Treat* **158:** 113–126. doi:10.1007/s10549-016-3847-3

Kos K, Aslam MA, van de Ven R, Wellenstein MD, Pieters W, van Weverwijk A, Duits DEM, van Pul K, Hau CS, Vrijland K, et al. 2022. Tumor-educated T$_{regs}$ drive organ-specific metastasis in breast cancer by impairing NK cells in the lymph node niche. *Cell Rep* **38:** 110447. doi:10.1016/j.celrep.2022.110447

Kowanetz M, Wu X, Lee J, Tan M, Hagenbeek T, Qu X, Yu L, Ross J, Korsisaari N, Cao T, et al. 2010. Granulocyte-colony stimulating factor promotes lung metastasis through mobilization of Ly6G$^+$Ly6C$^+$ granulocytes. *Proc Natl Acad Sci* **107:** 21248–21255. doi:10.1073/pnas.1015855107

Kuett L, Catena R, Özcan A, Plüss A, Ali HR, Al Sa'd M, Alon S, Aparicio S, Battistoni G, Balasubramanian S, et al. 2022. Three-dimensional imaging mass cytometry for highly multiplexed molecular and cellular mapping of tissues and the tumor microenvironment. *Nat Cancer* **3:** 122–133. doi:10.1038/s43018-021-00301-w

Kugel CH, Douglass SM, Webster MR, Kaur A, Liu Q, Yin X, Weiss SA, Darvishian F, Al-Rohil RN, Ndoye A, et al. 2018. Age correlates with response to Anti-PD1, reflecting age-related differences in intratumoral effector and regulatory T-cell populations. *Clin Cancer Res* **24:** 5347–5356. doi:10.1158/1078-0432.CCR-18-1116

Lai SCA, Gundlapalli H, Ekiz HA, Jiang A, Fernandez E, Welm AL. 2021. Blocking short-form Ron eliminates breast cancer metastases through accumulation of stem-like CD4$^+$ T cells that subvert immunosuppression. *Cancer Discov* **11:** 3178–3197. doi:10.1158/2159-8290.CD-20-1172

Lavin Y, Mortha A, Rahman A, Merad M. 2015. Regulation of macrophage development and function in peripheral tissues. *Nat Rev Immunol* **15:** 731–744. doi:10.1038/nri3920

Leach DR, Krummel MF, Allison JP. 1996. Enhancement of antitumor immunity by CTLA-4 blockade. *Science* **271:** 1734–1736. doi:10.1126/science.271.5256.1734

Lee J, Zhou YJ, Ma W, Zhang W, Aljoufi A, Luh T, Lucero K, Liang D, Thomsen M, Bhagat G, et al. 2017. Lineage specification of human dendritic cells is marked by IRF8 expression in hematopoietic stem cells and multi-

Cite this article as *Cold Spring Harb Perspect Med* doi: 10.1101/cshperspect.a041324

potent progenitors. *Nat Immunol* **18:** 877–888. doi:10
.1038/ni.3789

Le Mercier I, Poujol D, Sanlaville A, Sisirak V, Gobert M,
Durand I, Dubois B, Treilleux I, Marvel J, Vlach J, et al.
2013. Tumor promotion by intratumoral plasmacytoid
dendritic cells is reversed by TLR7 ligand treatment. *Cancer Res* **73:** 4629–4640. doi:10.1158/0008-5472.CAN-12-
3058

Ley K, Hoffman HM, Kubes P, Cassatella MA, Zychlinsky A,
Hedrick CC, Catz SD. 2018. Neutrophils: new insights
and open questions. *Sci Immunol* **3:** eaat4579. doi:10
.1126/sciimmunol.aat4579

Li CH, Karantza V, Aktan G, Lala M. 2019a. Current treatment landscape for patients with locally recurrent inoperable or metastatic triple-negative breast cancer: a systematic literature review. *Breast Cancer Res* **21:** 143. doi:10
.1186/s13058-019-1210-4

Li X, Gruosso T, Zuo D, Omeroglu A, Meterissian S, Guiot
MC, Salazar A, Park M, Levine H. 2019b. Infiltration of
CD8[+] T cells into tumor cell clusters in triple-negative
breast cancer. *Proc Natl Acad Sci* **116:** 3678–3687.
doi:10.1073/pnas.1817652116

Li P, Lu M, Shi J, Gong Z, Hua L, Li Q, Lim B, Zhang XH,
Chen X, Li S, et al. 2020. Lung mesenchymal cells elicit
lipid storage in neutrophils that fuel breast cancer lung
metastasis. *Nat Immunol* **21:** 1444–1455. doi:10.1038/
s41590-020-0783-5

Li K, Li T, Feng Z, Huang M, Wei L, Yan Z, Long M, Hu Q,
Wang J, Liu S, et al. 2021. CD8[+] T cell immunity blocks
the metastasis of carcinogen-exposed breast cancer. *Sci
Adv* **7:** eabd8936. doi:10.1126/sciadv.abd8936

Lobera M, Madauss KP, Pohlhaus DT, Wright QG, Trocha
M, Schmidt DR, Baloglu E, Trump RP, Head MS, Hofmann GA, et al. 2013. Selective class IIa histone deacetylase inhibition via a nonchelating zinc-binding group. *Nat
Chem Biol* **9:** 319–325. doi:10.1038/nchembio.1223

Loi S, Sirtaine N, Piette F, Salgado R, Viale G, van Eenoo F,
Rouas G, Francis P, Crown JPA, Hitre E, et al. 2013. Prognostic and predictive value of tumor-infiltrating lymphocytes in a phase III randomized adjuvant breast cancer
trial in node-positive breast cancer comparing the addition of docetaxel to doxorubicin with doxorubicin-based
chemotherapy: BIG 02-98. *J Clin Oncol* **31:** 860–867.
doi:10.1200/JCO.2011.41.0902

Loi S, Michiels S, Salgado R, Sirtaine N, Jose V, Fumagalli D,
Kellokumpu-Lehtinen PL, Bono P, Kataja V, Desmedt C,
et al. 2014. Tumor infiltrating lymphocytes are prognostic
in triple negative breast cancer and predictive for trastuzumab benefit in early breast cancer: results from the
FinHER trial. *Ann Oncol* **25:** 1544–1550. doi:10.1093/an
nonc/mdu112

Lu RJ, Taylor S, Contrepois K, Kim M, Bravo JI, Ellenberger
M, Sampathkumar NK, Benayoun BA. 2021. Multi-omic
profiling of primary mouse neutrophils predicts a pattern
of sex- and age-related functional regulation. *Nat Aging* **1:**
715–733. doi:10.1038/s43587-021-00086-8

Mahmoud SMA, Lee AHS, Paish EC, Macmillan RD, Ellis
IO, Green AR. 2012. Tumour-infiltrating macrophages
and clinical outcome in breast cancer. *J Clin Pathol* **65:**
159–163. doi:10.1136/jclinpath-2011-200355

Mamessier E, Sylvain A, Bertucci F, Castellano R, Finetti P,
Houvenaeghel G, Charaffe-Jaufret E, Birnbaum D, Mo-

retta A, Olive D. 2011. Human breast tumor cells induce
self-tolerance mechanisms to avoid NKG2D-mediated
and DNAM-mediated NK cell recognition. *Cancer Res*
71: 6621–6632. doi:10.1158/0008-5472.CAN-11-0792

Mamessier E, Pradel LC, Thibult ML, Drevet C, Zouine A,
Jacquemier J, Houvenaeghel G, Bertucci F, Birnbaum D,
Olive D. 2013. Peripheral blood NK cells from breast
cancer patients are tumor-induced composite subsets.
J Immunol **190:** 2424–2436. doi:10.4049/jimmunol.120
0140

Martinet L, Garrido I, Filleron T, Le Guellec S, Bellard E,
Fournie JJ, Rochaix P, Girard JP. 2011. Human solid tumors contain high endothelial venules: association with
T- and B-lymphocyte infiltration and favorable prognosis
in breast cancer. *Cancer Res* **71:** 5678–5687. doi:10.1158/
0008-5472.CAN-11-0431

Matson V, Fessler J, Bao R, Chongsuwat T, Zha Y, Alegre ML,
Luke JJ, Gajewski TF. 2018. The commensal microbiome
is associated with anti-PD-1 efficacy in metastatic melanoma patients. *Science* **359:** 104–108. doi:10.1126/science
.aao3290

McDowell SAC, Luo R, Arabzadeh A, Doré S, Bennett N,
Breton V, Karimi E, Rezanejad M, Yang R, Lach KD, et al.
2021. Neutrophil oxidative stress mediates obesity-associated vascular dysfunction and metastatic transmigration. *Nat Cancer* **2:** 545–562. doi:10.1038/s43018-021-
00194-9

McQuade JL, Daniel CR, Hess KR, Mak C, Wang DY, Rai RR,
Park JJ, Haydu LE, Spencer C, Wongchenko M, et al.
2018. Association of body-mass index and outcomes in
patients with metastatic melanoma treated with targeted
therapy, immunotherapy, or chemotherapy: a retrospective, multicohort analysis. *Lancet Oncol* **19:** 310–322.
doi:10.1016/S1470-2045(18)30078-0

Mellman I, Steinman RM. 2001. Dendritic cells: specialized
and regulated antigen processing machines. *Cell* **106:**
255–258. doi:10.1016/S0092-8674(01)00449-4

Michea P, Noël F, Zakine E, Czerwinska U, Sirven P, Abou-
zid O, Goudot C, Scholer-Dahirel A, Vincent-Salmon A,
Reyal F, et al. 2018. Adjustment of dendritic cells to the
breast-cancer microenvironment is subset specific. *Nat
Immunol* **19:** 885–897. doi:10.1038/s41590-018-0145-8

Mills CD, Kincaid K, Alt JM, Heilman MJ, Hill AM. 2000.
M-1/M-2 macrophages and the Th1/Th2 paradigm. *J Immunol* **164:** 6166–6173. doi:10.4049/jimmunol.164.12
.6166

Mitroulis I, Ruppova K, Wang B, Chen LS, Grzybek M,
Grinenko T, Eugster A, Troullinaki M, Palladini A, Kourtzelis I, et al. 2018. Modulation of myelopoiesis progenitors is an integral component of trained immunity. *Cell*
172: 147–161.e12. doi:10.1016/j.cell.2017.11.034

Mortara L, Castellani P, Meazza R, Tosi G, de Lerma Barbaro
A, Procopio FA, Comes A, Zardi L, Ferrini S, Accolla RS.
2006. CIITA-induced MHC class II expression in mammary adenocarcinoma leads to a Th1 polarization of the
tumor microenvironment, tumor rejection, and specific
antitumor memory. *Clin Cancer Res* **12:** 3435–3443.
doi:10.1158/1078-0432.CCR-06-0165

Mosmann TR, Cherwinski H, Bond MW, Giedlin MA, Coff-
man RL. 1986. Two types of murine helper T cell clone. I:
Definition according to profiles of lymphokine activities

and secreted proteins. *J Immunol* **136**: 2348–2357. doi:10 .4049/jimmunol.136.7.2348

Movahedi K, Laoui D, Gysemans C, Baeten M, Stangé G, Van den Bossche J, Mack M, Pipeleers D, In't Veld P, De Baetselier P, et al. 2010. Different tumor microenvironments contain functionally distinct subsets of macrophages derived from Ly6C^high monocytes. *Cancer Res* **70**: 5728–5739. doi:10.1158/0008-5472.CAN-09-4672

Müller P, Kreuzaler M, Khan T, Thommen DS, Martin K, Glatz K, Savic S, Harbeck N, Nitz U, Gluz O, et al. 2015. Trastuzumab emtansine (T-DM1) renders HER2⁺ breast cancer highly susceptible to CTLA-4/PD-1 blockade. *Sci Transl Med* **7**: 315ra188. doi:10.1126/scitranslmed.aac4925

Muraille E, Leo O. 1998. Revisiting the Th1/Th2 paradigm. *Scand J Immunol* **47**: 1–9. doi:10.1111/j.1365-3083.1998-47-1.00383.x

Murray PJ, Allen JE, Biswas SK, Fisher EA, Gilroy DW, Goerdt S, Gordon S, Hamilton JA, Ivashkiv LB, Lawrence T, et al. 2014. Macrophage activation and polarization: nomenclature and experimental guidelines. *Immunity* **41**: 14–20. doi:10.1016/j.immuni.2014.06.008

Murthy VH, Krumholz HM, Gross CP. 2004. Participation in cancer clinical trials: race-, sex-, and age-based disparities. *JAMA* **291**: 2720–2726. doi:10.1001/jama.291.22 .2720

Nagareddy PR, Kraakman M, Masters SL, Stirzaker RA, Gorman DJ, Grant RW, Dragoljevic D, Hong ES, Abdel-Latif A, Smyth SS, et al. 2014. Adipose tissue macrophages promote myelopoiesis and monocytosis in obesity. *Cell Metab* **19**: 821–835. doi:10.1016/j.cmet.2014.03.029

Nagayama A, Vidula N, Ellisen L, Bardia A. 2020. Novel antibody-drug conjugates for triple negative breast cancer. *Ther Adv Med Oncol* **12**: 175883592091598. doi:10 .1177/1758835920915980

Najmeh S, Cools-Lartigue J, Rayes RF, Gowing S, Vourtzoumis P, Bourdeau F, Giannias B, Berube J, Rousseau S, Ferri LE, et al. 2017. Neutrophil extracellular traps sequester circulating tumor cells via β1-integrin mediated interactions. *Int J Cancer* **140**: 2321–2330. doi:10.1002/ijc.30635

Nanda R, Chow LQM, Dees EC, Berger R, Gupta S, Geva R, Pusztai L, Pathiraja K, Aktan G, Cheng JD, et al. 2016. Pembrolizumab in patients with advanced triple-negative breast cancer: phase Ib keynote-012 study. *J Clin Oncol* **34**: 2460–2467. doi:10.1200/JCO.2015.64.8931

Netea MG, Domínguez-Andrés J, Barreiro LB, Chavakis T, Divangahi M, Fuchs E, Joosten LAB, van der Meer JWM, Mhlanga MM, Mulder WJM, et al. 2020. Defining trained immunity and its role in health and disease. *Nat Rev Immunol* **20**: 375–388. doi:10.1038/s41577-020-0285-6

Ng LG, Ostuni R, Hidalgo A. 2019. Heterogeneity of neutrophils. *Nat Rev Immunol* **19**: 255–265. doi:10.1038/s41577-019-0141-8

O'Brien J, Martinson H, Durand-Rougely C, Schedin P. 2012. Macrophages are crucial for epithelial cell death and adipocyte repopulation during mammary gland involution. *Development* **139**: 269–275. doi:10.1242/dev .071696

Ock CY, Hwang JE, Keam B, Kim SB, Shim JJ, Jang HJ, Park S, Sohn BH, Cha M, Ajani JA, et al. 2017. Genomic landscape associated with potential response to anti-CTLA-4

treatment in cancers. *Nat Commun* **8**: 1–12. doi:10.1038/ s41467-017-01018-0

Olson OC, Kim H, Quail DF, Foley EA, Joyce JA. 2017. Tumor-associated macrophages suppress the cytotoxic activity of antimitotic agents. *Cell Rep* **19**: 101–113. doi:10.1016/j.celrep.2017.03.038

Ombrato L, Nolan E, Kurelac I, Mavousian A, Bridgeman VL, Heinze I, Chakravarty P, Horswell S, Gonzalez-Gualda E, Matacchione G, et al. 2019. Metastatic-niche labelling reveals parenchymal cells with stem features. *Nature* **572**: 603–608. doi:10.1038/s41586-019-1487-6

Orecchioni M, Ghosheh Y, Pramod AB, Ley K. 2019. Macrophage polarization: different gene signatures in M1 (LPS⁺) vs. classically and M2(LPS⁻) vs. alternatively activated macrophages. *Front Immunol* **10**: 1084. doi:10 .3389/fimmu.2019.01084

Pal B, Chen Y, Vaillant F, Capaldo BD, Joyce R, Song X, Bryant VL, Penington JS, Di Stefano L, Tubau Ribera N, et al. 2021. A single-cell RNA expression atlas of normal, preneoplastic and tumorigenic states in the human breast. *EMBO J* **40**: e107333.

Park J, Wysocki RW, Amoozgar Z, Maiorino L, Fein MR, Jorns J, Schott AF, Kinugasa-Katayama Y, Lee Y, Won NH, et al. 2016. Cancer cells induce metastasis-supporting neutrophil extracellular DNA traps. *Sci Transl Med* **8**: 361ra138.

Patsouris D, Li PP, Thapar D, Chapman J, Olefsky JM, Neels JG. 2008. Ablation of CD11c-positive cells normalizes insulin sensitivity in obese insulin resistant animals. *Cell Metab* **8**: 301–309. doi:10.1016/j.cmet.2008.08.015

Perego M, Tyurin VA, Tyurina YY, Yellets J, Nacarelli T, Lin C, Nefedova Y, Kossenkov A, Liu Q, Sreedhar S, et al. 2020. Reactivation of dormant tumor cells by modified lipids derived from stress-activated neutrophils. *Sci Transl Med* **12**: eabb5817. doi:10.1126/scitranslmed.abb5817

Perou CM, Sørile T, Eisen MB, Van De Rijn M, Jeffrey SS, Rees CA, Pollack JR, Ross DT, Johnsen H, Akslen LA, et al. 2000. Molecular portraits of human breast tumours. *Nature* **406**: 747–752. doi:10.1038/35021093

Petroni G, Formenti SC, Chen-Kiang S, Galluzzi L. 2020. Immunomodulation by anticancer cell cycle inhibitors. *Nat Rev Immunol* **20**: 669–679. doi:10.1038/s41577-020-0300-y

Podojil JR, Miller SD. 2017. Potential targeting of B7-H4 for the treatment of cancer. *Immunol Rev* **276**: 40–51. doi:10 .1111/imr.12530

Reinfeld BI, Madden MZ, Wolf MM, Chytil A, Bader JE, Patterson AR, Sugiura A, Cohen AS, Ali A, Do BT, et al. 2021. Cell-programmed nutrient partitioning in the tumour microenvironment. *Nature* **593**: 282–288. doi:10 .1038/s41586-021-03442-1

Rice CM, Davies LC, Subleski JJ, Maio N, Gonzalez-Cotto M, Andrews C, Patel NL, Palmieri EM, Weiss JM, Lee JM, et al. 2018. Tumour-elicited neutrophils engage mitochondrial metabolism to circumvent nutrient limitations and maintain immune suppression. *Nat Commun* **9**: 5099. doi:10.1038/s41467-018-07505-2

Routy B, Le Chatelier E, Derosa L, Duong CPM, Alou MT, Daillère R, Fluckiger A, Messaoudene M, Rauber C, Roberti MP, et al. 2018. Gut microbiome influences efficacy of PD-1-based immunotherapy against epithelial tumors. *Science* **359**: 91–97. doi:10.1126/science.aan3706

Ruffell B, Au A, Rugo HS, Esserman LJ, Hwang ES, Coussens LM. 2012. Leukocyte composition of human breast cancer. *Proc Natl Acad Sci* 109: 2796–2801. doi:10.1073/pnas.1104303108

Ruffell B, Chang-Strachan D, Chan V, Rosenbusch A, Ho CM, Pryer N, Daniel D, Hwang ES, Rugo HS, Coussens LM. 2014. Macrophage IL-10 blocks CD8+ T cell-dependent responses to chemotherapy by suppressing IL-12 expression in intratumoral dendritic cells. *Cancer Cell* 26: 623–637. doi:10.1016/j.ccell.2014.09.006

Satthaporn S, Robins A, Vassanasiri W, El-Sheemy M, Jibril JA, Clark D, Valerio D, Eremin O. 2004. Dendritic cells are dysfunctional in patients with operable breast cancer. *Cancer Immunol Immunother* 53: 510–518. doi:10.1007/s00262-003-0485-5

Savas P, Virassamy B, Ye C, Salim A, Mintoff CP, Caramia F, Salgado R, Byrne DJ, Teo ZL, Dushyanthen S, et al. 2018. Single-cell profiling of breast cancer T cells reveals a tissue-resident memory subset associated with improved prognosis. *Nat Med* 24: 986–993. doi:10.1038/s41591-018-0078-7

Schaer DA, Beckmann RP, Dempsey JA, Huber L, Forest A, Amaladas N, Li Y, Wang YC, Rasmussen ER, Chin D, et al. 2018. The CDK4/6 inhibitor abemaciclib induces a T cell inflamed tumor microenvironment and enhances the efficacy of PD-L1 checkpoint blockade. *Cell Rep* 22: 2978–2994. doi:10.1016/j.celrep.2018.02.053

Scherer SD, Riggio AI, Haroun F, DeRose YS, Ekiz HA, Fujita M, Toner J, Zhao L, Li Z, Oesterreich S, et al. 2021. An immune-humanized patient-derived xenograft model of estrogen-independent, hormone receptor positive metastatic breast cancer. *Breast Cancer Res* 23: 100. doi:10.1186/s13058-021-01476-x

Schmidt M, Weyer-Elberich V, Hengstler JG, Heimes AS, Almstedt K, Gerhold-Ay A, Lebrecht A, Battista MJ, Hasenburg A, Sahin U, et al. 2018. Prognostic impact of CD4-positive T cell subsets in early breast cancer: a study based on the FinHer trial patient population. *Breast Cancer Res* 20: 1–10. doi:10.1186/s13058-018-0942-x

Schulz D, Zanotelli VRT, Fischer JR, Schapiro D, Engler S, Lun XK, Jackson HW, Bodenmiller B. 2018. Simultaneous multiplexed imaging of mRNA and proteins with subcellular resolution in breast cancer tissue samples by mass cytometry. *Cell Syst* 6: 25–36.e5. doi:10.1016/j.cels.2017.12.001

Shevchenko JA, Khristin AA, Kurilin VV, Kuznetsova MS, Blinova DD, Starostina NM, Sidorov S, Sennikov SV. 2020. Autologous dendritic cells and activated cytotoxic T-cells as combination therapy for breast cancer. *Oncol Rep* 43: 671–680.

Slattery K, Woods E, Zaiatz-Bittencourt V, Marks S, Chew S, Conroy M, Goggin C, Maceochagain C, Kennedy J, Lucas S, et al. 2021. TGFβ drives NK cell metabolic dysfunction in human metastatic breast cancer. *J Immunother Cancer* 9: e002044. doi:10.1136/jitc-2020-002044

Solinas C, Carbognin L, de Silva P, Criscitiello C, Lambertini M. 2017. Tumor-infiltrating lymphocytes in breast cancer according to tumor subtype: current state of the art. *Breast* 35: 142–150. doi:10.1016/j.breast.2017.07.005

Sørlie T, Tibshirani R, Parker J, Hastie T, Marron JS, Nobel A, Deng S, Johnsen H, Pesich R, Geisler S, et al. 2003. Repeated observation of breast tumor subtypes in independent gene expression data sets. *Proc Natl Acad Sci* 100: 8418–8423. doi:10.1073/pnas.0932692100

Spiegel A, Brooks MW, Houshyar S, Reinhardt F, Ardolino M, Fessler E, Chen MB, Krall JA, DeCock J, Zervantonakis IK, et al. 2016. Neutrophils suppress intraluminal NK cell–mediated tumor cell clearance and enhance extravasation of disseminated carcinoma cells. *Cancer Discov* 6: 630–649. doi:10.1158/2159-8290.CD-15-1157

Stagg J, Loi S, Divisekera U, Ngiow SF, Duret H, Yagita H, Teng MW, Smyth MJ. 2011. Anti-ErbB-2 mAb therapy requires type I and II interferons and synergizes with anti-PD-1 or anti-CD137 mAb therapy. *Proc Natl Acad Sci* 108: 7142–7147. doi:10.1073/pnas.1016569108

Strachan DC, Ruffell B, Oei Y, Bissell MJ, Coussens LM, Pryer N, Daniel D. 2013. CSF1R inhibition delays cervical and mammary tumor growth in murine models by attenuating the turnover of tumor-associated macrophages and enhancing infiltration by CD8+ T cells. *Oncoimmunology* 2: e26968. doi:10.4161/onci.26968

Su S, Liao J, Liu J, Huang D, He C, Chen F, Yang LB, Wu W, Chen J, Lin L, et al. 2017. Blocking the recruitment of naive CD4+ T cells reverses immunosuppression in breast cancer. *Cell Res* 27: 461–482. doi:10.1038/cr.2017.34

Szczerba BM, Castro-Giner F, Vetter M, Krol I, Gkountela S, Landin J, Scheidmann MC, Donato C, Scherrer R, Singer J, et al. 2019. Neutrophils escort circulating tumour cells to enable cell cycle progression. *Nature* 566: 553–557. doi:10.1038/s41586-019-0915-y

Szpor J, Streb J, Glajcar A, Frączek P, Winiarska A, Tyrak KE, Basta P, Okoń K, Jach R, Hodorowicz-Zaniewska D. 2021. Dendritic cells are associated with prognosis and survival in breast cancer. *Diagnostics* 11: 702. doi:10.3390/diagnostics11040702

Tallón de Lara P, Castañón H, Vermeer M, Núñez N, Silina K, Sobottka B, Urdinez J, Cecconi V, Yagita H, Movahedian Attar F, et al. 2021. CD39+PD-1+CD8+ T cells mediate metastatic dormancy in breast cancer. *Nat Commun* 12: 1–14. doi:10.1038/s41467-021-21045-2

Textor S, Fiegler N, Arnold A, Porgador A, Hofmann TG, Cerwenka A. 2011. Human NK cells are alerted to induction of p53 in cancer cells by upregulation of the NKG2D ligands ULBP1 and ULBP2. *Cancer Res* 71: 5998–6009. doi:10.1158/0008-5472.CAN-10-3211

Thorsson V, Gibbs DL, Brown SD, Wolf D, Bortone DS, Ou Yang TH, Porta-Pardo E, Gao GF, Plaisier CL, Eddy JA, et al. 2018. The immune landscape of cancer. *Immunity* 48: 812–830.e14. doi:10.1016/j.immuni.2018.03.023

Tofigh A, Suderman M, Paquet ER, Livingstone J, Bertos N, Saleh SM, Zhao H, Souleimanova M, Cory S, Lesurf R, et al. 2014. The prognostic ease and difficulty of invasive breast carcinoma. *Cell Rep* 9: 129–142. doi:10.1016/j.celrep.2014.08.073

Tohme S, Yazdani HO, Al-Khafaji AB, Chidi AP, Loughran P, Mowen K, Wang Y, Simmons RL, Huang H, Tsung A. 2016. Neutrophil extracellular traps promote the development and progression of liver metastases after surgical stress. *Cancer Res* 76: 1367–1380. doi:10.1158/0008-5472.CAN-15-1591

Tomay F, Wells K, Duong L, Tsu JW, Dye DE, Radley-Crabb HG, Grounds MD, Shvlakadze T, Metharom P, Nelson DJ, et al. 2018. Aged neutrophils accumulate in lymphoid tissues from healthy elderly mice and infiltrate T- and B-cell zones. *Immunol Cell Biol* **96**: 831–840. doi:10.1111/imcb.12046

Treilleux I, Blay JY, Bendriss-Vermare N, Ray-Coquard I, Bachelot T, Guastolla JP, Bremond A, Goddard S, Pin JJ, Bartfaelemy-Dubois C, et al. 2004. Dendritic cell infiltration and prognosis of early stage breast cancer. *Clin Cancer Res* **10**: 7466–7474. doi:10.1158/1078-0432.CCR-04-0684

Tyagi A, Sharma S, Wu K, Wu SY, Xing F, Liu Y, Zhao D, Deshpande RP, D'Agostino RB Jr, Watabe K. 2021. Nicotine promotes breast cancer metastasis by stimulating N2 neutrophils and generating pre-metastatic niche in lung. *Nat Commun* **12**: 474. doi:10.1038/s41467-020-20733-9

Verma C, Kaewkangsadan V, Eremin JM, Cowley GP, Ilyas M, El-Sheemy MA, Eremin O. 2015. Natural killer (NK) cell profiles in blood and tumour in women with large and locally advanced breast cancer (LLABC) and their contribution to a pathological complete response (PCR) in the tumour following neoadjuvant chemotherapy (NAC): differential restoration of blood profiles by NAC and surgery. *J Transl Med* **13**: 180. doi:10.1186/s12967-015-0535-8

Villacampa G, Tolosa P, Salvador F, Sánchez-Bayona R, Villanueva L, Dienstmann R, Ciruelos E, Pascual T. 2022. Addition of immune checkpoint inhibitors to chemotherapy versus chemotherapy alone in first-line metastatic triple-negative breast cancer: a systematic review and meta-analysis. *Cancer Treat Rev* **104**: 102352. doi:10.1016/j.ctrv.2022.102352

Vitale I, Manic G, Coussens LM, Kroemer G, Galluzzi L. 2019. Macrophages and metabolism in the tumor microenvironment. *Cell Metab* **30**: 36–50. doi:10.1016/j.cmet.2019.06.001

Vonderheide RH, Domchek SM, Clark AS. 2017. Immunotherapy for breast cancer: what are we missing? *Clin Cancer Res* **23**: 2640–2646. doi:10.1158/1078-0432.CCR-16-2569

Voorwerk L, Slagter M, Horlings HM, Sikorska K, van de Vijver KK, de Maaker M, Nederlof I, Kluin RJC, Warren S, Ong SF, et al. 2019. Immune induction strategies in metastatic triple-negative breast cancer to enhance the sensitivity to PD-1 blockade: the TONIC trial. *Nat Med* **25**: 920–928. doi:10.1038/s41591-019-0432-4

Vu Manh TP, Bertho N, Hosmalin A, Schwartz-Cornil I, Dalod M. 2015. Investigating evolutionary conservation of dendritic cell subset identity and functions. *Front Immunol* **6**: 260.

Wang ZX, Cao JX, Wang M, Li D, Cui YX, Zhang XY, Liu JL, Li JL. 2014. Adoptive cellular immunotherapy for the treatment of patients with breast cancer: a meta-analysis. *Cytotherapy* **16**: 934–945. doi:10.1016/j.jcyt.2014.02.011

Wang Q, Guldner IH, Golomb SM, Sun L, Harris JA, Lu X, Zhang S. 2019a. Single-cell profiling guided combinatorial immunotherapy for fast-evolving CDK4/6 inhibitor-resistant HER2-positive breast cancer. *Nat Commun* **10**: 3817. doi:10.1038/s41467-019-11729-1

Wang Z, Aguilar EG, Luna JI, Dunai C, Khuat LT, Le CT, Mirsoian A, Minnar CM, Stoffel KM, Sturgill IR, et al. 2019b. Paradoxical effects of obesity on T cell function during tumor progression and PD-1 checkpoint block-

ade. *Nat Med* **25**: 141–151. doi:10.1038/s41591-018-0221-5

Wang C, Barnoud C, Cenerenti M, Sun M, Caffa I, Kizil B, Bill R, Liu Y, Pick R, Garnier L, et al. 2023. Dendritic cells direct circadian anti-tumour immune responses. *Nature* **614**: 136–143. doi:10.1038/s41586-022-05605-0

Wculek SK, Malanchi I. 2015. Neutrophils support lung colonization of metastasis-initiating breast cancer cells. *Nature* **528**: 413–417. doi:10.1038/nature16140

Wculek SK, Cueto FJ, Mujal AM, Melero I, Krummel MF, Sancho D. 2020a. Dendritic cells in cancer immunology and immunotherapy. *Nat Rev Immunol* **20**: 7–24. doi:10.1038/s41577-019-0210-z

Wculek SK, Bridgeman VL, Peakman F, Malanchi I. 2020b. Early neutrophil responses to chemical carcinogenesis shape long-term lung cancer susceptibility. *iScience* **23**: 101277. doi:10.1016/j.isci.2020.101277

Wei SC, Duffy CR, Allison JP. 2018. Fundamental mechanisms of immune checkpoint blockade therapy. *Cancer Discov* **8**: 1069–1086. doi:10.1158/2159-8290.CD-18-0367

Wei WZ, Gibson HM, Jacob JB, Frelinger JA, Berzofsky JA, Maeng H, Dyson G, Reyes JD, Pilon-Thomas S, Ratner S, et al. 2020. Diversity outbred mice reveal the quantitative trait locus and regulatory cells of HER2 immunity. *J Immunol* **205**: 1554–1563. doi:10.4049/jimmunol.2000466

Wellenstein MD, Coffelt SB, Duits DEM, van Miltenburg MH, Slagter M, de Rink I, Henneman L, Kas SM, Prekovic S, Hau CS, et al. 2019. Loss of p53 triggers WNT-dependent systemic inflammation to drive breast cancer metastasis. *Nature* **572**: 538–542. doi:10.1038/s41586-019-1450-6

Wenisch C, Patruta S, Daxböck F, Krause R, Horl W. 2000. Effect of age on human neutrophil function. *J Leukoc Biol* **67**: 40–45. doi:10.1002/jlb.67.1.40

Wesolowski R, Sharma N, Reebel L, Rodal MB, Peck A, West BL, Marimuthu A, Severson P, Karlin DA, Dowlati A, et al. 2019. Phase Ib study of the combination of pexidartinib (PLX3397), a CSF-1R inhibitor, and paclitaxel in patients with advanced solid tumors. *Ther Adv Med Oncol* **11**: 175883591985423. doi:10.1177/1758835919854238

Wolchok JD, Chiarion-Sileni V, Gonzalez R, Rutkowski P, Grob JJ, Cowey CL, Lao CD, Wagstaff J, Schadendorf D, Ferrucci PF, et al. 2017. Overall survival with combined nivolumab and ipilimumab in advanced melanoma. *N Engl J Med* **377**: 1345–1356. doi:10.1056/NEJMoa1709684

Wu SZ, Al-Eryani G, Roden DL, Junankar S, Harvey K, Andersson A, Thennavan A, Wang C, Torpy JR, Bartonicek N, et al. 2021. A single-cell and spatially resolved atlas of human breast cancers. *Nat Genet* **53**: 1334–1347. doi:10.1038/s41588-021-00911-1

Xia S, Sha H, Yang L, Ji Y, Ostrand-Rosenberg S, Qi L. 2011. Gr-1$^+$ CD11b$^+$ myeloid-derived suppressor cells suppress inflammation and promote insulin sensitivity in obesity. *J Biol Chem* **286**: 23591–23599. doi:10.1074/jbc.M111.237123

Xie X, Shi Q, Wu P, Zhang X, Kambara H, Su J, Yu H, Park SY, Guo R, Ren Q, et al. 2020. Single-cell transcriptome profiling reveals neutrophil heterogeneity in homeostasis and infection. *Nat Immunol* **21**: 1119–1133. doi:10.1038/s41590-020-0736-z

Yam C, Yen EY, Chang JT, Bassett RL, Alatrash G, Garber H, Huo L, Yang F, Philips AV, Ding QQ, et al. 2021. Immune phenotype and response to neoadjuvant therapy in triple-negative breast cancer. *Clin Cancer Res* **27:** 5365–5375. doi:10.1158/1078-0432.CCR-21-0144

Yáñez A, Ng MY, Hassanzadeh-Kiabi N, Goodridge HS. 2015. IRF8 acts in lineage-committed rather than oligo-potent progenitors to control neutrophil vs monocyte production. *Blood* **125:** 1452–1459. doi:10.1182/blood-2014-09-600833

Yáñez A, Coetzee SG, Olsson A, Muench DE, Berman BP, Hazelett DJ, Salomonis N, Grimes HL, Goodridge HS. 2017. Granulocyte-monocyte progenitors and mono-cyte-dendritic cell progenitors independently produce functionally distinct monocytes. *Immunity* **47:** 890–902. e4. doi:10.1016/j.immuni.2017.10.021

Yuan X, Wang J, Huang Y, Shangguan D, Zhang P. 2021. Single-cell profiling to explore immunological heterogeneity of tumor microenvironment in breast cancer. *Front Immunol* **12:** 471.

Zhang D, Chen G, Manwani D, Mortha A, Xu C, Faith JJ, Burk RD, Kunisaki Y, Jang JE, Scheiermann C, et al. 2015. Neutrophil ageing is regulated by the microbiome. *Nature* **525:** 528–532. doi:10.1038/nature15367

Zhang J, Bu X, Wang H, Zhu Y, Geng Y, Nihira NT, Tan Y, Ci Y, Wu F, Dai X, et al. 2019. Author correction: cyclin D–CDK4 kinase destabilizes PD-L1 via cullin 3–SPOP to control cancer immune surveillance. *Nature* **571:** E10. doi:10.1038/s41586-019-1351-8

Zhu YP, Padgett L, Dinh HQ, Marcovecchio P, Blatchley A, Wu R, Ehinger E, Kim C, Mikulski Z, Seumois G, et al. 2018. Identification of an early unipotent neutrophil progenitor with pro-tumoral activity in mouse and human bone marrow. *Cell Rep* **24:** 2329–2341.e8. doi:10.1016/j.celrep.2018.07.097

Zilionis R, Engblom C, Pfirschke C, Savova V, Zemmour D, Saatcioglu HD, Krishnan I, Maroni G, Meyerovitz CV, Kerwin CM, et al. 2019. Single-cell transcriptomics of human and mouse lung cancers reveals conserved my-eloid populations across individuals and species. *Immunity* **50:** 1317–1334.e10. doi:10.1016/j.immuni.2019.03.009

Immunotherapy Approaches for Breast Cancer Patients in 2023

Leisha A. Emens[1,2] and Sherene Loi[3,4]

[1]Department of Medicine, University of Pittsburgh/UPMC Hillman Cancer Center, Pittsburgh, Pennsylvania 15232, USA

[2]Ankyra Therapeutics, Boston, Massachusetts 02116, USA

[3]The Sir Peter MacCallum Department of Medical Oncology, University of Melbourne, Parkville, Victoria 3010, Australia

[4]Division of Cancer Research, Peter MacCallum Cancer Centre, Melbourne, Victoria 3000, Australia

Correspondence: lemens@ankyratx.com, sherene.loi@petermac.org

Immunotherapy, particularly agents targeting the immunoregulatory PD-1/PD-L1 axis, harnesses the power of the immune system to treat cancer, with unique potential for a durable treatment effect due to immunologic memory. The PD-1 inhibitor pembrolizumab combined with neoadjuvant chemotherapy followed by adjuvant pembrolizumab improves event-free survival and is a new standard of care for high-risk, early-stage triple-negative breast cancer (TNBC), regardless of tumor PD-L1 expression. For metastatic TNBC, pembrolizumab combined with chemotherapy is a new standard of care for the first-line therapy of PD-L1$^+$ metastatic TNBC, with improvement in overall survival. The PD-L1 inhibitor atezolizumab combined with nab-paclitaxel is also approved outside the United States for the first-line treatment of metastatic PD-L1$^+$ TNBC. Current research focuses on refining the use of immunotherapy in TNBC by defining informative predictive biomarkers, developing immunotherapy in early and advanced HER2-driven and luminal breast cancers, and overcoming primary and secondary resistance to immunotherapy through unique immune-based strategies.

Breast cancer is often immunologically cold with low tumor mutational burden (TMB). However, tumor-infiltrating lymphocytes (TILs) are present in some breast cancers. TILs are enriched in more aggressive, highly proliferative breast cancers, particularly triple-negative breast cancers (TNBCs) and HER2$^+$ breast cancers (Stanton et al. 2016). Key questions are (1) Why do some patients develop immune infiltrates and others do not? and (2) How can we most effectively harness antitumor immunity for breast cancer treatment? Recent work demonstrates that immune checkpoint inhibitors (ICIs) targeting the programmed cell death 1 receptor (PD-1) or its ligand programmed death ligand 1 (PD-L1) is effective in the treatment of both early- and late-stage TNBC, resulting in multiple health authority approvals around the

Cite this article as *Cold Spring Harb Perspect Med* doi: 10.1101/cshperspect.a041332

world. Intense investigation into the potential efficacy of immunotherapy for other breast cancer subtypes is ongoing.

IMMUNE INFILTRATE IN BREAST CANCER

Although normal breast tissue does not typically contain high levels of immune cells, breast cancers sometimes harbor TILs. Immune infiltrates are composed primarily of T effector and effector memory cells (Savas et al. 2018). As the quantity of intratumoral immune infiltrates increases, the ratio of $CD8^+$ to $CD4^+$ T cells increases, reflecting an evolving cytotoxic T-cell response. Tissue-resident memory T cells have been identified in high-TIL TNBCs, signifying the importance of the local T-cell memory response. High levels of TILs present in breast cancer at diagnosis is strongly prognostic, particularly in TNBC. Although not yet incorporated into standard staging systems, the levels of TILs can effectively up- or down-stage anatomic staging (Loi et al. 2022a). Furthermore, intratumoral T cells most often express the immune checkpoints PD-1 and CTLA-4, but not TIM-3 or LAG-3, with tissue-resident memory T cells expressing PD-1, CTLA-4, TIM-3, LAG-3, and TIGIT in addition to cytotoxic T-cell markers (Savas et al. 2018). These observations underlie the rationale that immune checkpoint blockade can improve outcomes in patients with breast cancer. Notably, metastatic breast tumors have fewer TILs than primary tumors, suggesting increasing immune suppression with disease progression (Cimino-Mathews et al. 2016; Szekely et al. 2018).

EARLY-STAGE BREAST CANCER

Immunotherapy in Early TNBC

Several trials evaluating ICIs targeting PD-1/PD-L1 in early breast cancer have been reported. Given the high levels of TILs in triple-negative tumors, immunotherapy trials for early breast cancer have focused on TNBC. Importantly, preclinical data demonstrate improved responses to neoadjuvant relative to adjuvant ICI, suggesting that improved priming and ex-

pansion of T cells occurs when the primary lesion remains as a source of tumor antigens (Liu et al. 2016). For TNBC, a pathological complete response (pCR), the absence of invasive cancer at surgery after neoadjuvant treatment, strongly correlates with excellent clinical outcomes and is a common clinical end point in breast cancer clinical trials (Cortazar et al. 2014). Several trials have tested the combination of ICI and neoadjuvant chemotherapy in early TNBC, demonstrating significantly improved pCR and/or event-free survival (EFS) rates (Table 1).

KEYNOTE-522 is a randomized phase III trial that tests the addition of pembrolizumab or placebo to neoadjuvant chemotherapy with carboplatin and paclitaxel sequenced with doxorubicin and cyclophosphamide (AC), followed by surgery and adjuvant pembrolizumab (Schmid et al. 2020b, 2022). Adding pembrolizumab to neoadjuvant chemotherapy significantly increased the pCR rate, and more importantly reduced the risk of recurrence for all patients (hazard ratio [HR] 0.63, $P < 0.001$), regardless of PD-L1 expression. Based on these data, the U.S. Food and Drug Administration (FDA) has approved neoadjuvant pembrolizumab with chemotherapy for patients with early TNBC. Multiple important questions remain: (1) What is the optimum chemotherapy backbone? (2) Can we de-escalate therapy for appropriate patients? (3) What is the proper role of adjuvant ICI therapy? and (4) What is the best way to select patients for neoadjuvant immunotherapy? Other clinical trials illustrating these issues are discussed below.

Benefit of Neoadjuvant Carboplatin

Three randomized trials (IMpassion031, I-SPY2, and GeparNuevo, testing atezolizumab, pembrolizumab, and durvalumab, respectively) have yielded positive results without adding carboplatin to sequential AC and taxane (Mittendorf et al. 2020; Nanda et al. 2020; Loibl et al. 2022). These trials also allowed or mandated dose-dense (2-wk) AC, rather than the 3-wk AC used in KEYNOTE-522. GeparNuevo demonstrated improved invasive and distant disease-free survival (iDFS and DDFS) and overall

Table 1. Selected clinical trials in early triple-negative breast cancer (TNBC)

Clinical trial (sample size)	Design	Disease stage	ICI	Chemotherapy backbone	Adjuvant therapy	Results
KEYNOTE-522 (n = 1174)	Randomized phase III	T1N1-2, T2-4N0-2	Pembro	Carbo + paclitaxel, then AC	Pembro to complete 1 yr total (no cape allowed)	pCR in ITT 64.8% vs. 51.2%, $P < 0.001$; 3-yr EFS 84.5% vs. 76.8%, HR 0.63, $P < 0.001$
IMpassion031 (n = 333)	Randomized phase III	T2-4N0-3	Atezo	Nab-paclitaxel, then AC	Atezo to complete 1 yr total (cape allowed)	pCR in ITT 58% vs. 41%, $P = 0.0044$; pCR in PD-L1$^+$ 69% vs. 49%, $P = 0.021$; EFS immature
NeoTripaPDL1 (n = 280)	Randomized phase III	T1N1-3, T2-4N0-3	Atezo	Carbo + nab-paclitaxel	AC/EC/FEC	pCR 48.6% vs. 44.4%, $P = 0.48$; EFS immature
GeparNuevo (n = 174)	Randomized phase II	T2-4N0-3	Durva	2 wk durva alone, then nab-paclitaxel, then AC	None	pCR 53.4% vs. 44.2%, OR 1.45; 3-yr DDFS 91.4% vs. 79.5%, HR 0.37, $P = 0.0148$; 3-yr OS 95.1% vs. 83.1%, HR 0.26, $P = 0076$
I-SPY2 (n = 29 TNBC) (69 total)	Randomized phase II	cT2-4dN0-3	Pembro	Paclitaxel, then AC	Physician's discretion	pCR 60% vs. 22% in TNBC
NCI10013 (n = 67)	Randomized phase II	cT2-4Nany	Atezo	Carbo + paclitaxel	ddAC	pCR 55.6% vs. 18.8%, $P = 0.018$; pCR in mBRCA 80% vs. 50%
NeoPACT (n = 117)	Single-arm phase II	Stage I-III	Pembro	Carbo + docetaxel	Physician's discretion	pCR 60%; 2-yr EFS 89%
CHARIOT (n = 34)	Single-arm phase II	Stage III >15 mm RD after AC × 4	Ipi + nivo	Paclitaxel (AC prior to randomization)	Nivo	pCR 24.2%; 1-yr EFS 85%
I-SPY2 (n = 21 TNBC) (73 total)	Randomized phase II	Stage II-III	Durva + olaparib	Paclitaxel, then AC	None	pCR 47% vs. 27% in TNBC

(ICI) Immune checkpoint inhibitor, (pembro) pembrolizumab, (atezo) atezolizumab, (durva) durvalumab, (nivo) nivolumab, (ipi) ipilimumab, (carbo) carboplatin, (AC) doxorubicin and cyclophosphamide, (EC) epirubicin and cyclophosphamide, (dd) dose dense, (pCR) pathological complete response, (EFS) event-free survival, (DDFS) distant disease-free survival, (OS) overall survival, (ITT) intent to treat, (RD) residual disease, (cape) capecitabine, (mBRCA) mutant breast cancer gene, (PD-L1) programmed death ligand 1, (HR) hazard ratio, (OR) odds ratio.

survival (OS) with neoadjuvant durvalumab and chemotherapy independent of pCR, and survival data from IMpassion031 are awaited. The BrighTNess trial previously demonstrated EFS benefit with carboplatin added to paclitaxel followed by 3-wk AC, but the CALGB 40603 trial did not, so the utility of carboplatin for early TNBC remains a topic of debate (Geyer et al. 2022; Shepherd et al. 2022). Moreover, whether intensification of chemotherapy with carboplatin is needed when pembrolizumab is added to neoadjuvant chemotherapy is unknown.

Defining the short- and long-term impact of chemotherapy on antitumor immunity will inform the optimum chemotherapy backbone for ICI. Different chemotherapy classes induce distinct types of tumor cell death with varying mechanisms of potential immunologic synergy with ICIs (Emens and Middleton 2015; Park et al. 2020). For example, anthracyclines can induce immunogenic cell death, which may enhance dendritic cell uptake of tumor antigens (Mattarollo et al. 2011; Montico et al. 2018). Taxanes inhibit microtubule formation and may bind to Toll-like receptor 4 (TLR4) to promote innate immunity (Pfannenstiel et al. 2010). This immune-modulating activity enhances natural killer cell function and decreases both T regulatory cells and myeloid-derived suppressor cells (MDSCs) (Muraro et al. 2015). DNA-damaging therapies also stimulate immune surveillance through necrotic cell death, a source of type I interferons. During mitosis, cells with double-stranded DNA breaks accumulate micronuclei sensed by the cyclic GMP-AMP synthase stimulator of the interferon genes (cGAS-STING) pathway, culminating in increased interferon signaling (Zierhut et al. 2019; Serpico et al. 2020).

While chemotherapy may enhance tumor immunogenicity, the potential for long-term negative impact on adaptive immunity has also been recognized. A reduction in the naive T-cell pool occurs with more intensive chemotherapy regimens (Verma et al. 2016; Gustafson et al. 2020) and may be worsened by adjuvant radiotherapy (Mozaffari et al. 2007, 2009; Venkatesulu et al. 2018). Lymphopenia is associated with adverse clinical outcomes and a higher risk of future cancers (Ménétrier-Caux et al. 2019). Further defining the immunomodulatory activity of distinct chemotherapics will help optimize ICI efficacy and maintain patients' lymphocyte pool long term.

Less Cytotoxic Neoadjuvant Chemotherapy

It may be possible to reduce the intensity of the chemotherapy backbone with neoadjuvant ICI for two main reasons. First, attaining a pCR is associated with excellent outcomes regardless of the therapy used. Second, post-neoadjuvant treatment provides an opportunity to escalate or de-escalate subsequent therapy according to the pathologic response. Because adding ICI to neoadjuvant chemotherapy results in higher pCR rates, evaluating ICIs in combination with less toxic neoadjuvant chemotherapy regimens is an attractive strategy for reducing the risk of long-term cardiotoxicity and secondary hematological malignancies associated with anthracyclines in lower-risk patients with early TNBC. The single-arm phase 2 NeoPACT trial tested six cycles of neoadjuvant pembrolizumab, carboplatin, and docetaxel (Sharma et al. 2022), with a pCR rate of 60% and a 2-yr EFS rate of 88% without adjuvant pembrolizumab therapy. Patients with higher immune activation (TILs ≥30%) had higher pCR rates. The key to successful chemotherapy de-escalation to improve clinical outcomes and reduce toxicity lies in accurate selection of patients most likely to achieve pCR with ICIs.

Postoperative ICI Considerations

Post-neoadjuvant therapy provides an opportunity to escalate or de-escalate adjuvant therapy according to pathological response. The impact of continuing adjuvant pembrolizumab for patients who do not experience a pCR with neoadjuvant pembrolizumab and chemotherapy is unclear. The GeparNuevo trial demonstrated a statistically significant improvement in long-term clinical outcomes from neoadjuvant ICI despite not including an adjuvant ICI phase. In exploratory subgroup analysis, even patients who did not experience pCR had clinically meaningful improvements from durvalumab, including improved 3-yr iDFS (76.3% vs. 69.7%), DDFS (84.3% vs. 71.9%), and OS (92.0% vs. 78.8%), suggesting that clinical benefit can be derived from neoadjuvant-only administration of ICI even in the absence of pCR (Loibl et al. 2022). Notably, patients who do experience pCR with durvalumab in Gepar-Nuevo had a 3-yr DDFS and OS of 100%, suggesting adjuvant ICI can be omitted in these patients. While this question remains to be answered, in the future other trials without the adjuvant ICI component will also provide long-term clinical outcome data that may give confidence in this approach to those that achieve pCR. Notably, those patients who achieved pCR or residual cancer burden class I (RCB-I) in KEYNOTE-522 with or without neoadjuvant pembrolizumab also had excellent clinical outcomes. In patients with RCB-II, pembrolizumab significantly improved 3-yr EFS (75.5% in the pembrolizumab arm vs. 55.9% in the placebo arm, HR 0.52, 95% CI 0.32–0.82) (Pusztai et al. 2022), although it is unknown whether this is mediated by the neoadjuvant or adjuvant pembrolizumab phase, or both. Taken together, these data from KEYNOTE-522 and Gepar-Nuevo support a clinical trial to evaluate cessation versus continuation of adjuvant ICI in patients who achieve pCR and RCB-I. The Optimice-pCR trial is planned to address this question, randomizing patients with pCR following neoadjuvant chemo-immunotherapy to an additional 27 wk of pembrolizumab or observation.

Outside of these trials, patients with residual disease after neoadjuvant chemotherapy currently receive adjuvant capecitabine, or adjuvant olaparib when a germline BRCA1/2 alteration is present. Whether the incorporation of neoadjuvant or adjuvant pembrolizumab to these agents adds further benefit is unknown.

In KEYNOTE-522, patients with RCB-II had a partial pathological response to neoadjuvant treatment. The long-term improvements in this population imply that they have mounted meaningful antitumor immunity that results in improved clinical outcomes. In contrast, all patients with RCB-III in KEYNOTE-522 (5.1% and 6.7% of patients in the pembrolizumab and placebo arms, respectively) have very poor outcomes irrespective of treatment, implying primary resistance to both chemotherapy and immunotherapy (3-yr EFS 26.2% vs. 24.6% in the pembrolizumab and placebo arms, respectively [HR 1.24, 95% CI 0.69–2.23]). These patients are in critical need of novel non-cross-resistant therapies. New immunotherapeutic strategies that circumvent intrinsic immune resistance such as personalized cancer vaccines, adoptive T-cell therapy or immune agonists are particularly attractive.

Two trials (SWOG1418 and A-BRAVE) are evaluating the role of ICI in the adjuvant phase only specifically, with results pending (Conte et al. 2020). SWOG1418 is evaluating adjuvant pembrolizumab in patients who do not experience pCR with neoadjuvant chemotherapy (no prior ICI), while A-BRAVE is also evaluating adjuvant ICI in this group as well as those who receive no neoadjuvant therapy. These are important questions for patients who have not had the opportunity to access neoadjuvant ICI, as well as potentially tailoring treatment based on a pathological result post-neoadjuvant chemotherapy approach without ICI.

Toxicity Profiles

Relevant to breast cancer, women may be at higher risk of immune-related adverse events (irAEs) from immunotherapy than men (Özdemir et al. 2018; Miceli et al. 2021). For patients with curable disease, the risk of permanent or life-threatening treatment-related toxicity is set at a lower threshold than in the advanced disease setting, especially as some patients will be cured with chemotherapy alone. Although immunotherapy is generally well-tolerated and the most common irAEs are readily manageable, the idiosyncratic nature of permanent endocrinopathies such as primary adrenal insufficiency, hypophysitis, and autoimmune diabetes mellitus can be problematic. In KEYNOTE-522, any-grade irAE occurred in 33.5% of patients receiving pembrolizumab (compared to 11.3% in the placebo group), including hypothyroidism in 15.1%, primary adrenal insufficiency in 2.6%, and hypophysitis in 1.9% (Schmid et al. 2020a). Three immune-related deaths occurred in the pembrolizumab arm (pneumonitis, immune-related encephalitis, and pulmonary embolus). Adrenal insufficiency (due to hypophysitis or primary adrenal insufficiency) was reported in 8.7% of patients treated with pembrolizumab in I-SPY-2, higher that what has previously been reported for anti-PD-1 therapy (Nanda et al. 2020). Immunotherapy also has unknown survivorship implications. Recently, ICIs were reported to mediate ovarian inflammation and reduce oocyte reserves in a murine model (Winship et al. 2022). If confirmed in patients, this effect could have implications for women of reproductive age receiving immunotherapy in the curative setting, particularly given the detrimental impact of chemotherapy on fertility. Further clinical research assessing menstrual cycle patterns, menopausal status, fertility, and changes in sex steroid/gonadotropin levels with ICI treatment is needed.

Patient Selection

The major challenge to the optimal use of immunotherapy in early-stage TNBC is patient selection. PD-L1 protein expression is an established predictive biomarker of response to PD-1/PD-L1 inhibitors for metastatic TNBC, despite varying methods of PD-L1 assessment (Schmid et al. 2018; Cortes et al. 2020). In both KEYNOTE-522 and IMpassion031, improved pCR rates were associated with the use of pembrolizumab or atezolizumab, respectively, in both PD-L1$^+$ and PD-L1-negative patients. In KEYNOTE-522 and GeparNuevo, clinical benefit with pembrolizumab or durvalumab and chemotherapy was also observed regardless of PD-L1 status. Hence, PD-L1-negative patients should not be excluded from receiving ICIs with neoadjuvant chemotherapy, as patients can generate an antitumor immune response that translates into clinical benefit regardless of PD-L1 status. Thus, there is currently no indication for assessing PD-L1 expression in a newly diagnosed patient with early-stage TNBC.

Why PD-L1 expression predicts immunotherapy response in metastatic disease but not in early-stage disease is unclear. However, higher tumor burdens are associated with poorer outcomes from ICI in the advanced setting (Tarantino et al. 2020). The NeoTrip trial included higher-stage N2/N3 primary disease and suggested greater clinical benefit from ICI in patients with PD-L1$^+$ disease, although EFS outcomes are not yet reported (Gianni et al. 2022). In multivariate analysis, PD-L1 positivity was the most significant factor associated with pCR (odds ratio 2.08, $P < 0.0001$) in this trial. Data from patients presenting with stage 4 de novo disease treated with PD-1/PD-L1 agents may be helpful in understanding the impact of tumor burden relative to prior chemotherapy in the requirement for PD-L1 IHC expression.

Given long-term toxicity concerns, it is debated whether all early-stage patients need neoadjuvant ICI with their chemotherapy, particularly those presenting with node-negative disease. Clinical and preclinical data support that the addition of PD-1 targeting agents to chemotherapy as compared with chemotherapy alone significantly augments the antitumor T-cell response and results in long-term immune protection from recurrence. It is hoped that in the future these agents can be pharmacologically improved to just target tumor-specific T cells, which will hopefully alleviate many of the immune toxicity concerns.

Current Status and Future Directions

Adding ICI to neoadjuvant chemotherapy both achieves a pCR in more patients and improves long-term clinical outcomes even in patients with residual disease at surgery. High-risk patients (large tumor size or node-positive tumors) without a medical contraindication (active autoimmune disease, for example) should receive a PD-1/PD-L1 inhibitor in combination with neoadjuvant chemotherapy (Emens et al. 2021c). Future clinical trials aim to optimize the chemotherapy backbone and/or duration of immunotherapy according to the TIL level or PD-L1 expression. For example, it is plausible that some patients with stage I/II disease and high TIL or PD-L1$^+$ disease could benefit from neoadjuvant taxane with short course ICI.

Biomarkers that predict response to neoadjuvant ICI are needed. On-treatment change in TILs and PD-L1 protein expression, immune gene signatures, homologous recombination deficiency (HRD), CD274 amplification, and TMB (Karn et al. 2020; Bianchini et al. 2021) have been evaluated and a few have been found to be predictive of benefit from ICI specifically. In GeparNuevo, only an increase in intratumoral TIL from baseline to cycle 1, day 15 significantly predicted benefit from durvalumab and not placebo (Loibl et al. 2019). High TMB and immune gene-expression profile were associated with increased odds of pCR in both treatment arms (Karn et al. 2020). Stromal TILs (sTILs) and up-regulation of immune signaling pathways at cycle 2, day 1 was associated with pCR in both atezolizumab and chemo-alone arms of NeoTrip (Bianchini et al. 2021). Spatial connectivity of PD-L1-expressing immune cells relative to epithelial cells appears to be relevant to response to atezolizumab in NeoTrip (Bianchini et al. 2022), suggesting that categorical or continuous characterization of individual immune markers such as PD-L1 and TILs may be too simplistic in this disease.

The pCR end point is not a perfect surrogate end point for either neoadjuvant chemotherapy alone (Conforti et al. 2022) or combined with ICI. Assessment of pathologic response as a surrogate for long-term outcome may be strengthened by combining it with biomarkers such as circulating tumor cells, circulating tumor DNA, and/or TILs in residual disease (Luen et al. 2019). Finally, the clinical benefit of adjuvant immunotherapy used alone or continued after neoadjuvant therapy remains an open question that will be informed by the results of current clinical trials.

Immunotherapy in Hormone Receptor-Positive (HR$^+$) and HER2$^+$ Early Breast Cancer

Immunotherapy is being actively investigated in HR$^+$ and HER2$^+$ breast cancer (Table 2). Patients with HER-2-driven disease have high levels of TILs and high pCR rates with neoadjuvant chemotherapy plus dual HER2 blockade. IMpassion050 investigated the addition of atezolizumab or placebo to neoadjuvant AC followed by paclitaxel, trastuzumab, and pertuzumab (Huober et al. 2022). There was no significant difference in pCR between the two arms regardless of PD-L1 expression, and EFS outcomes remain immature. Moreover, there were four deaths in the atezolizumab arm and none in the placebo arm; two deaths were attributed to treatment. Atezolizumab plus trastuzumab emtansine patients with residual disease after standard neoadjuvant therapy for HER2$^+$ disease is being evaluated in the randomized phase III ASTEFANIA trial (NCT04873362).

HR$^+$ breast cancer overall is associated with a good prognosis. However, certain patient subgroups have a worse prognosis, including those who are premenopausal, and/or whose disease is high grade, has greater nodal involvement, or has low estrogen receptor (ER) expression. High-grade disease is associated with higher TIL levels, even in HR$^+$ disease. HR$^+$ HER2-negative (HER2neg) patients were included in I-SPY2, with an improvement in pCR rates from 13% to 30% (Table 2; Nanda et al. 2020). The GIADA trial enrolled 43 patients with stage II–III luminal B breast cancer (HR$^+$, HER2neg, Ki67 $\geq$ 20%, and/or grade 3) to receive three cycles of EC followed by eight cycles of 2-wk nivolumab, with a pCR rate of 16.3% (Dieci et al. 2022). Checkmate 7FL (NCT04109066) and KEYNOTE-756 (NCT03725059) are randomized phase III trials investigating either nivolumab or pembrolizumab, respectively, in combination with neoadjuvant AC- and taxane-based chemotherapy, followed by adjuvant ICI with endocrine therapy in high-grade HR$^+$ breast cancers. Checkmate 7FL closed to accrual early after the MonarchE trial showed significant benefit for adjuvant abemaciclib (Harbeck et al. 2021), which cannot be safely combined with ICI. KEYNOTE-756 has completed accrual; pCR and EFS analyses are planned.

Combination Strategies

Combinations of PD-1/PD-L1 agents with other immune-modulatory agents may improve pCR rates for patients unlikely to respond to ICI alone (Table 3). The single-arm CHARIOT trial recruited 34 patients with stage III TNBC with poor response to neoadjuvant AC to receive combination ipilimumab and nivolumab with neoadjuvant weekly paclitaxel, resulting in pCR rates of 24.4%, 37.5%, and 23% in the overall, PD-L1$^+$ and PD-L1-negative populations, respectively (Loi et al. 2022b). The combination of PARP inhibitors and immunotherapy is particularly attractive due to the prevalence of HRD in breast cancer. HRD is a genomic feature associated with increased immune activation via up-regulation of the cGAS/STING pathway (Parkes et al. 2022; van Vugt et al. 2022). I-SPY2 evaluated the PARP-inhibitor olaparib with durvalumab and paclitaxel, followed by AC, for patients with HER2neg early breast cancer, and demonstrated numerically improved pCR rates compared to chemotherapy alone in both the HR$^+$ and TNBC cohorts (Pusztai et al. 2021).

ADVANCED BREAST CANCER

Metastatic TNBC—First-Line ICIs

Metastatic TNBC has historically been treated with chemotherapy, with bevacizumab sometimes added ex-U.S. Recent clinical progress has expanded treatment options for patients, which now include targeted agents and immunotherapy (Huppert et al. 2022). The PARP inhibitors olaparib and niraparib (for germline BRCA-mutated breast cancer) and the antibody-drug conjugates sacituzumab govitecan (specific for the cell-surface antigen TROP2) and trastuzumab deruxtecan (specific for HER2 and active in patients with HER2-low TNBC) are targeted therapies available for routine use in the United States. Pembrolizumab with chemotherapy (based on KEYNOTE 355; Cortes et al. 2022) is a standard of care for the first-line treatment of metastatic PD-L1$^+$ TNBC in the United States and globally, and atezolizumab with nab-paclitaxel (based on IMpassion 130) is approved for the first-line treatment of metastatic PD-L1$^+$ TNBC (Schmid et al. 2018; Emens et al. 2021b) outside the United States. These trials are summarized in Table 3. In contrast to the neoadjuvant setting, PD-L1 expression is used to select patients with advanced TNBC for immunotherapy (the SP142 assay for atezolizumab and the 22C3 assay for pembrolizumab were used in the trials). These PD-L1 assays are not the same, as demonstrated in a systematic analysis of patient samples evaluating the association of PD-L1 status and clinical response from the IMpassion130 study (Rugo et al. 2021).

The combination of chemotherapy and pembrolizumab was evaluated in KEYNOTE 355, a global clinical study that randomized 847 patients with advanced TNBC to receive pembrolizumab or placebo and chemotherapy with nab-paclitaxel, paclitaxel, or gemcitabine and carboplatin (Cortes et al. 2020, 2022). Eligible patients had untreated advanced disease with a treatment-free interval from (neo)adjuvant therapy of ≥6 mo. Primary end points included PFS and OS in the overall patient population, and in patients with PD-L1 CPS scores of ≥10 or ≥1. Pembrolizumab given with chemotherapy improved both PFS (Δ4.1 mo, HR 0.65, $P = 0.0012$) and OS (Δ6.9 mo, HR 0.73, $P = 0.0093$) in PD-L1$^+$ CPS ≥10 patients. Based on these data, pembrolizumab with chemotherapy is approved in the United States and globally for the first-line treatment of patients with advanced PD-L1$^+$ TNBC.

IMpassion130 is a global clinical trial that enrolled 902 patients with advanced TNBC, randomizing them to nab-paclitaxel with atezolizumab or placebo (Schmid et al. 2018, 2020a; Emens et al. 2021b). Patients had treatment-naive metastatic disease, with a treatment-free interval of ≥12 mo from (neo)adjuvant therapy. Primary end points included PFS and OS in the overall and PD-L1$^+$ patient groups. The study showed an improvement in PFS with the addition of atezolizumab to nab-paclitaxel in both the overall (Δ2.1 mo, HR 0.69, $P = 0.0025$) and PD-L1$^+$ (Δ2.5 mo, HR 0.62, $P < 0.0001$) groups. A trend toward improved OS in the overall patient population was not statistically significant, but an exploratory OS analysis in PD-L1$^+$ patients revealed a clinically meaningful OS improvement that was not formally tested due to the hierarchical statistical analysis plan for OS. These results resulted in accelerated approval by the U.S. FDA. The OS for PD-L1$^+$ patients re-

Table 2. Immune checkpoint inhibitor (ICI) trials in early HER2$^+$ and HR$^+$ breast cancer

Clinical trial (sample size)	Design	Disease type and stage	ICI	Chemotherapy backbone	Adjuvant therapy	Results
IMpassion050 (n = 454)	Randomized phase III	HER2$^+$ =T2-4N1-3	Atezo	ddAC, then THP	Atezo + HP if pCR T-DM1 if RD	pCR in ITT 62.4% vs. 62.7%, $P = 0.9551$
I-SPY2 (n = 40 HR$^+$) (69 total)	Randomized phase II	=T2-4dN0-3	Pembro	Paclitaxel, then AC	Physician's discretion	pCR 30% vs. 13% in HR$^+$
GIADA (n = 43)	Single-arm phase II	Stage II–IIIA premenopausal luminal B	Nivo	EC × 3, OFS exemestane	OFS + exemestane	pCR 16.3%
I-SPY2 (n = 52) (73 total)	Randomized phase II	Stage II–III	Durva + olaparib	Paclitaxel, then AC	None	pCR 28% vs. 14% in HR$^+$
KEYNOTE-756	Randomized phase III	T1c-2N1-2 T3-4N0-2 Grade 3	Pembro	Paclitaxel, then AC	Pembro × 9 + ET	Primary end points pCR and EFS Completed accrual
Checkmate 7FL	Randomized phase III	T1c-2N1-2 T3-4N0-2 Grade 3 ER 1%–9%	Nivo	Paclitaxel, then AC	Nivo × 7 + ET	Terminated early due to a change in the standard of care

(HER2) Human epidermal growth factor 2, (HR) hormone receptor, (atezo) atezolizumab, (pembro) pembrolizumab, (nivo) nivolumab, (durva) durvalumab, (AC) adriamycin and doxorubicin, (THP) paclitaxel, trastuzumab, and pertuzumab, (HP) trastuzumab and pertuzumab, (EC) epirubicin and cyclophosphamide, (OFS) ovarian function suppression, (ET) endocrine therapy, (ER) estrogen receptor, (pCR) pathologic complete response, (RD) residual disease, (EFS) event-free survival, (ITT) intent to treat.

Table 3. Immune checkpoint inhibitor (ICI) trials for the first-line therapy of metastatic triple-negative breast cancer (TNBC)

Clinical trial (sample size)	Design	Patient population	ICI	Chemotherapy backbone	Results
KEYNOTE-355 (n = 847)	Randomized phase III	Untreated advanced TNBC TFI >6 mo	Pembro vs. placebo	Nab-paclitaxel Paclitaxel Gemcitabine + carboplatin	ITT: PFS 7.5 mo vs. 5.6 mo, HR 0.82, P = not tested ITT: OS 17.2 vs 15.5 mo, HR 0.89, P = not tested
					CPS >10: PFS 9.7 vs 5.6 mo, HR 0.65, P = 0.0012 CPS ≥10: OS 23.0 vs. 16.1 mo, HR 0.73, P = 0.0185
IMpassion130 (n = 902)	Randomized phase III	Untreated advanced TNBC TFI ≥12 mo	Atezo vs. placebo	Nab-paclitaxel	ITT: PFS 7.2 mo vs. 5.5 mo, HR 0.80, P = 0.002 ITT: OS 21.0 vs. 18.7 mo, HR 0.87, P = not tested
					PD-L1[+]: PFS 7.5 vs. 5.0 mo, HR 0.62, P < 0.001 PD-L1[+]: OS 25.4 vs. 17.9 mo, HR 0.67, P = not tested
IMpassion131 (n = 651)	Randomized phase III	Untreated advanced TNBC TFI ≥12 mo	Atezo vs. placebo	Paclitaxel	PD-L1[+]: PFS 5.7 mo vs. 6.0 mo, HR 0.82, P = 0.002 PD-L1[+]: OS 22.1 vs. 28.3 mo, HR 1.11 P = not tested

(TFI) Treatment-free interval, (pembro) pembrolizumab, (atezo) atezolizumab, (ITT) intent to treat, (PD-L1) programmed death ligand 1, (PFS) progression-free survival, (OS) overall survival, (HR) hazard ratio.

mained consistent over time, with improvements of 7.5 (HR 0.62), 7.0 (HR 0.71), and 7.5 (HR 0.67) mo at the first, second, and final analyses, respectively (median follow up 18.8 mo at final analysis). Extensive biomarker analyses of IMpassion130 tumor specimens revealed that enriching for PD-L1 was not associated with greater clinical benefit, and patients with PD-L1-negative disease had no treatment effect (Emens et al. 2021a). Only 9% of tumors expressed PD-L1 on tumor cells, and 2% of tumors expressed PD-L1 exclusively on tumor cells. Improved clinical outcomes were observed in tumors with CD8[+] T cells and sTILs only if the tumors were PD-L1[+]. BRCA1/2 mutations were present in 14.5% (89/612) of patients, and not associated with PD-L1 status. PD-L1[+] patients derived clinical benefit regardless of BRCA1/2 mutation status.

IMpassion 131, an independent companion trial designed to independently evaluate these findings, was a global clinical trial that enrolled a similar population of 651 patients (Miles et al. 2021). It randomized patients to receive atezolizumab or placebo combined with paclitaxel (Table 3). The percentages of PD-L1[+] patients in IMpassion 130 and IMpassion 131 were 40% and 45%, respectively. Adding atezolizumab to paclitaxel did not improve PFS or OS in PD-L1[+] patients (Δ0.3 mo, HR 0.82, P = 0.20, and Δ −6.2 mo, HR 1.11, respectively). The reasons for the discordant results between the trials are unclear. They may relate to the steroid exposure required for paclitaxel or unknown biological differences between the patient populations. Given the negative results of IMpassion 131 and the lack of statistically significant OS data from IMpassion 130, the approval of pembrolizumab left no clear regulatory path forward for atezolizumab combined with nab-paclitaxel in the United States, and the accelerated approval indication was withdrawn by the sponsor. Atezolizumab and nab-paclitaxel is approved by regulatory authorities outside the United States.

Metastatic TNBC—ICIs Second Line and Beyond

Immunotherapy has been less active in metastatic TNBC beyond the first-line setting (Table 4). The phase 3 KEYNOTE 119 study evaluated pembrolizumab monotherapy versus chemotherapy of physician's choice in a global, randomized, open label trial enrolling patients with TNBC after one to two prior systemic therapies for metastatic disease (Winer et al. 2021). Patients received either pembrolizumab or chemotherapy with capecitabine, eribulin, gemcitabine, or vinorelbine. Primary end points were OS in patients with PD-L1 CPS ≥ 10 and CPS ≥ 1 disease, and the overall patient population. Pembrolizumab did not improve OS relative to chemotherapy in any group, although exploratory analyses revealed a potential survival benefit in the CPS ≥ 20 subgroup. These findings indicate that novel treatment and/or patient-selection strategies are needed for patients with previously treated advanced TNBC.

The TONIC study is a noncomparative, adaptive phase 1/2 clinical trial that evaluated nivolumab alone or after 2 wk priming with immunomodulatory doses of cyclophosphamide (50 mg orally daily), cisplatin (2×40 mg/m^2 intravenously), doxorubicin (2×15 mg intravenously), or radiation therapy (3×8 Gy) in 67 patients with metastatic TNBC, 24% of whom had no prior treatment for metastatic disease (Voorwerk et al. 2019). Biopsies were collected at baseline, and after priming and 3 wk of nivolumab. The overall objective response rate (ORR) by iRECIST was 20%, with ORRs of 23% and 35% with cisplatin and doxorubicin priming, respectively. Correlative studies revealed up-regulation of immune-related genes in the PD-1 and T-cell cytotoxicity pathways with both drugs, as well as inflammation and both JAK/STAT and TNF-α signaling with doxorubicin. Induction with low-dose doxorubicin followed by nivolumab versus nivolumab alone is currently being evaluated.

The FUTURE-C-PLUS trial evaluated the antiangiogenic agent famitinib combined with the PD-1 antibody camrelizumab and nab-paclitaxel as first-line therapy in 48 advanced CD8$^+$ T-cell-enriched immunomodulatory TNBC patients (Wu et al. 2022). The ORR was 81.3%, with a median PFS of 13.6 mo. Clinical benefit was enhanced in CD8$^+$ T-cell-enriched and PD-L1$^+$ tumors, and the PKD1 somatic mutation was associated with resistance. The TOPACIO/KEYNOTE 162 trial is a single-arm trial that evaluated niraparib combined with pembrolizumab in 55 patients (47 efficacy evaluable) with metastatic TNBC (Vinayak et al. 2019). The ORR was 21%, with a disease control rate (DCR) of 49%. In 15 efficacy evaluable patients with a germline BRCA mutation, the ORR was 47% ($n = 7$) with a DCR of 80% ($n = 12$).

The randomized phase II SAFIR02 Breast Immuno trial tested maintenance therapy with chemotherapy versus durvalumab in patients with metastatic HER2neg breast cancer with stable disease after induction chemotherapy (Batchelot et al. 2021). There was no clinical activity of durvalumab in the overall population, but in 82 patients with TNBC, durvalumab improved OS (HR 0.54, $P = 0.0377$) and CD274 gain/amplification was identified as a potential biomarker of sensitivity.

Immunotherapy for Metastatic HER2$^+$ Breast Cancer

The major challenge of developing immunotherapy for metastatic HER2$^+$ breast cancer is the availability of multiple HER2-directed therapeutics with survival benefit. These highly effective drugs raise the bar for demonstrating enhanced clinical activity with immunotherapy combinations (Table 5), while simultaneously creating hope for curing patients with advanced HER2$^+$ breast cancer. Several HER2-specific therapeutics have intrinsic immune-modulating activity, creating an opportunity for therapeutic synergy. A phase 1b study evaluated adding atezolizumab to ado-trastuzumab emtansine (TDM1) or trastuzumab/pertuzumab with docetaxel in patients with metastatic HER2$^+$ breast cancer (Hamilton et al. 2021).

Table 4. Trials testing immune checkpoint inhibitors (ICIs) in metastatic triple-negative breast cancer (TNBC) beyond the first-line setting

Clinical trial (sample size)	Design	Patient population	ICI	Intervention	Results
KEYNOTE-119 (*n* = 622)	Randomized open label phase III	Advanced TNBC second or third line	Pembro	Pembro alone *vs.* physicians' choice chemo: capecitabine, eribulin, gemcitabine, vinorelbine	OS: CPS ≥10: 12.7 vs. 11.6 mo, HR 0.78, *P* = 0.057 CPS ≥1: 10.7 vs. 10.2 mo, HR 0.86, *P* = 0.073 ITT: 9.9 vs. 10.8 mo, HR 0.97, *P* = nt CPS ≥20: 14.9 vs. 12.5 mo, HR 0.58, *P* = nt
TONIC (*n* = 67)	Noncomparative adaptive phase I/II	Metastatic TNBC	Nivo	Nivo alone *vs.* nivo beginning after 2 wk priming with: CY 50 mg orally daily Cis 2 × 40 mg/m^2 IV Dox 2 × 15 mg IV XRT 3 × 8 Gy	ORR: Overall 20% Nivo 17% CY 8% Cis 23% Dox 35% XRT 8%
FUTURE-C-PLUS (*n* = 48)	Single-arm phase II	Advanced CD8$^+$ T-cell-enriched metastatic TNBC	Camre	Camrelizumab + famitinib, nab-paclitaxel	ORR 81.3%, PFS 13.6 mo
KEYNOTE-162 (TOPACIO) (*n* = 55)	Single-arm phase II	Metastatic TNBC	Pembro	Pembro + niraparib	ITT (*n* = 47 efficacy evaluable): ORR 21% DCR 49% PFS 2.3 mo mBRCA (*n* = 15): ORR 47% (*n* = 7) DCR 80% (*n* = 12) PFS 8.3 mo wtBRCA (*n* = 27): ORR 11% (*n* = 3) DCR 33% (*n* = 9) PFS 2.1 mo

Continued

Table 4. *Continued*

Clinical trial (sample size)	Design	Patient population	ICI	Intervention	Results
SAFIR02-BREAST IMMUNO (n = 199)	Randomized phase II	Metastatic HER2neg breast cancer, stable after six to eight cycles of chemotherapy	Durva	Maintenance therapy durva vs. chemo	Overall: PFS HR 1.40, $P = 0.047$ OS HR 0.84, $P = 0.423$ TNBC (n = 82) All OS HR 0.54, $P = 0.0377$, $P = $ nt PD-L1$^+$ (n = 32) OS HR 0.37, $P = $ nt PD-L1$^-$ (n = 24) OS HR 0.49, $P = $ nt CD274 gain/amplification (n = 23) OS HR 0.18, $P = 0.0059$ CD274 normal/loss (n = 32) OS HR 1.12, $P = 0.8139$

(HER2neg) Human epidermal growth factor 2-negative, (pembro) pembrolizumab, (nivo) nivolumab, (camre) camrelizumab, (durva) durvalumab, (chemo) chemotherapy, (CY) cyclophosphamide, (Cis) cisplatin, (Dox) doxorubicin, (XRT) radiotherapy, (ITT) intent to treat, (PD-L1) programmed death ligand 1, (CPS) combined positive score, (ORR) objective response rate, (DCR) disease control rate, (PFS) progression-free survival, (OS) overall survival, (HR) hazard ratio, (Gy) gray, (IV) intravenous, (mBRCA) mutated breast cancer–related gene, (wtBRCA) wild-type breast cancer–related gene, (nt) not tested.

The ORR with atezolizumab and TDM1 was 35% (7/20 patients) and with atezolizumab plus trastuzumab, pertuzumab, and docetaxel was 100% (6/6 patients). Immune cell PD-L1 expression increased in patients treated with atezolizumab with TDM1 independent of clinical response.

PANACEA evaluated the safety and clinical activity of trastuzumab and pembrolizumab in 55 patients with metastatic HER2[+] breast cancer who had previously progressed on trastuzumab (Loi et al. 2019). The ORRs were 15.2% (7/42) in PD-L1[+] and 0% (0/12) PD-L1-negative patients, respectively. Moreover, 12-mo PFS and OS rates were 13% versus 0% and 65% versus 12% in patients with PD-L1[+] versus PD-L1-negative disease, respectively. sTILs tracked with clinical benefit in PD-L1[+] patients.

KATE2 is a randomized phase 2 trial that added atezolizumab or placebo to TDM1 in 202 patients with HER2[+] metastatic breast cancer previously treated with trastuzumab and a taxane (Emens et al. 2020). There was no significant difference in PFS or OS in the overall population, but exploratory analyses suggested improved PFS and OS in PD-L1[+] patients. KATE3, a randomized phase 3 clinical trial is enrolling biomarker-selected patients with metastatic HER2[+] breast cancer that is also PD-L1[+] (NCT04740918).

Immunotherapy for Metastatic HR[+] Breast Cancer

HR-expressing breast cancer tends to lack T cells and instead contains a significant population of myeloid cells. Given the distinct immunobiology of metastatic HR[+] breast cancer, it is not surprising that ICIs alone have displayed limited clinical activity. Pembrolizumab alone in 25 patients with ER[+], HER2neg, PD-L1[+] metastatic breast cancer had an ORR of 12% (Rugo et al. 2018). Most trials for advanced HR[+] breast cancer have thus focused on combination strategies (Table 5).

The anti-CTLA4-specific antibody tremelimumab combined with exemestane was evaluated in 26 patients with advanced HR[+] breast cancer (Vonderheide et al. 2010). Peripheral

ICOS[+] T cells increased, with a best response of stable disease for ≥12 wk in 42% of patients. The combination of tremelimumab and durvalumab was tested in 18 patients with advanced HER2neg breast cancer, 11 of whom had HR[+] disease (Santa-Maria et al. 2018). Although no responses occurred in ER[+] disease, the ORR in TNBC was 17%.

Pembrolizumab was given with eribulin in 44 patients with advanced HR[+] breast cancer, with an ORR of 41% (Perez-Garcia et al. 2021). A randomized study enrolled 88 patients with advanced HR[+] breast cancer to receive eribulin with or without pembrolizumab, with crossover permitted (Tolaney et al. 2020). There was no improvement in ORR (27% vs. 34%, $P = 0.49$), PFS, or OS, and no association of PFS with PD-L1, sTILs, TMB, or genomic alterations. Computational analysis of 52 pretreatment tumors showed an association between immune infiltrates and antigen presentation pathways and clinical response, with resistant tumors characterized by heterogeneity and active estrogen signaling (Keenan et al. 2021).

Another study evaluated epigenetic modulation with pembrolizumab in metastatic ER[+] breast cancer, randomizing 34 patients to receive tamoxifen and vorinostat with concurrent pembrolizumab ($n = 18$), or with pembrolizumab beginning after one cycle of priming with tamoxifen and vorinostat ($n = 16$) (Terranova-Barberio et al. 2020). The ORR was 4%, with a clinical benefit rate of 19%. Another study evaluated entinostat with nivolumab and low-dose ipilimumab in 24 patients with advanced HER2neg breast cancer; 12 had TNBC and 12 had HR[+] breast cancer. The ORR was 30%, mostly in TNBC. The CD8[+] T cells increased with therapy.

Eftilagomad alpha (efti, IMP321) is a soluble LAG-3 Ig fusion that binds to MHC class II molecules, promoting immune priming. The AIPAC trial randomized 227 patients with metastatic HR[+] breast cancer to weekly paclitaxel with efti or placebo as first-line chemotherapy in endocrine-experienced patients (Wildiers et al. 2021). Paclitaxel was given with either efti or placebo for six cycles, and then maintenance efti or placebo was continued for 52 wk. There

Table 5. Selected trials testing immune checkpoint inhibitors (ICIs) in HER2+ and HR+ HER2-negative metastatic breast cancer

Clinical trial (sample size)	Design	Patient population	ICI	Intervention	Results
KATE2 (n = 202)	Randomized phase II	Metastatic HER2+ breast cancer previously treated with trastuzumab and taxane	Atezo	Atezo or placebo + T-DM1	ITT PFS 8.2 vs. 6.8 mo, HR 0.82, P = 0.33; PD-L1+ PFS 8.5 vs. 4.1 mo, HR 0.62; PD-L1− PFS 6.8 vs. 8.2, HR 1.06
PANACEA (n = 55)	Phase Ib/II parallel cohort	Metastatic HER2+ breast cancer progressed on trastuzumab	Pembro	Pembro + trastuzumab	PD-L1+ vs. PD-L1− cohorts: ORR 15% vs. 0%; 12-mo PFS 13% vs. 0%; 12-mo OS rate 65% vs. 12%
AIPAC (n = 227)	Randomized phase IIb	Metastatic HR+ breast cancer, first-line chemo	Efti	Paclitaxel with efti or placebo for six cycles, then maintenance efti vs. placebo	ORR 48.3% vs. 38.4%; PFS at 6 mo 63% vs. 54%, HR 0.93, P = 0.341
SAFIR02-BREAST IMMUNO (n = 199)	Randomized phase II	Metastatic HER2neg breast cancer, stable after six to eight cycles of chemo	Durva	Maintenance therapy durva vs. chemo	Overall (117/199 HR+) PFS HR 1.40, P = 0.047; OS HR 0.84, P = 0.423; *Potential benefit in TNBC subset (r = 82)
Pembro + eribulin (n = 88)	Randomized phase Ib/II	Metastatic HR+ breast cancer	Pembro	Pembro + eribulin vs. eribulin	ORR 27% vs. 34%, P = 0.49; PFS 4.1 vs. 4.2 mo, HR 0.80, P = 0.33
Treme + exemestane (n = 26)	Single-arm phase I	Advanced HR+ breast cancer	Treme	Treme + exemestane	Stable disease ≥12 wk in 42%
Pembro + tamoxifen + vorinostat (n = 34)	Randomized phase II	Advanced HR+ breast cancer	Pembro	Concurrent tamoxifen + vorinostat with pembro vs. one cycle of priming with tamoxifen + vorinostat followed by concurrent therapy with pembro	ORR 4%; CBF 19%
MEDIOLA (n = 30)	Single-arm phase II	gmBRCA HER2neg metastatic breast cancer	Durva	4 wk olaparib priming followed by concurrent olaparib with durva	Overall ORR 63%; HR− (n = 13) ORR 69%

(HER2) Human epidermal growth factor 2, (HER2neg) human epidermal growth factor 2-negative, (HR) hormone receptor, (TNBC) triple-negative breast cancer, (atezo) atezolizumab, (pembro) pembrolizumab, (treme) tremelimumab, (durva) durvalumab, (chemo) chemotherapy, (efti) eftilagomad alpha, (T-DM1) trastuzumab emtansine, (ITT) intent to treat, (PD-L1) programmed death ligand 1, (CPS) combined positive score, (ORR) objective response rate, (DCR) disease control rate, (PFS) progression-free survival, (OS) overall survival, (HR) hazard ratio, (ORR) objective response rate, (gmBRCA) germline mutated breast cancer–related gene, (wtBRCA) wild-type breast cancer–related gene.

was no PFS difference (HR 0.93, $P = 0.341$), but efti and paclitaxel improved OS in patients <65 yr with low monocytes and more aggressive disease by about 8 mo (22.3 vs. 14.8 mo, $P = 0.17$). A follow-up phase 3 clinical trial is planned.

Biomarker-Defined Metastatic Breast Cancer beyond Subtype

The MEDIOLA study enrolled 30 patients with germline BRCA-mutated HER2neg metastatic breast cancer to receive olaparib priming for 4 wk, followed by durvalumab with olaparib starting at week 5; 17 patients had TNBC and the remainder were HR[+] (Domchek et al. 2020). The overall ORR was 63%, with ORRs in TNBC and HR[+] disease of 59% and 69%, respectively. The median duration of response was 9.2 mo. Nine patients had early disease progression at ≤28 wk. Potential mechanisms of resistance identified included BRCA2 reversion, lack of BRCA2 LOH, p53 mutation, and PD-L1/PD-L2 gene amplification.

The TAPUR study tests commercially available targeted agents in patients with advanced cancers and a predictive molecular alteration (Alva et al. 2021). This study treated 28 patients with metastatic breast cancer (any subtype) and a high TMB of 9–37 Mut/Mb. The ORR and DCR were 21% and 37%, respectively, and median OS was 30.6 wk (95% CI 18.3–103.3). There was no association between PFS and TMB.

The NIMBUS study tested nivolumab with low-dose ipilimumab in 30 patients with advanced HER2neg breast cancer and a TMB ≥9 mut/Mb; 21 had HR[+] disease, and eight were treated first-line (Barroso-Sousa et al. 2022). The overall ORR was 17% (5/30 patients). Three patients had HR[+] disease with TMBs of 110, 38, and 17.5, whereas two TNBC responders had TMBs of 10.9 and 9.1, with the first also PD-L1[+]. Patients with TMB ≥14 had an ORR of 60%, suggesting prospective trials are needed to determine the optimal TMB cutoff in breast cancer.

NOVEL EMERGING IMMUNOTHERAPIES

To date, pembrolizumab is the only immunotherapy approved for breast cancer in the United States,

with its utility currently limited to TNBC. Key goals for the field are to enhance the activity of immunotherapy for TNBC, and to expand its use to other breast cancer subtypes. One major strategy is to combine ICIs with standard and novel breast cancer therapies. Attractive combinations include ICIs with newer antibody drug conjugates (sacituzumab govitecan, fam-trastuzumab deruxtecan, for example), radiotherapy, cyroablation, and new precision drugs targeting unique aspects of intrinsic tumor biology and/or the tumor immune microenvironment (key signaling pathways and metabolic circuits). Initial combinations have been discussed above. A second major strategy is the development of innovative immunotherapies, including novel immune checkpoint modulators, vaccines, adoptive cell therapies, and bispecific molecules.

Novel Immune Checkpoints

New immune checkpoints are under active investigation. The LAG-3-specific antibody relatlimab is now approved for melanoma in combination with nivolumab and represents the third type of ICI approved for clinical use (Tawbi et al. 2022). Clinical trials have tested antibodies specific for TIM-3 and TIGIT, but clinical activity has not been clearly demonstrated. Other promising novel immune checkpoints under active clinical investigation include CD40, OX-40, ICOS, B7-H3, and B7-H4.

Vaccines

Cancer vaccines have been tested with minimal clinical success. Therapeutic breast cancer vaccines were first tested in metastatic disease, and then in the adjuvant setting; all phase 3 clinical trials so far have been negative (Solinas et al. 2020). There is now increasing interest in applying cancer vaccines for disease interception and prevention, as disease burdens and immune suppression are both minimal.

Adoptive Cell Therapy

After ICIs, adoptive cellular therapy with CD19-specific CAR-T cells is the second major class of immunotherapy that has revolutionized cancer therapy. It is approved for patients with CD19-expressing hematologic malignancies. CAR-T cell therapy has been challenging to develop in solid tumors, due to on-target/off-tumor toxicity and lack of CAR-T cell trafficking into solid tumors. In contrast, therapy with autologous TILs is promising for solid tumors, including breast cancer (Zacharakis et al. 2022).

Bispecific Small Molecules

Bispecific small molecules are engineered to simultaneously recognize two distinct targets (Roussos Torres and Emens et al. 2022). The bispecific T-cell engager (CD3 × CD19) blinatumomab is a standard of care for the treatment of B-cell malignancies. Bispecific T-cell engagers under evaluation in clinical trials for breast cancer include CD3 × HER2 and CD3 × p-cadherin molecules. Dual affinity retargeting (DART) molecules simultaneously target two immune checkpoints, and DARTs tested in breast cancer so far are PD-1 × CTLA-4 and PD-1 × LAG-3 agents. Finally, bintrafusp alfa is a bifunctional fusion protein that simultaneously targets PD-L1 and transforming growth factor β. Early clinical trials demonstrated promising activity particularly in HPV-related cancers.

CONCLUDING REMARKS

We have made enormous progress in breast cancer immunotherapy over the last decade. Immunotherapy with pembrolizumab and chemotherapy is now approved for neoadjuvant therapy of high-risk early TNBC, regardless of PD-L1 expression, and for the first-line therapy of metastatic PD-L1[+] TNBC. We have significant work ahead to clarify how best to use immunotherapy in breast cancer, and to extend the impact of immunotherapy to patients with HER2[+] or HR[+] disease. For early-stage TNBC, remaining questions include predictive biomarkers of response and resistance, the optimal neoadjuvant chemotherapy backbone, and strategies for the tailored management of residual disease and pCR after neoadjuvant immunotherapy. For metastatic TNBC, remaining needs are better predictive biomarkers beyond PD-L1 for ICIs, defining primary and secondary mechanisms of immunotherapy resistance, and new immunotherapy strategies for metastatic TNBC independent of PD-L1 expression, and for other breast cancer subtypes. Given the innovative immune-based agents already under development, the future of breast cancer immunotherapy is bright.

ACKNOWLEDGMENTS

We thank Dr. Julia Dixon-Douglas for her contributions to manuscript writing. L.A.E. acknowledges research funding from the Breast Cancer Research Foundation, Stand Up to Cancer, National Institutes of Health, Department of Defense Breast Cancer Program, Abbie, AstraZeneca, Bristol Myers Squibb, Compugen, CytomX, EMD Serono, Roche/Genentech, Immune Onc, Merck, Next Cure, Silverback Therapeutics, Takeda, and Tempest, all to the institution. She acknowledges a consulting/advisory role for AstraZeneca, Chugai, CytomX, Roche/Genentech, Gilead, GPCR, Immune Onc, Immutep, Mersana, and Shionogi. She also acknowledges Roche/Genentech for medical writing support, and the potential for future stock options from Molecuvax. S.L. receives research funding to her institution from Novartis, Bristol Meyers Squibb, Merck, Puma Biotechnology, Eli Lilly, Nektar Therapeutics, AstraZeneca, Roche/Genentech, and Seattle Genetics. She has acted as consultant (not compensated) to Seattle Genetics, Novartis, Bristol Meyers Squibb, Merck, AstraZeneca, Eli Lilly, Pfizer, Gilead Therapeutics and Roche/Genentech. S.L. has acted as consultant (paid to her institution) to Aduro Biotech, Novartis, GlaxoSmithKline, Roche/Genentech, AstraZeneca, Silverback Therapeutics, G1 Therapeutics, PUMA Biotechnologies, Pfizer, Gilead Therapeutics, Seattle Genetics, Daiichi Sankyo, Merck, Amunix, Tallac Therapeutics, Eli Lilly, and Bristol Meyers Squibb.

REFERENCES

Alva AS, Mangat PK, Garrett-Mayer E, Halabi S, Hansra D, Calfa CJ, Khalil MF, Ahn ER, Cannon TL, Crilley P, et al.

2021. Pembrolizumab in patients with metastatic breast cancer with high tumor mutational burden: results from the targeted agent and profiling utilization registry (TAPUR) study. *J Clin Oncol* **39:** 2443–2451. doi:10.1200/JCO.20.02923

Barroso-Sousa R, Li T, Reddy S, Emens LA, Overmoyer B, Lange P, Dilullo MK, Attaya V, Kimmel J, Winer EP, et al. 2022. Abstract GS2-10: nimbus: a phase 2 trial of nivolumab plus ipilimumab for patients with hypermutated HER2-negative metastatic breast cancer (MBC). *Cancer Res* **82:** GS2-10. doi:10.1158/1538-7445.SABCS21-GS2-10

Batchelot T, Filleron T, Bieche I, Arnedos M, Campone M, Dalenc F, Coussy F, Sablin MP, Debled M, Lefeuvre-Plesse C, et al. 2021. Durvalumab compared to maintenance chemotherapy in metastatic breast cancer: the randomized phase II SAFIR02-BREAST IMMUNO trial. *Nat Med* **27:** 250–255. doi:10.1038/s41591-020-01189-2

Bianchini G, Dugo M, Huang C, Egle D, Bermejo B, Seitz RS, Nielsen TJJ, Zamagni C, Thill M, Anton A, et al. 2021. LBA-12 predictive value of gene-expression profiles (GEPs) and their dynamics during therapy in the NeoTRIPaPDL1 trial. *Ann Oncol* **32:** S1283–S1284. doi:10.1016/j.annonc.2021.08.2084

Bianchini G, Wang XQ, Danenberg E, Huang C-S, Egle D, Callari M, Bermejo B, Zamagni C, Thill M, Anton A. et al. 2022. Abstract GS1-00: single-cell spatial analysis by imaging mass cytometry and immunotherapy response in triple-negative breast cancer (TNBC) in the NeoTRIP-aPDL1 trial. *Cancer Res* **82:** GS1-00.

Cimino-Mathews A, Thompson E, Taube JM, Ye X, Lu Y, Meeker A, Xu H, Sharma R, Lecksell K, Cornish TC, et al. 2016. PD-L1 (B7-H1) expression and the immune tumor microenvironment in primary and metastatic breast carcinomas. *Hum Pathol* **47:** 52–63. doi:10.1016/j.humpath.2015.09.003

Conforti F, Pala L, Bagnardi V, De Pas T, Colleoni M, Buyse M, Hotrobagyi G, Gianni L, Winer E, Loibl S, et al. 2022. Surrogacy of pathologic complete response in trials of neoadjuvant therapy for early breast cancer: critical analysis of strengths, weaknesses, and misinterpretations. *JAMA Oncol* **8:** 1668–1675. doi:10.1001/jamaoncol.2022.3755

Conte PF, Dieci MV, Bisagni G, De Laurentiis M, Tondini CA, Schmid P, De Salvo GL, Moratello G, Guarneri V. 2020. Phase III randomized study of adjuvant treatment with the ANTI-PD-L1 antibody avelumab for high-risk triple negative breast cancer patients: the A-BRAVE trial. *J Clin Oncol* **38:** TPS598.

Cortazar P, Zhang L, Untch M, Mehta K, Costantino JP, Wolmark N, Bonnefoi H, Cameron D, Gianni L, Valagussa P, et al. 2014. Pathological complete response and long-term clinical benefit in breast cancer: the CTNeoBC pooled analysis. *Lancet* **384:** 164–172. doi:10.1016/S0140-6736(13)62422-8

Cortes J, Cescon DW, Nowecki Z, Im SA, Yusof MM, Gallardo C, Lipatov O, Barrios CH, Holgado E, Iwata H, et al. 2020. Pembrolizumab plus chemotherapy versus placebo plus chemotherapy for previously untreated locally recurrent inoperable or metastatic triple-negative breast cancer (KEYNOTE-355): a randomised, placebo-controlled, double-blind, phase 3 clinical trial. *Lancet* **396:** 1817–1828. doi:10.1016/S0140-6736(20)32531-9

Cortes J, Rugo HS, Cescon DW, Im SA, Yusof MM, Gallardo C, Lipatov O, Barrios CH, Peres-Garcia J, Iwata H, et al. 2022. Pembrolizumab plus chemotherapy in advanced triple negative-breast cancer. *N Engl J Med* **387:** 217–226. doi:10.1056/NEJMoa2202809

Dieci MV, Guarneri V, Tosi A, Bisagni G, Musolino A, Spazzapan S, Moretti E, Vernaci GM, Griguolo G, Giarratano T, et al. 2022. Neoadjuvant chemotherapy and immunotherapy in luminal B-like breast cancer: results of the phase II GIADA trial. *Clin Cancer Res* **28:** 308–317. doi:10.1158/1078-0432.CCR-21-2260

Domchek SM, Postel-Vinay S, Im SA, Park YH, Delord J-P, Italiano A, Alexandre J, You B, Bastian S, Krebs MG, et al. 2020. Olaparib and durvalumab in patients with germline BRCA-mutated metastatic breast cancer (MEDIOLA): an open-label, multicentre, phase 1/2 basket study. *Lancet Oncol* **21:** 1155–1164. doi:10.1016/S1470-2045(20)30324-7

Emens LA, Middleton G. 2015. The interplay of immunotherapy and chemotherapy: harnessing potential synergies. *Cancer Immunol Res* **3:** 436–443. doi:10.1158/2326-6066.CIR-15-0064

Emens LA, Esteva FJ, Beresford M, Saura C, De Laurentiis M, Kim SB, Im SA, Wang Y, Salgado R, Mani A, et al. 2020. Trastuzumab emtansine plus atezolizumab versus trastuzumab emtansine plus placebo in previously treated, HER2-positive advanced breast cancer (KATE2): a phase 2, multicentre, randomised, double-blind trial. *Lancet Oncol* **21:** 1283–1295. doi:10.1016/S1470-2045(20)30465-4

Emens LA, Molinero L, Loi S, Rugo HS, Schneeweis A, Diéras V, Iwata H, Barrios CH, Nechaeva M, Nguyen-Duc A, et al. 2021a. Atezolizumab and *nab*-paclitaxel in advanced triple-negative breast cancer: biomarker evaluation of the IMpassion130 study. *J Natl Cancer Inst* **113:** 1005–1016. doi:10.1093/jnci/djab004

Emens LA, Adams S, Barrios CH, Diéras V, Iwata H, Loi S, Rugo HS, Schneeweiss A, Winer EP, Patel S, et al. 2021b. First-line atezolizumab plus nab-paclitaxel for unresectable locally advanced or metastatic triple-negative breast cancer: IMpassion130 final overall survival analysis. *Ann Oncol* **32:** 983–993. doi:10.1016/j.annonc.2021.05.355

Emens LA, Adams S, Cimino-Mathews A, Disis ML, Gatti-Mays ME, Ho AY, Kalinsky K, McArthur HL, Mittendorf EA, Nanda R, et al. 2021c. Society for immunotherapy of cancer (SITC) clinical practice guideline on immunotherapy for the treatment of breast cancer. *J Immunother Cancer* **9:** e002597. doi:10.1136/jitc-2021-002597

Geyer CE, Sikov WM, Huober J, Rugo HS, Wolmark N, O'Shaughnessy J, Maag D, Untch M, Golshan M, Lorenzo JP, et al. 2022. Long-term efficacy and safety of addition of carboplatin with or without veliparib to standard neoadjuvant chemotherapy in triple-negative breast cancer: 4-year follow-up data from BrighTNess, a randomized phase III trial. *Ann Oncol* **33:** 384–394. doi:10.1016/j.annonc.2022.01.009

Gianni L, Huang CS, Egle D, Bermejo B, Zamagni C, Thill M, Anton A, Zambelli S, Bianchini G, Russo S, et al. 2022. Pathologic complete response (pCR) to neoadjuvant treatment with or without atezolizumab in triple-negative, early high-risk and locally advanced breast cancer: NeoTRIP michelangelo randomized study. *Ann Oncol* **33:** 534–543. doi:10.1016/j.annonc.2022.02.004

Cite this article as *Cold Spring Harb Perspect Med* doi: 10.1101/cshperspect.a041332

Gustafson CE, Jadhav R, Cao W, Qi Q, Pegram M, Tian L, Weyand CM, Goronzy JJ. 2020. Immune cell repertoires in breast cancer patients after adjuvant chemotherapy. *JCI Insight* 5: e134569. doi:10.1172/jci.insight.134569

Hamilton EP, Kaklamani V, Falkson C, Vidal GA, Ward PJ, Patre M, Chui SY, Rotmensch J, Gupta K, Molinero L, et al. 2021. Impact of anti-HER2 treatments combined with atezolizumab on the tumor immune microenvironment in early or metastatic breast cancer: results from phase Ib study. *Clin Breast Cancer* 21: 539–551. doi:10.1016/j.clbc.2021.04.011

Harbeck N, Rastogi P, Martin M, Tolaney SM, Shao ZM, Fasching PA, Huang CS, Jaliffe GG, Tryakin A, Goetz MP, et al. 2021. Adjuvant abemaciclib combined with endocrine therapy for high-risk early breast cancer: updated efficacy and Ki-67 analysis from the MonarchE study. *Ann Oncol* 32: 1571–1581. doi:10.1016/j.annonc.2021.09.015

Huober J, Barrios CH, Niikura N, Jarząb M, Chang YC, Huggins-Puhalla SL, Pedrini J, Zhukova L, Graupner V, Eiger D, et al. 2022. Atezolizumab with neoadjuvant anti–human epidermal growth factor receptor 2 therapy and chemotherapy in human epidermal growth factor receptor 2–positive early breast cancer: primary results of the randomized phase III IMpassion050 trial. *J Clin Oncol* 40: 2946–2956. doi:10.1200/JCO.21.02772

Huppert LA, Gumusay O, Rugo HS. 2022. Emerging treatment strategies for metastatic triple-negative breast cancer. *Ther Adv Med Oncol* 14: 175883592210869. doi:10.1177/17588359221086916

Karn T, Denkert C, Weber KE, Holtrich U, Hanusch C, Sinn BV, Higgs BW, Jank P, Sinn HP, Huober J, et al. 2020. Tumor mutational burden and immune infiltration as independent predictors of response to neoadjuvant immune checkpoint inhibition in early TNBC in GeparNuevo. *Ann Oncol* 31: 1216–1222. doi:10.1016/j.annonc.2020.05.015

Keenan TE, Guerriero JL, Barroso-Sousa R, Li T, O'Meara T, Giobbie-Hurder A, Tayob N, Hu J, Severgnini M, Agudo J, et al. 2021. Molecular correlates of response to eribulin and pembrolizumab in hormone receptor-positive metastatic breast cancer. *Nat Commun* 12: 5563. doi:10.1038/s41467-021-25769-z

Liu J, Blake SJ, Yong MC, Harjunpää H, Ngiow SF, Takeda K, Young A, O'Donnell JS, Allen S, Smyth MJ, et al. 2016. Improved efficacy of neoadjuvant compared to adjuvant immunotherapy to eradicate metastatic disease. *Cancer Discov* 6: 1382–1399. doi:10.1158/2159-8290.CD-16-0577

Loi S, Giobbie-Hurder A, Gombos A, Bachelot T, Hui R, Curigliano G, Campone M, Biganzoli L, Bonnefoi H, Jerusalem G, et al. 2019. Pembrolizumab plus trastuzumab in trastuzumab-resistant, advanced, HER2-positive breast cancer (PANACEA): a single-arm, multicentre, phase 1b-2 trial. *Lancet Oncol* 20: 371–382. doi:10.1016/S1470-2045(18)30812-X

Loi S, Salgado R, Adams S, Pruneri G, Francis PA, Lacroix-Triki M, Joensuu H, Dieci MV, Badve S, Demaria S, et al. 2022a. Tumor infiltrating lymphocyte stratification of prognostic staging of early-stage triple negative breast cancer. *NPJ Breast Cancer* 8: 3. doi:10.1038/s41523-021-00362-1

Loi S, Francis PA, Zdenkowski N, Gebski V, Fox SB, White M, Kiely BE, Woodward NE, Hui R, Redfern AD, et al. 2022b. Neoadjuvant ipilimumab and nivolumab in combination with paclitaxel following anthracycline-based chemotherapy in patients with treatment resistant early-stage triple-negative breast cancer (TNBC): a single-arm phase 2 trial. *J Clin Oncol* 40: 602. doi:10.1200/JCO.2022.40.16_suppl.602

Loibl S, Untch M, Burchardi N, Huober J, Sinn BV, Blohmer J-U, Grischke E-M, Furlanetto J, Tesch H, Hanusch C, et al. 2019. A randomised phase II study investigating durvalumab in addition to an anthracycline taxane-based neoadjuvant therapy in early triple-negative breast cancer: clinical results and biomarker analysis of GeparNuevo study. *Ann Oncol* 30: 1279–1288. doi:10.1093/annonc/mdz158

Loibl S, Schneeweiss A, Huober J, Braun M, Rey J, Blohmer J-U, Furlanetto J, Zahm D-M, Hanusch C, Thomalla J, et al. 2022. Neoadjuvant durvalumab improves survival in early triple-negative breast cancer independent of pathological complete response. *Ann Oncol Aug* 33: 1149–1158. doi:10.1016/j.annonc.2022.07.1940

Luen SJ, Salgado R, Dieci MV, Vingiani A, Curigliano G, Gould RE, Castaneda C, D'Alfonso T, Sanchez J, Cheng E, et al. 2019. Prognostic implications of residual disease tumor-infiltrating lymphocytes and residual cancer burden in triple-negative breast cancer patients after neoadjuvant chemotherapy. *Ann Oncol* 30: 236–242. doi:10.1093/annonc/mdy547

Mattarollo SR, Loi S, Duret H, Ma Y, Zitvogel L, Smyth MJ. 2011. Pivotal role of innate and adaptive immunity in anthracycline chemotherapy of established tumors. *Cancer Res* 71: 4809–4820. doi:10.1158/0008-5472.CAN-11-0753

Ménétrier-Caux C, Ray-Coquard I, Blay JY, Caux C. 2019. Lymphopenia in cancer patients and its effects on response to immunotherapy: an opportunity for combination with cytokines? *J Immunother Cancer* 7: 85. doi:10.1186/s40425-019-0549-5

Miceli R, Eriksson H, Eustace AJ, Lo Russo G, Ballot J, Bergamini C, Bjaanaes MM, Corti F, De Cecco L, Frisardi L, et al. 2021. 1795P gender difference in side effects of immunotherapy: a possible clue to optimize cancer treatment. *Ann Oncol* 32: S1223–S1224. doi:10.1016/j.annonc.2021.08.1737

Miles D, Gligorov J, André F, Cameron D, Schneeweiss A, Barrios C, Xu B, Wardley A, Kaen D, Andrade L, et al. 2021. Primary results from IMpassion131, a double-blind, placebo-controlled, randomised phase III trial of first-line paclitaxel with or without atezolizumab for unresectable locally advanced/metastatic triple-negative breast cancer. *Ann Oncol* 32: 994–1004. doi:10.1016/j.annonc.2021.05.801

Mittendorf EA, Zhang H, Barrios CH, Saji S, Jung KH, Hegg R, Koehler A, Sohn J, Iwata H, Telli ML, et al. 2020. Neoadjuvant atezolizumab in combination with sequential nab-paclitaxel and anthracycline-based chemotherapy versus placebo and chemotherapy in patients with early-stage triple-negative breast cancer (IMpassion031): a randomised, double-blind, phase 3 trial. *Lancet* 396: 1090–1100. doi:10.1016/S0140-6736(20)31953-X

Montico B, Nigro A, Casolaro V, Dal Col J. 2018. Immunogenic apoptosis as a novel tool for anticancer vaccine development. *Int J Mol Sci* **19**: 594. doi:10.3390/ijms19020594

Mozaffari F, Lindemalm C, Choudhury A, Granstam-Björneklett H, Helander I, Lekander M, Mikaelsson F, Nilsson B, Ojutkangas ML, Österborg A, et al. 2007. NK-cell and T-cell functions in patients with breast cancer: effects of surgery and adjuvant chemo- and radiotherapy. *Br J Cancer* **97**: 105–111. doi:10.1038/sj.bjc.6603840

Mozaffari F, Lindemalm C, Choudhury A, Granstam-Björneklett H, Lekander M, Nilsson B, Ojutkangas ML, Österborg A, Bergkvist L, Mellstedt H, et al. 2009. Systemic immune effects of adjuvant chemotherapy with 5-fluorouracil, epirubicin and cyclophosphamide and/or radiotherapy in breast cancer: a longitudinal study. *Cancer Immunol Immunother* **58**: 111–120. doi:10.1007/s00262-008-0530-5

Muraro E, Comaro E, Talamini R, Turchet E, Miolo G, Scalone S, Militello L, Lombardi D, Spazzapan S, Perin T, et al. 2015. Improved natural killer cell activity and retained anti-tumor CD8$^+$ T cell responses contribute to the induction of a pathological complete response in HER2-positive breast cancer patients undergoing neoadjuvant chemotherapy. *J Transl Med* **13**: 204. doi:10.1186/s12967-015-0567-0

Nanda R, Liu MC, Yau C, Shatsky R, Pusztai L, Wallace A, Chien AJ, Forero-Torres A, Ellis E, Han H, et al. 2020. Effect of pembrolizumab plus neoadjuvant chemotherapy on pathologic complete response in women with early-stage breast cancer: an analysis of the ongoing phase 2 adaptively randomized I-SPY2 trial. *JAMA Oncol* **6**: 676–684. doi:10.1001/jamaoncol.2019.6650

Özdemir BC, Coukos G, Wagner D. 2018. Immune-related adverse events of immune checkpoint inhibitors and the impact of sex—what we know and what we need to learn. *Ann Oncol* **29**: 1067. doi:10.1093/annonc/mdx818

Park YH, Lal S, Lee JE, Choi Y-L, Wen J, Ram S, Ding Y, Lee S-H, Powell E, Lee SK, et al. 2020. Chemotherapy induces dynamic immune responses in breast cancers that impact treatment outcome. *Nat Commun* **11**: 6175.

Parkes EE, Savage KI, Lioe T, Boyd C, Halliday S, Walker SM, Lowry K, Knight L, Buckley NE, Grogan A, et al. 2022. Activation of a cGAS-STING-mediated immune response predicts response to neoadjuvant chemotherapy in early breast cancer. *Br J Cancer* **126**: 247–258. doi:10.1038/s41416-021-01599-0

Perez-Garcia JM, Llombart-Cussac A, Cortes MG, Curigliano G, Lopez-Miranda E, Alonso JL, Bermejo B, Calvo L, Caranana V, de la Cruz Sanchez S, et al. 2021. Pembrolizumab plus eribulin in hormone-receptor-positive, HER2-negative, locally recurrent or metastatic breast cancer (KELLY): an open-label, multicentre, single-arm, phase II trial. *Eur J Cancer* **148**: 383–394.

Pfannenstiel LW, Lam SS, Emens LA, Jaffee EM, Armstrong TD. 2010. Paclitaxel enhances early dendritic cell maturation and function through TLR4 signaling in mice. *Cell Immunol* **263**: 79–87. doi:10.1016/j.cellimm.2010.03.001

Pusztai L, Yau C, Wolf DM, Han HS, Du L, Wallace AM, Stringer-Reasor E, Boughey JC, Chien AJ, Elias AD, et al. 2021. Durvalumab with olaparib and paclitaxel for high-risk HER2-negative stage II/III breast cancer: results from the adaptively randomized I-SPY2 trial. *Cancer Cell* **39**: 989–998.e5. doi:10.1016/j.ccell.2021.05.009

Pusztai L, Denkert C, O'Shaughnessy J, Cortes J, Dent RA, McArthur HL, Kuemmel S, Bergh JCS, Park YH, Hui RH, et al. 2022. Event-free survival by residual cancer burden after neoadjuvant pembrolizumab + chemotherapy versus placebo + chemotherapy for early TNBC: exploratory analysis from KEYNOTE-522. *J Clin Oncol* **40**: 503. doi:10.1200/JCO.2022.40.16_suppl.503

Roussos Torres ET, Emens LA. 2022. Emerging combination immunotherapy strategies for breast cancer: dual immune checkpoint modulation, antibody-drug conjugates, and bispecific antibodies. *Breast Cancer Res Treat* **191**: 291–302. doi:10.1007/s10549-021-06423-0

Rugo HS, Delord JP, Im SA, Ott PA, Piha-Paul SA, Bedard PL, Sachdev J, Le Tourneau C, van Brummelen EMJ, Varga A, et al. 2018. Safety and antitumor activity of pembrolizumab in patients with estrogen receptor–positive/human epidermal growth factor receptor 2–negative advanced breast cancer. *Clin Cancer Res* **24**: 2804–2811. doi:10.1158/1078-0432.CCR-17-3452

Rugo HS, Loi S, Adams S, Schmid P, Schneeweiss A, Barrios CH, Iwata H, Diéras V, Winer EP, Kockx MM, et al. 2021. PD-L1 immunohistochemistry assay comparison in atezolizumab plus *nab*-paclitaxel–treated advanced triple-negative breast cancer. *J Natl Cancer Inst* **113**: 1733–1743. doi:10.1093/jnci/djab108

Santa-Maria CA, Kato T, Park J-H, Kiyotani K, Rademaker A, Shah AN, Gross L, Blanco LZ, Jain S, Flaum L, et al. 2018. A pilot study of durvalumab and tremelimumab and immunogenomic dynamics in metastatic breast cancer. *Oncotarget* **9**: 18985–18996. doi:10.18632/oncotarget.24867

Savas P, Virassamy B, Ye C, Salim A, Mintoff CP, Caramia F, Salgado R, Byrne DJ, Teo ZL, Dushyanthen S, et al. 2018. Single-cell profiling of breast cancer T cells reveals a tissue-resident memory subset associated with improved prognosis. *Nat Med* **24**: 986–993. doi:10.1038/s41591-018-0078-7

Schmid P, Adams S, Rugo HS, Schneeweiss A, Barrios CH, Iwata H, Diéras V, Hegg R, Im SA, Shaw Wright G, et al. 2018. Atezolizumab and nab-paclitaxel in advanced triple-negative breast cancer. *N Engl J Med* **379**: 2108–2121. doi:10.1056/NEJMoa1809615

Schmid P, Cortes J, Pusztai L, McArthur H, Kümmel S, Bergh J, Denkert C, Park YH, Hui R, et al. 2020a. Pembrolizumab for early triple-negative breast cancer. *N Engl J Med* **382**: 810–821. doi:10.1056/NEJMoa1910549

Schmid P, Rugo HS, Adams S, Schneeweiss A, Barrios CH, Diéras V, Henschel V, Molinero L, Chui SY, Maiya V, et al. 2020b. Atezolizumab plus nab-paclitaxel as first-line treatment for unresectable, locally advanced or metastatic triple-negative breast cancer (IMpassion130): updated efficacy results from a randomised, double-blind, placebo-controlled, phase 3 trial. *Lancet Oncol* **21**: 44–59. doi:10.1016/S1470-2045(19)30689-8

Schmid P, Cortes J, Dent R, Pusztai L, McArthur H, Kümmel S, Bergh J, Denkert C, Park YH, Hui R, et al. 2022. Event-free survival with pembrolizumab in early triple-negative breast cancer. *N Engl J Med* **386**: 556–567. doi:10.1056/NEJMoa2112651

Serpico AF, Visconti R, Grieco D. 2020. Exploiting immune-dependent effects of microtubule-targeting agents to improve efficacy and tolerability of cancer treatment. *Cell Death Dis* **11**: 361. doi:10.1038/s41419-020-2567-0

Sharma P, Stecklein SR, Yoder R, Staley JM, Schwensen K, O'Dea A, Nye LE, Ella M, Satelli D, Crane G, et al. 2022. Clinical and biomarker results of neoadjuvant phase II study of pembrolizumab and carboplatin plus docetaxel in triple-negative breast cancer (TNBC) (Neo-PACT). *J Clin Oncol* **40**: 513. doi:10.1200/JCO.2022.40.16_suppl.513

Shepherd JH, Ballman K, Polley MC, Campbell JD, Fan C, Selitsky S, Fernandez-Martinez A, Parker JS, Hoadley KA, Hu Z, et al. 2022. CALGB 40603 (alliance): long-term outcomes and genomic correlates of response and survival after neoadjuvant chemotherapy with or without carboplatin and bevacizumab in triple-negative breast cancer. *J Clin Oncol* **40**: 1323–1334. doi:10.1200/JCO.21.01506

Solinas C, Aiello M, Migliori E, Willard-Gallo K, Emens LA. 2020. Breast cancer vaccines: heeding the lessons of the past to guide a path forward. *Cancer Treat Rev* **84**: 101947. doi:10.1016/j.ctrv.2019.101947

Stanton SE, Adams S, Disis ML. 2016. Variation in the incidence and magnitude of tumor-infiltrating lymphocytes in breast cancer subtypes: a systematic review. *JAMA Oncol* **2**: 1354–1360. doi:10.1001/jamaoncol.2016.1061

Szekely B, Bossuyt V, Li X, Wali VB, Patwardhan GA, Frederick C, Silber A, Park T, Harigopal M, Pelekanou V, et al. 2018. Immunological differences between primary and metastatic breast cancer. *Ann Oncol* **29**: 2232–2239. doi:10.1093/annonc/mdy399

Tarantino P, Marra A, Gandini S, Minotti M, Pricolo P, Signorelli G, Criscitiello C, Locatelli M, Belli C, Belloni M, et al. 2020. Association between baseline tumour burden and outcome in patients with cancer treated with next-generation immunooncology agents. *Eur J Cancer* **139**: 92–98. doi:10.1016/j.ejca.2020.08.026

Tawbi HA, Schadendorf D, Lipson EJ, Ascierto PA, Matamala L, Castillo Gutiérrez E, Rutkowski P, Gogas HJ, Lao CD, De Menezes JJ, et al. 2022. Relatlimab and nivolumab versus nivolumab in untreated advanced melanoma. *N Engl J Med* **386**: 24–34. doi:10.1056/NEJMoa2109970

Terranova-Barberio M, Pawlowska N, Dhawan M, Moasser M, Shien AJ, Melisko ME, Rugo H, Rahimi R, Deal T, Daud A, et al. 2020. Exhausted T cell signature predicts immunotherapy response in ER-positive breast cancer. *Nat Commun* **11**: 3584. doi:10.1038/s41467-020-17414-y

Tolaney SM, Barroso-Sousa R, Keenan T, Li T, Trippa L, Vaz-Luis I, Wulf G, Spring L, Sinclair NF, Andrews C, et al. 2020. Effect of eribulin with or without pembrolizumab on progression-free survival for patients with hormone receptor–positive, *ERBB2*-negative metastatic breast cancer: a randomized clinical trial. *JAMA Oncol* **6**: 1598–1605. doi:10.1001/jamaoncol.2020.3524

Van Vugt MATM, Parkes EE. 2022. When breaks get hot. inflammatory signaling in BRCA1/2 mutant cancers. *Trends Cancer* **8**: 174–189. doi:10.1016/j.trecan.2021.12.003

Venkatesulu BP, Mallick S, Lin SH, Krishnan S. 2018. A systematic review of the influence of radiation-induced lymphopenia on survival outcomes in solid tumors.

Crit Rev Oncol Hematol **123**: 42–51. doi:10.1016/j.critrevonc.2018.01.003

Verma R, Foster RE, Horgan K, Mounsey K, Nixon H, Smalle N, Hughes TA, Carter CRD. 2016. Lymphocyte depletion and repopulation after chemotherapy for primary breast cancer. *Breast Cancer Res* **18**: 10. doi:10.1186/s13058-015-0669-x

Vinayak S, Tolaney SM, Schwartzberg L, Mita M, McCann G, Tan AR, Wahner-Hendrickson AE, Forero A, Anders C, Wulf GM, et al. 2019. Open-label clinical trial of niraparib combined with pembrolizumab for treatment of advanced or metastatic triple-negative breast cancer. *JAMA Oncol* **5**: 1132–1140. doi:10.1001/jamaoncol.2019.1029

Vonderheide RH, LoRusso PM, Khalil M, Gartner EM, Khaira D, Soulieres D, Dorazio P, Trosko JA, Ruter J, Mariani GL, et al. 2010. Tremelimumab in combination with exemestane in patients with advanced breast cancer and treatment-associated modulation of inducible costimulator expression on patient T cells. *Clin Cancer Res* **16**: 3485–3494. doi:10.1158/1078-0432.CCR-10-0505

Voorwerk L, Slagter M, Horlings HM, Sikorska K, van de Vijver KK, de Maaker M, Nederlof I, Kluin RJC, Warren S, Ong S, et al. 2019. Immune induction strategies in metastatic triple-negative breast cancer to enhance the sensitivity to PD-1 blockade: the TONIC trial. *Nat Med* **25**: 920–928. doi:10.1038/s41591-019-0432-4

Wildiers H, Dirix L, Armstrong A, De Cuypere E, Dalenc F, Chan S, Marme F, Schroder CP, Huober J, Vuylsteke P, et al. 2021. Final results from AIPAC: a phase IIB comparing eftilagimod alpha (a soluble LAG-3 protein) vs. placebo in combination with weekly paclitaxel in HR$^+$ HER2$^-$ MBC. *J Immunother Cancer* **9**. doi:10.1136/jitc-2021-SITC2021.948

Winer EP, Lipatov O, Im SA, Goncalves A, Muñoz-Couselo E, Lee KS, Schmid P, Tamura K, Testa L, Witzel I, et al. 2021. Pembrolizumab versus investigator-choice chemotherapy for metastatic triple-negative breast cancer (KEYNOTE-119): a randomised, open-label, phase 3 trial. *Lancet Oncol* **22**: 499–511. doi:10.1016/S1470-2045(20)30754-3

Winship AL, Alesi LR, Sant S, Stringer JM, Cantavenera A, Hegarty T, Requesens CL, Liew SH, Sarma U, Griffiths MJ, et al. Checkpoint inhibitor immunotherapy diminishes oocyte number and quality in mice. *Nat Cancer* **3**: 1–13. doi:10.1038/s43018-022-00413-x

Wu SY, Xu Y, Chen L, Fan L, Ma XY, Zhao S, Song XQ, Hu X, Yang WT, Chai WJ, et al. 2022. Combined angiogenesis and PD-1 inhibition for immunomodulatory TNBC: concept exploration and biomarker analysis in the FUTURE-C-plus trial. *Mol Cancer* **21**: 84. doi:10.1186/s12943-022-01536-6

Zacharakis N, Huq LM, Seitter SJ, Kim SP, Gartner JJ, Sindri S, Hill VK, Li YF, Paria BC, Ray S, et al. 2022. Breast cancers are immunogenic: immunologic analyses and a phase II pilot clinical trial using mutation-reactive autologous lymphocytes. *J Clin Oncol* **40**: 1741–1754. doi:10.1200/JCO.21.02170

Zierhut C, Yamaguchi N, Paredes M, Luo JD, Carroll T, Funabiki H. 2019. The cytoplasmic DNA sensor cGAS promotes mitotic cell death. *Cell* **178**: 302–315.e23. doi:10.1016/j.cell.2019.05.035

Emerging Therapies for Breast Cancer

Shom Goel[1,2] and Sarat Chandarlapaty[3,4,5]

[1]Peter MacCallum Cancer Centre, Melbourne 3000, Australia

[2]The Sir Peter MacCallum Department of Oncology, University of Melbourne, Melbourne 3010, Australia

[3]Human Oncology and Pathogenesis Program (HOPP), Memorial Sloan Kettering Cancer Center, New York, New York 10021, USA

[4]Weill Cornell Medicine, New York, New York 10021, USA

[5]Breast Medicine Service, Department of Medicine, Memorial Sloan Kettering Cancer Center, New York, New York 10021, USA

Correspondence: Shom.goel@petermac.org; chandars@mskcc.org

The steady, incremental improvements in outcomes for both early-stage and advanced breast cancer patients are, in large part, attributable to the success of novel systemic therapies. In this review, we discuss key conceptual paradigms that have underpinned this success including (1) targeting the driver: the identification and targeting of major oncoproteins in breast cancers; (2) targeting the lineage pathway: inhibition of those pathways that drive normal mammary epithelial cell proliferation that retain importance in cancer; (3) targeting precisely: the application of molecular classifiers to refine therapy selection for specific cancers, and of antibody–drug conjugates to pinpoint tumor and tumor promoting cells for eradication; and (4) exploiting synthetic lethality: leveraging unique vulnerabilities that cancer-specific molecular alterations induce. We describe promising examples of novel therapies that have been discovered within each of these paradigms and suggest how future drug development efforts might benefit from the continued application of these principles.

Breast cancer has historically been treated with a focus on localized therapy including surgery and radiation, while systemic therapies have mainly served as "adjuvants" to suppress occult, micrometastatic disease. This approach remains a highly successful strategy for many cases; however, it proves inadequate for patients that present with overt distant disease and for a substantial portion of patients with localized breast cancer harboring features that imply a high likelihood of distant relapse. Indeed, the phenomenon of late recurrence has revealed the commonality of breast cancer as a systemic disease requiring systemic therapies for optimal long-term outcomes for most patients (Pan et al. 2017).

Among systemic therapies, major advances have emerged through the identifying and targeting of lineage and oncoprotein-specific targets to enable movement away from combination chemotherapy and toward endocrine and anti-HER2 therapies. These therapies have led to dramatic improvements in overall survival in both the early (Romond et al. 2005) and advanced disease (Slamon et al. 2001; Baselga et al. 2012a) settings and

Cite this article as *Cold Spring Harb Perspect Med* doi: 10.1101/cshperspect.a041333

serve as worthy exemplars for the new generations of systemic therapies that are needed for the disease subsets not effectively covered by these treatments. While there have been numerous drug approvals in breast cancer in the last two decades, we would suggest that there have been four paradigms that have both led to outstanding efficacy, and that these principles should provide guidance for breast cancer drug developers in the years ahead (Fig. 1).

1. Targeting the driver. The identification of HER2 amplification and its subsequent targeting with numerous anti-HER2 therapies is one of cancer medicine's most extraordinary success stories. This began with the seminal observation of HER2 overexpression via gene amplification altering the biology (Shih et al. 1981; Schechter et al. 1984, 1985) and out-

comes (Slamon et al. 1987, 1989) of these tumors. It was brought to fruition through the development of highly selective inhibitors of HER2 (e.g., trastuzumab) (Fendly et al. 1990) and continues through the ongoing development of regimens that maximize HER2 inhibition while simultaneously exploiting downstream vulnerabilities (e.g., combinations with antimicrotubular chemotherapies and dual anti-HER2 therapy).

2. Targeting the lineage pathway. Mechanistic studies revealing the role of steroid hormones and the cell-cycle machinery in shaping both mammary gland development and tumorigenesis (Sutherland et al. 1993; Wang et al. 1994) have been instrumental in the development of endocrine-based therapies for breast cancer. Specifically, understanding the basis

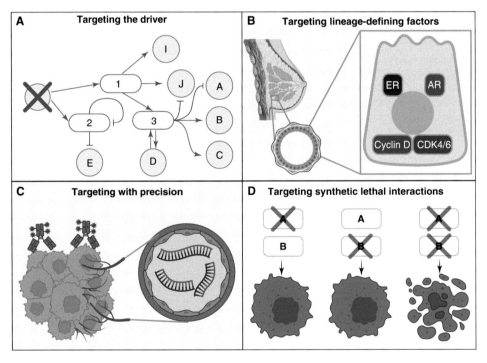

Figure 1. Approaches to the development of new breast cancer therapeutics. Drawing from recent successes in drug development, we propose four overarching strategies to guide future efforts in this field. (A) Targeting the driver: the identification, functional validation, and targeting of important oncoproteins in breast cancers. (B) Targeting lineage-defining factors: inhibition of pathways underpinning normal mammary epithelial proliferation that remain proliferative drivers in breast cancer. (C) Targeting with precision, including both the incorporation of molecular diagnostics to refine therapy selection and the development of novel antibody–drug conjugates. (D) Targeting synthetic lethal interactions: targeting the unique vulnerabilities of cancer cells that are conferred by their specific molecular alterations.

for how estrogens and the estrogen receptor (ER) regulate luminal breast cell proliferation (Eeckhoute et al. 2006) has directly informed the development of both antiestrogens and CDK4/6 inhibitors.

3. Targeting precisely. Treating breast cancer as a singular disease has led to vast overtreatment of breast cancers with agents such as chemotherapy. Moreover, the effects of these agents upon nontarget tissues have magnified the morbidity of this disease as patients incur toxicities. The use of molecular classifiers such as those based on gene expression (Perou et al. 2000; Sparano et al. 2015) have made the use of these drugs more exclusively for those with potential to derive benefit. Further, the development of antibody-directed chemotherapy has enabled a reduction in nontarget tissue exposure.

4. Exploiting synthetic lethalities. Fundamental research into the mechanisms by which BRCA1/2 loss induces genomic instability and thereby initiates breast cancer has also laid the foundation for therapeutic approaches that make use of this cancer-specific abnormality (Bryant et al. 2005; Farmer et al. 2005). The broad availability of CRISPR screening has unveiled new and unique vulnerabilities as a result of such changes akin to the efficacy of PARP inhibitors in BRCA1/2 mutant tumors.

In the following sections, we highlight several compelling examples of how these approaches are moving forward based on emerging biologic and chemical discoveries.

TARGETING THE DRIVER

PI3K-AKT-mTOR Pathway

Numerous profiling studies have demonstrated genetic alterations in components of the PI3K-AKT-mTOR signaling pathway are observed in upward of 70% of tumors (The Cancer Genome Atlas Network 2012). Activating mutations in phosphoinositide kinase, PIK3CA (PI3K), the downstream Ser/Thr kinase, AKT1, and inactivating mutations in phosphoinositide phosphatase, PTEN, comprise the majority. Beyond these, genetic alterations in many other components also induce this pathway including receptor tyrosine kinases (RTKs) (e.g., HER2), other kinases (PIK3CB, MTOR, S6K), or phosphatases (INPP4B). This has spurred the development of numerous PI3K pathway inhibitors for this disease that have ultimately disappointed, including drugs targeting PI3Ks (GDC0941, GDC0032, LY294002, BEZ235, BKM120), AKT (GDC0068, AZD5363, MK2206), and mTOR (AZD8055) (Table 1). However, initial generation inhibitors often lead to the development of second- and third-generation inhibitors such as the improvements in HER2 kinase inhibitors from lapatinib (Xia et al. 2002) to neratinib (Rabindran et al. 2004) and tucatinib (Dinkel et al. 2012). Two inhibitors of the PI3K pathway have led to sustained, if modest, improvements in progression-free survival (PFS) garnering drug approval: a p110α-selective inhibitor of PI3K (alpelisib) (André et al. 2019) and an allosteric inhibitor of mTOR (everolimus) (Baselga et al. 2012b). Both drugs were developed in combination with endocrine therapy (ET) in hormone receptor expressing (HR$^+$) metastatic breast cancer (MBC). Much work has gone into understanding the somewhat modest impact of these inhibitors and homed in on the challenges of cell-intrinsic (Chandarlapaty 2012) and systemic (Hopkins et al. 2018) feedback pathways, reducing the potency and efficacy of inhibitors as well as the toxicities incurred by through chronic PI3K pathway inhibition in normal tissues. New therapies have now been developed to overcome these challenges. (1) Mutant-selective inhibitors of PI3K are now being developed (Castel et al. 2021) that offer the opportunity to more potently inhibit the mutant oncoprotein in the tumor but not inhibit PI3K signaling in normal tissues. As a result, induction of hyperglycemia (toxicity) and reactivation of IGF1R/insulin receptor (reduced potency) may be ameliorated. It will be of interest to observe whether this class of inhibitors can still overcome the issue of tumor cell intrinsic feedback responses and thereby induce the drug-insensitive wild-type PI3K. (2) In addition to mutant-selective inhibitors, combinatorial approaches have been

Table 1. Emerging drugs and drug classes

Target (class)	Indication	Example(s) (stage in 2022)
PIK3CA (driver)—ATP competitive, selective inhibitors of wild-type (WT) and mutant α subunit	PI3K mutant; HER2 amplified	Alpelisib (approved/ phase 3*)
		Inavolisib (phase 3)
PIK3CA (driver)—allosteric, mutant selective inhibitors of α subunits that spare WT	PI3K mutant	RLY2608 (phase 1)
		LOXO783 (phase 1)
		STX-H1047-PI3Kα (phase 1)
mTOR (driver)—bisteric inhibitor of mTORC1	Pan-breast cancer	RMC5552 (phase 1)
MEK1/2 (driver)—allosteric inhibitors of MEK1/2 kinases	NF1 mutant and other MAPK pathway alterations	Mirdametinib (phase 1)
		Binimetinib (phase 1)
HER2 (driver)—ATP competitive inhibitors of HER2 kinase	HER2 somatic mutation; HER2 amplified	Neratinib (approved/ phase 3*)
		Pyrotinib (phase 3)
HER2 (driver)—HER2 antibody; antibody drug conjugates	HER2 somatic mutation; HER2 expressing; HER2 amplified	Trastuzumab deruxtecan (approved/phase 3*)
		Trastuzumab duocarmazine
ER (lineage)—ER antagonists including SERMs, SERDs, PROTACs	ER+	Elacestrant (phase 3)
		Lasofoxifene (phase 3)
		Giredestrant (phase 3)
		Imlunestrant (phase 3)
		ARV-471 (phase 3)
PR (lineage)—PR agonists, antagonists	ER+	Onapristone (phase 2)
		Prometrium (phase 2)
AR (lineage)—AR agonists, antagonists	ER+/TNBC	Enzalutamide (phase 2)
		Enobosarm (phase 2)
BCL-2 (lineage/cell cycle)—BH3 mimetics	ER+	Venetoclax (phase 2)
CDK2 (lineage/cell cycle)—ATP competitive CDK2 inhibitors	ER+/TNBC	PF-06873600 (phase 2)
CDK7 (lineage/cell cycle)—inhibitors of CDK7	ER+/TNBC	SY-5609 (phase ½)
		CT7001 (phase ½)
		LY3405105 (phase ½)
TROP2, Nectin, HER2, etc. (lineage/driver)—variety of antibody drug conjugates in testing/ approval	ER+/HER2 amp/TNBC	Sacituzumab govitecan (approved)
		Dato-DXd (phase 3)
		Enfortumab (approved)
		Trastuzumab DXd (approved)
PARP1 (synthetic lethality)—inhibitors of PARP1 enzymes	ER+/HER2 amp/TNBC	Olaparib (approved)
		Talazoparib (approved)
		AZD5305 (phase 1)
ATR—inhibitors of ATR kinase	ER+/TNBC	RP-3500 (phase 1)
Polθ—inhibitors of polymerase θ	ER+/TNBC	ART4215 (phase 1)
		Novobiocin (phase 1)

developed in an effort to block hyperglycemia including dietary changes (ketogenic diet) and SGLT2 inhibitors (Hopkins et al. 2018). (3) In addition to strategies that limit hyperglycemia, recent work has revealed the centrality of mTOR kinase as a key convergence of PI3K pathway signaling. Inhibition of PI3K/AKT signaling can lead to both induction of RTKs that can reactivate PI3K (Chandarlapaty et al. 2011) but also suppress the translation of PTEN (Mukherjee ct al. 2021) leading to induction of PI3Kβ. These have limited even efforts to target PI3Kα and PI3Kβ. However, newer inhibitors of mTOR kinase such as RAPA-LINK (Rodrik-Outmezguine et al. 2016) and selective inhibitors of mTORC1 such as RMC5552 (Lee et al. 2021) have shown promising activity in this context. Whether these will manifest a sufficient therapeutic index to allow long-term therapeutic efficacy will be investigated in early phase studies.

RTK-RAS-MAPK Pathway

While MAPK pathway activation is not a ubiquitous feature of primary breast cancer, recent studies sequencing ER^+ MBC have revealed the frequent acquisition of alterations that serve to induce MAPK signaling (Razavi et al. 2018). In particular, inactivating alterations in the RAS-GAP, NF1, and activating mutations in ERBB kinases, EGFR and ERBB2, are found and have been shown in preclinical models to emerge as oncogenic drivers that drive endocrine resistance (Zheng et al. 2020). (1) Targeting mutant, non-amplified HER2 has shown promising clinical activity using both irreversible kinase inhibitors such as neratinib in breast cancer (Smyth et al. 2020) as well as antibody drug conjugates such as TDM1 and TDXd in non-small-cell lung cancer (Li et al. 2020) (discussed further below). The utility of these approaches compared to or in combination with standard therapies will soon be evaluated in the clinic within the mutant population. (2) Germline mutation in NF1 has been previously linked to an increased risk of development of breast cancer among other malignancies; however, it was recent work in MBC that uncovered the wider prevalence of NF1 loss more commonly through sporadic mutation. Preclinical research

has suggested the efficacy of MAPK pathway inhibitors such as MEK inhibitors (Zheng et al. 2020; Smith et al. 2021); however, clinical trials are needed given the relatively low therapeutic index MEK inhibitors in this context as observed in other cancer types such as melanoma.

TARGETING THE LINEAGE PATHWAY

ER Inhibition

Antiestrogen therapy, given its widespread use in the early and advanced setting over many decades, has been one of the most effective and life-saving drug therapies in the history of oncology.

As discussed in the literature, ER may be targeted through the inhibition of estrogen production (aromatase inhibitors), antagonism of ER (ER antagonists), or degradation of ER (selective estrogen receptor degraders [SERDs]). In certain cases, an antiestrogen therapy may work through more than one of these mechanisms— for example, fulvestrant appears to act as both an ER antagonist and degrader. While these endocrine therapies work in the majority of patients, many tumors have been demonstrated to undergo evolution and acquire genetic alterations that mediate resistance. Principal among these alterations are highly recurrent (Robinson et al. 2013; Toy et al. 2013) somatic mutations in the ligand-binding domain (LBD) of the ER (ESR1) that cause estradiol-independent activity. While these mutant receptors are still susceptible to ER antagonists, they are observed to reduce binding and thus newer and/or more potent antagonists have been developed (Toy et al. 2017). These antagonists have taken a variety of approaches including enhanced bioavailability through different routes of administration (e.g., novel orally administered SERDS), enhanced degradation such as through PROTACS (Snyder et al. 2021), novel binding sites in the ligand-binding pocket (Puyang et al. 2018), and tissue-selective antagonism versus agonism (Lainé et al. 2021). While these have shown very promising anti-ESR1 mutant activity in preclinical studies (Liang et al. 2021; Shomali et al. 2021; Chen et al. 2022), clinical activity has been less pronounced. One speculation as to the basis for this is the interim development and adoption of CDK4/6 kinase inhibitors leading to tumor evolution beyond only antiestrogen resistance and toward antiestrogen plus CDK4/6i resistance. For instance, whether an Rb-null or CDK6 overexpressing tumor is likely to durably respond to antiestrogen monotherapy is unclear and perhaps a more refined biomarker than ESR1 mutant status will be required or else earlier use in combination with the kinase inhibitors.

Targeting Cooperative Steroid Receptors

In addition to the ER, most breast cancers are observed to express other ligand-induced nuclear receptors including androgen receptor (AR), progesterone receptor (PR), and glucocorticoid receptor (GR). As reviewed in the literature, genome-wide analyses of transcription factor binding and transcription have demonstrated the cooperativity between these receptors and the ER. In some instances, these cooperative receptors serve to redirect ER and thereby change the action of ER. For instance, activation of AR was shown to promote the movement of ER toward loci that favor cellular differentiation over proliferation (Hickey et al. 2021). Based on this and related work, analogs of testosterone for ER$^+$/AR$^+$ tumors and progestins (Mohammed et al. 2015) for ER$^+$/PR$^+$ tumors have been tested in the laboratory and moved forward to the clinic. Somewhat surprisingly, antagonists of AR (Cochrane et al. 2014) and PR (Singhal et al. 2016) have also been analyzed and shown preclinical activity on the basis of a different suite of transcripts that inhibition can potentially alter or a different receptor context (e.g., AR inhibition in AR$^+$/ER$^-$ breast cancer where AR has more of a driver role. As is apparent from these largely overlapping contexts in which antagonism versus agonism are both being proposed, further biomarker selection will likely be needed to clarify the most appropriate clinical context as protein expression alone proves insufficient to define the population that may benefit.

Targeting the Cell-Cycle Machinery

In addition to steroid hormones, specific components of the cell-cycle machinery play critical roles in mammary gland development. The G1/S transition is regulated by the D-type cyclins (cyclins D1/D2/D3) and cyclin-dependent kinases 4 and 6 and, remarkably, the cyclin D1-CDK4 holoenzyme plays a relatively selective role in driving proliferation of luminal mammary epithelial cells (Goel et al. 2022). As evidence for this, cyclin D1 knockout mice are viable, with a defect in mammary epithelial proliferation during pregnancy being one of their few notable phenotypes (Fantl et al. 1995; Sicinski et al. 1995). The importance for cyclin D1 in mammary gland biology is explained, at least in part, by the fact that *CCND1* is a direct ER-target gene (Prall et al. 1997).

Analogous to retained dependence on ER in luminal tumors, many breast cancers retain a dependence on cyclin D1 and CDK4 for growth. Transgenic mouse models have shown that the initiation of mammary tumors driven by certain oncoproteins (e.g., HER2, Ras) is both dependent on cyclin D1 and also on its binding partner CDK4 (Yu et al. 2001, 2006). Moreover, selective genetic ablation of cyclin D1 in already established mammary tumors results in tumor cell-cycle arrest and acquisition of a senescence-like phenotype (Choi et al. 2012; Goel et al. 2016).

Given the importance of cyclin D1 and CDK4 in mammary tumorigenesis, the rationale for developing CDK inhibitors has been clear for decades. However, this task has proved difficult for several reasons: (1) until recently, synthesizing potent and selective inhibitors of specific CDKs has presented a medicinal chemistry challenge (Sánchez-Martínez et al. 2019); (2) the cell-cycle machinery is inherently plastic, such that inhibition of a CDK can lead to cooption of other CDKs to maintain proliferation (Malumbres et al. 2004; Herrera-Abreu et al. 2016); (3) many cancers have mutations or deletions in cell-cycle machinery components, thereby rewriting the rules of orderly cell-cycle control (e.g., alterations in *RB1*, which encodes RB, the canonical CDK4/6 substrate) (Dean et al. 2012; Goel et al. 2022).

Despite these challenges, publications describing the first selective and potent CDK4/6 inhibitor—now known as palbociclib—emerged in the mid 2000s (Toogood et al. 2005), and subsequent work has transformed breast cancer therapy. Consistent with the preclinical phenotypes described above, palbociclib showed striking antiproliferative activity in breast cancer cell lines of luminal subtype (including HER2-amplified luminal lines) as well as a synergistic effect when combined with hormonal therapy (Finn et al. 2009). Similar results have been demonstrated with ribociclib and abemaciclib, two other FDA-approved CDK4/6 inhibitors (O'Brien et al. 2018).

The transit of CDK4/6 inhibitors from the preclinical to clinical arena has been swift, in no small part due to the strong and consistent efficacy signal for these drugs in the treatment of luminal cancers. A series of randomized phase 3 clinical trials has shown that the addition of CDK4/6 inhibition to ET improves the progression-free survival, and in some cases the overall survival, of patients with metastatic hormone receptor-positive, HER2-negative breast cancer (Spring et al. 2020). This has been observed in both pre- and postmenopausal women with both ET-naive and ET-refractory disease, and CDK4/6 inhibitors are now considered gold-standard treatment for hormone-receptor-positive breast cancer.

Given the ubiquitous use of CDK4/6 inhibitors in modern breast oncology, a deeper understanding of their effects is critical to inform the development of new CDK4/6 inhibitor-containing combinations as well as our understanding of resistance. Not surprisingly, several clinical studies have provided incontrovertible evidence that CDK4/6 inhibitors have an antiproliferative effect in luminal cancers, which is enhanced by concomitant ET (Ma et al. 2017; Johnston et al. 2019; Hurvitz et al. 2020). In addition, preclinical studies have uncovered numerous other mechanisms by which these agents exert their effects in breast cancers. In each case, the mechanistic insight has led to rational design of new CDK4/6 inhibitor-containing combination regimens:

1. By activating Rb, CDK4/6 inhibitors induce a senescence-like state in luminal breast cancer cells. This state remains to be deeply characterized, but is accompanied by certain hallmarks of classical cellular senescence including cellular enlargement, increased lysosomal content and β galactosidase activity, and a stereotypic pattern of chromatin remodeling (Choi et al. 2012; Goel et al. 2016; Yoshida et al. 2016; Watt et al. 2021). Of note, analysis of clinical specimens suggests that the addition of CDK4/6 inhibitors to ET reduces indices of apoptosis in ER-positive breast cancer cells, consistent with the notion that senescence is an antiapoptotic state (Childs et al. 2014; Johnston et al. 2019). This has, in turn, triggered examination of whether so-called "senolytic" therapies—drugs designed to specifically kill senescent cells—might be successfully combined with CDK4/6 inhibitors, and preclinical studies have suggested that this indeed might be the case. Specifically, like classically senescent cells, CDK4/6 inhibitor-treated breast cancer cells can develop specific dependencies on prosurvival members of the BH3 protein family, and this might in turn render them specifically vulnerable to BH3 mimetics including inhibitors of Bcl-2 and/or Bcl-xL (Whittle et al. 2020; Watt et al. 2021). In the case of Bcl-2, this hypothesis is currently the subject of clinical trials (NCT03900884). A key question with the use of senolytics relates to drug scheduling; in theory, a course of CDK4/6 inhibition to induce senescence followed by a short course of a senolytic (the so called "hit and run" approach) might be optimal but is not being formally tested in the clinical arena.

2. CDK4/6 inhibition can "rewire" kinase circuits in cancer cells, possibly through direct regulation of mTORC1 activity by CDK4/6 (Vora et al. 2014; Goel et al. 2016; Romero-Pozuelo et al. 2017, 2020). Importantly, studies have shown that cancer cells up-regulate various upstream oncogenic kinase pathways to drive proliferation in the face of CDK4/6 inhibition, and conversely that persistence of cyclin D–CDK4 activity maintains proliferation in response to inhibitors of upstream oncogenic kinases (Goel et al. 2016; Herrera-Abreu et al. 2016; Jansen et al. 2017; O'Brien et al. 2018; Zhao et al. 2021). In

breast cancer, potent synergy is observed when combining CDK4/6 inhibitors and HER2 kinase inhibitors (in luminal, HER2-amplified models) and PI3K inhibitors and in the case of HER2; this has led to the development of randomized trials, the final results of which are awaited (Metzger et al. 2019; Tolaney et al. 2020).

3. One surprising, yet consistent, preclinical observation has been that CDK4/6 inhibition can enhance antitumor immunity in breast and other cancers (Goel et al. 2017; Deng et al. 2018; Schaer et al. 2018). This might, in part, relate to the changes in DNA methylation and chromatin structure that accompany senescence, which in turn induce an interferon-driven gene expression within tumor cells (Goel et al. 2017; Watt et al. 2021). In addition, however, these agents exert effects on various T lymphocyte subsets (likely through the inhibition of CDK6 rather than CDK4) (Goel et al. 2017; Deng et al. 2018; Heckler et al. 2021; Lelliott et al. 2021). This has provoked the exciting question of whether CDK4/6 inhibitors might enhance the impact of immune checkpoint inhibitors in breast cancer, and preclinical studies provide strong evidence for this (Goel et al. 2017; Deng et al. 2018). Early trials have shown these combinations to be significantly toxic —ironically due to increased immune-related toxicity—but studies with newer agents given using different schedules and routes of administration are revisiting this question (Tan et al. 2019; Goel et al. 2021).

Given CDK4/6 inhibitors have cemented their position as a mainstay of breast cancer therapy and revigorated interest in cell-cycle inhibition, we enumerate some of the key questions for the field to address in years ahead: (1) Can CDK4/6 inhibitors cure early-stage breast cancer? Trials of adjuvant CDK4/6 inhibitors in moderate-to-high-risk luminal breast cancers have provided mixed results (Johnston et al. 2020; Mayer et al. 2021). Length of follow-up in trials where adjuvant therapy has proven to improve disease-free survival is short, and a burning question is whether ET-CDK4/6i combinations can eradicate mi-croscopic disease and thus effect cures, or merely delay the onset of clinical relapse until after adjuvant therapy is stopped. Limited preclinical data provide concerns that the latter might be true (Goel et al. 2016), and the answer will deeply inform our understanding of drug mechanism of action. (2) Are all CDK4/6 inhibitors the same? Although each of the approved agents is a potent CDK4/6 inhibitor, they differ with respect to their chemical structure, toxicity profiles, monotherapy response rates, and efficacy in the adjuvant setting (Hafner et al. 2019; Goel et al. 2022). This might in part relate to the different non-CDK4/6 targets inhibited by these agents, and understanding these differences remains a major unresolved issue. (3) Can CDK4/6 inhibitors be combined with chemotherapy or antibody–chemotherapy conjugates? Traditional dogma states that by inducing G1 arrest, CDK4/6 inhibitors antagonize cytotoxic chemotherapies, which often act upon cells in S phase or G2/M (Roberts et al. 2012). More recently, however, a sequence of inducing DNA damage with chemotherapy followed by CDK4/6 inhibition (which can impair homologous recombination-mediated DNA repair) might be highly effective in certain cases (Salvador-Barbero et al. 2020). This needs to be explored in breast cancer. (4) How do tumors acquire resistance to CDK4/6 inhibitors? In a subset of tumors, resistance to CDK4/6 inhibitors has been identified through genetic alterations that lead to reactivation of CDK4/6 activity (e.g., induction of CDK6 [Li et al. 2018; Yang et al. 2017]) and mechanisms that bypass the CDK4/6 requirement (Rb loss, CDK2-CCNE activation [Herrera-Abreu et al. 2016]). For this subset of tumors, new therapeutic strategies might involve more potent CDK4/6 inhibitors (Li et al. 2022) or inhibitors that are synthetically lethal with Rb1 loss (e.g., AURORA kinase inhibition [Oser et al. 2019]). Based on matched pre- and post-treatment data from randomized trials, however, most cases show no genetic alteration that clearly defines a conventional and binary sensitive versus resistant genotype (O'Leary et al. 2018). The fact that CDK4/6 inhibitors show some efficacy when used beyond progression (Kalinsky et al. 2022) suggests that resistance is not always genomically "hard-

Cite this article as *Cold Spring Harb Perspect Med* doi: 10.1101/cshperspect.a041333

wired," but rather that there may be an impact of plasticity of the cell-cycle machinery with cooption of other CDKs that can help drive the G1/S transition (e.g., CDK2, CDK7). To this end, CDK2 and CDK7 inhibitors are currently in clinical trials. Where these newer cell-cycle inhibitors might "fit" within the rapidly evolving landscape of therapies for advanced breast cancer (e.g., the arrival of antibody–drug conjugates) remains unclear.

TARGETING PRECISELY

Minimal Residual Disease

The efficacy of adjuvant therapy in breast cancer has lent strong credence to the notion that occult microscopic disease may be more vulnerable and subject to eradication than overt metastatic disease. This tenet is the foundation behind innumerable studies that move agents that have little curative role in the metastatic setting but have a strong one in the early stage setting. The main systemic therapies used for this approach have been chemotherapies or antiestrogen therapies largely given without the benefit of marker to track efficacy. The recent development of ultrasensitive measures of minimal residual disease using circulating tumor DNA as well as other markers (e.g., circulating tumor cells) holds promise to further, and even enhance, this paradigm (Cescon et al. 2021). First, the ultrasensitive detection of fragments of tumor DNA may allow narrowing of the population to those at risk versus not at risk for relapse. Second, the characterization of the particular mutational spectra present may provide insight on the types of systemic therapy that could be used. Third, the ease of sampling can allow tracking of response to therapy to ensure eradication of the clone or even guide rational therapy change. While there are clearly challenges to the widespread adoption of this approach, there is precedence from other malignancies such as the hematologic cancers. Importantly, these technologies could hasten the development of new therapies geared toward curing cancer rather than merely achieving temporary disease shrinkage in the advanced setting.

Antibody Drug Conjugates

Systemic chemotherapies have a strong stigma associated with their use largely due to their potential for disabling side effects that run the gamut from nausea/vomiting or hair loss to secondary leukemias or neuropathy. Weighed against this is the broad tumoricidal activity of many chemotherapies that can lead to rapid tumor shrinkage even among heterogenous tumors. The opportunity to see these effects can be limited specifically by the side effects that can reduce the intensity and density of dose that can be delivered. As a strategy to widen the therapeutic index of these agents, linking potent chemotherapies to tumor-specific antibodies has enabled more effective use of these drugs (Drago et al. 2021). The benefit has been most impressively realized in breast cancer through the use of anti-HER2 antibody–drug conjugates (ADCs), such as TDXd and TDM1, as well as those directed against Trop2, such as sacituzumab govitecan. In particular, the striking efficacy of TDXd in TDM1-refractory tumors and in HER2 low cancers has challenged several previously held dogmas and invigorated new concepts in cancer drug development (Cortés et al. 2022; Modi et al. 2022). First, the profound efficacy of TDXd in luminal ER+ breast cancers has challenged the notion of this subtype of disease being intrinsically chemoresistant and rather pointed to the need for greater potency to realize durable benefit. Second, the expanded benefit of TDXd beyond HER2-amplified breast cancer into most HER2-expressing breast cancers has forced a revisitation of the biomarker development strategy that may be useful in predicting drug efficacy. Similarly, the identification of Trop2 as a cell-surface protein that is up-regulated in the majority of breast cancers led to the development of sacituzumab govitecan, which has shown significant clinical activity in HER2-negative tumors (as defined by American Society of Clinical Oncology [ASCO]/College of American Pathologists [CAP] guidelines), regardless of hormone receptor status (Bardia et al. 2021; Rugo et al. 2022). Importantly, the modular design of ADCs in which there may be different types of antibodies (HER2, TROP2, Nectin, etc.) and payloads (antimicrotubule, topoisomerase inhibitor, immune stimulants), gives rise to the concept of personalizing ADC based on the tumor surface proteome and intrinsic sensitivities. The developments of this field are moving exceptionally fast and offer many opportunities for eradicating difficult to drug and heterogeneous tumor populations.

EXPLOITING SYNTHETIC LETHALITIES

PARP Inhibition as a Model for Exploiting Synthetic Lethal Interactions

The recognition that cancers with specific genetic (or nongenetic) aberrations display selective vulnerabilities to particular therapies defines the concept of synthetic lethality in cancer. In broad terms, synthetic lethalities can be identified either through (1) a rational application of knowledge acquired a priori, or (2) through screening approaches, in which models with or without a specific alteration are screened using yeast, compound libraries, siRNA or CRISPR libraries, or in silico–based machine learning approaches. In breast cancer, the concept of synthetic lethality has been most successfully exemplified using poly-(ADP)-ribose polymerase (PARP) inhibitors for cancers harboring mutations in BRCA1/2.

BRCA1/2-mutant cancers (most commonly due to a germline alteration) display a specific defect in homologous recombination (HR)—a mechanism that accurately repairs double-stranded DNA breaks (DSBs). PARP plays a critical role in base excision repair—a mechanism that repairs small base lesions on single-stranded DNA. When PARP is inhibited in cancer cells, single-stranded DNA lesions accumulate, ultimately resulting in DSBs that cannot be repaired in HR-deficient BRCA-mutant cancers (Bryant et al. 2005; Farmer et al. 2005; Fong et al. 2009). Early-phase trial data supported this molecular model, demonstrating tumor responses to PARP inhibitor monotherapy in BRCA1/2 germline carriers (Fong et al. 2009; Kaufman et al. 2015). Randomized trials in germline BRCA-mutant breast cancers have cemented a role for these agents in the treatment of metastatic disease, and more recently for early-stage disease where

adjuvant PARP inhibition improves overall survival (Tutt et al. 2021).

Given these results, a key question is whether the synthetic lethality framework underlying PARP inhibitor efficacy can be extended to include cancers other than those in individuals with pathogenic germline variants in BRCA1/2. For example, are tumors deficient in HR through other means also sensitive to PARP inhibition? In this light, a loose concept of "BRCAness" has arisen, intended to describe a molecular phenotype indicative of HR deficiency and hence sensitivity to PARP inhibitors (Lord and Ashworth 2016). A range of clinical studies aiming to address this issue have been conducted and, in sum, they have taught us that each biological hypothesis stemming from a putative synthetic lethal interaction needs to be rigorously validated in clinical trials. For example, it is likely that tumors harboring somatic BRCA1/2 mutations or germline mutations in other genes required for HR (most notably PALB2) will also benefit from PARP inhibitor therapy (Tung et al. 2020). On the other hand, combining PARP inhibitors with DNA-damaging chemotherapy in tumors not defined by a specific genetic defect has not proven highly effective (Loibl et al. 2018).

The Future of Synthetic Lethality-Based Drug Discovery

Following on from the success of PARP inhibitors, much of the focus for identification of other synthetic lethal interactions has been on agents targeting the DNA damage response (DDR), including inhibitors of ATR, DNA-PK, CHK1/2, and WEE1. One noteworthy case is that of DNA polymerase θ (POLQ) inhibitors. POLQ plays a role in numerous DNA repair mechanisms including θ-mediated end joining (TMEJ) to repair DSBs, and HR-deficient tumors often demonstrate up-regulation of POLQ and a dependence upon TMEJ-mediated DNA repair (Ceccaldi et al. 2015; Mateos-Gomez et al. 2015; Carvajal-Garcia et al. 2020). Although the precise mechanisms underlying the synthetic lethal relationship remains controversial (Schrempf et al. 2021), the efficacy of POLQ inhibitors in preclinical models of HR-deficient cancers (most notably BRCA1/2 mutant tumors (Zatreanu et al. 2021; Zhou et al. 2021) and their observed synergy with PARP inhibitors (Zatreanu et al. 2021) has triggered the development of several such agents.

It remains to be seen whether any of these DDR inhibitors will replicate the PARP inhibitor success story, and key challenges in the field include (1) the fact that cancer cells do not employ a single DNA repair pathway to survive, affording natural mechanisms of resistance; (2) defining the specific DNA repair defect present in any given tumor; and (3) the significant risk of toxicity to normal tissue when combining DDR inhibitors with DNA damaging cytotoxic chemotherapy. Looking forward, technologies that allow for high-throughput pharmacologic or genetic screening in cancer cells hold the greatest promise for the identification of novel synthetic lethal targets in breast cancer. Such efforts are well underway, with data defining cell line genotypes and their associated dependencies (derived from both CRISPR and compound screens) now publicly available.

REFERENCES

André F, Ciruelos E, Rubovszky G, Campone M, Loibl S, Rugo HS, Iwata H, Conte P, Mayer IA, Kaufman B, et al. 2019. Alpelisib for PIK3CA-mutated, hormone receptor–positive advanced breast cancer. N Engl J Med 380: 1929–1940. doi:10.1056/NEJMoa1813904

Bardia A, Mayer IA, Vahdat LT, Tolaney SM, Isakoff SJ, Diamond JR, O'Shaughnessy J, Moroose RL, Santin AD, Abramson VG, et al. 2021. Sacituzumab govitecan in metastatic triple-negative breast cancer. N Engl J Med 384: 1529–1541. doi:10.1056/NEJMoa2028485

Baselga J, Cortés J, Kim SB, Im SA, Hegg R, Im YH, Roman L, Pedrini JL, Pienkowski T, Knott A, et al. 2012a. Pertuzumab plus trastuzumab plus docetaxel for metastatic breast cancer. N Engl J Med 366: 109–119. doi:10.1056/NEJMoa1113216

Baselga J, Campone M, Piccart M, Burris HA III, Rugo HS, Sahmoud T, Noguchi S, Gnant M, Pritchard KI, Lebrun F, et al. 2012b. Everolimus in postmenopausal hormone-receptor–positive advanced breast cancer. N Engl J Med 366: 520–529. doi:10.1056/NEJMoa1109653

Bryant HE, Schultz N, Thomas HD, Parker KM, Flower D, Lopez E, Kyle S, Meuth M, Curtin NJ, Helleday T. 2005. Specific killing of BRCA2-deficient tumours with inhibitors of poly(ADP-ribose) polymerase. Nature 434: 913–917. doi:10.1038/nature03443

Carvajal-Garcia J, Cho JE, Carvajal-Garcia P, Feng W, Wood RD, Sekelsky J, Gupta GP, Roberts SA, Ramsden DA. 2020. Mechanistic basis for microhomology identification and genome scarring by polymerase θ. Proc Natl Acad Sci 117: 8476–8485. doi:10.1073/pnas.1921791117

Castel P, Toska E, Engelman JA, Scaltriti M. 2021. The present and future of PI3K inhibitors for cancer therapy. Nat Cancer 2: 587–597. doi:10.1038/s43018-021-00218-4

Ceccaldi R, Liu JC, Amunugama R, Hajdu I, Primack B, Petalcorin MIR, O'Connor KW, Konstantinopoulos PA, Elledge SJ, Soulton SJ, et al. 2015. Homologous-recombination-deficient tumours are dependent on Polθ-mediated repair. Nature 518: 258–262. doi:10.1038/nature14184

Cescon DW, Kalinsky K, Parsons HA, Smith KL, Spears PA, Thomas A, Zhao F, DeMichele A. 2022. Therapeutic targeting of minimal residual disease to prevent late recurrence in hormone-receptor positive breast cancer: challenges and new approaches. Front Oncol 11: 667397. doi:10.3389/fonc.2021.667397

Chandarlapaty S. 2012. Negative feedback and adaptive resistance to the targeted therapy of cancer. Cancer Discov 2: 311–319. doi:10.1158/2159-8290.CD-12-0018

Chandarlapaty S, Sawai A, Scaltriti M, Rodrik-Outmezguine V, Grbovic-Huezo O, Serra V, Majumder PK, Baselga J, Rosen N. 2011. AKT inhibition relieves feedback suppression of receptor tyrosine kinase expression and activity. Cancer Cell 19: 58–71. doi:10.1016/j.ccr.2010.10.031

Chen YC, Yu J, Metcalfe C, De Bruyn T, Gelzleichter T, Malhi V, Perez-Moreno PD, Wang X. 2022. Latest generation estrogen receptor degraders for the treatment of hormone receptor-positive breast cancer. Expert Opin Investig Drugs 31: 515–529. doi:10.1080/13543784.2021.1983542

Childs BG, Baker DJ, Kirkland JL, Campisi J, van Deursen JM. 2014. Senescence and apoptosis: dueling or complementary cell fates? EMBO Rep 15: 1139–1153. doi:10.15252/embr.201439245

Choi YJ, Li X, Hydbring P, Sanda T, Stefano J, Christie AL, Signoretti S, Look AT, Kung AL, von Boehmer H, et al. 2012. The requirement for cyclin D function in tumor maintenance. Cancer Cell 22: 438–451. doi:10.1016/j.ccr.2012.09.015

Cochrane DR, Bernales S, Jacobsen BM, Cittelly DM, Howe EN, D'Amato NC, Spoelstra NS, Edgerton SM, Jean A, Guerrero J, et al. 2014. Role of the androgen receptor in breast cancer and preclinical analysis of enzalutamide. Breast Cancer Res 16: R7. doi:10.1186/bcr3599

Cortés J, Kim SB, Chung WP, Im SA, Park YH, Hegg R, Kim MH, Tseng LM, Petry V, Chung CF, et al. 2022. Trastuzumab deruxtecan versus trastuzumab emtansine for breast cancer. N Engl J Med 386: 1143–1154. doi:10.1056/NEJMoa2115022

Dean JL, McClendon AK, Hickey TE, Butler LM, Tilley WD, Witkiewicz AK, Knudsen ES. 2012. Therapeutic response to CDK4/6 inhibition in breast cancer defined by ex vivo analyses of human tumors. Cell Cycle 11: 2756–2761. doi:10.4161/cc.21195

Deng J, Wang ES, Jenkins RW, Li S, Dries R, Yates K, Chhabra S, Huang W, Liu H, Aref AR, et al. 2018. CDK4/6 inhibition augments antitumor immunity by enhancing T-cell activation. Cancer Discov 8: 216–233. doi:10.1158/2159-8290.CD-17-0915

Dinkel V, Anderson D, Winski S, Winkler J, Koch K, Lee PA. 2012. ARRY-380, a potent, small molecule inhibitor of

ErbB2, increases survival in intracranial ErbB2$^+$ xenograft models in mice. *Cancer Res* 72: 852–852. doi:10.1158/1538-7445.Am2012-852

Drago JZ, Modi S, Chandarlapaty S. 2021. Unlocking the potential of antibody-drug conjugates for cancer therapy. *Nat Rev Clin Oncol* 18: 327–344. doi:10.1038/s41571-021-00470-8

Eeckhoute J, Carroll JS, Geistlinger TR, Torres-Arzayus MI, Brown M. 2006. A cell-type-specific transcriptional network required for estrogen regulation of cyclin D1 and cell cycle progression in breast cancer. *Genes Dev* 20: 2513–2526. doi:10.1101/gad.1446006

Fantl V, Stamp G, Andrews A, Rosewell I, Dickson C. 1995. Mice lacking cyclin D1 are small and show defects in eye and mammary gland development. *Genes Dev* 9: 2364–2372. doi:10.1101/gad.9.19.2364

Farmer H, McCabe N, Lord CJ, Tutt AN, Johnson DA, Richardson TB, Santarosa M, Dillon KJ, Hickson I, Knights C, et al. 2005. Targeting the DNA repair defect in BRCA mutant cells as a therapeutic strategy. *Nature* 434: 917–921. doi:10.1038/nature03445

Fendly BM, Winget M, Hudziak RM, Lipari MT, Napier MA, Ullrich A. 1990. Characterization of murine monoclonal antibodies reactive to either the human epidermal growth factor receptor or HER2/neu gene product. *Cancer Res* 50: 1550–1558.

Finn RS, Dering J, Conklin D, Kalous O, Cohen DJ, Desai AJ, Ginther C, Atefi M, Chen I, Fowst C, et al. 2009. PD 0332991, a selective cyclin D kinase 4/6 inhibitor, preferentially inhibits proliferation of luminal estrogen receptor-positive human breast cancer cell lines in vitro. *Breast Cancer Res* 11: R77. doi:10.1186/bcr2419

Fong PC, Boss DS, Yap TA, Tutt A, Wu P, Mergui-Roelvink MM, Mortimer P, Swaisland H, Lau A, O'Connor MJ, et al. 2009. Inhibition of poly(ADP-ribose) polymerase in tumors from BRCA mutation carriers. *N Engl J Med* 361: 123–134. doi:10.1056/NEJMoa0900212

Goel S, Wang Q, Watt AC, Tolaney SM, Dillon DA, Li W, Ramm S, Palmer AC, Yuzugullu H, Varadan V, et al. 2016. Overcoming therapeutic resistance in HER2-positive breast cancers with CDK4/6 inhibitors. *Cancer Cell* 29: 255–269. doi:10.1016/j.ccell.2016.02.006

Goel S, DeCristo MJ, Watt AC, BrinJones H, Sceneay J, Li BB, Khan N, Ubellacker JM, Xie S, Metzger-Filho O, et al. 2017. CDK4/6 inhibition triggers anti-tumour immunity. *Nature* 548: 471–475. doi:10.1038/nature23465

Goel S, O'Shaughnessy J, Tan AR, Krastev BM, Rugo HS, Aftimos PG, Yardley DA, Zoran A, Wolfgang CD, Sorrentino J, et al. 2021. Trial in progress: a phase 3, randomized, double-blind trial of trilaciclib versus placebo in patients receiving first- or second-line gemcitabine and carboplatin for locally advanced unresectable or metastatic triple-negative breast cancer (PRESERVE 2). *J Clin Oncol* 39: TPS1107. doi:10.1200/JCO.2021.39.15_suppl.TPS1107

Goel S, Bergholz JS, Zhao JJ. 2022. Targeting CDK4 and CDK6 in cancer. *Nat Rev Cancer* 22: 356–372. doi:10.1038/s41568-022-00456-3

Hafner M, Mills CE, Subramanian K, Chen C, Chung M, Boswell SA, Everley RA, Liu C, Walmsley CS, Juric D, et al. 2019. Multiomics profiling establishes the polypharmacology of FDA-approved CDK4/6 inhibitors and the potential for differential clinical activity. *Cell Chem Biol* 26: 1067–1080.e8. doi:10.1016/j.chembiol.2019.05.005

Heckler M, Ali LR, Clancy-Thompson E, Qiang L, Ventre KS, Lenehan P, Roehle K, Luona A, Boelaars K, Peters V, et al. 2021. Inhibition of CDK4/6 promotes CD8 T-cell memory formation. *Cancer Discov* 11: 2564–2581. doi:10.1158/2159-8290.Cd-20-1540

Herrera-Abreu MT, Palafox M, Asghar U, Rivas MA, Cutts RJ, Garcia-Murillas I, Pearson A, Guzman M, Rodriguez O, Grueso J, et al. 2016. Early adaptation and acquired resistance to CDK4/6 inhibition in estrogen receptor-positive breast cancer. *Cancer Res* 76: 2301–2313. doi:10.1158/0008-5472.CAN-15-0728

Hickey TE, Selth LA, Chia KM, Laven-Law G, Milioli HH, Roden D, Jindal S, Hui M, Finlay-Schultz J, Ebrahimie E, et al. 2021. The androgen receptor is a tumor suppressor in estrogen receptor-positive breast cancer. *Nat Med* 27: 310–320. doi:10.1038/s41591-020-01168-7

Hopkins BD, Pauli C, Du X, Wang DG, Li X, Wu D, Amadiume SC, Goncalves MD, Hodakoski C, Lundquist MR, et al. 2018. Suppression of insulin feedback enhances the efficacy of PI3K inhibitors. *Nature* 560: 499–503. doi:10.1038/s41586-018-0343-4

Hurvitz SA, Martin M, Press MF, Chan D, Fernandez-Abad M, Petru E, Rostorfer R, Guarneri V, Huang CS, Barriga S, et al. 2020. Potent cell-cycle inhibition and upregulation of immune response with abemaciclib and anastrozole in neoMONARCH, phase II neoadjuvant study in HR$^+$/HER2$^-$ breast cancer. *Clin Cancer Res* 26: 566–580. doi:10.1158/1078-0432.CCR-19-1425

Jansen VM, Bhola NE, Bauer JA, Formisano L, Lee KM, Hutchinson KE, Witkiewicz AK, Moore PD, Estrada MV, Sánchez V, et al. 2017. Kinome-wide RNA interference screen reveals a role for PDK1 in acquired resistance to CDK4/6 inhibition in ER-positive breast cancer. *Cancer Res* 77: 2488–2499. doi:10.1158/0008-5472.CAN-16-2653

Johnston S, Puhalla S, Wheatley D, Ring A, Barry P, Holcombe C, Boileau JF, Provencher L, Robidoux A, Rimawi M, et al. 2019. Randomized phase II study evaluating palbociclib in addition to letrozole as neoadjuvant therapy in estrogen receptor-positive early breast cancer: PALLET trial. *J Clin Oncol* 37: 178–189. doi:10.1200/jco.18.01624

Johnston SRD, Harbeck N, Hegg R, Toi M, Martin M, Shao ZM, Zhang QY, Martinez Rodriguez JL, Campone M, Hamilton E, et al. 2020. Abemaciclib combined with endocrine therapy for the adjuvant treatment of HR$^+$, HER2$^-$, node-positive, high-risk, early breast cancer (monarchE). *J Clin Oncol* 38: 3987–3998. doi:10.1200/jco.20.02514

Kalinsky K, Accordino MK, Chiuzan C, Mundi PS, Trivedi MS, Novik Y, Tiersten A, Raptis G, Baer LN, Oh SY, et al. 2022. A randomized, phase II trial of fulvestrant or exemestane with or without ribociclib after progression on anti-estrogen therapy plus cyclin-dependent kinase 4/6 inhibition (CDK 4/6i) in patients (pts) with unresectable or hormone receptor–positive (HR$^+$), HER2-negative metastatic breast cancer (MBC): MAINTAIN trial. *J Clin Oncol* 40: LBA1004. doi:10.1200/JCO.2022.40.17_suppl.LBA1004

Kaufman B, Shapira-Frommer R, Schmutzler RK, Audeh MW, Friedlander M, Balmaña J, Mitchell G, Fried G, Stemmer SM, Hubert A, et al. 2015. Olaparib monotherapy in patients with advanced cancer and a germline BRCA1/2 mutation. *J Clin Oncol* **33:** 244–250. doi:10.1200/JCO.2014.56.2728

Lainé M, Fanning SW, Chang YF, Green B, Greene ME, Komm B, Kurleto JD, Phung L, Greene GL. 2021. Lasofoxifene as a potential treatment for therapy-resistant ER-positive metastatic breast cancer. *Breast Cancer Res* **23:** 54. doi:10.1186/s13058-021-01431-w

Lee BJ, Boyer JA, Burnett GL, Thottumkara AP, Tibrewal N, Wilson SL, Hsieh T, Marquez A, Lorenzana EG, Evans JW, et al. 2021. Selective inhibitors of mTORC1 activate 4EBP1 and suppress tumor growth. *Nat Chem Biol* **17:** 1065–1074. doi:10.1038/s41589-021-00813-7

Lelliott EJ, Kong IY, Zethoven M, Ramsbottom KM, Martelotto LG, Meyran D, Zhu JJ, Costacurta M, Kirby L, Sandow JJ, et al. 2021. CDK4/6 inhibition promotes antitumor immunity through the induction of T-cell memory. *Cancer Discov* **11:** 2582–2601. doi:10.1158/2159-8290.Cd-20-1554

Li Z, Razavi P, Li Q, Toy W, Liu B, Ping C, Hsieh W, Sanchez-Vega F, Brown DN, Da Cruz Paula AF, et al. 2018. Loss of the FAT1 tumor suppressor promotes resistance to CDK4/6 inhibitors via the hippo pathway. *Cancer Cell* **34:** 893–905.e8. doi:10.1016/j.ccell.2018.11.006

Li BT, Michelini F, Misale S, Cocco E, Baldino L, Cai Y, Shifman S, Tu HY, Myers ML, Xu C, et al. 2020. HER2-mediated internalization of cytotoxic agents in *ERBB2* amplified or mutant lung cancers. *Cancer Discov* **10:** 674–687. doi:10.1158/2159-8290.CD-20-0215

Li Q, Jiang B, Guo J, Shao H, Del Priore IS, Chang Q, Kudo R, Li Z, Razavi P, Liu B, et al. 2022. INK4 tumor suppressor proteins mediate resistance to CDK4/6 kinase inhibitors. *Cancer Discov* **12:** 356–371. doi:10.1158/2159-8290.CD-20-1726

Liang J, Zbieg JR, Blake RA, Chang JH, Daly S, DiPasquale AG, Friedman LS, Gelzleichter T, Gill M, Giltnane JM, et al. 2021. GDC-9545 (Giredestrant): a potent and orally bioavailable selective estrogen receptor antagonist and degrader with an exceptional preclinical profile for ER⁺ breast cancer. *J Med Chem* **64:** 11841–11856. doi:10.1021/acs.jmedchem.1c00847

Loibl S, O'Shaughnessy J, Untch M, Sikov WM, Rugo HS, McKee MD, Huober J, Golshan M, von Minckwitz G, Maag D, et al. 2018. Addition of the PARP inhibitor veliparib plus carboplatin or carboplatin alone to standard neoadjuvant chemotherapy in triple-negative breast cancer (BrighTNess): a randomised, phase 3 trial. *Lancet Oncol* **19:** 497–509. doi:10.1016/S1470-2045(18)30111-6

Lord CJ, Ashworth A. 2016. BRCAness revisited. *Nat Rev Cancer* **16:** 110–120. doi:10.1038/nrc.2015.21

Ma CX, Gao F, Luo J, Northfelt DW, Goetz M, Forero A, Hoog J, Naughton M, Ademuyiwa F, Suresh R, et al. 2017. NeoPalAna: neoadjuvant palbociclib, a cyclin-dependent kinase 4/6 inhibitor, and anastrozole for clinical stage 2 or 3 estrogen receptor-positive breast cancer. *Clin Cancer Res* **23:** 4055–4065. doi:10.1158/1078-0432.CCR-16-3206

Malumbres M, Sotillo R, Santamaría D, Galán J, Cerezo A, Ortega S, Dubus P, Barbacid M. 2004. Mammalian cells cycle without the D-type cyclin-dependent kinases Cdk4 and Cdk6. *Cell* **118:** 493–504. doi:10.1016/j.cell.2004.08.002

Mateos-Gomez PA, Gong F, Nair N, Miller KM, Lazzerini-Denchi E, Sfeir A. 2015. Mammalian polymerase θ promotes alternative NHEJ and suppresses recombination. *Nature* **518:** 254–257. doi:10.1038/nature14157

Mayer EL, Dueck AC, Martin M, Rubovszky G, Burstein HJ, Bellet-Ezquerra M, Miller KD, Zdenkowski N, Winer EP, Pfeiler G, et al. 2021. Palbociclib with adjuvant endocrine therapy in early breast cancer (PALLAS): interim analysis of a multicentre, open-label, randomised, phase 3 study. *Lancet Oncol* **22:** 212–222. doi:10.1016/s1470-2045(20)30642-2

Metzger O, Mandrekar S, Loibl S, Mundhenke C, Seiler S, Valagussa P, Lim E, Tripathy D, Winer EP, Huang C, et al. 2019. PATINA: a randomized, open label, phase III trial to evaluate the efficacy and safety of palbociclib + anti-HER2 therapy + endocrine therapy (ET) vs. anti-HER2 therapy + ET after induction treatment for hormone receptor positive (HR⁺)/HER2-positive metastatic breast cancer (MBC). *Cancer Res* **79:** OT3-02-07. doi:10.1158/1538-7445.Sabcs18-ot3-02-07

Modi S, Jacot W, Yamashita T, Sohn J, Vidal M, Tokunaga E, Tsurutani J, Ueno NT, Prat A, Chae YS, et al. 2022. Trastuzumab deruxtecan in previously treated HER2-low advanced breast cancer. *N Engl J Med* **387:** 9–20. doi:10.1056/NEJMoa2203690

Mohammed H, Russell IA, Stark R, Rueda OM, Hickey TE, Tarulli GA, Serandour AA, Birrell SN, Bruna A, Saadi A, et al. 2015. Progesterone receptor modulates ERα action in breast cancer. *Nature* **523:** 313–317. doi:10.1038/nature14583

Mukherjee R, Vanaja KG, Boyer JA, Gadal S, Solomon H, Chandarlapaty S, Levchenko A, Rosen N. 2021. Regulation of PTEN translation by PI3K signaling maintains pathway homeostasis. *Mol Cell* **81:** 708–723.e5. doi:10.1016/j.molcel.2021.01.033

O'Brien N, Conklin D, Beckmann R, Luo T, Chau K, Thomas J, Mc Nulty A, Marchal C, Kalous O, von Euw E, et al. 2018. Preclinical activity of abemaciclib alone or in combination with antimitotic and targeted therapies in breast cancer. *Mol Cancer Ther* **17:** 897–907. doi:10.1158/1535-7163.Mct-17-0290

O'Leary B, Cutts RJ, Liu Y, Hrebien S, Huang X, Fenwick K, André F, Loibl S, Loi S, Garcia-Murillas I, et al. 2018. The genetic landscape and clonal evolution of breast cancer resistance to palbociclib plus fulvestrant in the PALOMA-3 trial. *Cancer Discov* **8:** 1390–1403. doi:10.1158/2159-8290.CD-18-0264

Oser MG, Fonseca R, Chakraborty AA, Brough R, Spektor A, Jennings RB, Flaifel F, Novak JS, Gulati A, Buss E, et al. 2019. Cells lacking the *RB1* tumor suppressor gene are hyperdependent on Aurora B Kinase for survival. *Cancer Discov* **9:** 230–247. doi:10.1158/2159-8290.CD-18-0389

Pan H, Gray R, Braybrooke J, Davies C, Taylor C, McGale P, Peto R, Pritchard KI, Bergh J, Dowsett M, et al. 2017. 20-year risks of breast-cancer recurrence after stopping endocrine therapy at 5 years. *N Engl J Med* **377:** 1836–1846. doi:10.1056/NEJMoa1701830

Perou CM, Sørlie T, Eisen MB, van de Rijn M, Jeffrey SS, Rees CA, Pollack JR, Ross DT, Johnsen H, Akslen LA, et al.

2000. Molecular portraits of human breast tumours. *Nature* **406:** 747–752. doi:10.1038/35021093

Prall OW, Sarcevic B, Musgrove EA, Watts CK, Sutherland RL. 1997. Estrogen-induced activation of Cdk4 and Cdk2 during G1-S phase progression is accompanied by increased cyclin D1 expression and decreased cyclin-dependent kinase inhibitor association with cyclin E-Cdk2. *J Biol Chem* **272:** 10882–10894. doi:10.1074/jbc.272.16 .10882

Puyang X, Furman C, Zheng GZ, Wu ZJ, Banka D, Aithal K, Agoulnik S, Bolduc DM, Buonamici S, Caleb B, et al. 2018. Discovery of selective estrogen receptor covalent antagonists for the treatment of ERα^{WT} and ERα^{MUT} breast cancer. *Cancer Discov* **8:** 1176–1193. doi:10.1158/ 2159-8290.CD-17-1229

Rabindran SK, Discafani CM, Rosfjord EC, Baxter M, Floyd MB, Golas J, Hallett WA, Johnson BD, Nilakantan R, Overbeek E, et al. 2004. Antitumor activity of HKI-272, an orally active, irreversible inhibitor of the HER-2 tyrosine kinase. *Cancer Res* **64:** 3958–3965. doi: 10.1158/ 0008-5472.CAN-03-2868

Razavi P, Chang MT, Xu G, Bandlamudi C, Ross DS, Vasan N, Cai Y, Bielski CM, Donoghue MTA, Jonsson P, et al. 2018. The genomic landscape of endocrine-resistant advanced breast cancers. *Cancer Cell* **34:** 427–438.e6. doi:10 .1016/j.ccell.2018.08.008

Roberts PJ, Bisi JE, Strum JC, Combest AJ, Darr DB, Usary JE, Zamboni WC, Wong KK, Perou CM, Sharpless NE. 2012. Multiple roles of cyclin-dependent kinase 4/6 inhibitors in cancer therapy. *J Natl Cancer Inst* **104:** 476– 487. doi:10.1093/jnci/djs002

Robinson DR, Wu YM, Vats P, Su F, Lonigro RJ, Cao X, Kalyana-Sundaram S, Wang R, Ning Y, Hodges L, et al. 2013. Activating ESR1 mutations in hormone-resistant metastatic breast cancer. *Nat Genet* **45:** 1446–1451. doi:10.1038/ng.2823

Rodrik-Outmezguine VS, Okaniwa M, Yao Z, Novotny CJ, McWhirter C, Banaji A, Won H, Wong W, Berger M, de Stanchina E, et al. 2016. Overcoming mTOR resistance mutations with a new-generation mTOR inhibitor. *Nature* **534:** 272–276. doi:10.1038/nature17963

Romero-Pozuelo J, Demetriades C, Schroeder P, Teleman AA. 2017. Cycd/Cdk4 and discontinuities in Dpp signaling activate TORC1 in the *Drosophila* wing disc. *Dev Cell* **42:** 376–387.e5. doi:10.1016/j.devcel.2017.07.019

Romero-Pozuelo J, Figlia G, Kaya O, Martin-Villalba A, Teleman AA. 2020. Cdk4 and Cdk6 couple the cell-cycle machinery to cell growth via mTORC1. *Cell Rep* **31:** 107504. doi:10.1016/j.celrep.2020.03.068

Romond EH, Perez EA, Bryant J, Suman VJ, Geyer CE Jr, Davidson NE, Tan-Chiu E, Martino S, Paik S, Kaufman PA, et al. 2005. Trastuzumab plus adjuvant chemotherapy for operable HER2-positive breast cancer. *N Engl J Med* **353:** 1673–1684. doi:10.1056/NEJMoa052122

Rugo HS, Bardia A, Marmé F, Cortes J, Schmid P, Loirat D, Trédan O, Ciruelos E, Dalenc F, Pardo PG, et al. 2022. Sacituzumab govitecan in hormone receptor-positive/human epidermal growth factor receptor-2-negative metastatic breast cancer. *J Clin Oncol* **40:** 3365–3376. doi:10 .1200/JCO.22.01002

Salvador-Barbero B, Álvarez-Fernández M, Zapatero-Solana E, El Bakkali A, Menéndez MDC, López-Casas PP, Di

Domenico T, Xie T, VanArsdale T, Shields DJ, et al. 2020. CDK4/6 inhibitors impair recovery from cytotoxic chemotherapy in pancreatic adenocarcinoma. *Cancer Cell* **37:** 340–353.e6. doi:10.1016/j.ccell.2020.01.007

Sánchez-Martínez C, Lallena MJ, Sanfeliciano SG, de Dios A. 2019. Cyclin dependent kinase (CDK) inhibitors as anticancer drugs: recent advances (2015–2019). *Bioorg Med Chem Lett* **29:** 126637. doi:10.1016/j.bmcl.2019 .126637

Schaer DA, Beckmann RP, Dempsey JA, Huber L, Forest A, Amaladas N, Li Y, Wang YC, Rasmussen ER, Chin D, et al. 2018. The CDK4/6 inhibitor abemaciclib induces a T cell inflamed tumor microenvironment and enhances the efficacy of PD-L1 checkpoint blockade. *Cell Rep* **22:** 2978– 2994. doi:10.1016/j.celrep.2018.02.053

Schechter AL, Stern DF, Vaidyanathan L, Decker SJ, Drebin JA, Greene MI, Weinberg RA. 1984. The neu oncogene: an erb-B-related gene encoding a 185,000-Mr tumour antigen. *Nature* **312:** 513–516. doi:10.1038/312513a0

Schechter AL, Hung MC, Vaidyanathan L, Weinberg RA, Yang-Feng TL, Francke U, Ullrich A, Coussens L. 1985. The neu gene: an erbB-homologous gene distinct from and unlinked to the gene encoding the EGF receptor. *Science* **229:** 976–978. doi:10.1126/science.2992090

Schrempf A, Slyskova J, Loizou JI. 2021. Targeting the DNA repair enzyme polymerase θ in cancer therapy. *Trends Cancer* **7:** 98–111. doi:10.1016/j.trecan.2020.09.007

Shih C, Padhy LC, Murray M, Weinberg RA. 1981. Transforming genes of carcinomas and neuroblastomas introduced into mouse fibroblasts. *Nature* **290:** 261–264. doi:10.1038/290261a0

Shomali M, Cheng J, Sun F, Koundinya M, Guo Z, Hebert AT, McManus J, Levit MN, Hoffmann D, Dourjaud A, et al. 2021. SAR439859, a novel selective estrogen receptor degrader (SERD), demonstrates effective and broad antitumor activity in wild-type and mutant ER-positive breast cancer models. *Mol Cancer Ther* **20:** 250–262. doi:10 .1158/1535-7163.MCT-20-0390

Sicinski P, Donaher JL, Parker SB, Li T, Fazeli A, Gardner H, Haslam SZ, Bronson RT, Elledge SJ, Weinberg RA. 1995. Cyclin D1 provides a link between development and oncogenesis in the retina and breast. *Cell* **82:** 621–630. doi:10 .1016/0092-8674(95)90034-9

Singhal H, Greene ME, Tarulli G, Zarnke AL, Bourgo RJ, Laine M, Chang YF, Ma S, Dembo AG, Raj GV, et al. 2016. Genomic agonism and phenotypic antagonism between estrogen and progesterone receptors in breast cancer. *Sci Adv* **2:** e1501924. doi:10.1126/sciadv.1501924

Slamon DJ, Clark GM, Wong SG, Levin WJ, Ullrich A, McGuire WL. 1987. Human breast cancer: correlation of relapse and survival with amplification of the HER-2/ neu oncogene. *Science* **235:** 177–182. doi:10.1126/science .3798106

Slamon DJ, Godolphin W, Jones LA, Holt JA, Wong SG, Keith DE, Levin WJ, Stuart SG, Udove J, Ullrich A, et al. 1989. Studies of the HER-2/neu proto-oncogene in human breast and ovarian cancer. *Science* **244:** 707– 712. doi:10.1126/science.2470152

Slamon DJ, Leyland-Jones B, Shak S, Fuchs H, Paton V, Bajamonde A, Fleming T, Eiermann W, Wolter J, Pegram M, et al. 2001. Use of chemotherapy plus a monoclonal antibody against HER2 for metastatic breast cancer that overexpresses HER2. *N Engl J Med* **344**: 783–792. doi:10.1056/NEJM200103153441101

Smith AE, Ferraro E, Safonov A, Morales CB, Arenas Lahuerta EJ, Li Q, Kulick A, Ross D, Solit DB, de Stanchina E, et al. 2021. HER2$^+$ breast cancers evade anti-HER2 therapy via a switch in driver pathway. *Nat Commun* **12**: 6667. doi:10.1038/s41467-021-27093-y

Smyth LM, Piha-Paul SA, Won HH, Schram AM, Saura C, Loi S, Lu J, Shapiro GI, Juric D, Mayer IA, et al. 2020. Efficacy and determinants of response to HER kinase inhibition in HER2-mutant metastatic breast cancer. *Cancer Discov* **10**: 198–213. doi:10.1158/2159-8290.CD-19-0966

Snyder LB, Flanagan JJ, Qian Y, Gough SM, Andreoli M, Bookbinder M, Cadelina G, Bradley J, Rousseau E, Chandler J, et al. 2021. The discovery of ARV-471, an orally bioavailable estrogen receptor degrading PROTAC for the treatment of patients with breast cancer. *Cancer Res* **81**: 44–44. doi:10.1158/1538-7445.Am2021-44

Sparano JA, Gray RJ, Makower DF, Pritchard KI, Albain KS, Hayes DF, Geyer CE Jr, Dees EC, Perez EA, Olson JA Jr, et al. 2015. Prospective validation of a 21-gene expression assay in breast cancer. *N Engl J Med* **373**: 2005–2014. doi:10.1056/NEJMoa1510764

Spring LM, Wander SA, Andre F, Moy B, Turner NC, Bardia A. 2020. Cyclin-dependent kinase 4 and 6 inhibitors for hormone receptor-positive breast cancer: past, present, and future. *Lancet* **395**: 817–827. doi:10.1016/S0140-6736(20)30165-3

Sutherland RL, Watts CK, Musgrove EA. 1993. Cyclin gene expression and growth control in normal and neoplastic human breast epithelium. *J Steroid Biochem Mol Biol* **47**: 99–106. doi:10.1016/0960-0760(93)90062-2

Tan AR, Wright GS, Thummala AR, Danso MA, Popovic L, Pluard TJ, Han HS, Vojnovic Z, Vasev N, Ma L, et al. 2019. Trilaciclib plus chemotherapy versus chemotherapy alone in patients with metastatic triple-negative breast cancer: a multicentre, randomised, open-label, phase 2 trial. *Lancet Oncol* **20**: 1587–1601. doi:10.1016/S1470-2045(19)30616-3

The Cancer Genome Atlas Network. 2012. Comprehensive molecular portraits of human breast tumours. *Nature* **490**: 61–70. doi:10.1038/nature11412

Tolaney SM, Wardley AM, Zambelli S, Hilton JF, Troso-Sandoval TA, Ricci F, Im SA, Kim SB, Johnston SR, Chan A, et al. 2020. Abemaciclib plus trastuzumab with or without fulvestrant versus trastuzumab plus standard-of-care chemotherapy in women with hormone receptor-positive, HER2-positive advanced breast cancer (monarcHER): a randomised, open-label, phase 2 trial. *Lancet Oncol* **21**: 763–775. doi:10.1016/S1470-2045(20)30112-1

Toogood PL, Harvey PJ, Repine JT, Sheehan DJ, VanderWel SN, Zhou H, Keller PR, McNamara DJ, Sherry D, Zhu T, et al. 2005. Discovery of a potent and selective inhibitor of cyclin-dependent kinase 4/6. *J Med Chem* **48**: 2388–2406. doi:10.1021/jm049354h

Toy W, Shen Y, Won H, Green B, Sakr RA, Will M, Li Z, Gala K, Fanning S, King TA, et al. 2013. ESR1 ligand-binding domain mutations in hormone-resistant breast cancer. *Nat Genet* **45**: 1439–1445. doi:10.1038/ng.2822

Toy W, Weir H, Razavi P, Lawson M, Goeppert AU, Mazzola AM, Smith A, Wilson J, Morrow C, Wong WL, et al. 2017. Activating *ESR1* mutations differentially affect the efficacy of ER antagonists. *Cancer Discov* **7**: 277–287. doi:10.1158/2159-8290.CD-15-1523

Tung NM, Robson ME, Ventz S, Santa-Maria CA, Nanda R, Marcom PK, Shah PD, Ballinger TJ, Yang ES, Vinayak S, et al. 2020. TBCRC 048: phase II study of olaparib for metastatic breast cancer and mutations in homologous recombination-related genes. *J Clin Oncol* **38**: 4274–4282. doi:10.1200/JCO.20.02151

Tutt ANJ, Garber JE, Kaufman B, Viale G, Fumagalli D, Rastogi P, Gelber RD, de Azambuja E, Fielding A, Balmaña J, et al. 2021. Adjuvant olaparib for patients with *BRCA1*- or *BRCA2*-mutated breast cancer. *N Engl J Med* **384**: 2394–2405. doi:10.1056/NEJMoa2105215

Vora SR, Juric D, Kim N, Mino-Kenudson M, Huynh T, Costa C, Lockerman EL, Pollack SF, Liu M, Li X, et al. 2014. CDK 4/6 inhibitors sensitize PIK3CA mutant breast cancer to PI3K inhibitors. *Cancer Cell* **26**: 136–149. doi:10.1016/j.ccr.2014.05.020

Wang TC, Cardiff RD, Zukerberg L, Lees E, Arnold A, Schmidt EV. 1994. Mammary hyperplasia and carcinoma in MMTV-cyclin D1 transgenic mice. *Nature* **369**: 669–671. doi:10.1038/369669a0

Watt AC, Cejas P, DeCristo MJ, Metzger-Filho O, Lam EYN, Qiu X, BrinJones H, Kesten N, Coulson R, Font-Tello A, et al. 2021. CDK4/6 inhibition reprograms the breast cancer enhancer landscape by stimulating AP-1 transcriptional activity. *Nat Cancer* **2**: 34–48. doi:10.1038/s43018-020-00135-y

Whittle JR, Vaillant F, Surgenor E, Policheni AN, Giner G, Capaldo BD, Chen HR, Liu HK, Dekkers JF, Sachs N, et al. 2020. Dual targeting of CDK4/6 and BCL2 pathways augments tumor response in estrogen receptor-positive breast cancer. *Clin Cancer Res* **26**: 4120–4134. doi:10.1158/1078-0432.CCR-19-1872

Xia W, Mullin RJ, Keith BR, Liu LH, Ma H, Rusnak DW, Owens G, Alligood KJ, Spector NL. 2002. Anti-tumor activity of GW572016: a dual tyrosine kinase inhibitor blocks EGF activation of EGFR/erbB2 and downstream Erk1/2 and AKT pathways. *Oncogene* **21**: 6255–6263. doi:10.1038/sj.onc.1205794

Yang C, Li Z, Bhatt T, Dickler M, Giri D, Scaltriti M, Baselga J, Rosen N, Chandarlapaty S. 2017. Acquired CDK6 amplification promotes breast cancer resistance to CDK4/6 inhibitors and loss of ER signaling and dependence. *Oncogene* **36**: 2255–2264. doi:10.1038/onc.2016.379

Yoshida A, Lee EK, Diehl JA. 2016. Induction of therapeutic senescence in vemurafenib-resistant melanoma by extended inhibition of CDK4/6. *Cancer Res* **76**: 2990–3002. doi:10.1158/0008-5472.CAN-15-2931

Yu Q, Geng Y, Sicinski P. 2001. Specific protection against breast cancers by cyclin D1 ablation. *Nature* **411**: 1017–1021. doi:10.1038/35082500

Yu Q, Sincinska E, Geng Y, Ahnström M, Zagozdzon A, Kong Y, Gardner H, Kiyokawa H, Harris LN, Stål O, et al. 2006. Requirement for CDK4 kinase function in breast cancer. *Cancer Cell* **9**: 23–32. doi:10.1016/j.ccr.2005.12.012

Zatreanu D, Robinson HMR, Alkhatib O, Boursier M, Finch H, Geo L, Grand D, Grinkevich V, Heald RA, Langdon S, et al. 2021. Polθ inhibitors elicit BRCA-gene synthetic lethality and target PARP inhibitor resistance. *Nat Commun* **12:** 3636. doi:10.1038/s41467-021-23463-8

Zhao M, Scott S, Evans KW, Yuca E, Saridogan T, Zheng X, Wang H, Korkut A, Cruz Pico CX, Demirhan M, et al. 2021. Combining neratinib with CDK4/6, mTOR and MEK inhibitors in models of HER2-positive cancer. *Clin Cancer Res* **27:** 1681–1694. doi:10.1158/1078-0432 .Ccr-20-3017

Zheng ZY, Anurag M, Lei JT, Cao J, Singh P, Peng J, Kennedy H, Nguyen NC, Chen Y, Lavere P, et al. 2020. Neurofibromin is an estrogen receptor-α transcriptional co-repressor in breast cancer. *Cancer Cell* **37:** 387–402.e7. doi:10.1016/j.ccell.2020.02.003

Zhou J, Gelot C, Pantelidou C, Li A, Yücel H, Davis RE, Färkkilä A, Kochupurakkal B, Syed A, Shapiro GI, et al. 2021. A first-in-class polymerase θ inhibitor selectively targets homologous-recombination-deficient tumors. *Nat Cancer* **2:** 598–610. doi:10.1038/s43018-021-00203-x

Index